SUSTAINABLE DESIGN FOR RENEWABLE PROCESSES

SUSTAINABLE DESIGN FOR RENEWABLE PROCESSES

Principles and Case Studies

Edited by

MARIANO MARTÍN

Department of Chemical Engineering, University of Salamanca, Salamanca, Spain

ELSEVIER

Elsevier
Radarweg 29, PO Box 211, 1000 AE Amsterdam, Netherlands
The Boulevard, Langford Lane, Kidlington, Oxford OX5 1GB, United Kingdom
50 Hampshire Street, 5th Floor, Cambridge, MA 02139, United States

MATLAB® is a trademark of The MathWorks, Inc. and is used with permission. The MathWorks does not warrant the accuracy of the text or exercises in this book. This book's use or discussion of MATLAB® software or related products does not constitute endorsement or sponsorship by The MathWorks of a particular pedagogical approach or particular use of the MATLAB® software.

Notices
Knowledge and best practice in this field are constantly changing. As new research and experience broaden our understanding, changes in research methods, professional practices, or medical treatment may become necessary.

Practitioners and researchers must always rely on their own experience and knowledge in evaluating and using any information, methods, compounds, or experiments described herein. In using such information or methods they should be mindful of their own safety and the safety of others, including parties for whom they have a professional responsibility.

To the fullest extent of the law, neither the Publisher nor the authors, contributors, or editors, assume any liability for any injury and/or damage to persons or property as a matter of products liability, negligence or otherwise, or from any use or operation of any methods, products, instructions, or ideas contained in the material herein.

British Library Cataloguing-in-Publication Data
A catalogue record for this book is available from the British Library

Library of Congress Cataloging-in-Publication Data
A catalog record for this book is available from the Library of Congress

ISBN: 978-0-12-824324-4

For Information on all Elsevier publications
visit our website at https://www.elsevier.com/books-and-journals

Publisher: Candice Janco
Acquisitions Editor: Peter Adamson
Editorial Project Manager: Ruby Gammell
Production Project Manager: Nirmala Arumugam
Cover Designer: Greg Harris

Typeset by MPS Limited, Chennai, India

Working together
to grow libraries in
developing countries

www.elsevier.com • www.bookaid.org

Dedication

A mi padre Mariano por todo su apoyo y consejo.

Contents

3

Biomass and waste based processes

4. Thermochemical processes

Antonio Sánchez and Mariano Martín

5. Biochemical-based processes

Mariano Martín and Guillermo Galán

6. Anaerobic digestion and nutrient recovery

Edgar Martín-Hernández and Mariano Martín

7. Basic concepts and elements in the design of thermally coupled distillation systems

Gabriel Contreras-Zarazúa, Juan Gabriel Segovia-Hernández and Salvador Hernández-Castro

8. Added-value products

Manuel Taifouris and Mariano Martín

4

Solar technologies

9. Solar thermal energy

Mariano Martín and Jose A. Luceño

10. Photovoltaic solar energy

César Ramírez-Márquez and Mariano Martín

5

Wind based processes

11. Wind energy: collection and transformation

Mariano Martín

6

Geothermal processes

12. Geothermal energy

Mariano Martín

7

Water as energy resource

13. Water as a resource: renewable energies and technologies for brine revalorization

Borja Hernández and Mariano Martín

8

Integration of resources

14. Renewable-based process integration

Salvador I. Pérez-Uresti, Ricardo M. Lima and
Arturo Jiménez-Gutiérrez

15. Energy storage

Mariana Corengia and Ana I. Torres

The codes pertaining to chapters 2 to 15 can be accessed through the ancillary site at https://www.elsevier.com/books-and-journals/book-companion/9780128243244

List of contributors

Brenda Cansino-Loeza Michoacan University of Saint Nicholas of Hidalgo, Morelia, Mexico

Gabriel Contreras-Zarazúa Department of Chemical Engineering, University of Guanajuato, Guanajuato, Mexico

Mariana Corengia Universidad de la República, Montevideo, Uruguay

Daniel Cortés-Borda Basic Sciences Faculty, University of the Atlantic, Puerto Colombia, Colombia

Guillermo Galán Department of Chemical Engineering, University of Salamanca, Salamanca, Spain

Ángel Galán-Martín Department of Chemical, Environmental and Materials Engineering, University of Jaén, Jaén, Spain; Center for Advanced Studies in Earth Sciences, Energy and Environment (CEACTEMA), University of Jaén, Jaén, Spain

Ignacio E. Grossmann Carnegie Mellon University, Pittsburgh, PA, United States

Borja Hernández Department of Chemical Engineering, University of Salamanca, Salamanca, Spain

Salvador Hernández-Castro Department of Chemical Engineering, University of Guanajuato, Guanajuato, Mexico

Arturo Jiménez-Gutiérrez Chemical Engineering Department, Tecnológico Nacional de México/Instituto Tecnológico de Celaya, Celaya, GTO, Mexico

Ricardo M. Lima Computer, Electrical and Mathematical Sciences & Engineering Division, King Abdullah University of Science and Technology (KAUST), Thuwal, Saudi Arabia

Jose A. Luceño Department of Chemical Engineering, University of Salamanca, Salamanca, Spain

Mariano Martín Department of Chemical Engineering, University of Salamanca, Salamanca, Spain

Edgar Martín-Hernández Department of Chemical Engineering, University of Salamanca, Salamanca, Spain

Salvador I. Pérez-Uresti Chemical Engineering Department, Tecnológico Nacional de México/Instituto Tecnológico de Celaya, Celaya, GTO, Mexico

José María Ponce-Ortega Michoacan University of Saint Nicholas of Hidalgo, Morelia, Mexico

César Ramírez-Márquez Universidad de Guanajuanto, División de Ciencias Exactas y Naturales, Guanajuato, Mexico

Antonio Sánchez Department of Chemical Engineering, University of Salamanca, Salamanca, Spain

Juan Gabriel Segovia-Hernández Department of Chemical Engineering, University of Guanajuato, Guanajuato, Mexico

Manuel Taifouris Department of Chemical Engineering, University of Salamanca, Salamanca, Spain

Ana I. Torres Universidad de la República, Montevideo, Uruguay

Carmen M. Torres Department of Chemical Engineering, University Rovira i Virgili, Tarragona, Spain; Technology Centre of Catalonia EURECAT, Sustainability Area - Water, Air and Soil, Tarragona, Spain

Javier Tovar-Facio Michoacan University of Saint Nicholas of Hidalgo, Morelia, Mexico

Preface

I find writing the preface of a book is, maybe, the most difficult part of the whole journey. It is a moment to declare the motivation and purpose behind it as well as the desired goals. It is also the moment to thank those who help put it together, not only as authors but also as reviewers and mentors. I hope to be up to the task with the next few lines.

Renewable resources are the key to the future of mankind. The world that we inherited from our parents and that we will leave to our sons, daughters, nephews, and nieces is the result of how we have made and make use of natural resources. We have transformed it, for better or worse. The sustainable use of resources is the only way to preserve the rights of our future generations, and sustainability is a concept that must grow on us as engineers so that we can build a better future or at least never a worse one. This book has been the result of more than 15 years of work in process modeling, design, and optimization of renewable-based processes, applied to and nurtured from working with companies and teaching systematic design courses at the University of Salamanca, as well as visiting professor at the University of Maribor, University of Leeds, University of Birmingham, Carnegie Mellon University, University of Wisconsin-Madison, University of Minnesota, Universidad de Concepción, Universidad de la República, Universidad Nacional del Sur—Plapiqui, Universidad de Guanajuato, and Universidad Michoacana de San Nicolás de Hidalgo. This book has also nurtured with the help of many friends and colleagues worldwide from ETH Zurich,

Rovira i Virgili, Universidad de Alcalá, Universidad de la República, Instituto Técnico de Celaya, Universidad de Guanajuato, Universidad Michoacana de San Nicolás de Hidalgo, King Abdullah University of Science and Technology, Carnegie Mellon University, as well as my current PhD students and past visitors at the University of Salamanca.

This book aims at providing the basics of process analysis and design using renewable resources, presenting, and addressing the challenges that they bring to the table. It is intended to be a textbook for the Master and PhD level students. It can be considered as a follow-up to the book "*Industrial chemical processes: Analysis and design,*" but it provides a step forward presenting systematic design tools and methods. This book focuses on the use of renewable raw materials and novel technologies within the Green Chemistry umbrella, beyond the classical chemical processes. It presents the resources evaluated in the book (Chapter 1: Management of Renewable Energy Sources) covering the principles in process modeling, simulation, synthesis, and optimization (Chapter 2: Mathematical Modeling for Renewable Process Design), and sustainable assessment (Chapter 3: Sustainability in Products and Process Design). Next, we go over each resource and the major processes to transform it to power and chemicals starting from biomass-based processes, considering thermochemical (Chapter 4: Thermochemical Processes), biochemical (Chapter 5: Biochemical-based Processes), and digestion processes (Chapter 6:

Anaerobic Digestion and Nutrient Recovery), the design of added-value products (Chapter 8: Added-Value Products), and process intensification principles (Chapter 7: Basic Concepts and Elements in the Design of Thermally Coupled Distillation Systems). Next, we move to solar-based power production, either thermal (Chapter 9: Solar Thermal Energy) or photochemical (Chapter 10: Photovoltaic Solar Energy), wind-based power and chemicals production (Chapter 11: Wind-based Processes), geothermal facilities and risk assessment (Chapter 2: Geothermal Energy), and water and seawater as resources for power and chemicals (Chapter 13: Water as a Resource: Renewable Energies and Technologies for Brine Revalorization). Finally, we analyze the issues and methodologies for the design of integrated processes based on variable resources (Chapter 14: Renewable-based Process Integration) and energy storage (Chapter 15: Energy Storage). The text does not only include the theory and concepts but also presents solved examples and case studies as well as the end-of-chapter problems that can be useful in teaching the materials and for the students to test their understanding of the topics. Many examples are solved in a variety of well-known and widely used software packages so that the students become familiar with them and can apply the learnings from other modules. This fact connects the book with a previous work of the group *"Introduction to software for chemical engineers."* The code is provided either in the text or as a supplementary material in the web of the editorial for reference and to support the learning process. Covering such a wide range of technologies and topics, I did not aim to present a deep analysis of all of them, but it pretends to provide a comprehensive overview of the principles and challenges of each technology.

I would like to thank my postdoc advisor, Prof. Ignacio E. Grossmann, coauthor of the second chapter, the one who developed most of the optimization concepts and promoted their extension for the design of renewable processes. My other professors at CMU specially Prof. Larry T. Biegler who also introduced me to many of the concepts also deserve big thanks. I would also like to thank my undergraduate, master, and previous PhD students, Lidia S. Guerras, Clara Montero, Sofía Núñez, Carlos Prieto, María Prieto, José Enrique Roldán, Elena Castellano, and Diego Santamaría, for their comments and suggestions on the manuscript. But above all those, I would like to thank them who accepted the invitation to contribute to this book, and without their contributions, it would not have been possible to cover such a wide range of concepts. I hope that this book will be helpful to instructors and students in providing the tools, concepts, and ideas to design the world of the future starting today.

This book has been prepared during the COVID-19 pandemic, an awful time in our lives that have changed them forever. I only hope that soon it becomes part of our history. I would like to remember those who are no longer with us, may they rest in peace, and extend my thanks to all the members of our society who have worked and fought against the virus from their positions at the health sector, the food production chain and logistics, policemen, army, teachers, and so many others. For us, working on this book has helped us overcome the difficult days full of hard news.

Mariano Martín

SECTION 1

Resources and raw materials

1

Management of renewable energy sources

Javier Tovar-Facio, Brenda Cansino-Loeza and
José María Ponce-Ortega
Michoacan University of Saint Nicholas of Hidalgo, Morelia, Mexico

1.1 Introduction

To achieve the 2°C goals of the Paris Agreement, fossil fuels need to be phased out and replaced by low-carbon sources of energy. This requires the nearly complete decarbonization of the power sector by 2050, and an accelerated shift toward electricity as a final energy carrier. The integration of energy efficiency and renewable energy technologies is key to develop a sustainable society. Significant efforts have been carried out to improve the efficiency of the current energy conversion systems, producing efficient energy conversion systems, and/or relying on renewable energy, such as wind energy, solar thermal, solar photovoltaic (PV), geothermal, hydro, and biomass energy (Rabaia et al., 2021). This sustainable energy transition, which seeks to transform the global energy sector from fossil-based to renewable energy sources, requires a massive amount of clean energy to decarbonize the power sector. Designing a power system with high renewable energy shares is challenging, mainly due to the temporal mismatch between the energy demand and the availability of renewable energy resources, price and demand fluctuations, technical limitations, technology innovations, and environmental policy.

Solar and wind energy resources vary through time and they are typically called variable renewable energy (VRE) sources (Lund et al., 2015), which on a large scale involves issues such as limited availability, economic obstacles, challenging ramping situations, periods of oversupply, as well as periods where the renewable sources are not able to meet the demand. Other renewable technologies based on sources such as biomass and hydro can vary the amount of energy they supply relatively quickly so that supply matches demand; however, these are also limited by weather conditions such as droughts. In this chapter, we introduce some of the renewable energy sources to show the limitations and advantages of each of them.

Sustainable Design for Renewable Processes
DOI: https://doi.org/10.1016/B978-0-12-824324-4.00004-4

1.2 Biomass

Biomass is an organic material used as an energy source derived mainly from plants, animals, and wastes. Biomass has the potential to store solar energy, which is known as biomass energy. During photosynthesis, green plants obtain the energy of sunlight to convert CO_2 and H_2O into simple sugars and oxygen that is released into the atmosphere, whereas the carbohydrates can be used as biomass energy that is burnt and back converted into CO_2 and H_2O. This way, it is considered that biomass is a CO_2 neutral resource because the CO_2 captured during photosynthesis is released in biomass combustion. Nevertheless, life cycle analysis studies have reported an important contribution to emissions associated with the application of fertilizers (650 g CO_2-eq/kWhe) (Amponsah et al., 2014).

There is a wide variety of biomass feedstocks composed mainly of cellulose, hemicellulose, and lignin, which can be converted into fuels, heat, electric power, and biobased products and chemicals. Biomass feedstocks can be originated from different sources that include forest residues, agricultural crops and residues, animal manure, human sewage, and municipal solid waste. In comparison with fossil fuels, biomass has low energy density and higher volatile matter content, which provides ignition stability. Biomass has lower heating values than fossil fuels, most of them ranging from 10 to 20 MJ/kg of dry matter. Higher heating value (HHV) and chemical composition of common biomass feedstock are presented in Table 1.1.

Biomass is characterized by the proximate and ultimate analysis to measure the most important biomass properties that determine the suitability of feedstocks for the conversion process. The proximate analysis considers the content of moisture, ash, and organic matter, and the ultimate analysis measures the elemental composition of biomass. The energy content or heating value of biomass feedstock can be estimated from several correlations on the basis of the proximate and ultimate analysis (Table 1.2).

Bioenergy, considering both the traditional use of biomass (energy for cooking and heating in simple and inefficient fires or stoves) and modern bioenergy, contributes around 12% of the global energy consumption. Modern bioenergy provides around 5.1% of total global demand, which accounts for about half of all renewable energy in final energy consumption (REN21, 2020).

Biomass has several positive impacts; carbon neutrality is one of the major advantages. In addition, biomass resources have abundant availability and can help in waste management and reduction. However, the main disadvantages of biomass energy are that it requires large amounts of water and land space. Furthermore, biomass energy is not completely clean because during biomass conversion additional fossil energy is used as heat or electricity, which results in CO_2 emissions.

1.2.1 Biomass conversion processes

Biomass can be converted into useful forms of energy such as heat, electricity, fuels, and chemicals, via different conversion pathways, which are classified into thermochemical, biochemical, and chemical processes. Generally, chemical methods are used to obtain

TABLE 1.1 Ultimate analysis and higher heating value (HHV) of typical biomass feedstocks.

Feedstock	Ultimate analysis (wt.%)					HHV (MJ/kg dry)
	C	H	O	N	S	
Oak wood	50.6	6.1	42.9	0.3	0.1	20.1
Pine chips	52.8	6.1	40.5	0.5	0.1	19.5
Olive tree pruning	49.9	6.0	43.4	0.7	0.0	18.9
Spruce wood	52.3	6.1	41.2	0.3	0.1	20.5
Willow	49.8	6.1	43.4	0.6	0.1	29.8
Switchgrass	49.7	6.1	43.4	0.7	0.1	19.8
Straw	48.8	5.6	44.5	1.0	0.1	17.1
Rice husk	49.3	6.1	43.7	0.8	0.1	16.0
Sugarcane bagasse	49.8	6.0	43.9	0.2	0.1	19.4
Corn cob	46.3	5.6	42.2	0.6	0.0	18.0
Wheat straw	47.2	6.2	42.0	0.8	0.1	17.5
Rice straw	41.8	4.6	36.6	0.7	0.1	16.3
Barley straw	47.5	6.3	41.7	0.6	0.1	17.1
Manure	50.2	6.5	34.6	5.2	0.9	14.2
Sewage sludge	50.9	7.3	33.4	6.1	2.3	10.7
Water hyacinth	53.0	6.8	37.2	2.5	0.5	16.0
Municipal solid waste	37.1	5.4	24.9	0.2	0.1	15.6

From Seitarides, T., Athanasiou, C., Zabaniotou, A., 2008. Modular biomass gasification-based solid oxide fuel cells (SOFC) for sustainable development. Renew. Sustain. Energy Rev. Pergamon. https://doi.org/10.1016/j.rser.2007.01.020; Ptasinski, K.J., 2016. Efficiency of Biomass Energy: An Exergy Approach to Biofuels, Power, and Biorefineries. John Wiley & Sons. <https://www.wiley.com/en-us/ Efficiency + of + Biomass + Energy%3A + An + Exergy + Approach + to + Biofuels%2C + Power%2C + and + Biorefineries-p-9781118702109; Han, J., Yao, X., Zhan, Y., Oh, S.-Y., Kim, L.-H., Kim, H.-J., 2017. A method for estimating higher heating value of biomass-plastic fuel. J. Energy Inst. 90 (2), 331–335. https://doi.org/10.1016/j.joei.2016.01.001.

TABLE 1.2 Correlations for estimating the higher heating value (HHV) of biomass.

Correlation	Reference
$HHV = 43.7C + 167.0$	Tillman (2012)
$HHV = 30.1C + 52.5H + 6.4O - 7.63$	Jenkins and Ebeling (1985)
$HHV = 35.2C + 116.2H + 6.3N + 10.5S - 11.1O$	Boie (1953)
$HHV = 31.37C + 70.09H + 3.18O - 136.75$	Sheng and Azevedo (2005)
$HHV = 36C + 120H - 16O$	Han et al. (2017)

more valuable products from biomass. Thermochemical processes are characterized for being performed at higher temperatures and conversion rates, whereas biochemical processes require low reaction time.

1.2.1.1 Thermochemical conversion processes

Thermochemical conversion processes use heat to promote chemical transformations of biomass into energy and chemical products. These processes have in common that they are carried out at high temperatures (450°C−1200°C). However, they substantially differ in the amount of applied oxygen. Biomass combustion involves complete fuel oxidation, gasification only partial oxidation, and pyrolysis is performed in the absence of oxygen. See Chapter 4 for process analysis.

Combustion: Combustion is the thermal conversion of organic matter that reacts with oxygen, during the process carbon and hydrogen are completely oxidized to carbon dioxide and water, which results in the release of a large amount of heat. This way, heat is used to raise steam in a boiler which in turn can drive a turbine to generate electricity. The operating temperature of combustion varies around 800°C−1000°C, and it is recommended that moisture content of biomass be less to 50%. Biomass combustion efficiencies are low, ranging between 20% and 40%. Combustion is the most widely used process for biomass conversion. It contributes to more than 90% of bioenergy production in the world (CTCN, 2020).

Gasification: Gasification is the thermal conversion of biomass into combustible gases by its partial oxidation at high temperatures, generally in the range of 800°C−900°C. Oxygen supply in gasification processes is commonly 35% of the oxygen demand for complete combustion. Products of biomass gasification are CO, H_2, CH_4, H_2O, and N_2. Synthesis gas (syngas) is the main product of gasification, which can be used for methanol and hydrogen production.

Pyrolysis: Pyrolysis is a thermal decomposition process, which occurs in the absence of oxygen for the conversion of biomass into solid charcoal, bio-oil, and gases at elevated temperatures. Conversion of biomass into solid, liquid, or gas products depends mainly on the temperature and reaction time. Temperature of pyrolysis ranges between 350°C and 500°C. At low temperature, the product is mainly charcoal, at high temperature the biomass will produce mainly gases, and a moderate temperature is optimum for producing liquids. Typically, pyrolysis processes can be classified into slow pyrolysis and fast pyrolysis. Fast pyrolysis occurs at high temperatures and short residence time, which results in a high yield of liquid product. On the other hand, the slow pyrolysis process takes place at low reaction temperatures and long residence time. Slow biomass pyrolysis has been widely applied for charcoal production.

Torrefaction: Torrefaction is a process for the pretreatment of biomass, it is used for the improvement of physical and chemical biomass properties. In this process, biomass is heated at temperatures of ∼250°C−350°C under an inert atmosphere that results in torrefied biomass with low moisture content (1%−3%), mass losses of about 30%, and energy losses of ∼10% (Ahmad, 2017). In addition, the fixed carbon content of torrefied biomass is high, between 25% and 40%, and energy density is improved by 10%−30% (Ptasinski, 2016). Consequently, torrefied biomass has better combustion properties, as it takes less time for ignition due to low moisture content and burns for a longer time due to a larger percentage of fixed carbon compared to raw biomass.

Liquefaction: Liquefaction is the thermochemical conversion of biomass into liquid bio-crude at moderate temperatures ranging between 300°C and 400°C and pressure from 4 to 22 MPa. Hydrothermal liquefaction is the most used liquefaction process and consists of the conversion of biomass into fuels by processing in a hot, pressurized water environment for sufficient time to break down the solid biopolymeric structure to mainly liquid components.

Hydrothermal processes are performed at different operating conditions, and depending on the temperatures of the processes, hydrothermal processing is divided into hydrothermal carbonization, hydrothermal liquefaction, and hydrothermal gasification. Hydrothermal carbonization is carried out at temperatures below 245°C and the main product of the process is hydrochar. At intermediate temperature ranges, between 245°C and 370°C, the process is called hydrothermal liquefaction, resulting in the production of biocrude, and at temperatures above 370°C, the process is defined as hydrothermal gasification, in this process the main product is synthetic fuel gas (Elliott et al., 2015).

1.2.1.2 Biochemical conversion processes

Biochemical conversion processes rely on microorganisms to convert biomass into bio-fuel. These processes require low energy consumption; they are carried out at low temperatures and are characterized by their low conversion rates due to the presence of microorganisms. The main biochemical conversion processes are fermentation (see Chapter 5 for process analysis) and anaerobic digestion (see Chapter 6 for process analysis).

Fermentation: Fermentation is the process where microorganisms metabolize plant sugar and produce ethanol, butanol, among others. The traditional ethanol fermentation uses sugar crops; nevertheless, there are various materials that can be used such as starchy crops or lignocellulosic biomass. Regularly, fermentation is carried out at atmospheric pressure and ambient temperature in the presence of bacteria, such as *Saccharomyces ceveresiae*. Ethanol concentration during fermentation is around 10%−18% by volume (Ahmad, 2017).

Anaerobic digestion: Anaerobic digestion is the process through which microorganisms break down biodegradable material in the absence of oxygen. Biomass feedstock in anaerobic digestion could be animal slurries, silage, food processing, and municipal solid wastes. During the conversion of biomass, microorganisms convert about 90% of the feedstock energy content into biogas, which in turn contains around 50%−70% methane (Naik et al., 2010; FAO, 2020). Biogas produced can be used directly in spark-ignition gas engines and gas turbines, and can be treated for CO_2 removal and increase its quality or used for chemicals production via dry reforming.

1.2.1.3 Chemical conversion processes

Chemical processes are used for the selective conversion of chemical compounds present in biomass into valuable products, which is known as a direct chemical conversion, or into intermediates for its process to obtain useful products known as an indirect chemical conversion, see Chapter 5 for details.

Esterification: Direct chemical conversion refers to the conversion of chemical compounds into valuable products. Esterification, mainly referred to as biodiesel production, is the most used direct chemical conversion and is given by the transesterification of triglycerides of fatty acids present in oilseed and animal tallow. Triglycerides (esters of glycerol) of fatty acids have high viscosity and, therefore, cannot be used as fuel in compression-ignition engines. However, during the transesterification of triglycerides with alcohol, methyl or ethyl esters of fatty acids are formed, these components constitute biodiesel.

Hydrolysis: The most common process of indirect chemical conversion is the hydrolysis of hemicellulose and cellulose. Ethanol production using lignocellulosic biomass is a common and efficient method. Hemicellulose and cellulose are complex carbohydrates that cannot be fermented to ethanol. However, the chemical hydrolysis of complex carbohydrates like polysaccharides is an efficient way to transform them into simple sugars that can be used for ethanol production

1.2.2 Biofuels

The term biofuel is referred to a solid, liquid, or gaseous fuel that is produced from biomass feedstocks. Depending on the origin and production technology of biofuels, they are classified into the following categories:

First-generation biofuels are biofuels derived from edible biomass such as sugar, starch, and oil crops. First-generation biofuels can help to improve domestic energy security but using edible feedstocks have a negative impact on biodiversity and food security.

Second-generation biofuels are fuels that can be derived from inedible biomass that is the case of lignocellulosic materials such as switchgrass, sawdust, low-priced woods, crop wastes, and municipal wastes. This type of feedstock makes them an attractive source for biofuel production because it is abundant and inexpensive biomass that does not affect the food supply. Despite the fact that second-generation lignocellulosic biomass can be cultivated on a large scale and involves a short rotation, several concerns remain about competitive land use.

Third-generation biofuels refer to biofuels derived from aquatic autotrophic organisms, where light, carbon dioxide, and nutrients are used to produce the feedstock. Third-generation biofuels are an attractive energy source as they do not compete with food and land use. Biomass used for this type of biofuels involves microalgae, macroalgae, and water plants.

Fourth-generation biofuels are based on the genetic modification of microorganisms, such as microalgae, yeast, fungi, and cyanobacteria, to create an artificial carbon sink to minimize carbon emissions. In addition to genetic modification, some fourth-generation technologies involve pyrolysis, gasification, and solar-to-fuel pathways.

1.3 Hydropower

Water energy resources include hydropower derived from the energy of moving water in rivers and lakes, and marine or ocean energy that takes advantage of the tides and waves and ocean temperature differential.

1.3.1 Hydropower

Hydropower is energy that harnesses the power of moving water. Hydroelectric energy is derived from the potential and kinetic energies of the movement of water between two points located at different altitudes. Then, the energy of water is transformed into mechanical energy in the turbine, which is connected to a generator for electricity production.

Among the types of renewable energy, hydroelectric energy is the largest source of renewable energy generation in the world (World Bank, 2020). Hydropower ranks third in gross electricity production (16.2%), only behind coal (38%) and natural gas (23%) (IEA, 2018). In 2019 it was estimated that 58% of electricity generation was generated by hydroelectric energy (REN21, 2020).

Hydropower is characterized by being the most efficient technology for the production of renewable energy, the efficiency of hydroelectric plants is around 90% (Bhatia, 2014) and also hydropower is the cheapest way to generate electricity (IRENA, 2020b).

Hydropower has many benefits in power generation; it is cost-effective, has a low generation of greenhouse gas emissions (75 g CO_2-eq/kWhe) (Amponsah et al., 2014), provides grid stability because it can respond immediately to fluctuations in electricity demand, can store energy which can be used for water supply, provides irrigation and flood control, and has flexibility and storage capacity for the use of intermittent renewable energy sources. However, the installation of hydroelectric schemes can also generate environmental and social concerns because the flow regimes of rivers can be modified, causing impacts on biodiversity ecosystems. In addition, the construction of hydroelectric plants can lead to the resettlement of people who previously lived near the area.

The amount of energy that a hydropower plant can generate is proportional to the hydraulic head and the flow rate (Letcher, 2018). Electrical power is calculated as follows:

$$P = \frac{Q \cdot g \cdot H \cdot \rho \cdot \eta}{10^6} \tag{1.1}$$

where P is the electrical power (MW), Q is the flow water (m^3/s), g is the gravitational constraint (9.81 m/s^2), H is the net head that refers to the elevation drop (m), ρ is the density of water (1000 kg/m^3), and η is the efficiency, referred to the product of all of the component efficiencies, which are normally the turbine, drive system, and generator. For rough estimation, 87% is used as typical overall plant efficiency.

The energy generated in the power plant (MWh) will be determined by the duration of the flow, t, in hours:

$$E = P \cdot t \tag{1.2}$$

1. Resources and raw materials

Further analysis can be found in Chapter 13. If the available water is much less than the capacity, the water flow could be diverted to prevent damage to the turbines. In cases where the water flow is lower but hydroelectric plants are operating, their efficiencies are very low because of their capacity factor. Usual capacity factors for hydroelectric plants are in the range of 0.2–0.7 (IFC, 2017), but in extreme cases, lower and higher values can also be found.

1.3.1.1 Classification of hydropower plants

Hydropower plants can be classified in different categories on the basis of their size and the type of scheme. There are different classifications of hydropower plants on the basis of their size that varies widely from one country to another. Facilities range in size from large power plants that supply energy for many consumers, to Pico plants that individuals operate for their own energy needs. However, it has been discussed that the classification of hydropower plants by size should be avoided because there is no clear connection between the size and impact that causes.

Therefore the following section presents the classification of hydropower plants by type. The main schemes of hydropower plants are conventional or impoundment hydropower plants, diversion or run-of-river hydropower plants, and pumped storage hydropower plants.

1.3.1.2 Conventional hydropower plants

The main component of a conventional hydropower plant is a dam, which raises the water level to create a reservoir to impound water (Fig. 1.1). In this way, energy is stored in the form of water in the reservoir and is released when needed according to the electric demand of the system. When electricity is needed, the gates of the dam are opened to conduct water at high pressure from the reservoir to a lower reservoir through the penstock. Flowing water in the penstock is conducted to the powerhouse where the kinetic energy of water spins the turbines to activate a generator that converts the mechanical energy of the turbine into electricity. The alternating current produced is sent to a transformer to increase the voltage and then is transmitted to the power line. Once the water has given up its energy, it returns to the river through a drainage channel.

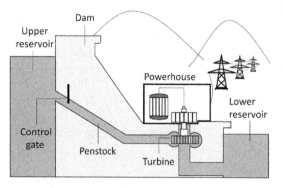

FIGURE 1.1 Impoundment hydroelectric power plant.

The main advantage of storage hydropower plants is their ability to store energy and generate electricity faster than other energy sources, maintaining the balance between supply and demand for electricity. In addition, water reservoirs can act as multipurpose systems that can be used for flood control, consumption, irrigation, and recreation.

1.3.1.3 Run-of-river hydropower plants

In run-of-river plants, the natural flow and elevation drop of a river are used to generate electricity. Run-of-river plants can be fed directly by a river or by a part of the river, which is separated by a canal. These hydropower plants are based on the natural fall of water from rivers with the regular flow that passes through very rugged terrain. Therefore a run-of-river plant requires sufficient hydrostatic head and a substantial flow rate. Its function is to divert the waterway from a river and guide it through a canal or penstock that leads to a powerhouse. Therefore the force of the moving water spins a turbine and drives a generator. The water is fed back into the main river further downstream.

The difference between run-of-river and conventional hydropower is that run-of-river systems do not make the river create a water reservoir. Most run-of-river facilities use a small dam or weir, to ensure enough water enters the penstock, and they have a small reservoir called pondage to store small amounts of water for same-day use. However, because of the absence of a major reservoir, large amounts of water cannot be stored for future use. Therefore it must be taken into account that the seasons of the year pass, the flow of the river also changes, so it is possible excess water causes water losses by the overflow of the dam. Otherwise, if the river water level is depleted due to water extraction, there will be no stored energy. Therefore run-of-river plants are only really feasible in rivers with large flow rates throughout the year.

1.3.1.4 Pumped storage hydropower plants

The operation in this type of plant allows regulating the production of energy according to the demand for electricity. Pumped storage plants are composed of two basins separated by a large difference in altitude and a turbine that can work as a pump. Pumped storage plants can be designed in places where a natural inflow in the higher basin could exist. Nevertheless, in most pumped storage plants, the basins do not have a natural inflow. When electric demand is high, the plant operates in turbine mode; water from the upper basin is conducted to the lower basin to activate the turbine and generate electricity. Otherwise, in the hours of less demand, generally at night, the plant works in pumping mode; the turbine pumps the water from the lower to the higher basin, this allows energy to be stored for electricity generation during peak hours.

1.3.2 Marine energy

Marine energy (or ocean energy) refers to a form of renewable energy that is harnessed from the ocean. More than 70% of Earth's surface is covered by oceans (Ressurreição et al., 2011); therefore, oceans represent an enormous source of renewable energy that is stored in the form of kinetic and thermal energy.

Marine energy is created by the rotation of the Earth that, in turn, creates wind that forms waves on the ocean surface, and by the gravitational pull of the Moon that creates tides and currents (Khare et al., 2020). In addition, oceans capture the thermal energy derived from the Sun creating a heat gradient from the surface to the depth.

Ocean energy system has many advantages; it is a type of energy environmentally friendly because it creates no harmful byproducts, has a low generation of emissions (50 g CO_2-eq/kWhe) (Amponsah et al., 2014), is abundant and widely available, and against most of the other alternative energy sources are easily predictable, and can be calculated the amount of energy that it can produce. Fig. 1.2 shows the global wave energy potential. Despite this, the main disadvantage of ocean energy is its location because it is not accessible to everyone. In addition, ocean energy disturbs the habitat of marine creatures and has the enormous cost of production.

1.3.2.1 *Tidal energy*

The gravitational pull of the Moon and Sun along with the rotation of the Earth create tides in the oceans. Each day, there are two high tides and two low tides. It takes about 12 h and 25 min between two consecutive high tides (Khan et al., 2017). The level of the ocean is constantly moving between high and low tide. Tidal power produces a variable amount of energy according to the position of the Earth, the Moon, and the Sun. When the Earth, the Sun, and the Moon are in a line, the gravitational pull of the Moon and the Sun are combined and creates high tides.

1.3.2.1.1 Tidal barrages

A tidal barrage is a system that consists of the construction of a low walled dam known as a "tidal barrage" in which a barrier is created between the sea and a tidal reservoir to

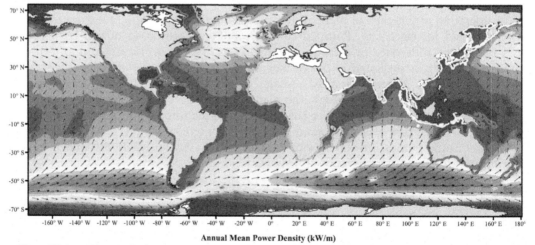

FIGURE 1.2 Global wave energy (Gunn and Stock-Williams, 2012). *Reprinted from Gunn, K., Stock-Williams, C., 2012. Quantifying the global wave power resource. Renew. Energy 44, 296–304, with permission from Elsevier.*

take advantage of the change in the tide levels to generate kinetic energy and produce power (Polis et al., 2017).

The bottom of the barrage dam is located on the seafloor with the top of the tidal barrage being just above the highest level that the water can get into at the highest annual tide. The barrage has a number of underwater tunnels that cut into its width allowing the seawater to flow through them in a controlled way using "sluice gates" on their entrance and exit points. Fixed within these tunnels are huge tidal turbine generators that spin as the seawater rushes past them either to fill or empty the tidal reservoir thereby generating electricity.

The base of the barrage dam is built on the seafloor and its height should exceed the level that the water can reach the highest annual tide. The barrage is sectioned into numerous underwater tunnels that allow seawater to flow through them using sluiced gates on their entry and exit points to control flow. Inside these tunnels, there are tidal turbine generators that rotate as the seawater passes; in this way, the kinetic energy of the water drives the turbines and then electrical energy is generated.

In tidal barrage, turbines can operate in one direction (i.e., Tidal Barrage Flood Generation and Tidal Barrage Ebb Generation) or in both flow directions during flood and ebb tides (Etemadi et al., 2011).

In the Tidal Barrage Flood Generation scheme, the tidal reservoir is empty during the low tide. When the tide begins to rise, the sluiced gates close to retain seawater and create a difference in level on both sides of the dam. Therefore the tidal reservoir is filled through the turbine tunnels that spin the turbines that generate electricity on the flood tide and then the reservoir is emptied through the open sluiced gate at low tide.

Conversely, the Tidal Barrage Ebb Generation scheme harnesses the ebb tide. During low tides, the gates open to fill the tidal reservoir, when the highest tide is reached the gates close.

Once the sea returns to its low tide level and there is a sufficient level difference for the electricity generation process, the gates connected to the turbine tunnels open allowing the water to flow. This rapid exit of the water through the tunnels with the outgoing tide causes the turbines to rotate at high speed generating electrical energy (Junejo et al., 2018).

A two-way tidal barrage scheme uses the energy over parts of both the rising tide and the falling tide to generate electricity (Khare et al., 2020). As the tide ebbs and flows, seawater flows in or out of the tidal reservoir through the same gate system. This flow of tidal water back and forth causes the turbine generators located within the tunnel to rotate in both directions producing electricity.

1.3.2.1.2 Tidal turbines

Tidal turbines use the kinetic energy of currents to generate electricity (Polis et al., 2017). Places, where the rise between the tides or their velocity produces strong currents, are potential sites for the installation of tidal turbines (Sangiuliano, 2017). The turbines are placed at the bottom of the sea so that the current that flows through the edges of the turbines drives a generator to produce electricity, which is supplied to the grid through the submarine wire.

1.3.2.1.3 Tidal fences

A tidal fence system is a scheme placed on the sea bed that is composed of vertical axis turbines mounted in a fence. It uses the kinetic energy of the water that passes through the turbines to generate electricity. Unlike the tidal turbines that only rotate around their own vertical axis, tidal fence harnesses the maximum available kinetic energy of the streams because it contains more than one vertical axis turbines that are mounted together in an only fence.

1.3.2.2 Wave energy

Wave energy consists in extracting energy from the movement of waves and converting it into electricity. Waves are generated by the interaction of wind with the surface of the ocean. The energy available for conversion mainly depends on the wind speed and the distance between wind and surface of the ocean. Wave energy can be extracted directly from surface waves or from pressure fluctuations below the surface (Melikoglu, 2018).

1.3.2.2.1 Oscillating water column

Oscillating water column devices are partially submerged hollow structures that form an air chamber with an opening underwater. As the waves rise and fall, the air trapped within the chamber compresses and expands, allowing the turbine to rotate and generate electricity (World Energy Council, 2004).

1.3.2.2.2 Wave Overtopping reservoir

Wave overtopping devices consist of a reservoir that stores wave water. Once a sufficient head of water is obtained between the level of the water in the reservoir and the level of the surrounding seawater, the energy from the reservoir is released into the sea by driving the turbines installed at the bottom of the reservoir to generate electricity.

1.3.2.2.3 Attenuator

An attenuator is a floating device made up of several hollow cylinders connected to a hydraulic pump by joints. The device captures the wave energy in different spatial orientations, causing hydraulic cylinders to pump oil to drive a hydraulic motor/generator through a power smoothing system.

1.3.2.2.4 Surface point absorber

Surface point absorbers are floating structures that absorb wave energy from all directions. Typical point absorbers are buoys. The buoys are fixed to the seabed and contain a linear generator, which consist of a set of magnets and a piston, and a stator formed by coils. When the buoy moves up and down due to the movement of the waves, the coils in turn move linearly around the piston generating electricity (Neill and Hashemi, 2018).

1.3.2.2.5 Ocean thermal energy conversion

Energy from the sun heats the surface water of the ocean. Ocean temperature varies from 24°C to 28°C on the surface to 4°C–6°C at 1 km depths (Khan et al., 2017). Ocean thermal energy conversion (OTEC) is a process that can produce electricity using the temperature difference between deep cold ocean waters and warm tropical surface waters. OTEC system works similar to a heat engine. The system consumes the thermal energy from the topmost layer of the sea and converts the portion of that energy into electrical energy. This type of system uses a temperature difference of at least 25°C to power a turbine to produce electricity. Warm surface water is pumped through an evaporator containing a working fluid. The vaporized fluid drives a turbine/generator and is turned back to a liquid in a condenser cooled with cold ocean water pumped from deeper into the ocean. OTEC systems using seawater as the working fluid can use condensed water to produce desalinated water (EIA, 2020a).

1.4 Geothermal power

Geothermal energy is the thermal energy that is generated and stored within the Earth, commonly associated with volcanic and tectonic activity and the decay process of radioactive elements (Fig. 1.3). Geothermal energy is available everywhere and is an unlimited resource because heat is continuously produced inside the Earth.

The rate of increasing temperature with respect to increasing depth in the Earth's interior is known as geothermal gradient and indicates that heat is continuously conducted to the Earth's surface. The geothermal gradient depends on the conductivity of rocks and the rate of heat production. In mid-oceanic ridges the gradients are highest (40−80 K/km),

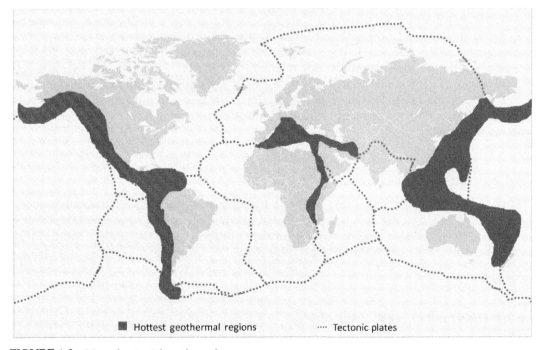

■ Hottest geothermal regions   ···· Tectonic plates

FIGURE 1.3   Map of potential geothermal resources.

whereas the lowest gradients (20−30 K/km) occur in stable continental areas (Arndt, 2011). Therefore the Earth's heat flow varies at different places in the world. It is estimated that the average heat flow of the Earth is 80 mW/m^2 (Morgan, 2006). However, it is not possible to use all the geothermal energy because it is so dispersed. Recovery factor of geothermal energy, which refers to the ratio of energy recovery to energy stored in the resources, ranges from 5% to 25% (Letcher, 2020).

Notwithstanding the Earth's surface temperature is affected by seasons and air temperature, with increasing depth the temperature below the Earth's surface is stable. On average, the temperature increases with a depth around 3°C/100 m but in potential geothermal areas, the temperature gradient could be greater than 7°C/100 m (Kruger et al., 1973).

Geothermal energy is manifested through different forms such as geysers, fumaroles, hot springs, hot pools, and steaming grounds. Geothermal sources are classified into hydrothermal, geopressured, and petrothermal. In hydrothermal systems, the geothermal fluid is heated by the hot rock, in this transition, if the fluid contains more vapor than liquid water, then it is called a vapor-dominated system. But if the liquid water content in the fluid is greater than the steam, then it is called a liquid-dominated system. On the other hand, geopressured zones are sedimentary basins where water is trapped at high pressures. Besides, water contains methane that can be used for electricity production. Petrothermal systems are based on the heat of the hot dry rock (HDR), heat is extracted by pumping water into the HDR to create steam that can be used for electricity production (Rashid, 2015).

One of the main advantages of using geothermal energy is that it is a constant source of energy, this means that geothermal plants can produce energy all the time and can be used for heat, cool, and power. In addition, geothermal power generation has lower life cycle greenhouse gas emissions (78 g CO_2-eq/kWhe) (Amponsah et al., 2014) and the land consumption for the surface installations is small (IRENA, 2017). Nonetheless, some of the disadvantages of geothermal energy are that energy cannot be transported over long distances, besides, the reinjection of streams into the Earth can result in minor seismic activity, and water conducted underground could leak toxic elements if the system is not properly insulated.

1.4.1 Geothermal uses

Low-temperature reservoirs can be used for domestic hot water, swimming pools, house heating among other applications. When geothermal energy uses low-temperature reservoirs, it is called *direct use* of geothermal energy. On the other hand, high-temperature reservoirs can be used to produce electricity using steam turbines and generators in power plants, then it is known as *indirect use* of geothermal energy. See Chapter 12 for process analysis.

1.4.1.1 Direct use of geothermal energy

Direct use of geothermal energy refers to the use of underground energy used directly in district heating systems. The main application of direct use of geothermal energy is the heat pump. These devices use low-temperature geothermal resources and leverage the nearly constant temperature below the Earth, which is warmer than the air that circulates above during winter, and during summer cooler than the air. This allows the heat pump to provide heating or cooling to an internal space taking advantage of the geothermal energy.

A geothermal heat pump system is made up of three main parts: the heat pump unit, the ground heat exchanger, and the distribution system (Fig. 1.4). The ground heat exchanger encompasses a series of pipes known as a ground loop, which is installed a few meters beneath the Earth close to the building. Through the ground, loop circulates an antifreeze solution to suck up or disseminate heat into the ground. The process in the heat pump system involves a cycle of evaporation, compression, condensation, and expansion. A refrigerant is used as the heat-transfer medium, which circulates within the heat pump. The cycle starts as the cold liquid refrigerant passes through the evaporator and absorbs heat from the fluid from the ground loop (1). The refrigerant evaporates into a gas as heat is absorbed. The gaseous refrigerant then passes through a compressor where the refrigerant is pressurized to raise its temperature (2). The hot gas then circulates through a condenser where heat is removed (3) and transferred to the building's distribution system. When it loses the heat, the refrigerant changes back to a liquid. The liquid is cooled as it passes through an expansion valve (4) and the process begins again. During summer, the system can run in reverse.

1.4.1.1.1 Types of geothermal heat pumps

Geothermal heat pump systems can be classified according to the climate, soil conditions, and available land that can be used for residential or commercial applications. The main types of geothermal heat pumps are closed-loop systems, pond or lake systems, open systems, and hybrid systems (EERE, 2017). Closed-loop systems are made of plastic tubing that is buried in the ground or submerged in water. Through the closed-loop, an antifreeze solution circulates; therefore, a heat exchanger transfers heat between the refrigerant in the heat pump and the antifreeze solution in the closed-loop. Closed-loop systems

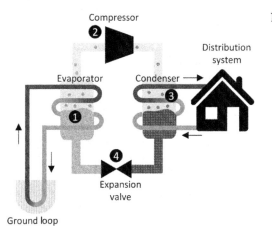

Ground loop

FIGURE 1.4   Geothermal heat pump system.

can be classified into horizontal and vertical. Horizontal closed loops are ideal for large areas of land. These systems are cost-effective and are commonly installed in residential areas. Vertical closed loops are often used in commercial buildings and schools where space is quite limited. Vertical loops are also applied in areas where the soil is too shallow for trenching, and they minimize the disturbance to existing landscaping. On the other hand, pond or lake systems are installed in places that have a water body. This way, a supply line pipe is run underground from the building to the water body and coiled into circles at least 2 m under the surface to prevent freezing. On the other hand, open-loop systems are ideal when the zones have a sufficient supply of groundwater. These systems use a well or surface water body as the heat exchange fluid. Once the fluid has circulated through the system, the water returns to the ground through the well, a recharge well, or surface discharge. This option is obviously practical only where there is an adequate supply of relatively clean water, and all local codes and regulations regarding groundwater discharge are met. In the case of hybrid systems, they use different geothermal resources or a combination of geothermal resources with outdoor air.

1.4.1.2 Indirect use of geothermal energy

Indirect use of geothermal energy refers to the use of geothermal energy for electricity production. The main schemes for electricity production are dry steam power plants, flash steam power plants, and binary cycle power plants (Manzella, 2019).

1.4.1.2.1 Dry steam power plants

Dry steam plants are used when the geothermal fluid is primary steam. Once the steam is pumped from underground reservoirs, it is sent directly to the turbine to produce kinetic energy which power the generator to produce electricity. After powering the turbine, the steam is sent to the condenser and is cooled by cooling towers. Then, water is reinjected to the Earth. The fluids used in these plants have temperatures above 250°C and the average size of dry steam plants is around 45 MW.

1.4.1.2.2 Flash steam power plants

In these systems the geothermal fluid pumped to the surface is partially vaporized. Therefore the geothermal fluid is sent to a flash tank that is at a much lower temperature causing the fluid to quickly flash into steam. The steam produced is sent to the turbines for electricity generation. After that, the steam is cooled and condensed to be reinjected back into the surface with the geothermal liquid separated in the flash tank. Flash steam plants reinject around 60%−90% of the geothermal fluid. Flash steam technology generally uses fluids with temperatures above 180°C. Plants have an average size of 30, 37, and 90 MW for single, double, and triple flash technologies, respectively.

1.4.1.2.3 Binary cycle power plants

In binary plants the geothermal fluid exchanges heat with a working fluid, usually an organic fluid with a low boiling point and high vapor pressure at low temperatures compared with steam. The main difference between binary power plants and the others is that the water or steam from below the Earth never comes in direct contact with the turbines. Instead, water from geothermal reservoirs is pumped through a heat exchanger where the organic fluid is heated. Then, the organic fluid enters the turbine for electricity generation after the fluid is cooled and condensed to start the cycle again. The hot water from the Earth is reinjected into the Earth through the injection well. The main disadvantage of these types of plants is its efficiency. The ratio between the net electricity produced and the energy input is lower than for other technologies, being around 12% for dry or flash steam plants and between 2% and 10% for binary systems.

1.5 Wind power

Eolic energy is the energy from the power of the wind, this energy is transformed into electric energy by turbines and systems of energy conversion (Fig. 1.5). Eolic energy is a renewable source of energy with a small environmental impact because it occurs naturally in the Earth's atmosphere.

When solar radiation is absorbed on Earth, the air above is heated so the air is expanded and therefore becomes less dense and then rises through the cold air. This thermal effect together with Earth's movements produces wind patterns with different speeds that also depend on the seasons of the year, region, and local variation. Extreme winds occur in hurricanes and cyclonic weather when the temperature difference between land and ocean is large; complex terrain of hills and mountains deflects and concentrates air movements due to uneven solar absorption and height differences (Twidell and Weir, 2015).

This technology can produce energy without producing greenhouse emissions directly. However, emissions associated with the construction of the type of energy are around 123 g CO_2-eq/kWhe (Amponsah et al., 2014). According to projections made by the Global Wind Energy Council (GWEC) in 2020, wind power alone could save 8.2 billion tons of CO_2 (Fried et al., 2017).

Wind energy is completely free and does not produce atmospheric emissions that cause acid rain or greenhouse gases (carbon dioxide or methane). Because wind is caused by the absorption of solar radiation and the sun shines, the energy will never run out. Besides,

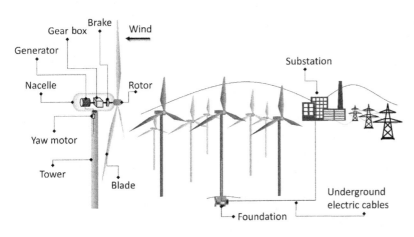

FIGURE 1.5 Simplified scheme of a horizontal axis wind turbine and a wind energy power system.

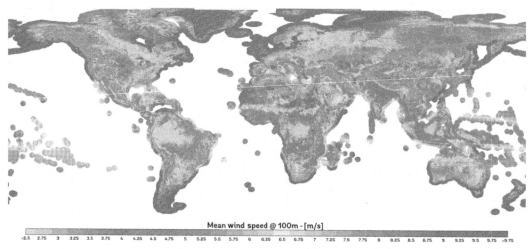

FIGURE 1.6   Wind resource map: 100 m mean wind speed (DTU, 2018). *Map obtained from the Global Wind Atlas 3.0, a free, web-based application developed, owned, and operated by the Technical University of Denmark (DTU). The Global Wind Atlas 3.0 is released in partnership with the World Bank Group, utilizing data provided by Vortex, using funding provided by the Energy Sector Management Assistance Program (ESMAP). For additional information: https://global-windatlas.info.*

turbines can be built directly on sites with good wind resources because they use only a part of the land which does not represent a problem for agriculture or land-use change. However, as many resources of energy, wind energy also has disadvantages, it is not a constant energy resource because the wind does not blow with the same intensity always and everywhere, so the location where it is installed must have an adequate supply of wind. Fig. 1.6 presents a wind resource map of the world and it is evident that some places have a much higher average speed that would make it more affordable to undertake a project of this nature. Something that should also be considered is that wind turbines may generate noise and visual pollution due to a simple turbine that can be seen from hundreds of meters away. On the other hand, turbines may be dangerous for wildlife; birds can die while flying into spinning turbine blades, and a wind farm requires roughly 0.1 km^2 of clear land per megawatt of turbine capacity (if they are sited closer together, the turbines start to interfere with each other's efficiency and they begin to harness less energy) (Towler, 2014).

1.5.1 Types of wind energy

There are three main types of wind energy:

Distributed or "small" wind: Single small wind turbines below 100 kW that are used to directly power a home, farm, or small business and are not connected to the grid.
Utility-scale wind: Wind turbines that range in size from 100 kW to several megawatts, where the electricity is delivered to the power grid and distributed to the end-user by electric utilities or power system operators.

Offshore wind: Wind turbines that are erected in large bodies of water, usually on the continental shelf. Offshore wind turbines are larger than land-based turbines and can generate more power.

Although onshore wind turbines are used in the majority of cases, offshore wind turbines are increasingly appealing. The reason is that the wind has a better quality offshore than onshore (it is stronger and more regular), as shown in Fig. 1.6. Also, people are in favor of wind energy but no one wants a wind turbine near urban areas. With offshore wind turbines, noise and visual pollution are mitigated due to location. In addition, there are also more available areas in the world to install wind turbines offshore than onshore. Nevertheless, floating offshore wind turbines have a lot of civil engineering challenges associated with design and control.

1.5.2 Wind turbines

The objective of turbines is to transform kinetic energy into rotating mechanical power of the turbine rotor blades (see Chapter 11). The formulation for the power in the wind in a specific location, perpendicular to the wind blowing directions is given by the formula:

$$P = \frac{1}{2}\rho A C_p v^3 \qquad (1.3)$$

where P is the power, ρ is the air density, v is the wind speed, and C_p is the power coefficient, which describes the fraction of the wind captured by a wind turbine. According to Betz rules, the value of the power coefficient features a theoretic limit connected with 59.7% (Sumathi et al., 2015). The reason for this efficiency is because wind power depends on a continuous flow of air in motion. If 100% of kinetic energy was extracted, then the flow of air would be zero and no velocity would remain available to sustain the flow through the extraction mechanism (Kalmikov, 2017).

There are two important types of turbines, named by the geometrical construction and the aerodynamics of the wind passing around the blades: horizontal axis wind turbines (HAWTs) and vertical axis wind turbines (VAWTs). See also Chapter 11 for further details. For a HAWT the value of C_p is between 0.40 and 0.50, while for VAWT, it is difficult to determine the exact value of C_p because the number of turbines operating is less (Eriksson et al., 2008).

1.6 Solar energy

Most of the energy plants use thermal conversion to generate steam and move a turbine, such as coal, natural gas, fuel oil, nuclear, and biomass technologies. On the other hand, wind and hydropower systems also use a generator that is moved using the kinetic energy of nature: either from the force of water in the case of hydro or other marine energies, or the air velocity for turning the wind turbine. However, the sun represents the biggest source of energy; more energy from the sunlight strikes the Earth in 1 h than all of the energy consumed by humans in an entire year (Zhang et al., 2013). Solar radiation often called the solar resource, is a general term for the electromagnetic radiation emitted by the sun. This energy can be transformed into heat and electricity (DOE, 2013). The solar radiation that reaches ground level on the Earth is of two types: direct radiation and diffuse radiation. Direct radiation is defined as radiation that has not experienced scattering in the atmosphere, so that it is directionally fixed, coming from the disk of the sun (Sørensen, 2017) but diffuse radiation is the solar radiation that reaches the Earth surface after having been scattered from the direct solar beam by molecules or particles in the atmosphere such as water vapor, air molecules, dust, pollutants, forest fires, and volcanoes. The sum of diffuse and direct solar radiation is called global solar radiation.

The technical feasibility and economic operation of solar technologies depend on the available solar resource, which is a function of several variables such as geographic location, time of day, season, topography, and the weather of the place. For example, in the early morning and late afternoon, the solar resource travels further through the atmosphere than at noon when the sun is at its highest point and the solar resource is better during summer than during winter due to the Earth's movements. Also, as shown in Fig. 1.7, the places with the highest levels of solar irradiation are found in Africa, the Middle East, parts of India, Australia, Mexico, the southwestern United States, and in some parts of South America such as Brazil and the west coast of Chile and Peru.

Solar resource has a lot of advantages that make it an excellent candidate to replace conventional fuels. For example, it is everywhere and can be utilized on-site without requiring transportation; solar energy has no waste residue, noise associated, and low emissions (150 g CO_2-eq/kWhe for solar thermal and 300 g CO_2-eq/kWhe for PV systems) (Amponsah et al., 2014); it has an inexhaustible supply and it can be transformed into

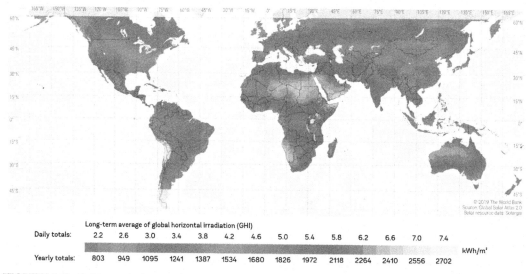

FIGURE 1.7   Solar resource map. Global Horizontal Irradiation in the world (Solargis, 2017). *GHI Solar Map © 2017 Solargis. The map is published by the World Bank Group, funded by ESMAP, and prepared by Solargis. For additional information: http://globalsolaratlas.info.*

other energy forms such as chemical, electrical, and thermal energy through the photosynthesis and different technologies [such as PV panels or power towers]. Nevertheless, the limitations of solar energy keep this renewable source of energy from being the ideal candidate in the sustainable energy transition (in some cases). The disadvantages of solar energy include a low power density that requires large power plants to produce electricity and consequently a potentially harmful land-use change, the instability in the solar energy supply because of the weather, the hour of the day, and the season impacts the electrical power generation. Furthermore, the biggest limitation, the discontinuity of solar radiation overnight which means that on its own solar energy cannot provide a continuous source of power anywhere in the world and it is necessary to use energy-storage systems to operate continuously (Wang, 2019).

As presented in Fig. 1.7, latitude is one of the major factors that affects solar resources received on a given surface area during a specific amount of time, and also climate and weather patterns. Locations in lower latitudes and in arid climates generally receive higher amounts of insolation than other locations (EIA, 2020b). Nonetheless, the regions with high energy consumption do not always match with the solar resource and they will be necessary to connect the electrical systems of different countries to make the most of the solar resource.

The type of solar collector also determines the type of solar radiation and level of insolation that a solar collector receives. Concentrating solar collector systems, such as those used in solar thermal-electric power plants, require *direct solar radiation*, which is generally greater in arid regions with few cloudy days. Flat-plate solar thermal and PV collectors can use global solar radiation (direct and diffuse solar radiation) (EIA, 2020b). This limits their application to specific regions.

1.6.1 Solar technologies

The concept of solar energy is based on taking advantage of the light and heat from the sun to generate useful energy for humans using different technologies. Solar energy technologies can be classified into passive solar technologies and active solar technologies. Passive technology involves the accumulation of solar energy without transforming thermal or light energy into any other form. For example, heat from sunlight can be used to cause air movement for ventilating to heat and cool houses or buildings without active mechanical or electrical devices. On the other hand, active solar systems collect solar radiation and use mechanical and electrical equipment for the conversion of solar energy to heat and electric power. In general, the latter can be grouped into two categories: PV (see also Chapter 10 for further details) and thermal technologies (see Chapter 9 for process analysis) (Kabir et al., 2018).

1.6.1.1 Solar photovoltaic

The only source of power generation that does not use a turbine is solar PV. This technology gets its name because it uses cells that can convert light (photons) to electricity (voltage), which is called the PV effect. These cells are constructed with materials that can generate this effect and produce direct current electricity (mostly silicon-based), and as

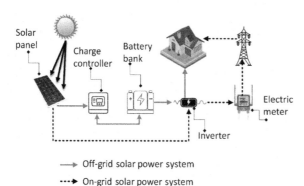

FIGURE 1.8 Simplified scheme of a photovoltaic power system.

→ Off-grid solar power system

⋯▸ On-grid solar power system

shown in Fig. 1.8, they can be part of an off-grid or on-grid power system. Electricity generated from solar PV panels is inverted into alternating current and injected either in the transmission grid, for large-scale plants, or into the distribution network for small PV units used by residential or small businesses.

Nowadays, electricity from solar cells has become cost-competitive in many regions and PV systems are being deployed at large scales to help power the electric grid. The current solar PV technologies are silicon solar cells, thin-film solar cells, dye-sensitized solar cells, perovskite solar cells, and quantum dot sensitized solar cells (Ludin et al., 2018).

The power output of a solar cell can be calculated in function of the area (m^2), the efficiency of the process, and the solar irradiance (W/m^2).

$$P = \eta \cdot A \cdot G \tag{1.4}$$

1.6.1.2 Solar thermal technology

Solar thermal technologies harness sunlight to produce thermal energy. This heat is then used directly or to generate electricity (Kantenbacher and Shirley, 2018). For example, solar water heating systems collect the thermal energy of the sun and use it to heat water in homes and businesses. On the other hand, concentrating solar thermal technologies can produce electricity on demand. In contrast with solar PV, concentrated solar power (CSP) systems use mirrors or lenses to focus solar energy, heating a working fluid (such as water or organic compounds) that can drive an electricity-generating turbine or a heating heat-transfer fluid (such as molten salt) which can then be used to produce electricity or stored for later use. Based on how they collect solar energy, there are three CSP design classification:

1.6.1.2.1 Linear concentrator systems

Linear concentrator systems collect the sun's energy using long rectangular, curved (U-shaped) mirrors. The mirrors are tilted toward the sun, focusing sunlight on tubes (or receivers) that run the length of the mirrors. The reflected sunlight heats a fluid flowing through the tubes. The hot fluid then is used to boil water in a conventional steam-turbine generator to produce electricity. There are two major types of linear concentrator systems: parabolic trough systems, where receiver tubes are positioned along the focal line of each parabolic mirror and linear Fresnel reflector systems, where one receiver tube is positioned above several mirrors to allow the mirrors greater mobility in tracking the sun.

1.6.1.2.2 Dish/engine systems

A dish/engine system uses a mirrored dish like a very large satellite dish, although to minimize costs, the mirrored dish is usually composed of many smaller flat mirrors formed into a dish shape. The dish-shaped surface directs and concentrates sunlight onto a thermal receiver, which absorbs and collects the heat and transfers it to the engine generator.

The most common type of heat engine used today in dish/engine systems is the Stirling engine. This system uses the fluid heated by the receiver to move pistons and creates mechanical power. The mechanical power is then used to run a generator or alternator to produce electricity.

1.6.1.2.3 Power tower systems

A power tower system uses a large field of flat, sun-tracking mirrors known as heliostats to focus and concentrate sunlight onto a receiver on the top of a tower (Fig. 1.9). A heat-transfer fluid heated in the receiver is used to generate steam, which, in turn, is used in a conventional turbine generator to produce electricity.

Some power towers use water/steam as heat-transfer fluid. Other advanced designs are experimenting with molten nitrate salt because of its superior heat transfer and energy-storage capabilities. The energy-storage capability, or thermal storage, allows the system to continue to dispatch electricity during cloudy weather or at night.

Compared with solar PV technologies, solar thermal technologies are relatively older, more mature, more space-efficient, and less complex. Both technologies share some of the environmental impacts such as water usage, possibly hazardous waste, land use, landscape fragmentation, possible extinction of local wildlife, and some microclimate changes.

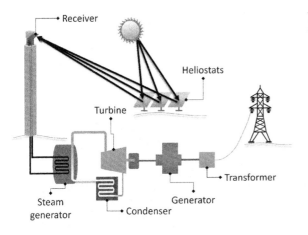

FIGURE 1.9 Simplified scheme of a power tower technology power system.

1. Resources and raw materials

1.7 Renewable energy integration and flexibility

According to the International Renewable Energy Agency (IRENA), increasing the power systems flexibility is one of the technology pillars for the future of energy (IRENA, 2020a). Flexibility is used to refer to the reliability of an energy system to cope with risks, threats, and adverse events that can jeopardize its capacity to satisfy the demand of the end-users (Gallo et al., 2016). In conventional power systems, flexibility could be ensured by providing reserves and generation planning. Nonetheless, major challenges must face with the increasing renewable energy penetration due to their variable capacity factors and uncertain generation (Impram et al., 2020). Renewable power generation capacity is commonly measured through the capacity factor, which represents the actual generation of a power plant compared with the maximum amount it could generate in a given period without any interruption (Morales Pedraza, 2019). The capacity factor of a power plant is a measure of availability and, therefore, an indirect measure of flexibility and startup times. Increasing the flexibility of power plants involves operational modifications to achieve the objectives of flexibilization. Generation flexibility in power system is based on parameters to characterize the operational flexibility (Hentschel et al., 2016; IRENA, 2019):

Lower minimum load. The minimum load is the lowest possible net load a generating unit can deliver under stable operating conditions. It is measured as a percentage of the nominal load.

Shorter startup time. It is defined as the period from starting plant operation unit reaching the minimum load. Therefore, with a shorter startup time, the plant can quickly reach full load.

Higher ramp rate. The ramp rate describes how fast a power plant can change its net power during operation. With higher ramp rates, the plant can quickly alter its production in line with the system's needs.

Minimum up/down time. Minimum length of time the plant must stay in an operational state taking offline and vice versa.

Before the widespread introduction of the VRE sources (such as wind and solar), the main sources for variability were changes in demand and failures in generators or disruptions of the energy network. Nonetheless, a disadvantage of VRE sources is their fluctuations in time and space with an associated uncertainty and lower capacity factors in comparison with conventional technologies. Furthermore, the effects of climate change such as temperature rise, changes in precipitation patterns, droughts, and/or the rise in sea level may limit the potential amount of energy from renewable sources such as biomass or hydro. Some sources of flexibility at the power systems are network expansion, energy storage, flexible generation, excess capacity of renewable energy technologies, and a well-planned mix of renewable energy technologies (type of technologies and geographical distribution) (Blanco and Faaij, 2018).

There is no one technology that can solve the problem of decarbonization on its own. Every country, city, or house has an optimal solution with a different percentage of participation of each of the renewable energies, see Chapter 14 for process integration analysis.

Therefore future power systems with high shares of renewable energy sources require to maximize its flexibility to avoid interruptions in the power supply.

The countries with the highest electricity generation using renewable energy sources are presented in Table 1.3, as well as the information corresponding to each type of technology. As can be seen, in 2018, China was the largest producer of electricity using renewable energy sources followed by the United States, both countries have a variety of renewable energy technologies to help meet demand.

In 2019 around 11% of global primary energy was generated by renewable technologies. Accelerating the transition to a renewable-based energy system brings benefits such as the reduction of greenhouse gas emissions, the creation of new employment opportunities, and the enhancement of human welfare. Fig. 1.10 shows the growth in the generation of renewable energy in the world. As can be seen, hydropower is the largest renewable energy source but wind and solar power are both growing rapidly.

It is expected that the share of renewable energy increases from 25% in 2017 to 85% by 2050, mainly through the rapid increase in the wind and solar power. This energy transition increases the need for more flexible generators in the power system.

TABLE 1.3 Country ranking of electricity generation from renewable energy in 2018 (IRENA, 2018).

Rank	Total (GWh)		Rank	Wind (GWh)	
1	China	1,884,073.561	1	China	366,452.204
2	United States	764,680.011	2	United States	275,834.000
3	Brazil	495,945.306	3	Germany	109,951.000
4	Canada	428,080.775	4	United Kingdom	56,903.961
5	India	240,671.895	5	India	55,008.662
Rank	**Bioenergy (GWh)**		**Rank**	**Geothermal (GWh)**	
1	United States	67,885.000	1	United States	18,773.000
2	China	67,300.603	2	Indonesia	13,295.810
3	Brazil	54,497.561	3	Philippines	10,435.305
4	Germany	50,880.000	4	New Zeeland	7,815.113
5	United Kingdom	32,870.111	5	Turkey	7,430.976
Rank	**Hydro (GWh)**		**Rank**	**Solar (GWh)**	
1	China	1,199,200.000	1	China	178,070.754
2	Brazil	388,971,076	2	United States	85,184.000
3	Canada	381,638.531	3	Japan	62,667.671
4	United States	295,501.000	4	Germany	45,784.000
5	Russian Federation	192,355.683	5	India	310,66.803

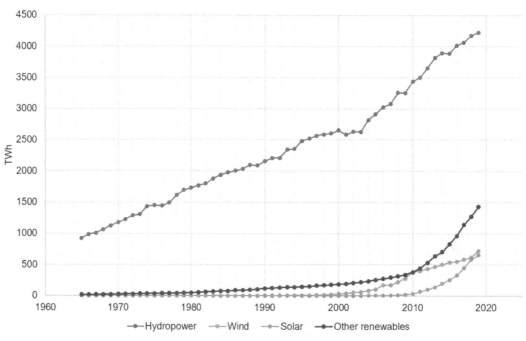

FIGURE 1.10   Global renewable energy generation in TWh (Ritchie and Roser, 2017).

References

Ahmad, M. (Ed.), 2017. Operation and Control of Renewable Energy Systems. Operation and Control of Renewable Energy Systems. John Wiley & Sons, Ltd, Chichester. Available from: https://doi.org/10.1002/9781119281733.

Amponsah, N.Y., Troldborg, M., Kington, B., Aalders, I., Hough, R.L., 2014. Greenhouse gas emissions from renewable energy sources: a review of lifecycle considerations. Renew. Sustain. Energy Rev. Available from: https://doi.org/10.1016/j.rser.2014.07.087.

Arndt, N., 2011. Geothermal gradient. Encyclopedia of Astrobiology. Springer, p. 662. Available from: https://doi.org/10.1007/978-3-642-11274-4_643.

Bhatia, S.C., 2014. Advanced Renewable Energy Systems. CRC PressEdited by CRS Press. Available from: https://www.sciencedirect.com/book/9781782422693/advanced-renewable-energy-systems.

Blanco, H., Faaij, A., 2018. A review at the role of storage in energy systems with a focus on power to gas and long-term storage. Renew. Sustain. Energy Rev. Available from: https://doi.org/10.1016/j.rser.2017.07.062.

Boie, W., 1953. Fuel technology calculations. Energietechnik 3, 309–316.

CTCN, 2020. Biomass Combustion and Co-Firing for electricity and Heat. Climate Technology Centre & Network. <https://www.ctc-n.org/technologies/biomass-combustion-and-co-firing-electricty-and-heat>.

DOE, 2013. Solar Radiation Basics. Department of Energy. <https://www.energy.gov/eere/solar/solar-radiation-basics>.

DTU, 2018. Global Wind Atlas. Wind Energy. <https://globalwindatlas.info/about/introduction>.

EERE, 2017. Geothermal Heat Pumps. Office of Energy Efficiency and Renewable Energy. <https://www.energy.gov/eere/geothermal/geothermal-heat-pumps>.

EIA, 2020a. Ocean Thermal Energy Conversion – United States Energy Information Administration. Energy Information Administration. <https://www.eia.gov/energyexplained/hydropower/ocean-thermal-energy-conversion.php>.

EIA, 2020b. Where Solar Is Found — United States Energy Information Administration. <https://www.eia.gov/energyexplained/solar/where-solar-is-found.php>.

Elliott, D.C., Biller, P., Ross, A.B., Schmidt, A.J., Jones, S.B., 2015. Hydrothermal liquefaction of biomass: developments from batch to continuous process. Bioresour. Technol. Available from: https://doi.org/10.1016/j.biortech.2014.09.132.

Eriksson, S., Bernhoff, H., Leijon, M., 2008. Evaluation of different turbine concepts for wind power. Renew. Sustain. Energy Rev. Available from: https://doi.org/10.1016/j.rser.2006.05.017.

Etemadi, A., Emami, Y., AsefAfshar, O., Emdadi, A., 2011. Electricity generation by the tidal barrages. Energy Proc. 12, 928—935. Available from: https://doi.org/10.1016/j.egypro.2011.10.122.

FAO, 2020. Bioenergy conversion technologies. <http://www.fao.org/3/T1804E/t1804e06.htm>.

Fried, L., Shukla, S., Sawyer, S., 2017. Growth trends and the future of wind energy. Wind Energy Engineering: A Handbook for Onshore and Offshore Wind Turbines. Elsevier Inc, pp. 559—586. Available from: https://doi.org/10.1016/B978-0-12-809451-8.00026-6.

Gallo, A.B., Simões-Moreira, J.R., Costa, H.K.M., Santos, M.M., Moutinho dos Santos, E., 2016. Energy storage in the energy transition context: a technology review. Renew. Sustain. Energy Rev. Available from: https://doi.org/10.1016/j.rser.2016.07.028.

Gunn, K., Stock-Williams, C., 2012. Quantifying the global wave power resource. Renew. Energy 44, 296—304. Available from: https://doi.org/10.1016/j.renene.2012.01.101.

Han, J., Yao, X., Zhan, Y., Oh, S.-Y., Kim, L.-H., Kim, H.-J., 2017. A method for estimating higher heating value of biomass-plastic fuel. J. Energy Inst. 90 (2), 331—335. Available from: https://doi.org/10.1016/j.joei.2016.01.001.

Hentschel, J., Babić, U., Spliethoff, H., 2016. A parametric approach for the valuation of power plant flexibility options. Energy Rep. 2 (November), 40—47. Available from: https://doi.org/10.1016/j.egyr.2016.03.002.

IEA, 2018. World Gross Electricity Production by Source. International Energy Agency. <https://www.iea.org/data-and-statistics/charts/world-gross-electricity-production-by-source-2018>.

IFC, 2017. Hydroelectric power. A guide for developers and investors. <http://www.fichtner.de>.

Impram, S., Varbak Nese, S., Oral, B., 2020. Challenges of renewable energy penetration on power system flexibility: a survey. Energy Strategy Rev. Available from: https://doi.org/10.1016/j.esr.2020.100539.

IRENA, 2017. Geothermal power: technology brief. International Renewable Energy Agency. Available from: http://www.irena.org.

IRENA, 2018. Country Rankings. /Statistics/View-Data-by-Topic/Capacity-and-Generation/Country-Rankings.

IRENA, 2019. Innovation landscape brief: flexibility in conventional power plants. <http://www.irena.org>.

IRENA, 2020a. Global Renewables Outlook: Energy Transformation 2050. /Publications/2020/Apr/Global-Renewables-Outlook-2020.

IRENA, 2020b. Hydropower. International Renewable Agency. /costs/Power-Generation-Costs/Hydropower.

Jenkins, B.M., Ebeling, J.M., 1985. Correlation of Physical and Chemical Properties of Terrestrial Biomass With Conversion. The Institute. <https://agris.fao.org/agris-search/search.do?recordID = United States8705549>.

Junejo, F., Saeed, A., Hameed, S., 2018. Energy management in ocean energy systems. Compr. Energy Syst. 5—5, 778—807. Available from: https://doi.org/10.1016/B978-0-12-809597-3.00539-3. Elsevier Inc.

Kabir, E., Kumar, P., Kumar, S., Adelodun, A.A., Kim, K.H., 2018. Solar energy: potential and future prospects. Renew. Sustain. Energy Rev. Available from: https://doi.org/10.1016/j.rser.2017.09.094.

Kalmikov, A., 2017. Wind power fundamentals. Wind Energy Engineering: A Handbook for Onshore and Offshore Wind Turbines. Elsevier Inc, pp. 17—24. Available from: https://doi.org/10.1016/B978-0-12-809451-8.00002-3.

Kantenbacher, J., Shirley, R., 2018. Renewable energy: scaling deployment in the United States and in developing economies. Sustainable Cities and Communities Design Handbook: Green Engineering, Architecture, and Technology. Elsevier Inc, pp. 89—109. Available from: https://doi.org/10.1016/B978-0-12-813964-6.00005-7.

Khan, N., Kalair, A., Abas, N., Haider, A., 2017. Review of ocean tidal, wave and thermal energy technologies. Renew. Sustain. Energy Rev. Available from: https://doi.org/10.1016/j.rser.2017.01.079.

Khare, V., Nema, S., Baredar, P., 2020. Fundamental and principles of the ocean energy system. Ocean Energy Modeling and Simulation with Big Data. Elsevier, pp. 1—48. Available from: https://doi.org/10.1016/b978-0-12-818904-7.00001-0.

Kruger, P, Otte, Mon, C., 1973. Geothermal energy. Resources, production, stimulation. Stanford Univ. Press.

Letcher, T.M., 2018. Managing Global Warming, first ed. Academic Press. Available from: https://www.elsevier.com/books/managing-global-warming/letcher/978-0-12-814104-5.

Letcher, 2020. Future Energy. Elsevier. Available from: https://doi.org/10.1016/C2018-0-01500-5.

Ludin, N.A., Mustafa, N.I., Hanafiah, M.M., Ibrahim, M.A., Mat Teridi, M.A., Sepeai, S., et al., 2018. Prospects of life cycle assessment of renewable energy from solar photovoltaic technologies: a review. Renew. Sustain. Energy Rev. Available from: https://doi.org/10.1016/j.rser.2018.07.048.

Lund, P.D., Lindgren, J., Mikkola, J., Salpakari, J., 2015. Review of energy system flexibility measures to enable high levels of variable renewable electricity. Renew. Sustain. Energy Rev. Available from: https://doi.org/10.1016/j.rser.2015.01.057.

Manzella, A., 2019. General introduction to geothermal energy, Lecture Notes in Energy, 67. Springer Verlag, pp. 1–18. Available from: https://doi.org/10.1007/978-3-319-78286-7_1.

Melikoglu, M., 2018. Current status and future of ocean energy sources: a global review. Ocean Engineering. Available from: https://doi.org/10.1016/j.oceaneng.2017.11.045.

Morales Pedraza, J., 2019. Current status and perspective in the use of coal for electricity generation in the North America region. Conventional Energy in North America. Elsevier, pp. 211–257. Available from: https://doi.org/10.1016/b978-0-12-814889-1.00004-8.

Morgan, P., 2006. Heat flow in the earth. Geophysics. Kluwer Academic Publishers, pp. 634–646. Available from: https://doi.org/10.1007/0-387-30752-4_79.

Naik, S.N., Goud, V.V., Rout, P.K., Dalai, A.K., 2010. Production of first and second generation biofuels: a comprehensive review. Renew. Sustain. Energy Rev. Available from: https://doi.org/10.1016/j.rser.2009.10.003.

Neill, S.P., Hashemi, M.R., 2018. Fundamentals of Ocean Renewable Energy: Generating Electricity from the Sea. Elsevier. Available from: https://doi.org/10.1016/C2016-0-00230-9.

Polis, H.J., Dreyer, S.J., Jenkins, L.D., 2017. Public willingness to pay and policy preferences for tidal energy research and development: a study of households in Washington State. Ecol. Econ. 136 (June), 213–225. Available from: https://doi.org/10.1016/j.ecolecon.2017.01.024.

Ptasinski, K.J., 2016. Efficiency of Biomass Energy: An Exergy Approach to Biofuels, Power, and Biorefineries. John Wiley & Sons. Available from: https://www.wiley.com/en-us/Efficiency + of + Biomass + Energy%3A + An + Exergy + Approach + to + Biofuels%2C + Power%2C + and + Biorefineries-p-9781118702109.

Rabaia, M.K.H., Abdelkareem, M.A., Sayed, E.T., Elsaid, K., Chae, K.J., Wilberforce, T., et al., 2021. Environmental impacts of solar energy systems: a review. Sci. Total. Environ. 754 (February), 141989. Available from: https://doi.org/10.1016/j.scitotenv.2020.141989.

Rashid, M.H., 2015. Electric Renewable Energy Systems. Elsevier Inc. Available from: https://doi.org/10.1016/C2013-0-14432-7.

REN21, 2020. Renewables 2020 Global Status Report. Secretariat, Paris.

Ressurreição, A., Gibbons, J., Dentinho, T.P., Kaiser, M., Santos, R.S., Edwards-Jones, G., 2011. Economic valuation of species loss in the open sea. Ecol. Econ. 70 (4), 729–739. Available from: https://doi.org/10.1016/j.ecolecon.2010.11.009.

Ritchie, H., Roser, M., 2017. Renewable Energy. Our World in Data. <https://ourworldindata.org/renewable-energy>.

Sangiuliano, S.J., 2017. Planning for tidal current turbine technology: a case study of the Gulf of St. Lawrence. Renew. Sustain. Energy Rev. Available from: https://doi.org/10.1016/j.rser.2016.11.261.

Seitarides, T., Athanasiou, C., Zabaniotou, A., 2008. Modular biomass gasification-based solid oxide fuel cells (SOFC) for sustainable development. Renew. Sustain. Energy Rev. Available from: https://doi.org/10.1016/j.rser.2007.01.020.

Sheng, C., Azevedo, J.L.T., 2005. Estimating the higher heating value of biomass fuels from basic analysis data. Biomass Bioenergy 28 (5), 499–507. Available from: https://doi.org/10.1016/j.biombioe.2004.11.008.

Solargis, 2017. Solar resource maps and GIS data for 200 + countries | Solargis. <https://solargis.com/maps-and-gis-data/download/world>.

Sørensen, B., 2017. Origin of renewable energy flows. Renew. Energy 39–218. Available from: https://doi.org/10.1016/b978-0-12-804567-1.00002-5.

Sumathi, S., Ashok Kumar, L., Surekha, P., 2015. Solar PV and wind energy conversion systems. Green Energy and Technology. Springer International Publishing. Available from: https://doi.org/10.1007/978-3-319-14941-7.

Tillman, D.A., 2012. Wood as an Energy Resource. Elsevier. Available from: https://www.elsevier.com/books/wood-as-an-energy-resource/tillman/978-0-12-691260-9.

Towler, B.F., 2014. The Future of Energy. Elsevier Inc. Available from: https://doi.org/10.1016/C2013-0-19049-6.

Twidell, J., Weir, T., 2015. Renewable Energy Resources. Routledge. Available from: https://doi.org/10.4324/9781315766416.

Wang, Z., 2019. The solar resource and meteorological parameters. Design of Solar Thermal Power Plants. Elsevier, pp. 47–115. Available from: https://doi.org/10.1016/b978-0-12-815613-1.00002-x.

World Bank, 2020. Operation and Maintenance Strategies for Hydropower: Handbook for Practitioners and Decision Makers. Operation and Maintenance Strategies for Hydropower. Available from: https://doi.org/10.1596/33313.

World Energy Council, 2004. In: Trinnaman, J., Clarke, A. (Eds.), 2004 Survey of Energy Resources. Elsevier. Available from: https://www.sciencedirect.com/book/9780080444109/2004-survey-of-energy-resources.

Zhang, H.L., Baeyens, J., Degrève, J., Cacères, G., 2013. Concentrated solar power plants: review and design methodology. Renew. Sustain. Energy Rev. Available from: https://doi.org/10.1016/j.rser.2013.01.032.

Design principles

Mathematical modeling for renewable process design

Mariano Martín[1] *and Ignacio E. Grossmann*[2]

[1]Department of Chemical Engineering, Universidad de Salamanca, Salamanca, Spain
[2]Carnegie Mellon University, Pittsburgh, PA, United States

Back in 1996 the CEO of DOW stated that modeling was the single technology that had the largest impact the previous decade (Popoff, 1996). The use of modeling, simulation, and optimization has become a key tool for the design and selection of technologies in many fields of engineering from aerospace to manufacturing, as well as other areas from sports to medicine. When novel technologies are becoming of age, the analysis of alternatives prior to pilot scale, the systematic analysis of alternative conditions, materials, or fluids can save money and time to market a product or system. In this chapter, we present the basis for the systematic analysis of alternatives starting from the modeling approaches to the systematic analysis and design of technologies. For a systematic analysis of a large number of alternatives, modeling is a key step, but it relies on experimental information on the operation of the system. Thus, the link between experimental data and modeling is strong (Rasmuson et al., 2014).

2.1 Modeling approaches for process synthesis

A model is a simplification or abstraction of a real item. Even though the word "model" can be used to represent from model cars, boats, or airplanes, in this chapter we focus on mathematical models. The advantages that a model provides can be summarized in the fact that they clarify existing relationships between variables, allow a systematic evaluation and the possibility of performing experiments, computational experiments in case of mathematical models. It is here where the strength and the importance of good models rise toward process synthesis and analysis. Modeling is more an art than a science. Experience in modeling and in the operation of the process and the unit that is being modeled are key for the model to be representative. Another issue is the needs that the model is trying to cover. In addition, different models can be developed for the same unit or process depending on the future use.

It is possible to classify the models following different criteria. For the purpose of this book, we follow the same structure as in (Martín and Grossmann, 2012) by establishing mechanistic versus empirical models and within each group, different levels of detail.

2.1.1 Mechanistic models

They are based on first principles such as conservation laws, thermodynamics, kinetics. While some models can be somehow independent of experimental data, most of them require inputs such as a conversion, the equilibrium, or kinetic constants that are to be experimentally validated or obtained. The level of detail depends on the complexity of the unit and the purpose of the model. Within this type of models, we consider:

Short cut models: These models are based on simple mass and energy balances, thermodynamics and can be used to provide a first approximation of units such as heat exchangers, compressors, splitters, mixers, or even separation units including distillation, adsorption, or absorption columns (Douglas, 1988; Martin, 2016a,b). For example,

Power from a compressor: It can be derived from thermodynamic laws, Eq. 2.1 (Moran and Shapiro, 1993) and we leave it as an exercise (Houghen et al., 1959).

$$W = - \int_{P_{ini}}^{P_f} p dV$$
$$W = (F) \cdot \frac{R \cdot k \cdot (T)}{((MW) \cdot (k-1))} \frac{1}{\eta_c} \left(\left(\frac{P_f}{P_{ini}} \right)^{(k-1)/k} - 1 \right); \tag{2.1}$$

Chemical Equilibrium model for reaction systems: Many chemical reactions are limited by chemical equilibrium, Eq. 2.2, (Sandler, 2017):

$$CH_4 + H_2O \rightarrow CO + 3H_2$$

$$CO_{(g)} + H_2O_{(g)} \leftrightarrow CO_{2(g)} + H_{2(g)}$$

$$kp = 10^{[\frac{11650}{T} + 13,076]} = \frac{P_{CO} \cdot P_{H_2}^3}{P_{CH_4} \cdot P_{H_2O}}$$

$$kp = 10^{[\frac{1910}{T} - 1,784]} = \frac{P_{CO_2} \cdot P_{H_2}}{P_{CO} \cdot P_{H_2O}} \tag{2.2}$$

$n_{C,in} = n_{C,out}, n_{O,in} = n_{O,out}, n_{H,in} = n_{H,out}$

In addition, process constraints to avoid C deposition on the catalyst can be added.

Phase equilibrium models: From vapor-liquid equilibrium, typical in flash operations as well as in distillation or absorption columns (Douglas, 1988; Poling et al., 2000), see Eq. (2.3),

$$F = V + L$$
$$F \cdot z_i = V \cdot y_i + L \cdot x_i$$
$$y_i = K_i \cdot x_i \tag{2.3}$$

to gas-solid equilibrium, Eq. 2.4, that take place is adsorption columns such as pressure swing adsorption systems.

$$\begin{aligned}
&\textit{Henry's law: } q_{eq} = kC\\
&\textit{Freundlinch: } q_{eq} = kC^{1/n}, 1 < n < 5\\
&\textit{Lagnmuir: } q_{eq} = \frac{q_m KC}{1 + KC}
\end{aligned} \tag{2.4}$$

Advantages: Simple representation of the units based on first principles.

Disadvantages: Limitations in modeling nonideal behavior of the mixtures involved.

Detailed models: This subtitle is broad by default. Within this category it is possible to include different levels of detail as well. In addition, it is possible to consider different scales of analysis from microscopic to macroscopic scale such as in the analysis of multiphase units where the dispersed phase, drops, bubbles or particles, are analyzed in detail and that model is integrated in the model of the entire bubble column or particle dryer. In this subsection a number of examples are presented that can be found in reactor engineering books as well as transport phenomena ones.

Kinetic models: A kinetic model for any system of reactions represents the mechanism of the transformation (Fogler, 1997). Eq. (2.5) shows the production of glycerol ethers from glycerol and i-butene as a system of three reactions in chemical equilibrium where the target is the production of di and tri tert butyl glycerol (DTBG and TTBG) (Martín and Grossmann, 2013).

$$\begin{aligned}
Glycerol + iButene &\underset{k_{-1}}{\overset{k_1}{\rightleftarrows}} MTBG\\
MTBG + iButene &\underset{k_{-2}}{\overset{k_2}{\rightleftarrows}} DTBG\\
DTBG + iButene &\underset{k_{-3}}{\overset{k_3}{\rightleftarrows}} TTBG
\end{aligned}$$

$$\frac{dC_{Glycerol}}{dt} = -k_1 C_{Glycerol} C_{Ibutene} + k_{-1} C_{MTBG}$$

$$\frac{dC_{MTBG}}{dt} = k_1 C_{Glycerol} C_{Ibutene} - k_{-1} C_{MTBG} - k_2 C_{MTBG} C_{Ibutene} + k_{-2} C_{DTBG}$$

$$\frac{dC_{DTBG}}{dt} = k_2 C_{MTBG} C_{Ibutene} - k_{-2} C_{DTBG} - k_3 C_{DTGB} C_{Ibutene} + k_{-3} C_{TTBG} \tag{2.5}$$

$$\frac{dC_{TTBG}}{dt} = k_3 C_{DTGB} C_{Ibutene} - k_{-3} C_{TTBG}$$

$$\frac{dC_{Ibutene}}{dt} = -k_1 C_{Glycerol} C_{Ibutene} + k_{-1} C_{MTBG} - k_2 C_{MTBG} C_{Ibutene} + k_{-2} C_{DTBG}$$
$$- k_3 C_{DTGB} C_{Ibutene} + k_{-3} C_{TTBG}$$

Transport phenomena-based models: The chemical kinetics can be a part of a more complex mechanism that involves mass and energy transport limited stages (Bird et al., 2006) including adsorption/desorption stages. They are based on three pillars:

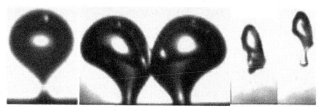

FIGURE 2.1   Transport phenomena in bubbles.

- Momentum transfer: Force balance (Newton law), Eq. 2.6.

$$\frac{d}{dt}\int\int\int_V \rho v\, dV = -\int\int_A (\rho v)(v \cdot n)dA + \int\int\int_V \rho g\, dV - \int\int_A pn\, dA + f\frac{\rho v^2}{2} + R \qquad (2.6)$$

- Mass transfer: Continuity equation, Eq. 2.7.

$$\frac{d}{dt}\int\int\int_V \rho\, dV = w_{i,in} - w_{i,out} + \int\int\int_V r_i\, dV \qquad (2.7)$$

- Heat transfer: Energy balance. Thermodynamic laws, Eq. 2.8.

$$\frac{d}{dt}\int\int\int_V \rho E\, dV = Q_{in} - \int\int_A \rho\left(H + \frac{v^2}{2} + gZ\right)(v \cdot n)dA + W_{shaft} - W_\mu \qquad (2.8)$$

Simple examples can be found in the laminar or axial flow in chemical reactors of the well-known shrinking core models just to mention a couple of typical examples.

Multiscale detailed models: They are based on transport phenomena and the physics of the system to represent units at industrial scale (Vlachos, 2005). Examples such as the computational fluid dynamics (CFD) and discrete element methods (DEM) models can be included in this category. Multiphase units are of particular interest where a discrete phase undergoes a number of phenomena within the general flow of a continuum phase. The analysis of the individual entities, such as particles, bubbles (Fig. 2.1), crystals is typically carried out including the phenomena that they are subjected to such as collisions, breakage, growing, and so on as well as mass and heat transfer from or to them. A population balance, Eq. 2.9, is the scheme used to analyze the discrete system (Himmelblau and Bischoff, 1968).

$$Accumulation = NetGeneration$$
$$\frac{d}{dt}\int_R \psi\, dR = \int_R (Birth - death)dR$$
$$dR = dx\,dy\,dz\,d\zeta_1 \ldots d\zeta_m$$
$$where$$
$$\zeta_i = \text{property of order i in entity } \psi \qquad (2.9)$$
$$\int_R = \int_x\int_y\int_z\int_{\zeta_1} \cdots \int_{\zeta_m}$$
$$resulting$$
$$\frac{\partial \mu_n}{\partial t} + \nabla(v\mu_n) - n\langle \zeta^{n-1} v_1 \rangle + \int \zeta^n (N - M)d\zeta = 0$$

However, the interaction with the continuum phase depends on the concentration of the discrete phase. Disperse systems are characterized by the negligible effect of the individual entities on the continuum phase. One-way coupling is enough to evaluate the effect of the continuum phase on the entities. Beyond certain concentration, two-way coupling between the continuum phase and the discrete phase is required because both affect each other. Finally, for dense systems four-way coupling, the effect between the entity and the continuum phase and the result of that is to be considered. Examples can be found in software books (Martín, 2019).

2.1.2 Empirical models

In this category we include different statistical based methods that have traditionally been used to predict the yield and operation of units and processes. These models rely on experimental data, and their lack of accuracy depends on the information available and the capability of the model to capture the relationship between the input and the output variables (Rasmuson et al., 2014). The better communication between the experimentalists and the modelers is, the more reliable and useful the model will be. For simplicity, several cases are presented.

Dimensionless correlations: These models are useful to capture and predict the operation of systems limited by mass and heat transfer (Martín et al., 2007; Branan, 2000), such as in gas-liquid-solid contact equipment or spray driers. They are based on first principles to determine the dimensionless numbers governing the particular phenomenon (Buckingham, 1914). The theorem states:

If there are n variables in a problem and these variables contain m primary dimensions (for example M, L, T) the equation relating all the variables will have (n-m) dimensionless groups.

Typical numbers such as Reynolds, Schmidt, Nusselt, Sherwood, Prandtl are well-known in chemical engineering (Perry and Green, 2018) for instance, Eq. 2.10:

$$Sh = k \cdot Re^{\alpha} Sc^{\beta} \tag{2.10}$$

Advantages: The models have physical meaning and scale-up issues are usually accounted for.

Disadvantages: The physics may not be fully captured if some of the variables are not included and the numerical issues that arise in optimization formulations.

Rules of thumb models: A number of common process units that have been operated in industry over the years such as carbon capture using MEA solutions, cooling towers, furnaces, molecular sieves, and so on have been experimentally studied at pilot plant and industrial scale, and typical operating values are available in the literature (Douglas, 1988; Baasel, 1989; Walas, 1990; Perry and Green, 2018; Branan, 2000; Towler and Sinnot, 2012). One particular example is the operation of CO_2 absorption columns. The regeneration process involves not only the desorption of a gas but also the breakage of bonds between the CO_2 and the alkaline liquid phase. Several studies allow estimating the duty of the different heat exchanger involved as well as the area (see Table 2.1).

Advantages: Models are based on experimental experience of the operation of the unit.

Disadvantages: Operational data are scarce. Their use is limited to the validated range.

TABLE 2.1   Rules of thumb for the design of the HX within amine absorption units (GPSA, 2012).

	Duty (BTU/h)
Reboiler	72,000 · F(gpm)
Rich-Lean amine	45,000 · F(gpm)
Amine cooler	15,000 · F(gpm)
Condenser	30,000 · F(gpm)

Empirical correlations: Sometimes the experimental studies are complete enough to present profiles of the output variables as a function of a few input operating variables. Parameter estimation is used to develop simple correlations out of the literature profiles. They can be used to compute the yield of a unit. For instance, Phillips et al. (2007) and Eggeman (2005) developed correlations for the composition of the gas resulting from the gasification of biomass as function of the operating temperature, or to produce models to estimate thermodynamic properties. As an example, we present here the correlations developed for the enthalpy and entropy of steam as a function of the pressure and temperature Eqs. (2.11) & (2.12) (León and Martín, 2016).

Superheated steam (Up to 10 bar)

$$H(kJ/kg) = \left(-6.3293 \cdot 10^{-6} \cdot (P(bar)) + 3.3179 \cdot 10^{-4}\right) \cdot (T)^2 + (0.0124 \cdot (P(bar)) + 1.8039)T + (-6.0707(P(bar)) + 2504.6)$$

(2.11)

$$S\left(\frac{kJ}{(kg\ K)}\right) = 9.42 \cdot 10^{-10}(T)^3 - 3.09 \cdot 10^{-6}(T)^2 + 5.24 \cdot 10^{-3} \cdot (T) + \left(6.8171 \cdot (P(bar))^{(-0.069455)}\right)$$

(2.12)

Advantages: The models involve physical meaning and experimental background.

Disadvantages: They are only valid for the range of experimental conditions. Sometimes the relationship between variables may not be easy to find when several variables are involved and, for optimization purposes, the models may present problems due to the mathematical functions used. It is here where Machine Learning can play a role to identify the mathematical terms that best fit the experimental data (i.e. ALAMO, Sahinidis (2016)).

2.1.2.1 Machine Learning-based models

They are based on statistics and within this category we consider the design of experiments, the artificial neural networks, and the Kriging modeling (Cozad et al., 2014; Venkatasubramanian, 2019).

Factorial design of experiments (DOE): It is typically used to study equipment whose performance is difficult to model using first principles due to the lack of physical understanding of the interaction of the different variables. DOE is a suitable technique (Montgomery,

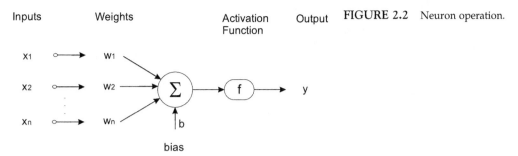

FIGURE 2.2   Neuron operation.

2001) for developing simplified models to evaluate the output of the process as function of a number of input variables. They typically are models of the form given by Eq. (2.13) where β are the adjustable parameters and apart from the effect of the individual variables, the second term in the right-hand side of the equation, the interactions between and among variables, third term in the right-hand side, are evaluated.

$$f(x) = \beta_0 + \sum_{i=1}^{n} \beta_i x_i + \sum_{i=1}^{n} \sum_{j \leq i}^{n} \beta_{ij} x_i x_j \tag{2.13}$$

For instance, the yield of the transesterification of oil is typically a function of the operating pressure and temperature, catalysis load, and alcohol ratio to oil and time. Simple models can be obtained using experimental data from the literature (Martín and Grossmann, 2012).

Artificial Neural Networks (ANN): In this methodology, the model tries to imitate the behavior of human neural networks in the brain (Hopfield, 1988). The input signal (x_i) from each previous neuron is weighted using different weight value (w_i). After the summation, the signal is sent to a transfer function f(x). This function is a sigmoid, see Fig. 2.2. The neuron works as follows, Eq. (2.14):

$$\begin{aligned} & \sum_i x_i w_1 + b \\ xw &= f\left(\sum_i x_i w_1 + b\right) \\ f(x) &= \frac{1}{1 + e^{-xw}} \end{aligned} \tag{2.14}$$

A collection of neuron nodes connected between them constitutes the neural network. The neurons are organized in layers. To establish the value of the weight for each neuron node, a training of the neural network is carried out using a collection of data (training set). It has been used in the context of process optimization for commercial modular packages (Henao and Maravelias, 2011).

Kriging modeling: This technique is originated in areas such as mining and geostatistics that involve spatially and temporally correlated data (Caballero and Grossmann, 2008). The idea is to interpolate a particular point considering the closest data available in the neighborhood assuming a linear correlation. The contribution of the different data points is related to their proximity. Thus, the spatial interpolation estimates the relationship between variables as in Eq. 2.15

$$y(x) = \sum_i w_i y_i \tag{2.15}$$

where the weights can be computed considering the importance of each data inversely to the distance to the one to be estimated, Eq. 2.16.

$$1 = \sum_i w_i$$

$$w_i = \frac{\dfrac{1}{d_{ik}}}{\sum\limits_{k=1}^{n} \dfrac{1}{d_{ik}}} \tag{2.16}$$

$$d_{ik} = \sqrt{\sum (x_i - x_k)^2}$$

However, the spatial variation is typically too irregular to estimate the values based on this approach. The interpolation based on geostatistics is known as Kriging. To account for that, the estimation is composed of two terms, the first one is a linear model (for example a polynomial) and the second one represents the fluctuations on the mean of the data set, Eq. 2.17.

$$y(x) = f(x) + Z(x) \tag{2.17}$$

where $Z(x)$ is a stochastic Gaussian process with a mean value of 0 and a covariance given by Eq. (2.18):

$$cov(Z(w), Z(x)) = \sigma^2 R(w, x) \tag{2.18}$$

where σ^2 is the process variance and $R(w, x)$ is a correlation function. One of the most popular structures for this correlation takes the following form (Eq. (2.19)):

$$R(w, x) = \exp\left(-\sum_{j=1}^{n} \theta_j \left(w_j - x_j\right)^{P_j}\right) \tag{2.19}$$

In process design, the first use was presented by Caballero and Grossmann (2008).

Advantages: Black box type of models. They allow the systematic study of the effect of a large number of variables.

Disadvantages: They can only predict within the range they have been obtained for. The models are subjected to scale up problems. Tight bounds for the variables are essential.

2.2 Process simulation

Once the process has been modeled unit by unit, the next step is to simulate the entire system. Two different approaches are used, equation based and sequential modular.

2.2.1 Equation-based simulation

The first one consists of formulating a problem including of all the equations involved in modeling each of the units. In case some of the units cannot be easily represented by simple equations, such as CFD, DEM models, surrogate models obtained based on DOE, ANN, or Kriging can be used. The simulation consists of solving such a system of equations. The type of equations determines the system to solve, a linear system, nonlinear system, system of ordinary differential equations (ODE), and differential algebraic system of equations (DEA). More specialized books can be found to address the solution (Westerberg et al., 1979; Kincaid and Cheney, 1991; Bielger, 2010). Here we just briefly comment some issues and solution approaches that will be useful for further reference. Additional details can be found in these references.

2.2.1.1 Linear problems

They are of the form $A \cdot x = b$ and can represent linear mass balances or the design of an absorption column assuming Henry's law. The typical methods for addressing these problems are presented in the following table (Kincaid and Cheney, 1991; Gentle, 2010).

Method	Structure	Solution
LU	$A = LU$	$Ly = b$
	L: Lower triangular	$Ux = y$
	U: Upper triangular	
Gaussian elimination	$A'x = b$	Direct substitution
	A' is an upper triangular obtained by operating rows and columns.	$a'_{(n,n)} = b_n$
	Note: large or small coefficients are responsible for numerical issues.	
QR	$A = QR$	$x = A^{-1}b = R^{-1}Q^T b$
	Q: Orthonormal ($Q^T = Q^{-1}$)	
	R: Upper triangular	
	Note: it is more stable for ill conditioned problems	

2.2.1.2 Nonlinear problems

Most problems in chemical engineering belong to this type to include detailed thermodynamic models that accurately represent the performance of the units. However, finding solutions to the problems is difficult due to numerical issues (Ortega and Rheinboldt, 1970). Here we present some of the difficulties and the attempts to address them. Note that it is easier to solve a problem with a physical meaning since the initialization and the bounds of the variables are easier to assign. We distinguish two main methods:

2.2.1.2.1 Methods based on divisions

Bolzano theorem states that if $sig(f(a)) \neq sig(f(b))$ and the function is continuous there exists $f(c) = 0$, $c \in [a,b]$.

It is possible to iterate to find a solution.

2.2.1.2.2 Methods based on fixed point

2.2.1.2.2.1 Newton method The function is approximated by a Taylor series around x_o, Eq. 2.20

$$0 = f(x) = f(x_o) + f'(x_o)(x_1 - x_o) + HOT$$
$$x_1 = x_o - \frac{f(x_o)}{f'(x_o)} \tag{2.20}$$

HOT stands for high order Taylor. By rearranging it becomes Eq. 2.21

$$0 = f(x) \approx f(x_o) + f'(x_o)(x_1 - x_o) + HOT$$
$$x^{t+1} \approx x^t - J^{-1}(x^t)f(x^t) \tag{2.21}$$

Instead of inverting a matrix we solve Eq. 2.22:

$$-f(x^t) \approx J(x^t)(x^{t+1} - x^t)$$
$$-f(x^t) \approx J(x^t)\Delta x \Rightarrow A\Delta x = b \tag{2.22}$$

We end up solving a linear system of equations, Eq. 2.23:

$$x^{t+1} = x^t + \Delta x \tag{2.23}$$

It is here where not only the algorithms presented earlier, but also the numerical issues of the algorithm play a role. Apart from these issues, the initial point is also of paramount importance to find a solution. A particular situation that can be encountered is given by Fig. 2.3 where iteration after iteration we cycle and never converge to the solution.

To avoid this problem there are a number of techniques that are implemented (Sioshansi and Conejo, 2017; Nocedal and Wright, 2006):

A. Line search: We control the step size to improve the performance, Eq. 2.24.

$$x^{t+1} = x^t + \alpha p^k \tag{2.24}$$

There are three conditions to modulate the step:

- Armijo: The step size is selected to provide sufficient decrease.
- Wolfe: It controls curvature so that the slope aims at a sufficient reduction of the function.
- Armijo-Goldstein: Both, step size and sufficient decrease are enforced.

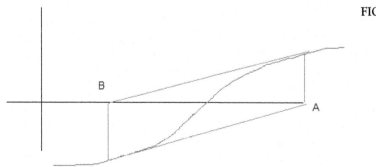

FIGURE 2.3 Cycled iteration.

FIGURE 2.4 Levenberg-Marquardt method.

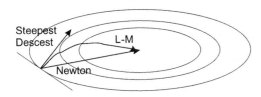

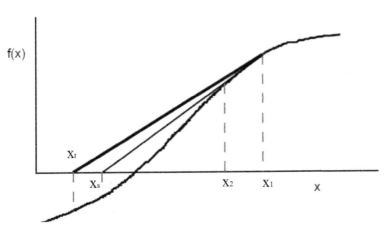

FIGURE 2.5 Broyden method. x_t: Tangent; x_s: Secant.

B. Trust region: We operate within a region where a quadratic approximation can be trusted. Within this region we compute the Newton step. Two approaches can be used to solve the problem that led to compute the step:

Exact Solution: Levenberg Marquardt. Combines the steepest descent with Newton's generating the Levenberg Marquardt progress curve, see Fig. 2.4.

Approximation: Dog Leg. The step is the linear combination between the Newton and the Cauchy steps.

C. Broyden method: It is a quasi-Newton method that instead of using the Jacobian (J) approximates it by a matrix B that corresponds to a secant in the one-dimensional case, Eq. 2.25, see Fig. 2.5, (Broyden, 1965).

$$-f\left(x^t\right) \approx \left(B^t\right)^{-1}\left(x^t\right)\left(x^{t+1} - x^t\right) \tag{2.25}$$

The basic idea of the method relies on the update of B so that in the next step it is as similar at the current one as possible. It is possible to achieve this by solving Eq. 2.26

$$
\begin{aligned}
&Min\left\|B^{t+1} - B^t\right\|_F \\
&s.t. \\
&B^{t+1}s = y \\
&where \\
&y = f\left(x^{t+1}\right) - f\left(x^t\right) \\
&s = x^{t+1} - x^t \\
&\|B\|_F = \left(\sum_i \sum_j B_{ij}^2\right)^{0.5}
\end{aligned}
\tag{2.26}
$$

2.2.1.3 Ordinary differential equations

Dynamics is an important part of process design, in particular the design of units such as chemical reactors that rely on the time it takes to achieve a certain conversion. In addition, process control is also a dynamic problem (Dorfman and Daoutidis, 2017). A two-level classification is presented, single versus multiple step and explicit versus implicit methods. It is important to evaluate the problem to address and decide between a simpler method, easier to implement and faster, and a more complex one depending on the properties of the dynamic processes at hand.

2.2.1.3.1 Single step

2.2.1.3.1.1 Explicit   *Euler:* It is based on a Taylor series expansion of the differential equation, Eq. 2.27.

$$\frac{dx}{dt} = f(x)$$

$$x(t + h) = x(t) + h\frac{dx(t)}{dt} \tag{2.27}$$

The error is due to truncation of the Taylor expansion and the error propagation. For instance, Eq. 2.28

$$\frac{dx}{dt} = -Kx \tag{2.28}$$

The approximation becomes Eq. 2.29

$$x_{i+1} = x_i - hKx_i = (1 - hK)x_i \tag{2.29}$$

Different cases arise, Eq. 2.30:

$$\begin{aligned}
|1 - hK| < 1 &\Rightarrow x_{i+1} = (1 - hK)x_i < x_i \Rightarrow x_\infty \to 0 \\
|1 - hK| > 1 &\Rightarrow x_{i+1} = (1 - hK)x_i > x_i \Rightarrow x_\infty \to \infty \\
|1 - hK| = 1 &\Rightarrow -hK = -2 \Rightarrow h = \frac{2}{K}
\end{aligned} \tag{2.30}$$

In Matlab the code is *eulers* ("function,"x_o,x_f,y_o, # steps)

Runge Kutta: It is not a method per se but a family of methods that can be both explicit or implicit, Eq. 2.31

$$x(t + h) = x(t) + h\phi_e(x(t))$$

$$x_{i+1} = x_i + \sum_{j=1}^{r} w_j k_j$$

$$k_j = hf\left(t_i + c_j h, x_i + \sum_{q=1}^{m} a_{jq} k_j\right) \tag{2.31}$$

w_j, c_j, and a_{jq} are constants. The method can be explicit ($m \leq j-1$), semi implicit ($m = j$), or implicit ($m = r$). The development of the methods follows a Taylor expansion. Let see a 2nd order for simplicity, Eq. 2.32

$$x_{i+1} = x_i + \sum_{j=1}^{r} w_j k_j$$

$$k_j = hf\left(t_i + c_j h, x_i + \sum_{q=1}^{m} a_{jq} k_j\right)$$

(2.32)

To obtain the constants a Taylor expansion is developed, Eq. 2.33:

$$x_{i+1} = x_i + x_i' h + x_i'' \frac{h^2}{2} + O(h^3)$$

$$x_i' = f(t_i, x_i)$$
$$x_i'' = \frac{df(t_i, x_i)}{dt} = \frac{\partial f}{\partial t} + \frac{\partial f}{\partial x} f_i$$

$$x_i''' = \frac{d^2 f(t_i, x_i)}{dt^2} = \frac{d}{dt}\left(\frac{\partial f}{\partial t} + \frac{\partial f}{\partial x} f_i\right)$$

(2.33)

$$x_{i+1} = x_i + (f(t_i, x_i))h + \frac{h^2}{2}\left(\frac{\partial f}{\partial t} + \frac{\partial f}{\partial x} f_i\right) + O(h^3)$$

For the 2nd order example, Eq. 2.34

$$x_{i+1} = x_i + w_1 k_1 + w_2 k_2$$
$$k_1 = hf(t_i, x_i), c_1 = 0; a_{11} = 0;$$
$$k_2 = hf(t_i + c_2 h, x_i + a_{21} k_1)$$
$$x_{i+1} = x_i + w_1(hf(t_i, x_i)) + w_2(hf(t_i + c_2 h, x_i + a_{21} k_1))$$

(2.34)

where, Eq. 2.35

$$f(t_i + h, x_i + k_1) = \sum_{i=0}^{n} \frac{1}{i!}\left(h\frac{\partial}{\partial x} + k\frac{\partial}{\partial y}\right)^i f(a, b)$$

$$\left(h\frac{\partial}{\partial x} + k\frac{\partial}{\partial y}\right)^0 f(a, b) = f(a, b)$$

$$\left(h\frac{\partial}{\partial x} + k\frac{\partial}{\partial y}\right)^1 f(a, b) = \left(h\frac{\partial f}{\partial x} + k\frac{\partial f}{\partial y}\right)(a, b)$$

(2.35)

$$\left(h\frac{\partial}{\partial x} + k\frac{\partial}{\partial y}\right)^2 f(a, b) = \left(h^2\frac{\partial^2 f}{\partial x^2} + 2hk\frac{\partial^2 f}{\partial x \partial y} + k^2\frac{\partial^2 f}{\partial y^2}\right)(a, b)$$

Resulting in Eq. 2.36

$$x_{i+1} = x_i + w_1(f(t_i, x_i))h + w_2\left(f(t_i, x_i) + \frac{\partial f}{\partial t} c_2 h + \frac{\partial f}{\partial x} a_{21} f_i h\right)h$$

(2.36)

2. Design principles

To compute the constants, w_1, w_2, c_2, a_{21}, this equation is compared with the Taylor expansion, Eq. 2.37

$$w_1 + w_2 = 1;$$
$$\frac{\partial f}{\partial t} h^2 \Rightarrow \frac{1}{2} = c_2 w_2$$
$$\frac{\partial f}{\partial x} f_i h^2 \Rightarrow \frac{1}{2} = a_{21} w_2$$

(2.37)

In Matlab, ode23 and ode45 correspond to the 2nd and 4th order RK with errors of 3rd and 5th order, respectively.

2.2.1.3.1.2 Implicit Implicit methods improve stability, but the iteration is computationally more expensive since a nonlinear problem is to be solved:

Euler: The formulation is given by Eq. 2.38

$$x(t + h) = x(t) + h\phi_i(x(t + h))$$

(2.38)

Starting at x_o, a nonlinear equation/system of equations is solved every time step, Eq. 2.39.

$$x(t_{i+1}) - x(t_i) - hf_i(x(t_{i+1})) = 0$$
$$x(t_o) = x_o$$

(2.39)

Example, Eq. 2.40:

$$\frac{dx}{dt} = -Kx \Rightarrow x_{i+1} = x_i - hKx_{i+1} \Rightarrow x_{i+1} = \frac{x_i}{1 + hK}$$

(2.40)

Runge Kutta: For a semi-implicit method, $m = j$, for an implicit one, $m = r$, Eq. 2.41

$$x_{i+1} = x_i + \sum_{j=1}^{r} w_j k_j$$
$$k_j = hf\left(t_i + c_j h, x_i + \sum_{q=1}^{m} a_{jq} k_j\right)$$

(2.41)

Rosenbrock: Stiffness is an interesting concept for ordinary differential equations (Marquardt, 1995). It takes place when the solution shows slow changing sections as well as rapidly changing ones. Non stiff methods can solve stiff problems, but they take longer. It is necessary to adjust the iteration step in that region. These problems can be found in chemical reaction systems where some species reach equilibrium real quick while others show slower kinetics, stiff. Explicit methods typically have problems with these cases. Rosenbrock method (i.e., ode23s in Matlab) is a stiff explicit method to deal with these problems.

2.2.1.3.2 Multistep methods

They can be in general be represented as follows, Eq. 2.42:

$$x_{i+1} = \sum_{j=0}^{k} \alpha_j x_{i-j} + h \sum_{j=-1}^{k} b_j f_{i-j}$$

(2.42)

where for b equal to 0 explicit methods are formulated. The main feature is that they need a number of previous points to compute the next iteration. These methods are appropriate for stiff problems. Here only two methods are discussed.

Adams:

The method or family of methods is derived from the fundamental theorem of Calculus (Bashforth and Adams, 1883) Eq. 2.43.

$$x_{k+1}^0 = x_k + \int_{t_k}^{t_{k+1}} f[t, x(t)] \tag{2.43}$$

The Adams-Bashforth-Moulton is a predictor corrector method that requires four points to compute the next one. It is typically initialized using RK.

- It uses an explicit formula to predict (Adams-Bashford)

At this stage the solution is fitted to a cubic polynomial using as data points (t_{k-3}, f_{k-3}), (t_{k-2}, f_{k-2}), (t_{k-1}, f_{k-1}) and (t_k, f_k). The polynomial is integrated in the interval (t_k, t_{k+1}), Eq. 2.44.

$$P_{k+1} = x_k + \frac{\Delta t}{24} \left(55 f_k - 59 f_{k-1} + 37 f_{k-2} - 9 f_{k-3}\right) \tag{2.44}$$

- It uses the implicit formula to correct (Adams−Moulton)

A cubic polynomial is fitted using (t_{k-2}, f_{k-2}), (t_{k-1}, f_{k-1}), (t_k, f_k), (t_{k+1}, f_{k+1}) where Eq. 2.45

$$f_{k+1} = f(t_{k+1}, P_{k+1}) \tag{2.45}$$

Integrating the polynomial in the interval (t_k, t_{k+1}), the corrector term Adams−Moulton becomes, Eq. 2.46:

$$x_{k+1} = x_k + \frac{\Delta t}{24} \left(9 f_{k+1} + 19 f_k - 5 f_{k-1} + f_{k-2}\right) \tag{2.46}$$

In Matlab this method uses the function ode113. It is therefore a variable order and variable step size method.

Gear (Gear's Backward difference formulas):

The general formulation is given by Eq. 2.47:

$$\sum_{j=0}^{k} \alpha_j x_{i+1-j} = h b_{-1} f_{i+1} \tag{2.47}$$

where coefficients b are 0 but for b_{-1}, m $= k-1$. They typically include a predictor to provide an initial point. Variable step size is used to reduce the computational cost. They require Newton's method to solve the iteration (Gear, 1971). In Matlab the method is ode15s.

EXAMPLE 2.1 Develop the 4th order RK method and code it in Matlab.

Solution

$$x_{i+1} = x_i + \frac{1}{6}(k_1 + 2k_2 + 2k_3 + k_4)$$

$$k_1 = f(t_i, x_i)h$$
$$k_2 = hf\left(t_i + \frac{h}{2}, x_i + \frac{k_1}{2}\right)$$
$$k_3 = hf\left(t_i + \frac{h}{2}, x_i + \frac{k_2}{2}\right)$$
$$k_4 = hf\left(t_i + \frac{h}{2}, x_i + k_3\right)$$

(2E1.1)

$$y_{k+1} = y_k + \frac{h}{6}\left(f_1 + 2f_2 + 2f_3 + f_4\right)$$

$$f_1 = f(t_k, y_k),$$
$$f_2 = f\left(t_k + \frac{h}{2}, y_k + \frac{h}{2}f_1\right),$$
$$f_3 = f\left(t_k + \frac{h}{2}, y_k + \frac{h}{2}f_2\right)$$
$$f_4 = f\left(t_k + h, y_k + hf_3\right)$$

(2E1.2)

```
function y1 = f1(x)
y1 = x(1)*3 + 2;
xo = 0;
xf = 10;
Pasos = 100;
h = (xf-xo)/steps;
n = 1;
y = [0];
eqs = n;
x(1) = xo;
y(1) = yo
for j = 1:steps;
        x1 = [x(j),y(j)];
        x2 = [x(j) + 0.5*h,y(j) + 0.5*f1(x1)];
        x3 = [x(j) + 0.5*h,y(j) + 0.5*f1(x2)];
        x4 = [x(j) + h,y(j) + 0.5*h*f1(x3)];
        y(j + 1) = y(j) + h*((1/6)*f1(x1) + (1/3)*f1(x2) + (1/3)*f(x3) + (1/6)*f1(x4));
            x(j + 1) = x(j) + h;
end
```

In summary

Explicit	Semiimplicit	Implicit
Computationally expensive for high order	High order	High order
They have stability limit	There is no stability limit	There is no stability limit
Easy to implement	Reasonable easy to implement	Harder to implement

2.2.1.4 Differential algebraic system of equations

Most of the problems involve simultaneous algebraic and differential equations (Pantelides et al., 1988). A process typically involves reactors and separation stages and while some of them can be modeled in steady state, to estimate the yield of reactors sometimes it is necessary to include the kinetics of the transformation. There are different approaches to deal with these problems:

2.2.1.4.1 Nested approach

At each time point the algebraic equations are solved

$$g(x_n, z_n) = 0 \rightarrow z_n(x_n)$$

Next, using an ODE method, Eq. 2.48.

$$x_{n+1} = \Phi(x_n, z_n(x_n), t_n) \tag{2.48}$$

2.2.1.4.2 Simultaneous

It consists of solving simultaneously the algebraic equations with the discretized differential ones. Commercial packages (i.e., gPROMS) typically use the Backward differentiation formulae (BDF, also: Gear's method) to represent the differential equations. The system becomes Eq. 2.49:

$$x_{n+1} = h\beta_{-1}f(x_{n+1}, z_{n+1}, t_n) + \sum_{j=0,k} a_j x_{n-j}$$
$$g(x_{n+1}, z_{n+1}, t_{n+1}) = 0 \tag{2.49}$$

Within this category there are special kinds of problems denoted as Boundary value problems (BVP) (Dorfman and Daoutidis, 2017), whose solution is relevant for chemical units operation, shooting methods and finite difference methods. The idea is that we know the solution to be achieved, i.e. the conversion of a reaction, but the initial conditions are to be computed or defined.

2.2.1.4.3 Shooting methods

The basic idea is to guess an initial condition and run an ODE simulation. If the target is achieved, convergence is achieved, else we guess another initial condition. To illustrate this, we consider a plug flow reactor where a first order kinetics reaction is carried out, see Example 2.2.

EXAMPLE 2.2 A plug flow reactor performs a reaction that follows first order kinetics, Eq. 2E2.1. The flow across is constant. The measured concentration at the exit is 0.05 M, but the initial concentration is unknown. The length of the reaction is 0.35 m and we assumed space velocity $v = 1$ m/s where the reaction rate $k = 1 \text{ s}^{-1}$.

$$v_z \frac{dc_A}{dz} = -kc_A \qquad (2E2.1)$$

Solution

Instead of trial and error we can write a simple code in Matlab to solve this problem as follows

```
function dc = tubular(t,c)
% Tubular es un reactor tubular con una cinética de primer orden
v = 1;
k = 10;
dc = -k/v*c
function error = funerror(ci)
cL = 0.05;
[z,cz] = ode23('tubular',[0 0.35], ci);
cLp = cz(max(size(cz)),1);
error = cL-cLp;
cOn = fzero('funerror',1)
```

2.2.1.4.4 Finite differences

The idea is to discretize the independent variable domain. We can have the boundary conditions,

$y(0) = \alpha - >$ Dirichlet conditions (Boundary values)
$y(1) = \beta - >$ Neumann conditions (Values of the derivative in the boundary)
Using a Taylor expansion, the derivatives are approximated by finite differences (Eq. 2.50).

$$y(t_{i+1}) = y(t_i) + y'(t_i)h + y''(t_i)\frac{h^2}{2} + y'''(t_i)\frac{h^3}{6} + y^{iv}(t_i)\frac{h^4}{24} + \dots$$

$$y(t_{i-1}) = y(t_i) - y'(t_i)h + y''(t_i)\frac{h^2}{2} - y'''(t_i)\frac{h^3}{6} + y^{iv}(t_i)\frac{h^4}{24} - \dots$$

(2.50)

The first derivative is given by Eq. (2.51)

$$\frac{y(t_{i+1}) - y(t_{i-1})}{h} = y'(t_i) + y'''(t_i)\frac{h^2}{6} + \dots \tag{2.51}$$

The second derivative is as follows, Eq. 2.52:

$$\frac{y(t_{i+1}) - 2y(t_i) + y(t_{i-1})}{h^2} = y''(t_i) + y^v(t_i)\frac{h^2}{12} + \dots \tag{2.52}$$

We substitute them as a function of the y_i to formulate the problem, Eq. 2.53

$$G(y(t_i), y(t_{i+1}), y(t_{i-1}), t_i) = 0 \tag{2.53}$$

It corresponds to a di or tridiagonal matrix, a matrix that has nonzero elements in the diagonal and ones above and/or below it, see Example 2.3.

EXAMPLE 2.3 A plug flow reactor performs a reaction that follows first order kinetics, Eq. 2E3.1. The flow across is constant. The measured concentration at the exit is 0.05 M, but the initial concentration is unknown. The length of the reactor is 0.35 m, and we assume space velocity $v = 1$ m/s where the reaction rate $k = 1\ s^{-1}$.

Solution

$$v_z\frac{dc_A}{dz} = -kc_A \tag{2E3.1}$$

$$v_z\frac{dc_A}{dz} = -kc_A \Rightarrow v\frac{c_i - c_{i-1}}{h} \approx -kc_i \tag{2E3.2}$$

which can be written as

$$Ax = b \tag{2E3.3}$$

Fig. 2E3.1

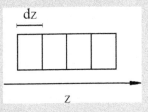

FIGURE 2E3.1   Discretization of the reactor.

$$\frac{1}{h}\begin{bmatrix} 1 & & & \\ -1 & 1 & & \\ & -1 & 1 & \\ & & -1 & 1 \end{bmatrix}\begin{bmatrix} c_1 \\ c_2 \\ c_3 \\ c_4 \end{bmatrix} + \frac{k}{v}\begin{bmatrix} 1 & & & \\ & 1 & & \\ & & 1 & \\ & & & 1 \end{bmatrix}\begin{bmatrix} c_1 \\ c_2 \\ c_3 \\ c_4 \end{bmatrix} = \frac{1}{h}\begin{bmatrix} c_o \\ 0 \\ 0 \\ 0 \end{bmatrix}$$

$$
\begin{aligned}
\frac{1}{h}(c_{A_1} - c_o) + \frac{k}{v}c_{A_1} &= 0 \\
\frac{1}{h}(c_{A_2} - c_{A_1}) + \frac{k}{v}c_{A_2} &= 0 \\
\frac{1}{h}(c_{A_3} - c_{A_2}) + \frac{k}{v}c_{A_3} &= 0 \\
\frac{1}{h}(c_{A_4} - c_{A_3}) + \frac{k}{v}c_{A_4} &= 0
\end{aligned}
\quad \rightarrow \quad
\begin{bmatrix}
1+\frac{hk}{v} & 0 & 0 & 0 \\
-1 & 1+\frac{hk}{v} & & \\
& -1 & 1+\frac{hk}{v} & \\
& & -1 & 1+\frac{hk}{v}
\end{bmatrix}
\begin{bmatrix} c_{A_1} \\ c_{A_2} \\ c_{A_3} \\ c_{A_4} \end{bmatrix}
=
\begin{bmatrix} c_0 \\ 0 \\ 0 \\ 0 \end{bmatrix}
\qquad (2E3.4)
$$

It can be seen that, in the end, a linear system of equations is solved requiring of the methods discussed above such as LU, QR, etc.

2.2.2 Sequential modular-based simulation

Sequential modular process simulation (Westerberg et al., 1979) started in industry by developing detailed models for particular equipment such as distillation columns. As more equipment were available, a process flowsheet is based on connecting the units one after the other (Martín, 2019). Thus, the name sequential modular comes from the fact that each of the units is a closed module where the equations of thermodynamics, the kinetics and chemical and phase equilibrium involving the species are solved. Furthermore, the modules are executed sequentially, and iteratively within recycle loops. The properties, that is, density, enthalpy, and so on, are computed by calling the appropriate thermodynamic package. The major advantage of this approach is the robust solutions that are obtained and validated over the years and with industrial input. This is the case for commercial software such as CHEMCAD (http://www.chemstations.com; https://www. chemstations.com/CHEMCAD-NXT ? gclid = Cj0KCQjwo-aCBhC -ARIsAAkNQiszfShgo1C MzmKeNzjqebf8evo4nyKShdVptS6I4ZFR00mHZrMPgqkaAogSEALw_wcB) or ASPEN Plus (http://www.aspentech.com; https://www.aspentech.com/en/products/engineering/aspen-plus). Further details can be found in Westerberg et al. (1979). Sometimes the drawback is how to address the modeling of novel technologies. The best performance for this software is the sequential solution of the flowsheet since the models for each unit are very efficient and connecting them is easy. However, most chemical processes involve recycles aiming at the efficient use of raw materials and energy. Dealing with recycles involves tearing of the streams in order to perform the calculations sequentially. The selection of the tearing streams requires the identification of loops (Westerberg et al., 1979; Westerberg and Piela, 1994). Different strategies have been developed to identify the tearing streams. Among them, two of the best well known are:

Minimum number of tear streams. Barkeley and Motard (1972). The algorithm is based on graph theory and allocates the streams as vertices and the units as edges. The loops are identified by evaluating the inlets that generate each of the outlets in a unit. By successive substitution of the outlet stream by the inlet in the cases, the unit has one inlet only, and the loops are found.

All recycles are teared a minimum number of times. Upadhye and Grens (1975). The algorithm identifies the loops by inspection of the flowsheet. By following the flow until a unit is repeated the units and the streams involved in each loop are identified. Once the loops are defined, a number of streams that allow tearing the loops efficiently are identified using the matrix of incidences that collects the information of the loops and the streams involved in each one.

Thus, the simultaneous solution of sequential modular based process simulations that address the recycling has been performed over four consecutive steps (Biegler, 1983).

A. Development of strategies for stream tearing: Among them different algorithms have been developed to minimize the number of tear streams (Barkeley and Motard, 1972), the minimum number of variables in the stream, aiming at the fact that all recycles are teared a minimum number of times (Upadhye and Grens, 1975) and aiming at convergence. These problems can be formulated as a set covering problem to decide which stream j to tear, Eq. 2.54:

$$Min \sum_{j=1}^{n} w_j y_j$$
$$s.t \sum_{j=1}^{n} a_{ij} y_j \geq 1 \tag{2.54}$$
$$i = 1, L$$
$$y_j = \{0, 1\}$$

where $w_j = 1$ for Barkeley and Motard (1972), $w_j = n_j$ to minimize the number of variables involved in a stream, and $w_j = \sum_i a_{ij}$, j corresponds to the streams and i to the loops. for Upadhye and Grens (1975).

B. Defining the recycle as a convergence variable and a design problem to solve.
C. The use of linear approximation of the units to help convergence.
D. Development on nonlinear models for the units to help solve the recycle design problem.

In the literature processes modeled using CHEMCAD or ASPEN are plentiful as conceptual analysis for technoeconomic analysis. The use of these packages is reported in Martín (2019).

2.3 Process integration and optimization

Simulation is an important stage toward the analysis of a process. However, the tight margins of benefit, the large number of alternative technologies that are being developed in particular for novel processes and the processing of renewable resources results in the need for advanced tools for systematic screening and comparison prior to installing pilot plants. It this at this stage where process simulators are not good enough, evaluating thousands of flowsheets and infinite number of sets of operating conditions is not feasible using simulation alone and optimization techniques play an important role. While modular simulators provide optimization capabilities, they are typically time consuming and can be considered for a second stage analysis once the most promising technologies and range of operating conditions are identified. For instance, the process optimization for the production of solar grade silicon takes 20 h of CPU time per variable using a differential evolution approach with tabu list (DETL) based optimization (Ramírez-Márquez et al., 2018). In addition, novel technologies are underrepresented in commercial modular process simulators while equation-based models are a flexible and powerful tool for their analysis. In this section the different optimization problems are presented and basic notes on the solution approaches are described. We leave more detailed analysis of the optimization strategies for specialized books and reports (Biegler et al., 1997; Bielger, 2010; Grossmann, 2021).

In a second stage the use of optimization toward basic integration of heat, water, and residues is discussed since it provides the background to understand the processes presented along this book.

2.3.1 Optimization problems

This section covers the four basic optimization problems per type of variables and equations involved including linear programming (LP), mixed integer linear programming (MILP), nonlinear programming (NLP) and mixed integer nonlinear programming (MINLP). In addition, very briefly we comment on dynamic optimization and global optimization for reference. We understand that complete books are available for each one of the topics, and it is not our intention to be a reference on them, but to present them for their use in process design (Grossmann, 2021).

2.3.1.1 Linear programming problems

Back in 1986 in an interview to George B Dantzig, the father of linear programming, he made the remark "Those in charge often do a hand-wave and say, 'I've considered all the alternatives,' but this is so much garbage. They couldn't possibly look at all possible combinations" (Albers and Reid, 1986). By planning the flights of airplanes and scheduling crews in the WWII the linear programming was developed.

Linear programs are of the form, Eq. 2.55:

$$\min Z = \sum_i c_i x_i = c^t x$$
$$s.t.$$
$$Ax \le b \qquad (2.55)$$
$$x \ge 0$$

where all the equations consist of linear products of coefficients and continuous variables. A feasible solution is the one that meets the constraints. The geometric interpretation of the problem consists of a convex, closed, region whose boundaries are the constraints. The mathematical definition for a convex region convex is as follows, Eq. 2.56:

$$x = \alpha x_1 + (1 - \alpha)x_2 \quad \in F \quad \forall \alpha \in [0, 1] \tag{2.56}$$

any point can be computed as a linear combination of two other points. A function is convex if, Eq. 2.57

$$f(\alpha x_1 + (1 - \alpha)x_2) \leq \alpha f(x_1) + (1 - \alpha)f(x_2) \quad \forall \alpha \in [0, 1] \tag{2.57}$$

$$x_1, x_2 \in R^n$$

Since an LP is a convex problem it only has one single optimum and it is solved in polynomial time. There are two main algorithms to solve linear problems, simplex and interior point methods (Vanderbei, 2020).

2.3.1.1.1 Simplex method

The mathematical proof of optimality of the method is left for more specialized books (Biegler et al., 1997; Sioshansi and Conejo, 2017; Grossmann, 2021), but it can be obtained using the KKT conditions. The solution method consists of the Simplex Tableau that is built as follows. For our original problem formulation, the inequalities are transformed into equalities by adding slack variables, s, Eq. 2.58.

$$Ax \leq b \rightarrow Ax + Ds = b \tag{2.58}$$

Variables now are defined as basic and nonbasic ones. The nonbasic ones, at the first iteration, x, take values of 0 and the problem is solved for the basic ones, s in the previous definition. We reformulate the problem as follows with x_B for basic variables and x_N for nonbasic ones as B and N for the coefficients of the augmented A matrix, Eq. 2.59.

$$\text{Min } Z = c_B^T x_B + c_N^T x_N$$

$$Bx_B + Nx_N = b \tag{2.59}$$

where, Eq. 2.60

$$x_B = -B^{-1}Nx_N + B^{-1}b \tag{2.60}$$

Using this definition in the objective function the problem becomes, Eq. 2.61.

$$\text{Min } Z = -\left[c_B^T B^{-1}N - c_N^t\right]x_N + c_B^T B^{-1}b$$

$$\text{s.t.} \quad x_N \geq 0 \tag{2.61}$$

Where the reduced gradient is as follows, Eq. 2.62:

$$g_R^T = -\left[c_B^T B^{-1}N - c_N^t\right] \tag{2.62}$$

The tableau becomes

Z	x_B^T	x_N^T	RHS	
1	$-c_B^T$	$-c_N^T$	0	Objective
0	B	N	b	Constraints

The idea is for the basic variables to set matrix $B = I$ and the coefficients, c_B^T, to be 0. By multiplying B^{-1} and adding to the objectives $c_B^T x$ the tableau becomes:

Z	x_B^T	x_N^T		RHS	
1	0	$c_B^T B^{-1} N - c_N^T$	$B^{-1} c_B^T b$		Objective
0	I	$B^{-1} N$	$B^{-1} b$		Constraints

The search terminates when

$$\left[c_B^T B^{-1} N - c_N^t \right] \leq 0$$

since the objective cannot be further minimized. Else, the variable x_{Nj} with the largest $c_B^T B^{-1} n_j - c_{nj}^T > 0$ enters the base and to decide which one abandons it, the min test ratio, b_i / n_{ij}, $i = 1, \ldots, m$ is computed. The row with the lower positive ratio corresponds to the variable that exits the base. See Example 2.4.

EXAMPLE 2.4 Solve using the simplex method.

Solution

$$
\begin{array}{ll}
\min & Z = -x_1 - 3x_2 \\
s.t. & x_1 + x_2 \leq 4 \\
& x_1 + 2x_2 \leq 6 \\
x_1, x_2 \geq 0
\end{array}
\quad \rightarrow \quad
\begin{array}{ll}
\min & Z = -x_1 - 3x_2 \\
s.t. & x_1 + x_2 + x_3 = 4 \\
& x_1 + 2x_2 + x_4 = 6 \\
x_1, x_2 \geq 0 \\
x_3, x_4 \geq 0
\end{array}
\qquad (2E4.1)
$$

		No basic		Basic		
Z	x_1	x_2		x_3	x_4	RHS
1	+1	+3		0	0	0
0	1	1		1	0	4
0	1	2		0	1	6

$$Z = 0, x_1 = x_2 = 0, x_3 = 4, x_4 = 6$$
$$-g_R^T > 0$$

x_2 shows the largest $c_B^T B^{-1} n_j - c_{nj}^T$. It enters the base. We now compute the min test ratio

$$\text{Min}(4/1, 6/2) = 3,$$

Row 3 corresponds to variable x_4 that exits the base activating the constraint.

		No basic	Basic	Basic	No basic	
	Z	x_1	x_2	x_3	x_4	RHS
$(1) - 3 \cdot (3)$	1	$-1/2$	0	0	$-3/2$	-9
$(2) - (3)$	0	$1/2$	0	1	$-1/2$	1
$x(1/2)$	0	$1/2$	1	0	$1/2$	3

$$Z = -9, x_1 = 0, x_2 = 3, x_3 = 1, x_4 = 0$$
$$c_1, c_4 < 0 \rightarrow \text{Finished}$$

Fig. 2E4.1

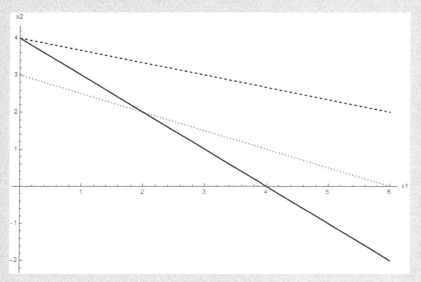

FIGURE 2E4.1 LP Simplex (.., _ constraints; − objective function).

2.3.1.1.2 Interior point methods

Starting with a feasible point, the Interior Point method iteratively approaches the solution by following a central path located far from the feasible region boundaries (Fig. 2.6). A particular case is the one that uses a barrier function, among them the most common is the log barrier, Eq. 2.63. Because the logarithm term forces the optimum away from zero, it rules out any small values of x_i. Reducing μ allows the optimal value to tend to 0. The solution alternates between taking a step in the direction that results in a sufficient decrease in the current barrier problem $B(x,\mu)$, and then decreasing the value of μ for the next iteration. A descent step rapidly ends near the feasible region boundary. By using steps which follow the central path, larger steps can be taken to avoid the feasible region boundary.

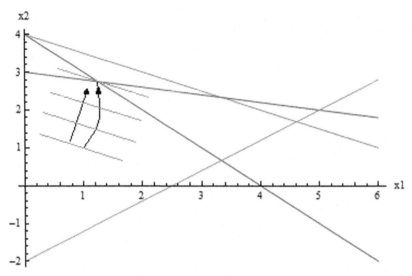

FIGURE 2.6   Interior point methods.

$$\min c^T x$$
$$st.Ax = b$$
$$x \geq 0$$
$$B(x, \mu) = c^T x - \mu \sum_{i=1}^{n} \ln x_i \qquad (2.63)$$
$$\min B(x, \mu)$$
$$st.Ax = b$$
$$x \in R^n$$

2.3.1.1.2.1 Bilevel optimization   When within the constraints there is another optimization problem, the problem becomes multilevel (Dempe, 2002). This kind of problems arise in the management on any company where each department has its goals, while the company as an entity has its own. The idea is for each department to be optimal but that should not jeopardize the entire company. Two different formulations can be distinguished, the optimistic and the pessimistic ones. In the optimistic formulation, the leader can choose a value, them the follower formulates the best response, and the leader develops an optimal strategy. In the pessimistic formulation, the leader minimizes the damage due to the follower's decision. This kind of problems may not have a solution. Let us present a 2D problem that is easy to follow and later we will present a mathematical formulation to solve the problem, Eq. 2.64, see Fig. 2.7a&b.

$$\min_{x,y} F(x,y) = x - 5y \qquad (2.64)$$
$$s.t. \min_{y} f(x,y) = y$$
$$s.t. -2x - y \leq -3$$
$$-2x + y \leq 0$$
$$2x + y \leq 12$$
$$3x - 2y \leq 4$$

By inspection, if y is to be minimized the lower bound of the feasible region is found.

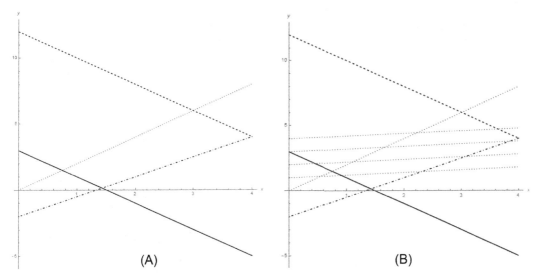

FIGURE 2.7   Bilevel formulation (A). Bilevel formulation (B).

If F is to be minimized, we find the optimum at the right corner of the feasible region
To address the solution of the bilevel optimization problems it is possible to use:

- Vertex enumeration
- Penalty methods.
- KKT conditions (see Section 2.3.1.3). It consists of substituting the second problem by its KKT conditions. It is only valid for the optimistic case (Eq. 2.65).

$$\min_y f(\hat{x}, y^*) \quad \nabla f\left(\hat{x}, y^*\right) + \sum_{j=1}^{r} \mu_j \nabla g_j(\hat{x}, y^*) = 0$$

$$s.t. \qquad \rightarrow \mu_j g_j(\hat{x}, y^*) = 0 \qquad \forall \quad j \qquad\qquad (2.65)$$

$$g(\hat{x}, y^*) \geq 0 \qquad \mu_j \geq 0$$

By using this approach, the reformulated problem becomes nonlinear.

EXAMPLE 2.5 Reformulate using the KKT conditions and solve using an MINLP optimization software (e.g. BARON) the following bilevel problem.

$$\min_{x,y} F(x,y) = x - 5y \qquad\qquad (2E5.1)$$
$$s.t. \min_{y} f(x,y) = y$$
$$s.t. - 2x - y \leq -3$$
$$-2x + y \leq 0$$
$$2x + y \leq 12$$
$$3x - 2y \leq 4$$

2. Design principles

Solution

$$\min_{x,y} F(x,y) = x - 5y$$
$$s.t. -2x - y \leq -3$$
$$-2x + y \leq 0$$
$$2x + y \leq 12$$
$$3x - 2y \leq 4$$
$$1 + \left(\mu_1(-1) + \mu_2(1) + \mu_3(1) + \mu_4(-2)\right) = 0$$
$$\mu_1\left(-2x - y + 3\right) = 0$$
$$\mu_2\left(-2x + y\right) = 0$$
$$\mu_3\left(2x + y - 12\right) = 0$$
$$\mu_4\left(3x - 2y - 4\right) = 0$$

(2E5.2)

Positive variables

x,y, u1, u2, u3, u4;

Equations

e1, e2, e3, e4, e5, e6, e7, e8, e9;

e1.. $-2{*}x - y = L = -3$;

e2.. $-2{*}x + y = L = 0$;

e3.. $2{*}x + y = L = 12$;

e4.. $3{*}x - 2{*}y = L = 4$;

e5.. $1 + (u1{*}(-1) + u2{*}(1) + u3{*}(1) + u4{*}(-2)) = E = 0$;

e6.. $u1{*}(-2{*}x - y + 3) = E = 0$;

e7.. $u2{*}(-2{*}x + y) = E = 0$;

e8.. $u3{*}(2{*}x + y - 12) = E = 0$;

e9.. $u4{*}(3{*}x - 2{*}y - 4) = E = 0$;

Variable

Z;

Equation

obj;

obj.. $Z = E = x - 5{*}y$;

model Bilevel /all/

Option MINLP = BARON;

Solve Bilevel Using MINLP Minimizing Z;

$x = 4$, $y = 4$, $\mu_4 = 0.5$

2.3.1.2 Mixed integer linear programming problems

In most cases in a design problem apart from selecting values of continuous variables, integer decisions, whether selecting a number of indivisible entities that have to be bought or installed, or binary decisions, whether to select or not kind of decisions must also be considered. As a result of the addition of this new type of variables the model becomes a mixed integer, since the new variable is an integer or binary, linear programming, all the equations are linear products of constants and the variable, problems, MILP. The general formulation is as follows, Eq. 2.66:

$$Min\ z = a^T x + b^T y$$
$$s.t.\quad Ax + By \leq d$$
$$x \geq 0;\ y \in \{0, 1\}^m$$

(2.66)

The binary or integer variables represent decisions or entities. It is common to use binary variables to help select technologies for chemical processes. For instance, if a selection between two reactors is to be make, once one of the units is selected, the cost, yield and mass and energy balances that model that particular unit are activated while the equations modeling the other must be silenced from the model. Note that as an optimization problem the entire set of alternatives, and equations that model them, is to be included in the problem formulation. Therefore, it is the job of the binary variables to help activate the equations and select the units. To selected among units the following equation is used, Eq. 2.67:

$$\sum_i y_i ? 1 \tag{2.67}$$

where ? is $\le$ if at most one unit is selected, $= 1$ in one decision must be made and $\ge$ if more that one unit can be selected. Once the unit is selected, we need to activate the model of that unit. Two different approaches can be used, either big $-$ M formulation or convex hull (Vecchietti et al., 2003; Grossmann and Trespalacios, 2013).

2.3.1.2.1 Big-M

Each alternative is assigned a binary variable to identify it, Eq. 2.68.

$$\sum_i y_i = 1 \tag{2.68}$$

And the equations that govern each alternative are activated if the technology is selected. To do that, a large parameter, M, is used so that in case that technology is not selected, the equations of that model become redundant, Eq. 2.69

$$f_i(x) \le M(1 - y_i) \tag{2.69}$$

If i is selected the equation must hold, otherwise since M is large enough it is irrelevant what value the function takes.

2.3.1.2.2 Convex hull

The variables are divided into as many as alternatives so that the original variables are written as a linear combination of new variable assigned to each disjunction. Only the one that corresponds to the solution is activated using a binary variable, Eq. 2.70

$$
\begin{aligned}
x &= \sum_{v \in Y} \beta_v \hat{x}^v \\
g(\hat{x}^v, v) &\le 0, \quad v \in Y \\
x^U &\le \hat{x} \le x^L \\
\sum_{v \in Y} \beta_v &= 1, \quad \beta_v \in [0, 1]
\end{aligned} \tag{2.70}
$$

Thus, the variable transformation is $x^v = \beta_v \hat{x}^v$
And the problem becomes, Eq. 2.71

$$
\begin{aligned}
x^v &= \sum_{v \in Y} \hat{x}^v \\
g\left(\frac{x^v}{\beta_v}, v\right) &\le 0, \quad v \in Y \\
\beta_v x^L &\le x^v \le \beta_v x^U, \quad v \in Y \\
\sum_{v \in Y} \beta_v &= 1, \quad \beta_v \in [0, 1]
\end{aligned} \tag{2.71}
$$

The problem becomes nonconvex. By multiplying the inequality g by beta, the resulting function is known as the perspective function, which is convex if the original function g is also convex, Eq. 2.72:

$$x^v = \sum_{v \in Y} \hat{x}^v$$

$$\beta_v g\left(\frac{x^v}{\beta_v}, v\right) \leq 0, \quad v \in Y$$

$$\beta_v x^L \leq x^v \leq \beta_v x^U, \quad v \in Y$$

$$\sum_{v \in Y} \beta_v = 1, \quad \beta_v \in [0, 1]$$

(2.72)

EXAMPLE 2.6 Let's assume the following disjunction:

$$\begin{bmatrix} (x_1 - 1.5)^2 + (x_2 - 1.5)^2 \leq 1.5 \end{bmatrix} \vee$$
$$\begin{bmatrix} (x_1 - 4)^2 + (x_2 - 2)^2 \leq 1 \end{bmatrix} \vee$$
$$\begin{bmatrix} (x_1 - 3)^2 + (x_2 - 4)^2 \leq 1 \end{bmatrix}$$

(2E6.1)

Reformulate the problem using a big-M and a convex − hull.

Solution

The big-M formulation of the problem becomes, Eq. 2E6.2

$$(x_1 - 1.5)^2 + (x_2 - 1.5)^2 \leq 1.5 + 15(1 - y_1)$$
$$(x_1 - 4)^2 + (x_2 - 2)^2 \leq 1 + 14(1 - y_2)$$
$$(x_1 - 3)^2 + (x_2 - 4)^2 \leq 1 + 17(1 - y_3)$$
$$y_1 + y_2 + y_3 = 1$$
$$0 \leq x_1, x_2 \leq 5, 0 \leq y_i \leq 1, i = 1, 2, 3$$

(2E6.2)

Fig. 2E6.1

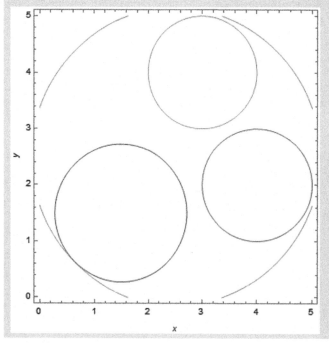

FIGURE 2E6.1 Big M reformulation.

The convex hull formulation generates as many new variables as possible disjunctions for each of the original variables.

$$x_1 = v_{11} + v_{12} + v_{13}$$
$$x_2 = v_{21} + v_{22} + v_{23}$$

And with this variable change the problem is reformulated as follows, Eq. 2E6.3

$$(y_1 + \varepsilon)\left[\left(\frac{v_{11}}{y_1+\varepsilon} - 1.5\right)^2 + \left(\frac{v_{21}}{y_1+\varepsilon} - 1.5\right)^2 - 1.5\right] \leq 0$$

$$(y_2 + \varepsilon)\left[\left(\frac{v_{12}}{y_2+\varepsilon} - 4\right)^2 + \left(\frac{v_{22}}{y_2+\varepsilon} - 2\right)^2 - 1\right] \leq 0$$

$$(y_3 + \varepsilon)\left[\left(\frac{v_{13}}{y_3+\varepsilon} - 3\right)^2 + \left(\frac{v_{23}}{y_3+\varepsilon} - 4\right)^2 - 1\right] \leq 0$$

(2E6.3)

$$y_1 + y_2 + y_3 = 1$$
$$y_i v_{ij}^{LO} \leq v_{ji} \leq y_i v_{ij}^{UP}$$

Fig. 2E6.2

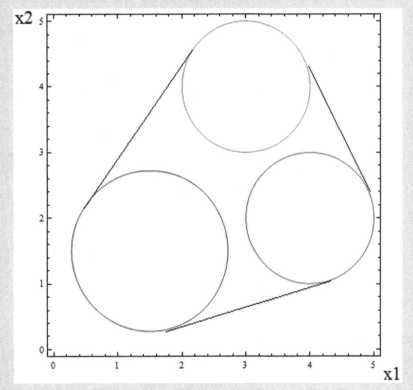

FIGURE 2E6.2 Figure -Convex Hull.

The big-M reformulation is less tight, but the problem size is smaller. An interesting note is that it is important to maintain linearity in the formulation since that will allow solving larger problems. Therefore, nonlinear equations are to be avoided or reformulated.

Piecewise linear approximations can be a valid approach. In addition, it is not wise to multiply binary variable by continuous ones. That happens in linear cost functions, Eq. 2.73, (Papoulias and Grossmann, 1983)

$$C = a + bx \qquad (2.73)$$

To cancel the contribution of the cost of a particular unit if it is not selected, the entire cost is not multiplied by the binary variable that identifies this alternative, but it is used to force x to be 0, Eq. 2.74

$$LB \cdot y < x < UPy \text{ and } C = ay + bx \qquad (2.74)$$

To address the solution of this type of problems several techniques can be used. It is possible to enumerate all the alternatives, a *brute force* kind of approach can be used where the objective function value is compared afterwards. It is also possible to solve the problem as an LP and approximate the values of the integer and binary solutions. However, commercial software typically uses a *relaxation approach*, as well as heuristics that study the problem upfront. Different strategies can be used (Sioshansi and Conejo, 2017; Grossmann, 2021).

Gomory cutting planes (Gomory, 1960)

The basic idea is to generate additional constraints, cuts, that will reduce the feasible region iteratively. It is possible to build those constraints using heuristics, by approximating the original constraints by the floor, as well as from the simplex tableau. It is an elegant method, but slow.

EXAMPLE 2.7 Consider the following example, Eq. 2E7.1:

$$\max z = x_1 + 3x_2$$
$$st.$$
$$3x_1 + 2x_2 \le 10 \qquad (2E7.1)$$
$$x_1 + 5x_2 \le 16$$
$$x_1, x_2 \ge 0;$$

Solution
We can generate cuts by taking the second constraint, Eq. 2E7.2.

$$1x_1 + 5x_2 \le 16 \qquad (2E7.2)$$

Divide all by the largest coefficient and approximate the resulting constraint by the floor, Eq. 2E7.3:

$$(1/5)x_1 + (5/5)x_2 \le 16/5$$

$$\lfloor 1/5 \rfloor x_1 + 1x_2 \le \lfloor 16/5 \rfloor$$
$$x_2 \le 3 \qquad (2E7.3)$$

And we have a cut to be added to the problem. By adding this we immediately solve the problem. Fig. 2E7.1

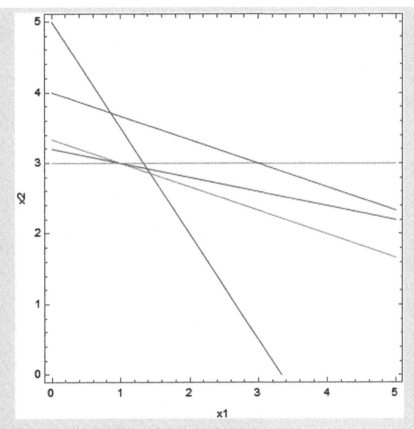

FIGURE 2E7.1 Gomory cutting plane.

Alternatively, we can use the simplex tableau. We define s_1 and s_2 as slack variables for each of the constraints.

Eq	x1	x2	s1	s2		Test ratio
Z	1	3				
1	3	2	1		10	5
2	1	5		1	16	3.2
Z	0.4	0	0	−0.6	−9.6	
1	2.6	0	1	-0.4	3.6	1.38461538
2	0.2	1	0	0.2	3.2	16
Z	0	0	−0.1538462	−0.5384615	−10.153846	
1	1	0	0.38461538	−0.1538462	1.38461538	
2	0	1	−0.0769231	0.23076923	2.92307692	

From the tableau we take any of the equations, either for x_1 or x_2, i.e. Eq. 2E7.4

$$x_2 - 0.077 \cdot s_1 + 0.23 \cdot s_2 = 0.92 \qquad\qquad (2E7.4)$$

It is possible to prove that by taking the floor of all the terms the equality becomes an inequality. We leave this as an exercise. Thus, a cut is generated as follows, Eq. 2E7.5

$$(1 - \lfloor 1 \rfloor)x_2 + (-0.077 - \lfloor -0.077 \rfloor) \cdot s_1 + (0.23 - \lfloor 0.23 \rfloor)s_2 \geq 2.92 - \lfloor 2.92 \rfloor$$
$$-0.077 s_1 + 0.23 \cdot s_2 \geq 0.92$$

With

$$3x_1 + 2x_1 + s_1 = 10$$
$$x_1 + 5x_2 + s_2 = 16 \qquad\qquad (2E7.5)$$

A cut

$$-0.077(10 - 3x_1 - 2x_2) + 0.23 \cdot (16 - x_1 - 5x_2) \geq 0.92$$
$$x_2 \geq 2$$

Similarly, from the other variable x_1, Eq. 2E7.6

$$0.385 \cdot s_1 - 0.154 \cdot s_2 = 1.385 \qquad\qquad (2E7.6)$$

We can obtain another cut, Eq. 2E7.6.

$$0.385(10 - 3x_1 - 2x_2) - 0.154 \cdot (16 - x_1 - 5x_2) \geq 0.385$$
$$x_1 \leq 1 \qquad\qquad (2E7.7)$$

We know add one and solve a new LP problem, Eq. 2E7.8

$$\begin{aligned}
\max z &= x_1 + 3x_2 \\
&st. \\
3x_1 + 2x_2 &\leq 10 \\
x_1 + 5x_2 &\leq 16 \\
x_1 &\leq 1 \\
x_1, x_2 &\geq 0;
\end{aligned} \qquad\qquad (2E7.8)$$

Solving this as an LP we already obtain an integer solution and we are done. $x_1 = 1$, $x_2 = 3$

Branch and Bound.

The idea is to avoid exhaustive enumeration by solving sequences of LPs and generate bounds to prune branches. At each node we solve an LP generating two problems where the binary (integer) variable is to the floor or ceiling of the continuous value solution. If the variable is integer, it is fixed to 0 or 1. The total number of nodes is $2^{n+1} - 1$. As we progress evaluating nodes, either deep or breath first, the algorithm does not explore every single node, but evaluates the solutions by Bound Updating and Branching. There are some rules for these two steps. For a minimization problem:

1. If p_i is not feasible, p_k (the next one) will not be either. The node can be removed.
2. If p_k is feasible, p_i is also feasible and $Z_i \leq Z_k$ (We are adding constraints)
3. If at node pk there is a feasible integer solution, $Z_k^* \geq Z^*$. We have an upper bound

Let's see an example.

EXAMPLE 2.8 Solve the following problem using a branch and bound approach.

$$\max z = 3x_1 + 5x_2 + 6x_3 + 3x_4$$
st
$$2x_1 + 6x_2 + 2x_3 + 3x_4 \leq 10$$
$$x_1 + x_2 \leq 1$$
$$x_1 - x_3 \leq 0 \qquad\qquad (2E8.1)$$
$$x_2 - x_4 \leq 0$$
$$x_i = binary$$

Solving the problem given by Eq. 2E8.1 as an LP, we get $x_1 = 0.25$, $x_2 = 0.75$, $x_3 = 1$; $x_4 = 1$.

We develop a tree search by starting with $(0,\_,\_,\_)$ allowing x_2-x_4 to be continuous. In the table below in bold we present the fixed values for the variables and below the objective function. Fig. 2E8.1

Thus, the solution is (1,0,1,1)

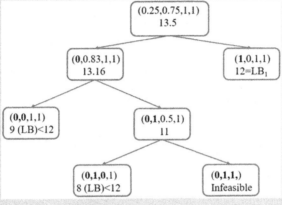

FIGURE 2E8.1   Solution tree.

Benders decomposition.

It is especially useful for problems with variables that complicate the solution, for instance, design problems where one variable remains constant over time. The structure of the problems is as follows:

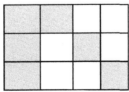

The procedure consists of:
1. Fixing the complex variables.
2. Generate cuts from the dual of this problem.
3. Reformulate the problem.

Adding all the cuts simultaneously, as in Example 2.9, is not practical for large problems and an iterative process is typically implemented to add the feasibility and the optimality cuts.

EXAMPLE 2.9

$$\min_{x,n} x + y_1 + y_3$$
$$st \quad 2x + y_1 - 6y_2 - 5y_3 \geq 1$$
$$3x - y_1 + y_2 + 2y_3 \geq 2 \tag{2E9.1}$$
$$x \geq 0 \in integer; \, y_i \geq 0$$

Solution

In Eq. 2E9.1, we fix the complicating variables, Eq. 2E9.2

$$\min_{x,n} y_1 + y_3$$
$$st \quad y_1 - 6y_2 - 5y_3 \geq 1 - 2x$$
$$-y_1 + y_2 + 2y_3 \geq 2 - 3x \tag{2E9.2}$$
$$y_i \geq 0$$

Formulate the dual (hint, see solution to Problem P2.13 in the Solutions Appendix)

$$\max(1 - 2x)u_1 + (2 - 3x)u_2$$
$$st \quad u_1 - u_2 \leq 1$$
$$-6u_1 + u_2 \leq 0 \tag{2E9.3}$$
$$-5u_1 + 2u_2 \leq 1 \quad .$$
$$u_1, u_2 \geq 0$$

The rays and the extreme points of the feasible region are computed. The **extreme points** are determined from the cut between each constraint of the dual:

$$u^1 = (1, 0); \ u^2 = (0, 0); \ u^3 = (1/7, 6/7)$$

the **Rays** correspond to the directions that bound the open feasible region.

$$v^1 = (1, 1); v^2 = (2/5, 1),$$

Fig. 2E9.1

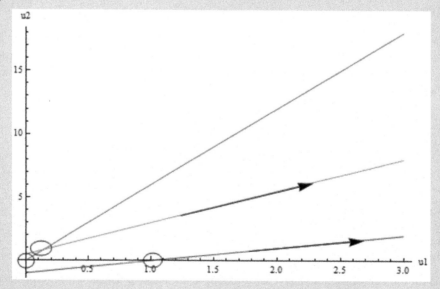

FIGURE 2E9.1   Benders cuts.

We reformulate the problem, Eq. 2E9.4, generating feasible cuts, from the rays.

$$v(b - By) \leq 0 \quad \forall j \in J$$

And optimality cuts from the extremes,

$$u^k(b - By) \leq n \quad \forall k \in K$$

$$
\begin{aligned}
&\min_{x,n} n \\
&st \quad n \geq (1 - 2x) \\
&n \geq 0 \\
&n \geq \frac{1}{7}(1 - 2x) + \frac{6}{7}(2 - 3x) = \frac{13}{7} - \frac{20}{7}x \\
&0 \geq (1 - 2x) + (2 - 3x) = 3 - 5x \\
&0 \geq \frac{5}{2}(1 - 2x) + (2 - 3x) = \frac{9}{2} - 8x \\
&x \geq 0
\end{aligned}
$$
(2E9.4)

By solving this reformulated problem, we have
$x^* = 1 \text{ y } z^* = 0$

Lagrangean relaxation.

This approach is of particular interest for problems with complicating constraints. The mathematical structure of the problem is as follows, Eq. 2.75:

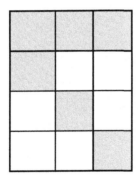

$$
\begin{aligned}
Z = &\max cx \\
&Ax \leq b \\
&Dx \leq e \\
&x \in Z_+^n
\end{aligned}
$$
(2.75)

The idea consists of dualizing the complex constraints so that they are evaluated within the objective function. As a result, the new objective will always present a gap with the original one. Only in the case of equality constraints to be dualized, at the solution the gap could be zero. In addition, a new variable appears, the Lagrange multipliers, u, corresponding to each of the dualized constraints, Eq. 2.76.

$$
\begin{aligned}
Z_{LR}(u) = &\max cx + u(b - Ax) \\
&Dx \leq e \\
&x \in Z_+^n
\end{aligned}
$$
(2.76)

Thus, the new problem, dual, that is to be solved is as follows Eq. 2.77:

$$Z_D = \min Z_{LR}(u)$$
$$u \geq 0$$
$$Z'_D = \max cx$$
$$Ax \leq b \tag{2.77}$$
$$x \in Conv\left(Dx \leq e, x \in Z^n_+\right)$$
$$x \geq 0$$

It is clear that the reformulation of the problem is a relaxation, we remove one or more constraints and therefore, for a maximization problem, the dual represents an upper bound. To solve the dual problem, it is critical to evaluate, initialize and update, the multipliers. Among the methods used, the subgradient method is presented in the example below.

$$Z(P) \leq Z_D \leq Z_{LR}(u) \leq Z_{LP}$$

EXAMPLE 2.10 Solve the following MILP problem, Eq. 2E10.1, Eq. 2E10.1, using the Lagrange relaxation and the subgradient method to update the multipliers.

min $2x_1 + 3x_2 + 4x_3 + 5x_4$

st $x_1 + x_3 \geq 1$

$x_1 + x_4 \geq 1$

$x_2 + x_3 + x_4 \geq 1$

$x_j \in (0, 1), j = 1, .., 4 \tag{2E10.1}$

Solution:

min $2x_1 + 3x_2 + 4x_3 + 5x_4 + \lambda_1(1 - x_1 - x_3) + \lambda_2(1 - x_1 - x_4) + \lambda_3(1 - x_2 - x_3 - x_4)$

st $x_j \in (0, 1), j = 1, .., 4$

$\lambda_i \geq 0 \quad \forall i \tag{2E10.2}$

min $(2 - \lambda_1 - \lambda_2)x_1 + (3 - \lambda_3)x_2 + (4 - \lambda_1 - \lambda_3)x_3 + (5 - \lambda_2 - \lambda_3)x_4 + \lambda_1 + \lambda_2 + \lambda_3$

st $x_j \in (0, 1), j = 1, .., 4$

Hence,

$C_1 = (2 - \lambda_1 - \lambda_2)$

$C_2 = (3 - \lambda_3)$

$C_3 = (4 - \lambda_1 - \lambda_3)$

$C_4 = (5 - \lambda_2 - \lambda_3) \tag{2E10.3}$

$x_j = 1$ si $C_j \leq 0$ o 0 otherwise

$$Z_{LB} = C_1 x_1 + C_2 x_2 + C_3 x_3 + C_4 x_4 + \lambda_1 + \lambda_2 + \lambda \tag{2E10.4}$$

The iteration consists of:

1. Initialize the multipliers.
2. Find a feasible solution to obtain a bound for the solution.
3. Evaluate the subgradients

$$G_i = b_i - \sum_j a_{ij} X_j \tag{2E10.5}$$

4. Compute the step size

$$T = \frac{\pi (Z_{UP} - Z_{LB})}{\sum_i G_i^2} \tag{2E10.6}$$

π is typically [0,2]

5. Update the multipliers as follows: $\lambda_i = \max(0, \lambda_i + T G_i)$

We assume $\pi = 2$

A feasible solution represents an upper bound. For instance,

$x_1 = 1 = x_2, x_3 = 0 = x_4$

Thus, we have an UP, $Z_{UP} = 5$

For $\lambda_1 = 1.5$, $\lambda_2 = 1.6$ and $\lambda_3 = 2.2$, only C1 ≤ 0, this $x_1 = 1$ and $Z_{LB} = 4.2$

The subgradients are given by Eq. 2E10.7

$$
\begin{aligned}
G_1 &= (1 - x_1 - x_3) = 1 - 1 - 0 = 0 \\
G_2 &= (1 - x_1 - x_4) = 1 - 1 - 0 = 0 \\
G_3 &= (1 - x_2 - x_3 - x_4) = 1 - 0 - 0 - 0 = 1
\end{aligned}
\tag{2E10.7}
$$

The step size is computed as follows, Eq. 2E10.8:

$$T = \frac{2(5 - 4.2)}{(0^2 + 0^2 + 1^2)} = 1.6 \tag{2E10.8}$$

We update the multipliers as follows, Eq. 2E10.9.

$$
\begin{aligned}
\lambda_1 &= \max(0, 1.5 + 1.6(0)) = 1.5 \\
\lambda_2 &= \max(0, 1.6 + 2.6(0)) = 1.6 \\
\lambda_3 &= \max(0, 2.2 + 1.6(1)) = 3.8
\end{aligned}
\tag{2E10.9}
$$

Solving again the Lagrange dual with these multipliers, $x_1 = x_2 = x_3 = x_4 = 1$ so that $Z_{LB} = 3.3$.

We continue with the iterations, see Fig. 2E10.1 so that in 11 iterations we reach the upper bound with $x_1 = x_2 = 1; x_3 = x_4 = 0$.

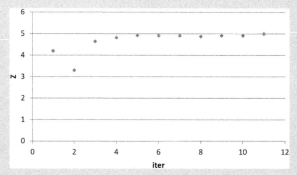

FIGURE 2E10.1   Lagrange decomposition.

A characteristic of this method is that there is no proof that at each step the solution improves. A particular case of the Lagrangean relaxation is the Lagrangean decomposition that allows splitting the original problem into two, duplicating variables so that it is possible to split the problems so that the same variable is not in two problems, only a copy, and solving the decomposed problems independently.

Typical solvers for MILP are, GUROBI, CPLEX, or XPRESS.

2.3.1.3 Nonlinear programming

Most of the problems in chemical process design involve nonlinear equations. Mass and energy balances include splitting fractions, heat capacities, thermodynamic and phase equilibrium, and so on, whose formulation involves nonlinear equations. This type of problems is particularly useful for the selection of the operating conditions of the units. As a result, the optimization of the operating conditions as well as the selection of technologies involving this type of models require a different formulation and solution algorithms. The general formulation of nonlinear programming (NLP) problems is as follows, Eq. 2.78:

$$
\begin{aligned}
&\min f(x), \\
&\text{s.t.} \\
&h(x) = 0 \\
&g(x) \le 0 \\
&x \in R^n; h(m - vector); g(r - vector) \\
&m \le n \\
&f(x), h(x), g(x) \quad \text{differentiables}
\end{aligned}
\tag{2.78}
$$

The necessary conditions to obtain a constrained minimum correspond to the KKT (Karusk-Kuhn-Tucker) conditions that consist of:

1. Linear dependency of the gradients: The physical meaning behind this equation is a force balance between the weight, the objective function drives you, constrained by the inequality constraints that do not allow you to go pass, and the equality constraints, that guide you. An interesting representation can be found in Biegler et al. (1997) where this condition is represented as a valley where you let a ball roll. The objective function takes you to the bottom, but normal forces due to fences, the inequality constrains, and railways, the equality constrains, limit your descent, Eq. 2.79.

$$\nabla f(x^*) + \sum_{j=1}^{m} \lambda_j \nabla h_j(x^*) + \sum_{j=1}^{r} \mu_j \nabla g_j(x^*) = 0 \tag{2.79}$$

2. Constraint feasibility: The constraints must hold since describe the problem, Eq. 2.80.

$$\begin{aligned} h_j(x^*) &= 0, j = 1, 2, \ldots, m; \\ g_j(x^*) &\leq 0, j = 1, 2, \ldots, r \end{aligned} \tag{2.80}$$

3. Complementarity constraint: Representing the active constraint, Eq. 2.81.

$$\begin{aligned} \mu_j g_j(x^*) &= 0, j = 1, 2, \ldots, r \\ \mu_j &\geq 0 \end{aligned} \tag{2.81}$$

In case $\mu_j \geq 0$ the gradient of the objective function and that of the constraint point to the same direction, and the objective function can still decrease.

One of the major issues with nonlinear problems is the convexity of the feasible region and of the objective function. Only in case of convexity, if there exist a local minimum.

a. x^* is global
b. Constraints qualification holds
c. KKT are necessary and sufficient for a global minimum.

However, in most cases neither the objective function nor the feasible region is convex, and we cannot prove global optimality with this approach alone. Note that solving the KKT conditions is only the solution of a nonlinear system of equations that can be addressed using a Newton-based method. There are a number of constraints that may not have an effect in the solution. Thus, one of the strategies to solve the NLP is by the active constraint strategy. The objective function is optimized together with the equality constraints. At this point, the inequalities are evaluated and the constraints that are violated are enforced as equality constraints. Otherwise, we have obtained our solution.

EXAMPLE 2.11 Compute the optimum for the NLP problem below, Eq. 2E11.1, using the active constraint method.

$$\min f(x) = \frac{(x_1+2)^2}{4} + \frac{(x_2-3)^2}{2}$$

s.t.

$$g_1 = -x_1 + x_2 \leq 6$$
$$g_2 = x_1 + \frac{1}{2}x_2 \leq 6 \qquad \text{(2E11.1)}$$

$$g_3 = x_1 - 3x_2 \leq 2$$
$$g_4 = -x_1 \leq 0$$
$$g_5 = -x_2 \leq 0$$

Solution

First, we compute the gradients

$$\nabla f = \begin{bmatrix} \frac{1}{2}(x_1+2) \\ x_2-3 \end{bmatrix}$$

$$\nabla g_1 = \begin{bmatrix} -1 \\ 1 \end{bmatrix}$$

$$\nabla g_2 = \begin{bmatrix} 1 \\ \frac{1}{2} \end{bmatrix} \qquad \text{(2E11.2)}$$

$$\nabla g_3 = \begin{bmatrix} 1 \\ -3 \end{bmatrix}$$

$$\nabla g_4 = \begin{bmatrix} -1 \\ 0 \end{bmatrix}$$

$$\nabla g_5 = \begin{bmatrix} 0 \\ -1 \end{bmatrix}$$

Fig. 2E11.1

Solving the unconstraint problem, we have Eq. 2E11.3

$$\begin{bmatrix} \frac{1}{2}(x_1+2) \\ x_2-3 \end{bmatrix} = 0 \qquad \text{(2E11.3)}$$

$$x_1 = -2; x_2 = 3$$

$$g_1 = -(-2) + 3 \leq 6 \Rightarrow OK$$
$$g_2 = -2 + \frac{1}{2}(3) \leq 6 \Rightarrow OK$$

$$g_3 = -2 - 3(3) \leq 2 \Rightarrow OK \qquad \text{(2E11.4)}$$
$$g_4 = 2 \leq 0 \Rightarrow X$$
$$g_5 = -3 \leq 0 \Rightarrow OK$$

We activate g_4 Eq. 2E11.5

$$\begin{bmatrix} \frac{1}{2}(x_1+2) \\ x_2-3 \end{bmatrix} + \mu_4 \begin{bmatrix} -1 \\ 0 \end{bmatrix} = \begin{bmatrix} 0 \\ 0 \end{bmatrix} \qquad \text{(2E11.5)}$$

$$x_1 = 0$$

$x_1 = 0$, $x_2 = 3$, $\mu_4 = 1$

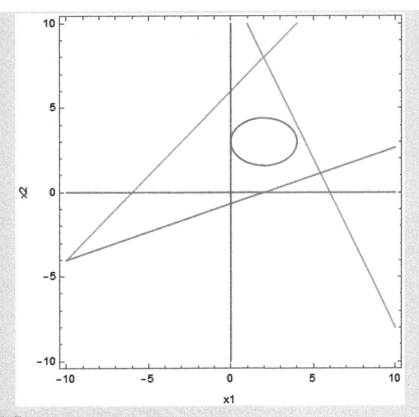

FIGURE 2E11.1 Active constraint method.

We check the constraints, Eq. 2E11.6:

$$g_1 = 0 + 3 \leq 6 \Rightarrow OK$$

$$g_2 = -0 + \frac{1}{2}(3) \leq 6 \Rightarrow OK$$

$$g_3 = -0 - 3(3) \leq 2 \Rightarrow OK$$ (2E11.6)

$$g_4 = 0 \leq 0 \Rightarrow X$$

$$g_5 = -3 \leq 0 \Rightarrow OK$$

The point is an optimal $Z = 1$.

Is the solution unique, Eq. 2E11.7

$$H = \begin{bmatrix} \dfrac{\partial^2 f}{\partial x_1 \partial x_1} & \dfrac{\partial^2 f}{\partial x_1 \partial x_2} \\ \dfrac{\partial^2 f}{\partial x_1 \partial x_2} & \dfrac{\partial^2 f}{\partial x_2 \partial x_2} \end{bmatrix} = \begin{bmatrix} 1/2 & 0 \\ 0 & 1 \end{bmatrix}$$ (2E11.7)

The eigenvalues are Eq. 2E11.8

$$\begin{bmatrix} 1/2 - \lambda & 0 \\ 0 & 1 - \lambda \end{bmatrix} = (1 - \lambda)(1/2 - \lambda)$$ (2E11.8)

$$\lambda_1 = 1/2; \lambda_2 = 1$$

Both positive, so unit and global.

2. Design principles

```
Variable Z;
Positive Variables x1,x2;
Equations obj,g1,g2,g3;
obj.. Z = E = (1/4)*(x1 + 2)*(x1 + 2) + (1/2)*(x2 – 3)*(x2 – 3);
g1.. – x1 + x2 = L = 6;
g2.. x1 + (1/2)*x2 = L = 6;
g3.. x1 – 3*x2 = L = 2;
model KKT /all/
Option NLP = conopt;
Solve KKT Using NLP Minimizing Z;
```

There are a number of commercial codes available for NLP problems. It is possible to classified them based on the solution approach (Bielger, 2010)

Quadratic programming (SQP) – SNOPT

Solve KKT with Newton's method for equality constraints and extension to inequalities using QP solvers. It has issues with large problems since relies on BFGS approximation for the Hessian.

Reduced gradient. MINOS, CONOPT, GRG

The basic idea is to use the equality constraints to reduce the dimensionality of the problem and use the Newton methods: Reduce Gradient.

Infeasible path: MINOS. Good when most of the constraints are linear.

Feasible path: CONOPT, GRG (EXCEL). Slower but more robust.

Interior point method: KNITRO, IPOPT

It follows similar principles as presented for the LP problems. The problem is reformulated using a penalty function in the objective function. The solution is forced to follow a path away from the constraints until the solution is found. It is very interesting to help initialize problems.

2.3.1.4 Mixed integer nonlinear programming

The more general class of problems is the one that involves at the same time linear and nonlinear equations and continuous and integer/binary variables. This type of problems receives the name of mixed integer nonlinear programming (MINLP) problems. They are used for technology selection, operating condition definition as well as unit design (Grossmann, 2021). The formulation is as follows, Eq. 2.82:

$$\min Z = c^T y + f(x)$$
$$s.t.$$
$$h(x) = 0$$
$$g(x) + By \le 0 \qquad (2.82)$$
$$Ay \le a$$

The solution approach follows two main trends, either a Branch and Bound or a decomposition algorithm.

Branch and Bound: It is a tree search similar to the one presented in the case of MILP above. For a minimization problem at any node, an NLP is solved where some binaries are fixed. As such it is a relaxation, since the rest can take continuous values. It is a lower bound. If at any point all the binaries take integer values, the solution is an upper bound, a feasible solution. The prune is carried out as in the case of MILP problems where either there is no feasible solution when a binary variable is fixed or if the lower bound is worse.

Decomposition methods: Within the decomposition methods, we can distinguish two different cases. Those which follow a master MILP and NLP problem based algorithm such as Generalized Benders decomposition (Geoffrion, 1972) or the Outer approximation (Duran and Grossmann, 1986a,b) and that which solves only linear problems, the Extended cutting plane (Westerlund and Pettersson, 1992).

a. Algorithm MILP-NLP

Within this class, we find two methods that differ in the way the MILP is obtained. In both cases they follow the algorithm presented in Fig. 2.8. For a minimization problem, the NLP is a feasible solution and represents an upper bound to the solution (Z_{UB}). Using the NLP, the master MILP is generated. The MILP is a relaxation and provides a lower bound (Z_{LB}). It computes the binary variables for the next iteration.

The difference is how the MILP is generated. Two cases are found:

Outer approximation (Duran and Grossmann, 1986a,b): It uses the NLP solution to linearize the nonlinear constraints using a Taylor expansion. The nonlinear constraints are supported by this approximation. It is an external one. At each iteration new hyperplanes are generated and added to the MILP that grows at each iteration. This algorithm requires an adaptation when equality constraints are present. Kocis and Grossmann (1987) proposed the equality relaxation so that the equality constraints are relaxed to become inequalities based on the Lagrange multipliers associated to them. Thus, the master MILP becomes Eq. 2.83:

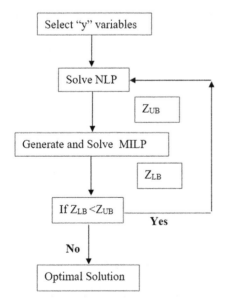

FIGURE 2.8 MINLP algorithm scheme.

$$minZ = c^T y + \alpha$$
$$s.t.$$
$$\alpha \geq f(x^k) + \nabla f(x^k)(x - x^k)$$
$$0 \geq g(x^k) + \nabla g(x^k)(x - x^k) + By \quad\quad \forall k \in T \quad\quad\quad (2.83)$$
$$0 \geq T^k[h(x^k) + \nabla h(x^k)(x - x^k)]$$
$$Cx + By \leq d$$

Generalized Benders decomposition Benders (1962), Geoffrion (1972): In this case the MILP is generated using only the constraints that involve binary variables where the nonlinear terms are evaluated at the solution provided by the NLP. The objective function is a new variable, α. The MILP also grows at each iteration by adding new cuts α, Eq. 2.84.

$$minZ = \alpha$$
$$s.t.$$
$$\alpha \geq c^T y + f(x^k) + (\mu^k)^T[g(x^k) + By] \quad \forall k \in K_F^k \quad\quad\quad (2.84)$$
$$(\mu^k)^T[g(x^k) + By] \leq 0$$

b. Extended cutting plane (Westerlund and Pettersson, 1992).

The main issue with the previous methods is the difficulty in solving the NLP in each iteration. The extended cutting plane consists of removing the nonlinear constraint from the problem and iteratively solve MILP where at each iteration the nonlinear constraints are evaluated. The most violated one is linearized using a Taylor approximation and added to the MILP until a tolerance is achieved. Typically, the number of iterations required is larger.

2.3.1.5 Optimization involving dynamic models

In chemical engineering is rather typical to have kinetics to represent the performance of reactors. Sometimes it is the easiest way to compute the conversion, but it requires including a set of ODE's within the optimization formulation. The basic idea is to discretize, convert the set of ODE's (or DEA's) into an algebraic problem as if a Runge Kutta is used to discretize the differential equations so that we solve an algebraic problem using Lagrange polynomials to fit the function, Eq. 2.85.

$$\frac{dy}{dt} = f(y(t), t) \approx f(y_k, t_k) = \sum_{j=0}^{K} l_j(t_k) y_j \quad\quad\quad (2.85)$$

For further details on the formulation and the discretization we refer to more detailed literature (Bielger, 2010).

2.3.1.6 Global optimization

It is possible to define two types of methods, deterministic and stochastic. The first one is typically used within equation-based optimization and allows proving optimality (Tawarmalani and Sahinidis, 2002). The second one, widely used in process simulators due to the lack of open models to evaluate the gradients, it is also being currently used also for equation-based models, but there is no mathematical proof of optimality.

2.3.1.6.1 Deterministic methods

Spatial branch and bound: This method consists of continuous approximation of the non-convex terms or binary variables to compute a lower bound while the upper bound is provided by a feasible solution to the problems. There are different approaches to approximate the nonconvex terms. The space is reduced by analyzing the regions following a tree search discarding region based on the objective function value in each one. Depending to the type of nonconvex term we have:

Bilinear terms: They are typical in mass and heat exchange networks due to the binary products between variables either flows and concentrations ($F \cdot C$) or flows and temperatures ($F \cdot T$). A Typical approach is the use of convex envelopes (McCormick, 1976). By using the bound to the variables, Eq. 2.86

$$x^L \leq x \leq x^U; y^L \leq y \leq y^U \tag{2.86}$$

We generate envelops based on the principle given by Eq. 2.87

$$g_1 \geq 0; g_2 \geq 0 \Rightarrow g_1 g_2 \geq 0 \tag{2.87}$$

where, Eq. 2.88

$$\begin{aligned} g_i &= x - x^L \\ g_k &= x - x^U \end{aligned} \tag{2.88}$$

For each of the variables of the product $x \cdot y$

Piecewise Linear Approximation (Algorithm branch and Refine):

The non-convex functions are approximated using a piecewise linear function given by Eq. 2.89:

$$p(x) = \frac{x - x_{k+1}}{x_k - x_{k+1}} f(x) + \frac{x - x_k}{x_{k+1} - x_k} f(x) \tag{2.89}$$

Different algorithms are available in the literature such as BARON, based on Branch and Reduce. It implements a spatial branch and bound algorithm that computes lower bounds for each subproblem utilizing linear relaxations and duality theory. It also relies on bound tightening techniques such as probing and violation transfer. Couenne, is the Convex Over-and Under-Envelopes for Nonlinear Estimation. It is an open-source branch and bound algorithm that obtains a lower bound through an LP relaxation using the reformulation techniques. SCIP: derived from Solving Constrained Integer Problems started as a MILP solver but evolved into a nonconvex MINLP solver. It follows the same approach of BARON and Couenne. Antigone: Branch and bound framework so that the solution of QCQPs benefits from dynamically generated cutting planes and piecewise linear and edge concave relaxations. Lindo Global: Uses a branch and bound/relax approach. It allows a wide range of mathematical functions, including nonsmooth, trigonometric, logical, and statistical. While earlier versions were based on mathematical bounding procedures, the most recent version contains additional heuristics that sacrifices any guarantee of global optimality. Or the Branch and Refine, where the feasible solution provides an upper bound while the lower bound is given by a piecewise linear approximation. An approximation with a larger number of pieces allows the refine stage (Misener and Floudas, 2014).

2.3.1.6.2 Stochastic methods

A number of alternatives are available such as Simulated annealing, Genetic algorithms, Particle swarm optimization. They are particularly useful for models that behave as black boxes, such as CFD or modular process simulators. The algorithm evaluated the progress of the objective function by changing the decision variables. To make it more efficient different techniques evaluate the progress of the objective function including tabu search (Ramírez-Márquez et al., 2018).

2.3.2 Integration

Renewable-based processes have big shoes to fill. With the competition against conventional processes, the optimization and integration are paramount for them to provide attractive alternatives. While conventional technologies have benefited over the decades from the improvement in technology, the advantage of novel facilities is the opportunity to create them from scratch instead of a continuous refinement. The disadvantage is the uncertainty of the operation of units and technologies with a low maturity level. Among the challenges that renewable processes pose on process design it is possible to identify several issues that result in the need for systematic methods to improve them. For instance, most biomass processing technologies operate at moderate to low temperature and show exothermic reactions such as fermentations where heat and water integration are a challenge (Grossmann and Martín, 2010). In addition, most of these processes are characterized by processing dilute mixtures, that is, ethanol synthesis (Martín and Grossmann, 2011) or algae growing (Martín and Grossmann, 2012) where not only energy demand is high to remove a large fraction of it, but also the use of water within the process is an opportunity for better integration. Therefore, concepts and methodologies for energy and water integration as well as the circular economy play an important role form the early design of renewable based processes. This section discusses these three concepts and their application in process design.

2.3.2.1 Energy integration

Pinch analysis has been one of the most well know technologies. It is a graphic method developed back in 1970s by Linnhoff and has had a major impact in improving the energy efficiency of refineries. We can see the method in traditional references (Martín, 2016a; Sinnott and Towler, 2019).

From a mathematical point of view, at process level we can identify the design of **heat exchanger networks** (Yee and Grossmann, 1990) that will allow using the excess of energy available in certain streams to provide thermal energy to cold ones. It is a posterior method once the flowsheet is already in place. The problem is formulated as an MINLP superstructure for the simultaneous selection of the number of heat exchangers and the energy transferred minimizing the cost of the network and that of the utilities. The model consists of

- Global energy balance per stream.
- An energy balance per stream and stage that assumed isothermal mixing.
- Feasibility of temperatures for heat transfer, heat is transferred from higher to lower temperature, and
- Logical constraints so that in case there exit heat exchange there is a temperature gradient.

To include the phase change, a disjunctive formulation was developed by Ponce-Ortega et al. (2008).

In addition, at process level we also find energy integration within distillation columns ether in the form of **multieffect columns** (Karuppiah et al., 2008) where a separation task is divided into a number of columns so that the condenser of the higher-pressure column provides energy to the reboiler of a lower pressure column. The operating pressures and the flows processed at each column allow for the perfect heat integration Eq. (2.90).

$$
\begin{aligned}
&F = \sum_k F_k \\
&Tb_k + dt \leq Tc_{k+1}, k \in COL \\
&Tb_k \geq Tc_k + dt, k \in COL \\
&Qb_k = Qw_{k+1} \\
&P_{k+1} \geq P_k \\
&DesignEqs.\, k
\end{aligned}
\tag{2.90}
$$

It is also possible to evaluate the heat transferred within the units involved in a **sequence of distillation columns**. The model for the sequence of the columns (Biegler et al., 1997) already provides a systematic selection of the separation stages based on yield ad costs. The model is an MILP superstructure that consist of:

- Mass balance to the different intermediates.
- Assumes sharp separation.
- Energy balances are precomputed to all the alternatives and are formulated as linear equations.
- The cost of the column is of the form $C = a \cdot F + b$. It has also been computed for the different columns beforehand.

On top of this model, it is possible to consider the heat transferred from a condenser to a reboiler so that a condenser at a higher temperature can be the hot source for the reboiler providing its temperature is above that of the operation of the reboiler. This fact can be controlled by the operating pressure of the columns and also depends on the composition of the mixture that is separated. Chapter 7 is devoted to process intensification where further details are presented on this technique (Stankiewicz and Moulijn, 2000).

When several processes are integrated, apart from the synergies in the use of raw materials, the basic idea is to be able to integrate energy among them to reduce the consumption of utilities. Examples can be found in the integration of solar and biomass where the energy provided by the biomass helps operate the facility (Vidal and Martín, 2015). In line with this process, not only thermal energy but also power is integrated. Power can be produced in gas or steam turbines and can be used to power pumps and compressors extending the concept of energy integration (Colmenares and Seider, 1987). In this case, both pressures and temperatures are key variables and compression, cooling cycles, heat pumps, and heat engines are in between the heat and the work exchange networks. Recent works on this topic evaluate the integration of heat and work networks (Yu et al., 2020). A larger scale corresponds to the integrated utility plants that will be evaluated along this text.

While these methods have typically been applied on top of the processes themselves, the idea of simultaneous optimization and integration is a well-known concept (Duran and Grossmann,

1986a,b). It allows playing with pressures, temperatures, and compositions of the different streams to make better use of the resources. It is based on pinch location; it allows for embedding the minimum utility target within the process optimization Eq. 2.91. It provides interesting results in case of processes involving recycles, as in the case of Biodiesel production and has been implemented in process optimizers such as MYPSIM (Kravanja and Grossmann, 1990; Kravanja, 2010).

$$\min C = f(x) + c_s Q_s + c_W Q_W$$
$$s.t.$$
$$h(x) = 0$$
$$g(x) \leq 0$$
$$Q_s \geq \sum_{j=1}^{n_c} f_j \left[\max\left\{ 0, t_j^{out} - (T^p - \Delta T_{\min}) \right\} - \max\left\{ 0, t_j^{in} - (T^p - \Delta T_{\min}) \right\} \right]$$
$$- \sum_{i=1}^{n_H} F_i \left[\max\left\{ 0, (T_i^{in} - T^p) \right\} - \max\left\{ 0, (T_i^{out} - T^p) \right\} \right]$$
$$Q_w = Q_s + \sum_{i=1}^{n_H} F_i (T_i^{in} - T_i^{out}) - \sum_{j=1}^{n_c} f_j \left(t_j^{out} - t_j^{in} \right)$$

(2.91)

Due to the complexity of identifying hot and cold streams, Quirante et al. (2018) extended the method to account for the automatic classification of streams based on a disjunctive formulation.

At complex level is important to mention **total site integration**. From the modeling point of view, the formulations are similar to the ones presented for heat exchanger networks and heat and power integration, but the scale of application increases to chemical complexes. In this case, the transportation of fluids and energy becomes important for the feasibility of the operation. The concept was established in 1997 (Klemeš et al., 1997) and has attracted a lot of attention lately.

2.3.2.1.1 Water integration

Water was a cheap abundant resource that rarely represented a limitation in process design. However, two issues have altered the paradigm. The current world situation where water scarcity is becoming an issue in many regions (IWIM, 2006) so that water is more expensive than crude oil and the expected scarcity over the following decades. The second is the use of renewable resources that require water for their production or operation such as biomass growing or solar energy. While biomass can be grown based on rainfall, it typically requires additional water (Elcock, 2008). In addition, the availability of solar is inversely related to that of water while it is required for the operation of such plants either as cooling agent in wet cooling towers or to clean the PV panels and mirrors. Water networks were presented back in 1980 (Takama et al., 1980), but the complexity of addressing the mathematical problem Eq. 2.92 led to the development of graphical methods, Wang and Smith (1994). The advances in global optimization allowed solving the mathematical formulation (Galan and Grossmann, 1998). Recently, efforts for the integration of water and energy simultaneously have been carried out in particular for biofuels facilities due to the challenges that they pose. They typically consume a fair amount of water and require to be as efficient as possible to be competitive.

$$
\begin{aligned}
\min \quad & \varphi = FW + \sum_{\substack{t \in TU \\ i \in t_{out}}} F^i \\
s.t. \quad & F^k = \sum_{i \in m_{in}} F^i \quad \forall m \in MU, k \in m_{out} \\
& F_k C_j^k = \sum_{i \in m_{in}} F_i C_j^i \quad \forall j, \ \forall m \in MU, k \in m_{out} \\
& F^k = \sum_{i \in s_{out}} F^i \quad \forall s \in SU, k \in s_{in} \\
& C_j^k = C_j^i \quad \forall j, \ \forall s \in SU, \ \forall i \in s_{out,} \ k \in s_{in} \\
& P_{in}^p C_j^k + L_j^p = P_{out}^p C_j^i \quad \forall j, \forall p \in PU, \ k \in p_{in}, i \in p_{out} \\
& F^k = F^i \quad \forall t \in TU, \ \forall i \in t_{out}, \ \forall k \in t_{in} \\
& C_j^i = \beta_j^t C_j^k \quad \forall j, \ \forall t \in TU, \ \forall i \in t_{out}, \forall k \in t_{in} \\
& F^{k,min} \le F^k \le F^{k,max} \quad \forall k \\
& C_j^{k,min} \le C_j^k \le C_j^{k,max} \quad \forall j, k
\end{aligned}
\tag{2.92}
$$

2.3.2.1.2 Circular economy

We finally arrive to the concept of circular economy that can be considered as the milestone for the efficiency of any industry and in particular can provide important benefits in the design of renewable based processes. The idea is to be able to integrate waste streams, not only mass but also energy streams, to provide further value out of them reducing the consumption of natural resources (Reh, 2013). While for many years the production system was linear, the large amount of residues generated coupled with the challenging handling and the limited availability of resources have paved the way towards a more sustainable process design. Even though the attention to circular economy is new, within the process community it is a concept that has taken many names such as the energy integration, water integration, recycle of streams, intensification, and so on. Solvay process is a very good example of this already back in 1864. Therefore, although the attention and the name are new, the ideas are already in the process engineering community as this section has summarized.

2.4 Economic evaluation

Estimating the investment and production costs for renewable-based processes is a challenge. This is the first metric used to evaluate a process that come to compete or substitute a well-established one. Apart from the social or environmental benefits that the use of renewable resources may bring, the first barrier toward its penetration in the market is cost. Therefore, the estimation of the cost is paramount in the analysis. Cost estimation is an art, but all design books present a methodology. Here we follow the one by (Sinnott and Towler, 2019), the factorial method. It is based on using a series of factors to estimate both, investment and production costs. Different levels of accuracy can be distinguished. With the guidelines we present below, the 2nd - 3rd level can be achieved, which means that an error of 20% must be assumed Table 2.2.

TABLE 2.2   Accuracy of cost estimation.

	Estimate	Error	Used to	Method
1	Order of magnitude	40%–50%	Profitability analysis	Comparison with similar plants
2	Study	25%–40%	Preliminary design	Preliminary engineering and equipment sizing
3	Preliminary	15%–25%	Budget approval	P&ID and equipment sizing
4	Definitive	10%–15%	Construction	Full equipment design.
5	Detailed	5%–10%	Turnkey contract	Detailed engineering. Quotation

TABLE 2.3   Factors for the estimation of the fixed capital cost (Sinnott and Towler, 2019).

Item	Processing		
	Fluids	Solids	Fluids and Solids
Equipment cost	C_e	C_e	C_e
f_1 Equipment erection	0.3	0.6	0.5
f_2 Piping	0.8	0.2	0.6
f_3 Instrumentation and control	0.3	0.2	0.3
f_4 Electrical	0.2	0.15	0.2
f_5 Civil engineering	0.3	0.2	0.3
f_6 Structures and buildings	0.2	0.1	0.2
f_7 Lagging and paint	0.1	0.05	0.1
Physical Plant cost(CP) = $\left(\sum_i f_i\right) C_e$	3.3	2.5	3.2
Offsites (OF)	0.3	0.4	0.4
Design and Engineering (D&E)	0.3	0.2	0.25
Contingency (X)	0.1	0.1	0.1
Total Fixed capital cost (CF) = CP(1 + OF) · (1 + D&E + X) = C_e · F	6.0	4.55	6.05

2.4.1 Investment cost estimation

Investment costs are based on the costs of the units. Table 2.3 shows the coefficients for processes that process fluids, solids or a mixture of them. The starting point is what is specific for the process, the estimation of the cost of the units, that in case of renewable-based processes, are particular such as fermenters, heliostat fields, wind turbines. Therefore, a modification is required to estimate these plants investment. Typically, the heliostat field, the PV panels or the wind farm cost is provided already as a block. Therefore, the chemical plant that uses the power from any of them is to be estimated as two sections where one is provided by the values in Table 2.3, while the power source is to be estimated as such since the data already refers to installed facility Eq. 2.93.

$$I = 1.1(I_{Chemical} + I_{Power}) = 1.1\left(F \cdot C_{eq} + I_{Power}\right) \tag{2.93}$$

Where the factors for the different items can be found in Table 2.3. On top of the fixed capital cost, the working capital is around 10% of the fixed capital.

Thus, all comes down to estimating the cost of the units. While heat exchangers, compressors, or vessels are widely known and studied, some of them are not that common such as gasifiers, fermenters, or digestion tanks (Almena and Martín, 2015). Table 2.4

TABLE 2.4 Equipment costs estimation.

Unit	Additional info	Cost (€)$_{2020}$
Steam turbine[a]		$633,000(W(MW))^{0.398}$
ORC Turbine[u]		$4,750(W(kW))^{0.75}$
Gas turbine[a]		$3,800(W(MW))^{0.754}$
Dryer[z]	Rotary	$11743\left(A\left(m^2\right)\right)^{0.45}$
	Steel	
	$< 450m^2$	
Dryer[z]	Spray	$49965A\left(m^2\right))^{0.28}$
	Steel	
	$450 < A(m^2) < 9000$	
Boiler[a]	Biomass	$1,340,000(W(MW))^{0.694}$
Gasifier[a]	Biomass	$1600M\left(kg/h\right)^{0.917}$
Gasifier[x]	Coal <150 psi	$1,164,817\left(F(kg/s)\right)^{0.698}$
	600 psi	$1,592,418\left(F(kg/s)\right)^{0.681}$
	1500 psi	$2,179,923\left(F(kg/s)\right)^{0.684}$
Cyclons[x]		$4,463\left(F(m^3/s)\right)^{0.8945}$
Electrostatic precipitator[x]		$5,111.6\left(F(m^3/s)\right) + 149,832$
Filter[x]		$5,497.4\left(F(m^3/s)\right) + 18,219$
Wet Scrubber[x]		$4796.1\left(F(m^3/s)\right) + 10,173$
Cooling towers[x]		$3229.2 \cdot (W(kW))^{0.5909}$
Anaerobic digester[b]	$<6000m^3$	$360\ (€/m^3)$
Centrifuges	Stainless Steel	$142,188\ (M(t/h)) + 20783$
	Up to 60 t/h	

(Continued)

2. Design principles

TABLE 2.4 (Continued)

Unit	Additional info	Cost (€)$_{2020}$
Fermentors[x]	Stainless Steel	$78,809\left(V(m^3)\right)^{0.53}$
	Typically below 500m^3	
Storage tanks[x]	cone roof, and flat bottom made of stainless steel	$6,839.8\left(V(m^3)\right)^{0.65}$
Vessels[x]	vertical vessel stainless steel	$Volume = \left(\frac{m\cdot}{\rho}\right)\tau; L = 4D_c; D_C = \sqrt[3]{\frac{6\cdot Volume}{7\cdot\pi}}$
		$W = \rho_{steel}\left(\pi\left[\left(\frac{D_c}{2}+e\right)^2 - \left(\frac{D_c}{2}\right)^2\right]\right.$
		$\left. L + \frac{4}{3}\pi\left[\left(\frac{D_c}{2}+e\right)^3 - \left(\frac{D_c}{2}\right)^3\right]\right)$
		$e = 0.0023 + 0.003D_c$
		$62379\left(W(kg)\right)^{0.878}$
Trays cost	Stainless steel	$315 \cdot e^{0.72\cdot D_C(m)}$
Compressors [x]	P < 125 psi	$335.27(W(kW)) + 36,211$
	P < 1000 psi	$2534.80(W(kW))^{0.8}$
Heat Exchangers	Stainless steel 316 working up to 150 psig	$815.55A(m^2) - 6601.9$
Reboiler	AKT	
	A$_{max}$ < 2000m^2	
Heat exchangers. including condensers	Stainless steel 316 working up to 150 psig	$293.59A(m^2)^{1.4915}; A < 25m^2$
		$1,593.8A(m^2) + 2,584.2; 25 < A < 140m^2$
	Shell and tubes	$22.234A(m^2)^{0.4671}; A > 140m^2$
	A$_{max}$ < 2000m^2	
Cristalizer[z]	Cooling evaporative	$165,394(M(t/h))^{0.6883}$
	Stainless Steel	
	< 100 t/h	
Membranes[m]	Gas separation	$50\ A(m^2)$
Membranes[n]	Reverse osmosis	$30\ A(m^2)$. Module of 40 m^2
Ponds[y]	Concrete	$0.62\ A(m^2)$
	1000m^2	
Wind turbines[c]		$1600\ W(kW_e)$
PV Panels[d]		$1050\ W\ (kW_e)$
Fuel cells[e]		$4600\ W\ (kW_e)$

(Continued)

TABLE 2.4 (Continued)

Unit	Additional info	Cost (€)$_{2020}$
Electrolyzer[f]		300−2000 W(kW$_e$)
Receiver [h]		200 W(kW$_t$)
Heliostats[g]		$120 \cdot A(m^2)$
Parabolic trough[p]		$170 \cdot A(m^2)$
Well [q]		$0.982355 \cdot e^{4.128 \cdot 10^{-04}(\text{Depth}(m))}$

[a]*Caputo et al. (2005).*
[b]*Taifouris and Martín (2018).*
[c]*Stehly et al. (2018).*
[d]*Fu et al. (2018).*
[e]*US DOE (2016).*
[f]*Saba et al. (2018).*
[g]*Sánchez and Martín (2018).*
[h]*Turchi and Heath (2013).*
[m]*He et al. (2018).*
[n]*Judd (2017).*
[p]*Kurip and Turchi (2015).*
[q]*Mansure and Blakenship (2008).*
[u]*Akrami et al. (2017).*
[x]*Almena and Martín (2015).*
[y]*http://www.aces.edu/pubs/docs/A/ANR-1114/ANR-1114.pdf.*
[z]*Garret (1989).*

shows a list of units' costs for different types of equipment. Design literature, process simulators, and web tools (http://www.matche.org) provide additional means to estimate cost for different units and, some cases, the results can vary from one source to the other. However, it is important to have correlations available to be used within modeling, simulation, and optimization formulations. Note that if one unit maximum size is reached, another one is to be bought. It is where economies of scale, observed in the exponents of the equations used, loose the advantage.

Equipment cost must be updated, for that inflation indexes are used such as the chemical price index, industrial energy price index, and so on. The indexes depend on the country of reference of the cost. To update any cost Eq. 2.94

$$C_{2020} = C_{year} \frac{I_{2020}}{I_{year}} \tag{2.94}$$

The material cost can be updated using the following factors, assumed 1 for carbon steel, 2 for stainless steel, 3.4 for Monel (http://www.matche.org; (Sinnott and Towler, 2019)).

2.4.2 Production cost estimation

They correspond to the costs related to the operation of the facility. It is possible to classify them into variable costs, those related to the use of materials, fixed costs, those related

TABLE 2.5 Estimation of the production costs.

Item	Range of values
Variable costs	
1. Raw materials	Market value and data from M&E balances
2. Utilities	Market value and data from M&E balances
3. Miscellaneous materials	10% of (5)
4. Shipping and packing	Variable. High for Consumer goods but negligible for bulk chemicals.
Total VC	
Fixed costs	
5. Maintenance	5%−10% of fixed capital
6. Operating labor	Personnel and salaries cost
7. Laboratory costs	20%−25% of (6)
8. Supervisions	20% of (6)
9. Plant overheads	50% of (6)
10. Capital Charges	15% of fixed capital
11. Insurance	1% of fixed capital
12. Taxes	2% of fixed capital
13. Royalties	1% of fixed capital
Total FC	
Direct Production costs (DPC)	VC + FC
14. Sales expense	20%−30% of DPC
15. General overheads	
16. Research and development	
Total EOR&D	EOR&D
Annual production costs	VC + FC + EOR&D

to maintenance, labor, taxes, and so on, and finally those related to additional overheads and research and development. Table 2.5 shows a scheme and some typical rages of values to estimate the production costs.

2.5 Multiscale modeling and simulation

While most of this book focuses on process scale, the use of renewable resources results in the need to consider a wider scope. We can distinguish two types of scales, size, and time. Although both can be considered simultaneously, for the sake of presenting them, we separate the effects and definitions.

2.5.1 Multiscale analysis: size dimension

Although this textbook deals with the level of the process analysis and design, this scale is in the middle of the unit and network levels (see Fig. 2.9). The macroscopic and strategic decision making takes place at network level, that provides a wide overview of the entire system. However, the decision on the selection of technologies and the cost relies on the performance of the processes themselves. It is not possible to use fully detailed process models at network level, the problems and evaluations would be intractable, but the information must be the proper one. Network problems model the different units/technologies as black boxes characterized by their yield and economics. However, a detailed process analysis is required for this information to be representative of the transformations evaluated. The development of these input-output models relies on the use of accurate process models and techniques like the ones presented in this chapter (i.e., DOE, correlations, etc.) can be used. In addition, the use of information from different sources is an issue. If the estimation methods are not the same, the network level comparison between technologies would be tricky and the results misleading. It is possible that the selection of one technology over the other may be due to the fact that the estimation reported a lower cost. Thus, the same basis must be provided for all processes. This is typically neglected at supply chain level analysis, due to the time-consuming step of process analysis. An example of good practice is the work by Martin and Grossmann (2018) where all the process model surrogates were developed individually over 10 years work within the same group using the same cost estimation procedure.

The other scale extreme is the detailed unit level. This is key to understand the operation and yield of the different units. It is typically based on the analysis of the physico-chemical phenomena that take place in a particular unit. Detailed CFD, DEM models including hydrodynamics, mass and heat transfer and multiphase analysis are included. However, the analysis of the entire process and or the selection of the particular technology for a transformation stage requires also a higher or wider view. As a result, simpler or surrogate models are required for the design of the process.

Typically, a MILP problem is formulated. However, behind this framework, process design and surrogate modeling is required. It is here where the **multiscale analysis** concept is important.

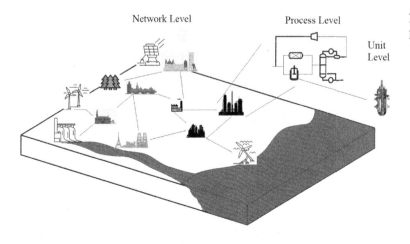

Network Level   Process Level

Unit
Level

FIGURE 2.9 Multiscale process modeling.

One important aspect related to plant size is the economies of scale. The traditional chemical industry has benefited from them over the decades. It is not twice as expensive to produce double capacity due to the savings in equipment and manpower. The typical six tenths rule has been used as a default rule to scale up the cost of an entire process or a particular unit as follows Eq. 2.95.

$$C_{Size2} = C_{Size1} \left(\frac{Size2}{Size1}\right)^{0.6} \tag{2.95}$$

Apart from this general rule, in Table 2.4 for particular units and the Appendix of Garret (1989) we can find coefficients for the scale up/down of typical chemical processes. However, this paradigm has changed with renewable processes. There are two important features. First, apart from discounts for large shipments, the price of a solar field or wind farm scales proportionally with the number of PV panels or wind turbines. In addition, for chemical reactors, fermenters or digesters, mixing is an important issue and maximum standard volumes are easily reached meaning that the unit must be duplicated to scale up a process. In addition, **distributed production** has been a hot topic due to the regional availability of renewable resources such as biomass, solar, or wind. As a result, instead of scaling up to benefit from the economies of scale, scale down has become a trend. Scale down is not as straightforward as it may be expected. In some cases, the technologies suggested for a particular unit change with the scale. Two simple examples can be highlighted here. Heat exchangers are shell and tube above 20 m^2 approximately. Below that, double pipe heat exchangers are to be used. In addition, air separation at large scale uses the Linde double column, while at smaller scale Pressure Swing Adsorption (PSA) or membrane units are recommended. A good example is presented in ammonia production in the recent paper (Sánchez and Martín, 2018).

Therefore, apart from the change in technology, scale up or down has to be properly carried out all the way up from the units. The cost of each stage, unit by unit, must be computed taking into account the number of units required based on the standard maximum capacity as follows, Eq. 2.96.

$$n = \left\lceil \frac{\text{Capacity}}{\text{Max Standard Capacity}} \right\rceil$$
$$C_{e,i} = (n_i - 1) \cdot C_e(\text{Max Standard Capacity}) + C_e(\text{Capacity}) \tag{2.96}$$

If the cost is linear, there is not much of a difference in the estimation but is there is some economy of scale the cost of the number of units required to perform such a stage differs. Using this consideration, we can proceed to evaluate the investment and production costs as presented in the previous section. For instance, for the fixed capital costs it becomes, Eq. 2.97:

$$I = F \sum_i C_{e,i} \tag{2.97}$$

For the production costs, note that variable costs are proportional to the production capacity, and therefore they do not present economies of scale. However, the fixed costs are affected since they depend on the investment.

2.5.2 Multiscale design: time dimension

Another important issue in process design involving renewables is the variability of the resources. This is concerned with the **time scale**. The design must be feasible over time, but the use of the capacity varies. Including time scales and uncertainty in the design of a process results in large MINLP problems whose solution is complex, see Chapter 14 for further details. While MINLP solution procedures based on the ones presented along the text (i.e. Branch and Bound, Benders decomposition) have been used to design particular units, such as air coolers for concentrated solar power plants (Luceño and Martín, 2018), the extension to process design requires additional modeling approaches.

Multiperiod design (Martin and Grossmann, 2017): We can address the solution of such a problem in several ways. The most computationally expensive requires the definition of all the mass and energy flows as a function of the period. Alternatively, we can take advantage of the fact that *intensive variables* are constant with the production capacity, such as temperature, pressures, or concentrations. *Extensive variables* are the ones which fluctuate over time as a function of the availability of resources. For the design of a process affected by the variability of the resources over time, we consider two groups of equations. Those whose operation is not directly affected by the environmental conditions, g^{II}, that is, processing steps, and those that depend on the input from the environment, g^I, either in the form of raw materials (water, air) or renewable energy (wind velocity, solar, or cooling systems). The variables belonging to the first group, g^I, cannot be reformulated since its evaluation is directly dependent on time dependent variables. However, for the second group we can decompose it as follows. We consider a reference time interval for which the model of the process is formulated. The operating variables of any other time period are computed as a function of that reference. We denote as α the scale-up factor with respect to the reference. It is related to the availability of resources. In addition, the parameter η is used to compute the fraction of capacity used with respect to the full capacity. The design variables, d, are defined to be larger or equal that the operating variable of each period of time. Thus, the formulation becomes, Eq. 2.98:

$$\min Z = C(d) + \sum_{t=1}^{\Gamma} f_t(d, x_t)$$

$$st.$$
$$g_t^I\left(d, w_t^{int,I}, w_t^{ext,I}\right) \leq 0 \ t = 1, \ldots, t_f$$
$$g_{ref}^{II}\left(d, w_{ref}^{int,II}, w_{ref}^{ext,II}, \theta_{ref}\right) \leq 0$$
$$w_t^{int,II} = w_{ref}^{int,II} \quad int \in \{P, T\} \tag{2.98}$$
$$\eta_t \alpha_t w_{ref}^{int,II} = w_t^{ext,II} \quad ext \in \{fc, F, Q, W\}$$
$$\alpha_t = f\left(t, d, w_{ref}^{int}, w_{ref}^{ext}, \theta_t\right)$$
$$\eta_t \in \{0, 1\}$$
$$\theta_t = f(\text{environmental } conditions) = f\left(w_t^{int,I}, w_t^{ext,I}\right), t = 1, \ldots, \Gamma$$
$$d \geq do_t$$

Multiperiod design under uncertainty (Martin, 2016b). This problem corresponds to an extension of the previous one. In this case in addition to multiperiod operation, there is uncertain availability of the resources, the variables add an addition dimension to their operation, the scenario under which they are evaluated. The size of the problem to consider all scenarios and periods is untractable if the level of process detail is large enough. Therefore, a two-stage procedure is presented, Eq. 2.99.

1. Optimization of the process to determine the yield and economics on a reference period. This corresponds to a superstructure optimization problem to select the technologies and the operating conditions of major units.
2. Development of input-output models assuming the fact that process operation will remain constant. This is not true for cooling using atmospheric water or other operations depending on the weather conditions. For this, a formulation similar to the one presented above would be needed.
3. Problem formulation including time and scenario dimensions for the extensive variables. The design variables, d, are defined as to be larger or equal that the operating variable, $do_{t,sc}$, of each scenario and time

$$
\begin{aligned}
&\min Z = C(d) + \sum_{t=1}^{\Gamma} f_t\left(d, x_{t,sc}\right) \\
&st.g_t\left(d, x_{t,sc}\right) \leq 0 \qquad t = 1, \ldots, \Gamma \\
&d \geq do_{t,sc}
\end{aligned}
\tag{2.99}
$$

Exercises

P2.1 Derive the equation for the power required to compress a gas with polytropic behavior.

$$
W = -\int_{P_{ini}}^{P_f} pdV
$$

P2.2 In a stirred tank reactor, the power required by the rotor is a function of the diameter of the impeller, the rotation speed, the liquid density and viscosity. Using the Pi Theorem, determine the relation between the power and the variables.

P2.3 Develop a design of experiments model from the data on the yield of biodiesel production (Canakci and Gerpen, 1999).

Temperature	Ratio	Cat	Conversion
25	6	3	10
45	6	3	55
60	6	3	85
60	3.3	3	77
60	3.9	3	80
60	6	3	87
60	20	3	95
60	30	3	98
60	6	1	72
60	6	3	88
60	6	5	95

P2.4 Estimate the height of the point (4,5) using the IDW kriging for the set of coordinates provided below. Compare with the actual value if the function is $Z = x^2 + 2xy - 3y + 2$.

Run	x	y	Z
1	1	2	7
2	2	3	18
3	2	6	30
4	3	5	41
5	4	8	82
6	5	9	117
7	4	1	26
8	5	3	57
9	3	9	65
10	6	6	110

P2.5 Using the Simplex tableau compute the optimum for the following problem.

$$\min \quad Z = -x_1 - 2x_2$$
$$s.t. \quad x_1 + x_2 \leq 4$$
$$x_1 + 5x_2 \leq 15$$
$$-4 \cdot x_1 + 5x_2 \leq -10$$
$$x_1, x_2 \geq 0$$

P2.6 Formulate the Big M and the convex hull of the disjunction.

$$[x_1 - x_2 \leq -1] \vee [-x_1 + x_2 \leq -1]$$
$$0 \leq x_i \leq 5$$

P2.7 Solve the follow LP problem and compare the solution to **P2.14**.

$$Min \quad y_1 + 2y_2$$
$$st. \quad (y_1 + 2y_2) \geq 5$$
$$(4y_1 - y_2) \geq 1$$
$$y_2 \geq 3$$
$$y_i \geq 0$$

$y_1 = 1; y_2 = 3.$

P2.8 Prove that by taking the floor of an equality generates a greater or equal inequality.

P2.9 Solve the NLP problem using the active constraint method.

$$\min f(x) = \left(\frac{(x_1 - 5)^2}{4} + \frac{(x_2 - 3)^2}{2} \right)$$

s.t.

$$g_1 = -x_1 + x_2 \leq 3$$
$$g_2 = x_1 + \frac{1}{2}x_2 - 6 \leq 0$$
$$g_3 = x_2 \geq 1$$

P2.10 Compute a Gomory cut for the following problem.

$$\max \quad z = x_1 + x_2$$
$$st.$$
$$10x_1 + 3x_2 \leq 55$$
$$4x_1 + 20x_2 \leq 55$$
$$x_1, x_2 \geq 0;$$

P2.11 Update the Lagrange multipliers. Assume $\lambda_1 = 1.9$; $\lambda_2 = 3.1$; $\lambda_3 = 1.1$.

$$\min \quad 4x_1 + 1x_2 + 3x_3 + 5x_4$$
$$st \quad x_1 + x_3 \geq 1$$
$$x_2 + x_4 \geq 1$$
$$x_1 + x_3 + x_4 \geq 1$$
$$x_j \in (0, 1), j = 1, .., 4$$

P2.12 Compute the feasibility and optimality cuts of the Benders decomposition for the following problem.

$$\min_{x,y} 3x_1 + 2x_2 - 4x_3 - 5y_1 + 3y_2 + 2y_3$$
$$st \quad 2x_1 - 4x_2 + x_3 - 3y_1 + 2y_2 - 3y_3 \geq 4$$
$$x_1 + 2x_2 - 4x_3 - 5y_1 - 4y_2 + 2y_3 \geq -2$$
$$x_i \leq 6; i = 1, 2, 3$$
$$x \in Z_+^3, y \in R_+^3$$

P2.13 Develop the concept of the dual for linear programs.
P2.14 Formulate the dual for the problem. See Problem P2.13 for reference.

$$\max \quad 5x_1 + 4x_2$$
$$st. \quad x_1 + x_2 \leq 2$$
$$3x_1 + x_2 \leq 3$$
$$x_1 \geq 0, x_2 \geq 0$$

References

Akrami, E., Chitsaz, A., Nami, H., Mahmoudi, S.M.S., 2017. Energetic and exergoeconomic assessment of a multi-generation energy system based on indirect use of geothermal energy. Energy 124, 625–639.

Albers, D.J., Reid, C., 1986. An interview with George B. Dantzig: the father of linear programming. Coll. Mathematics J. 17 (4), 292–314.

Almena, A., Martín, M., 2015. Techno-economic analysis of the production of epichlorohydrin from glycerol. Ind. Eng. Chem. Res 55 (12), 3226−3238.

Baasel, W.D., 1989. Preliminary Chemical Engineering Plant Design. van Nostrand Reinhold, Amsterdam.

Barkeley, R.W., Motard, R.L., 1972. Decomposition of nets. Chem. Eng. J. 3, 265.

Bashforth, F., Adams, J.C., 1883. Theories of Capillary Action. Cambridge University Press, London.

Benders, J.F., 1962. Partitioning procedures for solving mixed-variables programming problems. Numerische Mathematik 4.

Biegler L., 1983. Simultaneous Modular simulation and optimization. DRC-O6-JW-83.

Bielger L., 2010. Nonlinear Programming: Concepts, Algorithms, and Applications to Chemical Processes. MOS-SIAM Series on Optimization. Philadelphia PA.

Biegler, L., Grossmann, I.E., Westerberg, A.W., 1997. Systematic Methods of Chemical Process Design. Prentice Hall.

Bird, R., Stewart, W.E., Lightfoot, E.N., 2006. Transport Phenomena. John Wiley and Sons, New York.

Branan, C.R., 2000. McGraw Hill 2a Edición.

Broyden, C.G., 1965. A class of methods for solving nonlinear simultaneous equations. Maths. Comp. Am. Math. Soc. 19 (92), 577−593.

Buckingham, E., 1914. Phys. Rev. 4, 345−376.

Caballero, J.A., Grossmann, I.E., 2008. AIChE J. 54 (10), 2633−2650.

Canakci, M., Gerpen, J.V., 1999. Biodiesel Production via Acid Catalysis. Trans. ASAE 42 (5), 1203−1210.

Caputo, A.C., Palumbo, M., Pelagagge, P.M., Scacchia, F., 2005. Economics of biomass energy utilization in combustion and gasification plants:effects of logistic variables. Biomass Bioenergy 28, 35−51.

Colmenares, T.R., Seider, W.D., 1987. Heat and power integration of chemical processes. AIChE J. 33 (6), 898−915.

Cozad, A., Sahinidis, N., Miller, D.C., 2014. Learning surrogate models for simulation-based optimization. AIChE J. 60 (6), 2211−2227.

Dempe, S., 2002. Foundations of Bilevel Programming. Kluwer, Dordrecht.

Dorfman, K.D., Daoutidis, P., 2017. Numerical Methods with Chemical Engineering Applications (Cambridge Series in Chemical Engineering), 1st Edition Cambridge University Press, Cambridge UK.

Douglas, J.M., 1988. Conceptual Design of Chemical Processes. McGraw-Hill, New York.

Duran, M.A., Grossmann, I.E., 1986a. An outer-approximation algorithm for a class of mixed-integer nonlinear programs. Math. Programm. 36, 307−339.

Duran, M.A., Grossmann, I.E., 1986b. Simultaneous optimization and heat integration of chemical processes. AIChE, J. 32, 123−138.

Eggeman, T., 2005. Updated Correlations for GTI Gasifier WDYLD8. NREL, Golden, CO, June 27.

Elcock D. (2008) *Baseline and Projected Water Demand Data for Energy and Competing Water Use Sectors*, ANL/EVS/TM/08−8.

Fogler, S., 1997. Elements of Chemical Reaction Engineering, 3rd Ed. Prentice Hall, New York.

Fu R., Feldman D., Margolis R. United States Solar Photovoltaic System Cost Benchmark: Q1 2018. NREL/PR-6A20−72133.

Galan, B., Grossmann, I.E., 1998. Optimal design of distributed wastewater treatment networks. Ind. Eng. Chem. Res. 37, 4036−4048.

Garret, D.E., 1989. Chemical Engineering Economics. Springer, Switzerland.

Gear, C.W., 1971. Numerical Initial-Value Problems in Ordinary Differential Equations. Prentice Hall, Englewood Cliffs, New York.

Gentle, J.E., 2010. Matrix Algebra. Theory, Computations and applications in Statistics. Springer, Berlin.

Geoffrion, A.M., 1972. Generalized benders decomposition. J. Opt. Theor. Appli 19 (4), 237−260.

Gomory, R., 1960. An algorithm for the mixed integer problem, Technical Report RM-2597, The Rand Corporation (1960).

GPSA, 2012. Engineering Processing data book (Gas Processing) FTS version.

Grossmann, I.E., 2021. Advanced optimization for process systems engineering. Cambridge Series in Chemical Engineering, UK.

Grossmann, I.E., Martín, M., 2010. Energy and Water Optimization in Biofuel Plants. Chinese J. Chem. Eng, 18 (6), 914−922.

Grossmann, I.E., Trespalacios, F., 2013. Systematic modeling of discrete-continuous optimization models through generalized disjunctive programming. AIChE J. 59, 3276−3295.

2. Design principles

He, X., Chu, Y., Lindbrathen, A., Hillestad, M., Hagg, M.-B., 2018. Carbon molecular sieve membranes for biogas upgrading:techno-economic feasibility analysis. J. Clean. Prod. 194, 584–593.

Henao, C.A., Maravelias, C., 2011. Surrogate-based superstructure optimization framework. AIChE J. 57 (5), 1216–1232.

Himmelblau, D.M., Bischoff, K.B., 1968. Process Analysis and simulation. J Wiley and Sons, New York.

Hopfield, J.J., 1988. Artificial neural networks. IEEE Circuits Devices Mag. 4 (5), 3–10.

Houghen, O.A., Watson, K.M., Ragatz, R.A., 1959. Chemical Process Principles. Vol 1. Material and Energy Balances. Wiley, New York.

IWIM, 2006. Insights from the comprehensive assessment of water management in agriculture. Stockholm World Water Week, 2006. http://news.bbc.co.uk/2/shared/bsp/hi/pdfs/21_08_06_world_water_week.pdf.

Judd, S.J., 2017. Membrane technology costs and me. Water Res. 122, 1–9.

Karuppiah, R., Peschel, A., Grossmann, I.E., Martín, M., Martinson, W., Zullo, L., 2008. Energy optimization of an ethanol plant. AICHE J. 54 (6), 1499–1525.

Kincaid, D., Cheney, W., 1991. Numerical Analysis: Mathematics of Scientific computing. Brooks/Cole Publishing, California.

Klemeš, J., Dhole, V.R., Raissi, K., Perry, S.J., Puigjaner, L., 1997. Targeting and design methodology for reduction of fuel, power and co2 on Total Sites. Appl. Therm. Eng. 17 (8–10), 993–1003.

Kocis, G.R., Grossmann, I.E., 1987. Relaxation strategy for the structural optimization of process flow sheets. Ind. Eng. Chem. Res. 26, 1869–1880.

Kravanja, Z., 2010. Challenges in sustainable integrated process synthesis and the capabilities of an MINLP process synthesizer. MipSyn. Comp. Chem. Engng 34, 1831–1848.

Kravanja, Z., Grossmann, I.E., 1990. PROSYN-an MINLP process synthesizer. Comp. Chem. Eng. 14 (12), 1363–1378.

Kurip, P., Turchi, C.S., 2015. Parabolic Trough Collector Cost Update for the System Advisor Model (SAM) Technical Report NREL/TP-6A20–65228, November 2015.

León, E., Martín, M., 2016. Optimal production of power in a combined cycle from manure based biogas. Energ. Conv. Manag. 114, 89–99.

Luceño, J.A., Martín, M., 2018. Two-step optimization procedure for the conceptual design of A-frame systems for solar power plants. Energy 165, 483–500.

Mansure A.J., Blakenship D.A. Geothermal well cost analyses; 2008. SAND2008–3807C.

Marquardt, W., 1995. Numerical methods for the simulation of differential-algebraic process models. In: Berber, R. (Ed.), Methods of Model Based Process Control, vol 293. Springer, DordrechtNATO ASI Series (Series E: Applied Sciences). Available from: https://doi.org/10.1007/978-94-011-0135-6_2.

Martín, M., 2016a. Industrial Chemical Process. Analysis and Design. Elsevier, Oxford.

Martin, M., 2016b. Methodology for solar and wind based process design under uncertainty: methanol production from CO2 and hydrogen. Comp. Chem. Eng. 92, 43–54.

Martín, M., 2019. Introduction to Software for Chemical Engineers, 2nd (Ed.) CRC Press, Boca Raton USA.

Martín, M., Grossmann, I.E., 2011. Energy optimization of lignocellulosic bioethanol production via gasification. AIChE J. 57, 12. 3408, 3428.

Martín, M., Grossmann, I.E., 2012. Simultaneous optimization and heat integration for biodiesel production from cooking oil and algae. Ind. Eng. Chem. Res. 51 (23), 7998–8014.

Martín, M., Grossmann, I.E., 2013. On the systematic synthesis of sustainable biorefineries. Ind. Eng. Chem. Res. 52 (9), 3044–3064.

Martin, M., Grossmann, I.E., 2017. Optimal integration of a self-sustained algae based facility with solar and/or wind energy. J. Clean. Prod. 145 (1), 336–347.

Martin, M., Grossmann, I.E., 2018. Optimal integration of renewable based processes for fuels and power production: Spain case study. Appl. Energy 213, 595–610.

Martín, M., Montes, F.J., Galán, M.A., 2007. Oxygen transfer from growing bubbles: effect of the physical properties of the liquid. Chem. Eng. J. 128, 21–32.

McCormick, G., 1976. Computability of global solutions to factorable nonconvex programs: Part I - convex underestimating problems. Math. Program. 10 (1).

Misener, R., Floudas, C.A., 2014. ANTIGONE: algorithms for continuous/integer global optimization of nonlinear equations. J. Glob. Opt. 59, 503–526.

Montgomery, D.C., 2001. Design and Analysis of Experiments, 5th (Ed.) John Wiley & Sons. Inc, New York.

Moran, M.J., Shapiro, H.N., 1993. Fundamentals of Engineering thermodynamics, 2nd John Wiley and Sons.

Nocedal, J., Wright, S.J., 2006. Numerical Optimization, 2nd Ed Springer, New York.

Ortega, J.M., Rheinboldt, W.C., 1970. Iterative Solution of Nonlinear Equations in Several Variables. Elsevier. Academic Press, London.

Pantelides, C.C., Gritsis, D., Morison, K.R., Sargent, R.W.H., 1988. The mathematical-modeling of transient systems using differential algebraic equations. Comput. Chem. Eng. 12 (5), 449–454.

Papoulias, S.A., Grossmann, I.E., 1983. A structural optimization approach in process synthesis I: utility systems. Comput. Chem. Eng. 7 (6), 695–706.

Perry, R.H., Green, D.W., 2018. Perry's Chemical Engineer's Handbook, 8th (Ed.) McGraw-Hill, New York.

Phillips, S., Aden, A., Jechura, J., Dayton, D. Eggeman, T., 2007. Thermochemical ethanol via indirect gasification and mixed alcohol synthesis of lignocellulosic biomass. NREL/TP -510–41168 April 2007.

Poling, B.E., Prausnitz, J.M., O'Connell, J.P., 2000. The Properties of Gases and Liquids. McGraw-Hill, New York.

Popoff, F., 1996. Former CEO Dow Chemical April. 1996.

Ponce-Ortega, J.M., Jiménez-Gutiérrez, A., Grossmann, I.E., 2008. Optimal synthesis of heat exchanger networks involving isothermal process streams. Comp. Chem. Eng. 32 (8), 1918–1942.

Quirante, N., Grossmann, I.E., Caballero, J.A., 2018. Disjunctive model for the simultaneous optimization and heat integration with unclassified streams and area estimation. Comp. Chem. Eng. 108, 217–231.

Ramírez-Márquez, C., Vidal, M., Vázquez-Castillo, J.A., Martín, M., Segovia-Hernández, J.G., 2018. Process design and intensification for the production of solar grade silicon. J. Cleaner Prod. 170, 1579–1593.

Rasmuson, A., Andersson, B., Olsson, L., Andersoss, R., 2014. Mathematical Modelling in Chemical Engineering. Cambridge University Press, Cambridge UK. Available from: http://doi.org/10.1017/CBO9781107279124.

Reh, L., 2013. Process engineering in circular economy. Particuology 11, 119–133.

Saba, S.M., Muller, M., Robinius, M., Stolten, D., 2018. The investment costs of electrolysis -a comparison of cost studies from the past 30 years. Int. J. Hydrog. Energ. 43, 1209–1223.

Sahinidis, N., 2016. The ALAMO approach to machine learning. Comp. Aidede. Chem. Eng. 38, 2410.

Sánchez, A., Martín, M., 2018. Scale up and Scale down issues of renewable ammonia plants: towards modular design. Sust. Prod. Consump 16, 176–192.

Sandler, S.I., 2017. Chemical, Biochemical, and Engineering Thermodynamics, 5th (Ed.) Wiley, New York.

Sinnott, R., Towler, G., 2019. Chemical Engineering Design, 6th Elsevier.

Sioshansi, R., Conejo, A.J., 2017. Optimization in Engineering. Springer, Switzerland.

Stankiewicz, A., Moulijn, J.A., 2000. Process intensification: transforming chemical engineering. Chem. Eng. Prog. 96 (1), 22–34.

Stehly T., Beiter P., Heimiller D., Scott G. 2017 Cost of Wind Energy Review National Renewable Energy Laboratory Technical Report NREL/TP-6A20–72167. 2018.

Taifouris, M.R., Martín, M., 2018. Multiscale scheme for the optimal use of residues for the production of biogas across Castile and Leon. J. Clean. Prod. 185, 239–251.

Takama, N., Kuriyama, T., Shiroko, K., Umeda, T., 1980. Optimal water allocation in a petroleum refinery. Comput. Chem. Eng. 4, 251–258.

Tawarmalani, M., Sahinidis, N.V., 2002. Convexificaition and Global optimization in Continuous and Mixed-Integer Nonlinear programming. Theory, Algorithms, software and applications. Springer, Switzerland.

Towler, G., Sinnot, R.K., 2012. Chemical Engineering Design, Principles, Practice and Economics of Plant and Process Design, 2nd Edition Elsevier, Singapore.

Turchi C.S., Heath, G.A., 2013. Molten Salt Power Tower Cost Model for the System Advisor Model (SAM) NREL/TP-5500–57625. February 2013.

United States DOE, 2016. Combined Heat and Power Technology Fact Sheet Series. DOE/EE-1332.

Upadhye, R.S., Grens, A.E., 1975. Solution of decompositions for chemical process simulation. AIChE J. 21 (1), 136.

Vanderbei, R.J., 2020. Linear Programming: Foundations and Extension, 2nd (Ed.) Springer, Switzerland.

Vecchietti, A., Lee, S., Grossmann, I.E., 2003. Modeling of discrete/continuous optimization problems: characterization and formulation of disjunctions and their relaxations. Comput. Chem. Eng. 27, 433–448.

Venkatasubramanian, V., 2019. The promise of artificial intelligence in chemical engineering: is it here, finally? AIChE J. 65 (2), 466–478.

Vidal, M., Martín, M., 2015. Optimal coupling of biomass and solar energy for the production of electricity and chemicals. Comp. Chem. Eng 72, 273–283.

Vlachos, D.G., 2005. A review of multiscale analysis: examples from systems biology, materials engineering, and other fluid–surface interacting systems. Adv. Chem. Eng. 30, 1–61. Available from: https://doi.org/10.1016/S0065-2377(05)30001-9.

Walas, 1990. Chemical Process Equipment: Selection and Design Butterworth. Heinemann, Boston.

Wang, Y.P., Smith, R., 1994. Wastewater minimization. Chem. Eng. Sci. 49, 981–1006.

Westerberg, A.W., Piela, P.C., 1994. Equational-based process modelling.

Westerberg, A.W., Hutchinson, H.P., Motard, R.L., Winter, P., 1979. Process Flowsheeting. Cambridge University Press.

Westerlund, T., Pettersson, F., 1992. An extended cutting plane (ECP) method for the solution of MINLP problems, Process Design Laboratory, Åbo Akademi University (1992). Report 92–124-A.

Yee, T.F., Grossmann, I.E., 1990. Simultaneous optimization models for heat integration-II. Heat exchanger network synthesis. Comp. Chem. Eng. 14, *1165-l* 184.

Yu, H., Fu, C., Gundersen, T., 2020. Work exchange networks (WENs) and work and heat exchange networks (WHENs): a review of the current state of the art. Ind. Eng. Chem. Res. 59 (2), 507–525.

Further reading

Eia, 2016. Capital Cost Estimates for Utility Scale Electricity Generating Plants.

Jiménez, A., 2003. Diseño de procesos en Ingeniería Química. Reverté, Barcelona.

Rudd, D., Powers, G., Siirola, J., 1973. Process Synthesis. Prentice-Hall, Englewood Cliffs, NJ.

Seider, W.D., Seader, J.D., Lewin, D.R., 2004. Product and Process Design Principles. Wiley, New York.

Smith, R., 2005. Chemical Process: Design and Integration. Wiley, New York.

Yeomans, H., Grossmann, I.E., 1999. A systematic modeling framework for superstructure optimization in process systhesis. Comp. Chem. Eng. 23, 555–565.

Zlokarnik, M., 2006. Scale-up in Chemical Engineering, 2nd ed. Wiley-VCH, Germany.

Sustainability in products and process design

Daniel Cortés-Borda[1], Carmen M. Torres[2,3] and Ángel Galán-Martín[4,5]

[1]Basic Sciences Faculty, University of the Atlantic, Puerto Colombia, Colombia [2]Department of Chemical Engineering, University Rovira i Virgili, Tarragona, Spain [3]Technology Centre of Catalonia EURECAT, Sustainability Area - Water, Air and Soil, Tarragona, Spain [4]Department of Chemical, Environmental and Materials Engineering, University of Jaén, Jaén, Spain [5]Center for Advanced Studies in Earth Sciences, Energy and Environment (CEACTEMA), University of Jaén, Jaén, Spain

3.1 The quest for sustainability and the role of life cycle assessment

The world economies, together with population, have grown exponentially from the second half of the 20th century. Consequently, humanity's demands increased at an accelerated pace, often met under production and consumption patterns exceeding the carrying capacity of the planet. Beyond 1950, the human footprint has exerted a tremendous pressure on ecosystems and natural resources, driving unprecedented impacts on the environment that are intrinsically linked with the wellbeing and health of the human populations (Steffen et al., 2015a). Hence, recently, an international team of earth system scientists raised the alarm showing that four out of nine planetary boundaries (including climate change and loss of biosphere integrity) have been already transgressed; which impacts the stability of our planet threatening our future (Rockström et al., 2009; Steffen et al., 2015a,b).

Today, humanity struggles to recover from the damage caused so far by the excessive exploitation of resources. The main concerns of modern society stem from the fact that previous generations were focused solely on the single and main goal of economic growth, without being aware of the social and environmental consequences in the mid to long term. Experience taught us that wealth at all costs may lead to an irreversible situation.

Hence, substantial ongoing efforts attempt to decouple economic growth from environmental impacts and resource use. A development goal that is consistent with today's needs should aim at a balanced welfare of society, economy, and environment, that can be maintained indefinitely. In other words, today's global thinking requires a change toward achieving a sustainable development state.

Sustainability entails meeting our current needs (with proper life standards) without compromising future generations' wellness. Accordingly, sustainable processes and/or products are those meeting economic, environmental, and social standards that guarantee their long-term use. Our modern society is increasingly aware that the priority must be sustainability; however, the priority shifting will take decades and further efforts will be needed to ultimately achieve sustainable development.

The term sustainability was first introduced in 1987 in the report "Our common future," later known as the Brundtland Report, published by the World Commission on Environment and Development (WCED, 1987). This document placed environmental issues as a main concern of all nations and formulated innovative proposals to solve environmental issues and strengthen international cooperation. Later, in 1994, industries started to consider environment, wealth, and society as performance indicators. Building on the concept of sustainability, the traditional "bottom line" approach based only on the profit of industries, was replaced by a new approach known as the "Triple Bottom Line" (TBL) approach. TBL embraces the three pillars of sustainability, that is, the traditional economic dimension together with environmental and social dimensions. More recently, in 2015, the United Nations Member States committed to 17 Sustainable Development Goals (SDGs) (and their corresponding targets) as part of the 2030 Agenda for Sustainable Development (United Nations, 2015). In essence, these SDGs are an urgent call for actions in all countries and in all levels—from governments to business and all individuals—to make progress in particular sustainability problems such as ending poverty, protecting life on water and land, promoting a sustainable industry or combating climate change, among others.

Making progress on the SDGs requires comprehensive and robust tools to support decision-making by identifying solutions that best adhere to sustainability principles. These decisions span different scales from consumer choices to large-scale policy decisions, which must be taken following a systems perspective to avoid collateral damages to other parts of the system. Hence, assessing the potential impact of a process and/or product from the TBL framework requires a thorough study of the direct and indirect consequences taking place from along the supply chains, that is, material extraction (cradle) to the final disposal of products after their end-use (grave). This concept here introduced is known as life cycle thinking, which constitutes a philosophy aiming at making informed decisions based on the entire life cycle of the products or processes (see Fig. 3.1). The holistic focus of the life cycle thinking aids decision-making when comparing the sustainability level of different processes and/or products; and selecting the best alternatives. According to the United Nations Environment Programme (UNEP) and the Society of Environmental Toxicology and Chemistry (SETAC)), the main goals of life cycle thinking is to lower the amount of resource use and the emissions of a process/product, as well as improve its socioeconomic performance through its life cycle (Heiskanen, 2002; Jolliet et al., 2004).

Among the different approaches to life cycle thinking, Life Cycle Assessment (LCA) has emerged as the most popular tool over the last two decades for understanding the

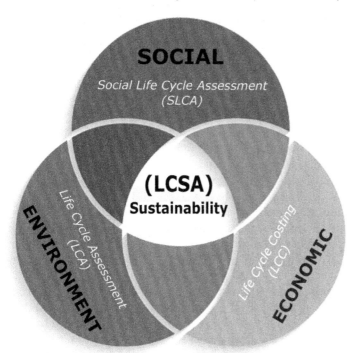

FIGURE   3.1 Life   Cycle Sustainability Assessment framework building on the Triple Bottom Line model. Sustainability is supported by the environmental, economic and social pillars.

environmental impacts of products and processes over their life cycle (Hellweg and Milà i Canals, 2014). In particular, LCA is a standardized methodology that allows assessing all the potential impacts of products/processes from a systems perspective, from "cradle to grave," that is, from the raw material acquisition to the end-use phase and disposal of wastes. Hence, ISO 14040 describes the principles and framework, while ISO 14044 provides an outline of the requirements and guidelines to carry out an LCA (ISO, 2006a,b). Notably, LCA is powerful in decision-making as it allows identifying opportunities for improvement (i.e., hotspots) without burden shifting (i.e., preventing improvements in one impact at the expense of adverse side effects in others), which enables selecting the best technological options entailing lower environmental impacts.

As applied for sustainability, LCA has recently broadened its scope into the so-called Life Cycle Sustainability Assessment (LCSA) (Onat et al., 2017). LCSA encompasses social and economic aspects besides the traditional environmental ones. LCSA is a powerful tool for assessing sustainability, since it simultaneously evaluates environmental, economic, and social aspects (Fig. 3.1) by the application of Environmental Life Cycle Assessment (E-LCA), Life Cycle Costing (LCC), and Social Life Cycle Assessment (S-LCA), respectively. Unlike E-LCA, where the alternatives are analyzed using indicators related to environmental impacts, S-LCA and LCC explore the socioeconomic consequences focused on indicators such as employment or health expenditure. Unfortunately, despite some initiatives toward the expansion to LCSA, its application is still immature (mostly due to the lack of data and consensus on how to perform it) (Ciroth et al., 2011; Fauzi et al., 2019). For this reason, in the chapter, we will focus on the environmental impacts (i.e., E-LCA), which will be here referred to as LCA. However, we advocate for the practical application of LCSA, whenever possible.

LCA has been widely applied during recent decades to support environmental decision-making in governments, policy initiatives, industry and business, and ultimately guiding the consumers' choices (Hellweg and Milà i Canals, 2014). Notably, LCA is very well-suited to support environmentally informed decisions in products and process design and, in particular, in the early stages of development where there is still room for improvements. LCA has been combined with Process Systems Engineering (PSE) tools such as simulation and mathematical programming for the eco-design of products; the process optimization for more sustainable chemicals and fuels; the design and operation of sustainable supply chains; or the design of sustainable energy systems, among other applications (Guillén-Gosálbez et al., 2019; Kleinekorte et al., 2020).

This book aims at providing a thorough and pedagogical introduction to the LCA methodology with a particular focus on its application to the sustainable design of renewable processes. After reading this chapter, the reader should be able to: (1) identify LCA as an analytical tool for sustainability, (2) carry out an LCA study, and (3) pose the questions that can be addressed by LCA.

The remainder of the chapter is organized as follows. Section 3.2 provides a brief introduction to the LCA methodology as a framework presented in the ISO 14040 and ISO 14044 standards; and introduces an illustrative case study to ease the understanding of the methodological LCA phases. In Section 3.3, the four phases of the LCA are described and the motivating example is developed step by step as each phase of the LCA is presented. Section 3.4 presents some of the latest applications of LCA in the context of sustainable renewable processes are reviewed. Finally, the fifth section includes an exercise based on LCA application for the evaluation and comparison of fuel production alternatives from fossil resources and bioresources, to enable the reader to practice and auto-evaluate the acquired knowledge.

3.2 The importance of life cycle assessment: motivating example

All actors in the society—for example, industries, governments, and consumers—are increasingly aware of environmental protection and sustainable development, which drives toward an environmentally friendly market. Shifting to eco-friendly products and processes is becoming a key element to solve existing sustainability problems at different scales. However, using raw data to decide whether a product or process should be promoted in an environmentally conscious market is challenging and could lead to counterproductive results and myopic decisions. Comparing the environmental performance of different alternatives should be based on the mid-long term effects of their life cycle, not only on the immediate effect of a particular phase.

When assessing the environmental impact of a product, we should start by inquiring about relevant data such as: how long will this good be used?; how much energy and resources were required to manufacture it?; is this good recyclable?; how efficient is its recycling process?; how long will it take to degrade its wastes?; among other relevant questions. Once all this information is clear, we should translate it into environmental indicators to evaluate its potential long-term impact and compare it with other alternatives. The later procedure can be done by carrying out an LCA study, a well-defined method for exploring products and systems' potential environmental impacts. Essentially, LCA guides how to quantify all the materials and energy inputs and releases to air, soil and water; over the entire life cycle of a process or product. Moreover, LCA allows translating these energy and material flows into relevant environmental impacts, providing crucial information to improve the systems and support decisions and policy-making.

To acquaint with the overall concept of LCA, let us consider an exemplary case study of a daily life product, a water bottle. Typically, water bottles are single-use containers made up of plastics, whose manufacturing requires oil-derived polymers and energy-intensive processes. The polymers used in plastic bottles are 100% recyclable; however, recycling plastic bottles has not yet been efficiently established. It takes thousands of years to degrade them naturally, which causes plastics accumulation in the environment and leads to serious threats to biodiversity. Considering that today plastic pollution is one of the most important environmental issues globally, moving towards eco-friendly alternatives for disposable plastic water bottles is increasingly becoming popular.

One appealing alternative to lower fossil fuel consumption and plastic pollution might be replacing all plastic bottles by reusable aluminum bottles. Aluminum bottles can be washed and refilled repeatedly, which should lower the dumping of plastics in landfills and oceans. In "the use phase," aluminum bottles seem to have a lower environmental impact than plastic bottles. However, we cannot overlook other phases of the supply chain of aluminum bottles, such as the upkeeping wash, energy-intensive production processes, and bauxite mining, which might ultimately penalize its environmental performance. Hence, aluminum manufacturing requires extreme temperatures to melt aluminum, which entails large energy consumption and CO_2 emissions. Moreover, aluminum bottles are heavier, leading to more emissions related to the transport of bottles and the washing of bottles is water intensive and could require cleaning agents that are dangerous for the environment.

Table 3.1 reflects aspects related to the environmental performance of the described water bottle alternatives. As mentioned above, the comparative analysis directly based on these raw data is very challenging and may lead to spurious conclusions. During the application of the different stages of LCA method to this illustrative example, the reader

TABLE 3.1 Environmental performance of two water bottle alternatives: a plastic disposable water bottle versus a refillable aluminum water bottle.

	PET bottle	Aluminum bottle
Capacity	500 mL	750 mL
Number of uses	1 (disposable)	not limited (refillable)
Mass	19,10 g	119 g
Washing	No	Daily
End of life	Incineration recycling	Recycling
Etc.	. . .	. . .

will observe how this general rough information is structured, refined and converted to impact indicators in themes of environmental damage. The resulting quantification of the environmental burdens through LCA will enable a softer and fairer comparison of the target alternatives.

Performing an LCA could provide guidance on which of the two previous alternatives is more environmentally friendly from the life cycle viewpoint. However, it is important to pose the right question within the LCA context. Hence, some relevant questions related to LCA might be:

- Which of the two alternatives (PET vs aluminum bottle) is greener? Comparison of specific goods and services providing the same function is one of the most frequently intended LCA applications. Note that reviewing and reporting of comparative assertions disclosed to the public have specific requirements under ISO 14040/14044 to ensure coherency and prevent bias.
- Where are the sources of environmental impact located in the life cycle of the aluminum bottle? LCA can answer this question by identifying the so-called hotspots as one of the main goals of LCA. Since the studied system is disaggregated in different units of potential damage (direct and indirect impacts), the contribution of each stage in the value chain of the product to the overall impact can be precisely determined.
- Are plastic bottles environmentally sustainable? LCA has not a straightforward answer to this type of question. According to ISO 14040, LCA cannot be used to predict absolute impacts due to its limited representativeness given by the constraints related to the reference unit selection, the geographical and temporal domain of the data and the uncertainty associated with modeling of environmental impacts, including future impacts. However, recent research aims at developing absolute indicators using environmental carrying capacity references. In this sense, global normalization values, such as the Planetary Boundaries (Steffen et al., 2015b), can reflect the product/process performance relative to absolute environmental boundaries or full safe operating space. The downscaling of this framework (Bjoern et al., 2020; Ryberg et al., 2018) to specific actors (systems, activities) competing for an environmental space (regions, countries) allow the quantification of their assigned share of the safe operating space based on their impact contribution with relative to other systems and activities.

In the next sections of this chapter, we will develop an LCA study relying on the illustrative example of the water bottles to shed light on the previous questions. All the results and information related to the illustrative example will be included presented in boxes to facilitate comprehension.

3.3 Overview of the life cycle assessment methodology

The interest in understanding and addressing environmental issues has motivated the development of impact assessment methods. In the previous section, we have demonstrated that the environmental impact of different goods should not only be compared during their use phase, but it requires assessing direct and indirect potential impacts embodied in their life cycle. Hence, LCA emerged as a comprehensive approach to overcome the limitations of the production site impact assessment by assessing the potential impact of each step of the supply chain of a process/product (see Fig. 3.2). The LCA methodology can help to identify opportunities to improve the environmental performance of processes and products; advising industries, governments and/or nongovernment organizations in decision-making related to priority setting; selecting appropriate environmental indicators and measuring techniques; and achieving product differentiation through ecolabels.

LCA methodology was introduced during the 90s and is still under continuous development and improvement. The LCA methodology was standardized and documented in the ISO 14040:2006 Environmental management—Life cycle assessment—Principles and framework (ISO, 2006a). Assessing the environmental impact systematically according to this overarching standard ensures the quality of the study and the meaningfulness of results. The standard procedure of a LCA consists of four phases (see Fig. 3.3): (1) Goal and scope definition: where the so-called functional unit and the boundaries of the study are defined; (2) Life Cycle Inventory (LCI): the accounting of all the material and energy inputs and outputs related to the functional unit in all the steps of the life cycle; (3) Life Cycle Impact Assessment (LCIA): assessment of the potential environmental impacts caused by the inventoried materials and energy streams; and (4) Interpretation: where the conclusions are drawn from the analysis and recommendations are made.

It is worth mentioning that LCA is an iterative approach that contributes to the continuous improvement of the reported results. After the goal definition, the initial scope of the assessment is derived and hence the minimum requirements of the later work. However,

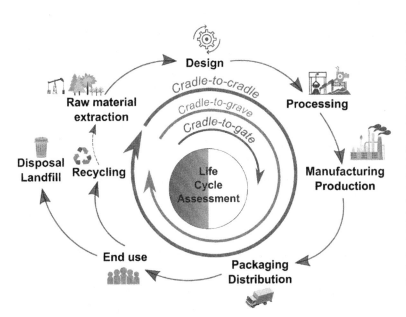

FIGURE 3.2 The life cycle of a process and/or product entails the use of materials and emission of wastes during extraction, manufacturing, distribution, use and final disposal. Resources are consumed and wastes are produced all along the supply chain of a product. The scope of LCA can be from cradle-to-grave (i.e., from raw materials extraction to the disposal phase), cradle-to-gate (i.e., from resource extraction to the factory gate) or from cradle-to-cradle (i.e., when the product is recycled after its use).

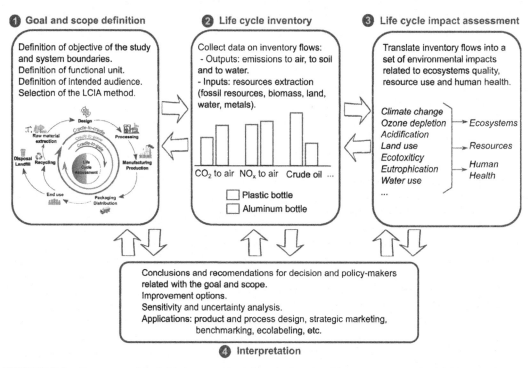

FIGURE 3.3   Four steps of the LCA framework as described in the ISO 14040:2006 standard (ISO, 2006a).

in most of the cases, more information becomes available during the subsequent phases of the LCA or when the author realizes that other aspects of the evaluation should be tackled, and so the initial scope usually needs to be updated and modified.

3.3.1 Goal and scope definition

Any LCA study begins with the definition of the purpose of the study. Before any data collection, an LCA should clearly state the main reason for carrying out the study and the intended use of the results (e.g., to compare multiple alternatives; to design or redesign a product system; strategic planning; marketing and/or brand positioning). Moreover, the intended audience and commissioner of the study and other influential actors should be also defined. Finally, this first step should also describe the limitations due to the method, assumptions, and impact coverage and whether the results will be disclosed to the public or not.

EXAMPLE 3E.1 Goal definition.

The LCA study aims to generate a quantitative environmental assessment of the production, use (1 year time-frame), and end-of-life of two packaging alternatives of drinking water during everyday outdoor activities, namely a plastic bottle (PET) and an aluminum bottle. The results drawn from this study are only instructional and will provide enough information to compare the environmental performance of both alternatives. The intended audience of this study is the readers of this chapter, who will have access to all the data of the LCA.

Once stated the goal of an LCA, the next step is to define the scope of the study. The level of detail of the defined scope should be enough to guarantee that the depth of the study is compatible with the stated goal. Details to be defined in the scope include: defining the product system of study by establishing the initial system boundaries; identifying the product system's function; selecting an appropriate functional unit; establishing the modeling framework to impact distribution and handling of product system multifunctionality; defining the impact indicators and their assessment methodology; specifying the data requirements, assumptions and limitations; among other relevant details. Since LCA is an iterative process, scope details are susceptible to modification as data is collected in order to meet the original goal of the LCA study.

Establishing the system boundaries entails defining the geographical borders and temporal frame of the study, that is, defining the accounting limits for energy and material flows. Standard variants of the delimitation of the LCA borders are known as cradle-to-grave, cradle-to-gate, gate-to-gate and cradle-to-cradle (Figure 2.2). Cradle-to-grave LCA studies include the whole life cycle of the process or product, that is, it considers all inputs and outputs from the materials extraction (cradle), the intermediate stages of the supply chain (e.g., manufacturing, transporting, storage, the use phase), until the final disposal at the end of life. Cradle-to-gate and gate-to-gate studies are partial variants of the LCA studies, omitting the use phase and final disposal. Partial LCA studies are of particular interest to businesses that aim to lower the environmental impact until the factory gate, before losing the control of their products when delivered to end-users. The cradle-to-cradle approach resembles the cradle-to-grave variant of the LCA study; however, the final disposal of materials is replaced by recycling. Hence, cradle-to-cradle LCA studies entail a closed loop of production that fits the "circular economy" targets. The definition of the LCA boundaries is suitable for the explicit definition of the assumptions and limitations of the study.

The function of a product (or process) refers to playing the role for which it is intended. The function of a system is often intuitive; however, some systems may have more than one function (multifunctional product systems). The scope of the LCA must clearly state the function of the study to avoid ambiguous interpretations. Once the function is stated, it is necessary to select the functional unit, a measurable reference unit that quantifies the function's performance. The functional unit is crucial for making LCA studies comparable. A properly defined functional unit places different systems on a common basis, enabling comparison of their environmental performance. The functional unit definition may not be straightforward, since comparable functional units should encompass the same quality, quantity, and duration of the function. Moreover, the functional unit should provide a direct link to the reference flow, for which all other input and output flows, i.e., the materials and/or energy inputs and outputs, are quantitatively related.

The scope definition of the LCA also specifies the data quality requirements for the study. The description of data quality provides details on how reliable will be the results of the LCA, how these results can be interpreted, and it allows evaluating whether the results are in accordance with the goal set.

EXAMPLE 3E.2 Scope definition.

The impacts of the whole life cycle of the bottles, including raw materials extraction, processing, use, and end-of-life options, will be accounted for according to a cradle-to-grave approach. Retail distribution, secondary and tertiary packaging, consumption during the use phase for on-way PET bottles (e.g., refrigeration) and infrastructure in the foreground system are excluded from the assessment. The resulting product system modeled through the primary data of material and energy exchanges is simplified based on own judgments to elaborate an illustrative example balancing realism and pragmatism.

In our example, the function of the two alternatives is to serve as a container for beverages, specifically water, for its outdoor consumption. To fully cover the functional features of each packaging option (mainly, capacity and number of uses), the selected functional unit is one year of use. For the calculation of the reference flows, the following assumptions for the 1-year period are taken: PET bottles of 500 mL capacity and aluminum bottles of 750 mL capacity; an average consumption of 1.5 L of water per day and adult; plastic bottles are for one-way use; and one aluminum bottle can be refilled throughout the year. Therefore, the reference flow of the PET product systems is 1095 bottles. The impacts generated in the manufacturing of the aluminum bottle corresponding to the functional unit are calculated considering a lifetime of 10 years as a baseline scenario. Therefore, the flows of materials and energy needed for the bottle production are amortized for one year of its lifetime.

The main sources of data for modeling the product systems are scientific literature and industrial reports, providing global average geographical representativeness and last decade temporal coverage.

3.3.2 Life cycle inventory

The second step of the LCA is the LCI phase which is documented in the ISO 14041 standard: Environmental management—life cycle assessment—goal and scope definition and inventory analysis (ISO, 2006a). This LCA phase involves data collection and calculation procedures to assess all the mass and energy inputs and outputs (related to the functional unit) across the boundaries defined in the scope. For instance, the LCI quantifies the input of water, raw materials, and fossil fuels entering the supply chain of a given product or process, as well as the emissions to air, soil, and water.

Collecting all the mass and energy inputs and outputs along the whole life cycle of a process or product is, at first glance, a daunting task. For this reason, the LCI adopts a product systems approach, which defines the flowsheet of the study and divides the problem into smaller sections called subsystems that facilitate the assessment of flows across the boundaries. A product system is the assembly of interconnected process units,

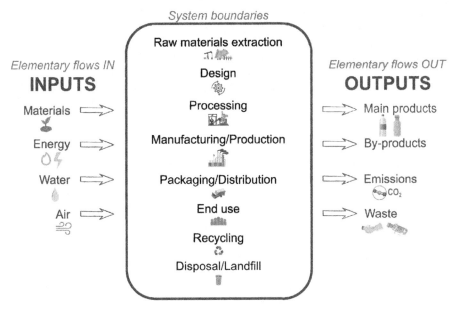

FIGURE 3.4 Example of the life cycle inventory applied to a product system.

which is intended to perform one or more defined functions (see Fig. 3.4). Unit processes are linked with each other by exchanging intermediate products and/or wastes; to the technosphere, through product flows and to the environment, by receiving and releasing elementary flows. According to the ISO standards, the term product flow refers to any product (as consumable good or commodity) entering from or leaving another product system. For instance, the electricity supply and raw materials that enter the system, or the waste to be treated leaving the system. These examples are typical nonfunctional product flows that come from the technosphere, while the target product and its coproducts are functional product flows. The elementary flows are those material and energy flows across the boundaries of the product system that go to or come from the environment without any human intervention, like direct pollutant releases or natural resources extraction.

When modeling the product system is essential to define its boundaries, so the foreground and the background systems are clearly differentiated. The foreground system comprises the processes for which primary data, i.e., elementary and product flows, are known and under the direct control of the decision-maker for which an LCA is carried out. In contrast, the background system refers to those processes whose impacts are indirect and for which the decision-making has indirect influence or no influence at all, for example, the production processes of nonfunctional product flows (e.g., electricity consumed in the process). Establishing the product system's flowsheet and dividing the whole life cycle into individual process units enables performing mass and energy balances to quantify the elementary flows.

EXAMPLE 3E.3 Flowsheet of the product system.

Figures 3E3.1 and 3E3.2 display the systems for both water bottle alternatives. The main flows and unit processes included in the cradle to grave scope are indicated. The flow diagrams show the unit processes and intermediate flows of the model that belong to the foreground system (blue line box) based on primary data of exchanging flows, that is, semi-finished product, manufacturing, and use (aluminum bottle). The background system activities are also represented by using secondary data from databases, that is, processes to produce the flows exchanged with the technosphere. As an example of elementary exchange, the freshwater extracted during the production of bottled mineral water is accounted for in the product system of the PET bottle, being a direct exchange with the environment.

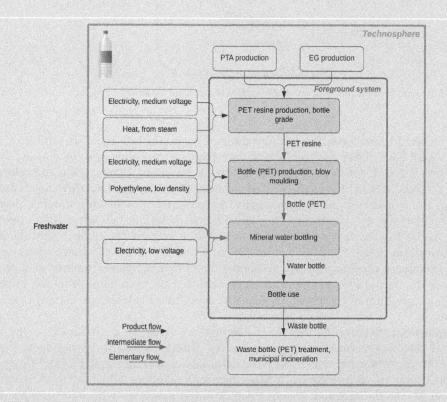

FIGURE 3E3.1   Flow diagram of PET bottle product system.

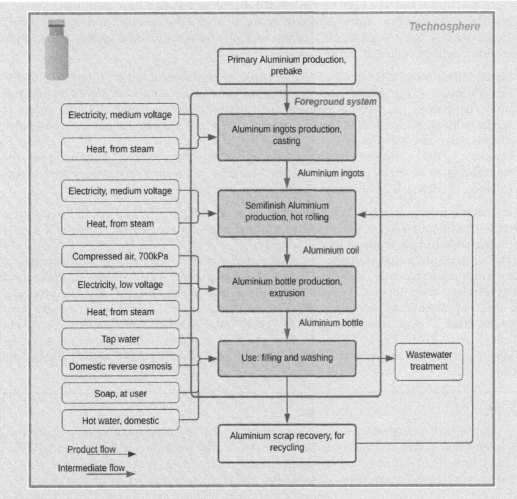

FIGURE 3E3.2 Flow diagram of aluminum bottle product system.

Once the limits and the unit processes of the product system are identified, the next step is the data collection. Data collection accounts for all the material and energy inputs and outputs that are related to each of the unit processes within the system boundaries. Relevant data to be collected are as follows:

- Inputs: These include (1) input of energy considering different types of fuels, electricity, the efficiency of conversion and distribution, and other inputs and outputs derived from the energy flow being assessed, (2) input of raw materials, (3) the input of ancillary materials (materials that are required for a process but are not raw materials), and (4) other physical inputs.
- Outputs: These include value-added products, byproducts involved in the manufacturing of main products, wastes (materials without a market) as well as the emission of materials (and energy) to air, water and soil.

It is worth mentioning that data collection is often challenging, LCI could be resource-intensive and time-consuming. Difficulties that may arise in accomplishing the LCI should be clearly stated in the scope of the LCA study.

The primary data that describe the foreground system often rely on experimental or real facility measurements, statistics, industrial reports and process design, mathematical programming, scale-up models, and simulations. In contrast, the background processes are often captured using secondary data of cumulative environmental loads already computed in other LCA studies. Therefore, inventory databases and software (e.g., Ecoinvent (Steubing et al., 2016), OpenLCA, GaBi, Simapro (Goedkoop et al., 2016)) provide a compilation of inventories for several processes and products in the most relevant industrial sectors.

EXAMPLE 3E.4 Data collection.

Considering the illustrative nature of the example and that the ultimate purpose is to show the accounting, calculation, and interpretation procedures, the primary data for the case study have been extracted from different sources. In both alternative bottles, the LCA model is representative of state-of-the-art technological options used globally. Notably, data for the PET bottle from resin production to blow-molding were collected from an LCA guideline report (Nessi et al., 2020) and an LCA study (Stefanini et al., 2020). Similarly, the inventory of flows for the aluminum bottle production was sourced from reports on aluminum in the packaging industry (EUA, 2013; Gesamtverband Der, 2019; International Aluminium Institute, 2015). Regarding the use stage of the aluminum bottle, the inputs were also adapted from literature (Garfí et al., 2016). Regarding the end-of-life phase, the most frequent options at the European level were chosen, that is, the incineration of PET waste and the recycling of aluminum scrap. With the didactic purpose in mind, the inventory here is limited to provide sufficient detail of the systems without being tedious, that is, the detail and depth of the models were established according to the objective set during the Goal and Scope of the LCA.

Once the flows of the study are fully characterized, the next step, data calculation, validates the collected data, relates the collected data with the unit processes and establishes the link between the collected data and the reference flow of the functional unit.

The LCI, that is, inventory of cumulative elementary exchanges, is calculated by converting the primary data, that is, in/output flows of materials and energy on the basis of the functional unit. At this stage, the use of LCA databases, like the Ecoinvent Database (Wernet et al., 2016), is crucial since often LCA practitioners only have information of elementary exchanges related to the unit processes of the foreground system. Hence, LCA databases provide emission factors for the product flows exchanged between the product system and the technosphere for the rest of the activities in the scope of the LCA. The emission factors include cumulative environmental loads of the product flows, before entering the product system (cradle-to-gate) or after exiting the product system (gate-to-grave). These emissions factors account for a huge variety of elementary exchanges, including emissions of polluting substances to air, water, and soil, and natural resources consumption including depletion of fossil, mineral, hydric and land resources. The resulting LCI is a vector composed of thousands of environmental loads expressed in the amount of elementary exchange (pollutant emission or natural resource consumption) per functional unit.

In mathematical terms, given a set of flows F (i.e., inputs and outputs), the calculation to obtain each of the elements of the LCI vector of environmental loads l can be done through the following general expression (Eq. 3.1):

$$lci_l = \sum_{f \in F} q_{f,l} + d_l = \sum_{f \in F} e_{f,l} \cdot x_f + d_l \quad \forall l \in L \tag{3.1}$$

where, lci_l is the elementary exchange for certain environmental load l generated in the studied system (e.g., kilograms of CO_2 emitted in the whole product life cycle of the PET bottle per year of bottle use) calculated as the summation of all flows of the same environmental load l; $q_{f,l}$ is the elementary exchange for certain environmental load l generated due to product flow f (e.g., kg of CO_2 emitted in the production and transport of the electricity that is used in the blow molding corresponding to a year of bottle use); d_l is the direct elementary exchange of environmental load l identified in the product foreground system (e.g., the direct CO_2 emissions released during the blow molding corresponding to a year of bottle use); $e_{f,l}$ is the emission factor for environmental load l generated due to product flow f (e.g., amount of CO_2 emitted per amount of electricity used in the blow molding); x_f is the product flow per functional unit (e.g., amount of electricity consumed in the blow molding corresponding to one year of bottle use).

The calculation using Eq. (3.1) must be performed for every type of environmental load, considering all unit processes and product flows identified in the scope. The final inventory obtained is enormous, making the analysis and derivation of conclusions very complex. Therefore, instead of working with the inventory of environmental loads, environmental indicators classified in areas of environmental damage (environmental categories) are calculated (see next section LCIA). Nevertheless, an analysis of the environmental profile of the studied system can already be carried out at this stage. Hence, the LCI studies, unlike the LCA studies, only go up to this stage of the calculation.

EXAMPLE 3E.5 Data calculation.

The primary data inventories for the PET and aluminum bottle systems for the selected functional unit are reported in Tables 3E5.1 and 3E5.2, respectively. All flows are referred to one year of bottle use, equivalent to 547.5 l of water. This also includes the depreciated amount of aluminum bottle that is allocated to one year of its lifetime (10 years of lifetime) to properly assign the corresponding impacts derived from the aluminum bottle production (amounts for 0.1 bottle units for a year).

Considering that no direct elementary exchanges were identified in the foreground system, the calculation of the LCI vectors for each bottle alternative through Eq. (3.1) implies multiplying the values in Tables 3E5.1 and 3E5.2 by emission factors for the list of environmental loads and for the specific product flows. In practice, this step is avoided, and LCIA is directly carried out from the primary data as is explained in Section 3.3.3

TABLE 3E5.1 Primary data for the PET bottle corresponding to one year of bottle use.

PET bottle system		
Reference flow	Value	Unit
PET bottle	1095	units
PET weight	20.91	kg
Flow	Value	Unit
Purified terephthalic acid	18.104	kg
Ethylene glycol	7.027	kg
Electricity, medium voltage, PET production	4.175	kWh
Heat, from steam, in chemical industry	8.139	MJ
Electricity, medium voltage, blow molding	35.662	kWh
Polyethylene, low density, granulate	0.627	kg
Freshwater	0.548	m^3
Electricity, low voltage, bottling	6.57	kWh
Waste treatment PET	20.915	kg

TABLE 3E5.2 Primary data for the aluminum bottle corresponding to one year of bottle use.

Aluminum bottle system		
Reference flows	Value	Unit
Aluminum bottle	0.1	units
Aluminum weight	0.011	kg
Product flows	*Value*	*Unit*
Aluminum, primary, liquid, prebake	3.30×10^{-4}	kg
Electricity, medium voltage, aluminum industry	3.14×10^{-5}	kWh
Heat, industrial, natural gas	4.95×10^{-4}	MJ
Electricity, medium voltage	6.02×10^{-3}	kWh
Heat, industrial, natural gas	0.021	MJ
Compressed air, 700 kPa	0.003	m³
Tap water	0.138	kg
Heat, industrial, natural gas	0.035	MJ
Electricity, low voltage	0.020	kWh
Tap water, refilling	0.548	m³
Electricity, low voltage, domestic osmosis	0.082	kWh
Tap water for cleaning	0.274	m³
Wastewater, for cleaning, sewage	0.274	m³
Soap, cleaning	0.730	kg
Heat for cleaning water, domestic	34.493	MJ
Aluminum scrap, for recycling	0.011	kg

Typically, product systems involve multiple products or services. The next step entails the allocation of flows and releases. The ISO 14044 standard (ISO, 2006a) states that the inputs and outputs of the product system should be allocated to the different products. Moreover, the allocation procedure should be clearly stated, documented and explained in the LCA. Hence, the inputs and outputs should be partitioned and fairly assigned to each product (or function) of the product system. Noteworthy, allocation should be based on the material balances so that the total inputs and outputs of a unit process should be equal to the allocated inputs and outputs of the unit process. The documented allocation procedure should be applied uniformly to similar inputs and outputs of the product system.

Allocation is of capital importance to establish the environmental responsibilities properly in multifunctional systems. Different allocation procedures are available, where the physical allocation is referred to as a physical magnitude (e.g., kg of product A, liters of product B). When a physical allocation is not suitable, the economic allocation is the last resort. Hence, the ISO 14044 provides the following guidelines for allocation: (1) avoid allocation (whenever is possible) by dividing the unit process into subprocesses or expanding the product system to include additional functions related to coproducts; (2) when it is not possible to avoid allocation, allocating inputs and outputs to each product or function should reflect physical relationships between them; (3) when the physical relationship is not possible to establish, another relationship (e.g., economic relationship) should be established.

EXAMPLE 3E.6 Allocation.

In the product systems established for the illustrative example, there are no other identified functions but only the use of the two bottle alternatives for outdoor drinking water. However, the method to solve the multifunctionality problem still has to be chosen because it affects the multioutput inventories in the background system. Hence, the chosen system model also applies to the secondary data because the cumulative LCA results obtained from other sources, such as LCA databases, were constructed following a particular approach. This method must remain consistent throughout the scope of the LCA.

Although not directly evaluated in the foreground system, the raw material ethylene glycol (EG) production is part of a multioutput process and is used here as a simple example of allocation procedure. The cumulative LCI corresponding to EG flow, namely its environmental footprint, has to be allocated (differentiated from its coproducts) to be used in the PET bottle production. In the production of EG, ethylene is directly oxidized with air or oxygen in the presence of a catalyst to ethylene oxide (EO). The EO is subsequently treated with water (hydrolyzed) and forms a variety of glycols, most notably mono ethylene EG, diethylene glycols (DEG), triethylene glycols, and heavy glycols (TEG). It is assumed that an average of 82.5% of the produced glycols consists of for example, accordingly, the following table shows the estimation of the environmental loads corresponding to the EG using two allocation approaches (mass and economic) and applied to the CO_2 emissions as environmental load (cumulative). Note that depending on the allocation method, the resulting impact assigned to the target product could significantly differ, see estimated CO_2 emissions for product EG in Table 3E6.1 in bold.

TABLE 3E6.1 Example of impact assignation using economic and mass allocation applied to CO_2 emissions of EG production.

	Units	EG	Other glycols
CO_2 emission	kg CO_2/t EO	2.23	
Mass production	%	82.50	17.50
Mass allocation	**kg CO_2/t glycol**	**1.84**	**0.39**
Price	$/lb	0.12	0.35
Revenue	$ / lb EO	9.9	6.125
Economic production	%	0.62	0.38
Economic allocation	**kg CO_2/kg glycol**	**1.38**	**0.85**

3.3.3 Life cycle impact assessment

The third phase of LCA, the LCIA, aims at evaluating the potential environmental impact of the studied system by using as input the data from the LCI. The LCIA associates the inventory data with particular impact categories and category indicators, calculated through different characterization models (see Fig. 3.5).

The LCIA consists of mandatory and optional elements. The mandatory elements include: (1) selecting the impact categories, category indicators, and characterization models; (2) the assignment of the inventory results to the selected impact categories (known as classification); and (3) calculating the category indicator results (characterization). Optional elements of the LCIA include normalizing the category indicators results with respect to a reference value; grouping, sorting and ranking impact categories; aggregating indicator results across impact categories in a weighted value; and analyzing the quality of data.

Next are broadly described the mandatory elements of the LCIA:

Impact categories represent the class where the environmental concerns identified in the LCI should be assigned. Impact categories are, for example, global warming, ozone depletion, acidification, and so on. According to the ISO 14044 standard, the impact categories should present accurate and descriptive names and be justified and consistent with the goal

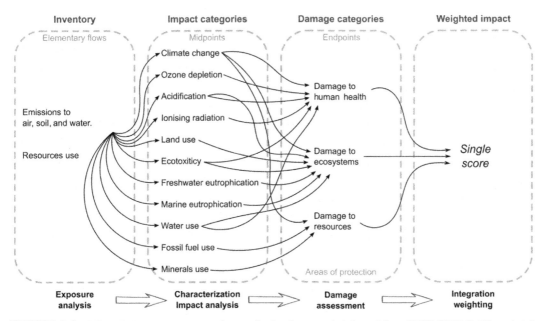

FIGURE 3.5   Life cycle impact assessment characterization framework (adapted from UNEP/SETAC, Life cycle initiative). Elementary flows are linked to eleven environmental indicators at the midpoint level and three endpoint levels related to areas of protection, which are ultimately aggregated into a single weighted impact score (Jolliet et al., 2004).

and scope of the LCA study. Moreover, among other recommendations, the categories should also reference the related information and sources and reflect a comprehensive set of environmental issues related to the system of study. Each impact category addressed in the LCIA requires identifying the category endpoint(s); defining the category indicators for given category endpoint(s); assigning LCI results properly in the impact category; and identifying the characterization models and characterization factors. The impact categories (also known as category indicators) and the characterization models used to quantify the intensity of the environmental impact should be internationally accepted. Ultimately, the impact categories should reflect the consequences of the LCI results on the category endpoint(s); add environmental data to the characterization model with respect to the category endpoint(s); and avoid double-accounting other recommendations found in the ISO 14044 standard.

The characterization model is a model describing the relationship between the elementary flows and their potential impact(s) on a particular impact category. Therefore, it allows converting the LCI results into a value that corresponds to the category indicator. The conversion is done through characterization factors, which translates the LCI to a common unit of the category indicator. The characterization models and characterization factors should be scientifically and technically valid, based on an identifiable environmental mechanism and reproducible. The LCA study should clearly describe the suitability of the characterization model used.

Considering the set C with impact categories c, the calculation of the indicator (I_c) in a certain category of impact (c) is formulated as in next Eq. (3.2):

$$I_c = \sum_{l \in L} lci_l \cdot cf_{c,l} \quad \forall c \in C \tag{3.2}$$

where, $cf_{c,l}$ is the characterization factor of environmental load l for the category of impact c. and lci_l is the amount of elementary exchange (emissions, resource extractions or land use, l). Previously to the use of this equation, selecting the category of impact and the classification of the environmental loads in the calculated LCI related to this category must be done. The procedure has to be repeated for every category of impact chosen in the study. The computation can be cumbersome depending on the number of impact categories selected and the number of environmental loads considered in the LCI.

Alternatively, the LCI calculation and characterization can be bypassed by translating the primary data directly to impact indicators. This can be done through impact factors (or equivalence factors) available in the LCA databases. These impact factors are LCIA results of the product flows expressed by the unit of product flow reference (e.g., kg of CO_2 equivalents per kWh of electricity consumed). Hence, by relying on the impact factors ($if_{c,f}$), the impacts generated in a certain category of impact based on the functional unit of the studied system would be calculated as shown in Eq. (3.3).

$$I_c = \sum_{f \in F} if_{c,f} \cdot x_f + \sum_{l \in L} d_l \cdot cf_{c,l} \quad \forall c \in C \tag{3.3}$$

It is worth mentioning that, in the LCIA step, it is possible to make specific choices that could introduce subjectivity to the study. Hence, the LCIA should report all the assumptions made to ensure transparency, and subjective value-choices should be avoided to the extent possible.

EXAMPLE 3E.7 LCIA.

The LCIA is conducted for the packaging alternatives for drinking water to calculate the impact indicator for the Climate Change (CC) category of impact. The indicator impact would actually represent the carbon footprint of both systems. Eq. (3.3) is applied by using the impact factors corresponding to the product flows in the primary data (Tables 3E7.1 and 3E7.2). In this case, the impact factors for the set of product flows were extracted from the Ecoinvent Database according to the Intergovernmental Panel of Climate Change (IPCC) method of characterization for which the characterization factors are the Global Warming Potentials for a 100 years horizon (GWP_{100}). The calculations for the PET and aluminum bottle systems are shown in Tables 3E7.1 and 3E7.2 respectively, where the values of Climate Change impact in the last column are the result of multiplying the values in the other two columns for the same row.

The CC impact indicators of both options are the bold values obtained as the summation of the scores in the last column of Tables 3E7.1 and 3E7.2. Based on these resulting indicators, the impact on CC for one-year use of the packaging options for outdoor water drinking is higher for PET bottles than for aluminum bottles. This is consistent with the hypothesis that one-way products would have higher environmental impacts than multiple-use products. However, given the assumptions made and the limitations of the simplified product systems, the comparison will be revised in detail in the *Interpretation* phase (Section 3.3.4).

Note that the freshwater flow in the PET bottle system represents the freshwater consumption from mineral water well in the bottle of water production. Since it is a direct consumption from nature, this consumed water is an elementary flow whose impact has to be calculated using Eq. (3.2). However, as it is not associated with an impact on CC, the *classification* step of the LCIA tells us that this flow does not have to be considered in the calculation. Hence, only greenhouse gas emissions are elementary flows with an effect on the category of impact CC, and so the characterization factors can be only found for these substances (see Table 3E7.3 for terminology related to LCIA using CC impact as example).

TABLE 3E7.1 LCIA calculation for the PET corresponding to one year of bottle use.

PET bottle Primary data	Value	Unit	CC Impact factor kg CO_2-Eq/ unit of flow	CC indicator kg CO_2-Eq/ FU
Purified terephthalic acid	18.104	kg	1.861	33.691
Ethylene glycol	7.027	kg	2.067	14.523
Electricity, medium voltage, PET production	4.175	kWh	0.046	0.192
Heat, from steam, in chemical industry	8.139	MJ	0.069	0.562
Electricity, medium voltage, blow molding	35.662	kWh	0.046	1.641
Polyethylene, low density, granulate	0.627	kg	2.051	1.286
Freshwater	0.548	m^3	—	—
Electricity, low voltage, bottling	6.570	kWh	0.066	0.432
Waste treatment PET	20.915	kg	1.086	22.709
Total Climate Change indicator (1 year of PET bottle use)				**75.036**

TABLE 3E7.2 LCIA calculation for the Aluminum bottle system corresponding to one year of bottle use.

Aluminum bottle Primary data	Value	Unit	CC impact factor kg CO_2-Eq/unit of flow	CC indicator kg CO_2-Eq/FU
Aluminum, primary, liquid, prebake	3.30E − 04	kg	9.781	0.003
Electricity, medium voltage, aluminum industry	3.14E − 05	kWh	0.160	0.000
Heat, industrial, natural gas	4.95E − 04	MJ	0.069	0.000
Electricity, medium voltage	0.006	kWh	0.160	0.001
Heat, industrial, natural gas	0.021	MJ	0.069	0.001
Compressed air, 700 kPa	0.003	m^3	0.034	0.000
Tap water	0.138	kg	0.001	0.000
Total heat extrusion	0.035	MJ	0.069	0.002
Total electricity extrusion	0.020	kWh	0.066	0.001
Tap water	0.548	m^3	0.001	0.000
Domestic reverse osmosis	0.082	kWh	0.066	0.005
Tap water for cleaning	0.274	m^3	0.001	0.000
Treatment of wastewater	0.274	m^3	0.475	0.130
Soap, cleaning	0.730	kg	2.704	1.974
Heat water, domestic	34.493	MJ	0.071	2.450
Recycling Al scrap	0.011	kg	− 16.914	− 0.186
Total climate change indicator (1 year of aluminum bottle use)				**4.384**

TABLE 3E7.3 Description of terms of the LCIA.

Term	Example
Impact category	Climate change
LCI results	Amount of a greenhouse gas per functional unit
Characterization model	Baseline model of 100 years of the Intergovermmental Panel on Climate Change
Characterization factor	Global warming potential (GWP_{100}) for each greenhouse gase (kg CO_2-equivalents/kg gas)
Category indicator result	Kilograms of CO_2-equivalents per functional unit
Areas of protection	Human health an ecosystem quality

Adapted from ISO 14044 standard.

3.3.4 Interpretation

The fourth phase of the LCA is the interpretation phase. Here, the results of the LCI and the LCIA are interpreted according to the goal and scope of the LCA. Please recall that the LCA is an iterative process, so this interpretation phase provides an opportunity to reconsider some previous decisions. For example, the suitability of the definitions of the system functions, functional unit and system boundaries, the quality of data collected and calculated, the sensitivity of the results upon certain variables of the study and the assumptions and choices during the preceding phases. A thorough revision of these aspects may lead to valuable conclusions and recommendations.

Nevertheless, the LCA interpretation could be ambiguous. Hence, the ISO 14040 and ISO 14044 standards simplify the LCA interpretation by the following elements: (1) identifying the significant issues, (2) evaluating the results, and (3) establishing conclusions, limitations and recommendations.

The "identification of significant issues" element interacts with the evaluation element to ultimately determine issues under the goal and scope definition. The interaction of these two elements aims at including the implications of the early stages of the LCA in the preceding phases. Significant issues refer to the main contributors of environmental impact (the so-called "hotspots"). Examples of significant issues are the inventory data (materials and energy inputs and outputs), the impact categories (e.g., resource use, climate change) and significant contributions from the LCI and LCIA results (such as unit processes with large shares of impact).

The purpose of the "evaluation of results" is to establish and enhance the LCA study's reliability. To this end, the evaluation element consists of: (1) the completeness check, which ensures that all the data and information required for the LCA is complete; (2) the sensitivity check, which assesses the reliability of the LCA outcomes, and how these are affected by the uncertainty of data, allocation methods, calculation of category indicators, etc; and (3) the consistency check, which evaluates if the goal and scope of the LCA are consistent with all the assumptions, methods and data of the study.

In the interpretation stage, sensitivity analysis is often used to assess the impact of uncertainty on the results, to provide robust conclusions and recommendations. Uncertainty arises from the assumptions, limitations and simplifications conducted in the methodological framework, the impact assessment data and the inherent variability of the flows collected as primary data (Heijungs and Lenzen, 2014). Moreover, other sources of uncertainty are related to the foreground and background data quality, including the characterization models, use of estimates, lacking verification, incompleteness and temporal, spatial and technological extrapolation. The uncertainty sources can be tackled in parametric uncertainty or scenario analysis (European Commission—Joint Research Centre—Institute for Environment and Sustainability, 2010). This is often done using sampling methods based on statistical distribution of the parameters which are used as input of Monte Carlo simulations (reference method in LCA for uncertainty propagation); while lacking of such data, the Pedigree matrix approach is used to relate quality indicators to uncertainty ranges (Weidema and Wesnaes, 1996). The resulting distribution functions allow establishing expected values and lower and upper bounds of a confidence interval; thus, an enhanced interpretation of the results and, most important, a fairer comparison between the process/product alternatives can be made.

After the evaluation element, the LCA interpretation phase establishes "conclusions, limitations and recommendations." The aim of this element is to advise the intended audience of the LCA with the main outcomes of the study. Conclusions drawn from the study are obtained from an iterative process with the other two elements of the LCA interpretation phase. The logical procedure for the interpretation phase as recommended by the ISO 14044 standard (ISO, 2006b) is as shown in Fig. 3.6.

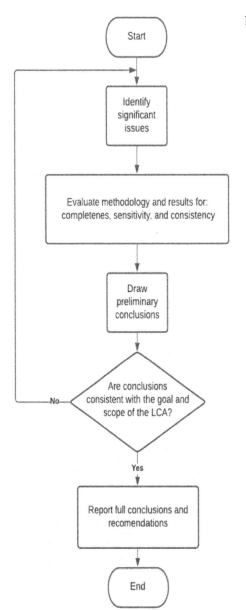

FIGURE 3.6   LCA interpretation logical procedure.

At this point, the LCA should provide specific recommendations to the intended audience (e.g., decision-makers) that reflect a reasonable consequence of the final conclusions. Direct applications of the LCA recommendations could be product design (or redesign), strategic planning, public policy-making or marketing, among other applications.

EXAMPLE 3E.8 Interpretation.

According to Table 3E7.1 the main contributors to environmental impact for the PET bottle are the production of chemical precursors, specifically, the purified terephthalate acid (45% of the total carbon footprint). In contrast, the significant issues in the aluminum bottle option are the impact derived from the use of materials needed for daily cleaning, mainly hot water and soap. Besides, the end-of-life activities in both bottle alternatives also have a significant impact. In the PET bottle case, the incineration has a footprint of about 1.1 kg CO_2eq/kg PET, which represents 30% of the equivalent CO_2 emitted in the whole PET bottle life cycle. Note that in the impact factor for the PET incineration credits (due to energy production) are subtracted because the municipal incineration of wastes is, in fact, a multifunctional process where an input (waste) is treated, and an output (energy) is produced. Since the substitution model was selected, the avoided impacts for the conventional production of heat and electricity are deducted from the total incineration impacts, which mainly involved combustion exhaust gases.

Given that the impacts related to the washing step are predominant in the total scores of the aluminum bottle, the assumptions made for the washing should be carefully revised through a sensitivity analysis. However, the frequency of washing is not expected to be higher than daily, and any reduction of washing activity will increase the advantage of the aluminum bottle in any case.

The life span of the aluminum bottle is found to be a critical parameter to which the impacts and final comparison are very sensitive. The environmental loads calculated for the aluminum bottle production are assigned to the functional unit (one-year of use) based on the number of years that the bottle is being reused. The longer? the bottle lasts, the lower is the impact per usage, as the impacts of the production are amortized along the bottle lifetime. In practice, the break-even number of uses for which the aluminum bottle starts to be more sustainable than the plastic bottle can be estimated. Considering the rest of the assumptions established in the first and second phases of the LCA, the use of an aluminum bottle would be environmentally worse if it is only used for six days or less in substitution of one-way PET bottles. If the bottle is dumped after this short usage, the Climate Change impact of its life cycle will be higher than for the PET bottle. In other words, the carbon footprint of the production of aluminum bottle unit is virtually 36 times (6 days multiplied by 3 uses a day) higher than the carbon footprint of the PET bottle production; however, its potential number of utilizations makes the difference.

3.4 Applications of LCA in the context of renewable processes

LCA was originally developed in business and industry to quantify the usage of energy and raw material in the design products. In the last decades, LCA has expanded its application to a much broader number of areas, and today it is widely recognized as a powerful support tool for improved sustainability. In the context of this book, LCA has been applied to aid decision-making in the design of more sustainable renewable processes. Hence, LCA has been combined with PSE tools, such as simulation and optimization, to incorporate environmental criteria besides the traditional economic and technical ones. This combined approach enables the systematic design of products, processes and services (Guillén-Gosálbez et al., 2019). For example, integrated LCA-optimization frameworks were developed to design more sustainable energy systems (Galán-Martín et al., 2018) and supply chains for biomass and fuels (Calvo-Serrano et al., 2019; Guillén et al., 2005). In the same context, data generated from simulation tools were used to carry out an LCA of a biorefinery coupled with carbon capture and storage technologies to investigate the potential of producing carbon-negative bioethanol from woody residues (Bello et al., 2020; Silva et al., 2018). Moreover, a growing number of studies apply LCA for evaluating the environmental performance of emerging renewable sources and promising renewable technologies at the earliest stages of development. These studies often compare conventional fossil fuel-based technologies with renewable alternatives to determine any environmental benefits and adverse side effects of their potential replacement. Most of them identify the hotspots where the efforts should be directed for improving environmental performance. Example of this type of LCA studies is the evaluation of different biomass feedstocks for electricity generation (Kadiyala et al., 2016) the study of biofuel production from microalgae following different conversion pathways (Sun et al., 2019; Torres et al., 2013), the environmental assessment of perovskite-silicon solar photovoltaic cells (Tian et al., 2020), the assessment of novel solar thermal technologies for calcination in the cement industry (Tomatis et al., 2020) or the assessment of onshore and offshore wind energy (Bonou et al., 2016), among others.

3.5 Case study of application of LCA in practice: carbon footprint of biodiesel

3.5.1 Problem setting

The proposed problem consists of the LCA application to the production of biodiesel from oil of vegetable origin, specifically from the extracted oil of thistle seeds (Cynara Cardunculus seeds). The thistle is a robust and wild perennial plant that does not compete for land with edible resources and grows spontaneously in nonirrigated lands. Its cultivation hardly needs fertilization, plowing, use of herbicide, and insecticide compared to other crops, such as sunflower or soybean.

The objective of the LCA is to assess the environmental performance of the proposed renewable source to allow its comparison with other bio-based fuels. The practical exercise includes the calculation of the CC indicator to identify the stages/unit operations that contribute most to the impact and establishing alternatives/changes in the balance of materials or energy or in the supply chain to reduce the impact.

3.5.2 Data and information

In the production of thistle oil-based biodiesel four subsystem are considered: cultivation, mechanical separation of seeds from biomass, oil extraction and biodiesel production through catalyzed transesterification of the oil's fatty acids to produce fatty acid methyl esters. The main flows of primary data for the product system are collected in Table 3.2 The impact factors to be used in the LCIA are indicated in Table 3.3.

Besides, the transport of the oil from the extraction facilities to the transesterification plant covers 80 km. The unit of reference for the good transport activities is usually the tkm, representing the transport of an amount of one metric ton of material by a certain means of transportation over one km of distance.

TABLE 3.2 Primary data for biodiesel production from thistle seed oil (each subsystem has its own units of reference flow).

Subsystem: C. Cardunculus cultivation		Value	Units
Subsystem output	*Biomass*	1	t
Inputs	Agricultural task (Diesel burned in agricultural machinery)	90	MJ
	Fertilizer	5	kg
Subsystem: seed separation from biomass		**Value**	**Units**
Subsystem output	*Seeds*	1	t
Inputs	Biomass	10.7	t
	Electricity	4	kWh
	Seed transport	80	km
Subsystem: oil extraction from seeds		**Value**	**Units**
Subsystem output	*Oil*	1	t
Inputs	Seeds	3.8	t
	Electricity	136	kWh
Subsystem: biodiesel production in transesterification plant		**Value**	**Units**
Subsystem output	*Biodiesel*	1	t
Inputs	Oil	1	t
	Methanol	118.5	kg
	Sodium hydroxide	14.8	kg
	Heat	4650	MJ
	Electricity	0.4	kWh

TABLE 3.3 Emission factors for the climate change impact through IPCC LCIA methodology (extracted from the Ecoinvent database).

Flow	IF value	Unit
Diesel burned in agricultural machinery	0.007	kg CO_2-eq/MJ
Fertilizer	1.4	kg CO_2-eq/kg
Electricity mix at consumer	0.47	kg CO_2-eq/kWh
Transport	0.52	kg CO_2-eq/tkm
Methanol	0.51	kg CO_2-eq/kWh

(*Continued*)

TABLE 3.3 (Continued)

Flow	IF value	Unit
Sodium hydroxide	1.18	kg CO_2-eq/kg
Heat, unspecific, in chemical plant	0.073	kg CO_2-eq/kg

Note that for the cultivation phase, in addition to the inputs of diesel and fertilizer, an elementary exchange is accounted: the direct emissions of N_2O due to the application of the fertilizer to soil. These N_2O emissions can be estimated, considering that the N_2O emitted represents 1% by weight of the total nitrogen content of the fertilizer. In this exercise, the nitrogen content is assumed to be 20% of the total weight of the NPK fertilizer. According to IPCC 2013, the global warming potential (GWP100) for N_2O is 265 kg CO_2-eq./kg of N_2O. For the separation of seeds and oil extraction, only electrical consumption will be considered.

Exercises

P3.1 Based on the established objective of the case study, what would be a suitable scope for the LCA?

P3.2 Build the inventory of the primary data referred to the functional unit: 1 metric ton of biodiesel

P3.3 What is the activity with a higher contribution to the climate change impact?

P3.4 A dataset for biodiesel production from soybean oil was found in an LCA database. The Climate Change impact indicator is 1.95 kgCO_2-eq per kilogram of biodiesel. Do you think that based on these results, the thistle oil would be a promising resource for biodiesel production in terms of sustainability?

P3.5 Indicate strategies of impact reduction:

References

Bello, S., Galán-Martín, Á., Feijoo, G., Moreira, M.T., Guillén-Gosálbez, G., 2020. BECCS based on bioethanol from wood residues: potential towards a carbon-negative transport and side-effects. Appl. Energy 279, 115884.

Bjoern, A., Chandrakumar, C., Boulay, A.-M., Doka, G., Fang, K., Gondran, N., et al., 2020. Review of life-cycle based methods for absolute environmental sustainability assessment and their applications. Environ. Res. Lett.

Bonou, A., Laurent, A., Olsen, S.I., 2016. Life cycle assessment of onshore and offshore wind energy-from theory to application. Appl. Energy 180, 327–337.

Calvo-Serrano, R., Guo, M., Pozo, C., Galán-Martín, A., Guillén-Gosálbez, G., 2019. Biomass conversion into fuels, chemicals, or electricity? A network-based life cycle optimization approach applied to the European Union. ACS Sustain. Chem. Eng. 7 (12), 10570–10582.

Ciroth, A., Finkbeiner, M., Hildenbrand, J., Klöpffer, W., Mazijn, B., Prakash, S., Sonnemann, G., Traverso, M., Ugaya, C., Valdivia, S., Vickery-Niederman, G., 2011. Towards a live cycle sustainability assessment: making informed choices on products. UNEP/SETAC Life Cycle Initiative.

EUA. (2013). Environmental Profile Report for the European Aluminium Industry. April 2013- Data for the year 2010. Life Cycle Inventory data for aluminium production and transformation processes in Europe. European Aluminium Association, (April), 1–78.

European Commission - Joint Research Centre - Institute for Environment and Sustainability. (2010). International Reference Life Cycle Data System (ILCD) Handbook – General guide for Life Cycle Assessment - Detailed guidance. First edition March 2010.

Fauzi, R.T., Lavoie, P., Sorelli, L., Heidari, M.D., Amor, B., 2019. Exploring the current challenges and opportunities of life cycle sustainability assessment. Sustainability 11 (3), 636.

Galán-Martín, A., Pozo, C., Azapagic, A., Grossmann, I.E., Mac Dowell, N., Guillén-Gosálbez, G., 2018. Time for global action: an optimised cooperative approach towards effective climate change mitigation. Energy Environ. Sci. 11 (3). Available from: https://doi.org/10.1039/c7ee02278f.

Garfí, M., Cadena, E., Sanchez-Ramos, D., Ferrer, I., 2016. Life cycle assessment of drinking water: comparing conventional water treatment, reverse osmosis and mineral water in glass and plastic bottles. J. Clean. Prod. 137, 997−1003.

Gesamtverband Der, A., 2019. Aluminium in the packaging industry. Aluminium-Zentrale 4−5, 12−17.

Goedkoop, M., Oele, M., Leijting, J., Ponsioen, T., & Meijer, E. (2016). Introduction to LCA with SimaPro. PRé.

Guillén, G., Mele, F.D., Bagajewicz, M.J., Espuna, A., Puigjaner, L., 2005. Multiobjective supply chain design under uncertainty. Chem. Eng. Sci. 60 (6), 1535−1553.

Guillén-Gosálbez, G., You, F., Galán-Martín, Á., Pozo, C., Grossmann, I.E., 2019. Process systems engineering thinking and tools applied to sustainability problems: current landscape and future opportunities. Curr. Opin. Chem. Eng. 26, 170−179.

Heijungs, R., Lenzen, M., 2014. Error propagation methods for LCA—a comparison. Int. J. Life Cycle Assess. 19 (7), 1445−1461.

Heiskanen, E., 2002. The institutional logic of life cycle thinking. J. Clean. Prod. 10 (5), 427−437. Available from: https://doi.org/10.1016/S0959-6526(02)00014-8.

Hellweg, S., Milà i Canals, L., 2014. Emerging approaches, challenges and opportunities in life cycle assessment. Science 344 (6188), 1109 LP-1113. Retrieved from. Available from: http://science.sciencemag.org/content/344/6188/1109.abstract.

International Aluminium Institute. (2015). Life Cycle Inventory Data and Environmental Metrics for the Primary Aluminium Industry - Data. 2017. Appendix A.

ISO, 2006a. Environmental Management: Life Cycle Assessment; Principles and Framework. ISO.

ISO. (2006b). ISO 14044. Environmental management—Life Cycle Assessment—Requirements and guidelines.

Jolliet, O., Müller-Wenk, R., Bare, J., Brent, A., Goedkoop, M., Heijungs, R., et al., 2004. The LCIA midpoint-damage framework of the UNEP/SETAC life cycle initiative. Int. J. Life Cycle Assess. 9 (6), 394.

Kadiyala, A., Kommalapati, R., Huque, Z., 2016. Evaluation of the life cycle greenhouse gas emissions from different biomass feedstock electricity generation systems. Sustainability 8 (11), 1181.

Kleinekorte, J., Fleitmann, L., Bachmann, M., Kätelhön, A., Barbosa-Póvoa, A., von der Assen, N., et al., 2020. Life cycle assessment for the design of chemical processes, products, and supply chains. Annu. Rev. Chem. Biomolecular Eng. 11.

Nessi S., Sinkko T.BC, Garcia-Gutierrez P., Giuntoli J.K.A., Sanye-Mengual E., Tonini D.P., et al. (2020). Comparative Life Cycle Assessment (LCA) of Alternative Feedstock for Plastics Production − Part 1, European Commission. https://doi.org/10.2760/XXXXX, JRCXXXXXX.

Onat, N.C., Kucukvar, M., Halog, A., Cloutier, S., 2017. Systems thinking for life cycle sustainability assessment: a review of recent developments, applications, and future perspectives. Sustainability 9 (5), 706.

Rockström, J., Steffen, W., Noone, K., Persson, A., Chapin, F.S., Lambin, E.F., et al., 2009. A safe operating space for humanity. Nature 461 (7263), 472−475. Available from: https://doi.org/10.1038/461472a.

Ryberg, M.W., Owsianiak, M., Richardson, K., Hauschild, M.Z., 2018. Development of a life-cycle impact assessment methodology linked to the planetary boundaries framework. Ecol. Indic. 88 (December 2017), 250−262. Available from: https://doi.org/10.1016/j.ecolind.2017.12.065.

Silva, R.O., Torres, C.M., Bonfim-Rocha, L., Lima, O.C.M., Coutu, A., Jiménez, L., et al., 2018. Multi-objective optimization of an industrial ethanol distillation system for vinasse reduction—a case study. J. Clean. Prod. 183, 956−963.

Stefanini, R., Borghesi, G., Ronzano, A., Vignali, G., 2020. Plastic or glass: a new environmental assessment with a marine litter indicator for the comparison of pasteurized milk bottles. Int. J. Life Cycle Assess. 1−18.

Steffen, W., Broadgate, W., Deutsch, L., Gaffney, O., Ludwig, C., 2015a. The trajectory of the Anthropocene: the great acceleration. Anthropocene Rev. 2 (1), 81−98.

Steffen, W., Richardson, K., Rockstrom, J., Cornell, S.E., Fetzer, I., Bennett, E.M., et al., 2015b. Planetary boundaries: guiding human development on a changing planet. Science 347 (6223), 1−10. Available from: https://doi.org/10.1126/science.1259855.

Steubing, B., Wernet, G., Reinhard, J., Bauer, C., Moreno-Ruiz, E., 2016. The ecoinvent database version 3 (part I): overview and methodology. Int. J. Life Cycle Assess. 21 (9), 1269−1281Retrieved from. Available from: http://link.springer.com/10.1007/s11367-016-1087-8.

2. Design principles

Sun, C.-H., Fu, Q., Liao, Q., Xia, A., Huang, Y., Zhu, X., et al., 2019. Life-cycle assessment of biofuel production from microalgae via various bioenergy conversion systems. Energy 171, 1033–1045.

Tian, X., Stranks, S.D., You, F., 2020. Life cycle energy use and environmental implications of high-performance perovskite tandem solar cells. Sci. Adv. 6 (31), eabb0055.

Tomatis, M., Jeswani, H.K., Stamford, L., Azapagic, A., 2020. Assessing the environmental sustainability of an emerging energy technology: solar thermal calcination for cement production. Sci. Total Environ. 742, 140510.

Torres, C.M., Ríos, S.D., Torras, C., Salvadó, J., Mateo-Sanz, J.M., Jiménez, L., 2013. Microalgae-based biodiesel: a multicriteria analysis of the production process using realistic scenarios. Bioresour. Technol. 147, 7–16.

United Nations. (2015). Transforming our world: The 2030 agenda for sustainable development. A/RES/70/1, 21 October.

WCED. (1987). Our Common Future – The Brundtland Report. World Commission on Environment and Development. Geneva.

Weidema, B.P., Wesnaes, M.S., 1996. Data quality management for life cycle inventories—an example of using data quality indicators. J. Clean. Prod. 4 (3–4), 167–174.

Wernet, G., Bauer, C., Steubing, B., Reinhard, J., Moreno-Ruiz, E., Weidema, B., 2016. The ecoinvent database version 3 (part I): overview and methodology. Int. J. Life Cycle Assess. 21 (9), 1218–1230. Available from: https://doi.org/10.1007/s11367-016-1087-8.

Biomass and waste based processes

Thermochemical processes

Antonio Sánchez and Mariano Martín

Department of Chemical Engineering, University of Salamanca, Salamanca, Spain

In this chapter, the thermochemical processes to valorize biomass are presented. They are technologies that operate over 400°C where the temperature, pressure, and residence time determine the product distribution. Four technologies are analyzed, based on physicochemical principles, such as gasification, pyrolysis, hydrothermal liquefaction (HTL), and combustion. Depending on the products obtained, upgrading and clean-up technologies such as reforming, carbon capture, hydrocracking, and fractionation are discussed as well as power cycles that will be described in more detail in Chapter 9.

4.1 Gasification

Biomass with low moisture content is the primary choice for the biomass gasification, as woody biomass or herbaceous plants. Prior to gasification, the biomass must be chopped off in particle sizes to 0.5−3.5 mm. The energy consumption is to be considered (Wright et al., 2010). The following equation calculates the energy consumption as a function of the final size of the biomass particle (size in mm).

$$Energy[kW/ton] = 5.31 size^2 - 30.86 size + 55.45 \qquad (4.1)$$

4.1.1 Gasification step

Gasification can be defined as the partial oxidation (less oxygen than necessary for the complete combustion) of a carbonaceous raw material, that is, biomass, at elevated temperature generating a gas that contains CO, CO_2, hydrogen, methane, water, and traces of higher hydrocarbons. It can be carried out using air, oxygen, steam, or a mixture as gasifying agent. Supercritical water is also proposed to be used in gasification, especially for wet biomasses. The major limitation of this alternative technology is the high pressure and temperature in the unit, increasing the investment cost, and the energy requirements

(Reddy et al., 2014; Okolie et al., 2019). Usually, biomass gasification takes place in three sequential steps (Bridgwater, 1995):

- Drying to evaporate moisture.
- Pyrolysis to generate gas, vaporized oils, tars, and a solid residue, char.
- Gasification or partial oxidation of the solid char, pyrolysis tars, and gases.

Syngas is obtained as a product of biomass gasification, whose common composition is 30%–60% CO, 25%–30% H_2, 5%–15% CO_2, and 0%–5% CH_4, and traces of H_2O, H_2S, NH_3, and others (Sikarwar et al., 2016). A general scheme of the gasification process is shown in Fig. 4.1.

Evaluating the performance of the gasification must consider the composition as well as the operation within the unit. First, the mechanisms of gasification are presented, namely the reactions that transform the biomass into a gas. Next, the different configurations for gasifiers are described. The configuration determines the design of the unit.

4.1.1.1 Principles and modeling

Although the list of reactions that take place is longer, the most important ones can be summarized in Fig. 4.2 according to the classification proposed by Sikarwar et al. (2016):

Others involve the production of higher hydrocarbons and pollutants (as ammonia or hydrogen sulfide). Among the reactions, there are those which are endothermic, such as reforming, Boudouard or water–gas shift reaction (WGSR), and exothermic, such as the oxidation or methanation reactions. Typically, the gasification is self-sustained, generating enough energy in the oxidation reactions to carry out the gasification process.

The biomass feedstock, which is mainly formed by cellulose, hemicellulose, and lignin, plays a key role in the gasification performance. As a general trend, the higher the cellulose and hemicellulose content in the inlet biomass, the larger the volume of gaseous products formed in the chamber. The typical range of biomass moisture to be introduced in the gasifier is about 15%–30% to avoid heat losses in the evaporation of this water in the gasifier.

The composition and yields of the gasification can be estimated from a theoretical point of view evaluating the equilibrium conditions. For this kind of modeling, the presence of higher hydrocarbons is neglected. Two approaches can be used, either the correlation for

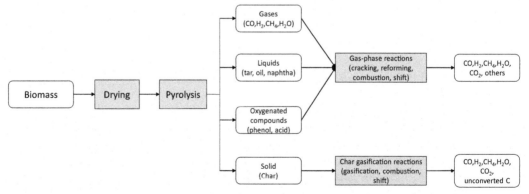

FIGURE 4.1 Routes for the biomass gasification. *Source: Adapted from Basu, P. (2010). Biomass Gasification and Pyrolysis: Practical Design and Theory. Academic Press.*

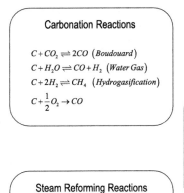

Carbonation Reactions

$C + CO_2 \rightleftharpoons 2CO$ (*Boudouard*)
$C + H_2O \rightleftharpoons CO + H_2$ (*Water Gas*)
$C + 2H_2 \rightleftharpoons CH_4$ (*Hydrogasification*)
$C + \frac{1}{2}O_2 \rightarrow CO$

Oxidation Reactions

$C + O_2 \rightarrow CO_2$
$CO + \frac{1}{2}O_2 \rightarrow CO_2$
$CH_4 + 2O_2 \rightleftharpoons CO_2 + 2H_2O$
$H_2 + \frac{1}{2}O_2 \rightarrow H_2O$

Water Gas Shift Reaction

$CO + H_2O \rightleftharpoons CO_2 + H_2$

Steam Reforming Reactions

$CH_4 + H_2O \rightleftharpoons CO + 3H_2$
$CH_4 + \frac{1}{2}O_2 \rightarrow CO + 2H_2$

Methanation Reactions

$2CO + 2H_2 \rightarrow CH_4 + CO_2$
$CO + 3H_2 \rightleftharpoons CH_4 + H_2O$
$CO_2 + 4H_2 \rightarrow CH_4 + 2H_2O$

FIGURE 4.2 Reactions involved in the gasification step. *Source: Adapted from Sikarwar, V.S., Zhao, M., Clough, P., Yao, J., Zhong, X., Memon, M.Z., et al. (2016). An overview of advances in biomass gasification. Energy Environ. Sci., 9, 2939.*

the gas equilibrium constants or a Gibbs free energy minimization, G, can be formulated (Jarungthammachote and Dutta, 2008). Below, the last of the two alternatives is described:

$$G = \sum_{i=1}^{N} n_i \mu_i \tag{4.2}$$

$$\mu_i = \overline{G_i^o} + RT \ln\left(\frac{f_i}{f_i^0}\right) = \overline{G_i^o} + RT \ln\left(\frac{\phi P_i}{P^o}\right) = \Delta\overline{G_{f,i}^o} + RT \ln(y_i) \tag{4.3}$$

$$G = \sum_{i=1}^{N} n_i \Delta\overline{G_{f,i}^o} + \sum_{i=1}^{N} n_i RT \ln(y_i) \tag{4.4}$$

where μ_i is the chemical potential of species i, n_i is the moles of species i, $\overline{G_i^o}$ is the standard Gibbs free energy, f_i is the fugacity of species i that is represented as a function of the partial pressure of the component, where ϕ and P_i are the fugacity coefficient and the partial pressure, respectively. At low pressure, f and P take the same value. Thus, if all the gases are assumed to be ideal, the molar fraction, y_i, is used for the ratio within the logarithm. $\overline{G_{f,i}^o}$ is the standard Gibbs free energy of formation of the species. To compute $\overline{G_{f,i}^o}$ as a function of the temperature, the following equations are used:

$$\Delta\overline{G_{f,i}^o} = \Delta\overline{H_{f,i}^o} - T\Delta\overline{S_{f,i}^o} \tag{4.5}$$

$$\Delta\overline{G_{f,i}^o} = \Delta\overline{H_{f,i}^o} - a'T\ln(T) + bT^2 - \left(\frac{c'}{2}\right)T^3 - \left(\frac{d'}{3}\right)T^4 + \left(\frac{e'}{2T}\right) + f' + g'T \tag{4.6}$$

NASA polynomials are typically used to compute these values (Turns, 2000). The nonidealities sometimes suggest the use of experimental data to compute the final gas composition (Jarungthammachote and Dutta, 2007) (Table 4.1).

Together with the equilibrium between species, the elemental balance must also hold:

$$\sum_{i=1}^{N} a_{ij} n_i = A_j \tag{4.7}$$

where a_{ij} is the number of atoms of the jth element in a mole of the ith species. A_j is defined as the total number of atoms of the jth element.

The equilibrium constant approach to compute the syngas composition from biomass gasification is explained through Example 4.1.

TABLE 4.1 Coefficients for the calculations of the Gibbs free energy of formation (Jarungthammachote and Dutta, 2007).

Compound	$\bar{h}_f^0$	a'	b'	c'	d'	e'	f'	g'
CO	-110.5	5.62E−03	−1.19E−05	6.38E−09	−1.85E−12	−4.89E+02	8.68E−01	−6.13E−02
CO$_2$	-393.5	−1.95E−02	3.12E−05	−2.45E−08	6.95E−12	−4.89E+02	5.27E+00	−1.21E−01
H$_2$O	-241.8	−8.95E−03	−3.67E−06	5.21E−09	−1.48E−12	0.00E+00	2.87E+00	−1.72E−02
CH$_4$	-74.8	−4.62E−02	1.13E−05	1.32E−08	−6.65E−12	−4.89E+02	1.41E+01	−2.23E−01

EXAMPLE 4.1 Biomass gasification: equilibrium conditions.

Wheat straw is processed in a downdraft gasifier. The moisture content of the wheat straw is equal to 7% and its ash content to 5.5% on a dry basis. The ultimate analysis of this biomass is: 46.32% C, 6.59% H, and 47.09% O. The gasification temperature is fixed to 900°C. Dry air is introduced as gasifying agent. Determine the composition of the outlet gases from the gasifier assuming equilibrium conditions.

Solution

To determine the gas composition, the equilibrium model proposed by Zainal et al. (2001) is used. Two equilibrium reactions are taken into account in this model: hydrogasification and the WGSR. The general reaction of the gasification process is as follows:

$$f_{\text{Biomass daf}} CH_\gamma O_\beta + \omega f_{\text{Biomass daf}} H_2O + f_{O_2} O_2 + \theta f_{O_2} N_2$$

$$\rightarrow f_{H_2} H_2 + f_{CO} CO + f_{CO_2} CO_2 + f_{H_2O} H_2O + f_{CH_4} CH_4 + \theta f_{O_2} N_2 \tag{4E1.1}$$

The following molar balances for each of the atom species involved (carbon, hydrogen, and oxygen) are formulated as well as the ratio between N_2 and O_2 in the feed stream. The ash from the biomass remains unchanged in the products.

$$f_{\text{Biomass daf}} = f_{CO} + f_{CO_2} + f_{CH_4} \tag{4E1.2}$$

$$\gamma f_{\text{Biomass daf}} + 2\omega f_{\text{Biomass daf}} = 2f_{H_2} + 2f_{H_2O} + 4f_{CH_4} \tag{4E1.3}$$

$$\beta f_{\text{Biomass daf}} + 2f_{O_2} + \omega f_{\text{Biomass daf}} = f_{CO} + 2f_{CO_2} + f_{H_2O} \tag{4E1.4}$$

$$f_{N_2} = f_{N_2} = \theta f_{O_2} \tag{4E1.5}$$

The equilibrium relationship for each of the reactions is presented in the next equations. The value of the equilibrium constant is expressed as a function of the partial pressure of the different gaseous components. The equilibrium constant is also calculated using the empirical correlation taking as variable the gasifier temperature.

$$K_1 = \frac{P_{CH_4}}{\left(P_{H_2}\right)^2} = \frac{f_{CH_4} f_{\text{total}}}{f_{H_2}^2 P_{\text{total}}} \tag{4E1.6}$$

$$\ln(K_1) = \frac{7082.848}{T} + (-6.567)\ln(T) + \frac{7.466 \cdot 10^{-3}}{2}T + \frac{-2.164 \cdot 10^{-6}}{6}T^2 + \frac{0.701 \cdot 10^{-5}}{2T^2} + 32.541 \tag{4E1.7}$$

$$K_2 = \frac{P_{CO_2} P_{H_2}}{P_{CO} P_{H_2O}} = \frac{f_{CO_2} f_{H_2}}{f_{CO} f_{H_2O}} \tag{4E1.8}$$

$$\ln(K_2) = \frac{5870.53}{T} + 1.86\ln(T) + 2.7 \cdot 10^{-4}T + \frac{58,200}{T^2} + 18.007 \tag{4E1.9}$$

Finally, the energy balance must hold. Adiabatic conditions are assumed for the biomass gasification process. The enthalpies of formation and heat capacities of the different components can be found in the Appendix B.

$$\begin{aligned}
f_{\text{Biomass daf}} H_f^0 {}_{\text{Biomass daf}} &+ f_{\text{Biomass daf}} w \left(H_f^0 {}_{H_2O(l)} + H_{\text{vap}} \right) + f_{O_2} H_f^0 {}_{O_2} + f_{N_2} H_f^0 {}_{N_2} \\
&= f_{H_2} H_f^0 {}_{H_2} + f_{CO} H_f^0 {}_{CO} + f_{CO_2} H_f^0 {}_{CO_2} + f_{H_2O} H_f^0 {}_{H_2O} + f_{CH_4} H_f^0 {}_{CH_4} + f_{N_2} H_f^0 {}_{N_2} \\
&\quad + \left(f_{H_2} C_{pH_2} + f_{CO} C_{pCO} + f_{CO_2} C_{pCO_2} + f_{H_2O} C_{pH_2O} + f_{CH_4} C_{pCH_4} + f_{N_2} C_{pN_2} \right) \Delta T
\end{aligned} \tag{4E1.10}$$

This simple model proposed for the biomass gasification assuming equilibrium conditions has been solved using Excel. The outlet stream from the gasifier is formed by two main phases: solid and gas. The solid phase contains the ashes from the biomass dragged by the gas stream. 0.05115 kg ashes/kg inlet biomass is generated in the process. The gas composition obtained for the given gasification temperature is collected in Table 4E1.1.

TABLE 4E1.1 Composition of the gases from the gasifier chamber.

Component	%	Component	%
H_2	21.04	H_2O	10.28
CO	23.28	CH_4	0.09
CO_2	9.17	N_2	36.14

4.1.1.2 Technologies

Gasifiers can be classified in four main groups (De et al., 2018): fixed bed gasifiers, fluidized-bed gasifiers, entrained-flow gasifiers, and plasma gasifiers. Fixed bed gasifiers correspond to the oldest technology proposed for gasification. There are three alternatives according to the direction of the fuel and gasifying agent: updraft, downdraft and cross-draft gasifiers. Commonly, fixed bed gasifiers have been used in small-scale applications (up to 10 MW). In updraft gasification, large particles of biomass are introduced from the top of the unit and the gasifying agent is introduced from the bottom, therefore, a counter-current contact between fuel and gasifier agent takes place. In this kind of gasifiers, four typical reaction zones can be distinguished: drying, pyrolysis, reduction, and oxidation. Significant production of tars appears in this type of unit due to the relatively low temperature in the outlet gases. In downdraft gasifiers (cocurrent contact), the biomass is fed at the top of the unit and the gasifier agent at an intermediate level (in a throat inside the oxidation zone). The product gas leaves the gasifier from the bottom. The exit temperature is higher than in the updraft; therefore less tar is produced. Finally, biomass is introduced from the top at cross-draft gasifiers, the gasifying agent from one side, and the products are extracted from the other side. The temperature of this unit is very high, involving a higher carbon monoxide concentration in the outlet gases and a reduced content in hydrogen and methane.

Fluidized-bed gasifiers require a small particle size to reach the fluidized-bed regime. The bed is more homogeneous; therefore it is not possible to differentiate the reaction zones. The gasifying agent is also the fluidizing agent in these units. Fluidized-bed gasifiers are used for a small-medium scale (500 kW to 50 MW) due to the easy scale-up and operation. Several advantages over the fixed bed can be identified: homogeneous and controllable temperature, low tar production, high conversion, excellent gas—solid contact, etc. Several alternatives have been proposed in this fluidized-bed gasifier category.

- Bubbling fluidized-bed gasifiers use a normal fluidization regime to gasify the biomass. An inert bed material is generally introduced in the chamber to improve the fluidization conditions. Biomass is introduced from the top of the unit or deep inside the fluidized bed. Most of the ash is removed from the bottom of the gasifier, nonetheless, a cyclone is also set up to eliminate the fly ashes from the outlet gases. Sulfur and chlorine can be removed in situ inside the bed. Typical inlet gas velocities used in this unit are about 1—2 m/s.
- Jetting fluidized-bed gasifier requires a high-velocity gas jet in the center of the bed. This fact improves the solid circulation, avoiding agglomeration, reaching a better carbon conversion.
- Circulating fluidized-bed (CFB) gasifiers consist in four main sections: a high-velocity riser, a cyclone, a downcomer, and a loop-seal valve. A high fluidization velocity is used and the particles are entrained by the gas. The particles are separated in the cyclone, returning to the chamber through the downcomer and the loop-seal valve. Typical gas velocities are about 5—10 m/s for this technology. An excellent gas—solid contact is reached in this chamber allowing high rates of mass and heat transfer.
- Dual fluidized-bed gasifiers introduce two chambers: the gasification and the combustion one. The gasification chamber is operated at bubbling fluidized-bed conditions; while, the

combustion one, at fast fluidized-bed conditions. The gasifying agent in the gasification chamber is, generally, steam generating an outlet gas stream rich in H_2. In the combustion chamber, char is burnt with air to heat up the bed material transferring the energy to the gasification chamber.
- Finally, vortex chamber fluidized-bed gasifiers generate a vortex inside the unit. A high-velocity gas is injected tangentially enhancing the mass and heat transfer. Very fine particles are required.

In entrained-flow gasifiers, biomass and gasifying agent are fed cocurrently from the top of the chamber. Very small particles are necessary in this technology, increasing the cost of grinding at the time of the pretreatment. The gas velocity here is higher than the CFB one, involving a really low residence time. Most of the facilities use oxygen as gasifying agent, obtaining a high temperature in the outlet gases (1200°C−1500°C). Due to the high temperature, a rapid conversion of the biomass is obtained with a low tar content. These units are mainly used in large-scale applications (>100 MW).

A more recent technology is plasma gasification. In this alternative, the energy is provided to the gasifier through a plasma system located near the bottom. Biomass is fed from the top of the chamber and the gasifying agent is introduced from the bottom, close to the plasma system. Very high temperatures are reached in this plasma gasifier allowing the treatment of hazardous and toxic materials. The main drawback of the plasma gasification is the high energy consumption and the large capital cost of the plasma system, essentially.

Several commercial gasifiers have been presented for different gasification technologies. For example, entrained-flow Siemens Fuel Gasifier, British Gas Lurgi gasifier, or Foster Wheeler CFB biomass gasifier. In particular, for biomass, the National Renewable Energy Laboratory (NREL) has presented a detailed evaluation of the performance of two fluidized-bed gasifiers: direct gasification, using RENUGAS gasifier (Eggeman, 2005), and indirect gasification, using a twin system, the FERCO BATTELLE gasifier (Phillips et al., 2007), see Fig. 4.3. In this chapter, a deeper analysis of these two gasifiers is performed.

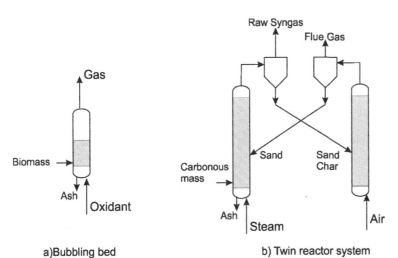

FIGURE 4.3 Typical gasifiers for biomass processing.

Fluidized-Bed Indirect Gasification: The low-pressure gasifier, Battelle Columbus (FERCO), is a twin reactor system. The first unit, Fig. 4.3B, is the gasifier. This unit is indirectly heated by means of a flow of hot sand, olivine. Biomass together with steam, 0.4 kg/kg$_{dry\ biomass}$ and olivine, 27 kg/kg$_{dry\ biomass}$ from the combustor, are fed to the gasifier which operates at 1.6 bar. The gasifier produces a raw syngas with low CO_2 content, but contains heavier hydrocarbons. The reactor is fast fluidized, allowing throughputs equal to the bubbling fluidized RENUGAS gasifier despite the nearly atmospheric operation (Phillips et al., 2007). The stream exiting the gasifier consists of solids, olivine and char, and raw syngas. The solids are recovered in a cyclone and sent to the combustor, while the gas is fed to the next stages. The composition of the gas has been experimentally evaluated as a function of the operating temperature, see Table 4.2. These empirical correlations are valid for a temperature range of 966K−1287K. The char is a carbonous residue that is comprised of the carbon that is left after the production of the gases as well as other residues such as oxygen, at least 4% of the oxygen of the biomass, the sulfur and nitrogen that do not generate gases (8.3% and 6.6%, respectively). Working at lower pressure conditions decreases the operating cost of the gasification system.

The char is used as fuel to heat up the sand, olivine, that is, the bed material used to provide the energy required for the endothermic reactions at the gasifier. In the second unit, the combustor, char is burned with 20% excess of preheated atmospheric air at 200°C. The gas can drag a fraction of the olivine, requiring a make-up of the sand. In the combustor, the char decomposes generating SO_2, CO_2, and N_2 and liberates the ash. The amount of energy generated is around 25,000 kJ/kg of char (Di Blasi, 2004). Typically, the combustor operates 100°C−200°C higher than the gasifier since the stream leaving it is

TABLE 4.2   Correlations for modeling the indirect low-pressure gasifier (Phillips et al., 2007).

Variable	Values and correlations (temperature ($T_{Fercogas}$) in °F)	
Mass of gas (kg)	$\left(28.993 - 0.043325 \cdot (T_{Fercogas}) + 0.000020966 \cdot (T_{Fercogas})^2\right)$	
	$*(Biomass_{maf}(kg)/0.454) * \rho_{gas}(kg/ft^3)$	(4.8)
Mass of tars (kg)	$(0.045494 - 0.000019759 \cdot (T_{Fercogas})) * (Biomass_{dry}(kg))$	(4.9)
Molar fraction		
CO_2	$0.01 * \left(-9.5251 + 3.77889 \cdot 10^{-2} \cdot (T_{Fercogas}) - 1.4927 \cdot 10^{-5}(T_{Fercogas})^2\right)$	(4.10)
CO	$0.01 * \left(133.46 - 1.029 \cdot 10^{-1}(T_{Fercogas}) + 2.8792 \cdot 10^{-5}(T_{Fercogas})^2\right)$	(4.11)
H_2	$0.01 * \left(17.996 - 2.6448 \cdot 10^{-2} \cdot (T_{Fercogas}) + 1.893 \cdot 10^{-5}(T_{Fercogas})^2\right)$	(4.12)
CH_4	$0.01 * \left(-13.82 + 4.4179 \cdot 10^{-2}(T_{Fercogas}) - 1.6167 \cdot 10^{-5}(T_{Fercogas})^2\right)$	(4.13)
C_2H_2	$0.01 * \left(-4.3114 + 5.4499 \cdot 10^{-3}(T_{Fercogas}) - 1.561 \cdot 10^{-6}(T_{Fercogas})^2\right)$	(4.14)
C_2H_4	$0.01 * \left(-38.258 + 5.8435 \cdot 10^{-2}(T_{Fercogas}) - 1.9868 \cdot 10^{-5}(T_{Fercogas})^2\right)$	(4.15)
C_2H_6	$0.01 * \left(11.114 - 1.1667 \cdot 10^{-2}(T_{Fercogas}) + 3.064 \cdot 10^{-6}(T_{Fercogas})^2\right)$	(4.16)

the one responsible for providing the energy to the gasifier in the form of sensible heat. The flue gas from the combustion of the char is to be treated to recover the sand in a cyclone, so that it is sent to the gasifier, and to remove at least 99% of the ash, using an electrostatic precipitator.

Fluidized-Bed Direct Gasification: The RENUGAS gasifier, also referred to as fluidized-bed direct gasification, is a pressurized direct oxygen-fired gasifier. It produces a gas rich in CO_2, with a smaller fraction of hydrocarbons compared to the FERCO Battelle design. The gasification takes place at higher pressure, 5–30 bar, allowing a large throughput per reactor volume, and reducing the need for pressurization downstream, so that less overall power is needed. However, the gasifier efficiency is lower and more steam is needed. Furthermore, the RENUGAS gasifier requires pure oxygen to reduce the equipment size or affect the catalysts downstream as well as to avoid diluting the syngas (Dutta and Phillips, 2009).

The gas composition from the RENUGAS gasifier can be estimated using the experimental correlations developed by Eggeman (2005). The general form of the correlations is given by Eq. (4.17) while the coefficients can be seen in Table 4.3 where the pressure, P, is given in psi, the temperature, T, in °F, X and Y are the ratios between the moles of oxygen or those of steam and the moles of carbon fed as biomass, $n_{O_2}/n_{C,feed}$ and $n_{H_2O}/n_{C,feed}$, respectively. The operational ranges for the variables in these equations are: for the temperature, between 1027K and 1255K; for the pressure, between 5.75 and 23.75 bar; for oxygen/inlet carbon, between 0.148 and 0.343; and, finally, for steam/inlet carbon ratio, between 0.24 and 1.97. The solid residue from the gasification, containing char and ash, is removed from the raw syngas in a cyclone.

$$f_i = A + B \cdot P + C \cdot T + D \cdot X + E \cdot Y \tag{4.17}$$

TABLE 4.3 High-pressure direct gasifier (Eggeman, 2005).

Molar ratio (f_i)	A	B	C	D	E
H_2/Feed H	− 3.830761E − 1	1.894350E − 4	2.666675E − 4	1.060088E − 1	7.880955E − 2
CO/Feed C	− 8.310017E − 2	− 3.340050E − 4	2.614482E − 4	1.495730E − 1	− 5.268367E − 2
CO_2/Feed C	7.157172E − 2	3.843454E − 4	1.286060E − 5	6.124545E − 1	9.980868E − 2
CH_4/Feed C	1.093589E − 2	1.388446E − 4	8.812765E − 5	− 2.274854E − 1	3.427825E − 2
C_2H_4/Feed C	5.301812E − 2	− 6.740399E − 5	− 1.372749E − 5	− 9.076286E − 3	− 4.854082E − 3
C_2H_6/Feed C	1.029750E − 1	− 5.440777E − 6	− 5.350103E − 5	− 3.377091E − 2	− 1.915339E − 3
C_6H_6/Feed C	4.676833E − 2	− 1.937444E − 5	− 1.270868E − 5	− 1.046762E − 2	− 8.459647E − 3
$C_{10}H_8$/Feed C	1.827359E − 2	− 2.328921E − 6	− 5.951746E − 6	− 1.936385E − 2	− 7.678310E − 4
Char					
% Feed N in Char	3.36	0.00	0.00	0.00	0.00
% Feed S in Char	8.45	0.00	0.00	0.00	0.00
% Feed O in Char	1.512040	1.582010E − 4	− 6.972612E − 4	1.573581E − 1	0.3332

4.1.2 Gas cleanup and conditioning

The syngas must be further processed to be used as synthesis gas or as fuel. Typically, four stages are required:

- The tars and other hydrocarbons are reformed producing additional CO and H_2.
- Particulates are removed by quenching.
- The gas composition can be adjusted depending on the needs using a water—gas shift reactor or a membrane system aiming the appropriate H_2 to CO ratio.
- Acid gases (CO_2 and H_2S) are removed from the gas stream.

4.1.2.1 Hydrocarbons reforming

While reformers can be furnaces of different designs (Martín, 2016), for the case of biomass processing they typically are bubbling fluidized-bed reactors where the hydrocarbons are converted to CO and H_2 while NH_3 is converted to N_2 and H_2 (Phillips et al., 2007).

Steam reforming: The chemical reactions taking place in steam reforming are of the form given by Eqs. (4.18) and (4.19). The process is endothermic and typically requires the use of fuel, either external or a fraction of the feed gas, to provide the heat necessary to get a certain level of conversion and to maintain isothermal operation. The mass balances for the different species are computed based on the stoichiometric relationships derived from them and the conversions provided in Table 4.4.

$$C_nH_m + nH_2O \rightarrow nCO + \left(\frac{m}{2} + n\right)H_2 \tag{4.18}$$

$$NH_3 \rightarrow \frac{1}{2}N_2 + \frac{3}{2}H_2 \tag{4.19}$$

For methane and ammonia, the composition highly depends on pressure and the yield of the steam reforming stage can be computed by the equilibrium (Martín, 2016). Two main equilibrium reactions are presented at this point: the methane decomposition and the WGSR. The equilibrium relationships for these two equilibriums are shown in Eqs. (4.20) and (4.21), respectively.

TABLE 4.4 Conversion of the reforming stage (Dutta and Phillips, 2009).

Compound	Conversion	
	1.6 bar	30 bar
CH_4	0.8	0.462
C_2H_6	0.99	0.99
C_2H_4	0.9	0.9
Tars	0.999	0.999
Benzene	0.99	0.99
NH_3	0.9	0.78

$$CH_4 + H_2O \leftrightarrow CO + 3H_2 \quad kp = 10^{[-\frac{11650}{T}+13.076]} \quad T[\ ] =]K \tag{4.20}$$

$$CO + H_2O \leftrightarrow CO_2 + H_2 \quad kp = 10^{[\frac{1910}{T}-1.784]} \quad T[\ ] =]K \tag{4.21}$$

While the use of steam reforming is widely used in biosyngas processing, aiming at increasing the H_2 to CO ratio, there are additional technologies to convert the hydrocarbons into CO and hydrogen. Here, two of the more mature ones are commented.

Partial oxidation: Pure oxygen is used, but in a limited amount to avoid the complete oxidation of the hydrocarbons, following the reaction given by Eq. (4.22). It is an exothermic process, so heat must be removed from the system.

$$C_nH_m + \frac{n}{2}O_2 \rightarrow nCO + \frac{m}{2}H_2 \tag{4.22}$$

Autothermal reforming: To make the most of the two alternatives presented above, the high yield to hydrogen obtained in the steam reforming and to self-sustain the process, autothermal reforming consists of combining both so that the operation is adiabatic. Therefore the consumption of steam and oxygen is adjusted for the reaction to self-sustain (Eq. 4.23).

$$C_mH_n + \frac{1}{2}mH_2O + \frac{1}{4}mO_2 \rightarrow mCO + \frac{1}{2}(m+n)H_2 \tag{4.23}$$

4.1.2.2 Gas cleanup

This stage includes particle and sour gases removal. However, note that, if the H_2:CO ratio is to be modified using a WGSR, CO_2 may be produced, in which case the sour gases removal stage must be postponed after the composition adjustment.

After the removal of hydrocarbons, *a scrubber* is typically used to remove particles and the remaining ash. The gas is treated with a stream of water so that the L/G ratio is 0.25 kg/m³ of gas (Martelli et al., 2009). It is assumed that solids (Ash, Char, Olivine) and NH_3 are eliminated. Typically, the amount of water added is such that the syngas is saturated after the contact and that helps control the steam to be added in a water–gas shift stage to adjust the composition of the H_2/CO ratio. As can be seen, there is a need to evaluate the process as a whole because the operation of one stage determines the next.

Sour gases removal: There are a number of technologies for sour gases, H_2S and CO_2, removal. They are applied not only to syngas, but as it will be shown in biogas upgrading, also to that operation. For the sake of the work, this section is focused on the use of amines, since they are capable of removing both at the same time. See Fig. 4.4, and we leave the use of PSA units and membranes for Chapter 6.

The CO_2 absorption systems using amines typically operate at low temperatures, around 25°C–30°C, and partial pressures above 0.05 bar, reaching removal yields of 90%–95% for partial pressures above 0.1 bar (Zhang and Chen, 2013). The amine absorption system is a scrubber that puts into contact a stream of absorbent with the gas phase containing the sour gas. To compute the flow of amines (Eq. 4.24), the pickup ratio of the solution is used. The

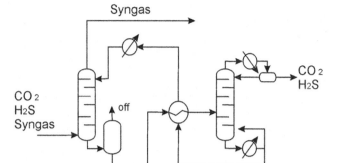

FIGURE 4.4   Sour gas removal.

TABLE 4.5   Amine properties for CO_2 capture GPSA (2004, 2012).

	MEA	DEA	MDEA
Gas pickup mol/mol amine	0.35	0.35–0.65	0.2–0.55
Solution concentration (wt.%)	20	35	45
Heat of reaction (BTU/lb CO_2)	620–700	580–650	570–600
GP	0.35	0.5	0.375
Density	1.01	1.05	1.045
Cost (€/kg)	1.3	1.32	3.09
Molecular weight	61	105	119

concentration of amine in the solution depends on the kind of amine used. A correction factor to secure the removal, GP, is also required, see Table 4.5.

$$fc_{amine} = \frac{MW_{amine}}{[amine]} \cdot \left(\frac{CO_{2eff} \cdot fc_{CO_2}}{MW_{CO_2}} + \frac{H_2S_{eff} \cdot fc_{H_2S}}{MW_{H_2S}} \right) \cdot \left(\frac{1}{GP} \right) \qquad (4.24)$$

This flow of amines results from the mixing between the recycled stream and the make-up needed due to the losses in the column that regenerates the amine solution. The absorption is an exothermic process. Since the absorption of gases reduces its efficiency with the temperature, this energy is to be refrigerated to maintain isothermal operation. The energy to be removed can be computed by the following equation:

$$Q_{Col1} = \Delta H_{react, amine} \cdot CO_{2eff} \cdot fc_{CO_2} + \Delta H_{react, amine} \cdot H_2S_{eff} \cdot fc_{H_2S} \qquad (4.25)$$

The acid-rich stream is sent to a stripping column to regenerate the amine. The liquid is to be heated to around 90°C for the operation of the column. The energy balance to the heat exchanger is as follows:

$$Q = F_{amine} \cdot q_{heat, amine} \qquad (4.26)$$

where F_{amine} is referred to the total mass flow of the amines stream, and q_{amine} the heat flow ratio based on the rules of thumb reported by GPSA (2012). The values are collected in Table 4.6. The solution is heated using the stream leaving the reboiler of the regeneration column with the aim of improving the energy efficiency of the desorption process.

The operation of column 2 is also based on rules of thumb (GPSA, 2004, 2012) including the estimation of the energy consumption in the reboiler and the cooler refrigeration requirements (Eqs. 4.27 and 4.28). The feed, the top and bottom temperatures for column 2 are 93°C, 54°C, and 125°C, respectively, and it operates at 1.7 bar.

$$Q_{Cond} = F_{Cond} \cdot q_{Cond, amine} \tag{4.27}$$

$$Q_{Reb} = F_{Reb} \cdot q_{Reb, amine} \tag{4.28}$$

From the reboiler, and after being used to heat up the amine-rich stream, the regenerated amine is cooled down heating up the feed to the column:

$$Q_{Cooling} = F_{Cooling} \cdot q_{Cooling, amine} \tag{4.29}$$

The gas leaving the regeneration column is saturated with the amines aqueous solution, producing the losses of amines which have to be replaced before being recirculated to the absorption column. The most typical amines are monoethanolamine (MEA), diethanolamine (DEA), and methyldiethanolamine (MDEA). Table 4.6 shows the parameters used in the amines absorption modeling for each amine considered (GPSA, 2004, 2012).

4.1.2.3 Gas composition adjustment

The syngas H_2 to CO ratio depends on the gasifier design. While the use of the indirect gasification yields a gas with a ratio of 0.6 prior to reforming and 1 almost after reforming (Phillips et al., 2007), the use of a direct gasifier results in a ratio close to 2 (Dutta and Phillips, 2009). The use of the syngas is what determines the further gas composition adjustment. The production of ethanol (Phillips et al., 2007) or dimethyl ether (DME) (Peral and Martín, 2015), requires a ratio of 1. Furthermore, the ratio for the production of Fischer–Tropsch liquids fuels is around 1.7 and for methanol around 2 (Martín, 2016). In the extreme case, the ammonia synthesis required the completed removal of carbon species (Sánchez et al., 2019). Therefore there may need to remove hydrogen or convert it into CO or, else, the composition does not need to be modified.

TABLE 4.6 Amine regeneration heat loads (GPSA, 2004).

	Duty (BTU/h)	Area (ft²)
Reboiler	72,000 GPM	11.3 GPM
Condenser	30,000 GPM	5.2 GPM
Amine feed to distillation	45,000 GPM	11.25 GPM
Amine cooler	15,000 GPM	10.20 GPM

Membrane separation: The high price of hydrogen may suggest the use of a membrane to recover the excess of hydrogen to be sold or used separately (Martín and Grossmann, 2011a). A palladium membrane can provide the purity required (Iyoha et al., 2007). It is not expected that the presence of H_2S affects the permeability of H_2 based on experimental results when H_2S is in the level of ppm (Emerson et al., 2012). The evaluation of the performance of a membrane can be modeled as follows, Eqs. (4.30)–(4.35), considering the entire unit. The control variable is the flux of the species that permeate, J_i. This flux is a function of the membrane and the species and is measured by its permeability. However, the gradient of concentrations along the membrane may require a discretization for a more accurate modeling.

$$F_{feed} = F_{permeate} + F_{retentate} \tag{4.30}$$

$$F_{feed} y_{i,\ feed} = F_{permeate} y_{i,\ permeate} + F_{retentate} y_{i,retentate} \tag{4.31}$$

$$J_i = \frac{F_{permeate} y_{i,permeate}}{A_{membrane}} \tag{4.32}$$

$$J_i = Q_i \left[\overline{y}_{feedside} P_{feed} - y_{i,permeate} P_{Permeate} \right] \tag{4.33}$$

$$\overline{y}_{feedside} = \frac{y_{i,\ feed} - y_{i,\ retentate}}{\ln\left(\frac{y_{i,\ feed}}{y_{i,\ retentate}}\right)} \tag{4.34}$$

$$Q_i = \frac{P_i}{\delta} \tag{4.35}$$

The units for the permeability are given in Barrer (1 barrer = 10^{-10} $cm^3_{STP} \cdot cm/(cm^2 \cdot s \cdot cmHg)$) or GPU (1 GPU = 3.35^{-10} mol/($m^2 \cdot s \cdot Pa$)). For hydrogen, values of 29,900 Barrer can be found (Yun and Ted Oyama, 2011)

Alternatively, *a sour WGSR can be used* (see Eq. 4.36). It is named as such due to the need to process gases containing hydrogen sulfide. For it to perform, a catalyst based on cobalt-molybdenum is used which can also contain Li, Na, Cs, or K. Furthermore, this catalyst can also convert carbonyl sulfide and other organic sulfur compounds into hydrogen sulfide (H_2S). The reactor operates adiabatically achieving CO conversions of around 97%. There are several designs for this reactor including traditional packed beds. However, the use of membrane reactors has attracted attention based on the high permeability of hydrogen across a palladium wall. It is possible to achieve purities of 99.99% and the membrane reactor combines both the equilibrium and the operation of the membrane, which also helps drive the equilibrium to the products (Iyoha et al., 2007). Fig. 4.5 shows a scheme of the operation of a membrane.

$$CO_{(g)} + H_2O_{(g)} \leftrightarrow CO_{2(g)} + H_{2(g)} \quad kp = 10^{\left[\frac{1910}{T} - 1.784\right]} \quad T[\ ]K \tag{4.36}$$

Once the syngas is cleaned, its use is similar than when it comes from any other source and different products can be obtained (Martín, 2016). The Example 4.2 evaluates the performance of a membrane reactor with hydrogen separation.

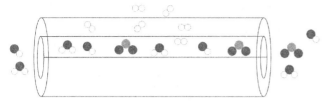

FIGURE 4.5   Membrane reactor.

EXAMPLE 4.2 Water–gas shift reactor with H_2 membrane.

Evaluate the performance of a membrane reactor for the in situ production of hydrogen. Based on the model by Ji et al. (2009). **Data:**

$K_{eq} = e^{\left(-4.946 + \frac{4897}{T}\right)}$; $P = 5$ bar; $d = 0.2$ m; $a_{H_2} = \pi d$; $km = 0.00015$ J/m $\cdot$ s $\cdot$ K

$\delta_{sp} = 0.005$ m; $\rho_s^{WGS} = 1200$ g/m^3; $\varepsilon_g = 0.05$; $Pm = 7.71 \cdot 10^{-4}$ mol m/(s m^{2}bar$^{0.5}$); $Ea = 29730$ kJ/kmol

$\delta_{H_2} = 5 \cdot 10^{-6}$; $k_o^{WGS} = e^{16.68}$; $Ea_{WGS} = 114600$ kJ/kmol

Solution

First, the differential equations to describe the behavior of hydrogen and CO inside the reactor are presented:

$$\frac{dF_{CO}}{dz} = -a_r\left(1 - \varepsilon_g\right)\rho_s^{WGS}\left(-r_{CO}^{WGS}\right) \tag{4E2.1}$$

$$\frac{dF_{H_2}}{dz} = a_r\left(1 - \varepsilon_g\right)\rho_s^{WGS}\left(-r_{CO}^{WGS}\right) - a_{H_2}N_{H_2} \tag{4E2.2}$$

It is assumed that only hydrogen can diffuse across the membrane. The kinetic expression for the reaction is as follows:

$$-r_{CO}^{WGS} = k^{WGS}(P_{CO})^{0.9}\left(P_{H_2O}\right)^{0.25}\left(P_{CO_2}\right)^{-0.6}(1 - \beta) \tag{4E2.3}$$

$$\beta = \frac{P_{CO_2}P_{H_2}}{P_{H_2O}P_{CO}K_{eq}} \tag{4E2.4}$$

$$k^{WGS} = k_0^{WGS}e^{\frac{-E_a}{RT}} \tag{4E2.5}$$

The hydrogen flow across the membrane can be expressed as a function of the difference in the hydrogen pressure on both sides of the membrane.

$$N_{H_2} = \frac{P_m e^{\frac{-E_a}{RT}}}{\delta_{H_2}}\left(\sqrt{P_{H_2}^{high}} - \sqrt{P_{H_2}^{low}}\right) \tag{4E2.6}$$

The mass balance for the permeate side of the membrane can be expressed as:

$$\frac{dG_{H_2}}{dz} = a_{H_2}N_{H_2} \tag{4E2.7}$$

The energy balance for both sides is represented with the next equations:

$$\frac{dT_r^{\text{WGS}}}{dz} = \frac{1}{\sum F_i Cp_i} \left(\Delta H_{\text{WGS}} a_r \left(1 - \varepsilon_g \right) \rho_s^{\text{WGS}} \left(-r_{\text{CO}}^{\text{WGS}} \right) - q - a_{H_2} N_{H_2} \Delta H_{H_2} \right) \qquad \text{(4E2.8)}$$

$$q = \frac{a_{H_2} k_m}{\delta_{sp}} \left(T_r^{\text{WGS}} - T_{nf}^{\text{WGS}} \right) \qquad \text{(4E2.9)}$$

$$\frac{dT_{nf}^{\text{WGS}}}{dz} = \frac{1}{\sum G_j Cp_j} \left(q + a_{H_2} N_{H_2} \Delta H_{H_2} \right) \qquad \text{(4E2.10)}$$

The set of algebraic and differential equations has been solved using MATLAB using the method ode45. The graphic results can be found in Fig. 4E2.1.

```
[a, b] = ode45(@Rmemh2,[0 0.75],[10 10 0 1 0 573 573]);
plot(a, b(:,1),'k−', a, b(:,2),'b−', a, b(:,3),'r−', a, b(:,4),'g−');
xlabel('Z(m)')
ylabel('F(mol/min)');
function Rmembrana = Rmemh2(t, F)
FH2O = F(1);
FCO = F(2);
FH2 = F(3);
FCO2 = F(4);
GH2 = F(5);
Tr = F(6);
Tnf = F(7);
%Constantes
%min − 1
```

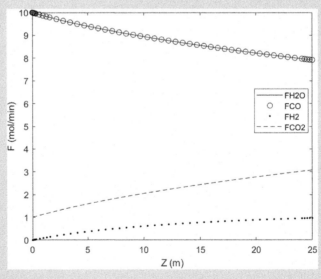

FIGURE 4E2.1   Flows profile.

```
Temp = 273 + 300;
Kceq = exp( - 4.946 + 4897/Temp);
Presion = 5;
Eg = 0.05;
diameter = 0.2;
area = (diameter/2)^2*3.14;
aH2 = 3.14*diameter;
rhowgs = 1200;
deltaH2 = 5e - 6;
ea2 = 29730;
Pm = 7.71e - 4;
Phigh = Presion;
Plow = 4;
kowgs = exp(16.68);
Ea = 114600;
kwgs = kowgs*exp( - Ea/(8.314*Temp));
NH2 = Pm*exp( - ea2/(8.314*Temp))*(Phigh^0.5 - Plow^0.5)/deltaH2;
km = 0.00015;
deltasp = 5e - 3;
Cpi = (1*28 + 1.1*44)*0.5;
%kJ/kgK*28 kg/kmol
DeltaHr = - 41000;
%kJ/kmol
Cph2 = (2*14);
%kJ/kmolK
DeltaH2 = Cph2*(Tnf - 298);
%kJ/kmol
%Variables
Ftotal = FH2O + FCO + FH2 + FCO2;
PCO = (FCO/Ftotal)*Presion;
PCO2 = (FCO2/Ftotal)*Presion;
PH2O = (FH2O/Ftotal)*Presion;
PH2 = (FH2/Ftotal)*Presion;
beta = PCO2*PH2/(PCO*PH2O*Kceq);
rCO = kwgs*PCO^0.9*PH2O^0.25*PCO2^( - 0.6)*(1 - beta);
q = aH2*km*(Tr - Tnf)/deltasp;
%Eqs diferenciales
Rmembrana(1,1) = - area*(1 - Eg)*rhowgs*rCO;
Rmembrana(2,1) = - area*(1 - Eg)*rhowgs*rCO;
Rmembrana(3,1) = area*(1 - Eg)*rhowgs*rCO - aH2*NH2;
Rmembrana(4,1) = area*(1 - Eg)*rhowgs*rCO;
Rmembrana(5,1) = aH2*NH2;
```

```
Rmembrana(6,1) = (1/(Ftotal*Cpi))*(DeltaHr*area*(1 − Eg)*rhowgs*rCO − q −
aH2*NH2*DeltaH2);
Rmembrana(7,1) = (1/(GH2*Cph2))*(q + aH2*NH2*DeltaH2);
```

4.1.2.4 Syngas usage

The use of syngas is common to the traditional industry in most aspects. The reactors are typically governed by the equilibrium, even though achieving the equilibrium is difficult (Martín, 2016). Most of the reactors are either multibed reactors or multiphase ones where the catalysis, the feed gas, and the products (liquid in case of heavier oils and gas for smaller molecules like methanol) are mixed. Table 4.7 shows the mechanisms that govern the reactions.

A detail for typical catalytic reactors can be seen in Martín (2016). Here,the most typical one is presented, the slurry one that is widely used in FT and methanol production, see Fig. 4.6; together with it, several designs for the fermentation of syngas are also included, such as a stirred tank aerated reactor, a bubble column, and an airlift (Stoll et al., 2020).

Companies, such as INEOS, Coskata, and LanzaTech, already operate syngas fermentation reactors. The main challenge is the fact that the reaction is exothermically operating at around 38°C which prevents from reusing the reactor as a hot stream, resulting in increasing the water consumption of the facilities.

In the case of catalyst synthesis, the limitation is the mass transfer from the gas to the catalyst particle and the removal of the heat produced. They are highly exothermic reactors. The cooling is typically carried out by the production of steam as seen in Fig. 4.6A. For the case of the fermentation of syngas, the limitation is the gas−liquid mass transfer. The mass transfer coefficient, $k_L a$, is a function of the mixing in the tank and the properties of the liquid (Martín et al., 2007). Values from 18 to 860 h^{-1} (Stoll et al., 2020) can be found for bubble columns (Fig. 4.6C) but for the stirred tanks and airlifts (Fig. 4.6B and D) the values depend on the flow pattern and energy input by the impeller. Example 4.3 computes the performance of a fixed-bed reactor for ethanol production from syngas. Additionaly, syngas fermentation is also evaluated in Example 4.4 using a bubble column reactor.

4.1.2.5 Product purification

The separation and purification of the products is based on typical phase separation units, including, flash and distillation columns. One particular issue is the absorption of CO_2 in DME. However, when using syngas, the presence of CO_2 is reduced. That will not be the case if CO_2 hydrogenation is the synthesis route as it will be presented later in the text. Table 4.8 summarized for the set of typical products the separation operations.

Water and energy integration allows reducing the needs of these processes. Energy integration can be integrated within the design of the process using Duran and Grossmann (1986) strategy, see Chapter 2, as well as the development of a heat exchanger network based on the models of Yee and Grossmann (1990). Thermochemical processes operating at high temperatures allow simple integration, but for the case of syngas fermentation. In addition, the possibility of using air cooling reduces the need for water. However, the

TABLE 4.7 Use of syngas: reaction mechanisms.

Product	Operating conditions	Synthesis stage	Reactions mechanism	
H$_2$ (Martín and Grossmann, 2013)	H$_2$/CO = inf	WGS	$CO_{(g)} + H_2O_{(g)} \leftrightarrow CO_{2(g)} + H_{2(g)}$ $kp = 10^{\left[\frac{1910}{T} - 1.784\right]}$	(4.37)
CH$_3$OH (Martín and Grossmann, 2017)	H$_2$/CO = 2–2.5	Catalysis	$CO + 2H_2 \rightleftarrows CH_3OH$	(4.38)
			$CO_2 + H_2 \rightleftarrows CO + H_2O$	(4.39)
	T = 200°C–300°C		They are governed by the equilibrium:	
	P = 50 bar		$\dfrac{[P_{CH_3OH}]}{[P_{CO}][P_{H_2}]^2} = 10^{\left[\frac{3971}{T} - 7.492\log T + 1.77\times10^{-3}T - 3.11\times10^{-8}T^2 + 9.218\right]}$	(4.40)
			$\dfrac{[P_{CO}][P_{H_2O}]}{[P_{CO_2}][P_{H_2}]} = Exp\left[13.148 - \dfrac{5639.5}{T} - 1.077\ln T - 5.44\times10^{-4}T + 1.125\times10^{-7}T^2 + \dfrac{49170}{T^2}\right]$	(4.41)
	H$_2$/CO = 1	Fermentation	$3CO + 3H_2 \rightarrow C_2H_5OH + CO_2; \quad X_{EtOH} = 0.7$	(4.42)
	T = 38°C			
	P = 1.2 bar			
Ethanol (Martín and Grossmann, 2011b)			$CO + H_2O \rightarrow H_2 + CO_2; \quad X_{CO_2} = 0.219$	(4.43)
			$CO + 2H_2 \rightarrow CH_3OH; \quad X_{MetOH} = 0.034$	(4.44)
			$CO + 3H_2 \rightarrow CH_4 + H_2O; \quad X_{CH_4} = 0.003$	(4.45)
	T = 300°C		$2CO + 4H_2 \rightarrow C_2H_5OH + H_2O; \quad X_{EtOH} = 0.282$	(4.46)
	P = 68 bar	Syngas	$2CO + 5H_2 \rightarrow C_2H_6 + 2H_2O; \quad X_{C_2H_6} = 0.003$	(4.47)
		Catalysis	$3CO + 6H_2 \rightarrow C_3H_7OH + 2H_2O; \quad X_{PropOH} = 0.046$	(4.48)
			$4CO + 8H_2 \rightarrow C_4H_9OH + 3H_2O; \quad X_{ButOH} = 0.006$	(4.49)
			$5CO + 10H_2 \rightarrow C_5H_{11}OH + 4H_2O; \quad X_{PentOH} = 0.001$	(4.50)
DME (Peral and Martín, 2015)	H$_2$/CO = 1	Catalysis	$CO_2 + 3H_2 \rightleftarrows CH_3OH + H_2O$	(4.51)
	T = 250°C		$CO + 2H_2 \rightleftarrows CH_3OH$	(4.52)
	P = 50 bar		$CO + H_2O \rightleftarrows H_2 + CO_2$	(4.53)
			$2CH_3OH \rightleftarrows CH_3OCH_3 + H_2O$	(4.54)

(Continued)

TABLE 4.7 (Continued)

Product	Operating conditions	Synthesis stage	Reactions mechanism	
			They are governed by:	
			$$Kp_1 = 10^{\left[\frac{3066}{T}-10.592\right]} = \frac{P_{CH_3OH}}{P_{CO}\cdot(P_{H_2})^2} = \frac{n_{CH_3OH}\cdot n_{Total}^2}{P_t^2\cdot n_{CO}\cdot(n_{H_2})^2}$$	(4.55)
			$$Kp_2 = \exp\left(\frac{2835.2}{T}+1.6775\ln T - 2.39\cdot10^{-4}T - 0.21\cdot10^{-6}T^2 - 13.36\right)$$ $$= \frac{P_{DME}\cdot P_{H_2O}}{(P_{CH_3OH})^2} = \frac{n_{DME}\cdot n_{H_2O}}{(n_{CH_3OH})^2}$$	(4.56)
			$$Kp_3 = 10^{\left[\frac{2073}{T}-2.029\right]} = \frac{P_{CO_2}\cdot P_{H_2}}{P_{CO}\cdot P_{H_2O}} = \frac{n_{CO_2}\cdot n_{H_2}}{n_{CO}\cdot n_{H_2O}}$$	(4.57)
FT (Hernandez and Martín, 2018)	$H_2/CO = 1.7$	Catalysis	$$nCO + \left(n+\frac{m}{2}\right)H_2 \rightarrow C_nH_m + nH_2O$$	(4.58)
	$T = 200°C$ (Heavier)		$$CO + 2H_2 \rightarrow -CH_2- + H_2O; \quad \Delta H_{FT} = -165\ kJ/mol$$	(4.59)
	$300°C$ (Gasolines)		The product distribution can be estimated as:	
	$P = 30$ bar		$$w_i = \alpha^{i-1}(1-\alpha)^2\cdot i$$	(4.60)
			where w_i is the weight percentage and α the chain length given by:	
			$$\alpha = \left(0.2332*\left(\frac{y_{co}}{y_{H_2}+y_{co}}\right)+0.633\right)*(1-0.0039*((T\_Synthesis+273)-533))$$	(4.61)

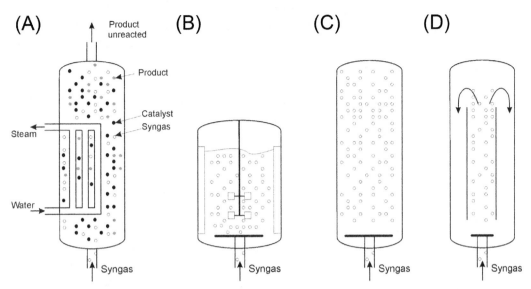

FIGURE 4.6 (A) Slurry reactor and (B) syngas fermenter (B) CSTR, (C) Bubble column, (D) Airlift.

EXAMPLE 4.3 Catalytic synthesis of ethanol from syngas.

Model the production of ethanol from syngas catalytic synthesis using the work in Portillo et al. (2016).

Solution

The reactions involved are the following:

$$CO + 2H_2 \rightarrow CH_3OH \tag{4E3.1}$$

$$CH_3OH + H_2 \rightarrow CH_4 + H_2O \tag{4E3.2}$$

$$CH_3OH + CO + 2H_2 \rightarrow C_2H_5OH + H_2O \tag{4E3.3}$$

$$C_2H_5OH + CO + 2H_2 \rightarrow C_3H_7OH + H_2O \tag{4E3.4}$$

$$CO + H_2O \leftrightarrow H_2 + CO_2 \tag{4E3.5}$$

For each one, the kinetics are given as follows:

$$k^i = k_0^i \, e^{\frac{-E_a}{RT}} \tag{4E3.6}$$

$$r_{MeOH} = k^{MeOH} P_{CO}^A P_{H_2}^B \tag{4E3.7}$$

$$r_{CH_4} = k^{CH_4} P_{H_2}^C P_{MeOH}^D \tag{4E3.8}$$

$$r_{EtOH} = k^{EtOH} P_{CO}^E P_{H_2}^F P_{MeOH}^G \tag{4E3.9}$$

$$r_{PrOH} = k^{PrOH} P_{CO}^H P_{H_2}^I P_{EtOH}^J \tag{4E3.10}$$

$$r_{CO_2} = k^{WGS}\left[P_{CO}P_{H_2O} - \frac{k_x}{K_{WS}}P_{CO_2}P_{H_2}\right] \tag{4E3.11}$$

$$K_{WS} = \exp\left(-\left(-8514 + 7.71T(K)[cal/mol]\right)/RT\right) \tag{4E3.12}$$

where the values for the reaction orders and kinetic constants are presented in Table 4E2.1. k_x is equal to 0.85.

TABLE 4E2.1 Values for the kinetic model.

	k_o'	Ea	CO order	H_2 order	MetOH order	EtOH order
Methanol	0.0036	83.16	1.93	0.44		
Ethanol	0.014	83.29	1.34	0.24	0.67	
Propanol	414.96	159.09	1.39	1.22		0.75
Methane	3152.07	114.58		0.06	0.53	
Carbon dioxide	10.65	57.18				

Thus the model becomes:

$$\frac{dF_{CO}}{dW} = -\left(r_{MeOH} + r_{EtOH} + r_{PrOH} + r_{CO_2}\right) \tag{4E3.13}$$

$$\frac{dF_{H_2}}{dW} = -\left(2r_{MeOH} + 2r_{EtOH} + 2r_{PrOH} + r_{CH_4}\right) + r_{CO_2} \tag{4E3.14}$$

$$\frac{dF_{H_2O}}{dW} = r_{EtOH} + r_{PrOH} + r_{CH_4} - r_{CO_2} \tag{4E3.15}$$

$$\frac{dF_{MetOH}}{dW} = r_{MeOH} - r_{EtOH} - r_{CH_4} \tag{4E3.16}$$

$$\frac{dF_{EtOH}}{dW} = r_{EtOH} - r_{PrOH} \tag{4E3.17}$$

$$\frac{dF_{PrOH}}{dW} = r_{PrOH} \tag{4E3.18}$$

$$\frac{dF_{CH_4}}{dW} = r_{CH_4} \tag{4E3.19}$$

$$\frac{dF_{CO_2}}{dW} = r_{CO_2} \tag{4E3.20}$$

```
[a, b] = ode15s(@ReacCatEtOH,[0 3000000],[4.5 4.5 0 0 0 0 0.2 0.4]);
plot(a, b(:,1),'k-', a, b(:,2),'b-', a, b(:,3),'r-', a, b(:,4),'g-', a, b
(:,5),'m-', a, b(:,6),'y-', a, b(:,7),'c-', a, b(:,8),'g-');
xlabel('W(kgcat)')
ylabel('F(kmol/s)')
legend('CO','H2','H2O','MeOH','EtOH','PrOH','CH4','CO2')
function Rcat = ReacCatEtOH(w, F)
FCO = F(1);
```

```
FH2 = F(2);
FH20 = F(3);
FMeOH = F(4);
FEtOH = F(5);
FPrOH = F(6);
FCH4 = F(7);
FCO2 = F(8);
%Constantes
%min - 1
Temp = 273 + 300;
Rgases = 0.008314;%kJ/molK
Rcal = 1.987; %cal/molK
Presion = 90;%bar
kx = 0.85;
kmeoh = 0.0036*exp( - 83.16/(Rgases*Temp));
ketoh = 0.014*exp( - 83.29/(Rgases*Temp));
kproh = 414.96*exp( - 159.09/(Rgases*Temp));
kch4 = 3152.07*exp( - 114.58/(Rgases*Temp));
kwgs = 10.65*exp( - 57.18/(Rgases*Temp));
KWS = exp( - ( - 8514 + 7.71*Temp)/(Rcal*Temp));
%Variables
Ftotal = FH20 + FCO + FH2 + FCO2 + FEtOH + FPrOH + FMeOH + FCH4;
PCO = (FCO/Ftotal)*Presion;
PCO2 = (FCO2/Ftotal)*Presion;
PH20 = (FH20/Ftotal)*Presion;
PH2 = (FH2/Ftotal)*Presion;
PEtOH = (FEtOH/Ftotal)*Presion;
PMeOH = (FMeOH/Ftotal)*Presion;
PPrOH = (FPrOH/Ftotal)*Presion;
PCH4 = (FCH4/Ftotal)*Presion;
%Rates
rMeOH = kmeoh*PCO^1.93*PH2^0.44;
rCH4 = kch4*PH2^0.06*PMeOH^0.53;
rEtOH = ketoh*PCO^1.34*PH2^0.24*PMeOH^0.67;
rPrOH = kproh*PCO^1.39*PH2^1.22*PEtOH^0.75;
rCO2 = kwgs*(PCO*PH20 - (kx/KWS)*PCO2*PH2);
%Eqs diferenciales
Rcat(1,1) = - (rMeOH + rEtOH + rPrOH + rCO2);
Rcat(2,1) = - (2*rMeOH + 2*rEtOH + 2*rPrOH + rCH4) + rCO2;
Rcat(3,1) = rEtOH + rPrOH + rCH4 - rCO2;
Rcat(4,1) = rMeOH - rEtOH - rCH4;
Rcat(5,1) = rEtOH - rPrOH;
Rcat(6,1) = rPrOH;
Rcat(7,1) = rCH4;
Rcat(8,1) = rCO2;
```

3. Biomass and waste based processes

The obtained results are shown in Fig. 4E3.1.

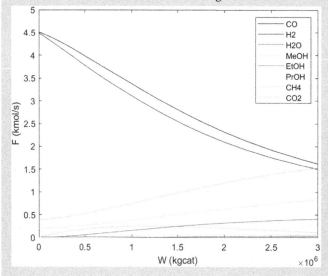

FIGURE 4E3.1   Profile of products for ethanol catalytic synthesis.

EXAMPLE 4.4 Syngas fermentation: bubble column reactor (BCR).

The kinetic model is taken from de Medeiros et al. (2019). This model assumes a Monod kinetics for the uptake rates of CO and H_2 with inhibition by substrate and product. Four main reactions were considered in this mechanics:

$$4CO + 2H_2O \rightarrow C_2H_4O_2 + 2CO_2 \tag{4E4.1}$$

$$4H_2 + 2CO_2 \rightarrow C_2H_4O_2 + 2H_2O \tag{4E4.2}$$

$$C_2H_4O_2 + 2CO + H_2O \rightarrow C_2H_6O + 2CO_2 \tag{4E4.3}$$

$$C_2H_4O_2 + 2H_2 \rightarrow C_2H_6O + H_2O \tag{4E4.4}$$

The following equations computed the kinetics rates:

$$v_i = - \frac{v_{\text{max},i} C_{l,i}}{K_{s,i} + C_{l,i}} I_E I_A I_{CO,i} \quad \forall i \in \{CO, H_2\} \tag{4E4.5}$$

$$I_i = \frac{1}{1 + \frac{C_{l,i}}{K_{l,i}}} \quad \forall i \in \{EtOH, HAc\} \tag{4E4.6}$$

$$I_{CO,H_2} = \frac{1}{1 + \frac{C_{l,CO}}{K_{l,CO}}} \quad ; \quad I_{CO,CO} = 1 \tag{4E4.7}$$

$$\mu = -v_{CO} Y_{X,CO} - v_{H_2} Y_{X,H_2} \tag{4E4.8}$$

$$v_k^R = \left(\frac{1}{2}\right)\left(\frac{2F_{\text{AcR},j}}{2F_{\text{AcR},j} + |v_j|}\right)|v_j| \quad \forall (i,k) \in \{(\text{CO},3),(\text{H}_2,4)\} \tag{4E4.9}$$

$$F_{\text{AcR},j} = \frac{v_{\max,\text{AcR}}^i C_{l,\text{HAc}}}{K_{S,\text{AcR}}^i + C_{l,\text{HAc}}} \quad \forall j \in \{\text{CO},\text{H}_2\} \tag{4E4.10}$$

$$v_k^R = -\frac{(v_i + 2v_{\text{AcR},i})}{4} \quad \forall (i,k) \in \{(\text{CO},1),(\text{H}_2,2)\} \tag{4E4.11}$$

$$v_{\text{CO}_2} = 2v_3^R - 2v_4^R + 2v_5^R \tag{4E4.12}$$

$$v_{\text{EtOH}} = v_5^R + v_6^R \tag{4E4.13}$$

$$v_{\text{HAc}} = v_3^R + v_4^R - v_5^R - v_6^R \tag{4E4.14}$$

$$v_{\text{H}_2\text{O}} = -2v_3^R + 2v_4^R - v_5^R + v_6^R \tag{4E4.15}$$

The model of the bubble column reactor is based on the following assumptions (Chen et al., 2015): ideal plug-flow for the gas phase, axial dispersion in the liquid phase, and an isothermal behavior of the reactor (with a temperature equal to 37°C). The biochemical reactions occur in the liquid phase according to the kinetic expressions presented previously. The equations to describe this model are as follows:

$$\frac{\partial C_{g,i}(x,t)}{\partial t} = -\frac{k_{m,i}}{\varepsilon_G}\left(C_i^*(x,t) - C_{l,i}(x,t)\right) - \frac{u_G}{\varepsilon_G}\frac{\partial C_{g,i}(x,t)}{\partial x} \quad \forall i \in \{\text{CO},\text{H}_2,\text{CO}_2\} \tag{4E4.16}$$

$$C_{g,i}(0,t) = C_{gf,i}; \qquad C_{g,i}(x,0) = C_{g0,i} \tag{4E4.17}$$

$$\frac{\partial C_{l,i}}{\partial t} = v_i X + \frac{k_{m,i}}{\varepsilon_L}\left(C_i^* - C_{l,i}\right) - \frac{u_L}{\varepsilon_L}\frac{\partial C_{l,i}}{\partial x} + D_A\frac{\partial^2 C_{l,i}}{\partial x^2} \quad \forall i \in \{\text{CO},\text{H}_2,\text{CO}_2\} \tag{4E4.18}$$

$$u_L C_{l,i}(0,t) - \varepsilon_L D_A\frac{\partial C_{l,i}(0,t)}{\partial x} = 0; \quad \frac{\partial C_{l,i}(L,t)}{\partial x} = 0; \quad C_{l,i}(z,0) = C_{l0,i} \tag{4E4.19}$$

$$\frac{\partial C_{l,i}}{\partial t} = M_i v_i X - \frac{u_L}{\varepsilon_L}\frac{\partial C_{l,i}}{\partial x} + D_A\frac{\partial^2 C_{l,i}}{\partial x^2} \quad \forall i \in \{\text{EtOH},\text{HAc}\} \tag{4E4.20}$$

$$u_L C_{l,i}(0,t) - \varepsilon_L D_A\frac{\partial C_{l,i}(0,t)}{\partial x} = 0; \quad \frac{\partial C_{l,i}(L,t)}{\partial x} = 0; \quad C_{l,i}(z,0) = C_{l0,i} \tag{4E4.21}$$

$$\frac{\partial X}{\partial t} = \mu X - \frac{u_L}{\varepsilon_L}\frac{\partial X}{\partial x} + D_A\frac{\partial^2 X}{\partial x^2} \tag{4E4.22}$$

$$u_L X(0,t) - \varepsilon_L D_A\frac{\partial X(0,t)}{\partial x} = 0; \quad \frac{\partial X(L,t)}{\partial x} = 0; \quad X(z,0) = X_0 \tag{4E4.23}$$

$$\varepsilon_G = \frac{\varepsilon_{G,\max} u_G}{K_{\varepsilon_G} + u_G} \tag{4E4.24}$$

$$C_i^*(x,t) = H_i P_{G,i} \tag{4E4.25}$$

$$P_{g,i}(x,t) = C_{g,i}(x,t)RT \tag{4E4.26}$$

3. Biomass and waste based processes

To run the model the parameters given in Table 4E4.1 are introduced into it.

TABLE 4E4.1 Parameters for the simulation of syngas fermentation in a BCR.

v_{max,H_2}	46.3 mmol/g/h	$k_{m,CO}$	80 h^{-1}
v_{max,H_2}	31.6 mmol/g/h	k_{m,H_2}	200 h^{-1}
$K_{s,CO}$	0.0115 mmol/L	k_{m,CO_2}	80 h^{-1}
K_{s,H_2}	0.675 mmol/L	$\varepsilon_{G,max}$	0.53
$K_{I,EtOH}$	217 mmol/L	K_{ε_G}	540 m/h
$K_{I,HAc}$	962 mmol/L	H_{CO}	8E − 4 mol/L/atm
$K_{I,CO}$	0.136 mmol/L	H_{H_2}	6.6E − 4 mol/L/atm
L_{total}	25 m	H_{CO_2}	2.5E − 2 mol/L/atm
$C_{g0,CO}$	80.64 mmol/L	C_{g0,H_2}	53.76 mmol/L
C_{g0,CO_2}	0 mmol/L	$C_{gf,CO}$	80.64 mmol/L
C_{gf,H_2}	53.76 mmol/L	C_{gf,CO_2}	0 mmol/L
$C_{l0,CO}$	1.642 mmol/L	C_{l0,H_2}	0.903 mmol/L
C_{l0,CO_2}	0 mmol/L	$C_{l0,EtOH}$	0 mmol/L
$C_{l0,HAc}$	0 mmol/L	D_A	0.25 m^2 h^{-1}
$v_{max,AcR}^{CO}$	37.6 mmol/g/h	$v_{max,AcR}^{H_2}$	22.2 mmol/g/h
$K_{S,AcR}^{CO}$	303 mmol/L	$K_{S,AcR}^{H_2}$	586 mmol/L
$Y_{X,CO}$	2.01E − 3 g/mmol	Y_{X,H_2}	0.23E − 3 g/mmol
u_G	75 m/h	u_L	0.25 m/h

The problem of syngas fermentation has been implemented in gPROMS ModelBuilder 6.0.4. The following codes include the model and process for this case study.

Model:

```
PARAMETER
   NR      as    Ordered_set    #Number of Reactions
   N       as    Ordered_set    #Number of species
   NN      as    Ordered_set
   NL      as    Ordered_set    #Liquid Species
   NG      as    Ordered_set    #Gas Species
   NLR     as    Ordered_Set
   vmax    as    Array(N) of    Real #kinetic vmax
   Ks      as    Array(N) of    Real #kinetic Ks
   KIE     as                   Real #Inhibition ethanol
   KIA     as                   Real #Inhibition HAc
   KICO    as                   Real #Inhibition CO
```

```
km              as      Array(N) of      Real
holdupmax       as                       Real
Kholdup         as                       Real
DA              as                       Real
vmaxAcR         as      Array(N) of      Real  #kinetic vmaxAcR
KsAcR           as      Array(N) of      Real  #kinetic KsAcR
YXCO            as                       Real  #yields
YXH2            as                       Real  #tields
length          as                       Real  #length
Henry           as      Array(N) of      Real  #Henry Law
R               as                       Real  #gas cosntant
T               as                       Real  #temperature
Cg_ini          as      Array(N)    of   Real
Cg0             as      Array(N)    of   Real
Cl0             as      Array(N)    of   Real
ClX0            as                       Real
Mj              as      Array(N) of      Real
DISTRIBUTION_DOMAIN
xdomain     as [0:length] # discretization array.
VARIABLE
  vR        as  DISTRIBUTION(NR, xdomain)  of  reaction_rate  #reaction rate per
                                                                   reaction
                                                              (mol/g h)
  vN        as  DISTRIBUTION(N, xdomain)  of  reaction_rate  #reaction rate per
                                                              specie (mol/g h)
  Cl        as  DISTRIBUTION(N, xdomain) of  liquid_concentration #concentration
                                                       in liquid phase (g/l)
  IE        as  DISTRIBUTION(xdomain)     of  I_factor  #Inhibition ethanol
  IA        as  DISTRIBUTION(xdomain)     of  I_factor  #Inhibition HAc
  ICOH2     as  DISTRIBUTION(xdomain)     of  I_factor  #Inhibition CO
  FAcRCO    as  DISTRIBUTION(xdomain)     of  F_factor  #Kinetic factor
  FAcRH2    as  DISTRIBUTION(xdomain)     of  F_factor  #Kinetic Factor
  grow_rate as  DISTRIBUTION(xdomain)     of  reaction_rate  #cell grownth rate
  holdup    as  DISTRIBUTION(xdomain)     of  hold_up  #gas hold up
  uG        as  DISTRIBUTION(xdomain)     of  velocity
  Cl_eq     as  DISTRIBUTION(N, xdomain)  of  liquid_concentration
  Cg        as  DISTRIBUTION(N, xdomain)  of  gas_concentration
  uL        as                               velocity
  ClX       as  DISTRIBUTION(xdomain)     of  liquid_concentration
  PG        as  DISTRIBUTION(N, xdomain) of  pressure
BOUNDARY
Cg(,0) = Cg_ini;
For j in NG DO
```

3. Biomass and waste based processes

```
    uL*Cl(j,0) - (1 - holdup(0))*DA*PARTIAL(Cl(j,0), xdomain) = 0;
    PARTIAL(Cl(j, length), xdomain) = 0;
end
For j in NLR DO
    uL*Cl(j,0) - (1 - holdup(0))*DA*PARTIAL(Cl(j,0), xdomain) = 0;
    PARTIAL(Cl(j, length), xdomain) = 0;
end
uL*ClX(0) - (1 - holdup(0))*DA*PARTIAL(ClX(0), xdomain) = 0;
PARTIAL(ClX(length), xdomain) = 0;
EQUATION
For x: = 0 TO length DO
    vN('CO', x) = - (vmax('CO')*Cl('CO', x)/(Ks('CO') + Cl('CO', x)))*IE(x)*IA(x);
    IE(x) = 1/(1 + Cl('EtOH', x)/(KIE*Mj('EtOH')));
    IA(x) = 1/(1 + Cl('HAc', x)/(KIA*Mj('HAc')));
    vN('H2', x) = - (vmax('H2')*Cl('H2', x)/(Ks('H2') + Cl('H2', x)))*IE(x)*IA(x)
     *ICOH2(x);
    ICOH2(x) = 1/(1 + Cl('CO', x)/KICO);
    vR('3', x) = (1/2)*(2*FAcRCO(x)/(1e - 20 + 2*FAcRCO(x) + ABS(vN('CO', x))))*ABS
     (vN('CO', x));
    FAcRCO(x) = vmaxAcR('CO')*Cl('HAc', x)/(KsAcR('CO')*Mj('HAc') + Cl('HAc', x));
    vR('4', x) = (1/2)*(2*FAcRH2(x)/(1e - 20 + 2*FAcRH2(x) + ABS(vN('H2', x))))*ABS(vN
     ('H2', x));
    FAcRH2(x) = vmaxAcR('H2')*Cl('HAc', x)/(KsAcR('H2')*Mj('HAc') + Cl('HAc', x));
    vR('1', x) = - (vN('CO', x) + 2*vR('3', x))/4;
    vR('2', x) = - (vN('H2', x) + 2*vR('4', x))/4;
    vN('CO2', x) = 2*vR('1', x) - 2*vR('2', x) + 2*vR('3', x);
    vN('EtOH', x) = (vR('3', x) + vR('4', x));
    vN('HAc', x) = vR('1', x) + vR('2', x) - vR('3', x) - vR('4', x);
    grow_rate(x) = ( - vN('CO', x)*YXCO - vN('H2', x)*YXH2)*1.5;
end
For j in NG DO
    For x: = 0| + TO length DO
        $Cg(j, x) = ( - km(j)/holdup(x))*(Cl_eq(j, x) - Cl(j, x)) - (uG(x)/holdup(x))
        *PARTIAL(Cg(j, x), xdomain);
    end
end
For x: = 0 TO length DO
    holdup(x) = holdupmax*uG(x)/(Kholdup + uG(x));
end
For j in NG DO
    For x: = 0 TO length DO
                PG(j, x) = Cg(j, x)*8.31*(37 + 273)/101325;
                Cl_eq(j, x) = Henry(j)*PG(j, x)*1000;
```

```
    end
  end
  #
  For j in NG DO
    For x: = 0| +  TO length| − DO
          $Cl(j, x) = vN(j, x)*ClX(x) + (km(j)/(1 − holdup(x)))*(Cl_eq(j, x)
            − Cl(j, x)) − (uL/(1 − holdup(x)))*PARTIAL(Cl(j, x), xdomain) + DA*PARTIAL
            (PARTIAL(Cl(j, x), xdomain), xdomain);
    end
  end
  For j in NLR DO
    For x: = 0| + TO length| − DO
                $Cl(j, x) = Mj(j)*vN(j, x)*ClX(x) − (uL/(1 − holdup(x)))*PARTIAL(Cl(j,
                x),
                xdomain) + DA*PARTIAL(PARTIAL(Cl(j, x), xdomain), xdomain);
    end
  end
  For x: = 0| + TO length| − DO
    $ClX(x) = grow_rate(x)*ClX(x) − (uL/(1 − holdup(x)))*PARTIAL(ClX(x), xdomain)
      + DA*PARTIAL(PARTIAL(ClX(x), xdomain), xdomain);
  end
```

Process:

```
UNIT
reactor1 AS Reactor_Ethanol
SET
reactor1.NR: = ['1','2','3','4'];
reactor1.N: = ['CO','H2','CO2','EtOH','HAc'];
reactor1.NN: = ['CO','H2','CO2','EtOH','HAc'];
reactor1.NG: = ['CO','H2','CO2'];
reactor1.NL: = ['EtOH','HAc'];
reactor1.NLR: = ['EtOH','HAc'];
within reactor1 DO
  vmax: = [46.3,31.6,0,0,0];#mmol/g hr
  Ks: = [0.0115,0.675,0,0,0];#mmol/l
  KIE: = 217;#mmol/l
  KIA: = 962;#mmol/l
  KICO: = 0.136;#mmol/l
  km: = [80,200,80,0,0];#h − 1
  holdupmax: = 0.53;
  Kholdup: = 540;#m/h
  Henry: = [8e − 4,6.6e − 4,2.5e − 2,0,0];#mol/l atm
  length: = 25;
```

3. Biomass and waste based processes

```
    Cg_ini: = [80.64,53.76,0,0,0];#mmol/l #Boundary
    Cg0: = [80.64,53.76,0,0,0];#mmol/l
    DA: = 0.25;#m2/hr
    Cl0: = [1.642,0.903,0,0,0];#mmol/l #Initial COnditiomn
    vmaxAcR: = [37.6,22.2,0,0,0];#mmol/g hr
    KsAcR: = [303,586,0,0,0]; #mmol/l
    YXCO: = 1.34/1000;
    YXH2: = 0.156/1000;
    R: = 8.31;#J/mol K
    T: = 37 + 273;
    xdomain: = [OCFEM,2,1000];
    ClX0: = 0.1;#g/l
    Mj: = [28E − 3,2E − 3,44E − 3,46E − 3,60E − 3];#g/mmmol
end
ASSIGN
reactor1.uG: = 75;
reactor1.uL: = 0.25;
For j IN reactor1.NL DO
  For x: = 0 TO reactor1.length DO
            reactor1.Cl_eq(j, x): = 0;
            reactor1.PG(j, x): = 0;
  end
end
For j IN reactor1.NL DO
  For x: = 0| + TO reactor1.length DO
            reactor1.Cg(j, x): = 0;
  end
end
INITIAL
Within reactor1 DO
  for i in NG DO
            for x: = 0| + TO length DO # No initial conditions needed at the boundary
                  Cg(i, x) = Cg0(i);
            end
  end
end
Within reactor1 DO
  for i in NG DO
            for x: = 0| + TO length| − DO # No initial conditions needed at the
boundary
                  Cl(i, x) = Cl0(i);
            end
  end
```

```
end
Within reactor1 DO
  for i in NLR DO
              for x: = 0| + TO length| − DO # No initial conditions needed at the
boundary
                  Cl(i, x) = ClO(i);
              end
  end
end
Within reactor1 DO
  for x: = 0| + TO length| − DO
              ClX(x) = ClXO;
  end
end
SCHEDULE
  Continue for 1000
```

Fig. 4E4.1 shows the main products (EtOh, HAc, and biomass) in the liquid phase of the bubble column reactor.

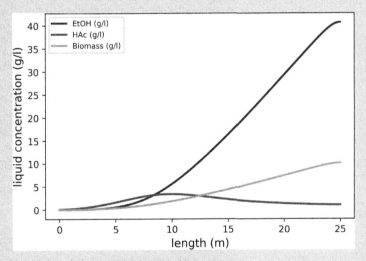

FIGURE 4E4.1   Profile of products for ethanol syngas fermentation.

excess of energy available in the system may result in larger consumption of water for thermochemical processes than for those based on hydrolysis (Martín et al., 2011). Only by integrating the production of chemicals via a thermochemical path with other facilities, total site integration will be possible to make better use of the energy available and reduce the water consumption of the complex, see Fig. 4.9.

TABLE 4.8 Product purification unit operations.

Product	Purification stage	Technology
Hydrogen	Gas separation	Membranes: see above.
Methanol	Gas–liquid	*Flash:* Recovery of unconverted CO and H_2.
	Liquid mixture	$F = V + L$ (4.62)
		$Fz_i = Vy_i + Lx_i$ (4.63)
		$y_i = K_i x_i$ (4.64)
		where K_i is computed (ideal case) as $P_{vap} (T)/P_{Total}$.
		Distillation: Water–methanol separation.
Ethanol	Water -ethanol mixture	*Distillation:* The limit is the azeotrope at 95.63% of ethanol by mass in the mixture. This is what determines the separation process. See Fig. 4.7. An additional step is required since the distillation can only reach this purity under ideal conditions. In addition, the large consumption of energy due to the dilute ethanol mixture, typically around 5%, makes process intensification a need to reduce production costs. The concept of the multieffect column presented in Chapter 2 is useful here to reduce the energy consumption by one-third (Karuppiah et al., 2008), see Fig. 4.8.
		Molecular sieves: They are the technology of choice to achieve fuel-grade ethanol. They are typically zeolite compounds that absorb water. The particular operation is that the area within the pores is only accessible to the small molecule of the water, with a size of 1.93 A, while other compounds like ethanol cannot enter the 3 A pore. That is why zeolite 3 A is appropriate for dehydration of ethanol. The adsorption isotherms are typically Langmuir type (Eq. 4.65).
		$$q \ (g/g) = \frac{Q \cdot K \cdot c(\%EtOH)}{(1 + K \cdot c(\%EtOH))} \qquad\qquad (4.65)$$
		For different adsorbent geometries, spheres or cylinders, the constants are found in Table 4.9 for zeolite 3 A (Carmo and Gubulin, 1997).
	Alcohols mixture	*Distillation sequence:* The sequence in which the mixture of products is separated affects the energy consumption. Biegler et al. (1997) present a formulation to determine the optimal sequence of distillation columns as an MILP problem.
DME	Gas–liquid	*Flash.* See methanol purification.
	Liquid–liquid	*Stripping column:* From the top, a saturated gas phase is obtained; while, from the bottom, the DME, water, and ethanol are recovered. CO_2 can be absorbed in the DME
		Distillation column: DME is separated from water and methanol.
FT	Water–oil	*Hydrocracking:* See Section 4.4.2.
	Separation	*Crude fractionation:* See Section 4.4.2.
	Oil fractionation	

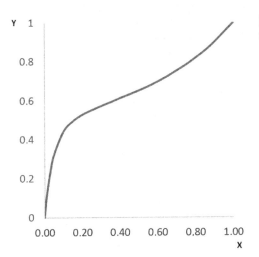

FIGURE 4.7 XY diagram for the water–ethanol mixture.

4.2 Pyrolysis

Pyrolysis consists of thermally breaking down of the biomass in the absence of air or oxygen generating three kinds of products: solid, liquid, and gas final compounds. Biomass pyrolysis is induced by the heat transfer from the surrounding to the biomass particle. Typical pyrolysis temperatures are between 400°C and 700°C, much lower than gasification (700°C–900°C) and combustion (~1000°C). Different stages can be found in the pyrolysis process (Basu, 2010). Up to approximately 100°C, biomass is dried, then, between 100°C and 300°C, dehydration of the biomass takes place releasing also some light components as CO or CO_2. Above 200°C, a progressive release of pyrolytic volatiles starts. Large molecules of the biomass are decomposed into char, tars, and gases. This stage is called primary pyrolysis and is completed below 500°C and 600°C. The last stage is called secondary pyrolysis and includes different cracking, reforming, polymerization reactions. The char can also be gasified in this step.

Three kinds of pyrolysis products are generated. Char is a solid product formed mainly by carbon (~85%) but also contains some amounts of oxygen and hydrogen. The lower heating value (LHV) of this char is about 32 MJ/kg. The liquid fraction is known as bio-oil. Two main groups are found in this bio-oil: water from biomass moisture or from the pyrolytic reactions (up to 20%) and the tars. Tars include a group of condensable organic components that, for the sake of simplicity, are included in a unique category. The LHV of this bio-oil is about 13–18 MJ/kg in wet basis. Finally, a mixture of gases is obtained from primary and secondary pyrolysis. Gases from the two steps are different and it is evident in the values of the LHV, 11 MJ/Nm^3 for the primary gases and 20 MJ/Nm^3 for the secondary gases.

According to Neves et al. (2011), the main factors governing the pyrolysis process are: particle size, temperature, gas dilution and residence time, and the amount of fuel. In terms of particle size, when the particle size increases, the char layer increases, promoting more intense intraparticle secondary reactions. Therefore the yields of liquids decrease when the

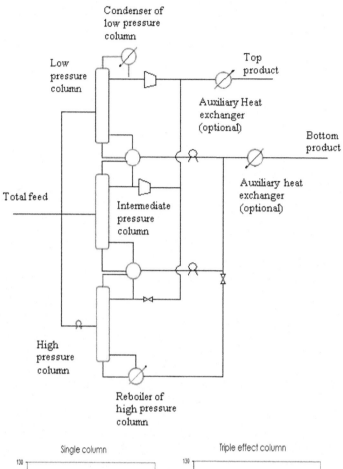

FIGURE 4.8 Multieffect column operation. *Source: Adapted from Karuppiah, R., Peschel, A., Grossmann, I.E., Martín, M., Martinson, W., Zullo, L. (2008). Energy optimization for the design of corn-based ethanol plants. AIChE J., 54(6), 1499–1525.*

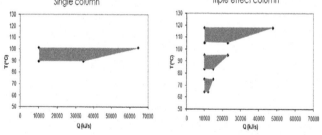

particle size increases. The yield of water from pyrolysis reactions also increases when the particle size increases. Temperature has a great importance on the final products of pyrolysis. High yields of liquids are obtained in the range of 400°C−450°C due to, below 500°C, it is not easy to crack the primary tars. Secondary reactions are enhanced when the temperature increases. The effect of the temperature on the yield of products can be seen in Fig. 4.10. The amount of fuel can influence in the secondary reaction because an increase in the amount of fuel involved an increase of the char holdup (Bridgwater et al., 1999).

TABLE 4.9 Langmuir coefficients for Zeolite 3A.

Temperature (°C)	Q_{esf}, Q_{cil} (g$_{água}$/g$_{ads}$)	K_{esf}, K_{cil} (g$_{soln}$/g$_{água}$)
25	0.249; 0.241	0.317; 0.307
40	0.238; 0.230	0.153; 0.148
50	0.220; 0.210	0.090; 0.086
60	0.200; 0.190	0.052; 0.049

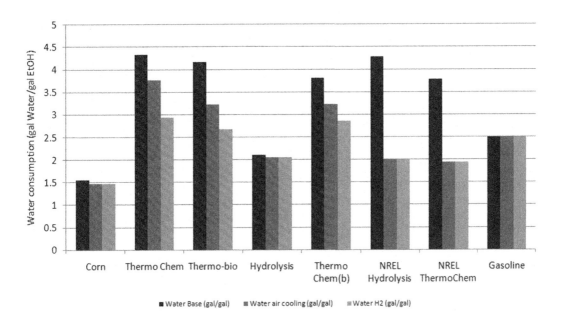

FIGURE 4.9 Water consumption in ethanol production. *Source: With permission from Martín, M., Ahmetovic, E., Grossmann, I.E. (2011). Optimization of water consumption in second generation bio-ethanol plants. Ind. Eng. Chem. Res., 50 (7), 3705–3721.*

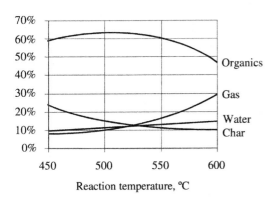

FIGURE 4.10 Yield to products via fast pyrolysis. *Source: With permission from Bridgwater, A.V., Meier, D., Radlein, D. (1999). An overview of fast pyrolysis of biomass. Org. Geochem. 30, 1479–1493.*

The pyrolysis process can be classified into two main categories: slow and fast pyrolysis. In the first one, the heating time is much longer than the reaction time. For the second one, it is the opposite. Slow pyrolysis uses lower temperature and higher reaction times, with a typical heating rate of $10^1-10^{2\circ}C/s$. Fast pyrolysis involves higher temperatures and the main objective is to increase the liquid fraction. The heating rate in this alternative is about $10^2-10^{3\circ}C/s$.

Pyrolysis has been used since ancient times to produce charcoal using a slow pyrolysis to increase the char yield. The reactor configuration is similar to the one presented for the gasification. For instance, fixed-bed, fluidized-bed, entrained-flow, and pyrolysis reactors have been proposed. In fixed bed configurations, the desired product is mainly char. In fluidized-bed reactors, an inert gas is used as fluidization agent and a high yield of liquids is obtained. A cyclone is responsible for recovering the char that can be used as a source of energy. The products are cooled to condense water and bio-oil, to be separated and fractionated and the gas is further used to provide energy and/or for the fluidization of the bed. The heat for fluidized-bed reactor can be provided by burning a fraction of the product gas inside the bed, using an external chamber or a CFB configuration, Fig. 4.11.

Different approaches have been proposed to model the pyrolysis process. Thermodynamic models (as equilibrium or Gibbs free energy minimization) are not accurate to represent the pyrolysis behavior. In general, these models try to overestimate the yields of hydrogen and carbon monoxide, underestimating the CO_2 yield. Furthermore, it is not possible to determine the concentration of methane and other hydrocarbons (Ranzi et al., 2008). Some empirical models have been evaluated for pyrolysis. One of them is illustrated in Example 4.5.

Some kinetic models have also been evaluated to describe the pyrolysis. The kinetic modeling of the reactor is based on the rate of decomposition of the biomass into the different products. Different models can be found in the literature. Hameed et al. (2019) presented a summary of mechanisms and models including (1) one step kinetic models, (2) three parallel reactions model, (3) competitive reaction models, detailed lumped kinetic

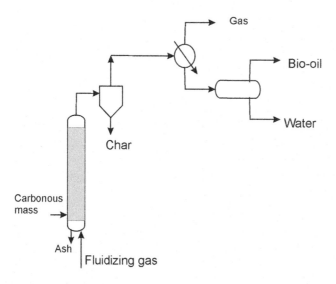

FIGURE 4.11 Scheme of the operation of a pyrolysis reactor.

EXAMPLE 4.5 Biomass pyrolysis—empirical model.

One hundred kilograms of wheat straw are introduced in a pyrolysis reactor at 500°C. Determine the amount and composition of the char, tar and gases produced. The solids from the reactor are separated using a cyclone and, after this solid separation, the bio-oil is separated cooling down up to 25°C. To calculate the pyrolysis yields, use the model proposed by Neves et al. (2011).

Solution

The ultimate analysis of wheat straw is the same as in Example 4.1. For every 100 kg of biomass, 7 kg of moisture, 5.1 kg of ash, and 87.9 kg of fixed carbon (FC) and volatiles are introduced.

The different mass balances for each of the atoms (C, O, and H) have been presented (Eq. 4E5.1, Eq. 4E5.2, Eq. 4E5.3). Ethylene, methane, carbon monoxide, and dioxide, water and hydrogen are considered as gas components in this example. Tar and char represent the liquid and solid fractions, respectively.

$$Y_{C,tar}Y_{tar,F} + Y_{C,C_2H_4}Y_{C_2H_4,F} + Y_{C,CH_4}Y_{CH_4,F} + Y_{C,CO}Y_{CO,F} + Y_{C,CO_2}Y_{CO_2,F} + Y_{C,char}Y_{char,F} = Y_{C,F}$$
(4E5.1)

$$Y_{O,tar}Y_{tar,F} + Y_{O,CO}Y_{CO,F} + Y_{O,CO_2}Y_{CO_2,F} + Y_{O,H_2O}Y_{H_2O,F} + Y_{O,char}Y_{char,F} = Y_{O,F}$$
(4E5.2)

$$Y_{H,tar}Y_{tar,F} + Y_{H,C_2H_4}Y_{C_2H_4,F} + Y_{H,CH_4}Y_{CH_4,F} + Y_{H,H_2O}Y_{H_2O,F} + Y_{H,H_2}Y_{H_2,F} + Y_{H,char}Y_{char,F} = Y_{H,F}$$
(4E5.3)

$Y_{j,i}$ is the mass fraction of element j in the product i (kg/kg) and $Y_{i,F}$ is the yield of product i (kg/kg daf biomass). Ashes do not suffer any transformation in the pyrolysis reactor and go to the solid fraction of the products.

The energy balance is expressed as follows:

$$\left(\sum_i Y_{i,F} - Y_{char,F}\sum_i Y_{i,char}\right)LHV_G = \left(Y_{tar,F} + Y_{H_2O,F}\right)LHV_G$$
$$+ Y_{C_2H_4,F}LHV_{C_2H_4} + Y_{CH_4,F}LHV_{CH_4} + Y_{CO,F}LHV_{CO} + Y_{CO_2,F}LHV_{CO_2} + Y_{H_2,F}LHV_{H_2}$$
(4E5.4)

Some of the yields are obtained from different empirical correlations:

$$\frac{Y_{H_2,F}}{Y_{CO,F}} = 3 \cdot 10^{-4} + \frac{0.0429}{1 + (T/632)^{-7.23}}$$
(4E5.5)

$$Y_{CH_4,F} = -2.18 \cdot 10^{-4} + 0.146 Y_{CO,F}$$
(4E5.6)

$$Y_{H_2,F} = 1.145\left(1 - \exp\left(-0.11 \cdot 10^{-2}T\right)\right)^{9.384}$$
(4E5.7)

$$Y_{char,F} = 0.106 + 2.43\exp\left(-0.66 \cdot 10^{-2}T\right)$$
(4E5.8)

$$Y_{C,char} = 0.93 - 0.92\exp\left(-0.42 \cdot 10^{-2}T\right)$$
(4E5.9)

$$Y_{O,char} = 0.07 + 0.85\exp\left(-0.48 \cdot 10^{-2}T\right)$$
(4E5.10)

$$\frac{Y_{C,tar}}{Y_{C,F}} = 1.05 + 1.9 \cdot 10^{-4}T$$
(4E5.11)

$$\frac{Y_{O,tar}}{Y_{O,F}} = 0.92 - 2.2 \cdot 10^{-4}T$$
(4E5.12)

$$\text{LHV}_G = -6.23 + 2.47 \cdot 10^{-2}T \tag{4E5.13}$$

where T is introduced in °C and LHV is expressed in MJ/kg.

The yields of hydrogen in the char and tar are calculated by difference:

$$Y_{H,char} = 1 - Y_{C,char} - Y_{O,char} \tag{4E5.14}$$

$$Y_{H,tar} = 1 - Y_{C,tar} - Y_{O,tar} \tag{4E5.15}$$

The char and ashes produced in the reactor constitute the solid fraction, the gas fraction, which includes the light gases, is saturated of water at 25°C, and the liquid fraction is formed by the tars and the remaining water. To determine the moisture content of the gases, the next equation has been used:

$$\omega = \frac{P_v^0(H_2O, T)}{P_{total} - P_v^0(H_2O, T)} \tag{4E5.16}$$

The vapor pressure of water at 25°C is equal to 23.76 mmHg.

If the mass and energy balances are solved using also the empirical correlations at pyrolysis temperature (500°C), the results shown in Table 4E5.1 are obtained:

TABLE 4E5.1   Results of the mass and energy balances.

Solid	Liquid	Gases
Total: 22.308 kg	Total: 58.442 kg	Total: 19.251 kg
Char: 17.193 kg (81.73% C, 14.71% O, 3.56% H)	Tar: 37.827 kg (53.04% C, 38.14% O, 8.82% H)	C_2H_4: 0.736 kg (4.72% v)
Ash: 5.115 kg		CH_4: 0.639 kg (7.18% v)
	H_2O: 20.615 kg (13.928 kg of pyrolytic water, 7 kg of moisture, −0.313 kg leaving with the gases)	CO: 4.509 kg (28.95% v)
		CO_2: 13.022 kg (53.20% v)
		H_2: 0.031 (2.82% v)
		H_2O: 0.313 (3.12% v)

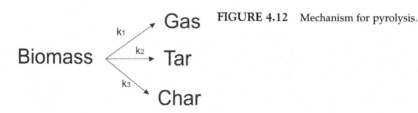

FIGURE 4.12   Mechanism for pyrolysis.

models, and (4) mechanisms with secondary interactions. Thurner and Mann (1981) proposed a simple model to determine the fraction of the different products only considering the primary pyrolysis, Fig. 4.12. A simple example is presented here using this model (Example 4.6).

EXAMPLE 4.6 Biomass pyrolysis—kinetic model.

Determine the gas, liquid, and char profiles versus time for the pyrolysis of a wood biomass at 350°C. The model of Thurner and Mann (1981) is used.

Solution

The mechanism proposed for Thurner and Mann (1981) is collected in Fig. 4.12. The following differential equations are introduced for each of the products and for the biomass:

$$\frac{dw_{\text{wood}}}{dt} = -(k_1 + k_2 + k_3)w_{\text{wood}} \tag{4E6.1}$$

$$\frac{dw_{\text{gas}}}{dt} = k_1 w_{\text{wood}} \tag{4E6.2}$$

$$\frac{dw_{\text{tar}}}{dt} = k_2 w_{\text{wood}} \tag{4E6.3}$$

$$\frac{dw_{\text{char}}}{dt} = k_3 w_{\text{wood}} \tag{4E6.4}$$

The variable w represents the mass fraction of each specie. The kinetic coefficients have been adjusted to an Arrhenius equation:

$$k_1\left(\text{min}^{-1}\right) = 8.607 \cdot 10^5 \exp\left(\frac{-88.6 \cdot 10^3}{8.31 T(\text{K})}\right) \tag{4E6.5}$$

$$k_2\left(\text{min}^{-1}\right) = 2.475 \cdot 10^8 \exp\left(\frac{-112.7 \cdot 10^3}{8.31 T(\text{K})}\right) \tag{4E6.6}$$

$$k_3\left(\text{min}^{-1}\right) = 4.426 \cdot 10^7 \exp\left(\frac{-106.5 \cdot 10^3}{8.31 T(\text{K})}\right) \tag{4E6.7}$$

The initial conditions for the set of differential equations are:

$$w_{\text{wood}} = 1; \quad w_{\text{gas}} = 0; \quad w_{\text{tar}} = 0; \quad w_{\text{char}} = 0 \tag{4E6.8}$$

This problem has been solved using MATLAB (the code is attached) and the following results have been obtained (represented in Fig. 4E6.1).

```
global T
T = 350 + 273;%rango 325 − 385
[a,b] = ode15s(@pirolisis,[0 30],[1,0,0,0]);
time = a;
ww = b(:,1);
wG = b(:,2);
wT = b(:,3);
wC = b(:,4);
hold on
plot(a,wC,'k − ')
```

3. Biomass and waste based processes

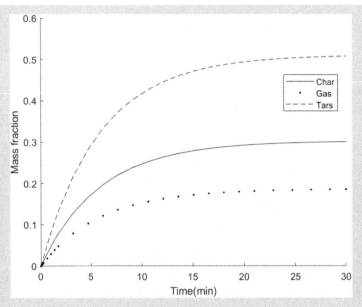

FIGURE 4E6.1   Results for the kinetic model of pyrolysis.

```
plot(a,wG,'k.')
plot(a,wT,'k−−')
legend('Char','Gas','Tars')
xlabel('Time(min)')
ylabel('Mass fraction')
hold off
function [Reac]=pirolisis(z,x)
%UNTITLED Summary of this function goes here
%      Detailed explanation goes here
global T
ww=x(1);
wG=x(2);
wT=x(3);
wC=x(4);
k1=8.607e5*exp(−88.6e3/(8.31*T));
k2=2.475e8*exp(−112.7e3/(8.31*T));
k3=4.426e7*exp(−106.5e3/(8.31*T));
Reac(1,1)=−(k1+k2+k3)*ww;
Reac(2,1)=k1*ww;
Reac(3,1)=k2*ww;
Reac(4,1)=k3*ww;

end
```

The typical composition of the products from pyrolysis of biomass, in particular corn stover (Wright et al., 2010), can be seen in Table 4.10.

The upgrading process considered is bio-oil hydrotreating and hydrocracking which is common for FT products as well as for the bio-oil obtained in the HTL. They are very well-known stages in the petroleum industry where they are used to remove undesired compounds such as sulfur and to break large hydrocarbon molecules into gasoline and diesel.

4.3 Hydrothermal liquefaction

HTL is a process that is based on applying moderate temperature ($>250°C$) and high pressure (>4 MPa) to convert wet biomass, with mass fractions of 5%−20% in the slurry, into a bio-crude oil as well as aqueous and gaseous byproducts (Amin, 2009). Unlike pyrolysis oil, the organic phase generated in the liquefaction of biomass presents lower oxygen content and higher heating value (Peterson et al., 2008) so that it has been classified as similar to heavy crude (Ross et al., 2010). As a result, the upgrading process is similar to crude oil fractionation.

4.3.1 Mechanism for hydrothermal liquefaction

The mechanism of biomass depolymerization is based on the work by Valdez and Savage (2013), Valdez et al. (2014). A network of reactions is considering for the transformation of biomass into an aqueous phase, a gas phase, and bio-oil, Fig. 4.13. Different modifications to this network can be seen in the literature, including light and heavy bio-crude. For this basic model,

TABLE 4.10 Product composition of pyrolysis reactor (Wright et al., 2010).

Gas compounds		Bio-oil		Other	
Species	kg/100 kg of dry biomass	Species	kg/100 kg of dry biomass	Species	kg/100 kg of dry biomass
CO_2	5.42	Acetic acid	5.93	Water	10.8
CO	6.56	Propionic acid	7.31	Char/ ash	16.39
CH_4	0.035	Methoxyphenol	0.61		
C_2H_6	0.142	Ethylphenol	3.80		
H_2	0.588	Formic acid	3.41		
C_3H_8	0.152	Propyl-benzonate	16.36		
NH_3	0.0121	Phenol	0.46		
		Toluene	2.27		
		Furfural	18.28		
		Benzene	0.77		

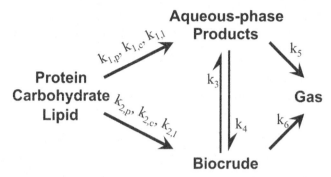

FIGURE 4.13   HTL decomposition mechanism.

$$\frac{dx_{1,P}}{dt} = -\left(k_{1,P} + k_{2,P}\right) \cdot x_{1,P} \tag{4.66}$$

$$\frac{dx_{1,C}}{dt} = -\left(k_{1,C} + k_{2,C}\right) \cdot x_{1,C} \tag{4.67}$$

$$\frac{dx_{1,L}}{dt} = -\left(k_{1,L} + k_{2,L}\right) \cdot x_{1,L} \tag{4.68}$$

$$\frac{dx_2}{dt} = -(k_4 + k_5) \cdot x_2 + k_{1,P}x_{1,P} + k_{1,C}x_{1,C} + k_{1,L}x_{1,L} + k_3 \cdot x_3 \tag{4.69}$$

$$\frac{dx_3}{dt} = -(k_3 + k_6) \cdot x_3 + k_{2,P}x_{2,P} + k_{2,C}x_{2,C} + k_{2,L}x_{2,L} + k_4 \cdot x_2 \tag{4.70}$$

$$\frac{dx_4}{dt} = k_5 \cdot x_2 + k_6 \cdot x_3 \tag{4.71}$$

the species involved include (1) the solids; (2) bio-crude; (3) the aqueous phase; and (4) the gas. In addition, bio-char, a solid residue, can also be produced. In Table 4.11, the values for the activation energies and preexponential factors are presented. The product composition of HTL using an algae with a specific composition as feedstock is presented in Example 4.7.

This process is typically catalyzed (López Barreiro et al., 2013) either using:

- Homogeneous catalysts: Such as alkali salts like Na_2CO_3 or KOH, they improve the liquefaction yield reducing the production of solid residues.
- Heterogeneous catalysts: Among them we find Pd, Pt, or Ru supported on C, CoMo, Ni, Pt, Ni/SiO_2 supported on Al_2O_3, and zeolite. However, they are reported to present drawbacks in the operation such as sintering, dissolution, poisoning, or intraparticle diffusion limitations.

4.3.2 Upgrading

In this section, two major unit operations are described, hydrotreating and hydrocracking. They are common for FT fluid production, HTL bio-crude processing, and pyrolysis

TABLE 4.11 Kinetic constants for HTL model (Valdez et al., 2014).

Constant	Path	K (min^{-1})			
		250	300	350	400
$k_{1,p}$	Protein → AP	0.095	0.2	0.28	0.3
$k_{1,c}$	Carbohydrates → AP	0.15	0.35	0.35	0.35
$k_{1,l}$	Lipids → AP	0.25	0.35	0.35	0.35
$k_{2,p}$	Protein → BC	0.13	0.13	0.28	0.32
$k_{2,c}$	Carbohydrates → BC	0.031	0.11	0.33	0.35
$k_{2,l}$	Lipids → BC	0.0001	0.0001	0.001	0.0032
k_3	BC → AP	0.0044	0.14	0.3	0.34
k_4	AP → BC	0.003	0.12	0.26	0.26
k_5	AP → Gas	0.0001	0.0004	0.0014	0.0014
k_6	BC → Gas	0.0001	0.0002	0.0009	0.0053

EXAMPLE 4.7 Determination of HTL product composition.

Determine the composition of the HTL for an algae with 11% protein, 29% carbohydrates, and 53% lipids operating at 300°C using the mechanism presented along with the text.

Solution

We implement the model presented above in Python to solve the system of differential Eqs. (4.66)–(4.71).

```
import numpy as np
import matplotlib.pyplot as plt
from scipy import integrate
#[kJ/mol]

#300°C      k en min-1

k1p = 0.2;
k1l = 0.35;
k1c = 0.35;
k2p = 0.13;
k2l = 0.11;
k2c = 0.001;
k3 = 0.14;
k4 = 0.12;
k5 = 0.0004;
k6 = 0.0002;
x1_p = 0.11;
x1_c = 0.29;
```

```
x1_l = 0.53;
solidsini = 1 − x1_p − x1_c − x1_l;
Xini = [x1_p, x1_c, x1_l,0,0,0];
#We create an empty list to store the concentrations data
x = [(),(),(),(),(),()]
tau = np.linspace(0,60,200)
#Equations: We define a function which collect the equations to solve
def    kinetics(x, tau):
        dxdtau = np.zeros(6)
        dxdtau[0] = − (k1p + k2p)*x[0]
        dxdtau[1] = − (k1c + k2c)*x[1]
        dxdtau[2] = − (k1l + k2l)*x[2]
        dxdtau[3] = − (k4 + k5)*x[3] + k1p*x[0] + k1c*x[1] + k1l*x[2] + k3*x[4]
        dxdtau[4] = − (k3 + k6)*x[4] + k2p*x[0] + k2c*x[1] + k2l*x[2] + k4*x[3]
        dxdtau[5] = k5*x[3] + k6*x[4]
        return dxdtau
#We collect the results in a list calles Res
res_x = integrate.odeint(kinetics, Xini, tau)
solids = 1 − res_x[:,0] − res_x[:,1] − res_x[:,2]
plt.plot(tau, res_x[:,0],'k − ')
plt.plot(tau, res_x[:,1],'k--')
plt.plot(tau, res_x[:,2],'k.')
plt.plot(tau, res_x[:,3],'k*')
plt.plot(tau, res_x[:,4],'k − <')
plt.plot(tau, res_x[:,5],'k. − ')
plt.xlabel('t(min)')
plt.ylabel('x')
plt.legend('PCLABG')
```

The product distribution is presented in Fig. 4E7.1.

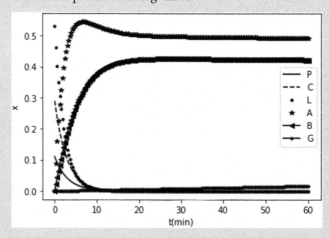

FIGURE 4E7.1 Profile of the HTL products.

bio-crude upgrading. However, the characteristics of the last one, due to the oxygen content of the hydrocarbons, may affect the process.

Hydrotreating: It consists of treating the oil in a hydrogen-rich environment (about 95% molar or 5% by weight) to remove impurities within the hydrocarbon mixture. It is a catalyzed stage using cobalt-molybdenum catalyst for the removal of sulfur and olefin saturation and nickel molybdenum for nitrogen removal and aromatic saturation, operating at 7–10 MPa and 300°C–400°C (CSM, 2019). A fixed bed reactor is used for this operation. The consumption of hydrogen accounts for 45–555 ft^3/bbl of treated crude (Watkins, 1979). The basic reactions are:

$$\text{Desulfuration: HBC-SH} + H_2 \rightarrow \text{HBC} + H_2S \tag{4.72}$$

$$\text{Denitrification: HBC-NH} + H_2 \rightarrow \text{HBC} + NH_3 \tag{4.73}$$

$$\text{Hydrocarbon saturation: RCH} = CH2 + H_2 \rightarrow \text{HBC} \tag{4.74}$$

$$\text{Oxygen removal: R-OH} + H_2 \rightarrow \text{RH} + H_2O \tag{4.75}$$

Hydrocracking: The idea is to break down long hydrocarbon molecules into shorter chains within diesel (C12) or gasoline (C8) range. Process conditions are a bit more severe than for hydrotreating, with pressures of 10–14 MPa and temperatures of 400°C–450°C using a nickel-molybdenum catalyst. The flow of hydrogen feed depends on the composition of the feed, the catalyst, and operating conditions. Values of 0.039 kg of H_2 consumed in the reactions per kg of crude can be found in the literature (CSM, 2019), representing from 7.5% to 23% of the total hydrogen fed to the cracker (Watkins, 1979). Thus the mass of crude oil to be processed downstream becomes:

$$\text{masshydro} = 1.039 \cdot fc_{(\text{Heavy})} \tag{4.76}$$

where the feed of hydrogen to the system is within:

$$0.15 \cdot fc_{(\text{Heavy})} \leq fc(H_2) \leq 0.52 \cdot fc_{(\text{Heavy})} \tag{4.77}$$

The heat of reaction is 77.53 kJ/kmol of hydrogen consumed (Coker, 2018).

The yield of the cracking step into gasoline, kerosene, and diesel is a complex calculation since each of the fractions is a mix of chemicals characterized by a range of distillation temperatures. Based on the typical ranges for gasoline, 40°C–200°C, kerosene, 170°C–220°C, and diesel, 180°C–360°C and the results by Bezergianni et al. (2009), correlations for the conversion of the process, X, and the selectivity, S_{Fuel}, to each one of the three fuels, Fig. 4.14, can be developed as follows:

$$X = 0.000185714 \cdot (T_{\text{HC}})^2 - 0.128829 \cdot T_{\text{HC}}3 + 22.6931 \tag{4.78}$$

$$S_D = -1.2232 \cdot 10^{-04} T_{\text{HC}}^2 + 8.2418 \cdot 10^{-02} T_{\text{HC}} - 1.2951 \cdot 10^1 \tag{4.79}$$

$$S_K = -2.5000 \cdot 10^{-05} T_{\text{HC}}^2 + 2.165010^{-02} T_{\text{HC}} - 4.465 \tag{4.80}$$

$$S_G = 1.4732 \cdot 10^{-04} T_{\text{HC}}^2 - 1.0407 \cdot 10^{-01} T_{\text{HC}} + 1.8416 \cdot 10^1 \tag{4.81}$$

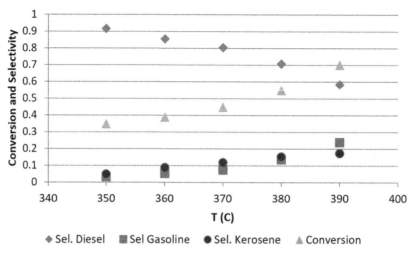

FIGURE 4.14 Conversion profiles for the hydrocracking reactor (Taimbú de la Cruz et al., 2019). *Source: With permission from Taimbú de la Cruz, C.A., Martín, M., Grossmann, I.E. (2019). Process optimization for the hydrothermal production of algae fuels. Ind. Eng. Chem. Res., 58(51), 23276–23283.*

Crude fractionation: Hydrocarbon fractionation is a typical operation in petroleum refineries. The closer the bio-oil to the petroleum the closer the operating conditions to the ones already known for refineries. A scheme of the column is presented in Fig. 4.15. A particular feature of these columns is the fact that they do not have reboiler and the energy for the separation, apart from heating up the feed using a furnace, is provided by a stream of steam directly fed to the column. The steam required is assumed to be from 0.043 kg/kg of residue (Gorak and Schoenmakers, 2014; Watkins, 1979) of which 40% goes directly to the atmospheric tower, 44% to the kerosene stripper, and the rest to the diesel stripper.

Gadalla et al. (2003) used a thermally equivalent series of distillation columns to represent the crude distillation unit. For four products, gasoline (G), kerosene (K), diesel (D), and heavy (H), a total of three columns are used, see Fig. 4.15. Chapter 7 details how to obtain thermally equivalent schemes from complex distillation columns. It is possible to assume sharp separation of each fraction. However, note that, since mixtures and nor pure components are separated, this approximation is valid. According to Speight (1991), the typical distillation towers for crude oil have 30 trays and the typical temperatures are 110°C for the top (Gasoline), 160°C for kerosene, 245°C for the diesel, and 372°C at the bottom for the heavier components (Riazi and Eser, 2013).

$$fc_{i,out} = fc_{i,in} \forall i = \{G, K, D, H, W\} \tag{4.82}$$

Based on the results by Gadalla et al. (2003), we formulate the energy balance to the reboilers and the condensers involved. Columns 2 and 3 have condenser and reboiler.

$$Q_{dist,3} = fc_{(Gasoline)}(R_3)\lambda_G + W\lambda_W \tag{4.83}$$

$$Q_{Reb,3} = fc_{(Gasoline)}(1 + R_3)\lambda_G + W\lambda_W \tag{4.84}$$

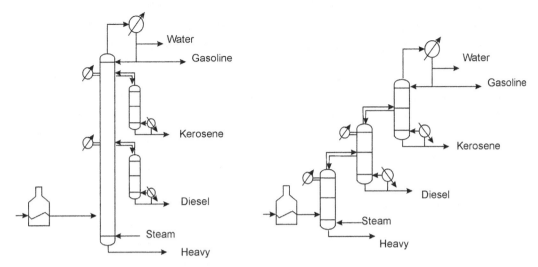

FIGURE 4.15 Scheme of the model for the bio-crude separation. *Source: With permission from Taimbú de la Cruz, C.A., Martín, M., Grossmann, I.E. (2019). Process optimization for the hydrothermal production of algae fuels. Ind. Eng. Chem. Res., 58(51), 23276–23283.*

$$Q_{dist,2} = \left[fc_{(Gasoline)}\lambda_G + fc_{(Kerosene)}\lambda_K + W\lambda_W \right](R_2) \tag{4.85}$$

$$Q_{Reb,2} = \left[fc_{(Gasoline)}\lambda_G + fc_{(Kerosene)}\lambda_K + W\lambda_W \right](1 + R_2) \tag{4.86}$$

$$Q_{dist,1} = \left[fc_{(Gasoline)}\lambda_G + fc_{(Kerosene)}\lambda_K + fc_{(Diesel)} \cdot \lambda_K + W\lambda_W \right](R_1) \tag{4.87}$$

The reflux ratios of columns 1, 2, and 3 (R_i) are taken from the literature as 0.41, 1.47, and 4.77 that have been obtained from the optimization of the separation of crude oil (Gadalla et al., 2003). The temperatures of the product streams are computed based on the number of carbons of the organics. The temperature of condensation of the water can be considered similar to that of gasoline.

4.4 Combustion of biomass

The combustion of biomass in different applications has been used since ancient times. The first uses correspond to small-scale applications, for instance, for heating in domestics uses (Van Loo and Koppejan, 2008). Technological development introduces the biomass for power generation at grid scale. Typical plants range up to 50 MW, much lower than the coal-based power facilities (Brown, 2011). The most common plant uses a boiler to generate steam with a Rankine cycle to produce power. Ideally, a complete combustion of the biomass is reached. The gases of the biomass combustion must be treated after being vented. Typical cleaning operation involves wet or dry scrubber to control sulfur and chlorine compounds and cyclones, filters, or electrostatic precipitators for particles removal. NO_x separation may be also necessary for complying with environmental legislation. The gas treatment must be adapted to the kind of biomass burnt. For example, municipal solid waste requires a more restrictive cleaning process than wood biomass. The combustion of the biomass can be divided in different steps (Demirbas, 2007): drying, pyrolysis (volatiles), primary gas-phase combustion, secondary combustion. The basic transport phenomena involved in this combustion process is as follows: oxygen has to diffuse to the biomass particle, oxygen has to be adsorbed in the particle surface, oxygen has to react with the solid generating the absorbed products, the products have to be desorbed and, finally, the desorbed products have to diffuse from the particle surface.

There are six main characteristics of the biomass influencing the combustion process (Demirbas, 2007): particle size and specific gravity, ash content, moisture content, extractive content, element (C, H, N, O) composition, and structural (cellulose, hemicellulose, and lignin) composition. Particle size is limited, in general, up to 0.6 cm. Ash is one of the main drawbacks in the combustion of biomass. Its presence in the biomass causes problems of fouling and corrosion in the equipment due to the alkaline nature of the ashes. The ash content in the wood biomass is about 0.5% but can reach up to 16% in certain kinds of biomass. Moisture content in the biomass is also an issue in the combustion because the heating value of the biomass decreases with it. Typical moisture contents for the biomass are in the range of 40%–70%. In general, it is preprocessed to reduce this value because it is difficult to maintain the combustion when the moisture content is higher than 55%.

Typically, biomass-based thermal power plants consist of a boiler followed by a steam turbine and a cooling system, see Fig. 4.16. The energy released in the combustion of the fuel is used to heat up, evaporate, and overheat steam to be used in a turbine for power production. The operation of the boiler involves multiple heat exchange stages to improve the efficiency of the system based on the temperature profile across the boiler. Two streams of air are used, one to the burner (primary) and another one (secondary) to the feed. Air, primary and secondary, is preheated using the colder gases before they are treated for particle removal. Water from the turbine, once condensed, is de-aerated, and preheated using the hot flue gases before they are used to heat up the air. The hot water enters the boiler at around 200°C and it is superheated in three different regions of the boiler before it is fed to the steam turbine. The analysis of the Rankine cycle is left for Chapter 9. Here, apart from this description, the energy that can be achieved from the biomass as a function of its composition is shown below in the proposed case of study.

Other alternatives have also been proposed to produce power from biomass. Mainly, the most developed is the use of gasification of biomass and then use the syngas in a combined cycle (De Kam et al., 2009). The whole process is called biomass integrated gasification combined cycle (Wu et al., 2008). The first step is the biomass gasification, using the units explained previously in this same chapter. The particles from the gasification are removed using a cyclone. Different alternatives to clean up the gases have been proposed: tar cracker, scrubber, venturi tubes, processes to remove the acid gases, and so on. Some works also proposed a shift reaction to adjust the hydrogen concentration (Corti and Lombardi, 2004). Then, the gases are burnt in the gas turbine and the remained heat is used in the steam turbine. A simplified example of this technology is also presented in the case of study section.

It is possible to compute the energy obtained, as high heating value (HHV), of different biomasses as a function of their composition (Demirbas, 2004).

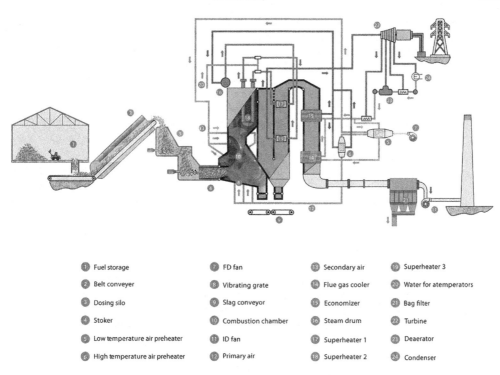

① Fuel storage	⑦ FD fan	⑬ Secondary air	⑲ Superheater 3
② Belt conveyer	⑧ Vibrating grate	⑭ Flue gas cooler	⑳ Water for atemperators
③ Dosing silo	⑨ Slag conveyor	⑮ Economizer	㉑ Bag filter
④ Stoker	⑩ Combustion chamber	⑯ Steam drum	㉒ Turbine
⑤ Low temperature air preheater	⑪ ID fan	⑰ Superheater 1	㉓ Deaerator
⑥ High temperature air preheater	⑫ Primary air	⑱ Superheater 2	㉔ Condenser

FIGURE 4.16 Biomass-based power plant. *Source: With permission from DP CleanTech.*

The HHV (MJ/kg) of lignocellulosic materials including C, H, O, and N (wt.%) can be calculated from:

$$HHV = (33.5[C] + 142.3[H] - 15.4[O] - 14.5[N]) \cdot 10^{-2} \qquad (4.88)$$

Alternatively, the HHV (MJ/kg) of the biomass samples as a function of FC (wt.%) and volatile matter (VM, wt.%) can be calculated from:

$$HHV = 0.312(FC) + 0.1534(VM) \qquad (4.89)$$

4.5 Case study

The case study evaluated in this chapter deals with the production of power using biomass as feedstock. Two different alternatives have been evaluated to transform biomass into power: direct combustion of the biomass with a Rankine cycle and biomass gasification with a combined cycle (gas turbine plus Rankine cycle). The raw material used for this case of study is wheat straw whose composition is presented in Example 4.1. The first alternative evaluated is the direct combustion of the biomass. For both cases, 100 kg/s of wheat straw are selected as a basis of calculation.

Direct combustion of biomass

The composition of the wheat straw is simplified considering only the three main components: carbon, oxygen, and hydrogen. With this composition, the empirical formula of the biomass can be calculated. For the combustion of the biomass, a complete combustion is assumed as it is represented in the next reaction:

$$CH_\gamma O_\beta + \left(2 + \frac{\gamma}{2} - \beta\right)O_2 \rightarrow CO_2 + \left(\frac{\gamma}{2}\right)H_2O \qquad (4E8.1)$$

For the case of the wheat straw, $\gamma = 1.71$ and $\beta = 0.76$. The combustion takes place with air at 25°C with a humidity (φ) of 30%. An excess of 10% of air is introduced in the chamber. Two different products are obtained from the combustion of the biomass. The final gases are formed by CO_2, H_2O, N_2, and O_2 and the solid fraction composed by the ashes presented in the inlet biomass. The mass balances are as follows:

$$\overline{H_2O}|_{biomass} = \left(\frac{7}{100}\right) \cdot \overline{Total}|_{biomass} \qquad (4E8.2)$$

$$\overline{Ash}|_{biomass} = \left(\frac{5.5}{100}\right) \cdot \left(\overline{Total}|_{biomass} - \overline{H_2O}|_{biomass}\right) \qquad (4E8.3)$$

$$O_2|_{air}^{est} = \left(2 + \frac{\gamma}{2} - \beta\right) \cdot \frac{\left(\overline{Total}|_{biomass} - \overline{H_2O}|_{biomass} - \overline{Ash}|_{biomass}\right)}{MW_{biomass\ daf}} \qquad (4E8.4)$$

$$O_2|_{air} = 1.1 \cdot O_2|_{air}^{est} \qquad (4E8.5)$$

$$N_2|_{air} = \left(\frac{0.79}{0.21}\right) \cdot O_2|_{air} \qquad (4E8.6)$$

$$\omega = \frac{\varphi P_v^0(H_2O, T)}{P_{total} - \varphi P_v^0(H_2O, T)} \qquad (4E8.7)$$

$$H_2O|_{air} = \omega \cdot (N_2|_{air} + O_2|_{air}) \qquad (4E8.8)$$

$$CO_2|_{gases} = \frac{\left(\overline{Total}|_{biomass} - \overline{H_2O}|_{biomass} - \overline{Ash}|_{biomass}\right)}{MW_{biomass\ daf}} \qquad (4E8.9)$$

$$H_2O|_{gases} = H_2O|_{biomass} + H_2O|_{air} + \left(\frac{\gamma}{2}\right) \cdot \frac{\left(\overline{Total}|_{biomass} - \overline{H_2O}|_{biomass} - \overline{Ash}|_{biomass}\right)}{MW_{biomass\ daf}} \qquad (4E8.10)$$

$$O_2|_{gases} = O_2|_{air} - O_2|_{air}^{est} \qquad (4E8.11)$$

$$N_2|_{gases} = N_2|_{air} \qquad (4E8.12)$$

The final temperature of the gases can be calculated using an adiabatic energy balance. It is assumed that the inlet biomass and air are introduced at 298K. Energy losses of 10% are adopted. Therefore the final equation for the energy balance is formulated as follows:

$$\left(\sum_{i \in biomass} n_i h_{f,i}^0 + \sum_{i \in air} n_i h_{f,i}^0 \right) \cdot \left(1 - \frac{\%losses}{100} \right) = \sum_{i \in gases} n_i \left(h_{f,i}^0 + \int_{T_{ref}}^{T_{out}} C_{p,i} dT \right) + \sum_{i \in ash} n_i \int_{T_{ref}}^{T_{out}} C_{p,i} dT$$

(4E8.13)

The enthalpies of formation and heat capacities of the gases are collected in the Appendix B. The heat capacity of the ashes is calculated using Eq. 4E8.14 (Brown, 2011):

$$C_{p,ash}(J/kg \ K) = 752 + 0.293 \cdot T(K)$$

(4E8.14)

The results from this first section are summarized in Table 4E8.1.

TABLE 4E8.1   Results of the biomass combustion.

Biomass	Air	Gases	Ash
T: 298.0K	T: 298.0 K	T: 2023.9 K	T: 2023.9 K
Total: 100.0 kg/s	Total: 539.0 kg/s	Total: 633.9 kg/s	Total: 5.1 kg/s
Moisture (H_2O): 7.0 kg/s	O_2: 124.8 kg/s	CO_2: 149.3 kg/s	Ash: 5.1 kg/s
Ash: 5.1 kg/s	N_2: 411.0 kg/s	H_2O: 62.3 kg/s	
Fixed carbon and volatiles: 87.9 kg/s (46.32% C, 6.59% H, 47.09% O)	Humidity (H_2O): 3.2 kg/s	O_2: 11.3 kg/s	
		N_2: 411.0 kg/s	

The next step in the transformation of biomass into power is to transfer the heat from the gases of biomass combustion to the steam inside the Rankine cycle. The gases from the combustion are cooled down from the outlet temperature (2023.9K) to the final temperature previous to release (fixed to 308K). The heat that it is necessary to remove is calculated with an energy balance:

$$Q_{boiler} = \sum_i n_i \left(\int_{T_{in}}^{T_{out}} C_{p,i} dT \right)$$

(4E8.15)

Some treatments are required to remove particles and other pollutants from the outlet gas stream; however, this is out of the scope of this example. The next stage is modeling the Rankine cycle coupled with the biomass combustion. For the sake of simplicity, a simple Rankine cycle is used, with only one expansion. The high and low pressures are also fixed at 130 and 0.1 bar. The boiler transforms a liquid into a superheated steam with no changes in the pressure. The heat involved in the boiler must be the same than the heat removed from the gases by biomass combustion.

$$Q_{boiler} = \dot{m}_{H_2O} \left(H_{SS} - H_{liquid} \right)$$

(4E8.16)

The expansion in the turbine is modeled as an isentropic process but an isentropic efficiency is also included. The value of this efficiency is fixed to 0.95. In the isentropic process, saturated steam leaves the turbine at a pressure equal to 0.1 bar.

$$\eta_s = \frac{H_{steam}^{out} - H_{steam}^{in}}{H_{steam}^{iso} - H_{steam}^{in}} \tag{4E8.17}$$

After the turbine, the steam is cooled down to transform the outlet steam in a liquid. To close the cycle, the liquid is pumped from the low pressure to the high one. The enthalpies and entropies of the different stream can be calculated according to the correlations presented in Appendix B. The final temperatures are presented in Fig. 4E8.1. For the gases produced with 100 kg/s of biomass, the following water flowrate is necessary:

$$\dot{m}_{H_2O} = 341.7 \text{ kg/s}$$

And, the total power produced in the Rankine cycle:

$$W_{total} = -499462.5 \text{ kW} = -5.0 \text{ MJ/kg biomass}$$

Biomass gasification with a combined cycle

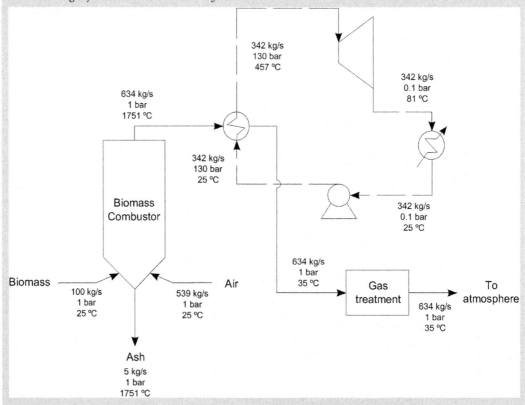

FIGURE 4E8.1 Flow diagram and main results of the direct combustion of biomass.

The second alternative analyzed here to transform biomass into power consists in gasifying the biomass and, then, to introduce the gases in a combined cycle (gas turbine and Rankine cycle). The first step is to gasify the biomass to convert it into a syngas. Direct gasification is used. The model presented in this chapter is used (Eggeman, 2005). To model the gasifier, the mass balances for each of the atoms with the empirical correlations have been used.

$$f_i = A + B \cdot P + C \cdot T + D \cdot \left(\frac{O_2}{FeedC}\right) + E \cdot \left(\frac{H_2O}{FeedC}\right) \tag{4E8.18}$$

$$\left(\frac{(\overline{Total}|_{biomass} - \overline{H_2O}|_{biomass} - \overline{Ash}|_{biomass})}{MW_{biomass\ daf}}\right) C_{biomass\ daf} = CO_2|_{syngas} + CO|_{syngas} + CH_4|_{syngas}$$

$$+ 6 \cdot C_6H_6|_{syngas} + 10 \cdot C_{10}H_8|_{syngas} + 2 \cdot C_2H_4|_{syngas} + 2 \cdot C_2H_6|_{syngas} + C|_{char} \tag{4E8.19}$$

$$\left(\frac{(\overline{Total}|_{biomass} - \overline{H_2O}|_{biomass} - \overline{Ash}|_{biomass})}{MW_{biomass\ daf}}\right) H_{biomass\ daf} + 2 \cdot \frac{\overline{H_2O}|_{biomass}}{MW_{H_2O}} + 2 \cdot H_2O|_{steam}$$

$$= 4 \cdot CH_4|_{syngas} + 6 \cdot C_6H_6|_{syngas} + 8 \cdot C_{10}H_8|_{syngas} + 4 \cdot C_2H_4|_{syngas}$$

$$+ 6 \cdot C_2H_6|_{syngas} + 2 \cdot H_2O|_{syngas} + 2 \cdot H_2|_{syngas} + H|_{char} \tag{4E8.20}$$

$$\left(\frac{(\overline{Total}|_{biomass} - \overline{H_2O}|_{biomass} - \overline{Ash}|_{biomass})}{MW_{biomass\ daf}}\right) O_{biomass\ daf} + \frac{\overline{H_2O}|_{biomass}}{MW_{H_2O}} + H_2O|_{steam} + 2 \cdot O_2|_{oxygen}$$

$$= 2 \cdot CO_2|_{syngas} + CO|_{syngas} + H_2O|_{syngas} + O|_{char} \tag{4E8.21}$$

In these equations, $C_{biomass\ daf}$, $H_{biomass\ daf}$, and $O_{biomass\ daf}$ correspond to the coefficient of the biomass in the empirical formula. For this case of study, the pressure, temperature, and the ratios between O_2/Feed C and H_2O/Feed C are fixed to the following values: 20 bar, 1050K, 0.2 (O_2/FeedC), and 0.7 (H_2O/FeedC). If the balances are solved (using a nonlinear solver, in this case in Excel), the results presented in Fig. 4E8.2 and Table 4E8.2 are obtained.

The next step is to feed this syngas in the combined cycle. The gas is previously cleaned up to remove particles and so on. Due to the high temperature of the gases, the first operation is to reduce the temperature of the outlet gases from the gasifier because it is not possible to introduce as such in the gas turbine. Steam is produced in this cooling down that is used later in the Rankine cycle. The gas is introduced in the heat exchanger at 1050K (outlet temperature of the gasifier) and the temperature is reduced up to 500K. Steam at 130 bar is generated.

$$\sum_i n_i \left(\int_{T_{in}}^{T_{out}} C_{p,i} dT\right) = \dot{m}_{H_2O}^1 (H_{SS} - H_{liquid}) \tag{4E8.22}$$

The cooled gases are introduced in the gas turbine. The gas is directly introduced in the combustion chamber at the pressure of 20 bar. The syngas is burnt with dry air. An excess of air

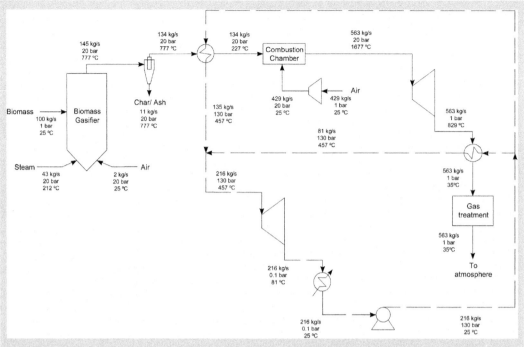

FIGURE 4E8.2   Flow diagram and main results of the biomass gasification and syngas combustion.

equal to 20% is fixed. A complete combustion of all the hydrocarbons presents in the syngas is assumed. An adiabatic combustion is considered to calculate the final temperature of the gases. The air must be compressed before being introduced in the combustion chamber. A three-step polytropic compression is assumed for the calculations with the same compression ratio.

$$O_2|_{air}^{est} = \left(\frac{1}{2}\right) CO|_{syngas} + 2 \times CH_4|_{syngas} + \left(\frac{15}{2}\right) \cdot C_6H_6|_{syngas} + 12 \cdot C_{10}H_8|_{syngas}$$

$$+ 3 \cdot C_2H_4|_{syngas} + \left(\frac{7}{2}\right) \cdot C_2H_6|_{syngas} + \left(\frac{1}{2}\right) \cdot H_2|_{syngas} \qquad (4E8.23)$$

$$O_2|_{air} = 1.2 \cdot O_2|_{air}^{est} \qquad (4E8.24)$$

$$N_2|_{air} = \left(\frac{0.79}{0.21}\right) \cdot O_2|_{air} \qquad (4E8.25)$$

$$W = \frac{F \cdot 8.314 \cdot k \cdot (T_{in} + 273.15)}{MW \cdot (k - 1)} \left(\left(\frac{P_{out}}{P_{in}}\right)^{\frac{k-1}{k}} - 1\right) \frac{1}{\eta_s} \qquad (4E8.26)$$

TABLE 4E8.2 Results of the biomass gasification.

Biomass	O_2	Steam	Gases	Char	Ash
Total: 100.0 kg/s	**Total: 42.7 kg/s**	**Total: 1.9 kg/s**	**Total: 133.4 kg/s**	**Total: 6.1 kg/s (51.99% C, 42.31% H, 5.70% O)**	**Total: 5.1 kg/s**
Moisture (H_2O): 7.0 kg/s	O_2: 42.7 kg/s	H_2O: 1.9 kg/s	CO_2: 59.0 kg/s		Ash: 5.1 kg/s
Ash: 5.1 kg/s			CO: 17.6 kg/s		
Fixed Carbon and volatiles: 87.9 kg/s (46.32% C, 6.59% H, 47.09% O)			CH_4: 8.5 kg/s		
			C_6H_6: 3.9 kg/s		
			$C_{10}H_8$: 2.0 kg/s		
			C_2H_4: 0.8 kg/s		
			C_2H_6: 1.7 kg/s		
			H_2O: 38.4 kg/s		
			H_2: 1.5 kg/s		

$$T_{out} = T_{in} + (T_{in} + 273.15)\left(\left(\frac{P_{out}}{P_{in}}\right)^{\frac{k-1}{k}} - 1\right)\frac{1}{\eta_s} \tag{4E8.27}$$

$$CO_2|_{gases} = CO_2|_{syngas} + CO|_{syngas} + CH_4|_{syngas} + 6 \cdot C_6H_6|_{syngas} + 10 \cdot C_{10}H_8|_{syngas}$$
$$+ 2 \cdot C_2H_4|_{syngas} + 2 \cdot C_2H_6|_{syngas} \tag{4E8.28}$$

$$H_2O|_{gases} = 2 \cdot CH_4|_{syngas} + 3 \cdot C_6H_6|_{syngas} + 4 \cdot C_{10}H_8|_{syngas}$$
$$+ 2 \cdot C_2H_4|_{syngas} + 3 \cdot C_2H_6|_{syngas} + H_2|_{syngas} + H_2O|_{syngas} \tag{4E8.29}$$

$$N_2|_{gases} = N_2|_{air} \tag{4E8.30}$$

$$O_2|_{gases} = O_2|_{air} - O_2|_{air}^{est} \tag{4E8.31}$$

$$\sum_{i \in syngas} n_i\left(h_{f,i}^0 + \int_{T_{ref}}^{T_{out}} C_{p,i}dT\right) + \sum_{i \in air} n_i h_{f,i}^0 = \sum_{i \in gases} n_i\left(h_{f,i}^0 + \int_{T_{ref}}^{T_{out}} C_{p,i}dT\right) \tag{4E8.32}$$

The power obtained in the gas turbine is calculated using the same equation as the compressor. The gas is expanded from 20 bar to ambient pressure. The total work involved in the compression of the air is:

$$W_{\text{air comp}} = 223740.8 \text{ kW}$$

The total work obtained from the expansion of the gases in the gas turbine is equal to:

$$W_{GT} = -923471.9 \text{ kW}$$

The gas leaves the combustion chamber at 1950.1K and, after the expansion in the gas turbine, the gas cools down up to 828.6K. The material flows for the combustion chamber and temperatures are summarized in Table 4E8.3.

TABLE 4E8.3   Results of the syngas combustion.

Syngas	Air	Gases
T: 500.0 K	T: 298.0 K	T: 1950.1K
Total: 133.4 kg/s	**Total: 429.4 kg/s**	**Total: 562.8 kg/s**
CO_2: 59.0 kg/s	O_2: 100.1 kg/s	CO_2: 137.8 kg/s
CO: 17.6 kg/s	N_2: 329.3 kg/s	H_2O: 79.0 kg/s
CH_4: 8.5 kg/s		O_2: 16.7 kg/s
C_6H_6: 3.9 kg/s		N_2: 329.3 kg/s
$C_{10}H_8$: 2.0 kg/s		
C_2H_4: 0.8 kg/s		
C_2H_6: 1.7 kg/s		
H_2O: 38.4 kg/s		
H_2: 1.5 kg/s		

The next stage in this gasification-based power production is the Rankine cycle. The gases leaving the gas turbine at 828.6K are cooling down up to 308K. This heat is transferred to the water of the Rankine cycle. This water is added to the vaporized water in the cooling just after the gasifier. The conditions of the cycle are the same than in the direct combustion of biomass. Only one expansion is modeled. The values of the high and low pressure are fixed at 130 and 0.1 bar. An isentropic efficiency of 0.95 is assumed. The conditions of the stream are presented in Fig. 4E8.2. The total flowrate of water in the Rankine cycle is equal to:

$$\dot{m}_{H_2O} = 216.1 \text{ kg}/s$$

And the power obtained from the steam turbine is:

$$W_{ST} = -315795.6 \text{ kW}$$

Therefore the total power of the combined cycle in the power production using the gasification of the biomass is equal to:

$$W_{total} = -1015526.7 \text{ kW} = -10.2 \text{ MJ/kg biomass}$$

Exercises

P4.1 Compute the energy required to regenerate MEA, DEA, and MDEA in a sour gases capture process that treats a syngas of 40 kmol/h of CO, 40 kmol/h of H_2, 19.5 kmol/h of CO_2, and 0.5 kmol/h of H_2S.

P4.2 Optimize the operating conditions toward H_2 production in a direct gasifier. The composition of the biomass is:

Water	15%
C	40.16%
H	4.73%
O	34.59%
S	0.07%
N	0.49%
Ash	4.96%

P4.3 Evaluate the effect of temperature of the bio-oil production in algae HTL.

P4.4 Based on Example 4.6, evaluate the effect of temperature on the product distribution.

P4.5 Following the model proposed in Example 4.5, determine the conditions to maximize the amount of gas in the pyrolysis of wheat straw.

References

Amin, S., 2009. Review on biofuel oil and gas production processes from microalgae. Energ. Convers. Manage. 50 (7), 1834–1840.

Basu, P., 2010. Biomass Gasification and Pyrolysis: Practical Design and Theory. Academic Press.

Bezergianni, S., Kalogianni, A., Vasalos, I.A., 2009. Hydrocracking of vacuum gas oil-vegetable oil mixtures for biofuels production. Bioresour. Technol. 100, 3036–3042.

Biegler, L.T., Grossmann, I.E., Westerberg, A.W., 1997. Systematic Methods of Chemical Process Design. Prentice Hall, Prentice Hall International Series in the Physical and Chemical Engineering Sciences.

Bridgwater, A.V., 1995. The technical and economic feasibility of biomass gasification for power generation. Fuel 14 (5), 631–653.

Bridgwater, A.V., Meier, D., Radlein, D., 1999. An overview of fast pyrolysis of biomass. Org. Geochem. 30, 1479–1493.

Brown, R.C., 2011. Thermochemical Processing of Biomass: Conversion Into Fuels, Chemicals and Power. Wiley.

Carmo, M.J., Gubulin, J.C., 1997. Ethanol-water adsorption on commercial 3a zeolites: kinetic and thermodynamic data. Braz. J. Chem. Eng. 14 (3).

Chen, J., Gomez, J.A., Hoffner, K., Barton, P.I., Henson, M.A., 2015. Metabolic modeling of synthesis gas fermentation in bubble column reactors. Biotechnol. Biofuels 8, 89.

Coker, A.K., 2018. Petroleum Refining Design and Applications Handbook. Willey:, New York.

Corti, A., Lombardi, L., 2004. Biomass integrated gasification combined cycle with reduced CO_2 emissions: performance analysis and life cycle assessment (LCA). Energy 29, 2109−2124.

CSM, 2019. Hydroprocessing: Hydrotreating & Hydrocracking. <https://inside.mines.edu/∼jjechura/Refining/08_Hydroprocessing.pdf> (last accessed July 2020).

De Kam, M.J., Morey, R.V., Tiffany, D.G., 2009. Biomass integrated gasification combined cycle for heat and power at ethanol plants. Energy Convers. Manag. 50, 1682−1690.

de Medeiros, E., Posada, J.A., Noorman, H., Filho, R.M., 2019. Dynamic modeling of syngas fermentation in a continuous stirred-tank reactor: multi-response parameter estimation and process optimization. Biotechnol. Bioeng. 116, 2473−2487.

De, S., Agarwal, A.K., Moholkar, V.S., Thallada, B., 2018. Coal and Biomass Gasification: Recent Advances and Future Challenges. Springer.

Demirbas, A., 2004. Combustion characteristics of different biomass fuels. Prog. Energy Combust. Sci. 30, 219−230.

Demirbas, A., 2007. Combustion of biomass. Energy Sources A: Recov. Util. Environ. Eff. 29 (6), 549−561.

Di Blasi, C., 2004. Modeling wood gasification in a countercurrent fixed-bed reactor. AIChE J. 50 (9), 2306−2319.

Duran, M.A., Grossmann, I.E., 1986. Simultaneous optimization and heat integration of chemical processes. AIChE, J. 32, 123−138.

Dutta, A. and Phillips, S.D., 2009. Thermochemical Ethanol via Direct Gasification and Mixed Alcohol Synthesis of Lignocellulosic Biomass. NREL/TP-510-45913.

Eggeman, T. (2005). Updated Correlations for GTI Gasifier WDYLD8. NREL, Golden, CO.

Emerson, S.C., Magdefrau, N.J., She Y. and Thibaud-Erkey, C. (2012). Advanced Palladium Membrane Scale-up for Hydrogen Separation. United Technologies Research Center. DE−FE0004967, PA.

Gadalla, M., Jobson, M., Smith, R., 2003. Shortcut models for retrofit design of distillation columns. Chem. Eng. Res. Des. 81, 971−986.

Gorak, A., Schoenmakers, H., 2014. Distillation: Operation and Applications. Elsevier, Oxford.

GPSA. (2004). Engineering Data Book. FPS Version.

GPSA (2012). Engineering Data Book. FPS Version.

Hameed, S., Sharma, A., Pareek, V., Wu, H., Yu, Y., 2019. A review on biomass pyrolysis models: kinetic, network and mechanistic models. Biomass Bioenergy 123, 104−122.

Hernandez, B., Martín, M., 2018. Optimization for biogas to chemicals via tri-reforming. analysis of Fischer-Tropsch fuels from biogas. Energy Convers. Manag. 174, 998−1013.

Iyoha, O., Enick, R., Killmeyer, R., Howard, B., Howard, B., Ciacco, M., et al., 2007. H_2 production from simulated coal syngas containing H_2S in multi-tubular Pd and 80 wt% Pd−20 wt% Cu membrane reactors at 1173 K. J. Membr. Sci. 306 (1−2), 103−115.

Jarungthammachote, S., Dutta, A., 2007. Thermodynamic equilibrium model and second law analysis of a downdraft waste gasifier. Energy 32, 1660−1669.

Jarungthammachote, S., Dutta, A., 2008. Equilibrium modeling of gasification: Gibbs free energy minimization approach and its application to spouted bed and spout-fluid bed gasifiers. Energ. Convers. Manage. 49, 1345−1356.

Ji, P., Feng, W., Chen, B., 2009. Production of ultrapure hydrogen from biomass gasification with air. Chem. Eng. Sci. 64, 582−592.

Karuppiah, R., Peschel, A., Grossmann, I.E., Martín, M., Martinson, W., Zullo, L., 2008. Energy optimization for the design of corn-based ethanol plants. AIChE J. 54 (6), 1499−1525.

López Barreiro, D., Prins, W., Ronsse, F., Brilman, W., 2013. Hydrothermal liquefaction (HTL) of microalgae for biofuel production: state of the art review and future prospects. Biomass Bioenergy 53, 113−127.

Martelli, E., Kreutz, T., Consonni, S., 2009. Comparison of coal IGCC with and without CO_2 capture and storage: shell gasification with standard vs. partial water quench. Energy Procedia 1, 607−614.

Martín, M., 2016. Industrial Chemical Processes. Analysis and Design. Elsevier, Oxford:.

Martín, M., Grossmann, I.E., 2011a. Energy optimization of hydrogen production from biomass. Computers Chem. Eng. 35 (9), 1798−1806.

Martín, M., Grossmann, I.E., 2011b. Energy optimization of bioethanol production via gasification of switchgrass. AIChE J. 57 (12), 3408–3428.

Martín, M., Grossmann, I.E., 2013. Optimal use of hybrid feedstock, switchgrass and shale gas for the simultaneous production of hydrogen and liquid fuels. Energy 55, 378–391.

Martín, M., Grossmann, I.E., 2017. Towards zero CO_2 emissions in the production of methanol from switchgrass. CO_2 to methanol. Comput. Chem. Eng. 105, 308–316.

Martín, M., Ahmetovic, E., Grossmann, I.E., 2011. Optimization of water consumption in second generation bioethanol plants. Ind. Eng. Chem. Res. 50 (7), 3705–3721.

Martín, M., Montes, F.J., Galán, M.A., 2007. Oxygen transfer from growing bubbles: effect of the physical properties of the liquid. Chem. Eng. J. 128, 21–32.

Neves, D., Thunman, H., Matos, A., Tartelho, L., Gómez-Barea, A., 2011. Characterization and prediction of biomass pyrolysis products. Prog. Energy Combust. Sci. 37, 611–630.

Okolie, J.A., Rana, R., Nanda, S., Dalai, A.K., Kozinski, J.A., 2019. Supercritical water gasification of biomass: a state-of-the-art review of process parameters, reaction mechanisms and catalysis. Sustain. Energy Fuels 3, 578.

Peral, E., Martín, M., 2015. Optimal production of dimethyl ether from switchgrass based syngas via direct synthesis. Ind. Eng. Chem. Res. 54 (30), 7465–7475.

Peterson, A.A., Vogel, F., Lachance, R.P., Fröling, M., Antal, M.J., Tester, J.W., 2008. Thermochemical biofuel production in hydrothermal media: a review of sub and supercritical water technologies. Energy Environ. Sci. 1, 32–65.

Phillips, S., Aden, A., Jechura, J., Dayton, D. and Eggeman, T. (2007). Thermochemical Ethanol via Indirect Gasification and Mixed Alcohol Synthesis of Lignocellulosic Biomass. NREL/TP -510-41168.

Portillo, M.A., Villanueva Perales, A.L., Vidal Barrero, F., Campoy, M., 2016. A kinetic model for the synthesis of ethanol from syngas and methanol over an alkali-codoped molybdenum sulfide catalyst: model building and validation at bench scale. Fuel Process. Technol., 151, 19–30.

Ranzi, E., Cuoci, A., Faravelli, T., Frassoldati, A., Migliavacca, G., Pierucci, S., et al., 2008. Chemical kinetics of biomass pyrolysis. Energy Fuels 22, 4292–4300.

Reddy, S.N., Nanda, S., Dalai, A.K., Kozinski, J.A., 2014. Supercritical water gasification of biomass for hydrogen production. Int. J. Hydrog. Energy 39, 6912–6926.

Riazi, M.R., Eser, S., 2013. Properties, specifications, and quality of crude oil and petroleum products. In: Riazi, M.R., Eser, S., Peña, J.L. (Eds.), Petroleum Refining and Natural Gas Processing. ASTM International, West Conshohocken, PA.

Ross, A.B., Biller, M.L., Kubacki, M.L., Li, H., Lea-Langton, A., Jones, J.M., 2010. Hydrothermal processing of microalgae using alkali and organic acids. Fuel 89 (9), 2234–2243.

Sánchez, A., Martín, M., Vega, P., 2019. Biomass based sustainable ammonia production: digestion vs gasification. ACS Sustain. Chem. Eng. 7 (11), 9995–10007.

Sikarwar, V.S., Zhao, M., Clough, P., Yao, J., Zhong, X., Memon, M.Z., et al., 2016. An overview of advances in biomass gasification. Energy Environ. Sci. 9, 2939.

Speight, J.G., 1991. The Chemistry and Technology of Petroleum, second ed. Marcel Dekker Inc., New York.

Stoll, K., Boukis, N., Sauer, J., 2020. Syngas fermentation to alcohols: reactor technology and application perspective. Chem. Ing. Tech. 92 (1–2), 125–136.

Taimbú de la Cruz, C.A., Martín, M., Grossmann, I.E., 2019. Process optimization for the hydrothermal production of algae fuels. Ind. Eng. Chem. Res. 58 (51), 23276–23283.

Thurner, F., Mann, U., 1981. Kinetic investigation of wood pyrolysis. Ind. Eng. Chem. Process. Des. Dev., 20, 482–488.

Turns, S.R., 2000. Introduction to Combustion, second ed. McGraw Hill, New York:.

Valdez, P.J., Savage, P.E., 2013. A reaction network for the hydrothermal liquefaction of Nannochloropsis sp. Algal Res. 2, 416–425.

Valdez, P.J., Tocco, V.J., Savage, P.E., 2014. A general kinetic model for the hydrothermal liquefaction of microalgae. Bioresour. Technol. 163, 123–127.

Van Loo, S., Koppejan, J., 2008. The Handbook of Biomass Combustion and Co-firing. Earthscan.

Watkins, R.M., 1979. Petroleum Refinery Distillation. Gulf. Pub. Co. Book Division, Houston, TX.

Wright, M.M., Satrio, J.A., Brown, R.C., Daugaard, D.E. and Hsu D.D. (2010). Technoeconomic Analysis of Biomass Fast Pyrolysis to Transportation Fuels. NREL/TP-6A20-46586.

Wu, C., Yin, X., Ma, L., Zhou, Z., Chen, H., 2008. Design and operation of a 5.5 MW biomass integrated gasification combined cycle demonstration plant. Energy Fuels 22, 4259–4264.

Yee, T.F., Grossmann, I.E., 1990. Simultaneous optimization models for heat integration-II. heat exchanger network synthesis. Comp. Chem. Eng. 14, 1165–1184.

Yun, S., Ted Oyama, S., 2011. Correlations in palladium membranes for hydrogen separation: a review. J. Membr. Sci. 375, 28–45.

Zainal, Z.A., Ali, R., Lean, C.H., Seetharamu, K.N., 2001. Prediction of performance of a downdraft gasifier using equilibrium modeling for different biomass materials. Energy Convers. Manag. 42, 1499–1515.

Zhang, Y., Chen, C.C., 2013. Modeling CO_2 absorption and desorption by aqueous monoethanolamine solution with Aspen rate-based model. Energy Procedia 37, 1584–1596.

Biochemical-based processes

Mariano Martín and Guillermo Galán

Department of Chemical Engineering, University of Salamanca, Salamanca, Spain

In this chapter, we focus on breaking down biomass into platform molecules such as sugars and oils. Compared with Chapter 4, where the biomass was broken into gases or a mix of chemicals in gas, liquid, and solid form, in this one the technologies aim to produce well known molecules to be used as platform for other chemicals. Therefore the operating temperature and pressure of the different stages is lower than in thermochemical processing while several stages allow controlling the depolymerization of the biomass. Two different types of biomass are analyzed, leaving out first-generation bioprocesses to avoid ethical issues related to the competence food versus chemicals, we consider lignocellulosic biomass and lipid sources, either nonedible seeds or algae. While biofuels are no longer a priority, bioethanol can be the raw material for renewable ethylene that is the starting point of organic synthesis and therefore this is considered as case study (Weissermel and Arpe, 1981; Rosales-Calderon and Arantes, 2019).

5.1 Sugar-based processes

The production of sugars can use grain, the carbohydrates within algae or lignocellulosic biomass as raw materials. The use of grain is classified as first-generation—based processes, and species such as corn or wheat have been among the most studied. The ethical issues on the use of biomass for food versus as a raw material for the chemical and fuel industry are displacing them. However, most of the facilities in operation still run first-generation processes. They are based on starch processing. Algae carbohydrates are also made of starch and can be an interesting raw material for further integrated processes as it will be presented along the chapter. Finally, lignocellulosic biomass consists of three major components, hemicellulose, cellulose, and lignin. Lignin provides the structure of the plant. It is a matrix of aromatic compounds. Within this structure, the cellulose is a polymer of mainly glucose monomers. Surrounding the cellulose, we find the hemicellulose, a polymer mainly made of xylose. The production of sugars out of lignocellulosic biomass consists of breaking down the protective structure of the lignin to release the cellulose and hemicellulose that will be later further broken down into the monomers.

5.1.1 Starch-based sugar production

The starch within the algae or from the grain is released via physical or chemical processes. The molecule is shown below. It is basically a polymer of glucose monomerts linked by α-glucosidic bonds, see Fig. 5.1.

Therefore, pretreatment of the starch consists of breaking the bonds to obtain sugar monomers of glucose. The process takes place in two stages. The first one is the liquefaction that breaks down the polymer into maltose, a dimer of glucose. It is carried out at 90°C and pH from 6 to 6.5 for 30 minutes. The reaction is as follows:

$$2(C_6H_{10}O_5)_n + nH_2O \xrightarrow{\alpha\text{-amylase}} nC_{12}H_{22}O_{11}$$

Next the maltose is further hydrolyzed to produce glucose at the saccharification step at 65°C and pH of 5.5 for 30 minutes. The reaction is as follows:

$$C_{12}H_{22}O_{11} + H_2O \xrightarrow{glucoamylase} 2C_6H_{12}O_6$$

The mechanism for the enzymatic hydrolysis is given by Michaelis Menten as follows, with E is the enzyme, S the substrate, and P is the product:

$$E + S \underset{k_{-1}}{\overset{k_1}{\rightleftarrows}} ES \overset{k_2}{\rightarrow} E + P \tag{5.1}$$

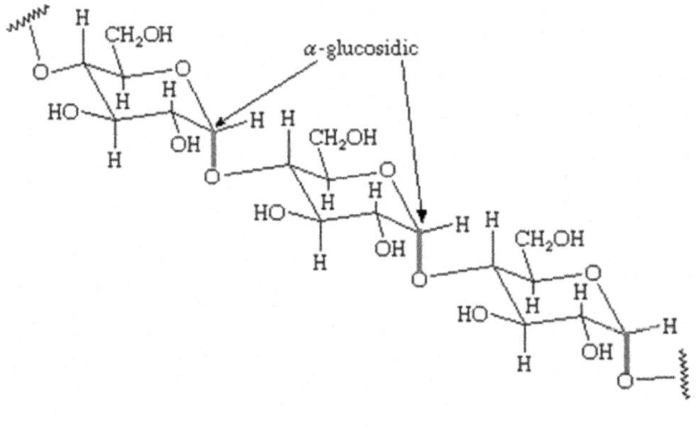

FIGURE 5.1   Starch molecule.

From the mechanism, it is possible to develop a kinetic rate. Assuming steady state for the enzyme−substrate (ES) complex:

$$\frac{d[ES]}{dt} = 0 = k_1[E][S] - k_{-1}[ES] - k_2[ES]$$

$$[ES] = \frac{k_1[E][S]}{k_{-1} + k_2}$$

$$K_m = \frac{k_{-1} + k_2}{k_1} \tag{5.2}$$

$$[ES] = \frac{[E][S]}{K_m}$$

$$\frac{d[P]}{dt} = k_2[ES]$$

$$[E_o] = [E] + [ES]$$
$$[E] = [E_o] - [ES]$$
Thus
$$[ES] = \frac{([E_o] - [ES])[S]}{K_m}$$

$$[ES] = [E_o]\frac{1}{1 + \frac{K_m}{[S]}}$$

$$\frac{d[P]}{dt} = k_2[E_o]\frac{[S]}{K_m + [S]} = V_{max}\frac{[S]}{K_m + [S]}$$

This basic mechanism has been extended to include the inhibitory effects:

Competitive inhibition: The inhibitor (I) binds to the enzyme active site, making up the enzyme−inhibitor complex, and preventing binding of the substrate.

$$E + S \underset{k_{-1}}{\overset{k_1}{\rightleftarrows}} ES \overset{k_2}{\rightarrow} E + P$$

$$E + I \underset{k_{-3}}{\overset{k_3}{\rightleftarrows}} EI \tag{5.3}$$

The kinetics equation is as follows:

$$-r_S = \frac{v_{max}C_s}{K_m\left(1 + \frac{C_I}{K_I}\right) + C_S} \quad \text{where} \quad K_I = \frac{k_{-3}}{k_3} \tag{5.4}$$

3. Biomass and waste based processes

Uncompetitive inhibition: The inhibitor binds to the ES complex, making up the enzyme–substrate–inhibitor complex, and preventing conversion to product (P).

$$E + S \underset{k_{-1}}{\overset{k_1}{\rightleftarrows}} ES \overset{k_2}{\rightarrow} E + P$$

$$ES + I \underset{k_{-4}}{\overset{k_4}{\rightleftarrows}} ESI \tag{5.5}$$

In this case, the global kinetics equation becomes:

$$-r_S = \frac{v_{max} C_s}{K_m + C_S\left(1 + \frac{C_I}{K_I}\right)} \quad \text{where} \quad K_I = \frac{k_{-4}}{k_4} \tag{5.6}$$

Noncompetitive inhibition: In this case, the inhibitor can bind to either free enzyme or ES complex, and likewise, the substrate can bind to free enzyme or the enzyme–inhibitor complex.

$$E + S \underset{k_{-1}}{\overset{k_1}{\rightleftarrows}} ES \overset{k_2}{\rightarrow} E + P$$

$$E + I \underset{k_{-3}}{\overset{k_3}{\rightleftarrows}} EI$$

$$ES + I \underset{k_{-4}}{\overset{k_4}{\rightleftarrows}} ESI \tag{5.7}$$

$$EI + S \underset{k_{-5}}{\overset{k_5}{\rightleftarrows}} ESI$$

The kinetic equation of this situation is given by Eq. (5.8)

$$-r_S = \frac{\dfrac{v_{max} C_s}{\left(1 + \frac{C_I}{K_I}\right)}}{K_m + C_S} \quad \text{where} \quad K_I = \frac{k_{-3}}{k_3} = \frac{k_{-4}}{k_4} \tag{5.8}$$

In this case, the inhibitor affinity is the same for the enzyme and the ES complex.

Mixed inhibition: It is the same as the noncompetitive inhibition but the inhibitor affinity is different for the enzyme and ES complex. The kinetic equation in this case is given by Eq. (5.9):

$$-r_S = \frac{v_{max}C_s}{K_m\left(1 + \frac{C_I}{K_{IE}}\right) + C_S\left(1 + \frac{C_I}{K_{IES}}\right)} \tag{5.9}$$

$$\text{where } K_{IE} = \frac{k_{-3}}{k_3}; K_{IES} = \frac{k_{-4}}{k_4}$$

Substrate inhibition: However, sometimes the substrate behaves as inhibitor binding to the ES complex and preventing conversion to product.

$$E + S \underset{k_{-1}}{\overset{k_1}{\rightleftarrows}} ES \overset{k_2}{\rightarrow} E + P$$

$$ES + S \underset{k_{-6}}{\overset{k_6}{\rightleftarrows}} ES_2 \tag{5.10}$$

In this case, the kinetics becomes:

$$-r_S = \frac{v_{max}C_s}{K_m + C_s + \frac{C_s^2}{K_I}} \quad \text{where} \quad K_I = \frac{k_{-6}}{k_6} \tag{5.11}$$

Thus, for the complete process from starch to glucose, the reactor model using Michaelis Menten mechanism with uncompetitive product inhibition would be given by Eq. (5.12) (Vrsalovic Presecki et al., 2008):

$$-\frac{dS}{dt} = r$$

$$\frac{dP_{Glucose}}{dt} = Y_{glucose/starch} \cdot r$$

$$\frac{dP_{maltose}}{dt} = Y_{maltose/starch} \cdot r$$

$$r = \frac{\varphi \cdot V \cdot S}{K_m + S\left(1 + \frac{P_{glucose}}{K_i^{glucose}} + \frac{P_{maltose}}{K_i^{maltose}}\right)} \tag{5.12}$$

$$\frac{dV}{dt} = -k_d^{\alpha-amylase}V$$

3. Biomass and waste based processes

where $k_m = 18.8$ g/L, $V = 6230.4$ mg/cm^3 min, $K_i^{glucose} = 101$ g/L, $K_i^{maltose} = 34.4$ g/L, $Y_{glucose/starch} = 0.12$ g/g, $Y_{maltose/sStarch} = 0.22$ g/g, $\varphi = 0.005$, $k_d = 0.0067$ min^{-1}.

5.1.2 Lignocellulosic sugar production

The depolymerization consists of a series of stages. First, a pretreatment breaks the physical structure of the biomass. It can be carried out following different pretreatments (Sun and Cheng, 2002). Among them, **chemical treatments** such as dilute acid, organosolv, **physicochemical treatments** such as ammonia fiber explosion, steam explosion, and **biological** ones have attracted most of the attention. Next, hydrolysis of the hemicellulose and cellulose releases the sugars. Depending on the pretreatment, the hemicellulose can already be broken down into its monomers. Cellulose is a polymer of glucoses. However, the hemicellulose is made of D-xylose, mannose, galactose, L-arabinose, and glucuronic acid. The structural units of the lignin are syringyl, guaiacyl, and para-hydroxy-phenyl.

5.1.2.1 Pretreatments

In this section, three major pretreatments are discussed, the ammonia fiber explosion (AFEX), the dilute acid with steam explosion, and organosolv because of their industrial use.

Physicochemical processes: They use carbon dioxide, steam, or ammonia under pressure so that when released, the biomass structure is broken. Among the different processes, AFEX is one of the few that have reached industrial development. In this case, the biomass is exposed to a solution of ammonia and water at moderate temperature and pressure for certain time. Pressure around 20 bar and temperatures from 90°C to 180°C are typically used. Then, pressure is reduced, and the biomass lignin fraction is reduced while the cellulose and hemicellulose fractions remain intact for hydrolysis. The advantage of this method is that it does not produce inhibitory by-products such as furans and shows high yield. The method was developed by Prof. Dale's Group (Holtzapple et al., 1992). The recovery of ammonia has changed over the years. In the beginning, after the expansion, the ammonia gas was recycled, recompressed, and condensed. However, gas compression (Fig. 5.2A) is expensive and alternatively, ammonia absorption in water was considered

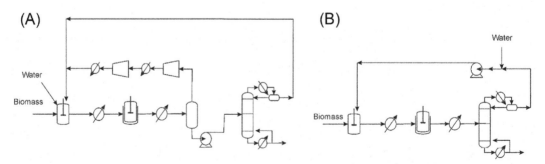

FIGURE 5.2   AFEX pretreatment. (A) Ammonia recompression and (B) ammonia absorption and liquid pumping.

(Fig. 5.2B). Water is needed for the pretreatment was fed to the reactor and that allowed the compression of a liquid using a pump.

The yield for sugar polymers recovery is a function of the operating temperature, the residence time (RT), and the ratios of ammonia and water added. Using the experimental results in Garlock et al. (2012) paper, a surface response model was developed to predict the yield as a function of the operating variables, Eq. (5.13) (Martín and Grossmann, 2012b). Table 5.1 presents the range of the variables involved. AFEX is more effective on biomasses with low-lignin content.

$$X = 0.01 \cdot (-88.7919 + 26.5272 \cdot AR - 13.6733 \cdot \text{water\_added} + 1.6561 \cdot T$$

$$+ 3.6793 \cdot t - 4.4631 \cdot AR^2 - 0.0057 \cdot T^2 + 0.0279 \cdot t^2 - 0.4064 \cdot AR \cdot t$$

$$+ 0.1239 \cdot \text{water\_added} \cdot T - 0.0132 \cdot T \cdot t; \tag{5.13}$$

Chemical pretreatment: This type of pretreatments uses alkali or acid solutions, peroxides, and ozone or solvents to degrade the biomass structure. Among them, two pretreatments have achieved industrial maturity such as dilute acid and organosolv.

- The National Energy Renewable Lab (NREL) has evaluated and scaled-up a pretreatment that is actually a combination between a physical one, due to the use of steam, and a chemical one, because a dilute solution of sulfuric acid pretreatment is also used (Aden and Foust, 2009; Kazi et al., 2010). The pretreatment consists of putting into contact the biomass with a solution of 0.5%–2% H_2SO_4, at 140°C–180°C for up to 1.5 hours and 12 atm. Next, the reactor is discharged so that a fraction of the water is evaporated, while the slurry containing the biomass depolymerized is further processed. The steam can be condensed and recycled. The slurry is sent to a centrifuge to separate the solids, cellulose, from the liquid phase containing the lignin and the hemicellulose. The liquid phase is neutralized with CaO, producing gypsum. In this treatment, a fraction of the hemicellulose can be lost, from 1% to 10%. Gypsum precipitates and is recovered in a filter. Finally, both streams can be either used separately or mixed, see Fig. 5.3

Based on experimental data, Martín and Grossmann (2014) developed models for the yield to cellulose and to hemicellulose as a function of the temperature of operation, the acid concentration (AC), the RT, and the enzyme added in the hydrolysis part. Table 5.2 presents the range for the operating variables.

TABLE 5.1 Range of operating variables for AFEX pretreatment.

	Lower bound	Upper bound
T (°C)	90	180
Ammonia added (g/g)	0.5	2
Water added (g/g)	0.5	2
Residence time (min)	5	30

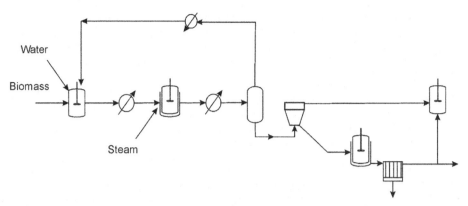

FIGURE 5.3   A basic flowsheet for dilute acid pretreatment.

TABLE 5.2   Range of operating variables for dilute acid pretreatment.

	Lower bound	Upper bound
T (°C)	140	180
Acid concentration (%)	0.5	2
Residence time (min)	1	80
Enzyme load (mg/g)	4.8	96.6

$$X_{cellulose} = -0.00055171 + 0.00355819 \cdot T + 0.00067402 \cdot AC$$
$$+ 0.00100531 \cdot RT - 0.0394809\ EL - 0.0186704 \cdot T \cdot AC$$
$$+ 0.00043556 \cdot T \cdot RT + 0.0002265 \cdot T \cdot EL$$
$$- 0.0013224 \cdot AC \cdot RT - 0.00083728 \cdot RT \cdot EL$$
$$+ 0.044353 \cdot AC \cdot EL + 0.000014412 \cdot T^2; \tag{5.14}$$

$$X_{hemicellulose} = -0.00015791 - 0.00056353 \cdot T + 0.000694361 \cdot AC$$
$$- 0.00014507 \cdot RT - 0.01059248 \cdot EL - 0.02142606 \cdot T \cdot AC$$
$$+ 0.000694055 \cdot T \cdot RT + 0.00013559 \cdot T \cdot EL$$
$$- 0.00145712 \cdot AC \cdot RT + 0.04769633 \cdot AC \cdot EL$$
$$- 0.00138362 \cdot RT \cdot EL + 0.0000059419 \cdot T^2 \tag{5.15}$$

Organosolv: It uses a solvent to break the biomass structure. Among them, alcohols such as ethanol or methanol, as well as acetone, ethylene, and glycol are the most used. The use of the solvent solubilizes the lignin and a fraction of the hemicellulose, while the cellulose remains in the solid phase. The liquid phase is further processed to recover the solvent generating a solid phase, mostly the lignin, and the liquid phase consists of the hemicellulose and other chemicals. The typical operating conditions are 160°C–200°C (Zhang et al., 2016). The demonstration and industrial scale facility was presented by Organocell, Alcell, or Lignol projects (Zhang et al., 2016; United States DOE, 2011). The yield to products depends on the solvent used, the AC, the temperature of operation, and the RT as well as on the biomass and operating conditions. An example for coffee pulp waste using ethanol as solvent with T (°C) and t (h) (Lini et al., 2018).

$$Yield = 1.912 + 7.595 \times 10^{-5}\text{pH} \cdot T \cdot t - 5 \times 10^{-4}\text{pH}^2\, T - 3.377 \times 10^{-4}\text{pH}^2\, t$$

$$- 1.6 \times 10^{-4}\text{pH} \cdot T^2 + 0.021\ \text{pH}^2 + 7.620 \times 10^{-4}\, T^2$$

$$+ 1.434 \times 10^{-4}\, t^2 + 0.018\ \text{pH} \cdot T - 1.513 \times 10^{-4}\text{pH} \cdot t$$

$$-3.550 \times 10^{-4}T \cdot t - 0.468\ \text{pH} - 0.074\, T + 3.583 \times 10^{-3}t \tag{5.16}$$

Note that in the most of the cases, the actual yield reported by the studies corresponds to the full process including hydrolysis because before it is not always possible to measure all the sugars since it has been described the cellulose remains as a solid after pretreatment. The economics (Eggeman and Elander, 2005; Martín and Grossmann, 2012a) are slightly in favor of the dilute acid.

5.1.2.2 Hydrolysis

The hydrolysis is the final stage by which the sugar polymers are broken into the monomers. It is an endothermic process that consumes water. Acid and enzymatic hydrolysis are presented later.

$$(C_6H_{10}O_5)_n + nH_2O \rightarrow nC_6H_{12}O_6 \quad \Delta H = 22.1\ n \quad \text{kJ/mol}$$

$$(C_5H_8O_4)_m + mH_2O \rightarrow mC_5H_{10}O_5 \quad \Delta H = 79.0\ m \quad \text{kJ/mol}$$

5.1.2.2.1 Dilute acid

The mechanism of the acid hydrolysis is that of a series reaction where the first stage is the desired one, the breakdown of the cellulose or hemicellulose into the sugars, while the second stage corresponds to the decomposition of the sugars into undesirable chemicals such as furans (Lenihan et al., 2011). It is important to mention that furfural and furan can be perfectly acceptable platform chemicals, see Chapter 7, but if the sugar is the target product, they must be minimized. It is typically carried out at moderate temperature but we risk the production of decomposition products.

$$Glucan \xrightarrow{k_1} glucose \xrightarrow{k} decomposition$$

As any other reaction in series, the mechanism and the kinetics are presented in Eq. (5.17). The value for the kinetic constants can be seen in Table 5.3.

$$(-r_{Sugar}) = k_2 C_{Sugar} - k_1 C_{Polymer}$$

$$(-r_{sugar}) = -\frac{1}{V}\frac{dN_{Sugar}}{dt}$$

$$\frac{dN_{Sugar}}{dt} = -k_2 N_{Sugar} + k_1 N_{Polymer}$$

$$N_{Polymer} = N_{Polyo} \cdot e^{-(k_1)t}$$

$$\frac{dN_{Sugar}}{dt} = -k_2 N_{Sugar} + k_1 N_{Polyo} \cdot e^{-(k_1)t} \tag{5.17}$$

$$\frac{dy}{dx} + P(x)y = Q(x)$$

$$N_{Sugar} = y; t = x; k_2 = P(x); k_1 N_{Poly0} \cdot e^{-k_1 t} = Q(x)$$

$$y = e^{-\int P(x)dx}\left[\int e^{\int P(x)dx} Q(x)dx + K\right]$$

$$N_{Sugar} = \left(\frac{k_1 N_{Poly0}}{k_2 - k_1}\right) \cdot e^{-k_1 t} + K \cdot e^{-k_2 t}$$

$$t = 0, N_{Sugar} = N_{Decomp} = 0 \text{ and } N_{Polymer} = N_{Poly0}$$

$$K = -\frac{k_1 N_{Poly0}}{k_2 - k_1}$$

$$N_{Sugar} = \left(\frac{k_1 N_{Poly0}}{k_2 - k_1}\right) \cdot \left[e^{-k_1 t} - e^{-k_2 t}\right]$$

An alternative model considers that only a fraction of the polymer reacts. Therefore a parameter α is used to correct the amount of monomer produced.

The optimization of the equation corresponding to the profile of glucose can be analytically computed as follows, see Example 5.1.

5.1.2.2.2 Enzymatic hydrolysis

The enzymatic typically takes place at 50°C and lasts 48−72 hours. In this case, the mechanism is shown in Fig. 5.4 (Obnamia, 2014):

TABLE 5.3 Kinetic constants for the acid hydrolysis (Lenihan et al., 2011).

Glucose	135°C	150°C	175°C	200°C
k_1 (min^{-1})	0.03731	0.0841	0.2772	0.5438
k_2 (min^{-1})	0.02002	0.0444	0.2054	0.5213

EXAMPLE 5.1 Optimize the operating time for the production of glucose.

$$\frac{d\left(\left[\frac{k_1 N_{A0}}{k_2 - k_1}\right] \cdot [e^{-k_1 t} - e^{-k_2 t}]\right)}{dt} = 0 \Rightarrow \left(\frac{k_1 N_{A0}}{k_2 - k_1}\right)(-k_1 e^{-k_1 t} + k_2 e^{-k_2 t}) = 0$$

$$k_1 e^{-k_1 t} = k_2 e^{-k_2 t} \Rightarrow \left(\frac{k_1}{k_2}\right) = \frac{e^{-k_2 t}}{e^{-k_1 t}} = e^{-(k_2 - k_1)t} \Rightarrow -(k_2 - k_1) \cdot t = \ln\frac{k_1}{k_2}$$

(5E1.1)

$$t = \frac{\ln(k_1/k_2)}{k_1 - k_2} =$$

$$N_{R,max} = \left(\frac{k_1 N_{A0}}{k_2 - k_1}\right) \cdot \left[e^{-k_1\left(\frac{\ln(k_1/k_2)}{k_1 - k_2}\right)} - e^{-k_2\left(\frac{\ln(k_1/k_2)}{k_1 - k_2}\right)}\right]$$

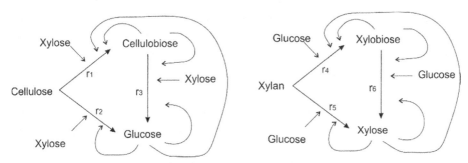

FIGURE 5.4 Reaction network for modeling cellulose and xylan hydrolysis.

Based on the mechanism shown in Fig. 5.4, the kinetics for the enzymatic hydrolysis of cellulose and xylan, the polymer for xylose is presented below (Kadam et al., 2004) where

E_a	activation energy (cal/mol); 1 cal = 4.184 J
E_{1T}	total enzyme concentration for CBH and EG (g protein/kg system)
E_{2T}	total enzyme concentration for βG (g protein/kg system)
E_{3T}	total enzyme concentration for EX (g protein/kg system)
E_{4T}	total enzyme concentration for βX (g protein/kg system)
E_{jB}	bound enzyme concentration (g protein/kg system); $j = 1, 2, 3,$ or 4
E_{jF}	free enzyme concentration (g protein/kg system); $j = 1, 2, 3,$ or 4
E_{jmax}	maximum enzyme that adsorbs to substrate (g protein/g substrate); $j = 1, 2, 3,$ or 4

E_{dose}	total enzymes added (g protein/g cellulose)
S, G_2, G	cellulose, cellobiose, and glucose concentration (g/kg system)
X_n, X_2, X	xylan, xylobiose, and xylose concentration (g/kg system)
S_0	initial cellulose concentration (g cellulose/kg system)
X_{n0}	initial xylan concentration (g xylan/kg system)
K_{jad}	ES dissociation constant (g protein/g substrate); $j = 1, 2, 3,$ or 4
k_{ir}	reaction rate constants (kg system/g protein h); i = rate Eq. 5.18
k_{ir}	reaction rate constants (g substrate/g protein h); i = rate Eq. 5.18
K_{iIG}	inhibition constants for glucose (g glucose/kg system); i = rate Eq. 5.18
K_{iIG2}	inhibition constants for cellobiose (g cellobiose/kg system); i = rate Eq. 5.18
K_{iIX}	inhibition constants for xylose (g xylose/kg system); i = rate Eq. 5.18
K_{iIX2}	inhibition constants for xylobiose (g xylobiose/kg system); i = rate Eq. 5.18
K_{3M}	cellobiose (substrate) saturation constant (g cellobiose/kg system)
K_{6M}	xylobiose (substrate) saturation constant (g xylobiose/kg system)
r_i	reaction rate equation (g substrate/kg system h); i = rate Eq. 5.18
R	universal gas constant (8.314 J/mol K)
R_S, R_X	substrate reactivity
T	temperature (K)
α, γ	substrate reactivity constant (dimensionless)

And the values obtained by fitting the model to experimental values result in Table 5.4 for the case of glucose.

TABLE 5.4 Values for the kinetics constants of the model (Obnamia, 2014).

Parameter	Value	Parameter	Value
K_{1ad}	0.4 (g protein/g cellulose)	K_{1I}	11.7×10^{-3} (g xylose/kg system)
K_{2ad}	0.1 (g protein/g cellobiose)	k_{2r}	4.8 (kg system/g protein h)
E_{1max}	0.06 (g protein/g cellulose)	K_{2G2}	0.612 (g cellobiose/kg system)
E_{2max}	0.01 (g protein/g cellobiose)	K_{2G}	0.140 (g glucose/kg system)
E_a	5540 (cal/mol)	K_{2X}	0.570 (g xylose /kg system)
α	1 (dimensionless)	k_{3r}	27.2 (g G2/g βG h)
k_{1r}	184.9 (kg system/g protein h)	K_{3M}	12.0×10^{-5} (g cellobiose/kg system)
K_{1G2}	98.0×10^{-5} (g cellobiose/kg system)	K_{3G}	13.0×10^{-5} (g glucose/kg system)
K_{1G}	0.00109 (g glucose/kg system)	K_{3X}	2.84×10^{-3} (g xylose/kg system)

$$MBCellulose: \frac{dS}{dt} = -r_1 - r_2$$

$$MBCellobiose: \frac{dG_2}{dt} = \left(\frac{342.3}{324.28}\right)r_1 - r_3$$

$$MBGlucose: \frac{dG}{dt} = \frac{180.16}{162.14}r_2 + \frac{360.31}{342.30}r_3$$

$$MBXylan: \frac{dX_n}{dt} = -r_4 - r_5$$

$$MBXylobiose\frac{dX_2}{dt} = \left(\frac{282.24}{264.22}\right)r_4 - r_6$$

$$MBLXylose: \frac{dX}{dt} = \frac{150.13}{132.11}r_5 + \frac{300.26}{282.24}r_6$$

$$Water: W = W_o - \frac{18.02}{342.30}G_2 - \frac{18.02}{180.16}G - \frac{18.02}{282.24}X_2 - \frac{18.02}{150.13}X$$

$$(5.18)$$

$$r_1 = \frac{k_{1r}E_{1B}R_SS}{1 + \frac{G_2}{K_{1/G2}} + \frac{G}{K_{1/G}} + \frac{X}{K_{1/X}}}; r_2 = \frac{k_{2r}(E_{1B} + E_{2B})R_SS}{1 + \frac{G_2}{K_{2/G2}} + \frac{G}{K_{2/G}} + \frac{X}{K_{2/X}}};$$

$$r_3 = \frac{k_{3r}(E_{2F})G_2}{G_2 + K_{3M}\left(1 + \frac{G}{K_{3/G}} + \frac{X}{K_{3/X}}\right)} r_4 = \frac{k_{4r}E_{3B}R_XX_n}{1 + \frac{X_2}{K_{4/X2}} + \frac{X}{K_{4/X}} + \frac{G}{K_{4/G}}};$$

$$r_5 = \frac{k_{5r}(E_{3B} + E_{4B})R_XX_n}{1 + \frac{X_2}{K_{5/X2}} + \frac{X}{K_{5/X}} + \frac{G}{K_{5/G}}}; r_6 = \frac{k_{6r}(E_{4F})X_2}{X_2 + K_{6M}\left(1 + \frac{X}{K_{6/X}} + \frac{G}{K_{6/G}}\right)}$$

$$E_{iB} = \frac{E_{imax}K_{iad}E_{iF}S}{1 + K_{iad}E_{1F}}$$

$$E_{iT} = E_{iF} + E_{iB}k_{ir(T2)} = k_{ir(T)} \cdot e^{-\frac{E_a}{R}\left(\frac{1}{T} - \frac{1}{T_2}\right)} R_S = \alpha\frac{S}{S_o} R_X = Y\frac{X}{X_{n0}}$$

5.1.3 Sugars fermentation

Once the sugars are available, they are an interesting raw material for the production of a number of chemicals. Biorefineries were based on the production of alcohols such as ethanol (Kazi et al., 2010) or butanol (Qureshi et al., 2013; Malmierca et al., 2017). However, platform chemicals such as furans can also be produced out of the sugars (Martin and Grossmann, 2015). In this chapter, we focus on the alcohols synthesis.

5.1.3.1 Ethanol production

The production of ethanol from the fermentation of sugars takes place under anerobic conditions operating at 28°C–38°C and 1.2 bar to avoid the entrance of air using *Zymomonas mobillis* bacterium (Wooley et al., 1999). The process is exothermic and its low operating temperature is a challenge resulting in the need for a large flow of cooling water. The main reactions are as follows for xylose and glucose, respectively. However, a large number of chemicals are produced together with ethanol. Table 5.5 presents the reactions including the growing of the bacterium that in its metabolism generates the ethanol. Nutrients, ammonia, are needed. This is an important issue since, in case of using AFEX, a certain amount of ammonia can be already available in the broth. The fermentation lasts 24–36 hours.

$$3C_5H_{10}O_5 \xrightarrow{yeast} 5C_2H_5OH + 5CO_2 \quad \Delta H = -74.986 \text{ kJ/mol}_{xylose}$$

$$C_6H_{12}O_6 \xrightarrow{yeast} 2C_2H_5OH + 2CO_2 \quad \Delta H = -84.394 \text{ kJ/mol}$$

The reactor is a jacketed stirred one that can also use a coil to remove the fermentation heat. The reactor is designed based on the time it takes the reaction to achieve the conversion. However, apart from the volume, the agitation and the heat transfer must be addressed. The reactor is a vertical vessel with two heads. Thus, the volume is that of the cylinder:

$$V_{Cylinder} = H\pi \left(\frac{D}{2}\right)^2 \tag{5.19}$$

TABLE 5.5   Stoichiometry and by-products in the production of ethanol (Wooley et al., 1999).

Reaction	Conversion
Glucose → 2 Ethanol + 2CO$_2$	Glucose : 0.92
Glucose + 1.2NH$_3$ → 6 Z. *mobilis* + 2.4 H$_2$O + 0.3O$_2$	Glucose : 0.035
Glucose + 2H$_2$O → glycerol + O$_2$	Glucose : 0.002
Glucose + 2CO$_2$ → 2 succinic acid + O$_2$	Glucose : 0.008
Glucose → 3 acetic acid	Glucose : 0.022
Glucose → 2 lactic acid	Glucose : 0.013
3 Xylose → 5 ethanol + 5CO$_2$	Xylose : 0.8
Xylose + NH$_3$ → 5 Z. *mobilis* + 2H$_2$O + 0.25O$_2$	Xylose : 0.03
3 Xylose + 5H$_2$O → 5 glycerol + 2.5O$_2$	Xylose : 0.02
3 Xylose + 5CO$_2$ → 5 succinic acid + 2.5O$_2$	Xylose : 0.03
2 Xylose → 5 acetic acid	Xylose : 0.01
3 Xylose → 5 lactic acid	Xylose : 0.01

where D and H are the diameter and the height, respectively. Hemisphere or elliptic heads can be used. In particular, we consider the elliptic one whose volume is as follows (Buthod, 1994):

$$V_{Head} = \frac{\pi}{24} \cdot D^3 \tag{5.20}$$

$$A_{Head} = 1.08 \cdot D^2 \tag{5.21}$$

where the height of the head, h, is computed as in Eq. (5.22):

$$h = 0.25 \cdot D \tag{5.22}$$

$$V_{Reactor} = \left(2 \cdot \frac{\pi}{24}D^3 + H\pi \cdot \left(\frac{D}{2}\right)^2\right) \tag{5.23}$$

We need to make sure that only two-thirds of the volume is filled due to the generation of gas along the reaction.

The stirring is typically scaled-up from experimental data based on the similarity. Even though the synthesis is anaerobic, the dissolved oxygen can affect the productivity of several compounds. About $20-60$ W/m^3 are typically required (Wooley et al., 1999).

The heat transfer is computed as any heat exchanger. Note that enough transfer area is required.

$$Q \cdot = U \cdot A_{Needed} \cdot \Delta T \tag{5.24}$$

where the area depends on the geometry of the vessel, Eq. (5.25):

$$A_{Geometry} = 1.08 \cdot D^2 + \pi \cdot D \cdot H_{liquid} \tag{5.25}$$

Alternatively is that of the coil. Logarithmic temperature gradient is used. Note that inside the reactor, a constant temperature is to be maintained around 38°C.

$$\Delta T = \frac{T_{W,out} - T_{W,in}}{\text{Ln}\left[\frac{T_{fermentor} - T_{W,in}}{T_{fermentor} - T_{W,out}}\right]} \tag{5.26}$$

The global heat transfer coefficient is computed considering five terms, the convective terms inside, h_i, the fermenter and at the coil or jacket, h_{out}, the fouling resistances, and the conduction across the wall as follows:

$$\frac{1}{U} = \frac{1}{h_i} + eff_{in} + \frac{x}{k} + eff_{out} + \frac{1}{h_{out}} \tag{5.27}$$

$eff_{in} = eff_{out} = 0.0005$ m^2 K/W (Branan, 2000).

For stainless steel, $k = 50$ W/(m K) (Sinnott, 1999).

The internal film coefficient can be computed as a function of the impeller and using correlations as the one presented in Eq. (5.31) (Sinnott, 1999).

$$h_i = \frac{k_f}{D}0.74Re^{0.67} \cdot Pr^{0.33} \cdot \left(\frac{\mu}{\mu_w}\right)^{0.14} \tag{5.28}$$

where Prandtl and Reynolds numbers are defined as

$$Pr = \frac{C_p \cdot \mu}{k_f} \tag{5.29}$$

$$Re = \frac{\rho \cdot N \cdot D_{impeller}^2}{\mu} \tag{5.30}$$

with N is the impeller revolutions, ρ, k_f, C_p, and μ are the density, conductivity, the heat capacity, and viscosity of the liquid respectively, and D is the diameter of the impeller.

The jacket film coefficient can be computed as (Sinnott, 1999):

$$Nu = \frac{h_{out} D_e}{k_f} = 0.023 \cdot Re^{0.8} \cdot Pr^{0.33} \tag{5.31}$$

where the cooling water velocity depends on the structure of the jacket and D_e is the equivalent diameter of the wetted perimeter. Alternatively, the diameter of a pipe is used if a coil is used (Wooley et al., 1999).

The time and the yield are computed from the kinetics. The following example presents the kinetics of ethanol production via xylose and glucose fermentation, Example 5.2.

EXAMPLE 5.2 Model the kinetics of xylose and glucose based on the mechanism Krishnan et al. (1999).

Solution

The kinetic model for ethanol production via a modified Monod type for expressions (Mulchandani and Loung, 1989):

$$E + S \underset{k_{-1}}{\overset{k_1}{\rightleftarrows}} ES \xrightarrow{k_2} E + P \tag{5E2.1}$$

In case of the production of ethanol, there is substrate and product inhibition.

$$EI + S \underset{k_{-3}}{\overset{k_3}{\rightleftarrows}} E + S + I \underset{k_{-1}}{\overset{k_1}{\rightleftarrows}} ES + I \xrightarrow{k_2} E + P + I \tag{5E2.2}$$

Assuming steady state for $[E]$, $[EI]$, and finally $[ES]$, the modified Monod kinetics is given as follows:

$$\mu = \frac{\mu_m \cdot S}{K_s + S + S^2/K_i} \tag{5E2.3}$$

Next, the product (P) inhibition is included as follows:

$$\frac{\mu}{\mu_0} = \left(1 - \left(\frac{P}{P_m}\right)^{\beta}\right) \tag{5E2.4}$$

Thus the models for the kinetics of the different species involved are as follows, where G represents glucose and X, xylose. The parameters for the fermentation are given in Table 5E2.1 (Krishnan et al., 1999).

Cells:

$$\mu_g = \frac{\mu_{m,g} \cdot S}{K_{s,g} + S + S^2/K_{i,g}} \left(1 - \left(\frac{P}{P_m}\right)^{\beta_g}\right)$$

$$\mu_x = \frac{\mu_{m,x} \cdot S}{K_{s,x} + S + S^2/K_{i,x}} \left(1 - \left(\frac{P}{P_m}\right)^{\beta_g}\right) \qquad (5E2.5)$$

$$\frac{1}{X}\frac{dX}{dt} = \frac{G}{G+X}\mu_g + \frac{X}{G+X}\mu_x$$

Product:

$$\nu_{E,g} = \frac{v_{m,g} \cdot S}{K_{s,g} + S + S^2/K_{i,g}} \left(1 - \left(\frac{P}{P_m}\right)^{\gamma_g}\right)$$

$$\nu_{E,x} = \frac{v_{m,x} \cdot S}{K_{s,x} + S + S^2/K_{i,x}} \left(1 - \left(\frac{P}{P_m}\right)^{\gamma_x}\right) \quad \text{with} \quad \begin{array}{l} \mu_{m,g} = 0.152 \cdot X^{-0.461} \\ \mu_{m,x} = 0.075 \cdot X^{-0.438} \\ v_{m,g} = 1.887 \cdot X^{-0.434} \\ v_{m,x} = 0.16 \cdot X^{-0.233} \end{array} \qquad (5E2.6)$$

$$\frac{1}{X}\frac{dP}{dt} = (\nu_{E,x} + \nu_{E,g})$$

Sustrate:

$$-\frac{dS}{dt} = \frac{1}{Y_{X/S}}\frac{dX}{dt} + mX = \frac{1}{Y_{P/S}}\frac{dP}{dt}$$

$$-\frac{dS}{dt} = \frac{1}{Y_{P/S}}\frac{dP}{dt}$$

$$-\frac{dxylo}{dt} = \frac{1}{Y_{P/S}}(\nu_{E,x}X) \qquad (5E2.7)$$

$$-\frac{dglu}{dt} = \frac{1}{Y_{P/S}}(\nu_{E,g}X)$$

The model can be written in any software to be solved see Fig. 5E2.1. The code in Python is presented below.

```python
import numpy as np
import matplotlib.pyplot as plt
from scipy import integrate
#[kJ/mol]
v_m_g = 2.005;
v_m_x = 0.250;
K_s_g = 0.565;
K_i_g = 283.7;
K_s_x = 3.4;
K_i_x = 18.1;
K_sp_g = 1.341;
```

TABLE 5E2.1   Kinetic parameters for fermentation.

Parameter	Glucose fermentation	Xylose fermentation
μ_m (h^{-1})	0.662	0.190
ν_m (h^{-1})	2.005	0.250
K_S (g/L)	0.565	3.400
K_S' (g/L)	1.342	3.400
K_i (g/L)	283.700	18.100
K_i' (g/L)	4890.000	81.300
P_m (g/L)	95.4 for $P \le 95.4$ g/L	
	129.9 for $95.4 \le P \le 129$ g/L	59.040
P_m' (g/L)	103 for $P \le 103$ g/L	
	136.4 for $103 \le P \le 136.4$ g/L	60.200
β	1.29 for $P \le 95.4$ g/L	
	0.25 for $95.4 \le P \le 129$ g/L	1.036
γ	1.42 for $P \le 95.4$ g/L	0.608
m (h^{-1})	0.097	0.067
$Y_{P/S}$ (g/g)	0.470	0.400
$Y_{X/S}$ (g/g)	0.115	0.162

```
K_ip_g = 4890;
K_sp_x = 3.4;
K_ip_x = 81.3;
m_g = 0.097;
m_x = 0.067;
Y_P_S_g = 0.470;
Y_P_S_x = 0.4;
P_m_x = 59.04;
P_mp_x = 60.2;
Beta_x = 1.036;
gamma_x = 0.608;
Xini = [200,50,0,1];
x = [(),(),(),()];
#We create an empty list to store the concentrations data
tau = np.linspace(0,36,250)
#Equations: We define a function which collect the equations to solve
def    kinetics(x,tau):
        Glucose = x[0];
        xylose = x[1];
        ethanol = x[2];
        cells = x[3];
        if ethanol <95.4:
```

```
                P_m_g = 95.4;
                Beta_g = 1.29;
                gamma_g = 1.42;
        else:
                P_m_g = 129;
                Beta_g = 0.25;
                gamma_g = 0;
        if ethanol <103:
                P_mp_g = 103;
        else:
                P_mp_g = 136;
        if cells < 5:
                mu_m_g = 0.152*(cells)**(-0.461);
                v_m_g = 1.887*(cells)**(-0.434);
                mu_m_x = 0.075*(cells)**(-0.438);
                v_m_x = 0.16*(cells)**(-0.233);
        else:
                mu_m_g = 0.662;
                v_m_g = 2.005;
                mu_m_x = 0.190;
                v_m_x = 0.25;
        mu_g = mu_m_g*Glucose*(1-(ethanol/P_m_g)**Beta_g)/
(K_s_g + Glucose + Glucose**2/K_i_g);
        mu_x = mu_m_x*xylose*(1-(ethanol/P_m_x)**Beta_x)/
(K_s_x + xylose + xylose**2/K_i_x);
        v_g = v_m_g*Glucose*(1-(ethanol/P_mp_g)**gamma_g)/
(K_sp_g + Glucose + Glucose**2/K_ip_g);
        v_x = v_m_x*xylose*(1-(ethanol/P_mp_x)**gamma_x)/(K_sp_x + xylose + xylose
**2/K_ip_x);
        dxdtau = np.zeros(4)
        dxdtau[0] = -(1/Y_P_S_g)*v_g*cells
        dxdtau[1] = -(1/Y_P_S_x)*v_x*cells
        dxdtau[2] = (v_g + v_x)*cells
        dxdtau[3] = cells*((Glucose)*mu_g/(Glucose + xylose) + (xylose)*mu_x/
(Glucose + xylose))
        return dxdtau
#We collect the results in a list calles Res
res_x = integrate.odeint(kinetics,Xini,tau)
plt.plot(tau,res_x[:,0],'k - ')
plt.plot(tau,res_x[:,1],'k.')
plt.plot(tau,res_x[:,2],'k - ')
plt.plot(tau,res_x[:,3],'k - .')
plt.xlabel('t(h)')
plt.ylabel('Concentration(g/L)')
plt.legend('GXEC')
```

3. Biomass and waste based processes

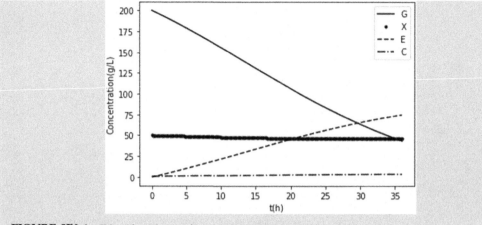

FIGURE 5E2.1   Ethanol production from hexoses and pentoses.

5.1.3.2 *Butanol production*

Butanol was deemed as an advanced fuel due to the higher energy density and its easy transportation, due to the lower vapor pressure, compared to ethanol. It has typically been produced following the acetone–butanol–ethanol (ABE) synthesis. The metabolism of the *Clostridium acetobutylicum* can be seen below (Fig. 5.5).

AB fermentations aim at improving the efficiency by producing mostly acetone and butanol (Qureshi and Blaschek, 1999) where the stoichiometry becomes (Malmierca et al., 2017):

$$100\ C_6H_{12}O_6 \rightarrow 82.68\ C_4H_{10}O + 23.15\ C_3H_6O + 2.03\ C_2H_6O + 195.76\ CO_2 + 10.44\ H_2 + 100.60\ H_2O$$

Different kinetic models can be found in the literature including complete ones such as those by Buehler and Mesbah (2016) and Raganati et al. (2015). We leave them for the reader to implement. Here we present a more recent effort but also simpler by Birgen et al. (2019) focused on the production of butanol alone.

$$Cells\,(X[g/L])$$

$$\frac{dX}{dt} = \mu_{net}X$$

$$\mu_{net} = \mu_g - k_d$$

$$\mu_g = (\mu_{SG} + \mu_{SX})\left(\frac{K_I}{K_I + SG + SX}\right)\left(1 - \frac{B}{B_{Max}}\right)^{i_B}$$

$$\mu_{SG} = \frac{\mu_{maxG} \cdot SG}{(K_{sG} + SG)\left(1 + \dfrac{SX}{K_{sX}}\right)}$$

$$\mu_{SX} = \frac{\mu_{maxX} \cdot SX}{(K_{sX} + SX)\left(1 + \dfrac{SG}{K_{sG}}\right)}$$

(5.32)

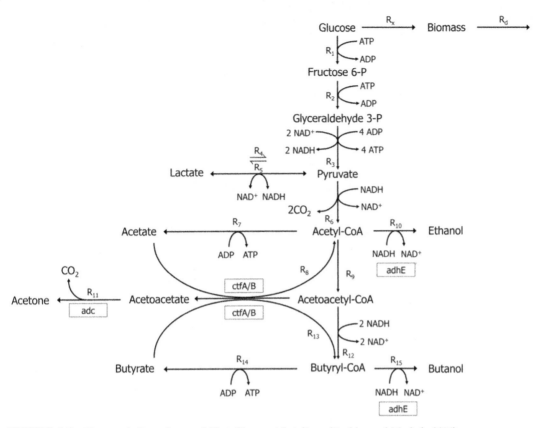

FIGURE 5.5 The metabolic pathway of *Clostridium acetobutylicum* (Buehler and Mesbah, 2016).

Net growth rate of cell mass, μ_{net}, is the difference between the specific growth rate and specific death rate, k_d. The specific growth rate is given by the consumption of glucose and xylose. Where μ_{maxG} and μ_{maxX} are maximum specific growth rates on glucose and xylose, and K_{sG} and K_{sX} are substrate affinity constants for glucose and xylose, respectively. μ_g is the specific growth rate of cell mass, K_I is the substrate inhibition constant, B_{max} is the concentration of butanol at which cell mass growth stops, and i_B is the butanol inhibition constant to cell mass growth.

The glucose and xylose uptakes are given in terms of the amounts of utilized cell mass growth.

Substrate Glucose $(SG [g/L])$

$$-\frac{dSG}{dt} = \mu_{SG}\left(\frac{1}{Y_{X/SG}} + \frac{1}{Y_{B/SG}}\right)X$$

Substrate Xylose $(SX [g/L])$

$$-\frac{dSX}{dt} = \mu_{SX}\left(\frac{1}{Y_{X/SX}} + \frac{1}{Y_{B/SX}}\right)X$$

(5.33)

where $Y_{X/SG}$ and $Y_{X/SX}$ are the cell yields on glucose and xylose and $Y_{B/SG}$ and $Y_{B/SX}$ are the butanol yield on glucose and xylose. The production of butanol is given as Eq. (5.34):

$$\frac{dB}{dt} = \left(\mu_{SG}Y_{B/XG} + \mu_{SX}Y_{B/XX}\right)X$$

$$Y_{B/XG} = \frac{Y_{B/SG}}{Y_{X/SG}}$$

$$Y_{B/XX} = \frac{Y_{B/SX}}{Y_{X/SX}}$$

(5.34)

Values for the parameters can be found in the literature. For our example we use the data in Table 5.6. Example 5.3 shows the kinetics.

TABLE 5.6   Kinetic parameters for the fermentation.

Parameter	Value
μ_{maxG} (h^{-1})	0.730
μ_{maxX} (h^{-1})	0.615
K_{SG} (g/L)	1.293
K_{SX} (g/L)	4.469
K_I (g/L)	171.492
i_B	0.616
B_{max} (g/L)	15.658
k_d (h^{-1})	0.076
$Y_{X/SG}$ (g/g)	0.523
$Y_{X/SX}$ (g/g)	0.058
$Y_{B/SG}$ (g/g)	0.196
$Y_{B/SX}$ (g/g)	0.232

EXAMPLE 5.3 Compute the concentration of butanol after an 8 hour fermentation of a broth of with initial concentrations of 25 g/L of glucose, 35 g/L of xylose, and 0.25 g/L of cells.

Solution

We implement the mode presented in the section above in Python. See the results in Fig. 5E3.1

```python
import numpy as np
import matplotlib.pyplot as plt
from scipy import integrate
#[kJ/mol]
#300°C       k en min⁻¹
mumaxG = 0.730;
mumaxX = 0.615;
Ksg = 1.293;
Ksx = 4.469;
Ki = 171.492;
iB = 0.616;
Bmax = 15.658;
kd = 0.076;
Yxsg = 0.523;
Yxsx = 0.058;
Ybsg = 0.196;
Ybsx = 0.232;
Ybxg = Ybsg/Yxsg;
Ybxx = Ybsx/Yxsx;
Xini = [0.25,25,30,0];
x = [(),(),(),()];
#We create an empty list to store the concentrations data
tau = np.linspace(0,100,500)
#Equations: We define a function which collect the equations to solve
def    kinetics(x,tau):
       muSG = (mumaxG*x[1])/((Ksg + x[1])*(1 + x[2]/Ksx));
       muSX = (mumaxX*x[2])/((Ksx + x[2])*(1 + x[1]/Ksg));
       mug = (muSG + muSX)*(Ki/(Ki + x[1] + x[2]))*(1 - x[3]/Bmax)**iB;
       munet = mug-kd;
       dxdtau = np.zeros(4)
       dxdtau[0] = munet*x[0]
       dxdtau[1] = - (muSG*(1/Yxsg + 1/Ybsg)*x[0])
       dxdtau[2] = - (muSX*(1/Yxsx + 1/Ybsx)*x[0])
       dxdtau[3] = (muSG*Ybxg + muSX*Ybxx)*x[0]
       return dxdtau
#We collect the results in a list calles Res
res_x = integrate.odeint(kinetics,Xini,tau)
plt.plot(tau,res_x[:,0],'k - ')
plt.plot(tau,res_x[:,1],'k - ')
plt.plot(tau,res_x[:,2],'k - .')
plt.plot(tau,res_x[:,3],'k - *')
plt.xlabel('t(h)')
plt.ylabel('x')
plt.legend('CGXB')
```

3. Biomass and waste based processes

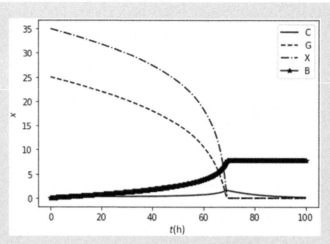

FIGURE 5E3.1 Profile of glucose (G), xylose (X), cells (C), and butanol (B).

5.1.4 Product recovery

In this chapter, we have focused on two products, biobutanol and bioethanol. In both cases, the mixture from the fermenter presents thermodynamic features that prevent from easy product purification in terms of nonidealities including azeotropes. We present the separation schemes for the purification of ethanol and butanol from the fermentation broth.

5.1.4.1 Ethanol recovery

In Chapter 4, we already presented the particular XY diagram for the water–ethanol mixture and the concept of multieffect column as well as the molecular sieves as final dehydration stage. Here we add to the description the pervaporation (Martín and Grossmann, 2012a; Wooley et al., 1999). Example 5.4 shows the performance of a multieffect column.

EXAMPLE 5.4 **Evaluate the steam consumption of a two-effect distillation column for the removal of water from a water–ethanol mixture 6%.**

Solution

For details on the modeling of the column, see Fig. 5E4.1, using CHEMCAD we refer to Martin (2019). We consider the NRTL thermodynamic model to represent the system.

100 kg/s of feed 6% ethanol is fed to the system. 52% of the feed goes to the low-pressure column, 0.4 bar, and the rest to the high-pressure column, at 1.3 bar see Fig. 5E4.1. The temperature profile along the columns is shown in Fig. 5E4.2. The major results for the operation of the low pressure column are $T_{LP, top} = 56.4°C$; $T_{LP, feed} = 70.3°C$; $T_{LP, Bottom} = 75.2°C$; $Q_{condenser} = 14.6$ MJ/s; $Q_{Reboiler} = 15.5$ MJ/s while for the high pressure column: $T_{HP, top} = 84.7°CT_{HP, feed} = 101.4°CT_{HP, Bottom} = 106.5°CQ_{condenser} = 15.4$ MJ/s$Q_{Reboiler} = 16.2$ MJ/s

---

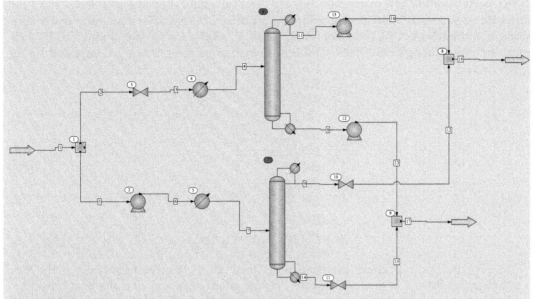

FIGURE 5E4.1 Scheme of the two-effect distillation column for ethanol purification.

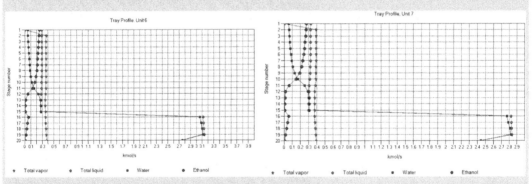

FIGURE 5E4.2 Profiles inside the columns.

We can see the temperature gradient along the columns and the match between the heat duty of the condenser of the higher-pressure column and the one of the reboiler of the lower-pressure one. See Chapter 7, for thermally coupled column design.

The final dehydration can be carried out using molecular sieves, as discussed in Chapter 4, where further details on the mechanism of water removal are presented or pervaporation. Here we discuss this technology.

3. Biomass and waste based processes

The pervaporation is a membrane, see Fig. 5.6. Typically, it operates at 50°C–90°C and permeate pressures of 3–30 mmHg achieving a purity of 99.8%. The maximum water composition at the inlet is 15% (Braisher, et al., 2006). A model for the membrane can be written as follows (Chang et al., 1998). The heat to evaporate the permeate is supplied by the liquid feed.

$$\frac{dF}{dA} = -J$$

$$\frac{dC_W^f}{dA} = J\frac{\left(C_W^f - C_W^p\right)}{F}$$

$$\frac{dT}{dA} = -J\frac{\lambda}{F \cdot cp}$$

(5.35)

$$\lambda = C_W^p \lambda_W + \left(1 - C_W^p\right)\lambda_e$$

$$cp = C_W^f cp_W + \left(1 - C_W^f\right)cp_e$$

The superindex p represents permeate and f feed. C_W is the water content, J is the flux, F is the species flow, T is the temperature, λ is the latent heat, and cp is the heat capacity. The flux has been modeled based on Fick's law as follows (Valentínyi et al., 2013):

$$J_i = \frac{1}{1 + \left(\dfrac{D_i}{Per \cdot p_{i,0}\gamma_i}\right)}\frac{D_i}{\gamma_i}\left(\frac{p_{i,f} - p_{i,p}}{p_{i,0}}\right) = K\left(p_{i,f} - p_{i,p}\right)$$

D: Difussion coefficient

(5.36)

Per: Permeability

$p_i \Rightarrow Antoine(T)$

Activity coefficients

$\gamma_i = \sqrt{\gamma_{i,f} \cdot \gamma_{i,p}}$

Average fluxes of 0.4 kg/m² h can be achieved with selectivities, α, of 200.

$$\alpha = \frac{y_{water}/y_{EtOH}|_{permeate}}{y_{water}/y_{EtOH}|_{feed}}$$

(5.37)

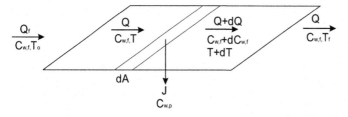

FIGURE 5.6 Membrane operation scheme.

5.1.4.2 *Butanol recovery*

The recovery of butanol from the fermentation broth is highly energy demanding due to the large content of water. Some studies using pure distillation process (Qureshi et al., 2005) show that the energy involved in that separation is often higher than the energy content of butanol. To reduce the energy consumption, complex hybrid processes involving both distillation and liquid—liquid extraction columns have been evaluated and proposed in the literature over the last 15 years. At laboratory scale, it has been possible to extract in situ the fermentation products. However, at industrial scale, this technology shows a number of challenges due to the slow mass transfer into solvent phase, the formation of emulsions while agitating, the possibility of cell inhibition by solvent and loss of cells at interphase, the fact that precipitates carry water into the solvent phase, scale up problems in terms of distribution coefficients and complex process control. As a result, external product removal still is considered the most promising alternative for large-scale operations. Fig. 5.7 shows the optimal flowsheet for the purification of an ABE mixture from the systematic work in Kraemer et al. (2011). A liquid—liquid extraction is used where water is mostly recovered from the bottoms. The solvent mesitylene is recovered in the atmospheric distillation column almost completely. The distillate is compressed to 2 bar and fed to a second distillation column operating at 2 bar in charge of recovering butanol from bottoms. The last distillation column operates at 0.5 bar and recovers acetone over the top. The model for the extraction column is based on liquid—liquid equilibrium on a tray-by-tray basis. Note that the feed is only present in one of the trays, for the rest it is zero.

Patrascu et al. (2017) developed intensification downstream process based on a dividing wall column for the separation of ABE mixtures. For further details on intensified schemes, we refer the reader to Chapter 7.

5.2 Lipid based

5.2.1 Algae growth

The increase in the demand for liquid fuels, and the limited yield and high land requirements for terrestrial biomass such as lignocellulosic species turned the focus into

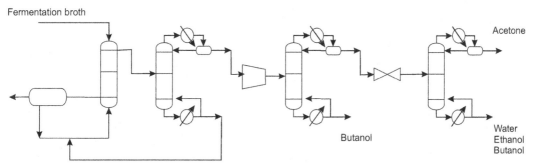

FIGURE 5.7   Extraction—distillation system for butanol recovery from ABE mixtures.

algae and microalgae, due to the far higher area yield, two orders of magnitude larger than other biomasses. Even though in the beginning, algae were considered a source for lipids, they can store up to 70% of their mass in the form of lipids, lately not only the starch, which can reach 50% by weight, but also other higher added value products such as carotenoids have been obtained in algae-based biorefineries.

Algae typically grow by absorbing CO_2 from the atmosphere performing photosynthesis, autotrophic. While this text focuses on this alternative, there are species that can use other sources of carbon and heterotrophic, and can grow on different carbon sources. The basic stoichiometry of the algae growing process is as follows:

$$106CO_2 + 16NO_3^- + HPO_4^{2-} + 122H_2O + 18H^+ \xrightarrow{\text{Solar radiation}} (CH_2O)_{106}(NH_3)_{16}H_3PO_4 + 138O_2$$

$$(5.38)$$

Apart from CO_2 and light, the growth of the algae depends on nutrients, phosphorous and nitrogen, as well as it is affected by the temperature and the pH of the medium. Although a Monod type of kinetics can be used to model its growth, Lee et al. (2015) present a review of kinetic models that are of the form:

$$\frac{dX}{dt} = (\mu_g - k_d)X$$

$$\mu_g = \frac{\mu_{max} \cdot S}{K_s + S}$$

$$(5.39)$$

For a typical algae species in biofuel production, Chlorella vulgaris, $\mu_{max} = 0.070 \text{ h}^{-1}$ and $K_s = 0.26 \text{ mg/L}$ when the growth is nutrient limited and under certain temperature and nutrient conditions. A detailed model including nutrient limitations can be found (Eze et al., 2018).

Accounting for all the variables within the specific growth rate, μ, is complex (Benson et al., 2016) and statistical analysis can be used to develop correlations for the growth rate as a function of nutrient concentration, total nitrogen (TotN, mg/L), total phosphorous (TotP, mg/L), weather conditions, temperature (T, °C) and light (μmol/m^2 s), and pH as follows (Taimbú de la Cruz et al., 2019):

$$\text{Growth}_N \left(\frac{g}{m^2 d} \right) = 0.418528 \cdot \text{TotP} + 0.52762 \cdot \text{TotN}$$

$$+ 0.225013 \cdot \text{TotN} \cdot \text{TotP} - 0.20754 \cdot \text{TotP}^2 - 0.03026 \cdot \text{TotN}^2 \quad (5.40)$$

$$\text{Growth}(g/m^2 \text{ d}) = \text{Growth}_N \frac{f(T, Light, pH)}{f_{ref}(25°C, 60 \, \mu mol/m^2/s, 7.4)}$$

$$f(T, Light, pH) = -156.77 + 0.21 \cdot T + 0.23 \cdot Light + 41.45 \cdot pH \quad (5.41)$$

$$+ 0.00000339 \cdot T \cdot Light + 0.022 \cdot pH + 0.0000181 \cdot pH \cdot Light$$

$$- 0.00863 \cdot T^2 - 0.0000509 \cdot Light^2 - 2.84 \cdot pH^2$$

The weather reports irradiance as kWh/m^2 d. Approximately, $4.57\ \mu mol/m^2$ s is equivalent to $1\ W/m^2$ and $1\ kWh/m^2$ day is equal to $41.7\ W/m^2$. Typical values of $20-50\ g/m^2$ d are optimistic but feasible.

The consumption of CO_2 is related to the algae growth as given by Eq. (5.42) (Sazdanoff, 2006):

$$CO_2\left(m^3/d\right) = 0.6565 \times Growth\left(g/m^2\ d\right) + 5.0784 \tag{5.42}$$

Algae are grown in ponds or photobioreactors (PBRs). The first ones are cheaper, civil engineering work only is needed. They consist of 20 cm deep channels where water is made to flow, and CO_2 and nutrients are injected. The standard size is $1000\ m^2$ with the geometry shown in Fig. 5.8. However, the contamination is likely to happen and the process control is difficult. In the case of operating ponds, $6.2\ m^3$ per day of water is lost by evaporation (Pate et al., 2007). The energy consumed by the pond system is 0.1 kW/ $(1000\ m^2)$ based on the result by Sazdanoff (2006).

On the other hand, the use of closed PBR systems allows a controlled environment such as tubular reactors, flat plane, air lift, and bubble columns but solar exposure is difficult, and the investment cost is higher. Fig. 5.8 shows examples of both systems, while air lifts and bubble columns are already presented in Chapter 4.

Once the algae are produced, the limiting stage for the production of oil in terms of cost is the harvesting of the algae. It typically consists of screening followed by thickening and dewatering and finally a drying stage so as to reach moistures below 10%, see Fig. 5.9 (Singh and Patidar, 2018; Barros et al., 2015). Table 5.7 presents the features of the different technologies for the harvesting and energy and yields (Vandamme, 2013; Fasaei et al., 2018).

The drying stage can be carried out in drum or spray dryers consuming 0.9 and $1.0-1.2\ kWh/kg$ of evaporated water, respectively.

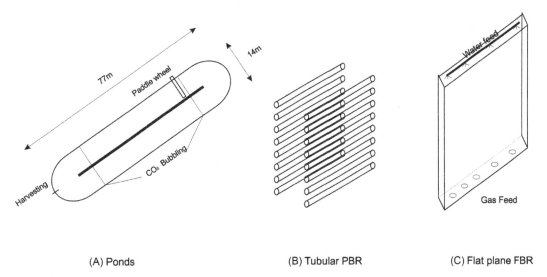

(A) Ponds   (B) Tubular PBR   (C) Flat plane FBR

FIGURE 5.8   Algae growth reactors. (A) Ponds, (B) tubular reactors, and (C) flat plane.

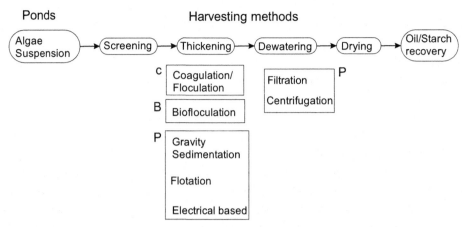

FIGURE 5.9   Harvesting stage. C, chemical; B, biochemical; P, physical.

TABLE 5.7   Summary of harvesting technologies (Vandamme, 2013; Fasaei et al., 2018).

Thickening		Dewatering	
Coagulation/ flocculation	*Advantages*:	Filtration	*Advantages*:
	Simple and fast		High efficiency
	Low energy requirements		Cost effective
	Easy for large scale		No need for chemicals
	Limited cell damage		Low energy consumption and shear stress
	Disadvantages:		Water recycle possible
	Expensive chemicals		*Disadvantages*:
	pH dependent		Slow requiring pressure o vacuum
	Limited recycle of culture medium		Not suitable for small algae
	Energy: 0.15 kWh/m^3 of feed		Membrane fouling and clogging resulting in high operating and maintenance costs
	Recovery: 80%–98% biomass		Pressure:
	Cationic:		Energy: 0.5–0.9 kWh/m^3 of feed
	Dosage: 165 mg/g of biomass		Recovery: 98% biomass
	Recovery: 80%–90% biomass		Vacuum:
			Energy 1.2–5.90 kWh/m^3 of feed
			Recovery: 98% biomass

(*Continued*)

3. Biomass and waste based processes

TABLE 5.7   (Continued)

Thickening		Dewatering	
Bioflocculation	*Advantages*:	Centrifugation	*Advantages*:
	Cheap		High efficiency
	Allows culture medium recycle		Fast method
	Nontoxic to algae		Suitable for all species of microalgae
	Disadvantages:		*Disadvantages*:
	Possible microbiological contamination		Expensive for large scale
	Changes in cellular composition		High energy consumption
	Chitosan		High operating and maintenance costs
	Dosage: 4–38 mg/g of biomass		Better suited to recover high added value products.
	Recovery: 85%–98% biomass		Time consuming
Sedimentation	*Advantages*:		Risk of cell destruction
	Simple and inexpensive		Energy: 0.7–1.3 kWh/m^3 of feed
	Disadvantages:		Recovery: 95%–99% biomass
	Time consuming		
	Possible biomass deterioration		
	Low concentration of the algal cake		
	Energy: 0.1–0.3 kWh/m^3 of feed		
Flotation	*Advantages*:		
	Flexible for large-scale operation		
	Low cost		
	Reduced space needs		
	Short operation time		
	Disadvantages:		
	Need for chemical flocculants		

(Continued)

3. Biomass and waste based processes

TABLE 5.7   (Continued)

Thickening	Dewatering
	Unfeasible for marine algae harvesting
	Energy: 0.015−1.5 kWh/m^3 of feed
Electrical-based processes	*Advantages*:
	Useful for all microalgae species
	No need for chemicals
	Disadvantages:
	Metal electrodes required
	Energy intense
	High units cost
	Metal contamination

Alternatively, capillarity-based process proposed by Univenture is supposed to obtain a biomass cake 5% moisture, with reduced energy consumption. The energy consumption is 0.288 kW/(kg/s of liquid stream processed) (Martín and Grossmann, 2012b).

5.2.2 Oil extraction

The dry biomass is processed to extract the oil. Typically, mechanical action and the use of solvents, that is, hexane, are used. It is possible to recover up to 95% of the oil. The mechanical action involves milling and grinding, similar to breakdown the biomass to make it accessible for further extraction. Values of 50.18−68 kWh/t for seeds are found in the literature (Kazmi, 2011). Next, the use of a solvent extracts the oil from the biomass. Typically 1:1 solvent to algae mass ratio is used (Martín and Grossmann, 2012b). Subsequently, the oil−solvent mixture is to be separated. It is important to mention that the oil cannot surpass 350°C to avoid decomposition. As a result, vacuum is typically required. To reduce the energy consumption of this stage, membranes are being evaluated to recover the solvent achieving fluxes of up to 6 kg/m^2 h and recoveries up to 93% (Weibin et al., 2011).

5.2.3 Synthesis

The oil has a high viscosity compared to diesel fuel. To reduce it, transesterification to transform the triglycerides into fatty acid esters is typically used. However, since the need is to break an organic molecule into smaller pieces, hydrotreating can also be used, see Chapter 4, for further details, as well as the production of emulsions. In this chapter, we focus on the transesterification route. It consists of the reaction of the oil with alcohols of short chain length obtaining the fatty acid esters and glycerol. The section of the alcohol has been based on economics and reaction kinetics. Methanol has been the most used one. It was cheap, and the kinetics was fast producing fatty acid methyl ester, biodiesel. However, the price was due to the source, fossil fuels. Lately, methanol has also been produced from renewable resources as presented in Chapter 4, and later in Chapter 11. The production of bioethanol within biorefineries has allowed the evaluation of the use of fatty acid ethyl esters. Regarding the reaction, we study its thermodynamics and kinetics.

$$
\begin{array}{ll}
\mathrm{CH_2-OOC-R_1} & \mathrm{R'-OOC-R_1} \quad \mathrm{CH_2-OH} \\
| & | \\
\mathrm{CH_2-OOC-R_2} + 3\mathrm{R'OH} \leftrightarrow & \mathrm{R'-OOC-R_2} + \quad \mathrm{CH-OH} \\
| & | \\
\mathrm{CH_2-OOC-R_3} & \mathrm{R'-OOC-R_3} + \quad \mathrm{CH_2-OH}
\end{array}
\tag{5.43}
$$

Thermodynamics of biodiesel production: The production of biodiesel is governed by a series of equilibria controlled by variables such as the alcohol to oil ratio, the operating temperature, pressure, and the RT in the case of supercritical noncatalyzed reactions, and the catalyst load is added to the list of variables for all the rest. We can classify the reaction between catalyzed and noncatalyzed under supercritical conditions. The last ones have not achieved industrial maturity. Among the catalyzed ones, it is possible to distinguish between homogeneous and heterogeneous whether the catalysis is in the same state of aggregation than the mixture or not (Martín and Grossmann, 2012b). At industrial scale, homogeneous alkali or acid catalysis has been used and in this text we focus on both for methanol and alkali for the use of ethanol. Further models for both alcohols and several catalysts, supercritical conditions can be found in the literature (Martín and Grossmann, 2012b; Severson et al., 2013).

Homogeneous alkali catalysts are the most used ones. However, in spite of the high conversion and short reaction times, they are sensitive to the presence of water and free fatty acids (FFAs). Therefore if the feedstock contains them, a pretreatment stage using an heterogeneous acid catalysis is required. The FFAs are esterified with an alcohol. The conversion of the FFAs can be computed as a function of the operating variables, temperature (T, °C), reaction time (time, hours), and methanol to oil ratio (RM, mol/mol) as follows:

$$
\begin{aligned}
X_{FFA} = {} & 31.03104 + 1.486403123 \cdot T - 6.97793097 \cdot RM + 19.77691899 \cdot \text{time} \\
& - 0.00018078 \cdot T^2 - 0.16677756 \cdot RM^2 - 1.6230585 \cdot \text{time}^2 \\
& + 0.02516368 \cdot T \cdot RM - 0.41625815 \cdot T \cdot \text{time} + 2.37322062 \cdot RM \cdot \text{time};
\end{aligned}
\tag{5.44}
$$

The feedstock either after pretreatment, if needed, or if it was free from FFAs is transesterified typically with NaOH or KOH. In most cases, NaOH has been used due to the lower price. However, the advantage of using KOH can be seen not in the reaction stage but in the biodiesel purification process. KOH can be neutralized with H_3PO_4 generating K_3PO_4 that is only slightly soluble in water. Therefore no cation traces are expected in the product.

$$
H_3PO_4 + 3KOH \rightarrow K_3PO_4 + 3H_2O
\tag{5.45}
$$

A response surface model to predict the conversion of the oil transesterification as a function of the methanol to oil ratio (*RM*), the temperature (*T*), and the catalyst load (Cat) can be seen in Eq. (5.46). Table 5.8 presents the typical range if the operating variables. The reactor operates at 4 bar (Martín and Grossmann, 2012b).

$$
\begin{aligned}
X = 74.6301 + 0.4209 \cdot T + 15.1582 \cdot Cat + 3.1561 \cdot RM - 0.0019 \cdot T^2 - 0.2022 \cdot T \cdot Cat \\
- 0.01925 \cdot T \cdot RM - 4.0143 \cdot Cat^2 - 0.3400 \cdot Cat \cdot RM - 0.1459 \cdot RM^2
\end{aligned} \tag{5.46}
$$

Homogeneous acid catalysts such as H_2SO_4 are not sensitive to the presence of FFAs, avoiding a pretreatment stage. However, the kinetics is slower. The conversion of the biodiesel production stage can be computed using Eq. (5.47) where the range of the variables including temperature (*T*), methanol to oil ratio (*RM*), and the catalyst load (Cat) can be seen in Table 5.9 (Martín and Grossmann, 2012b).

$$
X = 79.44 - 1.60 \cdot T - 3.13\, RM - 1.14\, Cat - 0.012 \cdot T
$$

$$
+ 0.75T \cdot Cat + 0.40 \cdot T \cdot RM + 0.18 Cat^2 - 6.56 CatRM - 0.025 RM \tag{5.47}
$$

The catalysts must be neutralized and removed from the products. Sulfuric acid can be neutralized using CaO to produce a low solubility product, $CaSO_4$ or gypsum, that precipitates and can be easily be separated.

$$
H_2SO_4 + CaO(\text{lime}) \rightarrow CaSO_4\,(\text{gypsum}) + H_2O \tag{5.48}
$$

Apart from the extended use of methanol due to its lower cost, either methanol is produced from renewable resources, which affects the price, or bioethanol is used as alcohol. Severson et al. (2013) evaluated the performance of several catalysis and among the KOH, based on the same reasoning as presented for the transesterification with methanol.

TABLE 5.8   Range of operation of the variables: alkali.

Variable	Lower bound	Upper bound
Temperature (°C)	45	65
Ratio methanol (mol/mol)	4.5	7.5
Cat (%)	0.5	1.5

TABLE 5.9   Range of operation of the variables: acid.

Variable	Lower bound	Upper bound
Temperature (°C)	25	65
Ratio methanol (mol/mol)	3.3	30
Cat (%)	1	5

The conversion of the reaction as a function of the operating variables and the range of operation can be seen below, Table 5.10:

$$X = 22.94293 + 113.88 \cdot Cat + 2.828881 \cdot RE - 1.02734 \cdot T - 1.44522 \cdot Cat \cdot RE + 0.250723 \cdot Cat \cdot T$$
$$+ 0.023375 \cdot RE \cdot T - 41.4402 \cdot Cat^2 - 0.07568 \cdot RE^2 + 0.006226 \cdot T^2;$$

(5.49)

Kinetics: The main issue related to biodiesel production is the large number of variables involved in predicting the yield. Response surface analysis has been widely used to obtained statistical-based models. However, remember that these models are only valid for the system used and even if there is a change in the scale, the models cannot longer be trusted. A more detailed analysis based on the reaction mechanism provides better insight. The drawback is the far larger number of experiments required to develop such a model so as to capture the effect of the catalysis as well as that of the temperature, time, and feedstock composition. The process to produce biodiesel is a set of three reactions in equilibrium. TG is the triglycerides, DG the diglycerides, MG the mono glycerides, ME the methyl ester, and GL is the glycerol.

$$TG + MeOH \xrightarrow{k1} DG + ME$$
$$DG + ME \xrightarrow{k2} TG + MeOH$$
$$DG + MeOH \xrightarrow{k3} MG + ME$$
$$MG + ME \xrightarrow{k4} DG + MeOH$$
$$MG + MeOH \xrightarrow{k5} GL + ME$$
$$GL + ME \xrightarrow{k6} MG + MeOH$$

(5.50)

The reactor design based on the kinetics is as follows:

$$\frac{dTG}{dt} = -k_1[TG][MeOH] + k_2[DG][ME]$$

$$\frac{dDG}{dt} = k_1[TG][MeOH] - k_2[DG][ME] - k_3[DG][MeOH] + k_4[MG][MeOH]$$

$$\frac{dMG}{dt} = k_3[DG][MeOH] - k_4[MG][MeOH] - k_5[MG][MeOH] + k_6[ME][GL]$$

TABLE 5.10 Range of operation of the variables: alkali–ethanol transesterification.

Variable	Lower bound	Upper bound
Temperature (°C)	25	80
Ratio ethanol (mol/mol)	3	20
Catalyst (%)	0.5	1.5

$$\frac{dME}{dt} = k_1[TG][MeOH] - k_2[DG][ME] + k_3[DG][MeOH] - k_4[MG][MeOH]$$

$$+ k_5[MG][MeOH] - k_6[ME][GL]$$

$$\frac{dGL}{dt} = + k_5[MG][MeOH] - k_6[ME][GL]$$

(5.51)

$$\frac{dMeOH}{dt} = - k_1[TG][MeOH] + k_2[DG][ME] - k_3[DG][MeOH] + k_4[MG][MeOH]$$

$$- k_5[MG][MeOH] + k_6[ME][GL]$$

$$k_i = k_{ic}C_{cat} + k_{in}$$

$$i = 1,\ldots,6$$

The rate constants depend on the oil, the catalyst, and the alcohol. For KOH 1%, we can see below some values from the literature in Table 5.11 (Issariyakul and Dalai, 2012). Example 5.5 shows the kinetics of biodiesel production.

TABLE 5.11   Kinetic constants for biodiesel production.

Reaction	Rate Constant	Palm oil 40°C	50°C	60°C
TG→DG	k_1	0.07	0.12	0.14
DG→TG	k_2	0.10	0.17	0.06
DG→MG	k_3	0.31	0.61	0.60
MG→DG	k_4	0.64	1.52	1.24
MG→GL	k_5	1.15	2.56	4.18
GL→MG	k_6	0.02	0.01	0.02

EXAMPLE 5.5 Compute the reaction time required to achieve almost 100% conversion of palm oil at 60°C.

Solution

We implement the model described above in Python, see Fig. 5E5.1.

```
import numpy as np
import matplotlib.pyplot as plt
from scipy import integrate
#[kJ/mol]

#60°C      k en h − 1

k1 = 0.14;
k_1 = 0.06;
```

```
k2 = 0.6;
k_2 = 1.24
k3 = 4.18;
k_3 = 0.02;
Cini = [1,6,0,0,0,0];
C = [(),(),(),(),(),()];
#We create an empty list to store the concentrations data
tau = np.linspace(0,4,100)
#Equations: We define a function which collect the equations to solve
def    kinetics(C,tau):
       Ctg = C[0];
       Cmetoh = C[1];
       Cmg = C[2]
       Cdg = C[3];
       Cgly = C[4];
       Cfame = C[5];
       r1 = k1*Ctg*Cmetoh;
       r_1 = k_1*Cdg*Cfame;
       r2 = k2*Cdg*Cmetoh;
       r_2 = k_2*Cmg*Cfame;
       r3 = k3*Cmg*Cmetoh;
       r_3 = k_3*Cgly*Cfame;
       dxdtau = np.zeros(6)
       dxdtau[0] = - r1 + r_1;
       dxdtau[1] = - r1 + r_1 - r2 + r_2 - r3 + r_3;
       dxdtau[2] = r2 - r_2 - r3 + r_3;
       dxdtau[3] = r1 - r_1 - r2 + r_2;
       dxdtau[4] = r3 - r_3;
       dxdtau[5] = r1 - r_1 + r2 - r_2 + r3 - r_3;
       return dxdtau
#We collect the results in a list calles Res
res_x = integrate.odeint(kinetics,Cini,tau)
plt.plot(tau,res_x[:,0],'k-')
plt.plot(tau,res_x[:,1],'k*')
plt.plot(tau,res_x[:,2],'k-.')
plt.plot(tau,res_x[:,3],'k-')
plt.plot(tau,res_x[:,4],'k.')
plt.plot(tau,res_x[:,5],'k>')
plt.xlabel('t(h)')
plt.ylabel('Mol')
plt.legend('OAMDGD')
```

3. Biomass and waste based processes

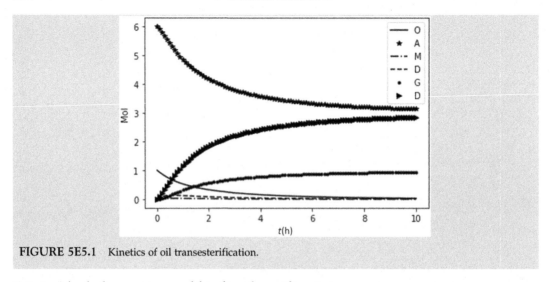

FIGURE 5E5.1 Kinetics of oil transesterification.

5.2.4 Alcohol recovery and biodiesel purification

As presented in the analysis of the equilibrium toward biodiesel, an excess of alcohol is typically used to drive the reaction into biodiesel. However, the decision on the excess of alcohol cannot me made just at the reactor level, aiming at higher conversions, but at the process level. The reason is that the larger the excess of alcohol, the conversion increases but the energy required to recover that excess is also larger. Therefore there is a tradeoff to be solved. In addition to avoid glycerol, biodiesel, and oil decomposition, typically above 150°C, 250°C, and 350°C, respectively, the operating pressures of the different columns must be carefully controlled. This fact provides an opportunity to heat integration within the flowsheet. Therefore simultaneous optimization and heat integration of the entire process, see Chapter 2, is the suggested process design procedure to address the operating conditions not only at the reactor, as seen in previous stage, but also at the distillation columns (Martín and Grossmann, 2012b). Fig. 5.10 shows the flowsheet for the production of biodiesel using heterogeneous catalysts.

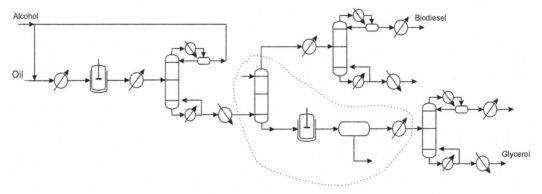

FIGURE 5.10 Biodiesel production using solid catalysts.

The distillation of the reaction mixture or the flash separation process involves two liquid phases consisting of the polar and the nonpolar species. The polar species include the glycerol and the alcohol, while the nonpolar ones contain the biodiesel and the oil. The homogeneous acid or alkali catalysis remains with the mixture. Thus the vapor pressure is that given by both phases as follows:

$$P_T = P_{v,polar} + P_{v,nonpolar} = \sum_{i = water, glycerol, alcohol} x_i Pv_i + \sum_{j = Biodiesel, oil} x_j Pv_j \qquad (5.52)$$

After distillation, a washing stage to remove the catalysis is required, typically adding a flow of water around 1% the mass of liquid product fed to the column (Zhang et al., 2003). Liquid−liquid separation is also aided by the addition of water yielding a polar phase to be neutralized and a nonpolar phase. However, if heterogeneous catalysts are used, no washing stage is required and a simple decanter operating at 40°C−60°C can be used to separate the phases. Thus, in Fig. 5.10, the section within the dotted line is substituted by a decanter. The nonpolar phase is distilled to purify the biodiesel operating at low pressure to avoid damage to the product. The polar phase is neutralized in case of using homogeneous catalysts. Next, either having used homogeneous or heterogeneous catalysts, the mixture is sent to a distillation column to purify the glycerol if needed. Glycerol can be an interesting raw material for the production of polyols as well as ethanol via fermentation, syngas via reforming or glycerol ethers (Almena et al., 2018). Example 5.6 shows the simulation of the process using CHEMCAD.

EXAMPLE 5.6 Model the production of 5 kg/s of biodiesel using al alkali (NaOH) catalyst using a process simulator. Assume complete neutralization of the catalysis and removal as a solid (using H_3PO_4). A stoichiometric reactor can be used for the transesterification stage as long as the conversion is computed off-line using the models in the text.

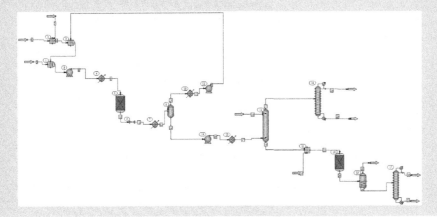

Reactor Summary

Equip. No.	5	16
Name		
Thermal mode	2	2
Temperature C	60.0000	25.0000
Heat duty MJ/s	2.3721	0.4125
Key Component	2	5
Frac. Conversion	0.9600	1.0000
Calc H of Reac. (J/kmol)	$-7.8359e+006$	$-3.7145e+007$
Stoichiometrics:		
Methyl Oleate	3.000	$0.000E+000$
Triacylgycerol	-1.000	$0.000E+000$
Methanol	-3.000	$0.000E+000$
Glycerol	1.000	$0.000E+000$
Sodium Hydroxid	$0.000E+000$	-3.000
TriNa Phosphate	$0.000E+000$	1.000
Phosphoric Acid	$0.000E+000$	-1.000
Water $0.000E+000$	3.000	

Extractor Summary

Equip. No.	11
Name	
No. of Stages:	8
1st feed stage	1
2nd feed stage	8
P Top bar	1.1000

SCDS Rigorous Distillation Summary

Equip. No.	17
Name	
No. of stages	4
1st feed stage	2
Condenser mode	1
Condenser spec	3.0000
Reboiler mode 5	
Reboiler spec.	0.0062
Reboiler comp i	4
Est. dist. rate (kmol/s)	0.0567
Est. reflux rate (kmol/s)	0.1702
Est. T top C	31.7711
Est. T bottom C	196.2380
Est. T 2C	102.1078

3. Biomass and waste based processes

Top pressure bar	0.0500		
Calc cond duty MJ/s	− 10.5125		
Calc rebr duty MJ/s	10.8209		
Initial flag 6			
Calc Reflux mole (kmol/s)	0.1702		
Calc Reflux ratio	3.0000		
Calc Reflux mass kg/s	3.1549		
Optimization flag	1		
Calc. tolerance	0.0012		
STREAM PROPERTIES			
Stream No.	1	2	3
Name			
− −Overall− −			
Molar flow kmol/s	0.0200	0.0016	0.0068
Mass flow kg/s	0.6408	0.0640	6.0000
Temp C	**25.0000**	**25.0000**	**25.0000**
Pres bar	**1.0000**	**1.0000**	**1.0000**
Vapor mole fraction	0.0000	0.0000	0.0000
Enth MJ/s	-4.7791	−1.2921	−16.834
Tc C	239.4900	2546.8501	1366.8500
Pc bar	80.9700	253.3101	4.7000
Std. sp gr. wtr = 1	0.801	1.934	0.909
Std. sp gr. air = 1	1.106	1.381	30.572
Degree API	45.2429	−58.3469	24.2280
Average mol wt	32.0420	39.9970	885.4322
Actual dens kg/m3	789.5790	1913.3403	909.0721
Actual vol m3/h	2.9218	0.1204	23.7605
Std liq m3/h	2.8816	0.1191	23.7719
Std vap OC m3/h	1613.7826	129.1123	546.7780
− −Liquid only− −			
Molar flow kmol/s	0.0200	0.0016	0.0068
Mass flow kg/s	0.6408	0.0640	6.0000
Average mol wt	32.0420	39.9970	885.4322
Actual dens kg/m3	789.5790	1913.3403	909.0721
Actual vol m3/h	2.9218	0.1204	23.7605
Std liq m3/h	2.8816	0.1191	23.7719
Std vap 0 C m3/h	1613.7826	129.1123	546.7780
Cp J/kmol − K	81119.5703	58877.1055	1494693.5000
Z factor	0.0022	0.0033	0.1027
Visc Pa-sec	0.0005380	3.590	0.001425
Th cond W/m − K	0.1999	0.3633	0.1713
Surf. tens. N/m	0.0222	0.5241	0.0061

3. Biomass and waste based processes

Flow rates in kg/s			
Methyl Oleate	0.0000	0.0000	0.0000
Triacylgycerol	0.0000	0.0000	6.0000
Methanol	0.6408	0.0000	0.0000
Glycerol	0.0000	0.0000	0.0000
Sodium Hydroxide	0.0000	0.0640	0.0000
Trisodium Phosphate	0.0000	0.0000	0.0000
Phosphoric Acid	0.0000	0.0000	0.0000
Water	0.0000	0.0000	0.0000

Problems

P5.1 Implement a model for enzymatic hydrolysis using the equations presented in the text.

P5.2 Implement Buehler and Mesbah (2016) kinetic model for the fermentation of sugars into acetone–butanol–ethanol fermentation.

P5.3 Compare the consumption of steam in the double pressure column with a single one operating under the same separation ratio per 100 kg/s feed.

P5.4 Evaluate using a process simulator the effect of the water content in the water–ethanol mixture on the energy consumption. Assume 4%, 6%, 8%, 12%, and 15% ethanol in water and represent the reboiler and condenser duties (kJ/kg of ethanol) aiming at a distillate with minimum 95% of the initial ethanol. Report the reflux ratio used.

P5.5 Evaluate the effect of temperature on the transesterification of palm oil.

P5.6 Develop McCormick envelops to the model for the conversion of the alkali catalyzed biodiesel production and globally optimized it (Table 5P6.1 shows the range of operation).

$$X = 74.6301 + 0.4209 \cdot T + 15.1582 \cdot Cat + 3.1561 \cdot RM - 0.0019 \cdot T^2 - 0.2022 \cdot T \cdot Cat$$
$$-0.01925 \cdot T \cdot RM - 4.0143 \cdot Cat^2 - 0.3400 \cdot Cat \cdot RM - 0.1459 \cdot RM^2$$

P5.7 For the solar and temperature data in Problem 10.4, evaluate the growth of algae assuming pH = 7.4 in all cases and a Growth N of 4. Which are the two best locations and their growth rates.

P5.8 Considering the cost of nitrogen €50/t and the cost for phosphorous at €65/t, determine the nutrients required to achieve Growth N of 4 g/m^2 d.

TABLE 5P6.1 Range of operation of the variables: alkali.

Variable	Lower bound	Upper bound
Temperature (°C)	45	65
Ratio methanol (mol/mol)	4.5	7.5
Cat (%)	0.5	1.5

P5.9 For the dilute acid, pretreatment determines a set of operating conditions that result in cellulose and hemicellulos yields above 90% and 60%, respectively. Evaluate the issues of the surface response model.

References

Aden, A., Foust, T., 2009. Technoeconomic analysis of the dilute sulfuric acid and enzymatic hydrolysis process for the conversion of corn stover to ethanol. Cellulose 16, 535–545.

Almena, A., Bueno, L., Díez, M., Martín, M., 2018. Integrated biodiesel facilities: review of transformation processes of glycerol based production of fuels and chemicals. Clean. Technol. Environ. Policy 20, 1639–1661.

Barros, A.I., Gonçalves, A.L., Simoes, M., Pires, J.C.M., 2015. Harvesting techniques applied to microalgae: a review. Renew. Sustain. Energy Rev. 41, 1489–1500.

Benson, B.C., Meyer, B.J., Bajpai, R.K., Gang, D.D., Dufreche, S.T., Zappi, M.E., 2016. Growth kinetics, light dynamics, and lipid production in microalgae from sugar-mill ponds. Chem. Biochem. Eng. Q. 30 (3), 331–339.

Birgen, C., Berglihn, O.T., Preisig, H.A., Wentzel, A., 2019. Kinetic study of butanol production from mixtures of glucose and xylose and investigation of different pre-growth strategies. Biochem. Eng. J. 147, 110–117.

Braisher, M., Gill, S., Treharne, W., Wallace, M., Winterburn, J., Cui, Z., et al., May 2006. Design proposal. In: Bioethanol Production Plant. Project Report.

Branan, C.R., 2000. Soluciones Prácticas Para el Ingeniero Químico, 2nd ed. McGraw Hill.

Buehler, E.A., Mesbah, A., 2016. Kinetic study of acetone-butanol-ethanol fermentation in continuous culture. PLoS One 11 (8), e0158243. Available from: https://doi.org/10.1371/journal.pone.0158243.

Buthod, P., 1994. Process Component Design. Universidad de Tulsa, Tulsa, OK.

Chang, J.H., Yoo, J.K., Ahn, S.-H., Lee, K.-H., Ko, S.-M., 1998. Simulation of pervaporation process for ethanol dehydration by using pilot test results. Korean J. Chem. Eng. 15 (1), 28–36.

Eggeman, T., Elander, R.T., 2005. Process and economic analysis of pretreatment technologies. Bioresour. Technol. 96, 2019–2025.

Eze, V.C., Velazquez-Orta, S.B., Hernández-García, A., Monje-Ramírez, I., Orta-Ledezma, M.T., 2018. Kinetic modelling of microalgae cultivation for wastewater treatment and carbon dioxide sequestration. Algal. Res. 32, 131–141.

Fasaei, F., Bitter, J.H., Slegers, P.M., van Boxtel, A.J.B., 2018. Technoeconomic evaluation of microalgae harvesting and dewatering systems. Algal Res. 31, 347–362.

Garlock, R.J., Balan, V., Dale, B.E., 2012. Optimization of AFEX™ pretreatment conditions and enzyme mixtures to maximize sugar release from upland and lowland switchgrass. Bioresour. Technol. 104, 757–768.

Holtzapple, M.T., Lundeen, J.E., Sturgis, R., Lewis, J.E., Dale, B.E., 1992. Pretreatment of lignocellulosic municipal solid waste by ammonia fiber explosion (AFEX). Appl. Biochem. Biotechnol. 34/35, 5–21.

Issariyakul, T., Dalai, A.K., 2012. Comparative kinetics of transesterification for biodiesel production from palm oil and mustard oil. The Canadian Journal of Chemical Engineering 90 (2), 342–350.

Kadam, K.L., Rydholm, E.C., McMillan, J.D., 2004. Development and validation of a kinetic model for enzymatic saccharification of lignocellulosic biomass. Biotechnol. Prog. 20, 698–705.

Kazi, F.K., Fortman, J.A., Anex, R.P., Hsu, D.D., Aden, A., Dutta, A., et al., 2010. Techno-economic comparison of process technologies for biochemical ethanol production from corn stover. Fuel 89 (Suppl. 1), S20–S28. Available from: https://doi.org/10.1016/j.fuel.2010.01.001.

Kazmi, A., 2011. Advanced Oil Crop Biorefinery. RSC, London.

Kraemer, K., Harwardt, A., Bronneberg, R., Marquardt, W., 2011. Separation of butanol from acetone–butanol–ethanol fermentation by hybrid extraction–distillation process. Comp. Chem. Eng. 35 (5), 949–963.

Krishnan, M.S., Ho, N.W.Y., Tsao, G.T., 1999. Fermentation kinetics of ethanol production from glucose and xylose by recombinant *Saccharomyces* 1400(pLNH33). Appl. Biochem. Biotechnol. 77–79. 373–388.

Lee, E., Jalalizadeh, M., Zhang, Q., 2015. Growth kinetic models for microalgae cultivation: a review. Algal Res. 12, 497–512.

Lenihan, P., Orozco, A., O'Neill, E., Ahmad, M.N.M., Rooney, D.W., Mangwandi, C., et al., 2011. Kinetic modelling of dilute acid hydrolysis of lignocellulosic biomass. In: dos Santoa Benardes, M.A. (Ed.), Biofuel Production: Recent Developments and Prospects. Intechopen, pp. 293–308.

Lini, F.Z., Widjaja, T., Hendrianie, N., Altway, A., Nurkhamidah, S., Tansil Y., 2018. The effect of organosolv pretreatment on optimization of hydrolysis process to produce the reducing sugar. In: MATEC Web of Conferences, vol. 154, p. 01022. Available from: https://doi.org/10.1051/matecconf/201815401022.

Malmierca, S., Díez-Antoínez, R., Paniagua, A.I., Martín, M., 2017. Technoeconomic study of AB biobutanol production. Part 2: Process design. Ind. Eng. Chem. Res. 56 (6), 1525–1533.

Martin, M, 2019. In: Introduction to software for chemical engineers, 2nd CRC Press, USA.

Martín, M., Grossmann, I.E., 2012a. Energy optimization of lignocellulosic bioethanol production via hydrolysis of switchgrass. AIChE J. 58 (5), 1538–1549.

Martín, M., Grossmann, I.E., 2012b. Simultaneous optimization and heat integration for biodiesel production from cooking oil and algae. Ind. Eng. Chem. Res. 51 (23), 7998–8014.

Martín, M., Grossmann, I.E., 2014. Optimization simultaneous production of ethanol and i-butene from switchgrass. Biomass Bioenergy 61, 93–103.

Martin, M., Grossmann, I.E., 2015. Optimal production of furfural and DMF from algae and switchgrass. Ind. Eng. Chem. Res. 55 (12), 3192–3202.

Mulchandani, A., Loung, J.H.T., 1989. Microbial inhibition kinetics revisited. Enzyme Microb. Technol. 11, 66–73.

Obnamia, J.A., 2014. Modeling the Reaction Kinetics of the Enzymatic Hydrolysis of Lignocellulosic Biomass (M. Sc. thesis). University of Toronto.

Pate, R., Hightower, M., Cameron, C., Einfeld. W., 2007. Overview of energy-water interdependencies and the emerging energy demands on water resources. In: Report SAND 2007–1349C. Los Alamos, NM: Sandia-National Laboratories.

Patrascu, I., Bildea, C.S., Kiss, A.A., 2017. Eco-efficient butanol separation in the ABE fermentation process. Sep. Purif. Technol. 1777, 49–61.

Qureshi, N., Blaschek, H.P., 1999. Production of acetone–butanol–ethanol (ABE) by a hyper-producing mutant strain of Clostridium beijerinckii BA101 and recovery by pervaporation. Biotechnol. Prog. 15, 594–602.

Qureshi, N., Hughes, S., Maddox, I.S., Cotta, M.A., 2005. Energy-efficient recovery of butanol from model solutions and fermentation broth by adsorption. Bioprocess. Biosyst. Eng. 27, 215–222.

Qureshi, N., Saha, B.C.;, Cotta, M.A., Singh, V., 2013. An economic evaluation of biological conversion of wheat straw to butanol: a biofuel. Energy Convers. Manage. 65, 456–462.

Raganati, F., Procentese, A., Olivieri, G., Götz, S.P., Marzocchella, A., 2015. Kinetic study of butanol production from various sugars by Clostridium acetobutylicum using a dynamic model. Biochem. Eng. J. 99, 156–166.

Rosales-Calderon, O., Arantes, V., 2019. A review on commercial-scale high-value products that can be produced alongside cellulosic ethanol. Biotechnol. Biofuels 12, 240.

Sazdanoff, N., 2006. Modeling and Simulation of the Algae to Biodiesel Fuel Cycle (Undegraduate thesis). The Ohio State University.

Severson, K., Martín, M., Grossmann, I.E., 2013. Optimal integration for biodiesel production using bioethanol. AICHE J. 59 (3), 834–844.

Singh, G., Patidar, S.K., 2018. Microalgae harvesting techniques: a review. J. Environ. Manage. 217, 499–508.

Sinnott, R.K., 1999. 3rd ed. Coulson and Richardson's Chemical Engineering, Vol. 6. Elsevier, Oxford.

Sun, Y., Cheng, J., 2002. Hydrolysis of lignocellulosic materials for ethanol production: a review. Bioresour. Technol. 83, 1–11.

Taimbú de la Cruz, C.A., Martín, M., Grossmann, I.E., 2019. Process optimization for the hydrothermal production of algae fuels. Ind. Eng. Chem. Res. 58 (51), 23276–23283. Available from: https://doi.org/10.1021/acs.iecr.9b05176.

United States DOE, 2011. Integrated biorefinery demonstration plant producing cellulosic ethanol and biochemicals from woody biomass. <https://www.energy.gov/sites/prod/files/2014/03/f14/ibr_demonstration_lignol.pdf>.

Valentínyi, N., Cséfalvay, E., Mizsey, P., 2013. Modelling of pervaporation: parameter estimation and model development. Chem. Eng. Res. Des. 91, 174–183.

Vandamme, D., 2013. Flocculation-Based Harvesting Processes for Microalgae Biomass Production (Doctoral dissertation). UGent.

Vrsalovic Presecki, A., Findrick, Z., Vasic-Racki, D., 2008. Mathematical modelling of amylase catalyzed starch hydrolysis. CHISA 2008, 1–8.

Weibin, C.A.I., Sun, Y., Piao, X., Li, J., Zhu, S., 2011. Solvent recovery from soybean oil/hexane miscella by PDMS composite membrane. Chin. J. Chem. Eng. 19 (4), 575–580.

3. Biomass and waste based processes

Weissermel, K., Arpe, H.J., 1981. Química Organica Industrial. Productos de partida e intermedios mas importantes. Reverte, Barcelona.

Wooley, R., Ruth, M., Sheehan, J., Ibsen, K., Majdeski, H., Galvez, A., 1999. Lignocellulosic biomass to ethanol process design and economics utilizing co-current dilute acid prehydrolysis and enzymatic hydrolysis current and futuristic scenarios. In: Technical Report. NREL/TP-580-26157.

Zhang, Y., Dube, M.A., McLean, D.D., Kates, M., 2003. Biodiesel production from waste cooking oil: 1. Process design and technological assessment. Bioresour. Technol. 89, 1−16.

Zhang, K., Pei, Z., Wang, D., 2016. Organic solvent pretreatment of lignocellulosic biomass for biofuels and biochemicals: a review. Bioresour. Technol. 199, 21−33.

Anaerobic digestion and nutrient recovery

Edgar Martín-Hernández and Mariano Martín

Department of Chemical Engineering, University of Salamanca, Salamanca, Spain

Anaerobic digestion (AD) is a microbiological process that decomposes organic matter in the absence of oxygen, resulting in a gas mixture and nutrient-rich residual stream, called biogas and digestate respectively. Biogas is mainly composed by CH_4 and CO_2, containing small amounts of impurities, particularly ammonia, hydrogen sulfide, and moisture. Although as a biochemical treatment of biomass, the anaerobic digestion could be part of the previous chapter, the extended use as a technology to process waste made us think that it was worth devoting a chapter to it. In particular, it takes place at lower temperatures compared to most of the processed described in the previous chapter. Anaerobic digestion can be a central part of waste management toward a circular economy (see Fig. 6.1) (Al Seadi, 2002).

6.1 Biogas production

The production of biogas is analyzed from the mechanistic perspective in this section. A general description is presented first, including the stages and dynamics of the process and the calculation of the biomass yield to biogas.

6.1.1 Anaerobic digestion mechanism

The decomposition of biomass through anaerobic digestion takes place in four stages, namely, hydrolysis, acidogenesis, acetogenesis, and methanogenesis, occurring in a sequence where the products of one stage are the feedstock for the next one (see Fig. 6.2) (Lauwers et al., 2013; Al Seadi et al., 2008).

Sustainable Design for Renewable Processes
DOI: https://doi.org/10.1016/B978-0-12-824324-4.00017-2

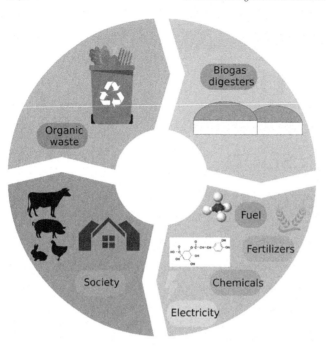

FIGURE 6.1 Valorization of organic waste through anaerobic digestion.

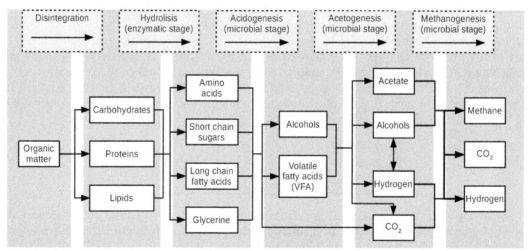

FIGURE 6.2 Mechanism of the anaerobic digestion of biomass.

1. Hydrolysis: The first stage consists of the hydrolysis of the biomass. It is an enzymatic governed process where the polymers constituting the organic matter are broken down into oligomers or monomers. In terms of biomolecules, this means that polysaccharides, proteins, and lipids are transformed into sugars, amino acids, and fatty acids. The enzymes are secreted by microorganisms present in the bulk liquid or attached to the organic matter particles.

2. Acidogenesis: In this stage the acidogenic microorganisms, fermentative bacteria, convert the products of the hydrolysis into alcohols and/or organic acids as well as volatile fatty acids (VFA), carbon dioxide, and hydrogen.
3. Acetogenesis: During this phase, the products from the acidogenesis that cannot be transformed into methane in the next step, such as long-chain VFA and alcohols, are converted into acetate, carbon dioxide, and hydrogen.
4. Methanogenesis: This step occurs in parallel to the acetogenesis. Acetic acid and the mixture between hydrogen and carbon dioxide from the previous stages are transformed into methane and either carbon dioxide or water, whether the raw material is acetic acid or the gas mixture.

$$\text{Acetic acid} \xrightarrow{\text{\textit{methanogenic bacteria}}} \text{Methane} + \text{Carbon dioxide}$$
$$\text{Hydrogen} + \text{Carbon dioxide} \xrightarrow{\text{\textit{methanogenic bacteria}}} \text{Methane} + \text{Water}$$

(6.1)

This stage is severely affected by the operating conditions including the feedstock composition, the temperature, pH among others.

The AD can take place at different temperatures with different retention times. Psychrophilic conditions require temperatures below 20°C, and long retention times, 70–80 days. Mesophilic conditions increase the growth rate reducing the retention time to 30–40 days by increasing the operating temperature to 30°C–42°C. Finally, thermophilic conditions require 43°C–55°C but only 15–20 days retention time. Most modern biogas plants operate at thermophilic conditions due to the advantages related to the better digestibility, higher growth rate, reduced retention time, and the effective destruction of pathogens. The thermophilic operating region has also some drawbacks that must be considered, such as the high risk of ammonia inhibition, the higher energy demand due to the higher operating temperature, and the fact that higher temperature increases the solubility of species in the liquid which may have an inhibitory effect on the process. On the contrary, the higher temperatures reduce the viscosity of the bulk media and enhances the diffusion of dissolved materials, increasing the reaction rates (Al Seadi et al., 2008). pH values from 5.5 to 8 are recommended for AD.

6.1.2 Anaerobic digestion yield estimation

The fraction of biomass behind biogas formation are the volatile solids contained in the volatile organic matter, while the remaining biomass is transformed into digestate.

The theoretical yield from biomass to biogas, as well as the gas fraction of methane, carbon dioxide, and impurities, can be estimated based on the waste composition through the following general stoichiometry by Eq. (6.2):

$$C_nH_aO_bN_cS_e + \left(n - \frac{a}{4} - \frac{b}{2} + \frac{3c}{4} + \frac{e}{2}\right)H_2O \leftrightarrow \left(\frac{n}{2} - \frac{a}{8} + \frac{b}{4} + \frac{3c}{8} + \frac{e}{4}\right)$$
$$CO_2 + \left(\frac{n}{2} + \frac{a}{8} - \frac{b}{4} - \frac{3c}{8} - \frac{e}{4}\right)CH_4 + cNH_3 + eH_2S$$

(6.2)

3. Biomass and waste based processes

This reaction accounts for the formation of biogas but does not take into consideration the degradation of organic matter for the bacteria metabolism. If required, a more detailed mass balance can be performed, although further information on the yield from the particular feedstock into biogas is required (León and Martín, 2016).

The moisture contained in the biogas corresponds to that of a saturated gas at the output conditions, usually the digestion temperature and atmospheric pressure, assuming ideal gas behavior, as shown in Eq. (6.3), where y denotes the specific saturated moisture of biogas, MW the molecular weight, T temperature, P the pressure, and Pv the vapor pressure.

$$y_{biogas} = \frac{MW_{H_2O}}{MW_{biogas\text{-}dry}} \cdot \frac{Pv(T)}{P - Pv(T)} \tag{6.3}$$

Therefore, the flow of biogas generated can be computed as per Eq. (6.4). Please note that the biogas molecular weight is defined in Eq. (6.7) from the mass fraction of the components, x_i:

$$F_{(Biogas)} = \frac{P_{atm} \cdot MW_{biogas}}{R \cdot (T_{digester} + 273)} \cdot \left[w_{VS/DM} \cdot w_{DM/Waste} \cdot V_{biogas/VS} \cdot F_{(Feedstock)} \right] \tag{6.4}$$

where R refers to the constant of ideal gases, F_j the total of the stream j, $w_{VS/DM}$ corresponds to the volatile solids per dry matter unit, $w_{DM/Waste}$ is the dry matter mass fraction of the waste, and $V_{biogas/VS}$ is the volume of biogas produced per mass unit of volatile solids. The flow of water contained in the biogas is calculated as shown in Eq. (6.5), where $fc_{i,j}$ denotes the mass flow of component i in stream j:

$$fc_{(H_2O,Biogas)} = y_{biogas} \cdot \sum_{a'} fc_{(a',Biogas)} \forall \, a' = \{CH_4, CO_2, NH_3, H_2S, N_2, O_2\} \tag{6.5}$$

The flow of each of the compounds is computed by the molar fraction, Y_i, of the dry biogas, where MW is the molecular weight, as given by Eq. (6.6).

$$\frac{fc_{(a',Biogas)}}{MW_{a'}} = \frac{Y_{a'/biogas\text{-}dry}}{MW_{biogas\text{-}dry}} \left(F_{(Biogas)} - fc_{(H_2O,Biogas)} \right) \tag{6.6}$$

$$MW_{biogas} \sum_{a} \frac{x_{a/biogas}}{MW_a} = \sum_{a} x_{a/biogas} \forall \, a = \{CH_4, CO_2, NH_3, H_2S, N_2, O_2, H_2O\} \tag{6.7}$$

The remaining biomass constitutes the digestate, whose composition is computed using mass balances as a function of the composition of the biomass as follows. $R_{C-N/k}$ refers to the carbon/nitrogen ratio and w_i is the mass ratio of component i in the dry matter (DM):

$$w_{C/k} = R_{C-N/k} \left(w_{N_{org}/k} + w_{N_{inorg}/k} \right); \; k = \{Waste\} \tag{6.8}$$

$$w_{C/DM} + w_{N_{org}/DM} + w_{N_{am}/DM} + w_{P/DM} + w_{K/DM} + w_{Rest/DM} = 1 \tag{6.9}$$

The mass balances to each one of the species in the digestate carbon, i.e., (C), organic and inorganic nitrogen (N_{org}, N_{am}), phosphorous (P), potassium (K), and others (rest) are computed as follows,

$$fc_{(C,Digestate)} = w_{C/DM} \cdot w_{DM/Waste} \cdot F_{(Feedstock)} - fc_{(CH_4,Biogas)} \cdot \frac{MW_C}{MW_{CH_4}} - fc_{(CO_2,Biogas)} \cdot \frac{MW_C}{MW_{CO_2}}$$
$$\tag{6.10}$$

$$fc_{(N_{org},Digestate)} = w_{N_{org}/DM} \cdot w_{DM/Waste} \cdot F_{(Feedstock)} \tag{6.11}$$

$$fc_{(N_{am},Digestate)} = w_{Nam/DM} \cdot w_{DM/Waste} \cdot F_{(Feedestock)} - fc_{(NH_3,Biogas)} \cdot \frac{MW_N}{MW_{NH_3}} \tag{6.12}$$

$$fc_{(P,Digestate)} = w_{P/DM} \cdot w_{DM/Waste} \cdot F_{(Feedstock)} \tag{6.13}$$

$$fc_{(K,Digestate)} = w_{K/DM} \cdot w_{DM/Waste} \cdot F_{(Feedstock)} \tag{6.14}$$

$$fc_{(Rest,Digestate)} = w_{Rest/DM} \cdot w_{DM/Waste} \cdot F_{(Feedstock)} - fc_{(CH_4,Biogas)} \cdot \frac{4 \cdot MW_H}{MW_{CH_4}} - fc_{(CO_2,Biogas)} \cdot \frac{2 \cdot MW_O}{MW_{CO_2}} -$$
$$fc_{(NH_3,Biogas)} \cdot \frac{3 \cdot MW_H}{MW_{NH_3}} - fc_{(H_2S,Biogas)} - fc_{(O_2,Biogas)} \tag{6.15}$$

$$fc_{(H_2O,Digesate)} = (1 - w_{DM/Waste}) \cdot F_{(Feedsctock)} - fc_{(H_2O,Biogas)} \tag{6.16}$$

The composition of the digestate is given in terms of the NPK index, as it is the standard metric to evaluate the quality of the fertilizer:

$$K_{index} = \frac{fc(K) \cdot \frac{MW_{K_2O}}{2 \cdot MW_K}}{fc(C) + fc(N_{org}) + fc(H_2O) + fc(Rest) + fc(K) \cdot \frac{MW_{K_2O}}{2 \cdot MW_K} + fc(P) \frac{MW_{P_2O_5}}{2 \cdot MW_P}} \cdot 100\% \tag{6.17}$$

$$P_{index} = \frac{fc(P) \frac{MW_{P_2O_5}}{2 \cdot MW_P}}{fc(C) + fc(N_{org}) + fc(H_2O) + fc(Rest) + fc(K) \cdot \frac{MW_{K_2O}}{2 \cdot MW_K} + fc(P) \frac{MW_{P_2O_5}}{2 \cdot MW_P}} \cdot 100\% \tag{6.18}$$

$$N_{index} = \frac{fc(N_{org})}{fc(C) + fc(N_{org}) + fc(H_2O) + fc(Rest) + fc(K) \cdot \frac{MW_{K_2O}}{2 \cdot MW_K} + fc(P) \frac{MW_{P_2O_5}}{2 \cdot MW_P}} \cdot 100\% \tag{6.19}$$

$$K_{index} + N_{index} + P_{index} \geq 6\%$$

$$0 \leq R_{C-N/fertilizer} = \frac{fc(C)}{fc(N_{org})} \leq 10 \tag{6.20}$$

The composition of the biomass can be fixed, or bounded between a lower bound and an upper bound (Defra, 2011a,b). Table 6.1 shows the typical composition of different of organic wastes.

6.1.3 Kinetics

The kinetics of the digestion includes multiphase mass transfer, biomass disintegration and hydrolysis, acidogenesis from sugars and amino acids, acetogenesis from long chain fatty acids, propionate and butyrate and valerate as well as aceticlastic and hydrogenotrophic methanogenesis. Fig. 6.2 shows the relationship between the different mechanisms involved in anaerobic digestion.

TABLE 6.1 Composition of different wastes.

	Cattle slurry		Pig slurry		Cattle manure	Pig manure	Sludge (wastewater)		Urban food waste	Urban green waste
	LOW	UP	LOW	UP			LOW	UP		
Vbiogas (m³/kg)	0.2	0.5	0.25	0.5	0.25	0.38	0.25	0.60	0.44	0.23
w_{DM}	0.1	0.2	0.03	0.08	0.25	0.25	0.08	0.25	0.31	0.25
w_{VS}	0.5	0.7	0.7	0.8	0.80	0.75	0.3	0.5	0.85	0.88
w_C						0.420			0.468	0.418
w_{Nam}	0.005	0.047	0.005	0.095	0.004	0.006	0.024	0.320	0.001	0.001
w_{Norg}	0.005	0.036	0.005	0.030	0.020	0.022	0.001	0.002	0.031	0.030
w_P	0.008	0.013	0.019	0.022	0.006	0.010	0.009	0.063	0.005	0.005
w_K	0.033	0.100	0.039	0.083	0.027	0.027	0.008	0.015	0.009	0.027
R_{C-N}	6	20	3	10	20	15	6	24	15	13
Cost (€/kg)	0.031		0.043		0.010	0.012	0.091		0.004	0.0084

One of the most detailed AD models developed is the one known as the Anaerobic Digestion Model no. 1 (ADM1) (Batstone et al., 2002). However, the complexity of the model leads to a number of more simple models to evaluate the effect of the waste composition. For instance, Weinrich and Nelles (2015) developed a model for the anaerobic digestion of agricultural wastes based on first order kinetics where the hydrolysis rate is the limiting stage, including also the fast kinetics of the methanogenesis, acetogenesis or acidogenesis stages. Similarly, models for manure (López et al., 2020) and municipal solid waste (Nopharatana et al., 2007) were developed under the assumption that the hydrolysis was the limiting stage.

The simplest AD modeling approach considers a first-order rate model for the substrate utilization, where k refers to the reaction rate coefficient, S the concentration of a degradable substrate s, and t the time:

$$\frac{dS}{dt} = -kS \tag{6.21}$$

Alternatively, Contois (1959) proposed a modified form of the Monod model for the first hydrolysis and acidogenesis stages (Nopharatana et al., 2007), and a Haldane function can be used for the methanogenesis step (Mairet et al., 2011), where μ_m and K_s are kinetic parameters, and X refers to the concentration of microorganisms. pH and the gas−liquid mass transfer must also be controlled. Example 6.1 shows the model for the kinetics of a digester.

$$\mu_k(S_k, X_k) = \frac{\mu_{m_k} \cdot S_k}{K_{s_k} X_k + S_k}; k \in \{\text{hydrolysis, acidogenesis}\} \tag{6.22}$$

$$\mu_k = \frac{\mu_{m_k} \cdot S_k}{K_{s_k} + S_k + S_k^2/K_k}; k \in \{\text{methanogenesis}\} \tag{6.23}$$

EXAMPLE 6.1 Model a digestor using the kinetics and the data from Weinrich and Nelles (2015).

Solution

Here a first-order kinetics based model is presented that allows the evaluation of the production of biogas a from carbohydrates, protein and lipids that are computed as follows (g/L):

$$x_{(ch-0)} = \frac{1}{V} \sum_{i=1}^{N} m_i \cdot DM_{(i)} \cdot DQ_{(ch-i)} \cdot (NfE_i + CF_i)$$

$$x_{(pr-0)} = \frac{1}{V} \sum_{i=1}^{N} m_i \cdot DM_{(i)} \cdot DQ_{(pr-i)} \cdot (CP_i) \qquad (6E1.1)$$

$$x_{(li-0)} = \frac{1}{V} \sum_{i=1}^{N} m_i \cdot DM_{(i)} \cdot DQ_{(li-i)} \cdot (CL_i)$$

where m_i is the mass of species i, DM is the dry mass, NfE_i is the extract free of nitrogen, CF_i CP_i CL_i corresponds to the crude fiber, protein, and lipids respectively, and DQ is the digestibility coefficient of each substrate. The bacteria are given by the following reaction:

$$x_{bac} \rightarrow 0.18 \cdot x_{pr} + 0.82 \cdot x_{ch} \qquad (6E1.2)$$

The kinetics is based on the stoichiometry and turns out to be:

$$\frac{dx_{ch}}{dt} = -k_{hyd-ch} \cdot x_{ch} + k_{dec-x_{bac}} \cdot 0.18 \cdot x_{bac}$$

$$\frac{dx_{li}}{dt} = -k_{hyd-li} \cdot x_{li}$$

$$\frac{dx_{pr}}{dt} = -k_{hyd-pr} \cdot x_{pr} + k_{dec-x_{bac}} \cdot 0.82 \cdot x_{bac}$$

$$\frac{dx_{bac}}{dt} = -k_{dec-x_{bac}} \cdot x_{bac} + k_{hyd-ch} \cdot 0.1509 \cdot x_{ch} + k_{hyd-pr} \cdot 0.1241 \cdot x_{pr} + k_{hyd-li} \cdot 0.1857 \cdot x_{li}$$

$$\frac{dx_{gas,CH_4}}{dt} = k_{hyd-ch} \cdot 0.2433 \cdot x_{ch} + k_{hyd-pr} \cdot 0.2383 \cdot x_{pr} + k_{hyd-li} \cdot 0.6588 \cdot x_{li} \qquad (6E1.3)$$

$$\frac{dx_{gas,CO_2}}{dt} = k_{hyd-ch} \cdot 0.6675 \cdot x_{ch} + k_{hyd-pr} \cdot 0.6822 \cdot x_{pr} + k_{hyd-li} \cdot 0.6644 \cdot x_{li}$$

$$\frac{dx_{gas,H_2S}}{dt} = k_{hyd-pr} \cdot 0.0012 \cdot x_{pr}$$

$$\frac{dx_{gas,NH_3}}{dt} = -k_{hyd-ch} \cdot 0.0227 \cdot x_{ch} + k_{hyd-pr} \cdot 0.1645 \cdot x_{pr} - k_{hyd-li} \cdot 0.0280 \cdot x_{li}$$

where k_{hyd-ch} is $0.4d^{-1}$, k_{hyd-pr} is $0.2d^{-1}$, k_{hyd-li} is $0.1d^{-1}$ and $k_{dec-xbac}$ is $0.02d^{-1}$. Actually, each substrate has its own hydrolysis constant. Thus, the models can be extended to capture the dynamics of the hydrolysis of each substrate (Table 6E1.1).

TABLE 6E1.1 Characterization of the substrates.

	Corn	Sugar beet	Grain
NfE g/kg$_{DM}$	640	872.3	717.9
CF g/kg$_{DM}$	205.3	47.8	30.6
CP g/kg$_{DM}$	72.7	32.7	107.2
CL g/kg$_{DM}$	41.9	27.5	42.9
DQ$_{carbohydrates}$ (−)	0.83	0.81	0.81
DQ$_{proteins}$ (−)	1	0.96	0.98
DQ$_{lipids}$ (−)	0.78	0.76	0.85
DM (%)	0.287	0.225	0.844

The model is written in MATLAB® as follows:

```
%Script%
NFE=[640,842.3,717,9];
CF=[205.3, 47.8, 30.6];
CP=[72.7,32.7, 107.2];
CL=[41.9, 27.5, 42.9];
DM=[0.287, 0.225, 0.844]
DQch=[0.83, 0.81, 0.81]:
DQpr=[1, 0.96, 0.98];
Dqli=[0.78, 0.76, 0.85];
Mass=[100,25,5];
Volume=37367;

For i=1:3
Xchoo(i)=Mass(i)*DM(i)*DQch(i)*(NFE(i)+CF(i)); Xproo(i)=Mass(i)*DM(i)*DQpr(i)
*(CP(i));
Xlioo(i)=Mass(i)*DM(i)*DQli(i)*(CL(i));
End

xch0=0;
xpr0=0;
xli0=0;

For i=1:3
xch0=xch0+Xchoo(i)/Volume;
xpr0=xpr0+Xproo(i)/Volume;
xli0=xli0+Xlioo(i)/Volume;
end
$Molecular weights of the biomass fractions
PMch=817.1;
PMpr=326.8;
```

```
PMli = 560.3;
[a,b] = ode45('Reac',[0 30],[xch0,xpr0,xli0,0,0,0,0,0.01])
subplot(2,2,1);plot(a,b(:,1));xlabel('Time (d)');ylabel('Carbohydrates(kg/d)')
subplot(2,2,2);plot(a,b(:,2));xlabel('Time (d)');ylabel('Protein (kg/d)')
subplot(2,2,3); plot(a,b(:,3));xlabel('Time (d)');ylabel('Lipids (kg/m3)')
subplot(2,2,4);plot(a,b(:,4));xlabel('Time (d)');ylabel('Cells (kg/m3)')
figure
subplot(2,2,1);plot(a,b(:,5));xlabel('Time (d)');ylabel('GasCH4 (kg/m3)')
subplot(2,2,2);plot(a,b(:,6));xlabel('Time (d)');ylabel('GasCO2 (kg/m3)')
subplot(2,2,3);plot(a,b(:,7));xlabel('Time (d)');ylabel('GasH2S (kg/m3)')
subplot(2,2,4);plot(a,b(:,8));xlabel('Time (d)');ylabel('GasNH3 (kg/m3)')
function Reactor = Reac(t,x)%Units, kg, mol,m^3 y s
xch = x(1); xpr = x(2); xli = x(3); xcell = x(4); xCH4 = x(5); xCO2 = x(6); xH2S = x(7);
xNH3 = x(8);
khyd1 = 0.4; khyd2 = 0.2; khyd3 = 0.1;kdeath = 0.02;
Reactor(1,1) = -khyd1*xch + 0.18*kdeath*xcell;
Reactor(2,1) = -khyd2*xpr + 0.82*kdeath*xcell;
Reactor(3,1) = -khyd3*xli;
Reactor(4,1) = -kdeath*xcell + (0.1506*khyd1*xch) + (0.1241*khyd2*xpr) +
(0.1857*khyd3*xpr);
Reactor(5,1) = ((0.2433*khyd1*xch) + (0.2383*khyd2*xpr) + (0.6588*khyd3*xli));
Reactor(6,1) = (0.6675*khyd1*xch) + (0.6822*khyd2*xpr) + (0.6644*khyd3*xli);
Reactor(7,1) = (0.0012*khyd2*xpr);
Reactor(8,1) = -(0.0227*khyd1*xch) + (0.1645*khyd2*xpr) -(0.0280*khyd3*xli);
end
```

See Figs. 6E1.1 and 6E1.2.

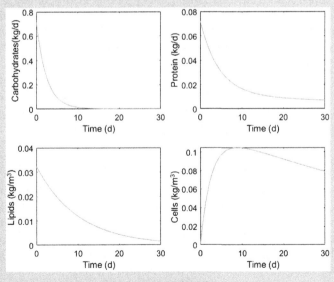

FIGURE 6E1.1 Degradation of products and cells kinetics.

3. Biomass and waste based processes

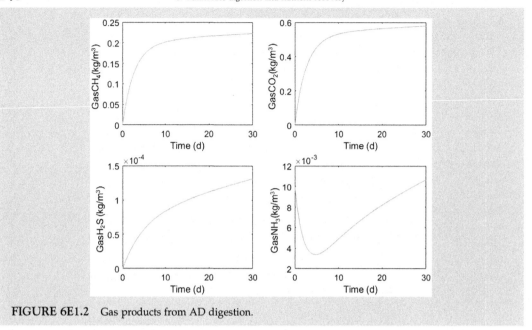

FIGURE 6E1.2 Gas products from AD digestion.

6.1.4 Reactor features

There are different types of biogas reactors that can be classified according to their operating mode: batch reactors, one-stage continuously fed systems, and two-stage (or even multistage) continuously fed systems where the hydrolysis/acidification and acetogenesis/methanogenesis stages are carried out in different compartments (Ward et al., 2008).

Regarding their construction, the most common digesters are continuous stirred tank reactors (CSTR), built either on the ground level or underground. Horizontal or vertical continuously feed plug flows are also used for AD. The digestion time can be easily controlled in these digesters, where the mixing is obtained through wheel agitators with pallets and/or the addition of baffles on the sides of the digester. Finally, some special digesters, with a more limited usage, with different compartments for carrying out the hydrolysis/ acidification and acetogenesis/ methanogenesis stages have been designed, although their usage is more limited than the other types of digesters (Fachagentur Nachwachsende Rohstoffe, 2010; Ward et al., 2008).

In addition to the physical unit, digesters can also be divided into wet or dry types. Wet digesters, used in applications such as manure processing, are those which operate with biomass containing up to 16% solids. On the other hand, dry digesters are fed with organic wastes containing 22%—40% of solids. These digesters are used in applications involving biomass with low moisture content, such as municipal solid waste. Another configuration in between of the conventional and dry digesters are semidry configurations for the decomposition of substrates with total solids content between 10% and 15% (Mata-Alvarez, 2015; Petracchini et al., 2018).

The energy required by the digester can be estimated through energy balances. The reactor energy balance can be computed by the heat of combustion of the species involved (Eq. 6.25). It is an endothermic process that requires energy to sustain the operating temperature during the digestion time. In addition, the feedstock must be preheated to the operating temperature (Eq. 6.24). Since AD is a long process carried out along several days, the energy losses must also be computed (Eq. 6.27).

$$Q_{\text{Feed}} = \sum_i m_i \cdot cp_i \cdot (T_{\text{AD}} - T_{\text{air}}) \tag{6.24}$$

$$Q = Q_R + Q_{\text{Losses}} \tag{6.25}$$

$$Q_R = \sum_P \Delta H_{\text{comb},i} - \sum_R \Delta H_{\text{comb},i} + m_{\text{evaporated}} \lambda \tag{6.26}$$

$$Q_{\text{Losses}} = \sum_{\text{areas}} A_i \cdot h_i \cdot (T - T_{\text{air}}) \tag{6.27}$$

The heat transfer coefficients for different walls are provided in Table 6.2 to compute the energy balance to the digester.

The energy required for the digestion process can be provided through direct and indirect heating:

- Direct heating: By injecting steam or hot water. However, this option presents major drawbacks, including high operating cost, potential local overheating, and variations of the composition of the digested biomass as a consequence of the injection of additional water in form of steam.
- Indirect heating: Among this alternative, four heating configurations are possible:
  - Floor heating: The use of this alternative can be hindered by the accumulation of solids that reduces the heat transfer.
  - In-vessels: The use of in-vessel heat exchangers is an interesting configuration as long as they can cope with the stress of the mixer and the recirculating pump.
  - On-vessel: Using heat conductors located in or on the walls of the digester. In spite of the large contact area, the main disadvantage of this technology is the loss of energy to the surroundings.
  - Ex-vessel: A heat exchanger is used to recirculate the slurry back to the digester and provide the energy required. It is the easier alternative to operate (see Fig. 6.3).

In addition to the thermal energy, pumping and mixing energy is consumed in the production. For digester with an ex-vessel heating system the piping and mixing energy requirements are 14 kW/500 m^3 (Zhang, 2003).

TABLE 6.2 Heat transfer coefficients for losses.

Structure	h (W/m$^2 \cdot$ °C)
Concrete wall 300 mm thick, not insulated (above ground)	4.7–5.1
Concrete wall 300 mm thick, insulated (above ground)	0.6–0.8
Concrete floor 300 mm thick (in contact with dry earth)	1.7
Fixed concrete cover 100 mm thick and covered, 25 insulation	1.2–1.6
Floating cover with 25 mm insulation	0.9–1.0
6 mm steel plate "sandwich" with 100 mm insulation	0.35

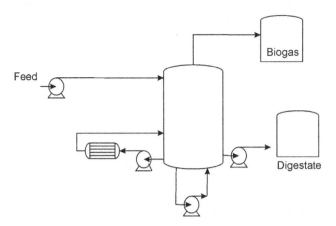

FIGURE 6.3 Scheme of ex-vessel heating for a digestor.

Standard sizes for the digesters of 1000, 3000, and 6000 m^3 can be found. Mixing and heat transfer across the superficial area can be a challenge for larger sizes (Fachagentur Nachwachsende Rohstoffe, 2010).

6.2 Biogas upgrading

Biogas upgrading refers to the purification of biogas into biomethane removing the impurities and the carbon dioxide. It can be divided in two stages, a purification stage where the biogas impurities are removed, that is, ammonia, sulfur hydride, and siloxanes, and a second stage where biomethane is produced through either the removal or the methanation of CO_2.

6.2.1 Biogas purification

6.2.1.1 Moisture removal

The biogas produced contains water saturating at operating temperature. The first stage for water removal is the moisture condensation, which can remove residual dust and oil as well. However, the biogas moisture can remain saturated at the cooled down temperature, and additional separation stage can be needed, using adsorbent agents such as silica gel in packed bed columns operating at pressures between 6 and 10 bar (Rykebosch et al., 2011).

6.2.1.2 H$_2$S removal

The removal of hydrogen sulfide can be carried out during digestion or from the biogas produced (Rykebosch et al., 2011).

6.2.1.2.1 In-digestion removal

1. H$_2$S can be biologically oxidized during the anaerobic digestion of the biomass through a controlled supply of air or oxygen, producing elemental sulfur. Removal yields up to

99% can be achieved using this process. The microorganisms involved in the process use the CO_2 from biogas as a source of C. However, sulfates can be produced during the biologival oxidation, and as a result, the concentration of sulfur compounds may be too large for the biogas to be used as substitute of natural gas.

$$O_2 + 2H_2S \rightarrow 2S + 2H_2O \tag{6.28}$$

2. Addition of iron chloride to form FeS that precipitates and is removed in solid form. This process faces significant drawbacks, including lower efficiency than biological oxidation, it is a process difficult to operate and expensive, and the addition of iron chloride affects the pH of the anaerobic digestion.

6.2.1.2.2 In-biogas removal

1. Precipitates formation: sulfur can be removed in form of Fe_2S_3 by treating the biogas in a fixed absorbent bed of Fe_2O_3 operating at 25°C–50°C. It is a highly efficient process, although it is sensitive to the presence of water. Therefore, the removal of moisture has to be carried out previously. The Fe_2S_3 is formed in the bed as follows:

$$Fe_2O_3 + 3H_2S \rightarrow Fe_2S_3 + 3H_2O \tag{6.29}$$

The regeneration of the bed is carried out using oxygen:

$$2Fe_2S_3 + 3O_2 \rightarrow 2Fe_2O_3 + 6S \tag{6.30}$$

2. Removal using liquids: it can be physical adsorption using compounds such as methanol, or chemical absorption through the use of amines scrubbing, diluted NaOH solutions forming solid Na_2S, $FeCl_2$ solutions that results in insoluble FeS, $Fe(OH)_3$ solutions generating Fe_2S_3, or just water scrubbing. The chemicals improve the absorption based on acid-base reactions.
3. Biological filters, similar to the operation of in-situ removal.
4. Adsorption on activated carbon.
5. Membrane separation: H_2S and CO_2 pass through the membrane while methane is retained in the biogas stream. This technology is also used for CO_2 removal.

6.2.1.3 NH₃ removal

Adsorption methods such as activated carbon or zeolites, as well as water scrubbing, are used for ammonia removal from biogas.

6.2.1.4 Siloxane removal

Siloxanes are organic components containing the Si-O-Si group linking organic chains. Although siloxanes are undesirable impurities in biogas, since they damage engines, siloxanes are valuable compounds used in cosmetics and pharma. The removal of siloxanes is carried out through absorption with organic solvents, reaching removal efficiencies of 97%, or through the use of adsorbent beds made by molecular sieves, silica gel or activated carbon, both of them with removal efficiencies up to 95%. The operation of adsorption systems require higher pressures than absorption-based processes.

6.2.2 Biogas to biomethane and power

After the removal of the impurities contained in the biogas, it can be upgraded to biomethane via the capture of CO_2, the largest species together with the methane (Martín-Hernández et al., 2020a). CO_2 capture is a sour gas removal stage that is common to other processes. Therefore, some of these processes have been already described in previous sections. In particular, we refer the reader to Chapter 4 for a detailed description about the use of amines for carbon dioxide removal.

6.2.2.1 Amines

See Chapter 4 for the details of this technology (GPSA, 2004).

6.2.2.2 Pressure swing adsorption

Pressure swing adsorption (PSA) is a gas removal system based on the adsorption of components on fixed beds at a certain pressure, while the regeneration of the bed is carried out through the desorption of the gas at a lower pressure. In the case of CO_2 capture, zeolites are one of the most common adsorbents. The PSA system consists of a compression train to raise the pressure of the biogas, and the zeolite beds, see Fig. 6.4. Two beds operate in parallel so that while one is in adsorbing mode the second in under regeneration.

Since the adsorption capacity of the zeolites is directly related to the partial pressure of the CO_2, a train of compressors with intermediate cooling may be needed to achieve the optimal operating pressure. The compressors can be modeled assuming polytropic behavior, considering a polytropic coefficient k of 1.4, and an efficiency of the compression stages of 0.85.

$$
\begin{aligned}
T_{\text{out/compresor}} &= T_{\text{in/compressor}} + T_{\text{in/compressor}} \left(\left(\frac{P_{\text{out/compressor}}}{P_{\text{in/compressor}}} \right)^{\frac{k-1}{k}} - 1 \right) \frac{1}{\eta_c} \\
W_{\text{(Compressor)}} &= (F) \cdot \frac{R \cdot z \cdot (T_{\text{in/compressor}})}{((MW) \cdot (k-1))} \frac{1}{\eta_c} \left(\left(\frac{P_{\text{out/compressor}}}{P_{\text{in/compressor}}} \right)^{\frac{k-1}{k}} - 1 \right)
\end{aligned}
\tag{6.31}
$$

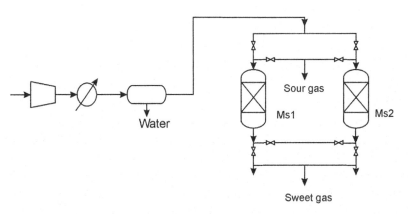

FIGURE 6.4 Zeolite system scheme.

The removal ratio is given by the breakthrough curve of the adsorbent bed. Based on experimental data (Hauchhum and Mahanta, 2014) for adsorption stages below 20 min, the CO_2 removal yield (η) is assumed to be 98%, containing below 2% CO_2 at the outlet stream, (Ferella et al., 2017). To compute the mass of bed, the adsorption capacity of the bed is calculated. Based on experimental data, the Langmuir isotherm is the adsorption model considered for this process, Eq. (6.32), since this is the one that best fits the performance of the zeolite 13X−CO_2 system (Hauchhum and Mahanta, 2014).

$$q = \frac{q_m \cdot K \cdot P_{CO_2}}{1 + K \cdot P_{CO_2}} \tag{6.32}$$

where the parameters q_m and K depend on the adsorbent material. Eqs. (6.33a) and (6.33b) show the correlations to estimate these parameters for zeolite 13X and 4A, respectively.

$$q_m(\text{mol/kg}) = -3.15551 \cdot 10^{-02} T(^\circ C) + 5.02915$$
$$K(\text{bar}^{-1}) = 1.63070 \cdot 10^{-03} \, T(^\circ C)^2 - 3.68662 \cdot 10^{-01} \, T(^\circ C) + 27.3737 \tag{6.33a}$$

$$q_m(\text{mol/kg}) = -1.82355 \cdot 10^{-02} \, T(^\circ C) + 3.72021$$
$$K(\text{bar}^{-1}) = 1.63070 \cdot 10^{-03} \, T(^\circ C)^2 - 3.68662 \cdot 10^{-01} \, T(^\circ C) + 27.3737 \tag{6.33b}$$

However, the adsorption capacity decays cycle after cycle until it stabilizes around 65% of the initial capacity (Hauchhum and Mahanta, 2014). Therefore, a corrected value for q is applied to compute the amount of zeolite used in the PSA system, as it can be shown in Eq. (6.34), assuming τ is equal to 20 min.

$$m_{Zeolite} = \frac{1}{q \cdot 0.65} \frac{fc_{CO_2} \cdot 1000}{MW_{(CO_2)}} \eta \cdot \tau \tag{6.34}$$

Typical lifetime of the zeolites bed is around 5 years, therefore that is the replacement time for the beds. The cost of the zeolites is around 5 \$/kg for both zeolite 13 X and zeolite 4A (Xiao et al., 2013). Example 6.2 shows the sizing of an adsorbent bed.

EXAMPLE 6.2 Compute the mass of zeolite 4A to remove 99% of the CO_2 in a stream of 0.1 kg/s of biogas, with a molar composition 0.59 CH_4 and the rest CO_2. Assume that the biomethane contains 2% CO_2 by mass and the bed operates at 25°C and 1 bar of total pressure neglecting the pressure drop across the bed.

Solution

CO_2 stream:
$fc_{CO_2}|_{out} = \eta fc_{CO_2}|_{in}$
$fc_{CH_4}|_{out} = fc_{CH_4}|_{in} - fc_{CH_4}|_{biomethane}$
CH_4 stream:
$fc_{CO_2}|_{biomethane} = (1-\eta)fc_{CO_2}|_{in}$
$0.02\left(fc_{CH_4}|_{out} + fc_{CO_2}|_{out}\right) = fc_{CO_2}|_{out}$

$$fc_{CH_4}|_{biomethane} = \frac{0.98}{0.02}fc_{CO_2}|_{out}$$

(6E2.1)

The composition across the PSA changes. Therefore, the driving force for absorption also changes. To compute an average partial pressure for the CO_2 we use the logarithmic mean where the final composition in CO_2, 0.007, is computed form the mass balance above:

$$y_{average} = \frac{y_{i,feed} - y_{i,final}}{\ln\left(\frac{y_{i,feed}}{y_{i,final}}\right)} = \frac{0.41 - 0.007}{\ln\left(\frac{0.41}{0.007}\right)} = 0.1 \tag{6E2.2}$$

$$q = \frac{q_m \cdot K \cdot P_{CO_2}}{1 + K \cdot P_{CO_2}} = \frac{3.26 \cdot 19.18 \cdot 0.1}{1 + 19.18 \cdot 0.1} = 2.15 \text{ mol/kg} \tag{6E2.3}$$

However, due to the losses in efficiency over the cycles q is corrected to compute the mass of bed:

$$m_{Zeolite} = \frac{1}{2.15\frac{mol}{kg} \cdot 0.65} \frac{0.065\frac{kg}{s} \cdot 1000\frac{g}{kg}}{44\frac{g}{mol}} (20 \cdot 60)s = 1270 \text{ kg} \tag{6E2.4}$$

6.2.2.3 Membranes

Membranes allows the separation of compounds that permeate the membrane material from the components that do not permeate. Different configurations of membranes and compression stages can be arranged (Deng and Hägg, 2010; Scholz et al., 2015), see Fig. 6.5, including single-stage configuration (a) and double-stage configuration. For the double-stage configuration, three different setups are possible: double-stage membranes with compression before each membrane and no recycle (b), double-stage membranes with compression before each membrane with recycle (c), and symmetric cascade and (d) asymmetric cascade. Among these alternatives, configuration (d) has proven to be the most economic process layout under wide ranges of feed conditions (Kim et al., 2017).

Each membrane module can be modeled through mass balances, considering the permeate and retentate streams (Eqs. 6.35 and 6.37), and the flux of the gases through the membrane, that is a function of the concentration gradient between both sides of the membrane (Eq. 6.38) (Fernandes Rodrigues, 2009). The flux is the parameter which allows computing the area of the membrane, as it is shown in Eq. (6.37), based on the permeability of the membrane (Eq. 6.40). As the driving force in the membrane separation process is the concentration gradient, the removal of CO_2 results in a change in the composition of the stream along the membrane, leading to a change in the driving force which controls the process. Therefore, an average molar fraction between the feed and the retentate composition is used to compute the separation driving force (Eq. 6.39). In these equations F denotes the total molar flow, $A_{membrane}$ and δ the membrane area thickness respectively, y_i the molar fraction, J_i the transversal flux ($kmol/(m^2 \cdot s)$), ε_i the permeance ($kmol/(kPa \cdot m^2)$), and $Perm_i$ the permeability of the component i ($mol/(kPa \cdot m)$).

$$F_{feed} = F_{permeate} + F_{retentate} \tag{6.35}$$

$$F_{feed} \cdot y_{i,feed} = F_{permeate} \cdot y_{i,permeate} + F_{retentate} \cdot y_{i,retentate}; \; i \in (CO_2, CH_4) \tag{6.36}$$

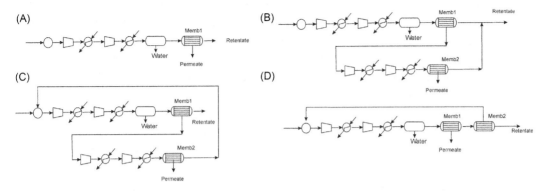

FIGURE 6.5 Membrane configurations. (A) Single module; (B) Two modules in series with permeate recompresion; (C) Two modules in series with retentate recompresion and recycle; (D) Two modules in series with a single compression stage and secondary permeate recycle.

TABLE 6.3 Gases permeability (Vrbová and Ciahotný, 2017).

	Permeability (Barrer)	
Polymer	CH$_4$	CO$_2$
Cellulose acetate	0.21	6.30
Polycarbonate	0.13	4.23
Polyimide	0.25	10.7

$$J_i = \frac{F_{\text{permeate}} \cdot y_{i,\text{permeate}}}{A_{\text{membrane}}}; \ i \in (CO_2, CH_4) \tag{6.37}$$

$$J_i = \varepsilon_i \left[y_{\text{feedside}} \cdot P_{\text{feed}} - y_{i,\text{permeate}} \cdot P_{\text{Permeate}} \right]; \ i \in (CO_2, CH_4) \tag{6.38}$$

$$y_{\text{feedside}} = \frac{y_{i,\text{feed}} - y_{i,\text{retentate}}}{\ln \left(\frac{y_{i,\text{feed}}}{y_{i,\text{retentate}}} \right)}; \ i \in (CO_2, CH_4) \tag{6.39}$$

$$\varepsilon_i = \frac{Perm_i}{\delta}; \ i \in (CO_2, CH_4) \tag{6.40}$$

The usual membrane thickness for industrial units is equal to 30 nm. The membrane materials more suitable for carbon dioxide removal are those one with large CO$_2$ permeability, low methane permeability, and therefore, high selectivity, including cellulose acetate, polyamide, and polycarbonate. Table 6.3 shows examples of several material permeabilities for CO$_2$ and CH$_4$ at 25°C (Vrbová and Ciahotný, 2017). A differential type of modeling to account for the variation of the concentration of the streams along the membrane can be seen in Fernandes Rodrigues (2009).

A membrane price of $50 \, \$/m^2$, based on the literature (Kim et al., 2017), can be used for economic assessment purposes. Considering a plant life equal to 20 years, the membranes, typical lifetime is 4 years, must be replaced 5 times during the plant life, $N_{Membranes}$ (Scholz et al., 2015). Example 6.3 presents the modelling of a membrane.

EXAMPLE 6.3 **For the production of bionatural gas from a stream 59% methane and the rest CH$_4$, minimize the area for maximum production of methane 98% pure using cellulose acetate membrane.**

Solution
By implementing the system of equations presented above in GAMS for instance we have:

```
Scalar
ConF        in kmol per s    m per(m2 atm)      /3.395E-14/
barrerCO2                               /6.3/
barrerCh4                               /0.21/
espesor     /30e-9/;

Positive variables
Ffeed, Freten, Fperm, Area, ych4feed, ych4reten, ych4perm, yco2feed, yco2reten,
yco2perm, Pfeed,Pperm, Jch4, JCO2, epsilonch4, epsilonCO2, Prel;

Variable
Z;

Ffeed.fx = 100;
yCH4feed.fx = 0.59;
yco2feed.fx = 0.41;
ych4reten.fx = 0.98;
Area.lo = 150;
Pfeed.LO = 35;
Pperm.fx = 1;

Equations
BalM1, BalM2, BalM3, BalM4, Flux1, Flux2, Flux3, Flux4, Yield, obj;
BalM1. Ffeed = E = Freten + Fperm;
BalM2.    Ffeed*ych4feed = E = Freten*ych4reten + Fperm*ych4perm;
BalM3.       ych4reten + yco2reten = E = 1;
BalM4.       ych4perm + yco2perm = E = 1;
Flux1.        Fperm*ych4perm = E = Jch4*Area;
Flux2.        Fperm*yco2perm = E = JCO2*Area;
```

```
Flux3.      Jch4*espesor = E = barrerCh4*conf*((ych4feed + ych4reten)/2*Pfeed-
ych4perm*Pperm);
Flux4.      JCO2*espesor = E = barrerCO2*conf*((yco2feed + yco2reten)/2*Pfeed-
yco2perm*Pperm);
Yield.    Freten*ych4reten = E = 0.75*Ffeed*ych4feed;

obj. Z = E = Area−Freten*ych4reten;

Model Membrane /ALL/;

Option NLP = conopt;
Solve Membrane Using NLP Minimizing Z;
```

The area is equal to 150 m^2.

6.2.2.4 Cryogenic

The difference in liquefaction temperature of methane and CO_2 allows their separation via compression and cooling. Water must be previously removed from biogas to avoid operating issues. In addition, the nature of the CO_2 as solvent separates the remaining impurities from biogas. Typically, 8 MPa are used and the mixture is cooled to $-45°C$ removing the CO_2. In a second step, the biogas is further cooled to $-55°C$ and expanded to 0.8–1 MPa. In the expansion, the gas cools down generating a solid phase consisting mainly of CO_2, and a gas phase consisting of high-purity methane. Example 6.4 presents the yield of the removal of CO_2 using a process simulator.

EXAMPLE 6.4 Simulate the cryogenic CO_2 removal from biogas considering a maximum pressure of 12.5 MPa, cooling to $-45°C$ and a using an expansion valve to 0.8 MPa. Compute the composition of the methane.

Solution

For details on how to model a flowsheet with CHEMCAD we refer to Martín (2019). Note that cryogenic conditions are used for two nonpolar species. Peng Robinson is the suggested by CHEMCAD, although SRK is typically an option too (Fig. 6E4.1).

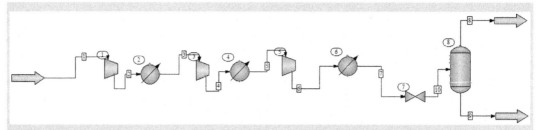

FIGURE 6E4.1 Flowsheet for cryogenic removal of CO_2.

	SRK	Peng-Robinson
CH_4 (% molar)	0.9	0.90
CO_2 (% molar)	0.1	0.1
T (°C)	−91.6	−92.8
Flow (kg/s)	4.46	4.38

6.3 Biogas uses

6.3.1 As fuel

Biogas, either with CO_2 or upgraded to biomethane, can be used as fuel for transportation and power generation. Clean biogas can be defined as the mixture of methane and carbon dioxide after impurities removal, while biomethane refers to upgraded biogas where the CO_2 is removed, resulting in high-purity methane.

Clean biogas can be burned in boilers or burners to generate utilities such as steam and hot water. In addition, clean biogas can be used to operate combined heat and power units (CHP), being a very widely used alternative. These engines have an overall efficiency of around 90%, which is divided into 35% for electricity generation and 65% for heat production. These facilities consist of a combustion engine (i.e., a gas-based Otto engine) coupled with a generator for power production. CHP units are usually used for facilities requiring power up to 2 MW. Alternatively, Stirling engines can be used. These engines do not use internal combustion, rather the pistons are moved when the gas is expanded as a result of its heating up using a burner. Their efficiency is lower than the gas-based Otto engine, 24%−28%, and are intended for power production ranges from 50 kW to 4 MW. Apart from engines, we can consider the use of gas micro-turbines for the generation of large amounts of power energy, which are gaining support due to the size of most biogas production facilities (Al Seadi et al., 2008). A gas turbine compresses air to burn the gas in a combustion chamber, increasing the temperature and pressure of the resulting flue gases, so that power is produced through expansion of these gases in the turbine. The residual heat contained in the flue gases leaving the turbine can still be used to produce steam, which can be used as a source of thermal energy, or to be used in a steam turbine within a combined cycle structure. León and Martín (2016) evaluated different configurations for the use of the hot flue gases to the production of steam in an integrated facility for the production of electricity from the anaerobic digestion of organic waste, either (a) the flue gas is split so that high temperature flue gas is used for both the overheating of the steam and for the regeneration step, or (b) the use of the flue gas to heat up in sequence first to overheat the steam before feeding the higher pressure turbine, then the reheating stage of the regenerative Rankine cycle, followed by the evaporation of water and finally the initial heating up of the compressed water (see Fig. 6.6). The results shown that this second option provided slightly better yield.

The gas turbine operates following a Brayton cycle (Moran et al., 2014). It consists of a multistep compression for the air and the biogas. An excess of air is typically used to control the temperature within the turbine so that it does not go over 1800°C. The polytropic coefficient and the compression efficiency can be assumed equal to 1.4 and 0.85, respectively. For the air stream, a maximum compression ratio of 40 can be considered based on typical achievements. The combustion of the biogas, assumed adiabatic, is used to heat up the air-biogas mixture. The turbine produces power through the expansion of the gases. Values for

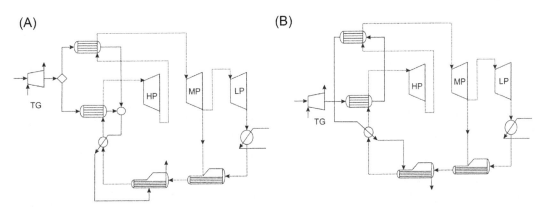

FIGURE 6.6 Electricity production using biogas turbines. (A) Hot flue gas split for the reheating and overheating stages; (B) Hot flue gas used in sequence at over and reheating.

the polytropic coefficient and the efficiency of 1.3 and 0.85 can typically be assumed. Open or closed Brayton cycles can be used whether the expanded gas is recompressed or exits the cycle. The detailed analysis of a gas turbine can be seen in Chapter 4 and Chapter 9.

6.3.2 Biogas to chemicals

Biogas contains all the components needed for the production of syngas via dry reforming. The mixture between methane and CO_2, with traces of water, is an interesting raw material for the synthesis of syngas. Several processes based on the use of methane and CO_2 are being studied including dry reforming, trireforming and superdry reforming. The difference between them is the mixture of CO_2 and CH_4, and the addition of water, CO_2, and/or oxygen.

6.3.2.1 Dry reforming

Dry reforming (DR) consists of an endothermic reaction between methane and carbon dioxide for the production of a mixture of CO and hydrogen, the syngas. This reaction is governed by the equilibrium presented below, where pressure P is given in bar and temperature T in Celsius (Luyben, 2014).

$$CH_4 + CO_2 \leftrightarrow 2CO + 2H_2$$

$$kp = e^{\left[31.447 - \frac{29580}{T}\right]} = \frac{P_{CO} \cdot P_{H_2}}{P_{CH_4} \cdot P_{CO_2}} \tag{6.41}$$

However, the presence of traces of water result in the occurrence of two well-known reactions, the steam reforming of methane (Eq. 6.42), and the water gas shift reaction (Eq. 6.43); both governed by thermodynamic equilibrium. In these correlations pressure P is given in bar and temperature T in Kelvin (Luyben, 2014; Roh et al., 2010).

$$CH_4 + H_2O \leftrightarrow CO + 3H_2$$

$$kp = 10^{\left[-\frac{11650}{T} + 13,076\right]} = \frac{P_{CO} \cdot P_{H_2}^3}{P_{CH_4} \cdot P_{H_2O}} \tag{6.42}$$

$$CO_{(g)} + H_2O_{(g)} \leftrightarrow CO_{2(g)} + H_{2(g)}$$

$$kp = 10^{\left[\frac{1910}{T} - 1,784\right]} = \frac{P_{CO_2} \cdot P_{H_2}}{P_{CO} \cdot P_{H_2O}} \tag{6.43}$$

In fact, among Eqs. (6.41)–(6.43), only two of them are linearly independent, while the remaining one can be computed as a linear combination of them. The analysis of the yield of the dry reforming is performed by the equilibrium presented above and the elementary mass balance (Hernández and Martin, 2018).

Two types of reactors are typically used, packed beds (Benguerba et al., 2015), or fluidized bed reactors (Zambrano et al., 2020) (see Fig. 6.7). In the first case, the reactors are similar to the reforming furnaces typically used in other methane reforming applications (Martín, 2016), while in the second reactor type the thermal fluid is in charge of controlling the temperature.

The kinetics of the dry reforming in packed beds, based on the work by Benguerba et al. (2015), are shown in Table 6.4. The values of the kinetic coefficients are collected in

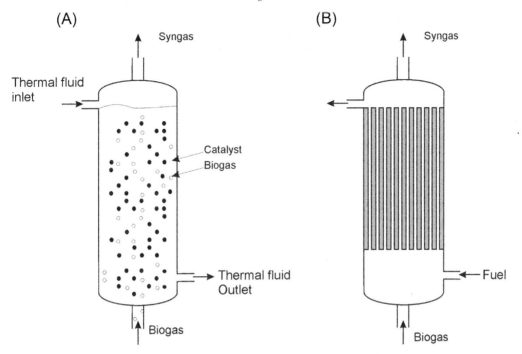

FIGURE 6.7 Reactor configuration for dry reforming. (A) Fluidized bed reactor; (B) Packed bed reactor.

TABLE 6.4 Kinetics for dry reforming in packed beds (Benguerba et al., 2015).

Reaction	Kinetics	ΔH (kJ/mol)
$CH_4 + CO_2 \leftrightarrow 2CO + 2H_2$	$r_1 = \dfrac{k_1 K_{CO_2,1} P_{CO_2} K_{CH_4,1} P_{CH_4}}{\left(1 + K_{CO_2,1} P_{CO_2} + K_{CH_4,1} P_{CH_4}\right)^2}\left(1 - \dfrac{\left(P_{CO} P_{H_2}\right)^2}{K_{P_1}\left(P_{CO_2} P_{CH_4}\right)}\right)$	247
$CO_{2(g)} + H_{2(g)} \leftrightarrow CO_{(g)} + H_2O_{(g)}$	$r_2 = \dfrac{k_2 K_{CO_2,2} P_{CO_2} K_{H_2,2} P_{H_2}}{\left(1 + K_{CO_2,2} P_{CO_2} + K_{H_2,2} P_{H_2}\right)^2}\left(1 - \dfrac{\left(P_{CO} P_{H_2O}\right)}{K_{P_2}\left(P_{CO_2} P_{H_2}\right)}\right)$	41.7
$CH_{4(g)} \leftrightarrow C + 2H_{2(g)}$	$r_3 = \dfrac{k_3 K_{CH_4,3}\left(P_{CH_4} - \dfrac{P_{H_2}^2}{K_{P_3}}\right)}{\left(1 + K_{CH_4,3} P_{CH_4} + \dfrac{1}{K_{2,3}} P_{H_2}^{1.5}\right)^2}$	74.87
$C + H_2O_{(g)} \leftrightarrow CO_{(g)} + H_{2(g)}$	$r_4 = \dfrac{\dfrac{k_4}{K_{H_2O,4}}\left(\dfrac{P_{H_2O}}{P_{H_2}} - \dfrac{P_{CO}}{K_{P_4}}\right)}{\left(1 + K_{CH_4,4} P_{CH_4} + \dfrac{P_{H_2O}}{K_{H_2O,4} P_{H_2}} + \dfrac{1}{K_{H_2,4}} P_{H_2}^{1.5}\right)^2}$	131.325
$C + CO_{2(g)} \leftrightarrow 2CO_{(g)}$	$r_5 = \dfrac{\dfrac{k_5}{K_{CO,5} K_{CO_2,5}}\left(\dfrac{P_{CO_2}}{P_{CO}} - \dfrac{P_{CO}}{K_{P_5}}\right)}{\left(1 + K_{CO,5} P_{CO} + \dfrac{P_{CO_2}}{K_{CO,5} K_{CO_2,5} P_{CO}}\right)^2}$	172

Table 6.5. Pressures (*P*) are in bar, the universal gas constant (*R*) in J/(mol · K), and temperature (*T*) in Kelvin. Example 6.5 shows the estimation of a kinetic model for the dry reforming of biogas.

TABLE 6.5 Kinetic and thermodynamic coefficients for dry reforming in packed beds. Reaction rates are in mol kg^{-1}s^{-1}, partial pressures are in bar, and temperatures are in Kelvin (Benguerba et al., 2015).

Parameter	Value	Parameter	Value
k_1	$1.29 \cdot 10^6 \exp\left(-\frac{102,065}{R \cdot T}\right)$	$K_{H_2O,4}$	$4.73 \cdot 10^{-6} \exp\left(\frac{97,770}{R \cdot T}\right)$
k_2	$0.35 \cdot 10^6 \exp\left(-\frac{81,030}{R \cdot T}\right)$	$K_{CH_4,4}$	$3.49 \exp\left(\frac{0}{R \cdot T}\right)$
k_3	$6.95 \cdot 10^3 \exp\left(-\frac{58,893}{R \cdot T}\right)$	$K_{H_2,4}$	$1.83 \cdot 10^{13} \exp\left(-\frac{216,145}{R \cdot T}\right)$
k_4	$5.55 \cdot 10^9 \exp\left(-\frac{166,397}{R \cdot T}\right)$	$K_{CO,5}$	$7.34 \cdot 10^{-6} \exp\left(\frac{100,395}{R \cdot T}\right)$
k_5	$1.34 \cdot 10^{15} \exp\left(-\frac{243,835}{R \cdot T}\right)$	$K_{CO_2,5}$	$2.81 \cdot 10^7 \exp\left(-\frac{104,085}{R \cdot T}\right)$
$K_{CO_2,1}$	$2.61 \cdot 10^{-2} \exp\left(\frac{37,641}{R \cdot T}\right)$	K_{P_1}	$6.78 \cdot 10^{14} \exp\left(-\frac{259,660}{R \cdot T}\right)$
$K_{CH_4,1}$	$2.60 \cdot 10^{-2} \exp\left(\frac{40,684}{R \cdot T}\right)$	K_{P_2}	$56.4971 \exp\left(-\frac{36,580}{R \cdot T}\right)$
$K_{CO_2,2}$	$0.5771 \exp\left(\frac{9,262}{R \cdot T}\right)$	K_{P_3}	$2.98 \cdot 10^5 \exp\left(-\frac{84,400}{R \cdot T}\right)$
$K_{H_2,2}$	$1.494 \exp\left(\frac{6,025}{R \cdot T}\right)$	K_{P_4}	$1.3827 \cdot 10^7 \exp\left(-\frac{125,916}{R \cdot T}\right)$
$K_{CH_4,3}$	$0.21 \exp\left(\frac{567}{R \cdot T}\right)$	K_{P_5}	$1.9393 \cdot 10^9 \exp\left(-\frac{168,527}{R \cdot T}\right)$
$K_{H_2,3}$	$5.18 \cdot 10^7 \exp\left(-\frac{133,210}{R \cdot T}\right)$		

EXAMPLE 6.5 Develop a kinetic model for the packed bed performing the dry reforming of biogas mixture. Assume 1:1 molar ratio of methane and CO_2 with a feed rate of 2, 2, 0.5, 0.1, and 0.1 for CO_2, CH_4, H_2O, H_2 and CO, respectively. The characteristics of the bed and the reactor are as follows:

Pressure = 1 bar; ρ = 800 kg/m^3; L = 20 m; D1 = 0.01 m; TW = 550 + 273 K; ω = (3.14*D1^2/4) m^2; U_{glo} = 4000; J/m^{2}K; cp$_{gases}$ = 2500; kJ/molK

$$\pm \frac{dF_i}{dz} = \rho \omega L \left(\sum_{j=1...5} \nu_i r_j \right)$$

$$\frac{dT}{dz} = \frac{\rho L}{u \rho_g cp} \left(\sum_{j=1...5} r_j \Delta H_j \right) - \frac{4 \cdot U_{glo} \cdot L}{D \cdot u \rho_g cp} (T - T_w) \qquad \text{(6E5.1)}$$

Solution

We write the model in MATLAB as follows:

```
FCOini = 0.1; FH2ini = 0.1; FH2Oini = 0.5; FCH4ini = 2; FCO2ini = 2; Tempini = 500 + 273;
[a,b] = ode15s('ReacDryref',[0 20],[FCH4ini,FCO2ini,FCOini,FH2ini,FH2Oini,Tempini]);
plot(a,b(:,1),'k-',a,b(:,2),'k-',a,b(:,3),'k-o',a,b(:,4),'k-*',a,b
(:,5),'k-<')
```

```
xlabel('Z (m)'); ylabel('Gas '); legend('CH4','CO2','CO','H2','H2O')
figure
plot(a,b(:,6),'k-'); xlabel('Z (m)'); ylabel('Temperature (K)')

function Rcat = ReacDryref(w,F)

FCH4 = F(1); FCO2 = F(2); FCO = F(3); FH2 = F(4); FH2O = F(5); Temp = F(6);

%Constantes
%min-1
Rgases = 8.314;%J/molK
Presion = 1;%Pa
rho = 800;%kg/m3
LLL = 20;%m
D1 = 0.01;%m
TW = 550 + 273;% K
omega = (3.14*D1^2/4);%m2
Uglo = 4000; %J/m2K
cp = 2500;%kJ/molK
DH1 = 247;%kJ/mol
DH2 = 41.7;%kJ/mol
DH3 = 74.87;%kJ/mol
DH4 = 131.325;%kJ/mol
DH5 = 172;%kJ/mol

k1 = 1.26e6*exp(-102065/(Temp*Rgases)); k2 = 0.35e6*exp(-81030/(Temp*Rgases));
k3 = 6.95e3*exp(-58893/(Temp*Rgases)); k4 = 5.55e9*exp(-166397/(Temp*Rgases));
k5 = 1.34e15*exp(-243835/(Temp*Rgases)); kco21 = 2.61e-2*exp(37641/(Temp*Rgases));
kch41 = 2.6e-2*exp(40684/(Temp*Rgases)); kco22 = 0.5771*exp(9262/(Temp*Rgases));
kh22 = 1.494*exp(6025/(Temp*Rgases)); kch43 = 0.21*exp(-567/(Temp*Rgases));
kh23 = 5.18e7*exp(-133210/(Temp*Rgases)); kh2o4 = 4.73e-6*exp(97770/
(Temp*Rgases));
kch44 = 3.49*(-0/(Temp*Rgases)); kh24 = 1.83e13*exp(-216145/(Temp*Rgases));
kco5 = 7.34e-6*exp(100395/(Temp*Rgases)); kco25 = 2.81e7*exp(-104085/
(Temp*Rgases));
kp1 = 6.78e14*exp(-259660/(Temp*Rgases)); kp2 = 56.4971*exp(-36580/(Temp*Rgases));
kp3 = 2.98e5*exp(-84400/(Temp*Rgases)); kp4 = 1.3827e7*exp(-125916/(Temp*Rgases));
kp5 = 1.9393e9*exp(-168527/(Temp*Rgases));

%Variables
Ftotal = FH2O + FCO + FH2 + FCO2 + FCH4;
PCO = (FCO/Ftotal)*Presion;
PCO2 = (FCO2/Ftotal)*Presion;
PH2O = (FH2O/Ftotal)*Presion;
PH2 = (FH2/Ftotal)*Presion;
PCH4 = (FCH4/Ftotal)*Presion;
```

```
Mol = (18*FH2O + 28*FCO + 2*FH2 + 44*FCO2 + 16*FCH4)/Ftotal;
rhog = Presion*101.325*Mol/(Temp*Rgases);
u = (Ftotal*Mol*0.001/rhog)/(3.14*D1^2/4);

%Rates
r1 = (k1*kco21*kch41*PCH4*PCO2)*(1-(PCO*PH2)^2/(kp1*PCH4*PCO2))/
(1 + kco21*PCO2 + kch41*PCH4)^2;
r2 = (k2*kco22*kh22*PCO*PH2)*(1-(PCO*PH2O)/(kp2*PCO2*PH2))/
(1 + kco22*PCO2 + kh22*PH2)^2;
r3 = (k3*kch43)*(PCH4*PH2^2/kp3)/(1 + kch43*PCH4 + PH2^1.5/kh23)^2;
r4 = (k4/kh2o4)*(PH2O/PH2 - PCO/kp4)/(1 + kch44*PCH4 + PH2O/(kh2o4*PH2) + PH2^1.5/kh24)^2;
r5 = (k5/(kco5*kco25))*(PCO2/PCO - PCO/kp5)/(1 + kco5*PCO + PCO2/(kco5*kco25*PCO))^2;

%Eqs diferenciales
Rcat(1,1) = -(rho*omega*LLL)*(r1 + r3);
Rcat(2,1) = -(rho*omega*LLL)*(r1 + r2 + r5);
Rcat(3,1) = (rho*omega*LLL)*(2*r1 + r2 + r4 + 2*r5);
Rcat(4,1) = (rho*omega*LLL)*(2*r1 - r2 + 2*r3 + r4);
Rcat(5,1) = (rho*omega*LLL)*(r2 - r4);
Rcat(6,1) = ((rho*LLL)/(u*rhog*cp))*(r1*DH1 + r2*DH2 + r3*DH3 + r4*DH4 + r5*DH5) -
((4*Uglo*LLL)/(D1*u*rhog*cp))*(Temp - TW);
```

See Fig. 6E5.1.

FIGURE 6E5.1   Profiles of the operation of the packed bed reactor

6.3.2.2 Trireforming

Since dry reforming is an endothermic process, energy needs to be provided to the reformer to reach a good conversion. This can be achieved through the combustion of a fraction of the organic feedstock to supply the necessary energy for the reforming. As a result, the addition of partial oxidation of the biomass has been considered to provide the energy required. This results in the trireforming (TR) technology, which combines the highly endothermic steam and dry reforming reactions with the exothermic partial oxidation of methane, improving the energy efficiency of the process. Therefore, in addition to control the hydrogen to CO ratio, the steam reforming is also included.

$$CH_4 + \frac{1}{2}O_2 \leftrightarrow CO + 2H_2 \qquad (6.44)$$

Kinetic studies showed that oxygen reacts with methane at the beginning of the reactor and then the components are distributed according to the equilibriums achieved. Thus, TR is modeled in two stages. First the oxygen reacts with 100% conversion with the methane producing energy to be used for the endothermic reforming reactions described in the dry reforming section. The reactors can also be either fixed bed reactors (Arab Aboosadi et al., 2011) or fluidized beds (Khajeh et al., 2015). The kinetics for the different designs is reported in the literature (Hernández and Martin, 2018). Note that in fluidized beds, the mass transfer from the bubble to the thermal fluid and the catalyst particle are the rate control stage. Example 6.6 illustrates the modelling of biogas trireforming process.

EXAMPLE 6.6 Evaluate the trireforming of biogas with a CO_2–CH_4 ratio of 1. Assume an addition of 0.1 mol of O_2 per mol of methane. Optimize the use of water and raw material usage for the production of a syngas with a H_2 to CO ratio of 1.

See the scheme in Fig. 6E6.1.

Solution

We formulate the model as follows, step by step and programing it in GAMS:

Balance to the splitter and compressor:

The mass balance to maintain the composition of the biogas as a fuel and as a raw material. We avoid the use of bilinear products in the model of this splitter:

$$
\begin{aligned}
fc_{(CH4,1)} &= fc_{(CH4,2)} + fc_{(CH4,3)} \\
mol_{(CH4,1)} &= Ratiomol \cdot mol_{(CO2,1)} \\
mol_{(CH4,2)} &= Ratiomol \cdot mol_{(CO2,2)} \\
mol_{(CH4,3)} &= Ratiomol \cdot mol_{(CO2,3)}
\end{aligned}
\qquad (6E6.1)
$$

The balance to the compressor assumes polytropic behavior. Note that we need to compute the biogas molecular weight.

$$W_{(Compres)} = (F) \cdot \frac{8.314 \cdot k \cdot (T + 273)}{((MW_{biogas}) \cdot (z - 1))} \frac{1}{\eta_c} \left(\left(\frac{P_{reform}}{P_{inlet}} \right)^{(z-1)/z} - 1 \right); \qquad (6E6.2)$$

And the temperature:

$$T = \left((T_{in} + 273) + \frac{1}{\eta_c}(T_{in} + 273) \cdot \left(\left(\frac{P_{refor}}{P_{inlet}} \right)^{(z-1)/z} - 1 \right) \right) - 273 \qquad (6E6.3)$$

Mass balance to the first stage

$$
\begin{aligned}
mol_{CO} &= 2 \cdot mol_{O_2,feed} \\
mol_{H_2} &= 4 \cdot mol_{O_2,feed} \\
R_{O_2} mol_{CH_4} &= mol_{O_2,feed} \\
mol_{CH_4}|_{toequilibrium} &= mol_{CH_4} - 2 \cdot mol_{O_2,feed}
\end{aligned}
\tag{6E6.4}
$$

Mass balance to second stage and equilibrium

$$
\begin{aligned}
mol_{CH_4} + mol_{CO} + mol_{CO_2}|_{in} &= mol_{CH_4} + mol_{CO} + mol_{CO_2}|_{out} \\
4 \cdot mol_{CH_4} + 2 \cdot mol_{H_2O} + 2 \cdot mol_{H_2}|_{in} &= 4 \cdot mol_{CH_4} + 2 \cdot mol_{H_2} + 2 \cdot mol_{H_2O}|_{out} \\
mol_{H_2O} + mol_{CO} + 2 \cdot mol_{CO_2}|_{in} &= mol_{H_2O} + mol_{CO} + 2 \cdot mol_{CO_2}|_{out}
\end{aligned}
\tag{6E6.5}
$$

And the equilibria. Only two are independent:

$$
kp = e^{[31.447 - \frac{29580}{T}]} = \frac{P_{CO} \cdot P_{H_2}}{P_{CH_4} \cdot P_{CO_2}}; kp = 10^{[-\frac{11650}{T} + 13,076]} = \frac{P_{CO} \cdot P_{H_2}^3}{P_{CH_4} \cdot P_{H_2O}}; kp = 10^{[\frac{1910}{T} - 1,784]} = \frac{P_{CO_2} \cdot P_{H_2}}{P_{CO} \cdot P_{H_2O}}
\tag{6E6.6}
$$

Global energy balance to the furnace

$$
Q('Furnance') = \sum_j \Delta H_f(j)|_{T(6)}|_{(6)} - \sum_j \Delta H_f(j)|_{T(5)}|_{(5)} - \sum_j \Delta H_f(j)|_{T(4)}|_{(4)} - \sum_j \Delta H_f(j)|_{T(7)}|_{(7)}
\tag{6E6.7}
$$

Objective function

$$
Z = (fc_{H_2} + fc_{O_2}) \cdot P_{Hydrogen} - P_{steam} \cdot fc_{Steam} - P_{O2} \cdot fc_{O_2}
\tag{6E6.8}
$$

```
$Title Biogas      20/05/20

Set
*               Define units
        unit    units
        /Furnance, Compres/
*               Define components
        J       components
        /Wa, CO2, CO, H2, CH4, O2/
        S       streams
        /1*8/
*               running variable     for cp vapor
        mar /1*4/;


*Data (For the sake of the length, the data is collected in the code file)


Scalars
*Define temperatures in C

        T_amb                           ambient temperature /20/;
```

```
Positive Variables
*               streams and mass fractions: all in kg/s
                F(s)                   total streams in kg s^ − 1
                fc(J,s)                individual components streams
                x(J,s)                 mass fraction of comp J in stream
                T(s)                   temperature of stream in C;
Variables
*               power
                W(unit)        power consumption of unit in kW (efficiency included)
*               heat
                Q(unit)      heat produced or consumed in unit (efficiency included);

*Define global bounds and fix specific variables

*mass fractions
x.UP(J,s) = 1;

*Total streams
F.UP(s) = 600;

*Component streams
fc.UP(J,s) = 600;

*Specifying heat and power consumption of certain units
Q.Fx('Compres') = 0;
W.Fx('Furnance') = 0;

*Temperature settings
*global temperature bounds—bounds get redefined for specific streams
T.LO(s) = 20;
T.UP(s) = 1000;

*Specifying temperatures
*Src3
fc.fx(J,'1')$((ord (J) ne 2) and (ord (J) ne 5)) = 0;
T.Fx('1') = T_amb;
*Src4
fc.fx(J,'5')$(ord (J) ne 1) = 0;
T.Fx('5') = 233;
*Src5
fc.fx(J,'7')$(ord (J) ne 6) = 0;
T.Fx('7') = 20;

*Global relationships
Equations

                Rel_1,Rel_2;
```

3. Biomass and waste based processes

```
*relationship between MT, m and x
Rel_1(J,s).     fc(J,s) = E = F(s)*x(J,s);
Rel_2(s). Sum(J,fc(J,s)) = E = F(s);


*                                             ****************
*                                             Rerforming Nat Gas
*                                             ***************

Variables

Q_prod
Q_reac;

Positive Variables

P_refor
K_met
K_dry
K_WGS
MW_biogas;

Scalar

PCI_gas             En kJ por kg              /39900/
Ratiobiogas         mol CH4 per mol CO2          /1/
Ratio02             mol 02 per mol CH4         /0.1/;

fc.fx('CH4','1') = 16;

Equations

Refor_1, Refor_2, Refor_3, Refor_4, Refor_5, Refor_6, Refor_7, Refor_8, Refor_9,
Refor_10,
Refor_11, Refor_12, Refor_13, Refor_14, Refor_15, Refor_16, Refor_17, Refor_18,
*Refor_17b,    Refor_19b,
 Refor_19, Refor_20,
Refor_21, Refor_22, Refor_23, Refor_24,
 Refor_25;

Refor_1.     fc('CH4','1') = E = fc('CH4','2') + fc('CH4','3');
Refor_2.     fc('CH4','1')/MW('CH4') = E = Ratiobiogas*fc('C02','1')/MW('C02');
Refor_3.       Ratiobiogas*fc('C02','2')/MW('C02') = E = fc('CH4','2')/MW('CH4');
Refor_4.       Ratiobiogas*fc('C02','3')/MW('C02') = E = fc('CH4','3')/MW('CH4');

fc.fx(J,'2')$((ord (J) ne 2) and (ord (J) ne 5)) = 0;
fc.fx(J,'3')$((ord (J) ne 2) and (ord (J) ne 5)) = 0;
*inicialiuzacion

Refor_5.     MW_biogas*(fc('CH4','1')/MW('CH4') + fc('C02','1')/MW('C02')) = E =
(fc('CH4','1') + fc('C02','1'));
```

3. Biomass and waste based processes

```
Refor_6.        W('Compres') = E = F('2')*(8.314*1.4*(T_amb + 273))*((P_refor)**
((0.4/1.4)) - 1)/((MW_biogas + 0.001)*(1.4-1));
P_refor.LO = 2;
P_refor.UP = 10;

Refor_7.        T('4') = E = (T_amb + 273) + (T_amb + 273)*((P_refor)**((0.4/1.4)) -
1) - 273;


Refor_8(J).              fc(J,'2') = E = fc(J,'4');
fc.fx(J,'4')$((ord (J) ne 2) and (ord (J) ne 5)) = 0;

*Prestage
Refor_9.        fc('CO','8')/MW('CO') = E = 2*fc('O2','7')/MW('O2');
Refor_10.       fc('H2','8')/MW('H2') = E = 4*fc('O2','7')/MW('O2');
Refor_11.       fc('O2','7')/MW('O2') = E = RatioO2*fc('CH4','4')/MW('CH4');
Refor_12.       fc('CH4','8')/MW('CH4') = E = fc('CH4','4')/MW('CH4') - 2*fc
('O2','7')/MW('O2');
Refor_13.       fc('CO2','8') = E = fc('CO2','4');
fc.fx('Wa','8') = 0;
fc.fx('O2','8') = 0;
T.fx('8') = 600;

*Equilibrium
*Balance al carbono
Refor_14.          fc('CH4','8')/MW('CH4') + fc('CO2','8')/MW('CO2') + fc
('CO','8')/MW('CO')
                        = E = fc('CH4','6')/MW('CH4')    + fc('CO','6')/MW
('CO') + fc('CO2','6')/MW('CO2');
*Balance al hidrogeno
Refor_15.          4*fc('CH4','8')/MW('CH4') + 2*fc('Wa','5')/MW('Wa') + 2*fc
('H2','8')/MW('H2') = E =
                        4*fc('CH4','6')/MW('CH4')    + 2*fc('H2','6')/MW
('H2') + 2*fc('Wa','6')/MW('Wa');
*Balance al oxigeno
Refor_16.          fc('Wa','5')/MW('Wa') + 2*fc('CO2','8')/MW('CO2') + fc
('CO','8')/MW('CO')
                        = E = fc('CO','6')/MW('CO')    + 2*fc('CO2','6')/MW
('CO2') + fc('Wa','6')/MW('Wa');

Refor_17.       K_met = E = 10**(-1650/(T('6') + 273) + 13.076);
*Refor_17b.     K_dry = E = exp(31.447 - 29580/(T('6') + 273));
Refor_18.       K_WGS = E = 10**(1910/(T('6') + 273) - 1.784);

*Refor_19.      (fc('CH4','6')/MW('CH4'))       *(fc('Wa','6')/MW('Wa'))
*(fc('CO2','6')/MW('CO2') + fc('CH4','6')/MW('CH4') + fc('Wa','6')/MW('Wa') + (fc
```

3. Biomass and waste based processes

```
('CO','6')/MW('CO'))+fc('H2','6')/MW('H2'))**2* K_met    = E = ((fc('CO','6')/
MW('CO'))*(fc('H2','6')/MW('H2'))**3)*(P_refor+eps1)**2;
Refor_19b.        (fc('CH4','6')/MW('CH4'))*(fc('CO2','6')/MW('CO2')) *
K_dry    = E = (fc('CO','6')/MW('CO'))*(fc('H2','6')/MW('H2'));
Refor_20.       (fc('CO2','6')/MW('CO2'))*(fc('H2','6')/MW('H2')) = E = K_WGS*(fc
('CO','6')/MW('CO'))*(fc('Wa','6')/MW('Wa'));
Refor_21.         Q_prod = E = sum(J,fc(J,'6')*(dH_f(J)+(1/MW(J))*(c_p_v
(J,'1')*(T('6')-25)+
(1/2)*c_p_v(J,'2')*((T('6')+273)**2-(25+273)**2)+
(1/3)*c_p_v(J,'3')*((T('6')+273)**3-(25+273)**3)+
(1/4)*c_p_v(J,'4')*((T('6')+273)**4-(25+273)**4))));

T.LO('6')=400;
T.UP('6')=800;

Refor_22.         Q_reac = E = sum(J,fc(J,'4')*(dH_f(J)+(1/MW(J))*(c_p_v
(J,'1')*(T('4')-25)+
(1/2)*c_p_v(J,'2')*((T('4')+273)**2-(25+273)**2)+
(1/3)*c_p_v(J,'3')*((T('4')+273)**3-(25+273)**3)+
(1/4)*c_p_v(J,'4')*((T('4')+273)**4-(25+273)**4))))+

sum(J,fc(J,'5')*(dH_f(J)+(1/MW(J))*(c_p_v(J,'1')*(T('5')-233)+
(1/2)*c_p_v(J,'2')*((T('5')+273)**2-(233+273)**2)+
(1/3)*c_p_v(J,'3')*((T('5')+273)**3-(233+273)**3)+
(1/4)*c_p_v(J,'4')*((T('5')+273)**4-(233+273)**4))))+

sum(J,fc(J,'7')*(dH_f(J)+(1/MW(J))*(c_p_v(J,'1')*(T('7')-25)+
(1/2)*c_p_v(J,'2')*((T('7')+273)**2-(25+273)**2)+
(1/3)*c_p_v(J,'3')*((T('7')+273)**3-(25+273)**3)+
(1/4)*c_p_v(J,'4')*((T('7')+273)**4-(25+273)**4))));

Refor_23.              Q('Furnance') = E = (Q_prod-Q_reac);
Refor_24.              fc('CH4','3') = E = Q('Furnance')/PCI_gas;
Refor_25.             fc('H2','6')/MW('H2') = E = fc('CO','6')/MW('CO');

Equations
              Obj;
Variables
              Z;

Scalar
P_steam /0.019/
P_o2       /0.021/;

fc.lo('H2','6')=0.25;

Obj.      Z = E = (fc('H2','6')+ fc('CO','6'))*1.6 - P_steam*fc('Wa','5') - P_o2*fc
('O2','7');
```

```
Model Natgas /ALL/;
option NLP = coinipopt;
Solve Natgas Using NLP Maximizing Z;
```

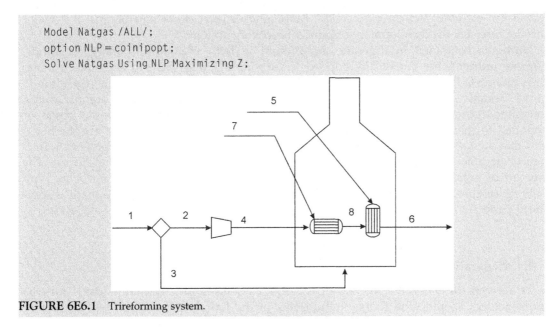

FIGURE 6E6.1   Trireforming system.

6.3.2.3 *Superdry reforming*

Superdry reforming was recently developed by intensifying Dr with the objective of increasing the conversion of CO_2 (Buelens et al., 2016). To achieve this improvement in the conversion of CO_2, the water gas shift reaction is avoided. A combined process composed by a $Ni/MgAl_2O_4$ catalyst for CH_4 reforming, a $Fe_2O_3/MgAl_2O_4$ support that acts as oxygen carrier and a CaO/Al_2O_3 sorbent for CO_2 has been designed for this purpose. Due to the different chemical equilibria involved in the superdry reforming process, eight parallel reactors that operate in two steps, four reactors carry out the oxidation of the methane and the remaining four reactors operate in CO_2 reduction mode, are needed to carry out the process in continuous mode. The global reaction taking into account both modes is presented in Eq. (6.42). This is a highly endothermic reaction.

$$CH_4 + 3CO_2 \leftrightarrow 4CO + 2H_2O \tag{6.45}$$

6.4 Recovery of nutrients from digestate

The remaining organic substrate after anaerobic digestion, called digestate, is mainly composed by stabilized organic carbon, and contains valuable nutrients, mainly phosphorus and nitrogen. Phosphorus and nitrogen are essential macronutrients for living organisms, and a key element for maintaining agricultural productivity. Ammonia can synthetically be produced through the Haber-Bosch process. However, phosphorus is a resource very sensitive to depletion since there is no known substitute or synthetic replacement for it and extractable deposits of phosphorus rock are limited. Projections estimate limited availability of phosphate over the next century (Cordell et al., 2009). Therefore, the search for nutrients recycling processes, especially for the case of phosphorus, is a major driving force for the development of nutrient recovery systems.

Nutrients contained in digestate are present in both organic and inorganic forms. Organic nutrients are chemically bonded to carbon, and they have to be converted into their inorganic forms through a mineralization process to be available for the vegetation to grow. Organic nutrients are mainly contained in the solid phase of the organic waste. Inorganic nutrients are water soluble, and they are mostly present in the liquid phase, or bounded to soluble minerals. They are immediately available to plants and are the nutrients that are feasible to be recovered. As a result of the anaerobic digestion process, a fraction of organic phosphorus and nitrogen is transformed into their inorganic forms. The amount of organic nutrients transformed into inorganic phosphorus and nitrogen, typically represents 24% and 16% over the original inorganic ammonia and phosphate, respectively (Smith et al., 2007; Martin, 2003; Alburquerque et al., 2012; Sørensen et al., 2011).

6.4.1 Separation of solid and liquid phases

To recover the inorganic fraction of nutrients, a solid-liquid separation stage must be implemented, keeping the inorganic nutrients in the liquid stage, which will be further processed, and the organic nutrients in the solid phase, which can be composted to mineralize nitrogen and phosphorus and be further used as fertilizers.

Different technologies can be used to carry out the separation of the solid-liquid phases of digestate, including tilted plane screens, screw press, belt press, and decanting centrifuges. Among these technologies, Møller et al. (2000) determined that screw press is the most cost-efficient liquid-solid separation system. The solid phase obtained contains most of the organic nutrients, which can be sent to composting, while the liquid phase containing most of the inorganic nutrients is the target stream for the recovery of nutrients.

6.4.2 Nutrient recovery technologies

The technologies to recover inorganic phosphorus can be classified in three categories: struvite-based phosphorus recovery, calcium precipitates-based phosphorus recovery, and modular separation systems (Fig. 6.8C). Struvite is a phosphate mineral with a chemical formula of $MgNH_4PO_4 \cdot 6H_2O$. The advantage of this technology is that struvite is a solid with a high nutrient's density, it is easy to transport, and it can be used as slow-release fertilizer without any postprocessing (Doyle and Parsons, 2002). Alternatively, phosphorus can also be recovered as calcium precipitates, although their performance as fertilizer is lower than struvite and the presence of calcium compounds can impact the pH level of soil. Finally, physical separation systems produce an organic solid rich in nutrients that can be used as nutrients supply for crops, but the exact amount of nutrients contained in it and the exact composition are not know, making difficult to determine the correct amount of fertilizer needed. In addition, this product has a lower nutrients density, making the transportation and redistribution of nutrients more expensive.

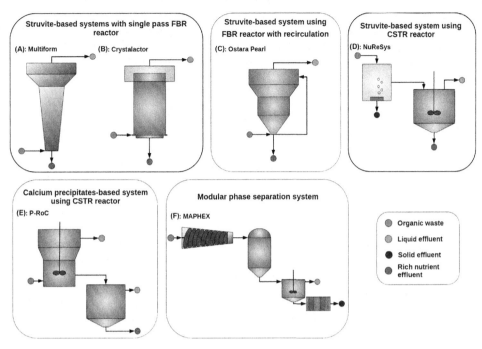

FIGURE 6.8   Phosphorus recovery technologies.

6.4.2.1 Struvite-based phosphorus recovery

Phosphorus can be recovered in form of struvite using fluidized bed reactors (FBR) (Fig. 6.8A and B), or CSTR (Fig. 6.8C). Struvite is formed by one molecule of ammonia, one molecule of phosphate, both of them present in significant amounts in wastewater sludge, livestock waste, and the digestates resulting from their anaerobic digestion decomposition, and one atom of magnesium. This element is supplied to the reactor by adding $MgCl_2$, reaching a molar ratio of 2 mol of Mg per mol of phosphate. pH is adjusted using sodium hydroxide at a value of 9 to optimize the production of struvite (Martín-Hernández et al., 2018).

Struvite is formed as a result of a chemical equilibrium. However, calcium-based precipitates are formed along with struvite as a result of the multiple chemical equilibria involving different species contained in the organic waste, mainly Mg^{2+}, Ca^{2+}, K^+, NH_4^+, PO_4^-, CO_3^{2-} and OH^-. It should be noted that the formation of different precipitates competes with struvite for the same ionic species. Therefore, operating conditions must be carefully optimized to promote the formation of struvite to the detriment of the other solid compounds. The main chemical equilibrium involved, and precipitates formed are shown in Table 6.6 (Martín-Hernández et al., 2020b).

In addition to the precipitates equilibrium, the equilibrium of the chemical systems in aqueous phase, that is, ammonia, water, phosphoric acid and carbonic acid must be considered, as shown in Table 6.7.

Since organic waste is a nonideal media, with a high concentration of dissolved ions, activities, Eq. (6.46) should be used for the calculation of chemical equilibria. The Debye–Hückel equation, Eq. (6.47), can be used to estimate the activity coefficients (γ) considering the effect of

TABLE 6.6 Solid species and chemical equilibrium involved in the formation of struvite from organic waste.

Name	Chemical system	pK$_{sp}$	Source
Struvite	$MgNH_4PO_4 \cdot 6H_2O \leftrightarrow Mg^{2+} + NH_4^+ + PO_4^{3-}$	13.26	Ohlinger et al. (1998)
K-struvite	$MgKPO_4 \cdot 6H_2O \leftrightarrow Mg^{2+} + K^+ + PO_4^{3-}$	10.6	Taylor et al. (1963)
Hydroxyapatite	$Ca_5(PO_4)_3OH \leftrightarrow 5Ca^{2+} + 3PO_4^{3-} + OH^-$	44.33	Brezonik and Arnold (2011)
Calcium carbonate	$CaCO_3 \leftrightarrow Ca^{2+} + CO_3^{2-}$	8.48	Morse et al. (2007)
Tricalcium phosphate	$Ca_3(PO_4)_2OH \leftrightarrow 3Ca^{2+} + 2PO_4^{3-}$	25.50	Fowler and Kuroda (1986)
Dicalcium phosphate	$CaHPO_4 \leftrightarrow Ca^{2+} + HPO_4^{2-}$	6.57	Gregory et al. (1970)
Calcium hydroxide	$Ca(OH)_2 \leftrightarrow Ca^{2+} + 2OH^-$	5.19	Skoog et al. (2014)
Magnesium hydroxide	$Mg(OH)_2 \leftrightarrow Mg^{2+} + 2OH^-$	11.15	Skoog et al. (2014)

TABLE 6.7 Aqueous chemical systems affecting the formation of struvite from organic waste.

Name	Chemical system	pK	Source
Ammonia	$NH_4^+ \leftrightarrow NH_3 + H^+$	9.2	Bates and Pinching (1949)
Water	$H_2O \leftrightarrow OH^- + H^+$	14	Skoog et al. (2014)
Phosphoric acid	$H_3PO_4 \leftrightarrow H_2PO_4^- + H^+$	2.1	Ohlinger et al. (1998)
	$H_2PO_4^- \leftrightarrow HPO_4^{2-} + H^+$	7.2	Ohlinger et al. (1998)
	$HPO_4^{2-} \leftrightarrow PO_4^{3-} + H^+$	12.35	Ohlinger et al. (1998)
Carbonic acid	$H_2CO_3 \leftrightarrow HCO_3^- + H^+$	6.35	Skoog et al. (2014)
	$HCO_3^- \leftrightarrow CO_3^{2-} + H^+$	10.33	Skoog et al. (2014)

temperature and ionic strength, Eqs. (6.48) and (6.49). z_x denotes the charge of ion x, EC the electrical conductivity of the waste, and A is a parameter Debye−Hückel equation.

$$\{x\} = [x] \cdot \gamma_x \tag{6.46}$$

$$\log_{10}(\gamma_x) = -A \cdot z_x^2 \left(\frac{\sqrt{I}}{1 + \sqrt{I}} \right) - 0.3 \cdot I \tag{6.47}$$

$$I(M) = 1.6 \cdot 10^5 \cdot EC \left(\frac{\mu S}{cm} \right) \tag{6.48}$$

$$A = 0.486 - 6.07 \cdot 10^{-4} \cdot T + 6.43 \cdot 10^{-6} \cdot T^2, T(K) \tag{6.49}$$

FBR reactors for struvite production can be configured as single pass FBR, with no recirculation, or with an internal recirculation loop used to recirculate liquid to the bottom of the reactor to increase the liquid flow in the reactor and achieve larger superficial velocities. This type of reactors has a conical or cylindrical design, where the organic waste is pumped from a side or the bottom of the reactor. In the reactor, the struvite particles grow until they reach a critical mass enough to overcome the drag force of the uplift liquid. The liquid stream is taken out from the top of the reactor (Rahaman et al., 2014). A final drying step is usually performed to remove the excess of moisture contained in the struvite particles. Struvite formation in continuous stirred tank reactors is carried out in units equipped with special impellers to minimize the breakage of struvite crystals. Similar to the FBR case, $MgCl_2$ is supplied to the reactor to increase struvite supersaturation. After struvite precipitation, both solid and liquid phases are extracted from the reactor in the same stream and it is injected in a settler where the separation of phases is carried out. Struvite fines are separated from the largest struvite particles through a hydrocyclone and they are recirculated to the process. The struvite particles are dried before their final collection.

6.4.2.2 Calcium precipitates-based phosphorus recovery

Phosphorus can be recovered as a variety of calcium precipitates. In some processes, the precipitation of phosphorus is carried out using calcium silicate hydrate (CSH), which is the support on which phosphorus is deposited forming a calcium precipitate. This process can be performed in a CSTR reactor, where the waste stream containing the nutrients to be recovered and the CSH are added, forming the calcium phosphate precipitates (Fig. 6.8D). The solid compounds and the liquid phase are taken out from the reactor through the same stream, and they are directed to a settler to carry out the solid-liquid separation stage. The calcium phosphate particles are dried following a two steps procedure composed by a belt filter and a conveyor dryer (Ehbrecht et al., 2011).

6.4.2.3 Modular separation systems for phosphorus recovery

Modular process for livestock waste treatment and nutrient recovery have been developed. These processes are modularized to fit them in semitrailers, so they can be easily transported and relocated. This feature makes the modular livestock waste treatment units interesting especially for small livestock facilities not large enough for the deployment of larger scale facilities. Among the modular processes developed for livestock waste treatment, MAPHEX is a nutrient recovery system based on physico-chemical separations developed by Penn State University and the USDA (Fig. 6.8E). It is composed by three stages, that is, liquid-solid separation with a screw press and a centrifuge, addition of iron sulfate to improve nutrients retention, and filtration with diatomaceous earth as filter media. It is conceived as a mobile modular system which can be set in two interconnected truck trailers. Each MAPHEX unit is able to process up to 18.54 kg of phosphorus as phosphate (P-PO_4) fed per day, with an associated operation cost of 110.8 USD per kg of P-PO_4 processed. Capital cost of a MAPHEX unit is 291,000 USD (Church et al., 2016, 2018). Example 6.7 estimates the amount of struvite that can be produced from cattle manure.

EXAMPLE 6.7 Determine the amount of struvite formed in the cattle waste described below. The implementation of this problem in Python can be found in the materials attached to the book.

Cattle waste composition and physico-chemical characteristics:

- $[PO_4^-] = 0.00341$ M
- $[NH_4^+] = 0.163$ M
- $[K^+] = 0.112$ M
- $[Ca^{2+}] = 0.00297$ M
- $[Mg^{2+}] = 2 \cdot [PO_4^-]$
- Alkalinity = 8770.5 mg $CaCO_3$
- Temperature = 298 K
- pH = 9
- Electrical conductivity (EC) = 18,800 μS/cm

Solution

Estimation of activity coefficients:

Ionic strength of the media has to be determined in order to compute activity coefficients:

$$I(M) = 1.6 \cdot 10^{-5} \cdot EC \tag{6E7.1}$$

Coefficient activities (γ) and activities ($\{x\}$) are calculated as follows:

$$A = 0.486 - 6.07 \cdot 10^{-4} \cdot T + 6.43 \cdot 10^{-6} \cdot T^2, \quad T(K) \tag{6E7.2}$$

$$\log_{10}(\gamma_x) = -A \cdot z_x^2 \cdot \left(\frac{I^{1/2}}{1 + I^{1/2}}\right) - 0.3 \cdot I \tag{6E7.3}$$

$$\{x\} = [x]\gamma_x \tag{6E7.4}$$

Distribution of species in aqueous phase:

The distribution of species for ammonia, water, phosphoric acid, and carbonate systems in cattle leachate is determined by chemical equilibria and the mass balance, as shown in Eqs. (6E7.5)–(6E7.7), where n_j and m_k are the stoichiometric coefficients of the reactants and products respectively, and J denotes the chemical systems described in Table 6.7:

$$\sum_j n_j \cdot \text{Reactant}_j \leftrightarrow \sum_k m_k \cdot \text{Product}_k \tag{6E7.5}$$

$$K_J = \frac{\left(\prod_k \{\text{Product}_k\}^{m_k}\right)_J}{\left(\prod_j \{\text{Reactant}_j\}^{n_j}\right)_J} \tag{6E7.6}$$

$$[i]_J^{\text{initial}} = \sum_J [\text{Compounds}]_J, \quad i \in \{NH^{4+}, Ca^{2+}, Mg^{2+}, PO_4^{3-}, CO_2^{-3}\} \tag{6E7.7}$$

Formation of precipitates:

Formation of solid species is calculated using solubility equilibrium and mass balances, Eqs. (6E7.9) and (6E7.10) respectively, n_a is the stoichiometric coefficients of the reactants, and L denotes the chemical systems described in Table 6.6. The supersaturation index (Ω), Eq. 6E7.11, is used to determine if a compound precipitate: $\Omega > 1$ indicates supersaturated conditions and precipitate may form, $\Omega = 1$ indicates equilibrium between solid and liquid phases, and $\Omega < 1$ indicates unsaturated conditions under which no precipitate can form. Additionally, since the higher the value of the supersaturation index, the larger the formation potential of a precipitate, Ω is used to determine the sequence of precipitation, i.e., what species precipitates first.

$$\sum_b m_b \cdot \text{Precipitate}_b \downarrow \leftrightarrow \sum_a n_a \cdot \text{Reactant}_a \tag{6E7.8}$$

$$K_{sp_L} = \left(\prod_a \{\text{Reactant}_a\}^{n_a} \right)_L \tag{6E7.9}$$

$$[i]_L^{\text{initial}} = \sum_L [\text{Compounds}]_L, \quad i \in \{NH^{4+}, Ca^{2+}, Mg^{2+}, PO_4^{3-}, CO_2^{-3}\} \tag{6E7.10}$$

$$\Omega_L = \frac{\left(\prod_a \{\text{Reactant}_a\}^{n_a} \right)_L}{K_{sp_L}} \tag{6E7.11}$$

Algorithm to calculate the formation of species in the equilibrium:

An iterative process is needed to calculate the concentration of all species when then thermodynamic equilibrium is reached. The algorithm used is shown in Fig. 6E7.1. First, the initial composition and physico-chemical characteristics of cattle waste are defined, box a. Secondly, box b, ionic strength and activity coefficients are computed. Next, boxes c and d, two parallel problems are solved, the equilibrium of the aqueous species, and the alkalinity problem to determine the distribution of carbonates. Once the concentration of all species is determined, the supersaturation index for all species is computed in box e. The compound with the maximum supersaturation index is assumed to precipitate first. The amount of formed precipitate is computed by solving the solubility equilibrium and the material balance. As a result of the precipitate formation, the concentration of some species in aqueous phase is reduced. Therefore, the equilibrium of the aqueous species and the alkalinity problem must be recalculated, to obtain the new concentration values of the different compounds in the waste, and the iterative process starts again. The iterative process runs until each component saturation index is equal or less than one, and the formation of the precipitates stops. The algorithm has been implemented in Python. Chemical equilibriums have been modeled as optimization problem using the Pyomo framework.

The results show that only struvite and hydroxyapatite are the only precipitates formed, resulting in 0.002013 and 0.000446 mol per liter of waste treated respectively.

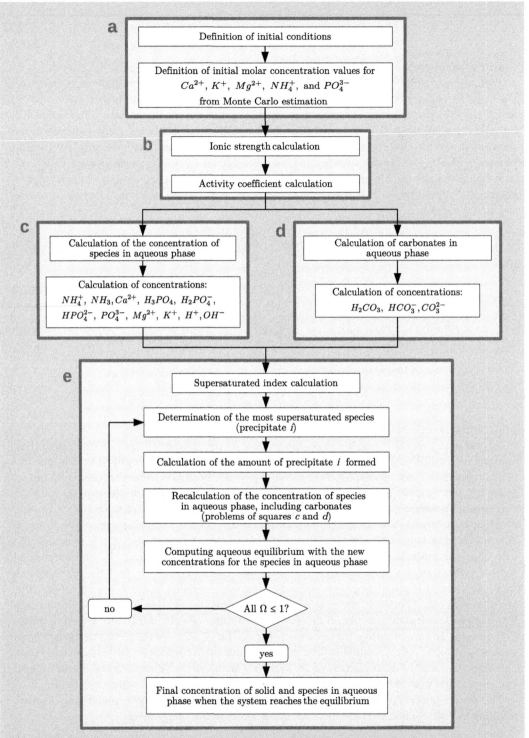

FIGURE 6E7.1 Algorithm to solve the thermodynamic model for the formation of precipitates in cattle organic waste. *With permission from Martín-Hernández, E., Ruiz-Mercado, G.J., Martín, M., 2020b. Model-driven spatial evaluation of nutrient recovery from livestock leachate for struvite production. J. Environ. Manag., 271, 110967.*

Exercises

P6.1 Evaluate the production of biogas for cattle slurry. See Composition in Table 6.1.

P6.2 A facility generates 0.1 kg/s of biogas, with a molar composition 0.59 CH_4 and the rest CO_2. A zeolite bed of 1000 kg capable of producing biomethane contains 2% CO_2 by mass, capturing 99% of the CO_2. The bed operates at 25°C and 1 bar of total pressure neglecting the pressure drop across the bed. Compute the fraction of the total flow capable of being processed.

P6.3 Compare the four configurations For the production of bionatural gas from a stream 59% methane and the rest $CH4$, minimizing the area for maximum production of methane 98% pure using cellulose acetate membrane.

P6.4 Perform a sensitivity analysis on the effect of the oxygen feed to the operation of the trireformer of Example 6.5.

P6.5 Evaluate the trireforming of biogas with a CO_2–CH_4 ratio of 1. Optimize the use of steam, oxygen and raw material usage for the production of a syngas with a H_2 to CO ratio of 2.

Ratio O_2 = 0.5; Steam = 31.3/60 (kg of biogas); Biogas as fuel 8.2/60, Pref: 8.2

P6.6 Determine the amount of struvite formed in the cattle waste with the following composition:

$[PO_4^-]$ = 0.00427 M; $[NH_4^+]$ = 0.241 M; $[K^+]$ = 0.0838 M; $[Ca^{2+}]$ = 0.00366 M; $[Mg^{2+}]$ = 4 · $[PO_4^-]$; Alkalinity = 8770.5 mg $CaCO_3$; Temperature = 298K; pH = 9; Electrical conductivity (EC) = 18,800 μS/cm

References

Al Seadi, T., 2002. Quality management of AD residues from biogas production. IEA Bioenergy, Task 24 – Energy from Biological Conversion of Organic Waste. January 2002. Available at: http://www.IEA-Biogas.net.

Al Seadi, T., Rutz, D., Prassl, H., Köttner, M., Finsterwalder, T., Volk, S., et al., 2008. Biogas Handbook. University of Southern Denmark Esbjerg, Esbjerg, Denmark.

Alburquerque, J., de la Fuente, C., Bernal, M., 2012. Chemical properties of anaerobic digestates affecting C and N dynamics in amended soils. Agric. Ecosyst. Environ. 160, 15–22.

Arab Aboosadi, Z., Jahanmiri, A.H., Rahimpour, M.R., 2011. Optimization of tri-reformer reactor to produce synthesis gas for methanol production using differential evolution (DE) method. Appl. Energy 88, 2691–2701.

Bates, R.G., Pinching, G., 1949. Acid dissociation constant of ammonium ion at 0–50°C, and the base strength of ammonia, and the base strength of ammonia. J. Res. Natl. Bur. Stand. 42, 419–430.

Batstone, D.J., Keller, J., Angelidaki, I., Kalyuzhnyi, S.V., Pavlostathis, S.G., Rozzi, A., et al., 2002. The IWA Anaerobic Digestion Model No 1 (ADM1). Water Sci. Technol. 45 (10), 65–73.

Benguerba, Y., Dehimi, L., Virginie, M., Dumas, C., Ernst, B., 2015. Modelling of methane dry reforming over Ni/Al_2O_3 catalyst in a fixed-bed catalytic reactor. Reac Kinet. Mech. Cat. 114, 109–119.

Brezonik, P., Arnold, W., 2011. Water Chemistry an Introduction to the Chemistry of Natural and Engineered -Aquatic Systems. Oxford University Press.

Buelens, L.C., Galvita, V.V., Poelman, H., Detavernier, C., Marin, G.B., 2016. Super-dry reforming of methane intensifies CO2 utilization via Le Chatelier's principle. Science 354, 449–452.

Church, C.D., Hristov, A.N., Bryant, R.B., Kleinman, P.J.A., Fishel, S.K., 2016. A novel treatment system to remove phosphorus from liquid manure. Appl. Eng. Agric. 32 (1), 103–112.

Church, C.D., Hristov, A.N., Kleinman, P.J., Fishel, S.K., Reiner, M.R., Bryant, R.B., 2018. Versatility of the MAnure PHosphorus EXtraction (MAPHEX) system in removing phosphorus, odor, microbes, and alkalinity from dairy manures: a four-farm case study. Appl. Eng. Agric. 34 (3), 567.

Contois, D.E., 1959. Kinetics of bacterial growth: relationship between population density and specific growth rate of continuous cultures. J. Gen. Microbiol. 21, 40–50.

Cordell, D., Drangert, J.O., White, S., 2009. The story of phosphorus: global food security and food for thought. Glob. Environ. Change 19 (2), 292–305.

Defra, 2011a. Cattle slurry and dirty water − total and available nutrients. Available at: http://adlib.everysite.co.uk/adlib/defra/content.aspx?id = 2RRVTHNXTS.88UF9N65FWLCJ.

Defra, 2011b. Fertilizer manual. Pig slurry − total and available nutrients. Available at: http://adlib.everysite.co.uk/adlib/defra/content.aspx?doc = 262994&id = 263068.

Deng, L., Hägg, M., 2010. Techno-economic evaluation of biogas upgrading process using CO_2 facilitated transport membrane. Int. J. Greenh. Gas. Control. 4, 638–646.

Doyle, J.D., Parsons, S.A., 2002. Struvite formation, control and recovery. Water Res. 36 (16), 3925–3940.

Ehbrecht, A., Schnauer, S., Fuderer, T., Schuhmann, R., 2011. P-Recovery from sewage by seeded crystallisation in a pilot plant in batch mode technology. Water Sci. Technol. 63 (2), 339.

Fachagentur Nachwachsende Rohstoffe, 2010. Guía sobre el Biogás. Desde la producción hasta el uso, fifth ed. FNR, Abt. Öffentlichkeitsarbeit, Gülzow, mediathek.fnr.de/media/downloadable/files/samples/l/e/leitfadenbiogas-es-2013.pdf (last accessed June 2017).

Ferella, F., Puca, A., taglieri, G., Rossi, L., Gallucci, K., 2017. Separation of carbon dioxide for biogas upgrading to biomethane. J. Clean. Prod. 164, 1205–1218. Available from: https://doi.org/10.1016/j.jclepro.2017.07.037.

Fernandes Rodrigues, D., 2009. Model Development of a Membrane Gas Permeation Unit for the Separation of Hydrogen and Carbon Dioxide (MSc thesis). Instituto Superior Technico, Lisbon.

Fowler, B., Kuroda, S., 1986. Changes in heated and in laser-irradiated human tooth enamel and their probable effects on solubility. Calcif. Tissue Int. 38, 197–208.

GPSA Engineering_Data_Book FPS VERSION 21−10, 2004.

Gregory, T., Moreno, E., Brown, W., 1970. Solubility of $CaHPO_4 \cdot 2H_2O$ in the system $Ca(OH)_2 − H_3PO_4 − H_2O$ at 5, 15, 25 and 37.5 °C. J. Res. Natl. Bur. Stand. 74, 461–475.

Hauchhum, L., Mahanta, P., 2014. Carbon dioxide adsorption on zeolites and activated carbon by pressure swing adsorption in a fixed bed. Int. J. Energy Environ. Eng. 5, 349–356. Available from: https://doi.org/10.1007/s40095-014-0131-3.

Hernández, B., Martin, M., 2018. Optimization for biogas to chemicals via tri-reforming analysis of fischer-tropsch fuels from biogas. Energy Convers. Manag. 174, 998–1013.

Khajeh, S., Arab Aboosadi, Z., Honarvar, B., 2015. Optimizing the fluidized-bed reactor for synthesis gas production by tri-reforing. Chem. Eng. Res. Des. 94, 407–416.

Kim, M., Kim, S., Kim, J., 2017. Optimization-based approach for design and integration of carbon dioxide separation processes using membrane technology. Energ. Proc. 136, 336–341. Available from: https://doi.org/10.1016/j.egypro.2017.10.284.

Lauwers, J., Appels, L., Thompson, I.P., Degreve, J., Van Impe, J.F., Dewil, R., 2013. Mathematical modelling of anaerobic digestion of biomass and waste: power and limitations. Prog. Energy Combust. Sci. 39, 383–402.

León, E., Martín, M., 2016. Optimal production of power in a combined cycle from manure based biogas. Energ. Conv. Manag. 114, 89–99.

López, I., Benzo, M., Passeggi, M., Borzacconi, L., 2020. A simple kinetic model applied to anaerobic digestion of cow manure. Environ. Technol. Available from: https://doi.org/10.1080/09593330.2020.1732473.

Luyben, W.L., 2014. Design and control of the dry methane reforming process. Ind. Eng. Chem. Res. 14423–14439.

Mairet, F., Bernard, O., Ras, M., Lardon, L., Steyer, J.-P., 2011. Modeling anaerobic digestion of microalgae using ADM1. Bioresour. Technol. 102, 6823–6829. Available from: https://doi.org/10.1016/j.biortech.2011.04.015.

Martin, J., 2003. A Comparison of Dairy Cattle Manure Management With and Without Anaerobic Digestion and Biogas Utilization. United States Environmental Protection Agency, AgSTAR Program.

Martín, M., 2016. Industrial Chemical Processes. Analysis and Design. Elsevier, Oxford.

Martín, M., 2019. Introduction to Software for Chemical Engineering, second ed. CRC Press, Boca Raton.

Martín-Hernández, E., Sampat, A.M., Zavala, V.M., Martín, M., 2018. Optimal integrated facility for waste processing. Chem. Eng. Res. Des. 131, 160–182.

Martín-Hernández, E., Guerras, L.S., Martín, M., 2020a. Optimal technology selection for the biogas upgrading to biomethane. J. Clean. Prod. 122032.

Martín-Hernández, E., Ruiz-Mercado, G.J., Martín, M., 2020b. Model-driven spatial evaluation of nutrient recovery from livestock leachate for struvite production. J. Environ. Manag. 271, 110967.

Mata-Alvarez, J., 2015. Biomethanization of the organic fraction of municipal solid wastes. Water Intell. Online 4 (0).

Møller, H., Lund, I., Sommer, S., 2000. Solid-liquid separation of livestock slurry: efficiency and cost. Bioresour. Technol. 74, 223−229.

Moran, M., Shapiro, H.N., Boettner, D.D., Bailey, M.B., 2014. Fundamentals of Engineering Thermodynamics, eighth ed. John Wiley & Sons, New Jersey.

Morse, J.W., Arvidson, R.S., Lüttge, A., 2007. Calcium carbonate formation and dissolution. Chem. Rev. 107, 342−381.

Nopharatana, A., Pullammanappallil, P.C., Clarke, W.P., 2007. Kinetics and dynamic modelling of batch anaerobic digestionof municipal solid waste in a stirred reactor. Waste Manag. 27, 595−603.

Ohlinger, K.N., Young, T., Schroeder, E., 1998. Predicting struvite formation in digestion. Wat. Res. 32 (12), 3607−3614.

Petracchini, F., Liotta, F., Paolini, V., Perilli, M., Cerioni, D., Gallucci, F., et al., 2018. A novel pilot scale multistage semidry anaerobic digestion reactor to treat food waste and cow manure. Int. J. Environ. Sci. Technol. 15 (9), 1999−2008.

Rahaman, M.S., Mavinic, D.S., Meikleham, A., Ellis, N., 2014. Modeling phosphorus removal and recovery from anaerobic digester supernatant through struvite crystallization in a fluidized bed reactor. Water Res. 51, 1−10.

Roh, H.-S., Lee, D.K., Koo, K.Y., Jung, U.H., Yoon, W.L., 2010. Natural gas steam reforming for hydrogen production over metal monolith catalyst with efficient heat-transfer. Int. J. Hydrog. 35 (3), 1613−1619.

Rykebosch, E., Brouillon, M., Vervaeren, H., 2011. Techniques for transformation of biogas to biomethane. Biomass Bioenergy 35, 1633−1645.

Scholz, M., Alders, M., Lohaus, T., Wessling, M., 2015. Structural optimization of membrane-based biogas upgrading processes. J. Memb. Sci. 474, 1−10. Available from: https://doi.org/10.1016/j.memsci.2014.08.032.

Skoog, D., West, D., Holler, F., Crouch, S., 2014. Fundamentals of Analytical Chemistry. Cengage Learning.

Smith, K., Grylls, J., Metcalfe, P., Jeffrey, B., Sinclair, A., 2007. Nutrient Value of Digestate from Farm-Based Biogas Plants in Scotland. Scottish Executive Environment and Rural Affairs Department.

Sørensen, P., Mejnertsen, P., Møller, H., 2011. Nitrogen fertilizer value of digestates from anaerobic digestion of animal manures and crops. NJF Rep. 7 (8), 42−44.

Taylor, A., Frazier, A., Gurneynium, E., 1963. Solubility products of magnesium ammonium and magnesium potassium phosphates. Trans. Faraday Soc. 59, 1580−1584.

Vrbová, V., Ciahotný, K., 2017. Upgrading biogas to biomethane using membrane separation. Energy Fuel 31 (9), 9393−9401. Available from: https://doi.org/10.1021/acs.energyfuels.7b00120.

Ward, A.J., Hobs, P.J., Holliman, P.J., Jones, D.L., 2008. Optimisation of the anaerobic digestion of agricultural resources. Bioresour.Technol. 99, 7928−7940.

Weinrich, S., Nelles, M., 2015. Critical comparison of different model structures for the applied simulation of the anaerobic digestion of agricultural energy crops. Biores. Technol. 178, 306−312.

Xiao, G., Webley, P., Hoadley, A., Ho, M., Wiley, D., 2013. Low Cost Hybrid Capture Technology Development: Final Report. ANLEC R&D Ref No: 3−0510-0046. Cooperative Research Centre for Greenhouse Gas Technologies, Canberra, Australia, CO2CRC Publication Number RPT13−4321.

Zambrano, D., Soler, J., Herguido, J., Menéndez, M., 2020. Conventional and improved fluidized bed reactors for dry reforming of methane: mathematical models. Chem. Eng. J. 393, 124775.

Zhang Y., 2003. Anaerobic Digestion System. Lecture 13 University of Southhampton.

Basic concepts and elements in the design of thermally coupled distillation systems

Gabriel Contreras-Zarazúa, Juan Gabriel Segovia-Hernández and Salvador Hernández-Castro

Department of Chemical Engineering, University of Guanajuato, Guanajuato, Mexico

7.1 Introduction: thermodynamic efficiency in distillation columns

The 21st century has been markedly characterized by increased environmental awareness and pressure from legislators to society to improve energy efficiency by adopting "greener technologies." In this context, the need for the chemical industry to develop processes that are more sustainable or eco-efficient has never been so vital. The successful delivery of green, sustainable chemical technologies at the industrial scale will inevitably require the development of innovative processing and engineering technologies that can transform industrial processes in a fundamental and radical manner.

Distillation is a thermal separation method for separating mixtures of two or more substances into its component fractions in a certain desired purity, based on differences in volatilities of components—which are in fact related to the boiling points of the components—by the application and removal of heat. Note that the term distillation refers to a physical separation process or a unit operation. At the commercial scale, distillation has many applications, such as the separation of crude oil into fractions (e.g., gasoline, diesel, kerosene), water purification and desalination, the splitting of air into its components (e.g., oxygen, nitrogen, argon), and the distillation of fermented solutions or the production of distilled beverages with high alcohol content (Kiss, 2013). Distillation underwent enormous development due to the petrochemical industry, and as such it is one of the most important technologies in the global energy supply system (Harmsen, 2010). Essentially, all transportation fuels go through at least one distillation column on its way from crude oil to readily usable fuel, with tens of thousands of distillation columns in operation worldwide. In view of the foreseen depletion of fossil fuels and the switch to renewable sources of energy such as biomass, the most likely transportation fuel will be ethanol, methanol, or derivatives. The synthesis of alternative fuels leads typically to aqueous mixtures that require distillation to separate ethanol or methanol from water. Consequently, distillation remains as the separation method of choice in the chemical process industry. The importance of distillation is unquestionable in providing most of the products required by our modern society (e.g., transportation fuel, heat, food, shelter, clothing). Lately, the application of this kind of designs to the separation and purification of platform chemicals has become key to the design of the production process (Contreras-Zarazúa et al., 2019).

Process intensification (PI) can provide such sought-after innovation of equipment design and processing to enhance process efficiency. PI aims to make dramatic reductions in plant volume, ideally between 100- and 1000-fold, by replacing the traditional unit operations with a novel, usually very compact designs, often by combining two or more traditional operations in one hybrid unit (Boodhoo and Harvey, 2013). Over the last two decades, the definition of PI has thus evolved from the simplistic statement of "the physical miniaturization of process equipment while retaining throughput and performance" to the complex definition provide recently by Tian et al. (2018) summarized activities that result in intensified processes, including the combination of multiple process tasks or equipment into a single unit (e.g., membrane reactors, reactive distillations), the miniaturization of process equipment (e.g., microreactors), the operation of equipment in a periodic manner (e.g., simulated moving bed, pressure adsorption swing), and tight process integration (e.g., dividing-wall distillation). Judging from the growth of research interest, it is clear that PI is a promising field that can enable a paradigm shift to the process industry, offering novel processing methods and equipment to achieve higher efficiency and safer operation (Segovia-Hernández and Bonilla-Petriciolet, 2015). Distillation is an energy-intensive process, which shares approximately 40% of the total energy consumed in process industries. Surprisingly, distillation offers a low thermodynamic efficiency of 5%−10% (Gómez-Castro et al., 2008). To win over these problems, heat integration came forth as an effective technique that leads to generating energy internally via a thermal coupling. There are typically two groups of heat integration, namely internal and external heat integration. The former type includes heat integrated distillation column (HIDiC) and thermally coupled distillation sequences (TCDS) and the latter one includes vapor recompression, bottoms flashing, closed-cycle compression, and absorption heat pump (Kiss and Infante Ferreira, 2016). This work is concerned with the TCDS.

The separation of multicomponent mixtures into three or more products is typically performed by operating sequences of simple distillation columns with one feed stream and two product streams. Some complex distillation arrangements, with or without thermal coupling, have been shown to provide significant energy and exergy savings. Such incentives have raised considerable scientific interest in the last 20 years. A good number of works have been reported on the design and analysis of TCDS for ternary separations (e.g., Tedder and Rudd, 1978; Glinos and Malone, 1988; Carlberg and Westerberg, 1989a,b; Yeomans and Grossmann, 2000; Rév et al., 2001). These studies have shown that the thermally coupled configurations can achieve energy savings of up to 30% in contrast to the conventional direct and indirect distillation sequences for the separation of feeds with low or high contents of the intermediate component. For zeotropic ternary, quaternary, and multicomponent mixtures, several arrangements for thermally coupled configurations have been proposed (Hernández and Jiménez, 1996, 1999; Agrawal and Fidkowski, 1999; Blancarte-Palacios et al., 2003; Calzon-McConville et al., 2006). Over the years, several heuristics and simple procedures have been approached to find the optimal thermally coupled sequence for several mixtures. In this way, two of the most classical works are the studies reported by Dejanović et al. (2010) and Asprion and Kaibel (2010), where they give a complete overview of the work done so far on the research and implementation of TCDS, from early ideas on thermal coupling of distillation columns to practical issues that needed to be solved for their successful implementation. Despite the potential benefits of thermally coupled columns and some reports of successful industrial applications (Kaibel and Schoenmakers, 2002), only a limited number of such columns has been implemented. The application of TCDS has been more noticeable in Europe and Japan, in part because of their dependence on imported crude oil (Kim, 2006). Finally, based on the aforementioned information, the aim of this chapter is illustrated to the reader about the basic concepts of thermally coupled distillation, current design strategies for designing the optimal thermally coupled sequences, and provide some simulation strategies for this kind of processes using common simulators like Aspen Plus.

7.2 Why do TCDS save energy? Remixing effect

Now, the multicomponent distillation applied to three component mixtures will be studied to exemplify thermodynamic inefficiencies of conventional distillation sequences.

When separating a three-product mixture using simple columns, there are only two possible sequences (see Fig. 7.1). Consider a system with the first characteristic of simple columns, a single feed is split into two products. As a first alternative to two simple

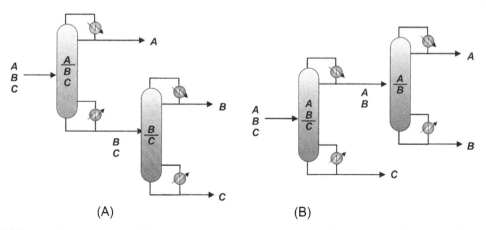

FIGURE 7.1 Conventional distillation sequences for three products. (A) Direct sequence and (B) indirect sequence.

columns, the possibilities shown in Fig. 7.2 can be considered. Here three products are taken from one column. The designs are in fact both feasible and cost-effective when compared to simple arrangements on a stand-alone basis (i.e., reboilers and condensers operating on utilities) for certain ranges of conditions.

If the feed is dominated by the middle product (typically >50% of the feed) and the heaviest product is present in small quantities (typically <5%), then the arrangements shown in Fig. 7.2A can be an attractive option. The heavy product must find its way down the column past the sidestream. Unless the heavy product has a small flow and the middle product has a high flow, a reasonably pure middle product cannot be achieved. In these circumstances, the sidestream is usually taken as a vapor product to obtain a reasonably pure sidestream. A large relative volatility between the sidestream Product B and the bottom Product C is also necessary to obtain a high purity sidestream. A practical difficulty with this arrangement should, however, be noted. While it is straightforward to split a liquid flow in a column, it is far less straightforward to split a vapor flow as shown.

The flow of vapor up the column must somehow be restricted to allow a split to be made between the vapor that must continue up the column and the vapor sidestream. This can be achieved by the use of a baffle across the column diameter, or vapor tunnels in the tray downcomers. However, taking a vapor sidestream is more problematic than taking a liquid sidestream. If the feed is dominated by the middle product (typically >50%) and the lightest product is present in small quantities (typically <5%), then the arrangement shown in Fig. 7.2B can be an attractive option. This time the light product

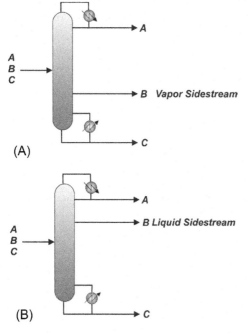

FIGURE 7.2 Distillation column with three products. (A) More than 50% middle component and <5% heaviest component. (B) More than 50% middle component and <5% lightest component.

7.2 Why do TCDS save energy? Remixing effect

287

must find its way up the column past the sidestream. Again, unless the light product is a small flow and the middle product is a high flow, a reasonably pure middle product cannot be achieved. This time the sidestream is taken as a liquid product to obtain a reasonably pure sidestream. A large relative volatility between the sidestream Product B and the overhead Product A is also necessary to obtain a high purity sidestream. In summary, single-column sidestream arrangements can be attractive when the middle product is in excess and one of the other components is present in only minor quantities. Thus the sidestream column only applies to special circumstances for the feed composition. More generally applicable arrangements are possible by relaxing the restriction that separations must be between adjacent key components.

Consider a three-product separation as shown in Fig. 7.3A in which the lightest and heaviest components are chosen to be the key separation in the first column. Two further columns are required to produce pure products, Fig. 7.3A. This arrangement is known as distributed distillation or sloppy distillation. The distillation sequence provides parallel flow paths for the separation of a product. At first sight, the arrangement in Fig. 7.3A seems to be inefficient in the use of equipment in that it requires three columns instead of two, with the bottoms and overheads of the second and third columns both producing pure B. However, it can be a useful arrangement in some circumstances. In the new design, the three columns can, in principle, all be operated at different pressures.

Also, the distribution of the middle Product B between the second and third columns is an additional degree of freedom in the design. The additional freedom to vary

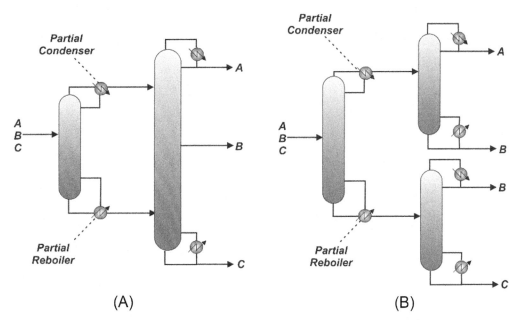

(A)   (B)

FIGURE 7.3   Configuration with prefactionator. (A) Distributed distillation for nonadjacent key components. (B) Prefractionator arrangement.

the pressures and the distribution of the middle product gives significant extra freedom to vary the loads and levels at which the heat is added to or rejected from the distillation. This might mean that the reboilers and condensers can be matched more cost-effectively against utilities, or heat integrated more effectively. If the second and third columns in Fig. 7.3A are operated at the same pressure, then the second and third columns could simply be connected, and the middle product taken as a sidestream as shown in Fig. 7.3B. The arrangement in Fig. 7.3B is known as a prefractionator arrangement. Note that the first column in Fig. 7.3B, the prefractionator, has a partial condenser to reduce the overall energy consumption. Comparing the distributed distillation in Fig. 7.3A and the prefractionator arrangement in Fig. 7.3B with the conventional arrangements in Fig. 7.1, the distributed and prefractionator arrangements typically require about 20%–30% less energy than conventional arrangements for the same separation duty. The reason for this difference is rooted in the fact that the distributed distillation and prefractionator arrangements are fundamentally thermodynamically more efficient than a simple arrangement. Based on the previous information, an important question arises, why does the prefractionator arrangement save energy? To response this question, consider the sequence of simple columns shown in Fig. 7.4. However, moving further down the column, the composition of Component B decreases again as the composition of the less-volatile Component C increases.

Thus the composition of B reaches a peak only to be remixed subsequently. Similarly, with the first column in the indirect sequence, the composition of Component B first increases above the feed as the less-volatile Component C decreases. It reaches a maximum only to decrease as the more volatile Component A increases. Again, the composition of Component B reaches a peak only to be remixed. This remixing that occurs in both sequences of simple distillation columns affect the separation; thus, it is considered

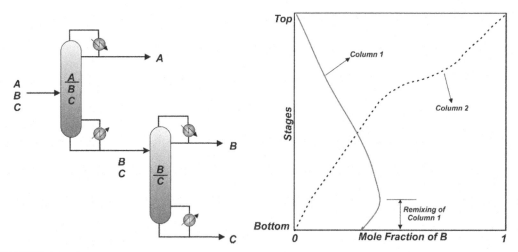

FIGURE 7.4   Composition profiles in direct sequence.

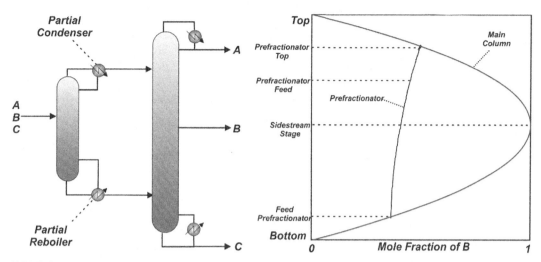

FIGURE 7.5   Remixing effect.

the main source of inefficiency in multicomponent separation. Other potential sources of inefficiencies are the inherent remixing phenomena in condensers and reboilers and the differences or mismatches between the composition of the column feed and the composition at the feed stage, even after optimizing the location of this tray, which promotes the remixing effects. In theory, for binary distillation in a simple column, a good match can be found between the feed composition and the feed stage. However, because the changes from stage to stage are finite, an exact match is not always possible. For multicomponent distillation in a simple column, except under extreme circumstances, it is not possible to match the feed composition and the feed stage. The solution to eliminate or mitigate the remixing effect in multicomponent distillation systems was proposed by Petlyuk (1965). Consider, again the first column of Fig. 7.4, as previously mentioned, this column has a maximum concentration of the intermediate volatility component and then this concentration decreases. Petlyuk et al. proposed that if instead of forcing this first column to completely separate the most volatile component, the middle component is distributed smoothly throughout the first column as in a prefractionation column (see Fig. 7.3B), it is possible to guarantee a good match between the feed stages and the compositions in the main column, which reduces the remixing losses in these trays. Additionally, a high concentration of the middle component can be obtained in a sidestream in the main column, by the reduction of remixing losses as shown in Fig. 7.5. The elimination of remixing losses in the prefractionator arrangement means that it is inherently more efficient than an arrangement using simple columns. The same basic arguments apply to both distributed distillation and prefractionator arrangements, with the additional degree of freedom in the case of distributed distillation to vary the pressures of the second and third columns independently. Finally, the Example 7.1 illustrates the effect of feed composition on remixing effect.

EXAMPLE 7.1 Effects of feed composition on remixing.

A mixture of pentane hexane and heptane is fed to a distillation column, the top product of this column is pentane with a molar purity of 98%. A diagram and design specifications of this column are shown inFig. 7E1.1, Fig. 7E1.2, Fig. 7E1.3, Fig. 7E1.4, Fig. 7E1.5, Fig. 7E1.6, Fig. 7E1.7. Use a process simulator to determinate the effect of the feed composition on remixing effect.

1. An equimolar mixture of 50 mol of pentane, 50 mol of hexane, and 50 mol of heptane.
2. A mixture of 50 mol of pentane, 25 mol of hexane, and 50 mol of heptane
3. A mixture of 25 mol of pentane, 50 mol of hexane, and 25 mol of heptane

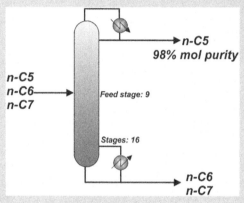

FIGURE 7E1.1   Distillation column Example 7.1.

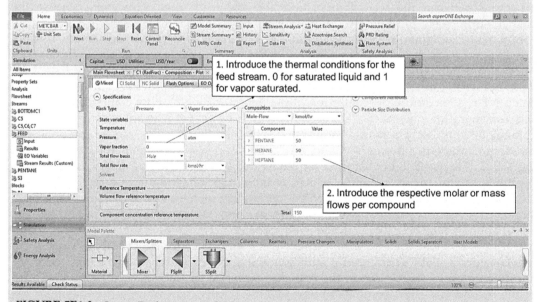

FIGURE 7E1.2   Step 1. Feed stream conditions.

7.2 Why do TCDS save energy? Remixing effect

291

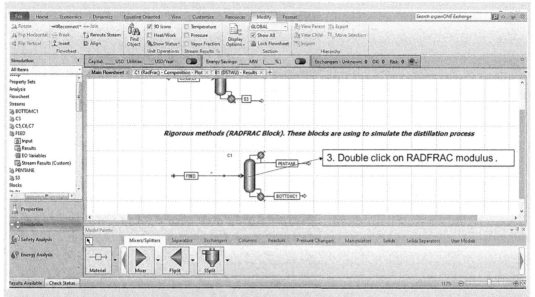

FIGURE 7E1.3 Step 2. RADFRAC modulus.

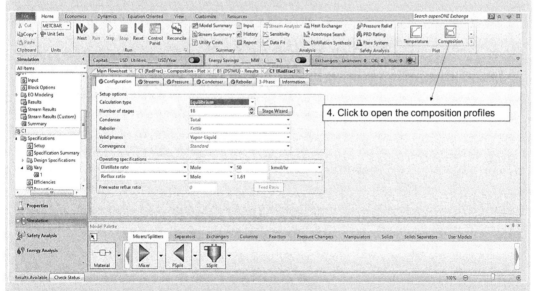

FIGURE 7E1.4 Step 3 Determine the composition profiles.

3. Biomass and waste based processes

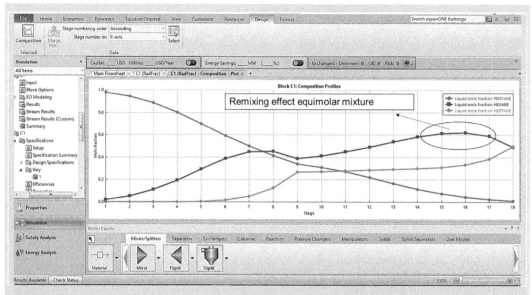

FIGURE 7E1.5   Composition profiles equimolar mixture.

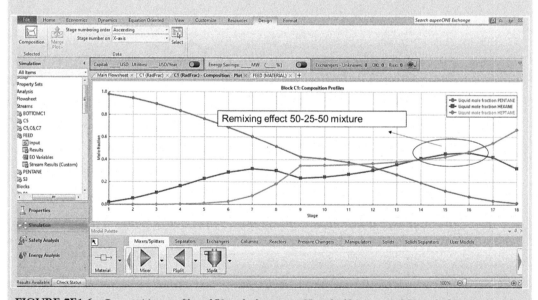

FIGURE 7E1.6   Composition profiles of 50 mol of pentane, 25 mol of hexane, and 50 mol of heptane mixture.

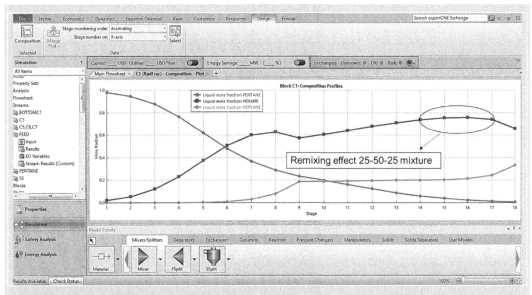

FIGURE 7E1.7 Composition profiles of 25 mol of pentane, 50 mol of hexane, and 25 mol of heptane mixture.

Which feed composition shows the most noticeable remix effect? Consider, the thermal conditions of feed stream as 1 atm and saturated liquid. (The ASPEN simulation file of this example is available on the website of this book.)

Solution:
Based on simulation results, the mixture of 25 mol of pentane, 50 mol of hexane, and 25 mol of heptane shows the most notorious remixing effect. This is an expected result due to the greater composition if the intermediate component.

7.3 Thermally coupled columns

An important restriction of simple columns mentioned earlier was that they should have a reboiler and a condenser. However, it is possible to use material flows to provide some of the necessary heat transfer by direct contact, which also mitigates the remixing phenomena caused by the condensers and reboilers. This transfer of heat via direct contact is known as thermal coupling.

First consider thermal coupling of the simple direct sequence shown in Fig. 7.1. Fig. 7.6B shows a thermally coupled direct sequence. The reboiler of the first column is replaced by a thermal coupling. Liquid from the bottom of the first column is transferred to the second as before but now the vapor required by the first column is supplied by the second column, instead of a reboiler on the first column. The four column sections are marked as 1, 2, 3, and 4 in Fig. 7.6A. It is important to mention that a section is a column part among condenser and feed or sidestream, reboiler and feed or sidestream, or between two streams. In Fig. 7.6C, the four column sections from Fig. 7.6B

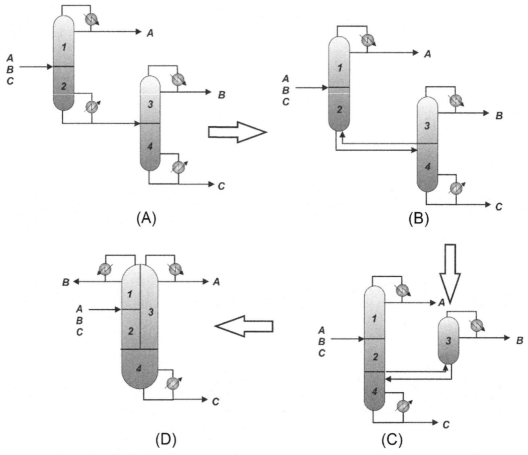

FIGURE 7.6 Thermally coupling in direct sequence. (A) Direct sequence, (B) direct thermally coupled sequence, (C) side-rectifier arrangement, and (D) partitioned side-rectifier arrangement.

are rearranged to form a side-rectifier arrangement, this configuration is called thermally coupled equivalent. There is a practical difficulty in engineering a side-rectifier arrangement. This is the difficulty in taking a vapor sidestream from the main column. As already noted for vapor sidestream columns, while it is straightforward to split a liquid flow in a column, it is not straightforward to split a vapor flow as required in Fig. 7.6C. This problem can be avoided by constructing the side rectifier in a single shell with a partition wall as shown in Fig. 7.6D. The partition wall in Fig. 7.6D should be insulated to avoid heat transfer across the wall as different separations are carried out on each side of the wall and the temperatures on each side will differ. Heat transfer across the wall will have an overall detrimental effect on column performance.

Similarly, Fig. 7.7A shows an indirect sequence and Fig. 7.7B the corresponding thermally coupled indirect sequence. The condenser of the first column is replaced by thermal coupling. The four column sections are again marked as 1, 2, 3 and 4 in Fig. 7.7B. In Fig. 7.7C, the four column sections are rearranged to form a side-stripper arrangement. As

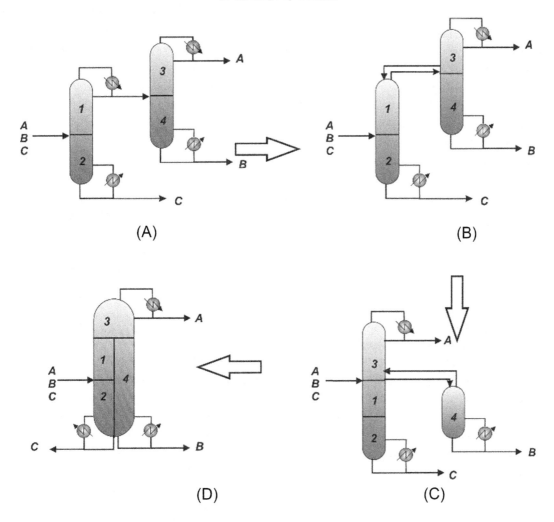

FIGURE 7.7 Thermally coupling in indirect sequence. (A) Indirect sequence, (B) indirect thermally coupled sequence, (C) side-stripper arrangement, and (D) partitioned side-stripper arrangement.

with the side rectifier, the side stripper can be arranged in a single shell with a partition wall, as shown in Fig. 7.7D. Again, the partition wall should be insulated, otherwise heat transfer across the wall will have an overall detrimental effect on the separation. Both the side-rectifier and side-stripper arrangements have been shown to reduce the energy consumption compared with simple two-column arrangements. This results from reduced mixing losses in the first (main) column. As with the first column of the simple sequence, a peak in composition occurs with the middle product. But now, advantage of the peak is taken by transferring material to the side rectifier or side stripper.

Now consider thermal coupling of the prefractionator arrangement from Fig. 7.3B. Fig. 7.8A shows a prefractionator arrangement with partial condenser and reboiler on the prefractionator.

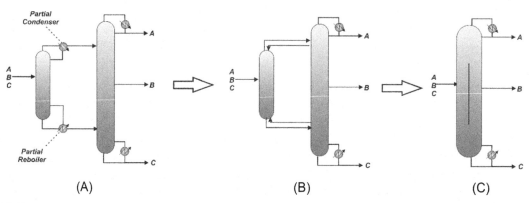

FIGURE 7.8 Thermally coupling in three columns arrangement. (A) Prefractionator sequence, (B) Petlyuk sequence, (C) dividing-wall column.

Fig. 7.8B shows the equivalent thermally coupled prefractionator arrangement, sometimes known as the Petlyuk column. To make the two arrangements in Fig. 7.8A and B equivalent, the thermally coupled prefractionator requires extra plates to substitute for the prefractionator condenser and reboiler. The prefractionator arrangement in Fig. 7.8A and the thermally coupled prefractionator (Petlyuk column) in Fig. 7.8B are almost the same in terms of total heating and cooling duties. There are differences between the vapor and liquid flows in the top and bottom sections of the main column of the designs in Fig. 7.8A and B, resulting from the presence of the partial reboiler and condenser in Fig. 7.8A. However, although the total heating and cooling duties are almost the same, there are greater differences in the temperatures at which the heat is supplied and rejected. In the case of the prefractionator in Fig. 7.8A, the heat load is supplied at two points and two different temperatures and rejected from two points and at two different temperatures. Fig. 7.8C shows an alternative configuration for the thermally coupled prefractionator that uses a single shell with a vertical partition dividing the central section of the shell into two parts, known as a dividing-wall column or partition column. The arrangements in Fig. 7.8B and C are equivalent if there is no heat transfer across the partition. As with side rectifiers and side strippers, the partition wall should be insulated to avoid heat transfer across the wall as different separations are carried out on each side of the wall and the temperatures on each side will differ. Heat transfer across the wall will have an overall detrimental effect on column performance. Partition columns offer several advantages over conventional arrangements:

1. Various studies have compared the thermally coupled arrangement in Fig. 7.8B and C, with a conventional arrangement using simple columns on a stand-alone basis. These studies show that the prefractionator arrangement in Fig. 7.8 requires typically to 30% less energy than the best conventional arrangement using simple columns (Fig. 7.1).
2. In addition, the partition column in Fig. 7.8C requires typically 20%−30% less capital cost than a two-column arrangement of simple columns.
3. The partition column has one further advantage over the conventional arrangements in Fig. 7.1. In partitioned columns, the material is only reboiled once and its residence time in the high-temperature zones is minimized. This can be important if distilling heat sensitive materials.

Example 7.2 illustrates how to design a thermally equivalent column using a commercial simulator such as Aspen Plus. The procedure is analogs for thermally equivalent columns and Petlyuk columns.

EXAMPLE 7.2 Simulation of a thermally equivalent configuration in Aspen Plus.

An equimolar mixture of hexane, 2-methylhexane, and heptane must be separated using a direct thermally equivalent distillation sequence. The feed molar flow rate is 75 kmol/h and its thermal conditions are 1 atm and saturated liquid. The scheme of the conventional distillation arrangement is shown in Fig. 7E2.1 and the design parameters are presented in Table 7E2.1. Use a simulation software to simulate the thermically equivalent process. (The ASPEN simulation file is available on the website of this book).

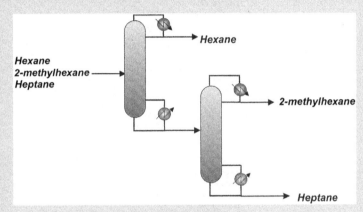

FIGURE 7E2.1 Conventional sequence Example 7.2.

TABLE 7E2.1 Design conventional distillation arrangement.

Design specifications	Column 1	Column 2
Stages (including condenser and reboiler)	28	59
Feed stage	13	30
Reflux ratio	3	10
Reboiler duty (kW)	821	2335
Operating pressure (atm)	1	1

Use the Chao-Seader thermodynamic model to perform the simulation.

The simulation of the thermally equivalent configuration using the Aspen Plus software is shown below. It is important to highlight that Aspen Plus considers the first and the last stage as condenser and reboiler, respectively. For this reason, the number of stages for the columns 1 and 2 must be specified as 27 and 61, these values include the column trays, the condenser and reboiler (Fig. 7E2.2, Fig. 7E2.3, Fig. 7E2.4, Fig. 7E2.5, Fig. 7E2.6, Fig. 7E2.7, Fig. 7E2.8, Fig. 7E2.9, Fig. 7E2.10, Fig. 7E2.11).

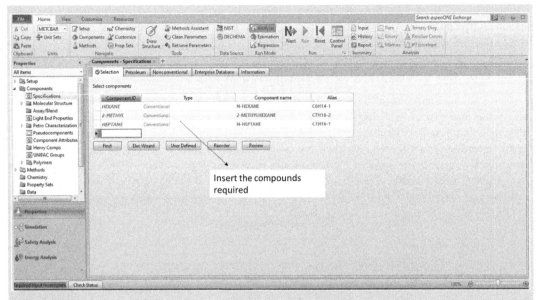

FIGURE 7E2.2   Step 1. Selection of compounds.

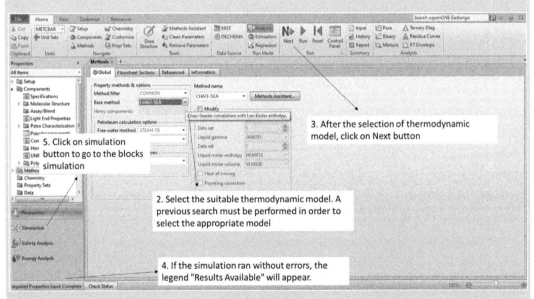

FIGURE 7E2.3   Step 2. Selection of thermodynamic model.

3. Biomass and waste based processes

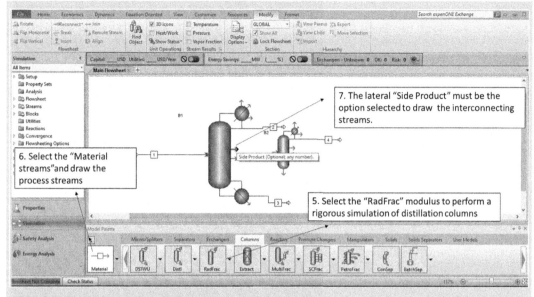

FIGURE 7E2.4   Step 3. Process flowsheet.

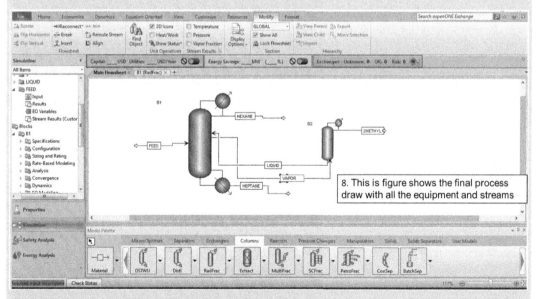

FIGURE 7E2.5   Step 4. Complete process flowsheet.

3. Biomass and waste based processes

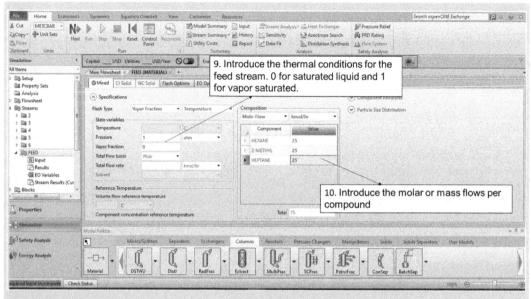

FIGURE 7E2.6   Step 5. Feed stream conditions.

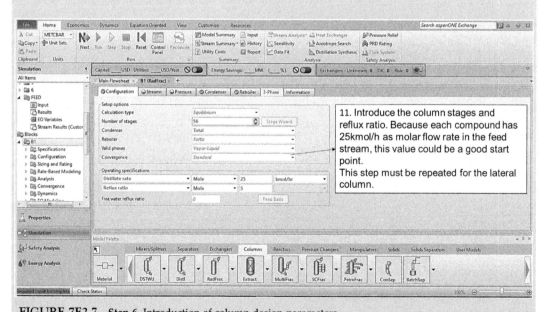

FIGURE 7E2.7   Step 6. Introduction of column design parameters.

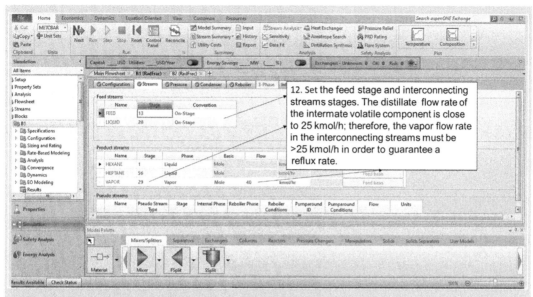

FIGURE 7E2.8 **Step 7.** Introduction of interconnecting and feed streams parameters.

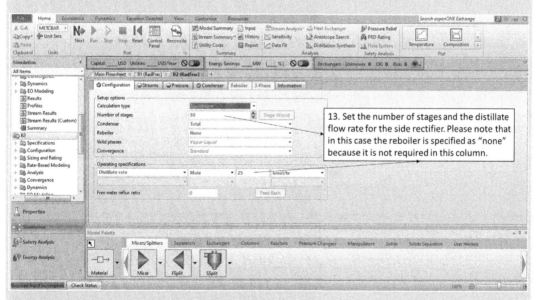

FIGURE 7E2.9 **Step 8.** Specifying the side rectifier.

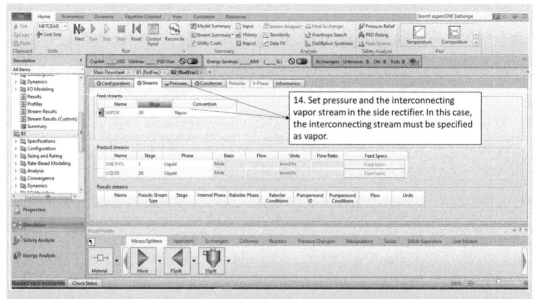

FIGURE 7E2.10　　**Step 9.** Set the feed stages side rectifier.

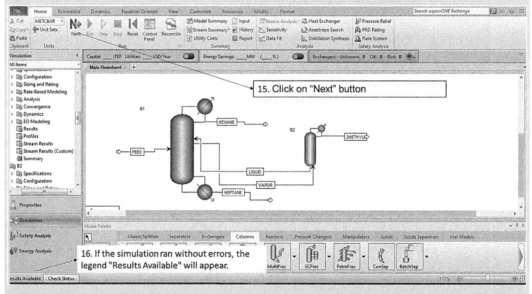

FIGURE 7E2.11　　**Step 10.** Simulation complete.

7.4 Design of thermally coupled columns and effect of interconnection mass flows

During several years, the industrial implementation of thermally coupled systems was limited due to the lack of design procedures, in fact the development of a suitable design method was one important challenge for engineers and researchers. The main problem associated with the design of thermally coupled systems is their inherently greater complexity. This superior complexity is directly related with many more degrees of freedom in contrast with the conventional distillation schemes (Dünnebier and Pantelides, 1999). Fig. 7.9 shows a comparison of the degrees of freedom for a conventional distillation and Petlyuk arrangements.

Most of the researchers have been focused on the development of short cut methods with the aim to get a fast screening of the possible alternatives and a good initialization of the degrees of freedom. However, most of the design methods have a restricted application and they usually solve the design problem assuming some assumptions such as constant relative volatility or constant molal overflow. The solution to the design problem was provided by Hernández and Jiménez (1996), they suggested to use the design of conventional distillation systems to generate the thermally coupled arrangements. The design methodologies for conventional distillation systems are well known, these schemes are easily designed using conventional techniques such as Fenske-Underwood-Gilliland that provides the number or stages, feed stages, reflux ratios, and reboiler duties to initialize the degrees of freedom. In this sense, once the conventional systems have been designed, Hernández and Jiménez (1996) proposed to divide the column in sections to generate the

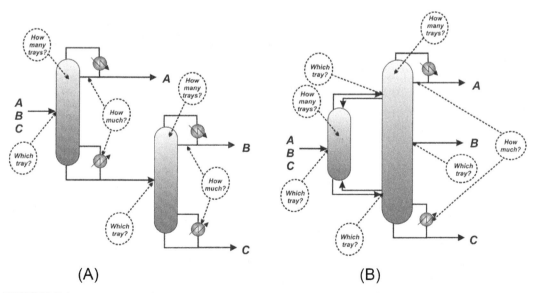

FIGURE 7.9 Degrees of Freedom for conventional and Petlyuk arrangements. (A) Direct column sequence and (B) Petlyuk column.

design of the thermally coupled arrangements. It is important to remember that a section is a column part between a condenser or reboiler and feed or sidestream (see Fig. 7.6 and Fig. 7.7). Once the column sections have been identified and due to utilization of the liquid and vapor interconnection streams, it is quite simple to generate the thermodynamic equivalent arrangements moving the section between columns as shown in Fig. 7.6 and Fig. 7.7, for this reason, the method proposed by Hernández and Jiménez (1996) is also called moving section method.

7.4.1 Columns with lateral equipment

Despite this method has been considered one of the most successful for designing thermally coupled systems, it does not provide an initial value for the interconnecting liquid or vapor streams; hence, a particular question arises about of how to specify and optimize those streams. Several previous works have determined that the interconnecting flows are one of the most important design variables because their values have direct effects on the energy requirements. Therefore bad values for those parameters can provoke poor energy savings for thermally coupled arrangements or even more energy consumption compared to their conventional counterparts (Hernández et al., 2003; Blancarte-Palacios et al., 2003; Dünnebier and Pantelides, 1999). As a representative case, the effect of the interconnecting liquid on energy requirements for an indirect thermally coupled distillation scheme used to separate an equimolar mixture of n-pentane, n-hexane, and n-heptane is shown in Fig. 7.10 (Hernández et al., 2003). Note the presence of a minimum energy consumption for an interconnecting liquid flow around 25 kmol/h, according to the data provided by Hernández et al. (2003), this minimum energy consumption corresponding to an energy saving of 21% in comparison with the indirect conventional sequence for separating the same mixture. To start the optimization procedure of the interconnection liquid shown in Fig. 7.10, the reflux rate of column C1 in the conventional indirect arrangement can be an initial value. In this sense, good initial values for the interconnecting streams can be obtained from the conventional schemes analyzing the flow rates or reflux ratios of the streams that connect both columns, these values can be easily calculated or obtained in a simulation software. In this sense, Example 7.3 illustrates the optimization of the interconnecting streams using a process simulation such as Aspen Plus.

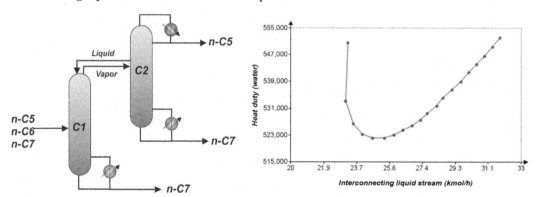

FIGURE 7.10   Effect of interconnecting flows on the energy consumption for an indirect thermally coupled sequence.

EXAMPLE 7.3 Optimization of the interconnecting streams for a direct thermally equivalent sequence.

Consider the thermally equivalent distillation sequence studied in Example 7.2. Perform the optimization of interconnecting streams using a simulation software. Consider that the hexane and 2-methylhexane must be recovered with a molar purity of 99% and 93%, respectively.

Solution:

Energy requirements

Firstly, before performing the optimization of the interconnection flows, it is necessary to set the desired compositions for hexane and 2-methylhexanes using a design specification. It is important to highlight that the optimization of the interconnecting flows must be done considering the same products compositions to achieve a representative comparison of the energy requirements for different interconnecting flows. The steps required to perform the interconnecting optimization are shown below (Fig. 7E3.1, Fig. 7E3.2, Fig. 7E3.3, Fig. 7E3.4, Fig. 7E3.5, Fig. 7E3.6):

Please note that in this case, the plot of energy requirement versus interconnecting flows is a line and not a curve. This is a common shape but it is important to highlight that the shape of the plot is specific for each case and it is not possible generalize just one shape.

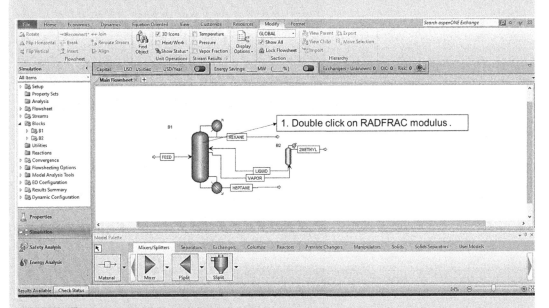

FIGURE 7E3.1   Step 1. Set the purities.

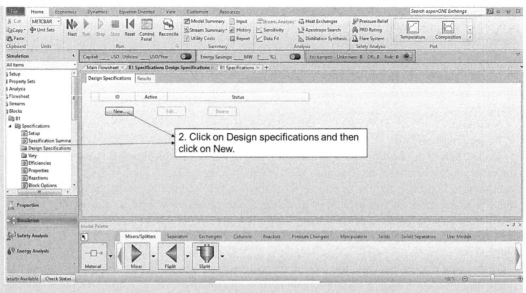

FIGURE 7E3.2   Step 2. Create a new design specification.

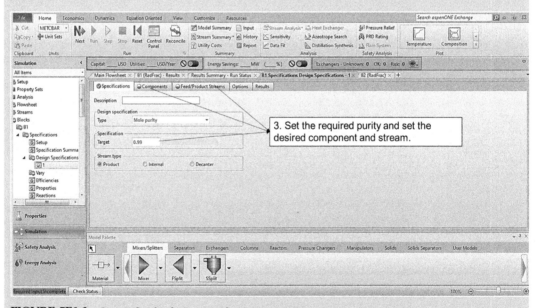

FIGURE 7E3.3   Step 3. Set the design specification parameters.

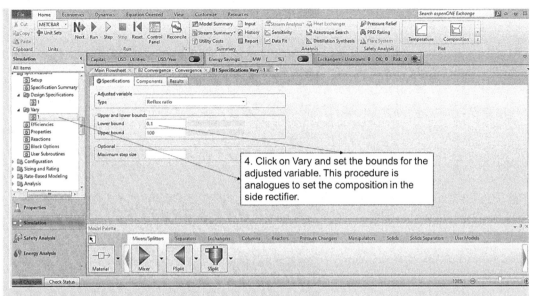

FIGURE 7E3.4 Step 4. Set the adjusted variables and its bounds.

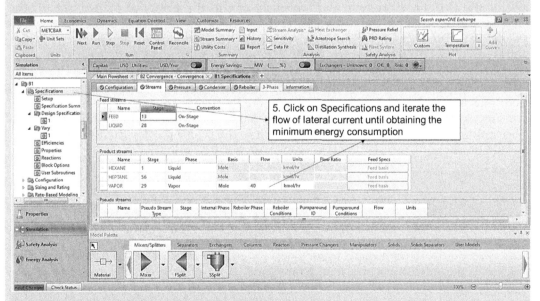

FIGURE 7E3.5 Step 5. Iterate the interconnecting flow.

3. Biomass and waste based processes

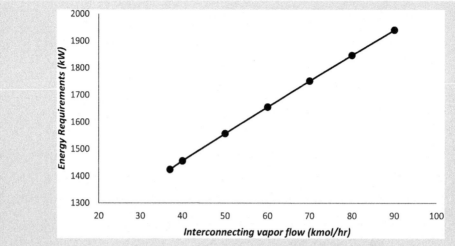

FIGURE 7E3.6   **Step 6.** Plot the energy requirement as a function of interconnection flow.

7.4.2 Columns with total integration

Similarly, the interconnecting streams show a strong effect on the energy consumption for the Petlyuk arrangements; however, due to the presence of two pair of the interconnecting streams, at least one liquid and one vapor streams must be specified generating a three-dimensional surface. However, it is preferred for the Petlyuk configuration to specify the vapor or liquid fractions of interconnecting streams in order to facilitate the convergence during simulation and design of this column. The vapor fractions can be obtained from the conventional distributed scheme using the internal vapor and liquid flows of the stages where the thermally couplings are implemented (see Fig. 7.11). These internal mass flows can be obtained from simulation data (Hernández and Jiménez, 1999). The vapor fraction can be calculated using the following equations:

$$\eta_V = \frac{V_B}{V_{NS2}} \tag{7.1}$$

$$\eta_L = \frac{L_T}{L_{NR}} \tag{7.2}$$

where V_B is the interconnecting vapor flow, V_{NS2} is the internal vapor flow of the stage NS2, L_T is the interconnecting liquid flow, and L_{NR} is the internal liquid flow of the stage NR. Based on the previous work reported by Hernández and Jiménez (1999), it is quite simple to set firstly the vapor fraction based on data from conventional arrangements. At the same time, Hernández and Jiménez (1999) found that usually the liquid fraction value is lower than the vapor fraction, which significantly reduces the search space to achieve a Petlyuk scheme with minimum energy consumption. Fig. 7.11 shows the effect

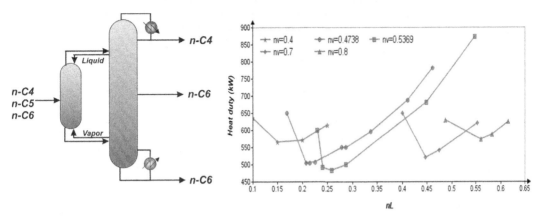

FIGURE 7.11   Effect of interconnecting flows on the energy consumption for an Petlyuk sequence.

of the interconnecting liquid and vapor streams on the energy consumption of a Petlyuk column used to separate an equimolar mixture of *n*-butane, *n*-pentane, and *n*-hexane (Hernández and Jiménez, 1999).

7.5 Bidirectionality of Petlyuk column: generation of a dividing-wall column

The method proposed by Hernández and Jiménez provides a powerful tool to generate different thermally coupled designs considering only the different sections of distillation columns, this can be achieved because this method takes advantage of flow direction of interconnecting streams and the internal liquid and vapor flows. Formally, two configurations can be considered thermodynamically equivalent if one arrangement can be rearrangement simply by moving sections from another scheme, which generates a changing on direction of interconnecting streams (see Fig. 7.6 and Fig. 7.7). In these cases, the change on interconnecting flow directions does affect thermodynamic advantages of thermally coupling, this can be easily corroborated analyzing an enthalpy diagram for the thermally schemes. Firstly, consider a single column in which the flow of the feed stream enters the column first as saturated liquid and then as superheated steam. The enthalpy–temperature diagrams for this case are shown in Fig. 7.12. Based on the enthalpy diagrams, the energy requirements for a saturated liquid feed stream are greater than for an overheating vapor feed because the reboiler requires overcoming the latent heat barrier to vaporize the feed stream.

Now consider the direct distillation sequences and its respective thermally coupled arrangement, the enthalpy diagrams for both sequences are shown in Fig. 7.13. Note, in Fig. 7.13, that because of the interconnection streams contain liquid and vapor flows at their respective saturation point, the compiled behavior of the two interconnecting streams that generate the thermal coupling among both columns is equivalent to a net current of superheat steam, which reduces the energy requirements used for heating in the thermally coupled schemes (Carlberg and Westerberg, 1989a). This compiled behavior of feed streams persists in the thermally equivalent

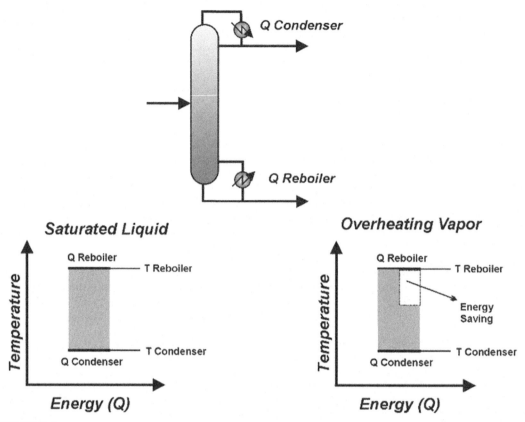

FIGURE 7.12   Enthalpy diagrams for saturated liquid and overheating vapor feeds.

columns and it is independent of the direction of the interconnecting liquid and vapor flows; therefore, it is also called as bidirectionality. A similar analysis can be performed for the indirect arrangements (Caballero and Grossmann, 2013; Caballero and Grossmann, 2004).

Carlberg and Westerberg (1989b) demonstrated that the energy savings in the Petlyuk column can be represented by a system where the direct and indirect thermally coupled schemes occur simultaneously in the prefractionator column, for this reason the Petlyuk arrangement is also known as distillation with fully thermally coupled. Therefore the Petlyuk column can be represented by an equivalent three-column scheme; however, to represent those three columns alternative, it is necessary to adjust the energy provided in the boiler and the condenser energy load to match the flows at the stage of extraction of intermediate volatility component. Fig. 7.14 exemplifies this equivalence.

Because the Petlyuk column only uses one condenser and reboiler instead of four heat exchanges of the conventional distillation sequences, it is possible to obtain important investment savings. Finally, the bidirectional streams can be rearranged for generating a dividing-wall column, where the prefractionator and the main column are integrated in a single shell and divided by an internal wall (see Fig. 7.15).

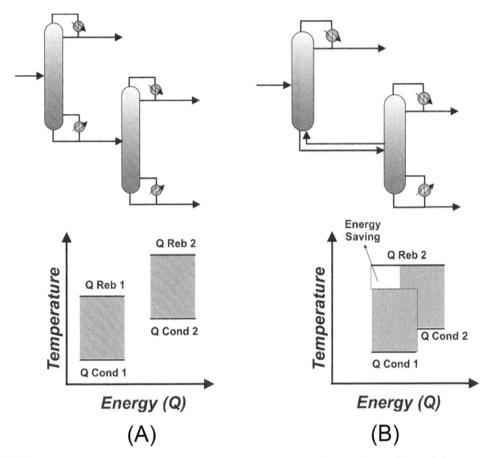

FIGURE 7.13   Enthalpy diagrams (A) Conventional direct sequence (B) Direct thermally coupled sequence.

7.6 Implementation in mixtures with four or more components

The thermally coupled distillation schemes can also be implemented for systems with four or more components to reduce the energy consumption of the separation. However, when the number of components is higher than three components, the set of possible thermally coupled schemes increases dramatically, most of those alternatives do not appear in three component systems. For this reason, some important structural considerations should be taken into account, to generate as much as possible of alternatives and determinate the best scheme. In conventional distillation arrangements, N components can be separated using $N-1$ columns, condensers, and reboilers, these $N-1$ columns are equivalent to $2(N-1)$ column sections (Agrawal, 1999, 2000). Based on the aforementioned information, it is possible to replace condenser and reboilers by interconnection columns, without modifying the process structure. In this case, the number of heat exchangers is the same when compared to the number of components as shown in Fig. 7.16.

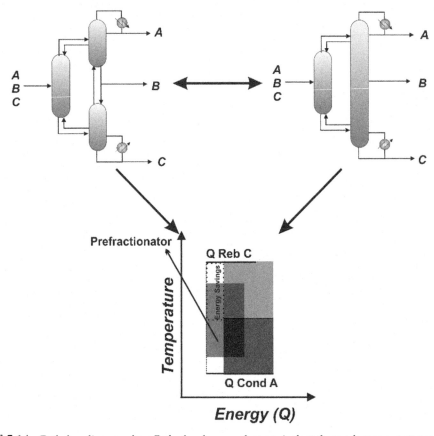

FIGURE 7.14 Enthalpy diagrams for a Petlyuk column and an equivalent three columns arrangement.

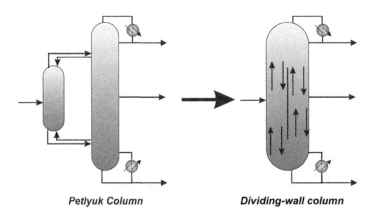

Petlyuk Column   *Dividing-wall column*

FIGURE 7.15 Enthalpy diagrams for a Petlyuk column and an equivalent three columns arrangement.

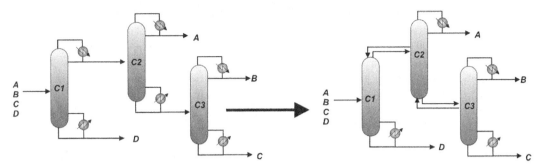

FIGURE 7.16 Thermally coupled arrangement for a four components separation.

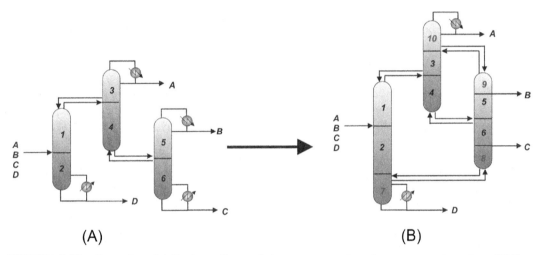

FIGURE 7.17 Generation of fully thermally coupled arrangements in a four components system: (A) thermally coupled scheme using 6 sections and (B) full thermally coupled scheme using 10 sections.

The replacement of condensers and reboilers for the unassociated product streams generates direct and indirect thermally coupled schemes. Note that in this case, the indirect and direct arrangements coexist in a single distillation sequence, which is very common in mixtures with four or more components. On the other hand, the thermally coupled sequence showed in Fig. 7.16 can be easily modified to generate a fully thermally coupled schemes, by replacing the heat exchangers associated with streams of intermediate volatility pure components by interconnecting streams. Nevertheless, to achieve the full thermally coupling and the desired purities, two extra column sections must be added for each exchanger removed. The number of stages of those new sections corresponds to the minimum theoretical stages required to separate the corresponding adjacent components. The minimum stages are calculated using the Fenske-Underwood-Gilliland method. Therefore the minimum number of column sections required to generate a full thermally coupled arrangement of mixtures with four or more components is $4N-6$ (Agrawal, 1999). The addition of this extra stages is presented in Fig. 7.17 in red numbers.

Note that although the minimum number of column section is $4N-6$ (without considering possible internal walls shames), the number of columns required to perform the separation is kept at $N-1$ columns.

It is important to highlight that thermo equivalent arrangements for systems with four or more components can be generated in analog way to three components systems using the moving section method. However, due to the numerous possible alternatives, many thermo equivalent configurations could be omitted with the methodology. Additionally, when a multicomponent thermally coupled arrangement is considered, it is not always easy to discern whether two sequences are thermodynamically equivalent or not. For example, Fig. 7.18 shows two thermodynamically equivalent sequences, although both alternatives look as two different configurations. The structure of fully thermally coupled sequence of Fig. 7.18A is called satellite column arrangement, whereas the scheme of Fig. 7.18B is known as sequential column arrangement or general thermally coupled system for four components (Sargent and Gaminibandara, 1976; Agrawal, 2000).

When two or more thermally coupled sequences are "thermodynamically equivalent," those configurations will have the same number of stages, column sections, product concentrations, and energy requirements, whereas the main differences between those schemes are in the vapor and liquid flows between sections, and mainly in the order of separation among different components. The number of column sections, stages, compositions, and energy requirements can be easily estimated using as reference the respective conventional distillation sequence. For this reason, a simple way of reducing or enclosing the number of possible thermally coupled configurations, and at the same time, selected the most adequate set of possible schemes, is selected the best conventional distillation sequence. The selection of the best conventional alternative is usually carried out taking into account a set of different metrics such as minimum energy requirements, minimum cost, best controllability criteria, safety properties, and among others. However, the

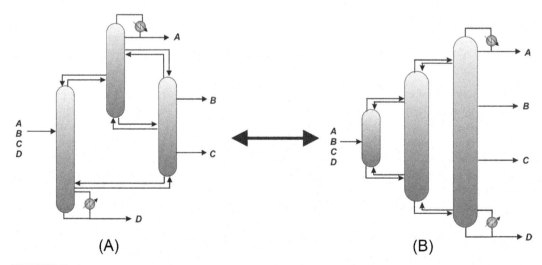

FIGURE 7.18 Thermally coupled equivalent configurations for a four components system: (A) satellite column arrangement and (B) sequential column arrangement.

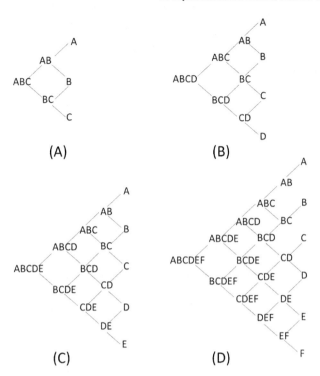

FIGURE 7.19 Possible separation alternatives for (A) three-, (B) four-, (C) five-, and (D) six-component components.

difficulty for selecting the best conventional alternative increases with the number of components involved in the mixture. Fig. 7.19 presents the superstructure of possible separation routes for three-, four-, five-, and six-component mixtures. Components in a mixture are ranked according to their relative volatility, where A is the most volatile component and volatility decreases in successive order.

Based on Fig. 7.19 note that the number of possible separation alternatives increases with the number of components. The selection of a specific separation alternative depends on many factors such as thermodynamic properties of mixture; component compositions; energy requirements; economic, control, safety, and environmental issues; and among others. Once the best alternative is conventional sequence selected, the set of thermally coupled schemes is derivate. Different heuristic rules have been developed to achieve the best conventional configuration. In this sense, Agrawal (2003) proposed six points for selecting basic distillation configurations, these points are mentioned as follows:

1. The lightest component (A) is always recovered at the top of a distillation column and has an associated condenser. In contrast, the heaviest component is always recovered at the bottom of a distillation column and it has an associated reboiler.
2. A component of intermediate volatility may be recovered with an associated condenser or a reboiler or neither of the two.
3. Whenever a submixture is associated with a condenser or a reboiler, this submixture is always removed from the distillation column.

4. If a submixture is transferred from one distillation column to another distillation column, then both the sections emanating from it are needed and they exist in the distillation column receiving the submixture.

5. When a submixture on the upper branch is transferred from one distillation column to another, then it has a condenser associated with it at the withdrawal location. In contrast, when a submixture on the lower branch is transferred from one distillation column to another, then it has a reboiler associated with it at the remotion location. In addition, an internal submixture, which is neither on the upper nor the lower branch, can be transferred between the columns with or without an associated reboiler or condenser.

6. When a condenser is associated with an internal submixture at a depth m in a network, then transfers of all submixtures are eliminated at depths earlier than m that contain the internal submixture as the heavy subgroup within the submixture. On the other hand, when a reboiler is associated with an internal submixture at depth m, transfers of all submixtures at depths earlier than m that contain the internal submixture as the light subgroup within the submixture are eliminated. Finally, for the recovery of an internal submixture at depth m from a distillation column without an associated reboiler or a condenser, there must be distillation sections both above and below the withdrawal point.

7.7 Industrial application of thermally coupled and divided wall column configurations

The first known application of thermally coupled distillation is due to Wright and Elizabeth (1949), who in 1949 patented an apparatus for separating air components, nowadays, this process is known as an internal wall column. Despite of the thermally coupled schemes have been proposed since 1969 and provide important energy savings about 30%−50% in contrast to the conventional configurations, it has not been until the last years that the thermally coupled schemes started to be implemented with more frequency in the chemical industry. The reason is quite simple, and it is associated with the implementation of the interconnecting streams required to achieve energy reductions. These interconnecting streams rise the complexity of the thermally coupled arrangements, during many years the engineers and researchers thought that the energy savings were being achieved at expenses of designing a more difficult process to control. Due to this the controllability of thermally coupled processes started to be investigated, reaching the conclusion that the thermally coupled processes have similar control properties and, in many cases, they may even have better control properties than the conventional configurations.

Currently, the BASF company is the world leader in the implementation of thermally coupled and divided wall columns, with more than 50 processes operating these technologies for the year 2002 (Schultz et al., 2002; Yildirim et al., 2011). On other hand, the thermally coupled columns have been applied mainly of a broad spectrum of mixtures such as hydrocarbons, alcohols, aldehydes, ketones, acetals, amines, and others. Additionally, these technologies have been implemented in complex distillation systems, for example, azeotropic, extractive, and reactive distillation (Yildirim et al., 2011). Table 7.1 presents the number of implemented thermally coupled columns reported in the literature. Note that most of them (116 columns) were implemented for ternary separations. Finally, Fig. 7.20 shows the number of thermally

TABLE 7.1 Thermally coupled columns implemented in the industry (Yildirim et al., 2011).

Systems	Number of thermally coupled columns
Ternary mixture	116
Mixture with four or more components	2
Reactive systems	–
Azeotropic mixture	1
Extractive systems	2
Revamps	4

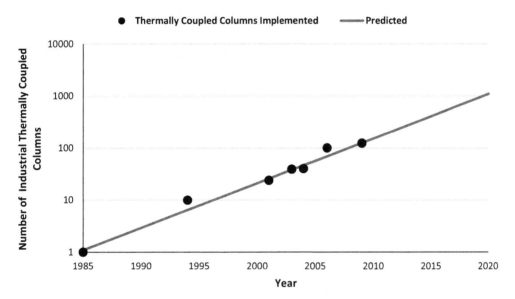

FIGURE 7.20 Thermally coupled columns implemented in the industry (Yildirim et al., 2011).

coupled columns implemented over the years and the corresponding implementation predictions according to the data provided by Yildirim et al. (2011). Note that for the year 2020, around 1100 thermally coupled should be implemented; however, these data are difficult to corroborate because the implementations of these technologies are commonly industrial secrets.

7.8 Conclusions

The energy used in the chemical process industry (particularly related to distillation) is essential for improving the sustainability of the chemical sector. Some concluding remarks can be established:

1. Despite the maturity of distillation, it still presents an enormous scope to reduce its energy consumption by significantly improving its low thermodynamic efficiency using thermally couplings, for example.
2. Thermally coupled distillation technologies should not be considered in isolation as solution to a problem but in the context of the overall system to find the optimal solution.
3. Optimization of TCDS requires all of the degrees of freedom to be manipulated simultaneously but this is challenging.

Sustainable success TCDS requires consistent cooperation between research and industry, certainly in view of the grand challenges of the next decades in energy area. Complex distillation is a key enabler for the process and product innovation, and has a bright future with sustainable impact on the chemical engineering area. TCDS must consider new opportunities in other areas, while maintaining its identity in an ever-changing context.

Problems

P7.1 Perform the simulation of the thermally coupled scheme of Example 7.2.
P7.2 If the separation process of Example 7.2 is carried out using an indirect sequence, perform the simulation of the indirect thermally coupled sequence and its respective thermally equivalent sequence. The design parameters and the process flowsheets are presented in Table 7P2.1 and Fig. 7P2.1, respectively.

TABLE 7P2.1 Design parameters example.

Design specifications	Column 1	Column 2
Stages	61	25
Feed stage	31	13
Reflux ratio	4.89	3.11
Reboiler duty (kW)	2500	871
Operating pressure (atm)	1	1

Use the Chao-Seader thermodynamic model and the previous feed steer conditions.

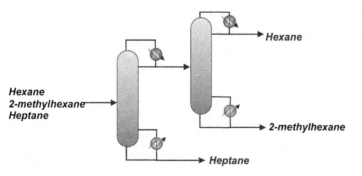

FIGURE 7P2.1 Conventional sequence problem 2.

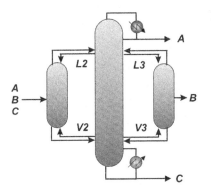

FIGURE 7P3.1 Thermally coupled system studied by Kim (2006).

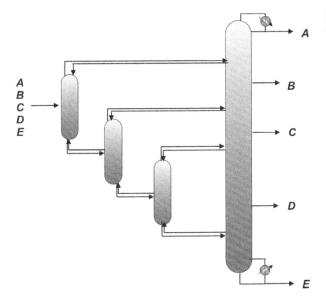

FIGURE 7P4.1 Thermally coupled system for five components.

P7.3 Kim (2006) has proposed the following column (Fig. 7P3.1) with a lateral postsplitter, which has a higher energy consumption than the Petlyuk column, under certain conditions.

Explain as clearly as possible according to the method of Hernández and Jiménez a possible design technique (number of plates, operating pressure, etc.) and optimization for this sequence.

P7.4 Considerer the following alternative thermally coupled sequence (Fig. 7P4.1):

Using the methodology proposed by Hernández and Jiménez, determine the conventional sequence.

P7.5 Rong (2011) studied the separation of multicomponent mixtures using thermally coupled systems. Fig. 7P5.1 shows two of the schemes studied by Rong. Draw at least two the thermally coupled schemes for both conventional sequences.

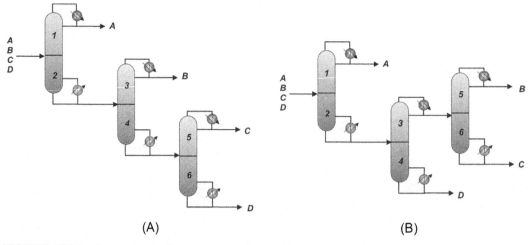

(A)   (B)

FIGURE 7P5.1   Conventional schemes studied by Rong (2011). (A) Conventional scheme 1 and (B) conventional scheme 2.

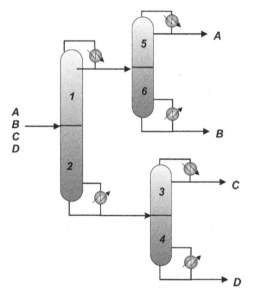

FIGURE 7P7.1   Sloppy distillation for a four-component mixture.

P7.6 Draw two thermodynamically equivalent sequences to thermally coupled system studied in problem 7.3.

P7.7 The sloppy distillation scheme for the separation of four-component mixture is shown in Fig. 7P7.1. Using the moving sections methodology proposed by Hernández and Jiménez (1996). Draw a Petlyuk column for this scheme.

References

Agrawal, R., 1999. More operable fully thermally coupled distillation column configurations for multicomponent distillation. Chem. Eng. Res. Des. 77 (6), 543–553.

Agrawal, R., 2000. A method to draw fully thermally coupled distillation column configurations for multicomponent distillation. Chem. Eng. Res. Des. 78 (3), 454–464.

Agrawal, R., 2003. Synthesis of multicomponent distillation column configurations. AIChE J. 49 (2), 379–401.

Agrawal, R., Fidkowski, Z.T., 1999. New thermally coupled schemes for ternary distillation. AIChE J. 45 (3), 485–496.

Asprion, N., Kaibel, G., 2010. Dividing wall columns: fundamentals and recent advances. Chem. Eng. Process. Process Intensif. 49 (2), 139–146.

Blancarte-Palacios, J.L., Bautista-Valdés, M.N., Hernández, S., Rico-Ramírez, V., Jiménez, A., 2003. Energy-efficient designs of thermally coupled distillation sequences for four-component mixtures. Ind. Eng. Chem. Res. 42 (21), 5157–5164.

Boodhoo, K., Harvey, A. (Eds.), 2013. Process Intensification Technologies for Green Chemistry: Engineering Solutions for Sustainable Chemical Processing. John Wiley & Sons.

Caballero, J.A., Grossmann, I.E., 2004. Design of distillation sequences: from conventional to fully thermally coupled distillation systems. Comput. Chem. Eng. 28 (11), 2307–2329.

Caballero, J.A., Grossmann, I.E., 2013. Synthesis of complex thermally coupled distillation systems including divided wall columns. AIChE J. 59 (4), 1139–1159.

Calzon-McConville, C.J., Rosales-Zamora, M.B., Segovia-Hernández, J.G., Hernández, S., Rico-Ramírez, V., 2006. Design and optimization of thermally coupled distillation schemes for the separation of multicomponent mixtures. Ind. Eng. Chem. Res. 45 (2), 724–732.

Carlberg, N.A., Westerberg, A.W., 1989a. Temperature-heat diagrams for complex columns. 2. Underwood's method for side strippers and enrichers. Ind. Eng. Chem. Res. 28 (9), 1379–1386.

Carlberg, N.A., Westerberg, A.W., 1989b. Temperature-heat diagrams for complex columns. 3. Underwood's method for the Petlyuk configuration. Ind. Eng. Chem. Res. 28 (9), 1386–1397.

Contreras-Zarazúa, G., Sánchez-Ramírez, E., Vázquez-Castillo, J.A., Ponce-Ortega, J.M., Errico, M., Kiss, A.A., et al., 2019. Inherently safer design and optimization of intensified separation processes for furfural production. Ind. Eng. Chem. Res. 58 (15), 6105–6120.

Dejanović, I., Matijašević, L., Olujić, Ž., 2010. Dividing wall column—a breakthrough towards sustainable distilling. Chem. Eng. Process. Process Intensif. 49 (6), 559–580.

Dünnebier, G., Pantelides, C.C., 1999. Optimal design of thermally coupled distillation columns. Ind. Eng. Chem. Res. 38 (1), 162–176.

Glinos, K., Malone, M.F., 1988. Optimality regions for complex column alternatives in distillation systems. Chem. Eng. Res. Des. 66 (3), 229–240.

Gómez-Castro, F.I., Segovia-Hernández, J.G., Hernandez, S., Gutiérrez-Antonio, C., Briones-Ramírez, A., 2008. Dividing wall distillation columns: optimization and control properties. Chem. Eng. Technol. 31 (9), 1246–1260.

Harmsen, J., 2010. Process intensification in the petrochemicals industry: drivers and hurdles for commercial implementation. Chem. Eng. Process. Process. Intensif. 49 (1), 70–73.

Hernández, S., Jiménez, A., 1996. Design of optimal thermally-coupled distillation systems using a dynamic model. Chem. Eng. Res. Des. 74 (3), 357–362.

Hernández, S., Jiménez, A., 1999. Design of energy-efficient Petlyuk systems. Comput. Chem. Eng. 23 (8), 1005–1010.

Hernández, S., Pereira-Pech, S., Jiménez, A., Rico-Ramírez, V., 2003. Energy efficiency of an indirect thermally coupled distillation sequence. Can. J. Chem. Eng. 81 (5), 1087–1091.

Kaibel, G., Schoenmakers, H., 2002. Process synthesis and design in industrial practice, Computer Aided Chemical Engineering, Vol. 10. Elsevier, pp. 9–22.

Kim, Y.H., 2006. A new fully thermally coupled distillation column with postfractionator. Chem. Eng. Process. Process Intensif. 45 (4), 254–263.

Kiss, A.A., 2013. Novel applications of dividing-wall column technology to biofuel production processes. J. Chem. Technol. Biotechnol. 88 (8), 1387–1404.

Kiss, Anton, Infante Ferreira, Carlos A., 2016. Heat Pumps in Chemical Process Industry, 1st ed. CRC Press, Boca Raton, pp. 1–442.

3. Biomass and waste based processes

Petlyuk, F.B., 1965. Thermodynamically optimal method for separating multicomponent mixtures. Int. Chem. Eng. 5, 555–561.

Rév, E., Emtir, M., Szitkai, Z., Mizsey, P., Fonyo, Z., 2001. Energy savings of integrated and coupled distillation systems. Comput. Chem. Eng. 25 (1), 119–140.

Rong, B.G., 2011. Synthesis of dividing-wall columns (DWC) for multicomponent distillations—a systematic approach. Chem. Eng. Res. Des. 89 (8), 1281–1294.

Sargent, R.W.H., Gaminibandara, K., 1976. Optimum design of plate distillation columns. Optimization in Action 1, 217–314.

Schultz, M.A., Stewart, D.G., Harris, J.M., Rosenblum, S.P., Shakur, M.S., 2002. Reduce costs with dividing-wall columns. Chem. Eng. Prog. 98 (5), 64–71.

Segovia-Hernández, J.G., Bonilla-Petriciolet, A., 2015. Process Intensication in Chemical Engineering. Springer.

Tedder, D.W., Rudd, D.F., 1978. Parametric studies in industrial distillation: Part I. Design comparisons. AIChE J. 24 (2), 303–315.

Tian, Y., Demirel, S.E., Hasan, M.F., Pistikopoulos, E.N., 2018. An overview of process systems engineering approaches for process intensification: state of the art. Chem. Eng. Process. Process Intensif. 133, 160–210.

Wright, R.O., Elizabeth, N.J., 1949.

Yeomans, H., Grossmann, I.E., 2000. Optimal design of complex distillation columns using rigorous tray-by-tray disjunctive programming models. Ind. Eng. Chem. Res. 39 (11), 4326–4335.

Yildirim, Ö., Kiss, A.A., Kenig, E.Y., 2011. Dividing wall columns in chemical process industry: a review on current activities. Sep. Purif. Technol. 80 (3), 403–417.

Added-value products

Manuel Taifouris and Mariano Martín

Department of Chemical Engineering, University of Salamanca, Salamanca, Spain

8.1 What are value-added products?

Value-added products can be defined as all those products obtained after one or more stages of processing, with increased value with respect to the initial raw material. Some examples of value-added products are all petroleum derivatives, leading to the design of refineries. However, oil is not the only complex raw material from which a large variety of products can be obtained, biomass is another example.

The concept of a biorefinery was established as a parallelism with crude oil processing facilities where from a raw material consisting of a mixture of species, a portfolio of products is obtained. Similarly, biomass as a complex material can be processed to obtain a number of valuable products. The need to control CO_2 emissions provided the first push toward the design of facilities that produced biofuels, bioethanol, and biodiesel, as substitutes for crude-based gasoline and diesel. They were the main products of the biorefineries. The low margins presented an opportunity to transform a single product facility into a refinery so that byproducts such as distiller's dry grains with solubles (DDGS) in the case of first-generation bioethanol (Karuppiah et al., 2008), or glycerol in biodiesel facilities (Martín and Grossmann, 2012), a valuable product in the cosmetics industry, were used to improve the tight economics. These facilities evolved into integrated plants that using biomass of different types were capable of self-producing the intermediates they needed for the production of advanced products. Examples can be a diesel substitutes facility that produces biodiesel and glycerol esters from algae. The required i-butene to obtain the glycerol ethers is obtained from glucose fermentation (De la Cruz et al., 2014), as well as integrated multiproduct facilities from lignocellulosic biomass producing platform chemicals and fuels (Bond et al., 2014). The use of biomass as raw material is suffering another revolution. Biomass is too valuable to be used for the production of energy, fuels, and bulk chemicals. In addition, the competition between biomass and crude for the production of chemicals is unbalanced since the use of biomass as raw material requires a large number of steps to decompose the raw material into the useful intermediate, sugars, or syngas. Chemicals and fuels (Martín and Grossmann, 2013a,b) and platform chemicals, levulinic acid (Martín-Alonso et al.,

2013), furfural and either hydroxymethylfurfural (Torres et al., 2010) and/or dimethyl furfural (Martin and Grossmann, 2016), and diols (Huang et al., 2017) can still be part of the portfolio; however, the new trend is aiming at achieving a more efficient use of this precious natural resource (Fig. 8.1).

Social expectations have a lot to do with this turn in the goal. High added-value products focus on a series of specialty chemicals demanded by the cosmetics, pharma, and food industry toward a better quality of life. Within this class of products, it is possible to identify carotenoids (Psycha et al., 2014), food supplements and polymers (García Prieto et al., 2017), essential oils (Dávila et al., 2015; Criado and Martín, 2020) and active principles for drugs (Xu and Arancon, 2014; Paek et al., 2014) so that they have entered the portfolio of the biorefinery. Specialty chemicals are in very small quantities and require a number of processing stages to be recovered and purified. However, after their extraction from biomass, it is still possible to further use the waste as a source of bulk chemicals or power for self-providing thermal and electrical energy to the facility (Criado and Martín, 2020) or as a service. The biomass can follow thermal treatment (Chapter 3), hydrolysis (Chapter 4), anaerobic digestion (Chapter 6), or a separation process to obtain some specific chemical. In case of using digestion, a digestate rich in nutrients is also produced. The design of such integrated refineries involves concepts like circular economy, integration and intensification, and process system engineering tools that have been developed to aid in the synthesis.

Several integrated facilities have been latterly proposed in the literature based on the newly coined process and product design concept together with the ideas of circular economy. The biomass composition can be a variable of the process to design a biomass that will be more efficient for the operation of the facility or the selection of the biomass among the ones that can be produced in a region. Examples of these are the production of glycerol ethers and biodiesel from algae, that used the starch in the algae to produce ethanol, for oil transesterification, and i-butene, for glycerol ethers production and the oil was the base for

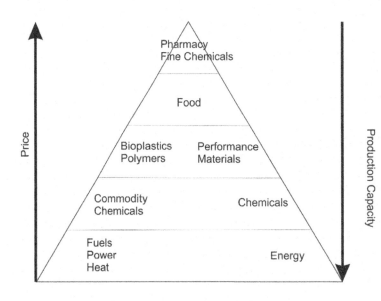

FIGURE 8.1 Pyramid of biomass-related products.

the biodiesel and glycerol production (De la Cruz et al., 2014). The production of biodiesel from waste is another example, where after anaerobic digestion the biogas was used to produce methanol and the digestate provided the nutrients for algae growing (Hernández and Martin, 2017). Apart from the use of algae as raw material, lignocellulosic biomass consisting of hemicellulose, cellulose, and lignin can be the raw material for the production of ETBE where hemicellulose was the raw material for the production of ethanol and cellulose via glucose to produce i-butene that together allowed the production of Ethyl tert-Butyl Ether (ETBE) (Galán et al., 2019). In addition, using orange peels it is possible to produce limonene and from the residue still produce fertilizers, to help in the next year harvest, thermal and electrical energy for self-consumption as well as product (Criado and Martín, 2020).

8.2 Value-added products design formulations

Following the discussion presented above, it is possible to identify two different but related problems, the blending/pooling problem by which a mixture of ingredients provides a final product with interesting properties and the refining problem, where from a formulated raw material it is possible to obtain a full portfolio of products (Fig. 8.2). These problems can be extended to integrate product design. The integrated design of processes and products has been applied to the design of solvents (Brand et al., 2016), heat transfer fluids (Peña-Lamas et al., 2018), catalysts and materials design such as zeolites or polymers (Gounaris et al., 2006a,b; Kang et al., 2019) but also biomass (Galán et al., 2019; Martín and Grossmann, 2013a) and consumer products (Martín and Martínez, 2013). Only by addressing the integrated problem, it is possible to address the trade-offs between the processing needs and costs and the product features resulting in a large, multi-objective, multidisciplinary problem.

8.2.1 Formulated product design

8.2.1.1 Problem definition: pooling problem

The design of a product from a number of intermediates or ingredients is a well-known problem in the chemical industry in particular for the production of gasoline or coal blends. The aim is to select the mixture so that the properties of the final product satisfy environmental and cost requirements (Audet et al., 2004). The problem can be as simple as a linear problem (Dantzig, 1963) or being extended to NLP or MINLP problems. These last problem formulations include the model of the entire process that uses the resulting coal mixture and includes the flue gas treatment (Guerras and Martín, 2019) as well as the use

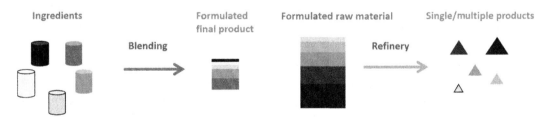

FIGURE 8.2 Formulated product design problems.

of the mixture of wastes for the production of a syngas with the right composition for the production of different chemicals (Hernández et al., 2017). A particular case of blending problems is the "pooling-problem." They were, as the previous one, originated in refineries aiming at adjusting the sulfur composition of the final products. Based on this formulation, extended pooling problems have been developed for fuel emissions control (Misener et al., 2010) or consumer products (Martín and Martínez, 2013). The main mathematical feature of the pooling problems is the presence of bilinear products that define the composition of the resulting mixture. This fact brings into scene global optimization techniques to discard the multiple local optima that the formulation gives rise.

8.2.1.2 Formulation

The mathematical formulation of the problem must include product performance and quality, environmental process, and economics constraints (Martín and Martínez, 2013). As such, in the extended version, it is a multi-objective optimization problem where the conflicting objectives typically are the cost, the environmental performance, and the quality or customer acceptance.

$$\text{Min Cost Production (raw materials)} + \text{Pools} \tag{8.1}$$

S.t.
Feed availability

$$A_i^L \leq \sum_{Tx} x_{i,l} + \sum_{T_z} z_{i,l} \leq A_i^U \quad \forall i \tag{8.2}$$

Pool capacity

$$\sum_{Tx} x_{i,l} \leq S_l \quad \forall l \tag{8.3}$$

Product demand

$$D_j^L \leq \sum_{T_y} y_{l,j} + \sum_{T_z} z_{i,j} \leq D_j^U \forall j \tag{8.4}$$

Material balance

$$\sum_{T_x} x_{i,l} - \sum_{T_y} y_{l,j} \leq 0 \; \forall l \tag{8.5}$$

Quality balance

$$\sum_{T_x} C_{i,k} \cdot x_{i,l} - p_{k,l} \sum_{T_y} y_{l,j} \leq 0 \quad \forall l, k \tag{8.6}$$

Product quality

$$\sum_{T_z} z_{i,j} - \sum_{T_y} p_{k,l} \cdot y_{l,j} \leq P_{j,k}^U \quad \forall l, k$$
$$\sum_{T_z} z_{i,j} - \sum_{T_y} p_{k,l} \cdot y_{l,j} \geq P_{j,k}^L \quad \forall l, k \tag{8.7}$$

Hard bounds

$$0 \leq x_{i,l} \leq \min\left\{A_i^U, S_l, \sum_{T_y} D_j^U\right\} \quad \forall T_x$$

$$0 \leq y_{i,l} \leq \min\left\{S_l, D_j^U, \sum_{T_x} A_l^U\right\} \quad \forall T_y \quad (8.8)$$

$$0 \leq z_{i,l} \leq \min\left\{D_j^U, A_l^U\right\} \forall T_z$$

where, A_j is the availability of ingredient j, (Upper (U) and lower (L)), $CC_{i,k}$ the composition of ingredient k in the flow i, D_k the demand of ingredient j (Upper (U) and lower (L)), $x_{(i,\,l)}$ the flows from raw material (i) to intermediate pool(l), $y_{(l,\,j)}$ corresponds to the flow from pool (l) to product (j), and $z_{(i,\,j)}$ the flow from raw material (i) to product (j). $p_{(l,\,k)}$ is the composition in component (k) of pool (l) and $PQ_{(j,\,k)}$ the composition in component (k) of product (i). Finally, S (l) is the Maximum tank size (kg).

8.2.1.2.1 Process constraints

Units conversion/yield.
Limits of operation of the units.
    Flows.
    Properties of the mixture.
    Properties of the product.

8.2.1.2.2 Product performance/quality

To relate the performance of the product with the ingredients that constitute it, surrogate models are usually developed. These models require not only experimental data but also market studies to evaluate the effect of the different ingredients on the final product performance.

8.2.1.2.3 Environmental Burden

The environmental impact associated with the production of a product can be estimated using different indexes (Chapter 3). The product life cycle analyses all the different types of environmental impact that a product can produce from its production process to its release into the environment. The ecolabel analyses the environmental impact produced by the release of a product in an aquatic environment, while the carbon footprint analyses the environmental impact produced in the atmosphere. Normally, the environmental impact is usually associated with the ingredients that make up the formulated product. But transportation and production process must also be taken into account when determining the environmental impact of a product.

The mathematical complexity of the model relies on the equations that represent the constraints of the product quality, the mixture of the ingredients, the process, and product performance. To draw conclusions from a number of indicators that are not aligned, several mathematical approaches are available. Marler and Arora (2004) presented a review describing multi-objective optimization methods including:

ε-*constraint method*: The second objective is added as a constraint that is forced to take lower values obtaining a PARETO CURVE (see Example 8.1).

EXAMPLE 8.1 **Consider the preparation of animal feed. Per kg of mixture, it should contain a minimum of nutrients as given below:**

Nutrient	A	B	C	D
g/kg	75	50	25	2

Each ingredient has its cost and nutritional value:

	A	B	C	D	Cost(€/kg)	Environmental impact (pt/kg)
Ingredient 1 (g/kg)	100	60	40	10	35	5
Ingredient 2 (g/kg)	150	175	20	—	65	2
Ingredient 3 (g/kg)	200	60	50	15	20	10
Ingredient 4 (g/kg)	250	150	25	2	30	8
Ingredient 5 (g/kg)	—	—	—	—	0	0

Optimize the mixture for different environmental burdens, computing the Pareto curve. Formulate the problem and solve it using Python

Solution

Constraints

Mass balance

$$x_1 + x_2 + x_3 + x_4 + x_5 = 1$$

Nutritional balance

$$100x_1 + 150x_2 + 200x_3 + 250x_4 \geq 75 \text{ (nutrient A)}$$
$$60x_1 + 175x_2 + 60x_3 + 150x_4 \geq 50 \text{ (nutrient B)}$$
$$40x_1 + 20x_2 + 50x_3 + 25x_4 \geq 25 \text{ (nutrient C)}$$
$$10x_1 + 15x_3 + 2x_4 \geq 2 \text{ (nutrient D)}$$

Environmental impact

$$\varepsilon \geq 5x_1 + 2x_2 + 10x_3 + 8x_4 = \text{Impact}$$

Objective function:

$$\text{Minimize: } F = 35x_1 + 65x_2 + 20x_3 + 65x_4$$

Using pyomo to formulate and solve the problem:

```
from pyomo.environ import *
from pyomo.opt import SolverFactory
import matplotlib.pyplot as plt
```

```
environ = [6,5.5,5,4.5,4,3.5,3]
coste = []
for ep in environ:
# Optimization model
        model = ConcreteModel(name = 'Nutrition')

#Variables
        model.x1 = Var(within = PositiveReals)# Create a variable x1 > = 0
        model.x2 = Var(within = PositiveReals)# Create a variable x2 > = 0
        model.x3 = Var(within = PositiveReals) # Create a variable x3 > = 0
        model.x4 = Var(within = PositiveReals) # Create a variable x4 > = 0
        model.x5 = Var(within = PositiveReals) # Create a variable x5 > = 0
        model.environ = Var(within = PositiveReals)

#CONSTRAINTS
        model.constraint1 = Constraint(expr = 100*model.x1 + 150*model.
x2 + 200*model.x3 + 250*model.x4 > = 75)
        model.constraint2 = Constraint(expr = 60*model.x1 + 175*model.x2 + 60*model.
x3 + 150*model.x4 > = 50)
        model.constraint3 = Constraint(expr = 40*model.x1 + 20*model.x2 + 50*model.
x3 + 25*model.x4 > = 25)
        model.constraint4 = Constraint(expr = 10*model.x1 + 15*model.x3 + 2*model.
x4 > = 2)
        model.constraint5 = Constraint(expr = model.x1 + model.x2 + model.x3 + model.
x4 + model.x5 == 1)
        model.constraint6 = Constraint(expr = model.environ == 5*model.x1 + 2*model.
x2 + 10*model.x3 + 8*model.x4)
        model.constraint7 = Constraint(expr = model.environ < = ep)

# OBJECTIVE FUNCTION
        model.obj = Objective(expr = 35*model.x1 + 65*model.x2 + 20*model.
x3 + 30*model.x4, sense = minimize)
#SOLVE
        solver = SolverFactory('glpk')
        Res = solver.solve(model)
        print('x1 = ',model.x1(),'x2 = ',model.x2(),'x3 = ',model.x3(),'x4 = ',model.
x4(),'x5 = ',model.x5())
        print('Environmental = ',model.environ())
        print('Cost = ',model.obj())
        coste.append(model.obj())

plt.plot(environ,coste,'k - ')
plt.xlabel('Impact(Pt/kg)')

plt.ylabel('Cost (€/kg)')
```

3. Biomass and waste based processes

Impact/Fraction	6	5	4	3.5	3
X1	0	0.152	0.455	0.563	0.500
X2	0	0	0	0.025	0.250
X3	0.417	0.303	0.076	0	0
X4	0.167	0.152	0.121	0.079	0
X5	0.417	0.394	0.348	0.333	0.25
Cost (€/kg)	13.3	15.9	21.1	23.7	33.7

See Fig. 8E1.1.

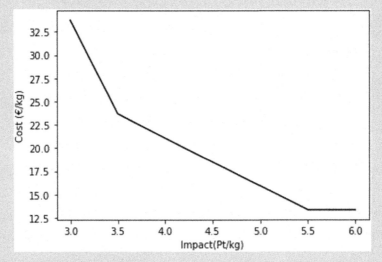

FIGURE 8E1.1   Pareto plot blending problem.

Weighted averages: Use weights to formulate a new objective function:

$$F = w_1 \cdot Obj_1 + w_2 \cdot OBj_2 + \dots \dots \tag{8.9}$$

Normalized objectives: Use a reference value to weight each objective. It is required to optimize the different objectives independently first to compute the reference values.

$$\varphi_i = \frac{Obj_i - Obj_i^{min}}{Obj_i^{max} - Obj_i^{min}} \tag{8.10}$$

8.2.2 Formulated raw materials

8.2.2.1 Refinery problem

Another type of problem is that where the raw material consists of a mixture of species that can be processed into a portfolio of products. Typically, there are major components and others in smaller amounts. This is the case of refineries and by extension of biorefineries. The feature of biorefineries is the fact that different biomasses show different compositions, but that composition within a species may be modulated. For instance, it is possible to modify the biomass composition during the growth by controlling the nutrients. An example is the algae. The accumulation of lipids can be modulated not only from one species to another but also by limiting the nutrients during the growth (Martín and Grossmann, 2013a,b). By modifying the composition of the biomass, a production facility that processes and requires different amounts of the various constituents may show an optimal operation for a particular biomass if a portfolio of products is fixed (Martín and Grossmann, 2013a,b; Galán et al., 2019) or the other way around, the portfolio of products depends on that composition. Thus, by solving this type of problem, it is possible to select the type of biomass that is appropriate for a particular use or design the biomass that is best for that use so that the plant is engineered to achieve that particular composition.

8.2.2.2 Formulation

The problem formulation corresponds to a superstructure optimization problem involving the selection of technologies and operating conditions of the various units as well as the design of the raw material to be processed in terms of its composition. It corresponds to an MINLP problem. Fig. 8.3 shows the superstructure for the production of ETBE from lignocellulosic biomass (Galán et al., 2019). Example 8.2 shows the optimization of the product, biomass, and the process for the production of ETBE.

Max: Product portfolio

St.
Mass & Energy balances processing facility
Process constraints.
          Technologies and operating conditions
Raw material composition

8.2.2.3 Added-value products from waste

A possible source of value-added products may be waste from different industries. Among them, the food industry stands out as one of the largest generators of organic waste. These residues, due to their composition, can be used to obtain a large number of products that can be used by the food industry itself, such as food supplements, favoring the economy, but also by other industries, such as pharmaceuticals, obtaining active principles for drugs, and the industry in general, that can benefit from pigments, resins and other materials of interest. This

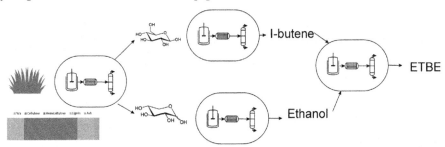

FIGURE 8.3   Process and product design for the production of ETBE from switchgrass.

EXAMPLE 8.2 In a second-generation biorefinery processing, lignocellulosic feedstock for the production of ETBE different biomasses are available with the composition in cellulose, hemicellulose, and lignin, to be within 35–50; 20–35 and 15–25, respectively. The conversion of cellulose to i-butene is 0.9 kg/kg and the conversion of hemicellulose to ethanol is 0.7 kg/kg. The molar ratio ethanol: i-butene has to be 1. Optimize the biomass composition for the facility to operate considering that lignin represents up to 18% of the biomass dry weight and the energy required must be provided from it as follows, 0.4 kg of lignin per kg of ethanol and 0.05 kg of lignin per kg of i-butene. The ethanol or the i-butene could be sold as products if an excess. Assume a selling price of the unused ethanol of 0.2€/kg and 0.4€/kg for the i-butene in excess. Note: this is a very simplified version of the work Galán et al. (2019) that has inspired the example. For the complete process, we refer the reader to that work where the process model is included in the formulation of the biomass.

   Solution

$$Max\ Etbe + P_{Et} \cdot Et_{nonused} + P_{Ib} \cdot Ib_{nonused}$$
$$s.t$$
$$Et = Et_{nonused} + Et_{used}$$
$$Et = \eta_{Hemi} \cdot Hemicellulose$$
$$Ibutene = Ib_{nonused} + Ib_{used}$$
$$Ibutene = \eta_{Cellulose} \cdot Cellulose$$
$$Etbe = Et_{used} + Ib_{used}$$

$$Hemicellulose + Cellulose + Lignin = Biomass$$
$$\frac{Et_{used}}{M_{Et}} = \frac{Ib_{used}}{M_{ib}}$$
$$20 \leq Hemicellulose \leq 35$$
$$35 \leq Cellulose \leq 50$$
$$15 \leq Lignin \leq 25$$

```
from pyomo.environ import *
from pyomo.opt import SolverFactory

# Optimization model
model = ConcreteModel(name = 'Biorefinery')

#Variables
model.x1 = Var(within = PositiveReals,bounds = (35,50))# Cellulose
model.x2 = Var(within = PositiveReals,bounds = (20,35))# Hemicellulose
model.x3 = Var(within = PositiveReals,bounds = (15,25))# Lignin
model.ethanol = Var(within = PositiveReals)
model.ibutene = Var(within = PositiveReals)
model.ethanolused = Var(within = PositiveReals)
model.ethanolnonused = Var(within = PositiveReals)
```

```
model.ibuteneused = Var(within = PositiveReals)
model.ibutenenonused = Var(within = PositiveReals)
model.etbe = Var(within = PositiveReals)
model.energy = Var(within = PositiveReals)

#CONSTRAINTS
model.constraint1 = Constraint(expr = model.ethanol == model.x2*0.9)
model.constraint2 = Constraint(expr = model.ethanol == model.ethanolused + model.
ethanolnonused)
model.constraint3 = Constraint(expr = model.ibutene == model.x1*0.8)
model.constraint4 = Constraint(expr = model.ibutene == model.ibuteneused + model.
ibutenenonused)
model.constraint5 = Constraint(expr = model.etbe == model.ibuteneused + model.
ethanolused)
model.constraint6 = Constraint(expr = model.ethanolused/46 == model.ibuteneused/56)
model.constraint7 = Constraint(expr = model.x1 + model.x2 + model.x3 == 100)
model.constraint8 = Constraint(expr = 0.4*model.ethanol + 0.05*model.ibutene
<= model.x3)

# OBJECTIVE FUNCTION
model.obj = Objective(expr = 1*model.etbe + 0.2*model.ethanolnonused +0.4*model.
ibutenenonused, sense = maximize)

#SOLVE
solver = SolverFactory('glpk')
Res = solver.solve(model)
print('x1', model.x1(),'x2', model.x2(),'x3', model.x3())
print('Ethanol', model.ethanol(), 'Ethanol non used', model.ethanolnonused())
print('Ibutene', model.ibutene(), 'Ibutene non used', model.ibutenenonused())
```

x1 = 50.0; x2 = 35.0; x3 = 15.0
Ethanol 31.5 Ethanol non used 0.0
Ibutene 40.0 Ibutene non used 1.652
objective 70.5

does not only favor the circular economy of the process, reducing the production of waste, but also does increase the profitability of the process, reducing the energy dependence of the producing company, and improving the sustainability of the process.

As in the previous cases, the products that can be obtained depend on the composition of the raw material and therefore, once the production process is established, the composition of the biomass should not change too much. In waste management, this is a rather complex task due to the fact that many factors affect its composition, such as the type of industry, the production processes or environmental regulations, among others. For example, in the food industry, depending on the type of food, one type of waste or the other will be produced, as well as its quantity and environmental impact it generates. Even in the same type of industry, a different production process can generate organic waste with a different composition that will affect the development of the processes for its recovery. This feature makes it very complex to develop general waste treatment and valorization processes that can be applied to a broad spectrum of waste, having to address each in a case-by-case basis.

Nevertheless, it is possible to follow several strategies to avoid excessive variation in the composition of the waste. Achieving the desired composition from the mixture of different types of waste or design waste treatment plants as an additional line in the production process of the industry that generates that waste.

In the first case, the idea would be similar to the case of the blending problem discussed in previous sections but applied to waste. The procedure to follow would be:

- The products to be obtained from the biomass are established
- The composition that the biomass must have to obtain this product is analyzed.
- The different types of biomass are mixed until obtaining that composition.

Note that this procedure is the same as the previous case (blending/pooling problem) but using wastes.

In some cases, mixing different types of biomass waste may not result in just a mass balance of ingredients. For example, in cases where anaerobic digestion is used to generate biogas and digestate, mixing different types of biomass can lead to a decrease in the yield of the transformation process due to the interaction between microorganisms and different types of substrates (inhibition due to excess substrate or the presence of toxic substances for microorganisms, etc.). In addition, some strains of microorganisms work best with specific substrates and begin to fail if these substrates are mixed with others. The effect of the interaction of different wastes is to be carefully analyzed to unveil the synergies and inhibitory effects. Therefore, the way to approach this type of problem is to create a superstructure of alternatives that will select the amount of raw material to be used in each process to adjust the mixture of the products and obtain the final product that meets the established objectives, which may be of quality, environmental impact or benefit. Fig. 8.4 shows a superstructure derived from the work of Taifouris et al. (2020a,b) in which 4 different types of waste are considered to obtain biogas and digestate.

In the previous example, the target product is previously selected and there are a number of raw materials available to obtain it. However, the recovery of waste can be approached from another point of view, starting from a particular type of waste, in order to obtain a product portfolio from it. In this case, the formulated product is the raw material, and each of the ingredients that compose it may be the final products to be marketed.

For this problem, it is necessary to evaluate the possible products that can be obtained out of the raw material and their final uses to estimate their value and the final consumer. This type of study is usually done at laboratory level, focusing on the characterization of the raw material and evaluating the physicochemical properties of the products obtained, so as to select the potential products that could be marketed are analyzed. However, in some cases, the number of products that can be used is enormous, but not all are profitable. In order to select which product has the largest potential from an economic or environmental point of view, it is necessary to carry out a techno-economic analysis of each of the products and compare them. To systematically evaluate the trade-offs and the alternatives, superstructures are developed that

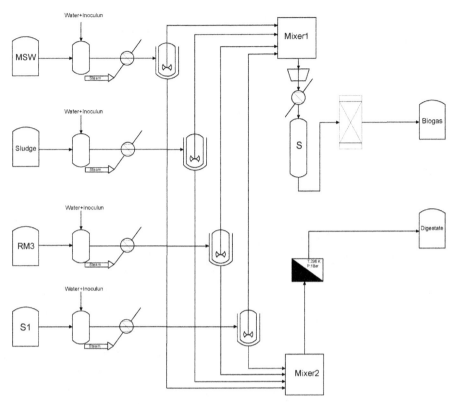

FIGURE 8.4 Example of a pooling problem applied to products.

consider all possible processes, while the amount of raw material sent to each process is a variable of the optimization model formulated. In this way, depending on the objectives set, one process or another and a particular portfolio of products will be chosen. These superstructures are the basis of the integrated biorefineries that use one or more residues as raw material. These can be designed as an additional line to a larger process or constitute an independent process itself oriented to favor the circular economy and the philosophy of processes with zero emissions. Furthermore, a part of the raw material can be used to produce energy that supplies the rest of the process, making it self-sustaining energetically.

An example of the application of the concept of integrated biorefineries could be the coffee industry. During the production of the coffee, several wastes are generated, but there is one that stands out among them, the spent coffee ground (*SCG*). It is possible to obtain a series of value-added product from *SCG* such as biofuels—that is, biodiesel, bioethanol (Kondamudi et al., 2008; Kwon et al., 2013), biogas (Vítěz et al., 2016), bio-oil (Ktori et al., 2018) and pellets (Kamil et al., 2019), food supplements and biocomponents for the pharmaceutical and cosmetic industries (caffeine, antioxidants, and phenolics (Shi, 2016)) natural extracts (Brazinha et al., 2015; Ribeiro et al., 2013), additives for industry (tannins (Low et al., 2015)), and fertilizer production for some types of crops (Liu and Price, 2011), and energy production (Ciesielczuk et al., 2015). The *SCG* can also be used to produce energy using a combustion or gasification process.

Techno-economic analysis of a series of products that can be obtained of this residue is carried out in the work of Taifouris et al. (2021a). Another example is the forest industry. Forest residues can be used to produce value-added products or the production of polymers as well as energy through combustion, either thermal energy or electrical energy (Roldan et al., 2021). This type of waste has tannins in its composition that can be used as a natural pigment for the textile industry. The agroindustry residues are also interesting resources of products. For instance, in juice production, the fruit peel is obtained as a waste. However, it can be used to produce energy through combustion or obtain value-added products, such as essential oils with a high concentration of limonene and pectin. This product can be sold to pharmaceutical or food industries, favoring the circular economy and improving the economic performance of the process (Criado and Martín, 2020). See Example 8.3.

EXAMPLE 8.3 Three processes are considered for the recovery of a residue from coffee production (Spent coffee ground): an extraction, filtration, and drying process to obtain a natural extract and natural pigment; anaerobic digestion to obtain biogas (which is used to produce electrical energy) and digestate; and a gasification process to produce power. Part of the waste generated by the first process can be recovered through anaerobic digestion and, therefore, the first process also produces biogas and digestate. A diagram of the processes considered can be seen in Fig. 8E3.1. The production yields of each product (with respect to the raw material fed to the process) are indicated in Table 8E3.1, as well as the price of the products, the fixed costs and variables of each process, and maximum demand.

The cost of the process is calculated with the following expression:

$$Process\ costs = A + b \cdot F_{SCG} \tag{8E3.1}$$

where, F_{SCG} is the Flow of raw material that is sent to each process (Table 8E3.1).

TABLE 8E3.1 Information on the technical economic analysis of the processes.

Process	Product	η	Price(€/t)	Process cost A	Process cost B	Demand (t/year)
1	Natural extract	0.01	4500	65,000	200	5
	Natural pigment	0.08	2300			30
	Biogas	0.05	500			—
	Digestate	0.11	1500			50
2	Biogas	0.06	500	32,000	80	—
	Digestate	0.12	1500			50
3	Power	2.2	100	30,000	100	—
AU	Steam	4	—	20,000	60	5000
	Hot air	21.10	—			—

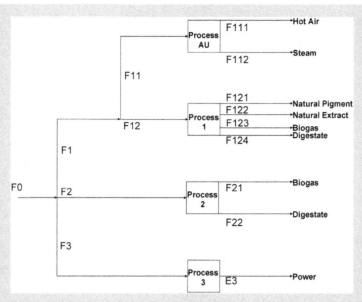

FIGURE 8E3.1 Superstructure that integrates all processes.

If process 1 is selected, part of the raw material must be used to supply the energy that this process requires. It will be necessary to send 0.173 kg of SCG per kg of *SCG* that sent to process 1. If the production of soluble coffee is 200 tons (400 tons of *SCG*). Determine which is the process that generates the greatest economic benefit.

Solution

The three processes are presented in the same process diagram and each of the possible streams is defined. Fig. 8E3.1 is obtained:

From the scheme, the mathematical model is elaborated:

Mass balances between raw material and products:

$$
\begin{aligned}
F111 &= F12 \cdot \eta_{Hot\ Air} \\
F112 &= F12 \cdot \eta_{Steam} \\
F13 &= F12 \cdot \eta_{NaturalPigment} \\
F14 &= F12 \cdot \eta_{NaturalExtract} \\
F15 &= F12 \cdot \eta_{Biogas1} \\
F16 &= F12 \cdot \eta_{Digestate1} \\
F21 &= F2 \cdot \eta_{Biogas2} \\
F22 &= F2 \cdot \eta_{Digestate2}
\end{aligned}
\tag{8E3.2}
$$

Constraints

$$
\begin{aligned}
F11 &= F1 \cdot 0.173 \\
F0 &= F1 + F2 + F3 \\
F1 &= F11 + F12
\end{aligned}
\tag{8E3.3}
$$

3. Biomass and waste based processes

Logic constraints

$$F1 - y1 \cdot U \leq 0$$
$$F2 - y2 \cdot U \leq 0$$
$$F3 - y3 \cdot U \leq 0 \qquad (8E3.4)$$
$$y1 + y2 + y3 = 1;$$

where, U can be 40000.

Costs

$$CostP1 = AP1 * y1 + BP1 \cdot F12$$
$$CostP2 = AP2 * y2 + BP2 \cdot F2$$
$$CostP3 = AP3 * y3 + BP3 \cdot F3 \qquad (8E3.5)$$
$$CostPAU = APAU * y1 + BPAU * F11$$

Income

$$B_{NaturalPigment} = F13 \cdot P_{NP}$$
$$B_{NaturalExtract} = F14 \cdot P_{NE}$$
$$B_{Biogas1} = F15 \cdot P_{B1}$$
$$B_{Digestate1} = F16 \cdot P_{D1} \qquad (8E3.6)$$
$$B_{Biogas2} = F21 \cdot P_{B2}$$
$$B_{Digestate2} = F22 \cdot P_{D2}$$
$$B_{Power} = E31 \cdot P_{POWER}$$

TABLE 8E3.2   Solution of the Example 8.3.

Process	Operation Cost	Income	Profit
1	155,320	178,010	22,690
2	64,000	84,000	20,000
3	88,000	70,000	18,000

Objective Function

$$Profit = \sum Income - \sum costs \qquad (8E3.7)$$

The model is solved in GAMS using a commercial solver and the results are shown in Table 8E3.2. As you can see, the most profitable process is 1.

For the sets and variable definitions we refer to the online supplementary file where the.gms file can be found.

```
*Mass balances:
eq1..F0 = E = F1 + F2 + F3;
eq2..F1 = E = F11 + F12;
eq4..F111 = E - F11*nhotair;
eq5..F112 = e = F11*nsteam;
```

```
eq7..F121 = E = F12*nNaturalPigment;
eq8..F122 = E = F12*nNaturalExtract;
eq9..F123 = E = F12*nBiogas1;
eq10..F124 = E = F12*nDigestate1;
eq12..F21 = E = F2*nBiogas2;
eq13..F22 = E = F2*nDigestate2;
eq14..E3 = e = F3*npower;

*Constraint
eq15..F11 = E = F1*0.1728;

*Costs
eq16..CostP1 = E = AP1*y1 + BP1*F12;
eq17..CostP2 = E = AP2*y2 + BP2*F2;
eq18..CostP3 = E = AP3*y3 + BP3*F3;
eq19..CostPAU = e = APAU*y1 + BPAU*F11;

eq20..totalcost = e = CostP1 + CostP2 + CostP3 + CostPAU;
*Income

eq21..BNaturalPigment = E = F121*PNP;
eq22..BNaturalExtract = E = F122*PNE;
eq23..BBiogas1       = E = F123*PB1;
eq24..BDigestate1    = E = F124*PD1;
eq25..BBiogas2       = E = F21*PB2;
eq26..BDigestate2    = E = F22*PD2;
eq27..BPower         = E = E3*PPOWER;

eq28..totalbenefit = e = BNaturalPigment + BNaturalExtract + BBiogas1 + BDigestate1 +
BBiogas2 + BDigestate2 + BPower;

*logic equations
eq29..F1 − y1*U = 1 = 0;
eq30..F2 − y2*U = 1 = 0;
eq31..F3 − y3*U = 1 = 0;
eq32..y1 + y2 + y3 = 1 = 1;

* objetive fuction
eq33..Profit = e = BNaturalPigment +
BNaturalExtract +
BBiogas1 +
BDigestate1 +
BBiogas2 +
BDigestate2 +
BPower −
```

```
CostP1 -
CostP2 -
CostP3 -
CostPAU;
model example/all/;
solve example Maximizing profit using miP
```

8.3 Integration of product, process, and supply chain design

Introducing a new product in an increasingly competitive market is a very complicated process that requires a deep understanding of the customer needs and the entire market chain. These needs are increasingly complex. It is no longer enough that there exists a demand for the product, but it must also show high quality and an adjusted price. In addition, lately, consumer looks for eco-friendly products whose traceability can be ensured. Therefore, integrated product design approaches are gaining great importance. Designing the product, the production process, and the supply chain simultaneously has the following advantages:

- It allows to discard product designs whose large-scale production is not feasible from an economic or environmental point of view, reducing time to market and allowing savings in the R&D stages by discarding many designs that will not have to be experimentally evaluated.
- It allows evaluating the environmental impact and costs of the entire process, from the conception of the product idea to its delivery to the consumer, being able to identify bottlenecks or designs that balance all objectives, coordinating each and every one of the development phases. In addition, it allows evaluating the availability of raw materials, adjusting transport costs by choosing ingredients that can be purchased locally.
- It allows adapting the design of the product to the demands of the consumer, up to the level or personalized products and reducing delivery times. In addition, it facilitates the complete traceability of the product.

Integrated design is especially important in the case of formulated products, since its design depends on its formulation, which can be easily changed by adjusting the conditions of the production process. Similarly, when designing a new formulated product (detergents, fertilizers, food, perfumes, etc.), the set of feasible products to investigate is large. The integrated design can act as a filter to substantially reduce this wide spectrum of possibilities. The set of stages that constitutes the integrated design of formulated products can be seen in Fig. 8.5 (Taifouris et al., 2020a).

The problem begins by designing and/or selecting the ingredients that will be part of the final product. The design of the final product is carried out in a mixing process that can be modeled as a pooling problem. Under the concept of integrated product design, the fraction of each ingredient in the final product will be influenced by the characteristics of the supply chain, the process restrictions, the objectives to be met (economic, environmental, and social), and the demands of the consumers. Therefore, all these factors must be

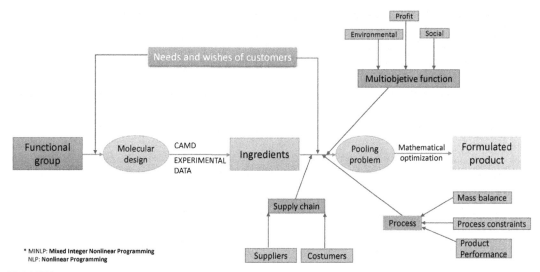

FIGURE 8.5   Extended design of formulated products. *With permission from Taifouris et al. (2020b).*

integrated into the mathematical model that will be used to obtain the optimal formulation of the product. This approach is applied in the same way to any product obtained by mixing others, whether it is a mixture of ingredients (in the case of detergents) or a mixture of products (biogas and digestate obtained from waste).

The integrated design applied to formulated raw materials is aimed at obtaining a series of products from the ingredients that make up that raw material but considering not only the techno-economic analysis of the production process and the sale of the products, but also the costs of supplying raw materials and distributing products. By integrating the design of the supply chain, the decision of the most profitable process to value a waste provides a wider vision of the problem, taking into account not only all the costs involved, but also the total environmental impact (process and transport), so the result will be closer to the actual optimum of the complete problem (Taifouris et al., 2021b).

In the cases previously proposed, the waste treatment systems were proposed as an additional extension to the industry that generated them, but most of the waste is generated at homes. For the treatment of this type of waste, a correct design of the supply chain and its integration in the design of the production process and the final product is key for the process to be profitable, because the correct location of the plant would be used to balance the costs of transporting raw materials and the costs of product delivery. This affects the selection of the process and the product, since it is possible that it is economically more profitable to obtain product A because the sales benefit is greater than product B; however, the clients who demand product A are decentralized and there is no ideal location to reduce transport costs, while in the case of product B, all customers are centralized, being able to place the factory at a point that allows reducing total costs. In this case, it would be better to choose product B, but it has been possible to reach this conclusion using the integrated design approach.

Even if there are three stages or problem levels, to evaluate the trade-offs properly it is necessary that all of them are evaluated simultaneously (product, process, and supply chain design). The mathematical models to address this kind of problem are usually hard to solve. To model production processes, it is necessary to use nonlinear and non-convex equations (material balances, process restrictions, price policies, etc.) as well as bilinear terms. Supply chains involve decision variables regarding the selection of suppliers, locations, customers, etc. that are usually binary variables, which transform nonlinear models (NLP) into mixed nonlinear integer models (MINLP). When these models reach sizes corresponding to a supply chain at the continental level (i.e., European level), commercial solvers cannot find the optimal values of the variables because the feasible region to search is too large and fractional (discontinuous problems). In these cases, it is necessary to develop decomposition techniques or algorithmic solutions, such as the Bender or Lagrange decomposition, or linearization methods, such as the piecewise linear approximation or the McCormick envelopes.

8.3.1 Formulation

Since different locations will be considered in the problems that integrate the supply chain design, the dimension 'location' can be added to the variables indicated in Section 8.2.3. In this way, the material balances, as well as the rest of the equations, will depend on each location considered. In addition, it will be necessary to introduce two new variables:

$Ccp_{i,\,sup,\,loc}$: That represents the amount purchased by the location 'loc' from the supplier 'sup' of the ingredients 'i'. The relationship between this variable and the internal flows of the factories is given by:

$$\sum_{sup} ccp_{i,sup,loc} = \sum_{l} x_{i,l,loc} + \sum_{j} z_{i,j,loc} \tag{8.11}$$

$Cv_{j,cust,loc}$: Amount sold of the product "j" to the customer "$cust$" by the location "loc." This variable must also be related to customer demands:

$$\sum_{sup} ccp_{i,sup,loc} = \sum_{l} x_{i,l,loc} + \sum_{j} z_{i,j,loc} \tag{8.12}$$

$$\sum_{loc} cv_{j,cust,loc} \geq D^{L}_{j,cust} \tag{8.13}$$

$$\sum_{loc} cv_{j,cust,loc} \leq D^{U}_{j,cust} \tag{8.14}$$

And to avoid overproduction, the following equality constraint is necessary:

$$\sum_{cust} cv_{j,cust,loc} = Massprodc_{j,loc} \tag{8.15}$$

The income of each location will depend on the variable $Cv_{j,cust,loc}$ and price of each product.

$$Income = \sum_{j,cust,loc} cv_{j,cust,loc} \cdot price_{j} \tag{8.16}$$

The selection of suppliers or customers, as well as the quantities of ingredients and products, do not require the use of binary variables to be selected, since the two previous variables can be used for this purpose. However, to determine the fixed costs for the manufacturing of the selected plant, it is necessary to introduce a binary that determines the place where the plant is going to be built. Therefore, it is necessary to introduce the following restriction, since only the built plants can manufacture products (Taifouris et al., 2020b)

$$Massprodc_{j,loc} - bi(loc) \cdot U \leq 0 \tag{8.17}$$

where, U is a large enough value, a big M type of constraint.

And finally, the transportation costs of both ingredients and products will have to be integrated. In Example 8.4 a case study is developed.

EXAMPLE 8.4 The case of Example 8.3 is applied to four soluble coffee factories located at different distances from the customers that demand the products derived from the spent coffee ground treatment. In this case, the cost of delivery of products is considered in the total costs of the plant and is calculated as follows:

$$Delivery\ cost = \frac{Minimum\ distance(km) \cdot Transport\ cost(€/km)}{Loading\ Capacity(t)} \cdot Product\ Amount(t) \tag{8E4.1}$$

The distance between the factory and the customers is shown in Table 8E4.1. The cost of transporting the biogas will be equal to 0 since it is converted into electrical energy and sent to the electrical grid to which the plant is connected. The transport cost is € 1/km and the loading capacity is 1 t.

TABLE 8E4.1 Distances between the factories and customers.

		Distance Factory			
Process	Product	1	2	3	4
1	Natural extract	300	150	170	480
	Natural Pigment	350	100	70	80
	Biogas	—	—	—	—
	Digestate	170	130	240	120
2	Biogas	—	—	—	—
	Digestate	170	120	240	120
3	Power	—	—	—	—

In this case, what is the best process to integrate into each of the factories?
Solution
It is only necessary to expand the dimensions of the material flows including the 'location' dimension, since the mass flows may be different in each of the factories considered:
For example:

$$F0 = F1(\text{loc}) + F2(\text{loc}) + F3(\text{loc}) \tag{8E4.2}$$

In addition, it will be necessary to add the delivery costs:

$$
\begin{aligned}
&\text{Delivery cost13(loc)} = M\text{í}nimum\ Distance \cdot 0.25 \cdot F13(\text{loc}) \\
&\text{Delivery cost14(loc)} = M\text{í}nimum\ Distance \cdot 0.25 \cdot F14(\text{loc}) \\
&\text{Delivery cost15(loc)} = M\text{í}nimum\ Distance \cdot 0.25 \cdot F15(\text{loc}) \\
&\text{Delivery cost16(loc)} = M\text{í}nimum\ Distance \cdot 0.25 \cdot F16(\text{loc}) \qquad (8E4.3) \\
&\text{Delivery cost21(loc)} = M\text{í}nimum\ Distance \cdot 0.25 \cdot F21(\text{loc}) \\
&\text{Delivery cost22(loc)} = M\text{í}nimum\ Distance \cdot 0.25 \cdot F22(\text{loc}) \\
&\text{Delivery cost31(loc)} = M\text{í}nimum\ Distance \cdot 0.25 \cdot F31(\text{loc})
\end{aligned}
$$

Therefore, the delivery costs are included in the objective function:
Objective Function

$$Profit = \sum Incomes - \sum Costs - \sum Delivery\ costs \tag{8E4.3}$$

The processes selected in each of the factories considered are shown in Table 8E4.2. When entering the distribution costs, in cases where the distances between customers and the factory were greater (Factory 1 and 4), the process selected in the Example 8.3 (process 1) was changed to process 3 (without transport cost). However, in factories that are closer to customers, delivery costs do not offset the extra benefit of process 1 compared to other processes, so the selection from the previous example is maintained.

TABLE 8E4.2   Solution of the Example 8.4.

Process	Factory 1	Factory 2	Factory 3	Factory 4
1		X	X	
2				
3	X			X

For the scalar, parameter, tables and variable definitions we refer to the.gms file in the editorial webpage of the book.

```
*Mass balances:
eq1(loc)..F0 = E = F1(loc) + F2(loc) + F3(loc);
eq2(loc)..F1(loc) = E = F11(loc) + F12(loc);
eq4(loc)..F111(loc) = E = F11(loc)*nhotair;
eq5(loc)..F112(loc) = e = F11(loc)*nsteam;
eq7(loc)..F121(loc) = E = F12(loc)*nNaturalPigment;
eq8(loc)..F122(loc) = E = F12(loc)*nNaturalExtract;
```

```
eq9(loc)..F123(loc) = E = F12(loc)*nBiogas1;
eq10(loc)..F124(loc) = E = F12(loc)*nDigestate1;
eq12(loc)..F21(loc) = E = F2(loc)*nBiogas2;
eq13(loc)..F22(loc) = E = F2(loc)*nDigestate2;
eq14(loc)..E3(loc) = e = F3(loc)*npower;

*Constraint
eq15(loc)..F11(loc) = E = F1(loc)*0.1728;

*Operational Costs
eq16(loc)..CostP1(loc) = E = AP1*y1(loc) + BP1*F12(loc);
eq17(loc)..CostP2(loc) = E = AP2*y2(loc) + BP2*F2(loc);
eq18(loc)..CostP3(loc) = E = AP3*y3(loc) + BP3*F3(loc);
eq19(loc)..CostPAU(loc) = e = APAU*y1(loc) + BPAU*F11(loc);
eq20(loc)..totalcost(loc) = e = CostP1(loc) + CostP2(loc) + CostP3(loc) + CostPAU(loc);
*Delivery costs

eq21(loc)..DeliveryCost121(loc) = e = MinimumDistancia121(loc)*(1)*F121(loc);
eq22(loc)..DeliveryCost122(loc) = e = MinimumDistancia122(loc)*(1)*F122(loc);
eq23(loc)..DeliveryCost124(loc) = e = MinimumDistancia124(loc)*(1)*F124(loc);
eq24(loc)..DeliveryCost22(loc) = e = MinimumDistancia22(loc)*(1)*F22(loc);

eq25(loc)..Totaldeliverycost(loc) = e = DeliveryCost121(loc) + DeliveryCost122(loc) +
DeliveryCost124(loc) + DeliveryCost22(loc);

*Incomes
eq26(loc)..BNaturalPigment(loc) = E = F121(loc)*PNP;
eq27(loc)..BNaturalExtract(loc) = E = F122(loc)*PNE;
eq28(loc)..BBiogas1(loc) = E = F123(loc)*PB1;
eq29(loc)..BDigestate1(loc) = E = F124(loc)*PD1;
eq30(loc)..BBiogas2(loc) = E = F21(loc)*PB2;
eq31(loc)..BDigestate2(loc) = E = F22(loc)*PD2;
eq32(loc)..BPower(loc) = E = E3(loc)*PPOWER;

eq33(loc)..totalbenefit(loc) = e =
BNaturalPigment(loc) + BNaturalExtract(loc) +
BBiogas1(loc) + BDigestate1(loc) +
BBiogas2(loc) + BDigestate2(loc) + BPower(loc);

*logic equations

eq34(loc)..F1(loc) − y1(loc)*U = l = 0;
eq35(loc)..F2(loc) − y2(loc)*U = l = 0;
eq36(loc)..F3(loc) − y3(loc)*U = l = 0;
eq37(loc)..y1(loc) + y2(loc) + y3(loc) = l = 1;
```

3. Biomass and waste based processes

```
* objetive fuction

eq38..Profit = e = sum(loc.BNaturalPigment(loc) + BNaturalExtract(loc) + BBiogas1
(loc) +
BDigestate1(loc) + BBiogas2(loc) + BDigestate2(loc) + BPower(loc)-CostP1(loc) −
CostP2(loc) − CostP3(loc) − CostPAU(loc) − DeliveryCost121(loc)-
DeliveryCost122(loc) − DeliveryCost124(loc) − DeliveryCost22(loc));

model example/all/;
solve example maximizing profit using miP
```

$$Delivery\ cost = \frac{distance(km) \cdot Transport\ cost\left(\frac{€}{km}\right)}{Loading\ Capacity(t)} \cdot Product\ amount(t) \qquad (8.18)$$

$$Supply\ transport\ cost = \frac{distance(km) \cdot Transport\ cost\left(\frac{€}{km}\right)}{Loading\ Capacity(t)} \cdot Ingredient\ amount(t) \qquad (8.19)$$

In this way, the objective function would become as follows:

$$objval = \sum_{j,loc,cli} cv_{j,cli,loc} \cdot priceP(j) - \sum_{l,j,loc} c_{pool} \cdot y(l,j,loc) - ccp(i,sup,loc) \cdot priceI(i)$$
$$- Delivery\ cost - Supply\ transport\ cost - \sum_{loc} bi_{loc} * Fixcosts_{loc} \qquad (8.20)$$

Problems

P8.1 The production of different crops requires a supply of nitrogen, phosphorous, potassium. Magnesium and calcium as presented in Table 8P1.1. Up to 5 fertilizers are available with costs and compositions as presented in Table 8P1.1. Optimize the purchase of fertilizers for the production of each crop individually.

TABLE 8P1.1 Nutrients and minerals needed for the crops.

Crop	Corn	Wheat	Sunflower
N (kg)	26	28	26
P (kg)	11	12	18
K (kg)	28	25	63
Mg (g)	2.5	2	22
Ca (g)	2.5	3	4

TABLE 8P1.2 Fertilizers composition and cost.

	F1	F2	F3	F4	F5
Cost (€/100 kg)	20	15	12	5	2
N (kg/100 kg)	15	10	15	0	0
P (kg/100 kg)	15	10	2	0	0
K (kg/100 kg)	15	20	15	0	0
Mg (g/100 kg)	0	0	0	1	0
Ca (g/100 kg)	0	0	0	0	1

P8.2 Select the best biomass for the production of ETBE following the process information provided in Example 8.2.

Biomasses (cellulose/hemicellulose/lignin)

Corn stover: 38/26/19
Switchgrass: 42/32/12
Wheat straw: 35/22/17

P8.3 Consider the preparation of animal feed. Per kg of mixture, it should contain a minimum of nutrients as given below:

Nutrient	A	B	C	D
g/kg	75	50	25	2

Each ingredient has its cost and nutritional value, see Table 8P3.1:

TABLE 8P3.1 Information on the ingredients

	A	B	C	D	Cost(€)/kg	Envirn. impact (pt/kg)	μ(cp)
Ingredient 1 (g/kg)	100	60	40	10	35	5	0.15
Ingredient 2 (g/kg)	150	175	20	—	65	2	0.2
Ingredient 3 (g/kg)	200	60	50	15	20	10	0.25
Ingredient 4 (g/kg)	250	150	25	2	30	8	0.2
Ingredient 5 (g/kg)	—	—	—	—	0	0	0.1

The environmental burden cannot be over 4 points per kg and the viscosity lower than 0.2 cp for the mixture to be processed.

P8.4 Value-added products from waste.

The aim is to design a process to produce biogas and digestate from 4 different residues, MSW, sludge, manure, and lignocellulosic. These residues cannot be mixed, but the biogas and digestate that are obtained from them can be mixed (see Fig. 8.3).

The composition of the final biogas must be 60% CH4 and 40% CO2 (by mass), while the digestate must have a minimum of 15% Nitrogen. The biogas and digestate production yields, in addition to their composition, can be seen in Table 8P4.1. The sale price of biogas is € 100/t, while the sale price of digestate would be € 200/t. Taking into account the compositional restrictions of each product and the price, determine the amount of each residue to maximize benefits (Table 8P4.1).

P8.5 Integration of product, process, and supply chain design: simple location problem

You want to produce three types of detergents from eight different ingredients for three clients. There are two different locations to locate a factory. Three suppliers at different distances from locations and with different prices are considered. Furthermore, the quantity of each ingredient that the plant can process is limited by the capacity of the factory as indicated in table Table 8P5.1. The raw material fed to the plant only contains a single ingredient, that is, the indexes of i and k coincide, so

TABLE 8P4.1 Information on the technical economic analysis of the processes.

Waste	Availability(t)	Yield per product ($t_{waste}/t_{product}$)		Yield per compund ($t_{waste}/t_{compund}$)				Process cost	
		Biogas	Digestate	CH4	CO2	N	nonN	A	B
MSW	1000	0.2	0.8	0.14	0.06	0.2	0.6	600	0.5
Sludge	1000	0.3	0.7	0.20	0.05	0.05	0.65	200	0.3
Manure	1000	0.1	0.90	0.03	0.07	0.25	0.65	300	0.35
Ligno	1000	0.3	0.7	0.2	0.1	0.2	0.5	500	0.4

TABLE 8P5.1 Information of the case of study of the problem 5 (part 1).

Suppliers/ingredient	Price (€/kg)			Availability (tn)			Process limit (tn)	Maximum composition(t_k/tn_j) Product			Minimal composition (t_k/tn_j) Product		
	1	2	3	1	2	3	All locations	1	2	3	1	2	3
1(Surfactant)	0.112	0.093	0.074	400	400	400	200	0.25	0.25	0.25	0	0	0
2(Builder)	0.013	0.011	0.008	500	500	500	250	0.6	0.6	0.6	0.1	0.1	0.1
3(Bleach)	0.120	0.100	0.080	250	250	250	175	0.25	0.25	0.25	0.05	0.05	0.05
4(Filler)	0.012	0.010	0.008	200	200	200	100	0.5	0.5	0.6	0.1	0.1	0.1
5(Antifoam)	0.576	0.500	0.389	50	50	50	25	0.05	0.05	0.05	0.001	0.001	0.001
6(Enzymes)	1.308	1.090	0.872	30	30	30	15	0.05	0.01	0.01	0.001	0.001	0.001
7(Polymers)	0.129	0.107	0.086	20	20	20	10	0.05	0.05	0.05	0.001	0.001	0.001
8(Water)	0.012	0.010	0.008	500	500	500	250	0.5	0.5	0.5	0.005	0.005	0.005

the parameter $C\ (i,\ k)$ is formed by a diagonal matrix where the main diagonal is formed by 1. Do not consider either the fixed or variable costs of the pools. The minimum demand is 0 for all products. Process restrictions will not be considered, but product quality will be considered. The relationship between the performance of the products and their composition is given by the following expression:

$$Performance(j, loc) = E = (107 * PQ(j, surfactants) + 1872 * PQ(j, enzyme)$$
$$+ 53.9 * (PQ(j, builder) + PQ(j, filler)) + 134 * PQ(j, polymer)$$
$$+ 119 * PQ(j, bleach))$$

While the performance required for each type of product is as follows:

$95 < Product1 < 100$
$80 < Product2 < 90$
$70 < Product3 < 80$

Note that it is an adaptation of the work of Martin and Martinez (Martín and Martínez, 2013), which has included the design of a supply chain. The fixed costs of location are the same in the two considered locations (100,000€).

Combine the formulation indicated in Sections 8.2.1.2 and 8.3.1 to select the suppliers, the formulation of the detergents, the amount of product sold to each customer, and the location to optimize the economic benefit of the factory. Note the data in Tables 8P5.1 and 8P5.2.

TABLE 8P5.2 Information of the case of study of the problem 5 (part 2).

Customers/products	Maximum demands (t)			Product Price (€/kg)	
	1	2	3	Product	Price
1	300	300	300	1	4
2	400	400	400	2	3.2
3	100	100	100	3	2.5
	Distances (km)				
Location/Supplier	1	2			
1	100	500			
2	50	300			
3	300	200			
Location/Customers	1	2			
1	100	400			
2	75	300			
3	100	200			

References

Audet, C., Brimberg, J., Hansen, P., Le Digabel, S., Mladenovi, N., 2004. Pooling problem: alternate formulations and solution methods. Manag. Sci. 50, 761−776.

Bond, J.Q., Upadhye, A.A., Olcay, H., Tompsett, G.A., Jae, J., Xing, R., et al., 2014. Production of renewable jet fuel range alkanes and commodity chemicals from integrated catalytic processing of biomass. Energy Environ. Sci. 7, 1500−1523.

Brand, C.V., Graham, E., Rodríguez, J., Galindo, A., Jackson, G., Adjiman, C.S., 2016. On the use of molecular-based thermodynamic models to assess the performance of solvents for CO2 capture processes: monoethanolamine solutions. Faraday Discuss. 192, 337−390.

Brazinha, C., Cadima, M., Crespo, J.G, 2015. Valorisation of spent coffee through membrane processing. J. Food Eng. 149, 123−130. Available from: https://doi.org/10.1016/j.jfoodeng.2014.07.016.

Ciesielczuk, T., Karwaczynska, U., Sporek, M., 2015. The possibility of disposing of spent cofee ground with energy recycling. J. Ecol. Eng 16 (4), 133−138. Available from: https://doi.org/10.12911/22998993/59361.

Criado, A., Martín, M., 2020. Integrated multiproduct facility for the production of chemicals, food, and utilities from oranges. Ind. Eng. Chem. Res. 59 (16), 7722−7731.

Dantzig, G., 1963. Linear Programming and Extensions. Princeton University Press.

Dávila, J.A., Rosenberg, M., Cardona, C.A., 2015. Techno-economic and environmental assessment of p-cymene and pectin production from orange peel. Waste Biomass Valor 6, 253−261.

De la Cruz, V., Hernandez, S., Martin, M., Grossmann, I.E., 2014. Integrated Synthesis of Biodiesel, Bioethanol, Isobutene, and Glycerol Ethers from Algae. Industrial & Engineering Chemistry Research 53 (37), 14397−14407. Available from: https://doi.org/10.1021/ie5022738.

Galán, G., Martín, M., Grossmann, I., 2019. Integrated renewable production of ETBE from switchgrass. ACS Sust. Chem. Eng. 7 (9), 8943−8953.

García Prieto, C.V., Daniel Ramos, F., Estrada, V., Villar, M.A., Díaz, M.S., 2017. Optimization of an integrated algae-based biorefinery for the production of biodiesel, astaxanthin and PHB. Energy 139, 1159−1172.

Gounaris, C.E., Floudas, C.A., Wei, J., 2006a. Rational design of shape selective separation and catalysis I: concepts and analysis. Chem. Eng. Sci. 61 (24), 7933−7948.

Gounaris, C.E., Floudas, C.A., Wei, J., 2006b. Rational design of shape selective separation and catalysis II: mathematical model and computational studies. Chem. Eng. Sci. 24, 7949−7962.

Guerras, L.S., Martín, M., 2019. Optimal gas treatment and coal blending for reduced emissions in power plants: a case study in Northwest Spain. Energy 169, 739−749.

Hernández, B., Martin, M., 2017. Optimal integrated plant for production of biodiesel from waste. ACS Sust. Chem. Eng. 5 (8), 6756−6767.

Hernández, B., León, E., Martín, M., 2017. Bio-waste selection and blending for the optimal production of power and fuels via anaerobic digestion. Chem. Eng. Res. Des. 121, 163−172.

Huang, K., Brentzel, Z.J., Barnett, K.J., Dumesic, J.A., Huber, G.W., Maravelias, C.T., 2017. Conversion of furfural to 1,5-pentanediol: process synthesis and analysis. ACS Sustain. Chem. Eng. 5 (6), 4699−4706.

Kamil, M., Ramadan, K., Olabi, A.G., Shanableh, A., Ghenai, C., Naqbi, A. K.Al, et al., 2019. Comprehensive evaluation of the life cycle of liquid and solid fuels derived from recycled coffee waste. Resour. Conserv. Recycl. 150, 104446. Available from: https://doi.org/10.1016/j.resconrec.2019.104446.

Kang, J., Shao, Z., Chen, X., Biegler, L.T., 2019. Reduced order models for dynamic molecular weight distribution in polymerization processes. Comput. Chem. Eng. 126, 280−291.

Karuppiah, R., Peschel, A., Grossmann, I.E., Martín, M., Martinson, W., Zullo, L., 2008. Energy optimization of an ethanol plant. AICHE J. 54 (6), 1499−1525.

Kondamudi, N., Mohapatra, S., Misra, M., 2008. Spent Coffee Grounds as a Versatile Source of Green Energy. Bioresour. Technol. 56 (24), 11757−11760. Available from: https://doi.org/10.1021/jf802487s.

Ktori, R., Kamaterou, P., Zabaniotou, A., 2018. Spent coffee grounds valorization through pyrolysis for energy and materials production in the concept of circular economy. Mater. Today: Proc. 5 (14), 27582−27588. Available from: https://doi.org/10.1016/j.matpr.2018.09.078.

Kwon, E.E., Yi, H., Jeon, Y.J., 2013. Sequential co-production of biodiesel and bioethanol with spent coffee grounds. Bioresour. Technol. 136, 475−480. Available from: https://doi.org/10.1016/j.biortech.2013.03.052.

Liu, K., Price, G.W., 2011. Evaluation of three composting systems for the management of spent coffee grounds. Bioresour. Technol. 102 (17), 7966−7974. Available from: https://doi.org/10.1016/j.biortech.2011.05.073.

Low, J.H., Rahman, W.A.W.A., Jamaluddin, J., 2015. The influence of extraction parameters on spent coffee grounds as a renewable tannin resource. J. Clean. Prod. 15, 222–228. Available from: https://doi.org/10.1016/j.jclepro.2015.03.094.

Marler, R.T., Arora, J.S., 2004. Survey of multi-objective optimization methods for engineering. Struct. Multidisc Optim. 26, 369–395.

Martín, M., Grossmann, I.E., 2012. Simultaneous optimization and heat integration for biodiesel production from cooking oil and algae. Ind. Eng. Chem. Res. 51 (23), 7998–8014.

Martín, M., Grossmann, I.E., 2013a. On the systematic synthesis of sustainable biorefineries. Ind. Eng. Chem. Res. 52 (9), 3044–3064.

Martín, M., Grossmann, I.E., 2013b. Optimal engineered algae composition for the integrated simultaneous production of bioethanol and biodiesel. AIChE J. 59 (8), 2872–2883.

Martin, M., Grossmann, I.E., 2016. Optimal Production of Furfural and DMF from Algae and Switchgrass. Ind. Eng. Chem. Res. 55 (12), 3192–3202. Available from: https://doi.org/10.1021/acs.iecr.5b03038.

Martín, M., Martínez, 2013. A Methodology for simultaneous process and product design in the consumer products industry: the case study of the laundry business. Chem. Eng. Res. Des. (91), 795–809.

Martín-Alonso, D., Wettstein, S.G., Mellmer, M.A., Gurbuz, E.L., Dumesic, J.A., 2013. Integrated conversion of hemicellulose and cellulose from lignocellulosic biomass. Energy Environ. Sci. 6 (1), 76–80.

Misener, R., Gounaris, C.E., Floudas, C.A., 2010. Mathematical modeling and global optimization of large-scale extended pooling problems with the (EPA) complex emissions constraints. Comput. Chem. Eng. 34, 1432–1456.

Paek, K.Y., Hosakatte, N.M., Zhong, J.J., 2014. Production of Biomass and Bioactive Compounds Using Bioreactor Technology. Springer, Berlin.

Peña-Lamas, J., Martinez-Gomez, J., Martín, M., Ponce-Ortega, J.M., 2018. Optimal production of power from mid-temperature geothermal sources: scale and safety issues. Energy. Convers. Manag. 165, 172–182.

Psycha, M., Pyrgakis, K., Harvey, P., Ben-Amotz, A., Cowan, A.K., Kokossis, A., 2014. Design analysis of integrated microalgae biorefineries. Comp. Aided Chem. Eng. 34, 591–596.

Ribeiro, H., Marto, J., Raposo, S., Agapito, M., Isaac, V., Chiari, B.G., et al., 2013. From coffee industry waste materials to skin-friendly products with improved skin fat levels. Eur. J. Lipid Sci. Technol. 115 (3), 330–336. Available from: https://doi.org/10.1002/ejlt.201200239.

Roldan, J.E., Martin Hernandez, E., Briones, R. and, Martín, M., 2021. Process design and scale-up study for the production of polyol-based biopolymer from sawdust. Sust. Consump. Prod. 27, 462–470.

Taifouris, M., Martín, M., Martinez, A., Esquejo, N., 2020a. Challenges in the design of formulated products: multiscale process and product design. Curr. Opin. Chem. Eng. 27, 1–9.

Taifouris, M., Martín, M., Martinez, A., Esquejo, N., 2020b. On the effect of the selection of suppliers on the design of formulated products. Comp. Chem. Eng. 141, 106980.

Shi, L., 2016. Bioactivities, isolation and purification methods of polysaccharides from natural products: A review. Int. J. Biol. Macromol. 92, 37–48. Available from: https://doi.org/10.1016/j.ijbiomac.2016.06.100.

Taifouris, M., Corazza, M., Martín, M., 2021a. Integrated design of biorefineries based on spent coffee ground. Ind. Eng. Chem. Res. 60 (1), 494–506.

Taifouris, M., Martín, M., Martinez, A., Esquejo, N., 2021b. Simultaneous optimization of the design of the product, process, and supply chain for formulated product. Comp. Chem. Eng. 152, 107384.

Torres, A.I., Daoutidis, P., Tsapatsis, M., 2010. Continuous production of 5-hydroxymethylfurfural from fructose: a design case study. Energy Environ. Sci. 3 (10), 1560–1572.

Vítěz, T., Koutný, T., Šotnar, M., Jan, C., 2016. On the Spent Coffee Grounds Biogas Production. Acta Univ. Agric. Silvic. Mendel. Brun. 64, 1279–1282. Available from: https://doi.org/10.11118/actaun201664041279.

Xu, C.P., Arancon, R.D.A., 2014. Bioactive compounds from biomass. RSC Green. Chem. (27), 176–186.

Solar technologies

Solar thermal energy

Mariano Martín and Jose A. Luceño

Department of Chemical Engineering, University of Salamanca, Salamanca, Spain

In this chapter we discuss solar thermal based processes, the solar fields design, the thermodynamic cycles as well as the cooling technologies. Two case studies are presented divided into three examples to show the performance of major technologies.

9.1 Process description

Concentrated solar technologies typically consist of the solar field where collectors or mirrors reflect the solar radiation to concentrate and direct it to a small surface to reach high temperatures. The receiver is so-call since it receives the solar energy from the previous device to heat up a heat transfer fluid that is responsible for transferring the solar energy into thermal energy to produce steam fed to a Rankine cycle to produce energy and a cooling system is required to close the cycle (IEA, 2010). Solar energy can be used in a different technology, by heating up a gas within a gas cycle, a Brayton cycle. Fig. 9.1a shows a scheme of a Rankine based concentrated solar power plant, and Fig. 9.1b shows a Brayton based CSP plant. One typical classification of CSP plants is based on the receiver design. 59% of the facilities use parabolic troughs, 27% solar towers, 10% Fresnel reflectors and the rest parabolic discs (Irena, 2012; IEA, 2010).

9.2 Solar field

The solar field is responsible for harvesting the solar energy that is received on the Earth surface as radiation. This section of the facility comprises the field of heliostats, the receiver and, in the case of Rankine cycles, the thermal storage that allows for CSP plants to be able to handle clouds and operate overnight. There are four main types of CSP plants defined by the receiver as shown in Table 9.1 (Irena, 2012; IEA, 2010; NREL, 2010). The

Sustainable Design for Renewable Processes
DOI: https://doi.org/10.1016/B978-0-12-824324-4.00009-3

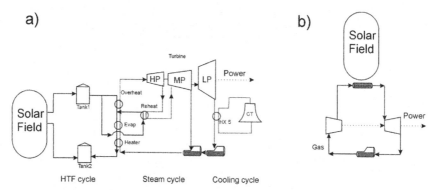

FIGURE 9.1　Schemes of CSP plants using different thermodynamic cycles. a) Regenerative Rankine. b) Closed Brayton.

two that have received more attention reaching commercial maturity are the parabolic troughs and the solar tower.

The energy that the solar field can harvest depends on the solar radiation received by the Earth surface. Apart from measured values that are typically reported by the meteorological services, it is possible to estimate it as a function of the position of the point (Zhuang et al., 2019)

$$DNI_t = 1.353\left((1-0.14h)0.7^{AM^{0.678}} + 0.14h\right)$$

$$AM = \frac{1}{\sin\alpha + 0.50572(\alpha+6.07995)^{-1.6364}} \tag{9.1}$$

$$\alpha = \sin^{-1}(\sin\delta\sin\phi + \cos\delta\cos\phi\cos\gamma)$$

Where δ is the solar declination angle, ϕ the local latitude, γ the solar hour angle and h is altitude above sea level (m). The annual DNI is computed integrating DNI_t

$$DNI = \int_{t_1}^{t_2} DNI_t dt \tag{9.2}$$

The design point corresponds with the highest solar incidence at midday at the location of choice. Values of 850–1000 W/m² are typical for appropriate locations.

9.2.1 Mirror/Heliostat field

The energy harvested is a fraction of that the Earth receives as radiation due to the loses and the area of the heliostats. The area is to be computed for a particular power production capacity.

$$Q_{(collector)} = DNI_i \cdot A_{field} \cdot \eta_{field} \tag{9.3}$$

TABLE 9.1 Comparison of CSP capture technologies (IEA, 2010; Irena, 2012; Zhou et al., 2019; NREL, 2010) Licence: www.iea.org/t&c/termsandconditions.

	Parabolic Troughs	Solar Tower	Parabolic Dishes	Linear Fresnel Reflectors
Size (MW)	30–320	10–200	$5-25 \cdot 10^{-3}$	1.5–5
Operating temperature (°C)	350–550	250–565	550.750	390
Concentration (suns) ratio	30–80	500–1000	>2000	25–40
Peak efficiency (%)	14–20	23–35	30	18
Annual efficiency (conversion Solar energy to power)	15–16	15–17	20–25	8–10
Thermodynamic efficiency	Low	High	High	Low
Tracking	One axis	Two axes	Two axes	One axis

(Continued)

TABLE 9.1 (Continued)

	Parabolic Troughs	Solar Tower	Parabolic Dishes	Linear Fresnel Reflectors
Maturity of technology	Mature	Recent	Mature	Recent
Commercial status	Available	Available	Pilot plant	Pilot plant
Technological risk	Low	Medium	High	Low
Storage possibility	Yes (HTF's)	Yes (HTF's)	Batteries	Yes (HTF's)
Receptor	Mobile	Fixed	Mobile	Fixed
Application type	On-grid	On-grid	On-grid/off-grid	On-grid
Cost (Investment) ($/kW)	8000 (6 h storage)	6300–7700 (6–9 h storage) 9000–10500 (12–15 h storage)		
LCOE ($/kWh)	0.15–0.3	0.18–0.32	0.14–0.28	0.21–0.657

9.2.1.1 Parabolic troughs

The area does not correspond to the mirrors as such but to the projected area due to the parabolic shape of the mirrors. The area that the field provides for collecting solar radiation is computed as in Eq. (9.4) (Desai et al., 2014; Kelly, 2006a,b,c):

$$A_{field} = n_h \cdot A_{proyected}$$
$$A_{Proyected} = a \cdot l \tag{9.4}$$

Where "a" is the aperture of the parabola. However, the area projected can be geometrically computed from the shape of the parabola as follows, Eq. (9.5):

$$A_{Proyected} = \left(\frac{a}{2} \sqrt{1 + \frac{a^2}{16f^2}} + 2f \ln \left(\frac{a}{4f} \sqrt{1 + \frac{a^2}{16f^2}} \right) \right) \cdot l \tag{9.5}$$

$$y = \frac{1}{4f} x^2$$

Different types of mirrors are available, however silver coated glass mirrors are the ones used by commercial scale facilities. They are made of a number of coats that end up with copper, silver and finally glass. The reflectivity of the mirrors reaches 93.5%. (Günther et al., 2011)

The efficiency of the field is computed as the optical efficiency, $\eta_o = 0.7-0.75$, minus the losses, convective and radiative, 2nd and 3rd terms in the right-hand side of Eq. (9.6).

$$\eta_{field} = \eta_o - U_L \left(\frac{T_m - T_a}{DNI \cdot \cos\theta} \right) - \sigma\varepsilon \left(\frac{T_m^4 - T_a^4}{DNI \cdot \cos\theta} \right) \tag{9.6}$$

Where the convective heat transfer coefficient is $U_L = 0.01-0.1$ (W/m^2 K), T_m is the collector outlet temperature, and T_a the atmospheric temperature, while θ is the incidence angle, the Stefan-Boltzmann constant σ is $5.670373 \cdot 10^{-8}$ W/m^{2}K^4 and ε is the emissivity, equal to 0.5. The annual collector efficiency is around 55 % (El-Gharbi et al., 2011)

9.2.1.2 Solar towers

In this case the area provided by the field is computed as follows, Eq. (9.7):

$$A_{field} = n_h \cdot A_h \cdot \eta_h \tag{9.7}$$

Where A_h is the area of a heliostat, from 40 to 150 m^2, n_h is the number of heliostats and η_h the efficiency of each one. Typical values for the heliostat efficiency (η_h) of 90% are used. However, the efficiency of the field, η_{field}, depends on the location of the facility and the design of the field of heliostats due to losses as a result of several phenomena, and can be computed as in Eqs. (9.8)-(9.9), see Fig. 9.2.

$$\eta_{field} = \eta_{optical,field} - U_L \left(\frac{T_m - T_a}{DNI \cdot \cos\theta} \right) - \sigma\varepsilon \left(\frac{T_m^4 - T_a^4}{DNI \cdot \cos\theta} \right) \tag{9.8}$$

$$\eta_{optical,field} = \eta_{cos} \cdot \eta_{shade\&block} \cdot \eta_{reflectivity} \cdot \eta_{atm} \cdot \eta_{int} \tag{9.9}$$

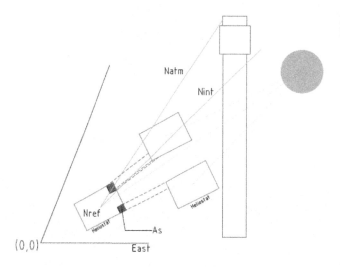

FIGURE 9.2   Scheme of the solar field losses.

Typically, the convective term is negligible.

Cosine losses, η_{cos}, (15−20%): They correspond to the largest loss of efficiency. They are due to the angle between the incident solar beam radiation and a vector normal to the surface of the heliostat. It can be computed using Eq. (9.10):

$$\eta_{cos} = \frac{\sqrt{2}}{2}(\sin\alpha \cdot \cos\lambda - \cos(\theta_H - A)\cos\alpha \cdot \sin\lambda + 1)^{0.5} \tag{9.10}$$

Where α is the solar altitude angle, λ is tilt angle, A is the Solar azimuth angle between the reflected ray incident on the center of the heliostat and the vertical direction, θ_H is the azimuth angle of the heliostat relative to the tower base, see Fig. 9.3.

Shading and blocking (2−8% losses): Due to other reflectors block or produce shades on the surrounding ones.

Heliostat reflectivity: from 0.90 to 0.95.

Interception efficiency: accounts for the fraction of the reflected rays that hit the target.

Transmission losses through the atmosphere (5% loses): The air between the reflector and the receiver due to the presence of particles difficult the passage of the rays, Eq. (9.11).

$$\begin{aligned}\eta_{atm} &= 0.99321 - 0.0001176 \cdot S_0 + 1.97 \cdot 10^{-8} \cdot S_0^2 \quad (S_0 \leq 1000\,\text{m}) \\ \eta_{atm} &= e^{-0.0001106 \cdot S_0}(S_0 > 1000\,\text{m})\end{aligned} \tag{9.11}$$

The field efficiency (η_{field}) shows with typical values of 55%−65% (Noone et al., 2012). This efficiency can be effectively computed in the design of the lay out. The lay out of the mirrors on the ground has been a matter of research towards the optimal location of each mirror maximizing the efficiency of the field reducing the cosine losses as well as the shading and blocking. There are a number of efforts to optimize the heliostat field including UHC-RCELL, DELSOL3 (winDELSOL), HFLCAL, MIRVAL, TieSOL, SCTHGM, and HFLD and Campo. They can be classified by the approach followed for the design of the layout such as the cellwise method, Campo code, heliostat minimum radial spacing

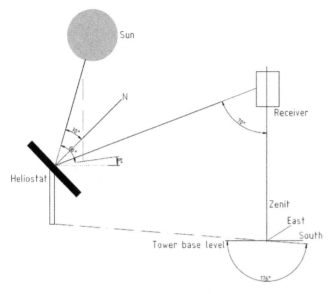

FIGURE 9.3 Cosine losses.

method, MUUEN algorithm, HGM methodology, and Greedy algorithm and biomimetic (Wei et al., 2010). The three most common are: DELSOL (Lipps and Vant-Hull, 1980), an extended version of the MUEEN (Siala and Elayeb, 2001) algorithm, and BIOMIMETIC (Noone et al., 2012):

DELSOL algorithm is based on the code RCEL that provides the initial layout spacings. The field is divided into zones according to radial and azimuth directions forming circular sectors. The heliostats are arranged following a radial pattern. They are located on iso-azimuthal and iso-radial lines. The spacings applied to each zone depend on the size of the heliostats, the tower height and the zone in the field to avoid contact between heliostats and reduce blocking and shading, see Fig. 9.4. For a North field, the radial spacing, ΔR, is given as Eq. (9.12):

$$\Delta R = \left(63.0093 - 0.5873130\theta_L + 0.0184239\theta_L^2 + \cos\phi\left(2.80873 - 0.1480\theta_L + 0.0014892\theta_L^2\right)\right)H_{Helio}$$

(9.12)

The azimuthal spacing, ΔAz, is given by Eq. (9.13):

$$\Delta Az = \left(2.46812 - 0.040105\theta_L + 9.2359 \cdot 10^{-4}\theta_L^2 + \cos\phi\left(0.17345 - 0.009113\theta_L + 1.2761 \cdot 10^{-4}\theta_L^2\right)\right)W_{Helio}$$

(9.13)

Where θ_L is the loft angle, defined as the complementary angle with respect to the vector that links the heliostat with the tower and the vertical with the ground, ϕ is the azimuthal angle, $0°$ corresponds to south, H_{helio} and W_{helio} are the height and the width of the heliostat.

MUEEN algorithm generates field layouts aiming at avoiding blocking between heliostats. It is a radial staggered algorithm to arrange the heliostats.

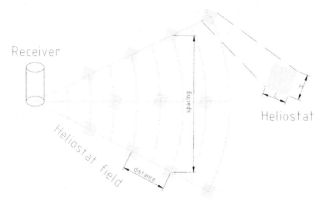

FIGURE 9.4 Heliostat field (Spacing, ΔAz; Distance; ΔRz).

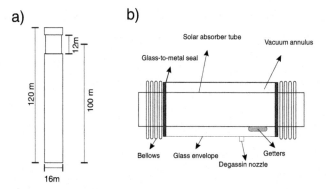

FIGURE 9.5 Scheme of a) solar tower and b) parabolic trough receiver.

BIOMIMETIC patters are generated following the Fermat's spiral algorithm. It uses only two parameters to generate the solar field without the need to subdivide it into zones.

For CSP plants using central receivers, the average land coverage ratio is typically 0.20−0.25 (Siala and Elayeb, 2001). For parabolic troughs, the capacity weighted area requirements are similar to those of the tower in terms of total area (Ong et al., 2013).

9.2.2 Solar receiver

The two most common designs are the tower and the pipes that are located in the focus of the parabolic thought, see Fig. 9.5

The **height of the tower** can be estimated as a function of the capacity of the facility, Eq. (9.14), (Jebamalai, 2016):

$$H_{tower} = 82.60 + 0.2552 \cdot (Q_{Collector}(MW)) \tag{9.14}$$

For the **Parabolic trough**, the pipes are designed for a flow velocity ranging from 2 to 4 m/s (Mohamed et al., 2019). If synthetic organic heat transfer fluids (HTF's) are used, hydrogen can be generated. It is allowed to permeate the pipe towards the vacuum annulus to avoid losses in the solar field efficiency.

9.2.3 Thermal storage

The current thermal energy storage types can be divided into four main groups including those using sensible heat, latent heat, steam accumulators and thermo-chemical storage (Sattler et al., 2011)

1. Thermal Energy Storage Systems for Sensible Heat

Within the group it is possible to distinguish between indirect and direct storage system. The main difference is the fact that indirect storage systems used different fluids as heat-transfer and storage fluids.

a. Indirect Storage Systems.

In this category, four technologies are used:

i. 2-Tank HTF Indirect Storage

The storage fluid from the low-temperature storage tank flows through a heat exchanger where it is heated by the high-temperature heat-transfer fluid. The high-temperature storage fluid then flows back to the high-temperature storage tank. This system is being used in many of the parabolic power plants in Spain and has also been proposed for several U.S. parabolic plants.

ii. Packed-Bed Thermal Energy Storage (Regenerator).

It is a vessel filled with a bed of packed material with a high thermal capacity. It typically uses air as HTF. Charging consists of feeding hot air across from the top of the bed to heat it up. The discharge changes the direction of the air circulation. Cold air is fed through the bottom duct and it is heated up as it crosses the bed exiting from the top

iii. Sand Storage

It can be the material of choice when the CSP is built in desert areas. It is similar in design an operation to the previous case. However, in this case the bed is not still but the grains flow across a duct and hot air is circulated across so that the hot sand is stored. In the discharge process, the hot sand is fluidized from the storage to come into contact with pipes carrying water that absorb the energy producing steam.

iv. Concrete Storage

The storage is a block of concrete across which pipes where the HTF flows heating up the concrete.

b. Direct Storage Systems

In this case the same fluid is heated up in the receiver and it is used as the heat-transfer and storage fluid.

i. 2-Tank HTF (molten salts/oil) Direct Storage

Different Heat transfer fluids can be circulated through the receiver being heated up with the solar energy captured. The maximum operating temperature is a feature of each of the fluids to avoid solidification at the cold end and decomposition at the hot end. Table 9.2 shows typical HTF and their properties.

ii. Single-Tank Thermocline Storage

TABLE 9.2   Molten salts and synthetic fluids properties (García and Martín, 2021).

HTF	Range of temperatures (°C)	Heat capacity (°C)
Molten Salts	290–565	$Cp = 1.443 + 0.000172\ T$
60%w/w $NaNO_3$–40% w/w KNO_3		
DTA	200–400	$Cp = 1.493 + 0.00274\ T$
Syltherm	200–400	$Cp = 1.5742 + 0.0017076\ T$
Therminol VP1	200–400	$Cp = 1.4864 + 0.0028229\ T$
Malotherm	200–350	$Cp = 1.52 + 0.0031872\ T$
HELISOL	200–425	$Cp = 1.609 + 0.00188\ T$

These systems store thermal energy in a solid medium such as silica sand in a single tank. During the operation, a fraction of the medium is at high temperature while the rest is at low temperature. Both fractions are separated by a thermocline, a temperature gradient. Hot HTF flows into the top and exit from the bottom so that the thermocline moves downwards storing energy. The discharge of the system is carried out by reversing the flow so that the thermocline moves upward, and the energy is used to generate steam and electricity. The stratification of the fluid in the tank is due to buoyancy of the material, helping stabilize and maintain the thermocline.

2. Latent-Heat Storage

Latent-heat storage systems use a phase change material (PCM) as storage medium. In principle there are three possible phase changes such as solid–solid, not typically used, solid–liquid and liquid–vapour, which is seldom used. In most cases a solid PCM is melted in the charging cycle to store the energy and is solidifies in the discharge. The process can be isothermal for the PMC, which is desired for a stable operation of the facility. Typical PCMs include nitrate and nitrite salts. Corrosivity is key in the selection of the material to reduce the high maintenance and investment cost.

3. Steam Accumulator

Steam accumulators are pressure vessels in which a charging system feeds steam in the hot water (well distributed) heating up the hot water. It can be considered as a direct storage system. For solar plants, sliding pressure systems (Ruth's storage) can be used. In this storage system, saturated steam is withdrawn during discharging reducing the pressure within the vessel resulting in the evaporation of the hot water. In the charging cycle, steam is fed superheated.

4. Thermo-Chemical Storage System

In thermo-chemical energy storage systems, reversible reactions are used to store energy, which are endothermic when charging and exothermic when discharging. Examples such as steam methane reforming and methanation can be used.

Two important concepts must be defined related not only with the storage system but also with the general design and operation of CSP plants.

- **The solar multiple** (SM) is a design parameter for the CSP plants. It estimates the oversize of the solar field as compared to the rest of the system. In other words, it is the real size of the solar field in comparison to a field size required to run the plant at design capacity under solar reference conditions. It is defined as *"the ratio of the receiver power at design point to the nominal cycle inlet power"*, Eq. (9.15).

$$SM = \frac{Q_{Solar\_field}}{Q_{power\_block}} \qquad (9.15)$$

The SM depends on the design point conditions, typically the solar noon at summer solstice or equinox. Higher multiples allow power plants to operate at full output even when the solar input is less than rated. This allows for an increased capacity factor value, increased annual solar share and a better overall utilization of the power block. Facilities without storage tend to have optimal SM's of 1.1 to 1.5, while storage integrated systems can have a SM up to 3 to 5. Solar multiple of 1.3 means that the heliostat field delivers a 30% greater thermal power capacity than is required to generate the nominal power for the design day. That 30% excess thermal energy can be delivered to the thermal energy storage.

- Another important concept is the **capacity factor, CF**, defined as the ratio of the hours of solar rated output power operation to the total hours per year (8760 h).

9.3 Thermodynamic cycle

9.3.1 Thermodynamic basics

The energy harvested is used to produce power using a thermodynamic cycle as a function of the use of that thermal energy. CSP facilities can use (Kuravi et al., 2013).

- Rankine cycle. It is the most commonly used when the temperatures of at the receptor range from 573–823 K. Two different cycles are used, a basic one characterized by steam generation, expansion of the steam at the turbine, condensing the exhaust steam and recycle and a regenerative one where the steam from the turbine at medium pressure is reheated up before feeding the turbine. While parabolic trough and solar towers (Martín and Martín, 2013) have implemented superheated steam Rankine cycles operating from 100 to 160 bar respectively, Linear Fresnel reflectors used saturated steam Rankine cycles at 533 K and 50 bar.
- Brayton cycle. It is the cycle used for gases and gas turbines. The gas is compressed, heated up, in this case using solar energy, and expanded. It becomes interesting for higher operating temperatures, 873−1473 K.
- Stirling Engine: Uses helium or hydrogen as HTF. It is only applicable for parabolic dish collectors.
- Combined cycle. It consists of using the hot flue gas from the gas turbine to produce steam for a steam turbine. Examples have been presented in Chapter 6 using biogas.

9.3.2 Rankine cycle

A Rankine cycle consists of four stages, fluid heating, evaporation and superheating, isentropic expansion at the turbine, exhaust steam condensation and pumping up the liquid. Two alternatives can be used, subcritical, Fig. 9.6a, and supercritical, Fig. 9.6b, that is required to operate above 250 bar. To improve the efficiency of the cycle, a reheating Rankine cycle is typically implemented such as the one presented in Fig. 9.6a (Moore et al., 2010). While ideally the expansions at the turbine are isentropic, there is an isentropic efficiency associated due to irreversibility that reduces the power production, dotted line. In addition, there are a number of expansions at the medium and finally at the low-pressure turbines. The extractions are used to heat up the condensed phase before using the energy from the hot source. Up to 7 extractions are used at industrial level (Martín and Martín, 2017). The extractions are referred to regeneration cycle, Fig. 9.6c (Montes et al., 2009).

Description of the cycle: The compressed liquid 6 is heated up using the HTF. The operating pressure at the high-pressure turbine ranges from 40 to 126 bar for CSP plants. Three heat exchangers are used in a subcritical Rankine cycle to heat up first, to evaporate (3) and finally to superheat the steam (4). Only a fraction of the HTF is used for the reheating stage, while the entire flow, is used in the evaporation and the heating up (from 6-3) see Fig. 9.1a. In the case of the supercritical there is no phase change, see Fig. 9.6b. The high-pressure high temperature steam is expanded into a medium pressure, from 5 to 35 bar, and is it reheated with a fraction of the HTF at the highest temperature coming from the storage tanks, see Fig. 9.1a. The reheated steam (4) is expanded at the medium and low-pressure turbine. The exhaust pressure is in the range of 0.05−0.19 bar, typically around 0.08 bar and a fraction of liquid is allowed by the turbines without mechanical issues, around 8% or less (5). The condensation takes place using cooling water that it is later cooled in the cooling system as it will be presented in the following section.

Analysis: The fraction of HTF to be used for steam superheating and for the reheating stage can be an optimization variable (Martín and Martín, 2013). The mass and energy balances to each of the heat exchangers is performed as follows, Eqs. (9.16)-(9.17).

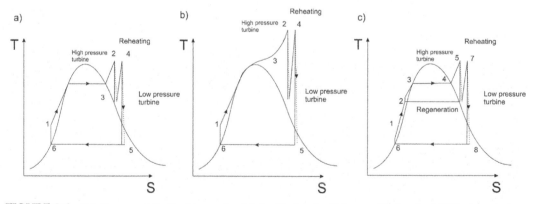

FIGURE 9.6 T-S diagram of the Rankine cycle. a) Subcritical cycle; b) Supercritical cycle c) With regeneration.

$$Q_{avail} = - fc_{(HTF)} \int_{T_{HTF,in}}^{T_{HTF,out}} cp_{HTF} dT \qquad (9.16)$$

$$Q_{Avail} = Q_{toCycle} = fc_{(Wa)} \cdot \left(H_{Out}(p, T_{S,out}) - H_{in}(p, T_{S,in}) \right) \qquad (9.17)$$

The enthalpies of the steam and water can be read in Tables (Moran et al., 2014) or computed using correlations like the ones developed by Martín and Martín (2013) that can be found in the appendix of the book. To avoid temperature crosses, considering countercurrent flow of the HTF and the water/steam it is important to establish the following constraint, Eq. (9.18):

$$T_{S,out} + \delta \leq T_{HTF,in}$$
$$T_{S,in} + \delta \leq T_{HTF,out} \qquad (9.18)$$
$$\delta \in [5, 10]$$

To ensure saturation conditions, the temperature at the inlet of the evaporator can be computed using Antoine equation for the operating pressure, Eq. (9.19):

$$P_{turb1} \cdot 760 = e^{\left(A_{(Wa)} - \frac{B_{(Wa)}}{\left(C_{(Wa)} + T_{Evaporator} \right)} \right)} \qquad (9.19)$$

The expansion of the steam in the different turbines is assumed to have an isentropic efficiency, η_s, Eq. (9.20), (Nezammahalleh et al., 2010). Therefore, the stream exiting the high-pressure turbine can be calculated using Eqs. (9.20)-(9.22)

$$\eta_{real} = \frac{H_{steam,out} - H_{steam,in}}{H_{steam,(isoentropy)} - H_{steam,in}} \qquad (9.20)$$

Where

$$H_{steam,(isoentropy)} = f\left(p_{out}, T_{out}^* \right) \qquad (9.21)$$

T^* represents the isoentropic temperature after the expansion. Thus, the entropy, s, remains constant:

$$s_{steam} = f\left(p_{in}, T_{in} \right) = f\left(p_{out}, T_{out}^* \right) \qquad (9.22)$$

To enforce that the output of the turbine is superheated steam, the Antoine equation can be used to secure that the temperature is above that of saturation.

However, one of the mayor issues of the turbines in CSP facilities is the fact that they do not operate at the full load the entire year due to the variability in the solar energy available. The efficiency of each expansion also depends on the load and several correlations have been developed to estimate that loss of efficiency, φ, as follows, Eqs. (9.23)-(9.25):

$$\varphi = \left(-1.0176 \cdot (Load)^4 + 2.4443 \cdot (Load)^3 - 2.1812 \cdot (Load)^2 + 1.0535 \cdot Load + 0.701 \right) \quad (9.23)$$

(Jüdes et al., 2009)

$$\varphi = 1 - \left(0.191 - 0.409 \cdot Load + 0.218 \cdot (Load)^2 \right) \qquad (9.24)$$

(NREL, 2019)

$$\varphi = 0.4097 \cdot Load + 0.5903 \tag{9.25}$$

(Erhart et al., 2011)
Therefore the actual efficiency is givem by Eq. (9.26):

$$\eta_{real} = \eta_s \varphi \tag{9.26}$$

The energy that is obtained at each expansion is given by Eq. (9.27):

$$W = fc_{(Wa)} \cdot \left(H_{steam,in} - H_{steam,out} \right) \tag{9.27}$$

The pumps used for the compression of the water also have an isentropic efficiency defined as a ratio with a designed reference, Eq. (9.28):

$$\frac{\eta_{s,pump}}{\eta_{s\_o,pump}} = e_0 + 2 \cdot (1 - e_0) \frac{fc_{(wa)}}{fc_{(wa),ref}} - (1 - e_0) \left(\frac{fc_{(wa)}}{fc_{(wa),ref}} \right)^2 \tag{9.28}$$

e_o defines the shape of the efficiency parabola and takes the value of 0 for constrant speed pumps.

While this is the energy that the thermodynamic predicts, the generator also presentes an efficiency due to the load. The efficieny is typically in the range of 92–98% and depends on the type of generator. The following correlation, Eq. (9.29) (Patnode, 2006), can be used.

$$\eta_{gen} = 0.908 + 0.258 \cdot Load - 0.3 \cdot Load^2 + 0.12 \cdot Load^3 \tag{9.29}$$

Example 9.1 shows the performance of a Rankine cycle for a CSP plant.

EXAMPLE 9.1 A parabolic trough CSP plant allocated in the South of Spain operates a Rankine cycle with regeneration, see Fig. 9E1.1. Determine the power generated by the facility, the cooling needs and the fraction required in the regeneration stage. Assume a design DNI of 850 W/m^2. Mixing is to be carried out at the same pressure. The maximum pressure at the turbine is 100 bar and the steam is overheated at 555°C. The HTF operates between a minimum of 290°C and a maximum of 565°C, assume molten salts. The first expansion reduces the pressure at 11 bar, 25% is used in the regeneration of the cycle, and the exhaust pressure is 0.08 bar. The isentropic efficiency of each turbine is equal to 0.8. The projected area is 100000 m^2 and the yield of the solar field is equal to 0.6.

Solution

$$Q_{(collector)} = DNI_i \cdot A_{field} \cdot \eta_{field} \tag{9E1.1}$$

$$A_{field} = n_h \cdot A_{proyected}$$
$$A_{Proyected} = a \cdot l \tag{9E1.2}$$

We formulate a global mass balance to produce steam for the turbine. Note, we use the subindexes to represent the streams in Fig. 9E1.1

$$Q_{(Collector)} = -fc_{(HTF)} \int_{T_{HTF,in}}^{T_{HTF,out}} cp_{HTF} dT \tag{9E1.3}$$

$$Q_{(Collector)} = fwa_2 \cdot (H_2(p_2, T_2) - H_1(p_1, T_1)) \tag{9E1.4}$$

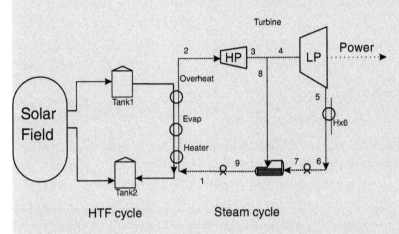

FIGURE 9E1.1 Scheme of the thermal plant.

The expansion of the first turbine yields an overheated steam at 11 bar. We assume full load for the design conditions. The isentropic efficiency ot the turbines is 0.8.

$$\eta_{real} = \frac{H_3 - H_2}{H_{3,(isoentropy)} - H_2} \tag{9E1.5}$$

$$\varphi = \left(-1.0176 \cdot (Load)^4 + 2.4443 \cdot (Load)^3 - 2.1812 \cdot (Load)^2 + 1.0535 \cdot Load + 0.701\right) \tag{9E1.6}$$

$$\eta_{real} = \eta_s \varphi \tag{9E1.7}$$

Where

$$H_3 = f(p_3, T_3)$$
$$H_2 = f(p_2, T_2) \tag{9E1.8}$$
$$H_{3(isoentropy)} = f(p_3, T_{3,iso})$$

$$W_1 = fwa_2 \cdot (H_2 - H_3) \tag{9E1.9}$$

By solving this system we can compute the actual temperature of stream 3. A fraction is extracted and used for the regeneration. The aim is to saturate the exhaust stream at 11 bar. The rest of the flow is fed to the second turbine at the same pressure and temperature as those leaving the first expansion stage.

$$fwa_2 = fwa_3 = fwa_4 + fwa_8 \tag{9E1.10}$$

The second turbine results in a vapor-liquid mixture at the exhaust pressure of 0.08 bar. Thus the enthalply is to be obtained as

$$H_5 = f_L \cdot H_{lsat}(p_5, T_5) + (1 - f_L) \cdot H_{vsat}(p_5, T_5) \tag{9E1.11}$$

As in the precious case, the expansion is not isentropic and therefore:

$$H_{5,iso} = f_{L,iso} \cdot H_{lsat}(p_5, T_{5,iso}) + (1 - f_{L,iso}) \cdot H_{vsat}(p_5, T_{5,iso}) \tag{9E1.12}$$

At saturation, the pressure and the temperature are related and can be computed using Antoine correlation:

$$p_5 \cdot 760 = e^{\left(A_{(Wa)} - \frac{B_{(Wa)}}{(C_{(Wa)} + T_5)}\right)} \tag{9E1.13}$$

The energy obtained in this section stage is computed as follows:

$$W_2 = fwa_4 \cdot (H_5 - H_4) \tag{9E1.14}$$

The exahust is to be condensed in HX6. The enthaly of the liquid, H_6, is that of saturation at 0.08 bar. Thus, the cooling needs of the cycle are computed as Eq. (9E1.15):

$$Q_{HX6} = fwa_5 \cdot (H_6 - H_5) \tag{9E1.15}$$

The liquid is to be compressed at 11 bar for the mixing. The energy consumed by the pump is computed as the following equations. We assume that the temperature of both streams across the pump remain the same and the excess is lost.

$$H_7 = H_6 + \nu_6(p_7 - p_6)$$
$$W_{pump1} = fwa_6 \cdot (H_7 - H_6) \tag{9E1.16}$$

Where the specific volume of water is assumed constand at 0.001 m^3/kg. The mixing is isobaric. The condition to be achieved is that it is saturated at 11 bar. That allows computing the fraction of steam stracted from the turbine

$$fwa_9 = fwa_7 + fwa_8$$
$$fwa_9 \cdot H_9 = fwa_7 \cdot H_7 + fwa_8 \cdot H_8$$
$$H_9 = H_{lsat}(11 \, bar)$$
$$H_8 = H_3 \tag{9E1.17}$$

The second pump is computed as the previous one and the cycle closes. By solving the system, the GAMS code can be seen in the supplementary material, Table 9E1.1 shows the flows, enthalpies, and entropies. The flow of steam/water is 18.8 kg/s and 22% of it is extracted for the regeneration, 4.1 kg/s. The flow of HTF, is 122.3 kg/s. Table 9E1.2 shows the power produced by the system 19,163 kW, and the system requires 31,840 kW of cooling.

TABLE 9E1.1 Thermodyanmic properties of the streams.

Stream	H (kJ/kg)	T (°C)	S(kJ/kg K)	Stream	H (kJ/kg)	T (°C)	S(kJ/kg K)
1	805.3	184.5		6	147.3	41.8	
2	3519.9	555	6.7	7	175.4	41.8	
3	3011.5	284.3	6.9	8	3011.5	284.3	6.9
4	3011.5	284.3		9	796.3	184.5	
5	2344.0	41.8					

TABLE 9E1.2 Electrical energy involved in the cycle.

W	(kW)
Turbine1	9553
Turbine2	9795
Pump1	16
Pump2	169

9.3.3 Brayton cycle

The Brayton cycle is the basis for jet engines and is the one used in gas turbine power plants. While in the case of Biogas, or natural gas, burning the gas using air is what heats up the mixture before the expansion, in the case of CSP plants the energy comes from the concentrated solar technology. Open and close Brayton cycles are available as well as different fluids. Open cycles, Fig. 9.7a, mean that the flue gas once expanded is not recycled and recompressed. This is typically in combustion-based cycles and in combined cycles that use that hot stream to produce steam. Closed cycles, Fig. 9.7b, uses a fluid that follows the cycle. The close cycle can be used in reverse for cooling. Pressure ratios up to 40 have been reached lately

a)

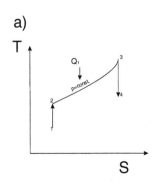

b)

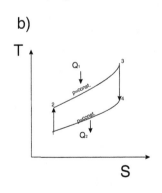

FIGURE 9.7 T-S diagram of a) open and b) closed Brayton cycle.

(General Electric, 2001). To reduce the energy consumption, the compression is carried out in a multistage configuration with intercooling. Similarly, using a multistage expansion with reheating also improves the efficiency of the cycle.

For CSP facilities the use of supercritical CO_2 (sCO_2) is being evaluated. The aim is to take advantage of the high density of the fluid to reduce the pumping power improving the efficiency of the cycle as long as the CO_2 is kept above 78.9 bar and 30.98°C. Among the advantages to be expected are the high efficiency, over 50%, good matches with the temperature of the solar tower (850°C), compactness of the turbine, low thermal mass (International Energy Agency, IEA, 2010).

Example 9.2 shows the performance of a Brayton cycle for a CSP plant.

EXAMPLE 9.2 A solar tower CSP plant allocated in the South of Spain operates a close Brayton cycle. Assuming an average DNI of 700 W/m², optimize the number of stages required to produce 25 MW. The gas flow rate is 50 kg/s of air. The yield, compressor and expansion performance, assumed polytropic, are found below.

Yield_field: 0.6; Solar radiation kW per m²/0.7; polytropic efficiency 0.80; polytropic coefficient compression: 1.4; expansion polytropic coefficient: 1.3; Heliostat efficiency: 0.9; Area heliostat (m²)/120;

Compute the tower height and the number of heliostats.
Solution.

To model the gas turbine, we assume multistage compression of the gas with intercooling with a number of stages from 1 to 3 to be optimized, Eqs. (9E2.1)-(9E2.2)

$$T_{out/compresor} = T_{in/compressor} + T_{in/compressor}\left(\left(\frac{P_{out/compressor}}{P_{in/compressor}}\right)^{\frac{z-1}{z}} - 1\right)\frac{1}{\eta_c} \quad (9E2.1)$$

$$W_{(Compressor)} = n(F) \cdot \frac{R \cdot z \cdot (T_{in/compressor})}{((MW) \cdot (z-1))}\frac{1}{\eta_c}\left(\left(\frac{P_{out/compressor}}{P_{in/compressor}}\right)^{\frac{1}{n}\frac{z-1}{z}} - 1\right) \quad (9E2.2)$$

Where T corresponds to the temperature, P the pressure, F the mass flow, n the number of compression stages, z the polytropic coefficient, MW the molecular weight of the stream compressed, and η_c the efficiency of the compression. After each compression stage, but before the heating, the stream is cooled down. The total cooling needs are computed as follows, Eq. (9E2.3):

$$Q(HX) = (n-1)\sum_i fc_i \cdot \int_{T_{out/compressor}}^{T_{in/compressor}} cpdt \quad (9E2.3)$$

The gas exiting the compression stage is heated up with the energy from the collector; we assume that it is at a high temperature enough to heat the gas from that exiting the compression stage, Eq. (9E2.4).

$$\begin{aligned}Q_{(collector)} &= DNI_i \cdot A_{field} \cdot \eta_{field}\\ A_{field} &= N_{Heliostat} \cdot A_{Heliostat} \cdot \eta_{Heliostat}\\ H_{Tower} &= 82.6 + 0.2552 \cdot 10^{-3} \cdot Q_{(collector)}(kW)\\ Q_{Prod} &= Q_{feed} + Q_{(collector)}\end{aligned} \quad (9E2.4)$$

The hot gas is expanded assuming one stage only, that is modeled as Eqs. (9E2.5)–(9E2.6):

$$T_{out/turbine} = T_{in/turbine} + T_{in/turbine}\eta_t \left(\left(\frac{P_{out/turbine}}{P_{in/turbine}} \right)^{\frac{z-1}{z}} - 1 \right) \qquad (9E2.5)$$

$$W_{(turbine)} = (F) \cdot \frac{R \cdot z \cdot (T_{in/turbine})}{((MW) \cdot (z-1))}\eta_t \left(\left(\frac{P_{out/turbine}}{P_{in/turbine}} \right)^{\frac{z-1}{z}} - 1 \right) \qquad (9E2.6)$$

For a closed cycle the $P_{out/turbine}$ is the same as that of $P_{in/compressor}$ and the $P_{out/compressor}$ is equal to $P_{in/turbine}$. The facility will consist of 1615 Heliostats with a tower of 101 m. A 3-stage compression system with intercooling is selected with a maximum pressure of 18.6 bar and an expansion down to 1 bar. The cooling required adds up to 12160 + 2762 kW for the intercooling and the cold extreme of the cycle.

The GAMS code can be seen below regarding equation writing. For the parameters' definition including the heat capacities, and molecular weights we refer to the code uploaded in the web.

```
*                                            ***************
*                                              Gas turbine
*                                            ***************
Positive Variable
mgas, Pressure_GT_in, Pressure_GT_out, MW_gas, Q_Sun, Area, Q_react, Q_prod, Power,
H_Tower, NHelios, Number;

Equations

Gasturb_1, Gasturb_2, Gasturb_3, Gasturb_4, Gasturb_5, Gasturb_6, Gasturb_7,
Gasturb_8, Gasturb_9, Gasturb_10, Gasturb_11, Gasturb_12, Gasturb_13, Gasturb_14,
Gasturb_15, Gasturb_16, Gasturb_17, Gasturb_18;

mgas.LO = 50;
mgas.UP = 50;

Gasturb_1(J).. fc(J,'HX3','Compres1') = E = mgas*Composition(J);

Gasturb_2.. mgas = E = MW_gas*SUM(J, fc(J,'HX3','Compres1') /MW(J));

Number.LO = 1;
Number.UP = 3;
Gasturb_3.. W('Compres1')*((MW_gas + 0.001)*(k_p - 1)) = E = Number*(1/nu_p)*mgas

*(8.314*k_p*(T_cooldown + 273))*(((Pressure_GT_in + eps1)/(Pressure_GT_out + eps1))
**((1/(Number + eps1))*((k_p - 1)/k_p)) - 1);

Gasturb_4.. T('Compres1','Furnace') = E = ((T_cooldown + 273) + (1/nu_p)*
(T_cooldown + 273)*(((Pressure_GT_in + eps1)/(Pressure_GT_out + eps1))**((1/
(Number + eps1))*((k_p - 1)/k_p)) - 1)) - 273;
```

```
Gasturb_5.. Q('HX1')=E= -(Number-1)*sum(J,fc(J,'Compres1','Furnace')*
((1/MW(J))*(c_p_v(J,'1')*(T('Compres1','Furnace')-25)+(1/2)*c_p_v(J,'2')*
((T('Compres1','Furnace')+273)**2-(298)**2)+(1/3)*c_p_v(J,'3')*
((T('Compres1','Furnace')+273)**3-(298)**3)+(1/4)*c_p_v(J,'4')*
((T('Compres1','Furnace')+273)**4-(298)**4))));

T.LO('Compres1','Furnace')=20;
T.UP('Compres1','Furnace')=450;

Pressure_GT_in.LO=5;
Pressure_GT_in.UP=40;

Gasturb_6.. Q_Sun =E= Area*Yield_field*Rad;

Area.LO=120;
Area.UP=350000;

Gasturb_7.. Q_react=E= sum(J,fc(J,'Compres1','Furnace')*((1/MW(J))*
(c_p_v(J,'1')*(T('Compres1','Furnace')-25)+(1/2)*c_p_v(J,'2')*
((T('Compres1','Furnace')+273)**2-(298)**2)+(1/3)*c_p_v(J,'3')*
((T('Compres1','Furnace')+273)**3-(298)**3)+(1/4)*c_p_v(J,'4')*
((T('Compres1','Furnace')+273)**4-(298)**4))));

Gasturb_8.. Q_prod=E= sum(J,fc(J,'Furnace','TurbGas')*((1/MW(J))*(c_p_v(J,'1')*
(T('Furnace','TurbGas')*10-25)+(1/2)*c_p_v(J,'2')*((T('Furnace','TurbGas')
*10+273)**2-(298)**2)+(1/3)*c_p_v(J,'3')*((T('Furnace','TurbGas')*10+273)**3-
(298)**3)+(1/4)*c_p_v(J,'4')*((T('Furnace','TurbGas')*10+273)**4-(298)**4))));

Gasturb_9(J).. fc(J,'HX3','Compres1') =E= fc(J,'Compres1','Furnace');
Gasturb_10(J).. fc(J,'Compres1','Furnace')=E=fc(J,'Furnace','TurbGas');

Gasturb_11.. Q_prod=E= Q_react+Q_Sun;
T.LO('Furnace','TurbGas')=100;
T.UP('Furnace','TurbGas')=160;

Gasturb_12.. T('TurbGas','HX3')*10 =E= ((T('Furnace','TurbGas')*10+273)+
(nu_p)*(T('Furnace','TurbGas')*10+273)*(((Pressure_GT_out+eps1)/
(Pressure_GT_in+eps1))**(0.3/1.3)-1))-273;

T.LO('TurbGas','HX4')=50;
T.UP('TurbGas','HX4')=150;

Gasturb_13.. W('TurbGas')*((MW_gas+0.001)*(k_p2-1)) =E= -nu_p*mgas*
(8.314*k_p2*(T('Furnace','TurbGas')*10+273))*(((Pressure_GT_out+eps1)/
(Pressure_GT_in+eps1))**((k_p2-1)/k_p2)-1);
```

```
Pressure_GT_out.LO = 1;
Pressure_GT_out.UP = 5;

Gasturb_14(J).. fc(J,'Furnace','TurbGas') = E = fc(J,'TurbGas','HX3');

Gasturb_15.. Q('HX3') = E = - sum(J,fc(J,'TurbGas','HX3')*((1/MW(J))*(c_p_v(J,'1')
*(T('TurbGas','HX3') - 25) + (1/2)*c_p_v(J,'2')*((T('TurbGas','HX3') + 273)**2 -
(298)**2) + (1/3)*c_p_v(J,'3')*((T('TurbGas','HX3') + 273)**3 - (298)**3) + (1/4)*c_p_v
(J,'4')*((T('TurbGas','HX3') + 273)**4 - (298)**4))));
Gasturb_16.. Power = E = W('TurbGas') - W('Compres1');
Gasturb_17.. H_tower = E = 82.6 + 0.2552*Q_Sun/1000;
Gasturb_18.. Area = E = NHelios*rd_helio*A_helio;
Power.fx = 25000;
Equations
                Obj;
Variables
                Z;
Obj..       Z = E = W('TurbGas') - W('Compres1');
Model CSPBrayton /ALL/;
Option NLP = conopt;
Solve CSPBrayton using nlp maximizing Z;
```

9.3.4 Stirling Engine

The Stirling Engine, developed by Robert Stirling in 1816, consists of reciprocating pistons that use a gas working fluid, typically hydrogen, helium or air, that are externally heated. The cycle consists of four stages: 1–2) Isothermal compression; 2–3) isochoric heating, constant volume; 3–4) isothermal expansion; 4-1) isochoric heat removal. Fig. 9.8 shows the T-S scheme of the thermodynamic cycle. The CSP provides the energy from the Sun at stage 2.

FIGURE 9.8   T-S diagram of a Stirling Engine.

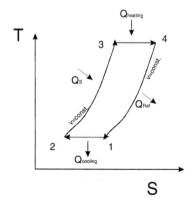

These cycles have only been implemented at small scale facilities but with high conversion efficiencies when the inlet temperatures are around 800°C. The system achieves AC efficiencies of 30%. However, so far it has not possible to implement storage systems. An alternative is its hybridization with natural gas.

9.3.5 Combined cycle

It consists in the integration of an open Brayton cycle and a steam cycle where the hot flue gas from the gas turbine is used to produce steam. See Chapter 6 for the flow diagram.

9.4 Cooling systems

Cooling is a critical stage of the cycle. It does not only determine the efficiency, it represents the cold extreme of the thermodynamic cycle, but it is responsible for the consumption of water of these facilities. While water consumption may not be a critical issue in fossil or biomass-based power plants, since the availability of water in the region is not limiting, CSP facilities are located in regions where the relative availability of water is low and the consumption of water may result in infeasible operation. In this section we comment on two different cooling systems, natural draft wet cooling towers (NDWCT) and A-frame air coolers, which are gaining attention. However, the analysis of the NDWCT can be easily used as a basis to evaluate natural draft dry cooling systems, that substitute the evaporative cooling by the use of convective cooling with air close to the A-frame systems or the forced towers, where there is no superstructure to generate the draft of air since a fan is used to provide the pressure drop.

9.4.1 Introduction: types

The main trade-off between the different cooling systems can be summarized into power vs water consumption. Natural draft systems do not consume power, they require the paraboloid structure representing a large investment cost, to generate the draft, see below. Dry systems rely on convective transfer reducing the efficiency while wet systems are typically used due to the higher efficiency at the expense of presenting water consumptions of around 2–3 L/kWh. The actual value is highly dependent on the allocation of the facility (Guerras and Martín, 2020). The weather conditions affect not only the size of the paraboloid but also the water consumed. Alternatively, forced draft can be used. In particular, air coolers are widely used, and for power plants the use of A-frame is common. Their operation is based on convective heat transfer generated by the flow of air induced by the fans. The consumption of water is limited to cleaning, but the energy efficiency of the entire facility reduces by 5–10% to power the fans (Luceño and Martín, 2018; Martín, 2015). Fig. 9.9 shows parabolic cooling towers and A-Frame structures for further discussion along this section. In addition to the cooling system, the efficiency in the solar harvesting affects the water consumption per kWh. Table 9.3 shows the values for the four main receivers. Solar towers show better water efficiency among those using Rankine cycles.

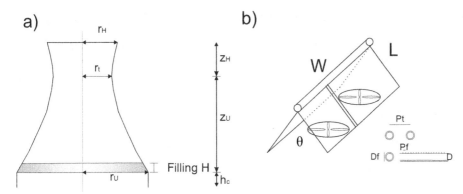

FIGURE 9.9   Cooling technologies: a) ND(WorD)CT; b) A-Frame.

TABLE 9.3   Sumarry of thermodynamic cycles for CSP technologies (EIA, 2010; Irena, 2012).

	Parabolic Trough	Solar Tower	Linear Fresnel Reflectors	Dish Parabolic
Cycle	Superheated steam Rankine	Superheated steam Rankine	Saturated steam Rankine	Stirling
Water requirement (L/kWh)	3-wet cooling	2–3 (wet cooling)	3-wet cooling	0.05–0.1 (mirror washing)
	0.3-dry cooling	0.25 (dry cooling)	0.2-dry cooling	

9.4.2 Wet cooling: Cooling tower

In this section both the packing design and the geometry structure of the unit.

9.4.2.1 Packing design

Cooling towers packing is designed based on the mass and energy transfer required for cooling the water flow (Geankoplis, 1993). The analysis of those processes is performed by formulating a differential energy balance to the packing including latent and sensible heat. Subindex "i" is used for the interphase, "g" for the gas phase and "l" for the liquid phase. T represents the air temperature and "h_g" is the heat transfer coefficient.

The sensible energy balance to the air bulk is given by Eq. (9.30):

$$h_g \, A \, S \, dP_{Height} (T_i - T_g) = f_{air} \, c_h \, dT_g \tag{9.30}$$

where S is the specific contact area, A is the cross-sectional area and dP_{Height} is the differential length along the packed bed. f_{air} is the flow of air across the tower with c_h as the heat capacity, that is given by Eq. (9.31):

$$c_h = h_g / k_y \tag{9.31}$$

k_y is the mass transfer coefficient. Therefore, Eq. (9.32) becomes:

$$k_y \, A \, S \, dP_{Height}\left(c_h \, T_i - c_h \, T_g\right) = f_{air} \, c_h \, dT_g \tag{9.32}$$

Latent energy balance to the air bulk is formulated as Eq. (9.33):

$$k_y \, A \, S \, dP_{Height}(Y_i - Y)\lambda_o = f_{air} \, \lambda_o \, dY \tag{9.33}$$

where λ_o is the latent heat, Y is the air moisture and dY is the moisture gradient.
The global energy balance by adding both, Eq. (9.32) and Eq. (9.33), results in:

$$k_y \, A \, S \, dP_{Height}\left[(c_h \, T_i + \lambda_o \, Y_i) - (c_h \, T_g - \lambda_o \, Y)\right] = f_{air}\left[c_h \, dT_g + \lambda_o \, dY\right] \tag{9.34}$$

Using the definition for gas enthalpy, H, we have Eq. (9.35):

$$k_y \, A \, S \, dP_{Height}[H_i - H] = f_{air}[dH] \tag{9.35}$$

so that the height of the packing is computed as follows, Eq. (9.36):

$$P_{Height} = \frac{f_{air}}{k_y A \, S} \int_{H_1}^{H_2} \frac{dH}{H_i - H} = HTU \cdot NTU \tag{9.36}$$

To compute the NTU, the driving force must be computed, the difference between the enthalpy at the interphase, H_i, and that of the gas, H. Here two methods are presented, a simplified one given by rules of thumb (Brannan, 2005) and the unit operations based one.

9.4.2.1.1 Rules of thumb packing sizing

The simpler way to compute it is by using correlations in the literature, Eqs. (9.37)-(9.40), (Brannan, 2005):

$$H_i(BTU/lb) = \frac{-665.432 + 13.4608 \cdot t - 0.784152 \cdot t^2}{t - 212} \tag{9.37}$$

Where t is in (°F) and it can be used for temperatures above 10 °C

$$NTU = \left(\frac{L}{G}\right)\left[\left(\frac{1}{2 \cdot a}\right) Ln\left[\frac{a \cdot t_2^2 + b \cdot t_2 + c}{a \cdot t_1^2 + b \cdot t_1 + c}\right] - \left(212 + \frac{b}{2a}\right)e\right] \tag{9.38}$$

Where

$$a = -\left(\frac{L}{G}\right) - 0.784152;$$

$$b = -H_1 + \left(\frac{L}{G}\right)(t_1 + 212) + 13.4608$$

$$c = -665.432 + 212 \cdot \left[H_1 - \left(\frac{L}{G}\right) \cdot t_1\right] \tag{9.39}$$

$$d = 4 \cdot a \cdot c - b^2$$

For d > 0

$$e = \frac{2}{\sqrt{d}} \cdot \left[\arctan\left(\frac{2 \cdot a \cdot t_2 + b}{\sqrt{d}}\right) - \arctan\left(\frac{2 \cdot a \cdot t_1 + b}{\sqrt{d}}\right) \right]$$

The height of the transfer unit can be computed as follows (Walas, 1990)

$$HTU = 5.51 \cdot \left(\frac{L}{G}\right)^{0.59} \tag{9.40}$$

9.4.2.1.2 Unit operation packing sizing

Alternatively, a more detailed analysis can be carried out. The equilibrium line for the air-water system is computed for the total pressure of the place where the cooling tower is located, Eq. (9.41).

$$H_i = f(t_i, Y_{i,sat})$$
$$Y_{i,sat} = 0.62 \frac{P_{v,sat}(t_i)}{P - P_{v,sat}(t_i)} \tag{9.41}$$

The operating line is determined by an energy balance between the air and the water flows, f_{air} and f_{Wa} respectively, where h is the water enthalpy, t_L and c_L are water temperature and heat capacity, Eq. (9.42).

$$f_{air} \, dH = f_{Wa} \, dh = f_{Wa} \, c_L \, dt_L = f_{Wa} \, dt_L \tag{9.42}$$

It is convenient to use kcal as unit for energy so that $c_L = 1 \, kcal/(kg°C)$. Thus, the energy balance to the packing section of the tower becomes Eq. (9.43):

$$\frac{H_2 - H_1}{t_{L,2} - t_{L,1}} = \frac{f_{Wa}}{f_{air}} = \frac{L}{V} \tag{9.43}$$

The liquid flow, f_{Wa}, is calculated from the energy balance to the heat exchanger that is used to condense the exhaust steam from the turbine. The rules of thumb recommend the cooling water to heat up 8–10 °C across the heat exchanger from $t_{L,in}$ to $t_{L,out}$ (Perry and Green, 1997), Eq. (9.44):

$$8 \leq \left(t_{L,in} - t_{L,out}\right) \leq 10 \tag{9.44}$$

The operation of the cooling tower requires a minimum flow of air, typically from 1.3 to 1.5 times the minimum given by the maximum slope, computed from the liquid inlet temperature and the tangent with the H_i curve.

The high efficiency of these towers is due to the evaporative heat transfer. In this operation water evaporates, $f_{(wa,Evaporated)}$. A lower bound of the evaporation can be established using rules of thumbs, Eq. (9.45), (Perry and Green, 1997).

$$f_{(Wa,Evaporated)} \geq 1.8 \cdot 0.00085 \cdot f_{Wa} \left(t_{L,in} - t_{L,out}\right) \tag{9.45}$$

The temperature of the air leaving the tower is calculated as the final temperature of the profile along the cooling tower in red in Fig. 9.10. The Lewis relationship,

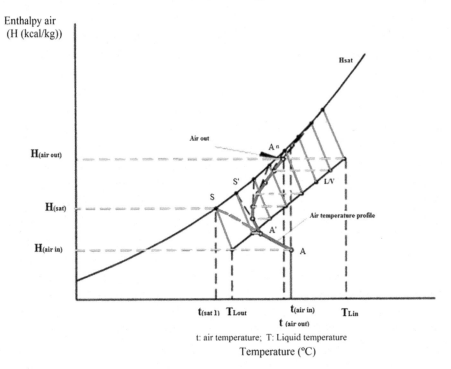

FIGURE 9.10 Mickley method for the cooling tower operation. *Martín and Martín (2017) with permission.*

Eqs. (9.46)-(9.47) based on an energy balance, relates the operation line with the equilibrium line determining the temperature of the air

$$h_L\ A\ S\ dP_{Height}(t_L - T_i) = f_{Wa}\ dh = f_{air}\ dH = k_y\ A\ S\ dP_{Height}(H_i - H) \qquad (9.46)$$

$$\frac{H_i - H}{T_i - t_L} = -\frac{h_L}{k_y} \qquad (9.47)$$

An ideal operation would consider $-h_L/k_y$ to be $-\inf$. However, at industrial scale it typically takes values from -3 and -10, in green in Fig. 9.10.

The profile of the air temperature is computed as the ratio of the change on enthalpy and temperature as it exchanges moisture with the liquid flow Eqs. (9.48)–(9.49)

$$\frac{k_y\ A\ S\ P_{Height}(H_i - H)}{k_y\ c_h\ A\ S\ P_{Height}(T_i - T_g)} = \frac{f_{air}dH}{f_{air}c_h dT_g} \qquad (9.48)$$

$$\frac{(H_i - H)}{(T_i - T_g)} = \frac{dH}{dT_g} \qquad (9.49)$$

To solve the temperature profile, it is possible to solve the system of differential equations. Alternatively, the profile can be discretized. At each step, the Lewis equation determines the saturation, S. The air temperature at each step is computed at the intersection between the line A-S in blue, and the horizontal line, corresponding to the enthalpy of the are, in orange resulting in A'.

We obtain a series of points, A-A',A'', ... A^n that constitute the temperature profile of the air along the column. Solving this as a set of equations can be mathematically tricky. The lines may present large slopes. It is more efficient to combine both lines to compute the next temperature as follows, Eq. (9.50), (Martín and Martín, 2017):

$$\frac{(H_{j+1} - H_j)}{(T_{j+1} - T_j)} = \frac{(H_{i,j} - H_j)}{(T_{i,j} - T_j)} \tag{9.50}$$

Where subindex i represents the saturation conditions and subindex j corresponds to the discretization point. Finally, the enthalpy of the exiting air, its moisture and temperature must be consistent. Typically, the air does not leave saturated but at 95%

The **HTU** is computed based on the mass transfer. Depending on the type of packing different correlations can be used. k_y, is computed using the correlation given by Coulson and Richardson (1999), Eq. (9.51), assuming that the contact area provided by the fillings (a) is constant and equal to 250 m^2/m^3, where Gflux (kg s^{-1} m^2) and Lflux (kg s^{-1} m^2) are the cross sectional flows and "a" the specific contact area ($k_y a[=]$ (kg s^{-1} m^2))

$$k_y = \left(\frac{2.95}{a}\right) \cdot (\text{Gflux})^{0.72} \cdot (\text{Lflux})^{0.26} \tag{9.51}$$

The specific flows across the contact region are computed per the gas and liquid flows and the actual area, Eqs. (9.52)–(9.54):

$$A = \left(\frac{\pi}{4}\right) \cdot D_{base}^2 \tag{9.52}$$

$$\text{Gflux} = \frac{f_{air}}{A} \tag{9.53}$$

$$\text{Lflux} = \frac{fc_{Wa}}{A} \tag{9.54}$$

The cost of the fill is around 25 €/m^3 (Guerras and Martín, 2020).

9.4.2.2 Paraboloid design

One of the main issues in the design of the paraboloid is to provide enough draft for the air to flow across as well as to secure the mass transfer in case of wet cooling towers. The driving force, $\Delta P_{Generated}$, is given by buoyancy, Eq. (9.55):

$$\Delta P_{Generated} = CT_{Height} \cdot g \cdot \left(\rho_{g,in} - \rho_{g,out}\right) \tag{9.55}$$

Where CT_{height} is the total height of the tower, g is gravity and ρ are air densities. Assuming ideal gases, has densities are computed as follows, Eq. (9.56):

$$\rho_g = \left(\frac{1}{M_{air}} + \frac{Y}{M_{water}}\right)\frac{RT_g}{P} \tag{9.56}$$

The air that flows suffers pressure drop due to a number of elements such as the mist eliminator, the support of the tower, the contraction to the air inlet, the pressure drop generated by the water spray used to cooldown the water and the packing itselft (EPA, 1970; Kreith and Goswami, 2005), Eq. (9.57):

$$\Delta P_{Loss} = \left(N_{Mist} + N_{Support} + N_{Contraction} + N_{Spray} + N_{Packing}\right)\rho_{air}\frac{v_{air}^2}{2}$$

where:

$$N_{Contraction} = 0.167\left(\frac{D_{base}}{h_c}\right)^2$$

$$0.02 \cdot D_{base} \leq h_c \leq 0.12 \cdot D_{base}$$

$$N_{Spray} = 0.16 \cdot h_c \cdot \left(\frac{f_{Wa,in}}{f_{air,in}}\right)^{1.32} \tag{9.57}$$

$$N_{Packing} = Height_{packing} \cdot 2.36\left(\frac{L_{flux}}{3.391}\right)^{1.1} \cdot \left(\frac{G_{flux}}{3.391}\right)^{-0.64}$$

Where L_{flux} and G_{flux} are defined in Eqs. (9.49) and (9.50). N_{mist} and $N_{Support}$ take values from 0.5–2 m typically. The total height of the tower is given by Eq. (9.58):

$$CT_{Height} = Z_H + Z_u + h_c \tag{9.58}$$

The geometry of the cooling tower is that of a hyperboloid (Gould and Kratzig, 1999). Each section of the hyperboloid, Z_H and Z_u, is computed using the mathematical expression for that particular geometry, Eq. (9.59), where d_i is the diameter at Z_i, d_t is the largest diameter and h_c is the height of the opening for the air fed to the CT, Eq. (9.59):

$$\frac{d_i^2}{d_t^2} - \frac{Z_i^2}{b^2} = 1 \quad \forall i \in \{u, H\} \tag{9.59}$$

For the mechanical structure to maintain stability, additional rules of thumb are imposed (Almási, 1981), Eq. (9.60)

$$0.55 \cdot D_{base} \leq 2 \cdot r_t \leq 0.65 \cdot D_{base}$$
$$0.61 \cdot D_{base} \leq 2 \cdot r_H \leq 0.73 \cdot D_{base}$$
$$1.25 \cdot D_{base} \leq CT_{Height} \leq 1.5 \cdot D_{base} \tag{9.60}$$
$$1.1 \cdot D_{base} \leq Z_u + Z_h \leq 1.3 \cdot D_{base}$$
$$0.92 \cdot D_{base} \leq Z_u \leq 1.02 \cdot D_{base}$$

The concrete used to build the tower shell can be estimated as Eq. (9.61)

$$V_{Shell} = \frac{\pi}{2}(2 \cdot r_u + 2 \cdot r_H) \cdot t_s \cdot T_{Height} \tag{9.61}$$

where t_s is the thickness of the shell varies across the tower., Eq. (9.62) For a 100 m tower at the bottom 65 cm while the minimum is 15 cm (Lang and Straub, 2011). A useful value from actual columns where t_{ratio} is 0.0023.

$$t_s = t_{ratio} D_{Base} \tag{9.62}$$

The cost of the concrete is 200 €/m^3 (Guerras and Martín, 2020)

9.4.2.3 *Water consumption*

The water lost by the operation of a cooling tower includes of the evaporation loss, F_E, the drift loss, F_D, and the blowdown, F_B, Eq. (9.63):

$$F_M = F_E + F_D + F_B \tag{9.63}$$

Drift is expected to be negligible in newly designed towers. F_B losses is computed as a function of the F_E as in Eq. (9.64):

$$F_B = \frac{1}{COC - 1} F_E \tag{9.64}$$

COC is the ratio of the concentration of salts or dissolved solids in the circulating water/blowdown c_B (ppm) to that in the makeup water c_M (ppm). Industrial practice uses COC's from 3 to 7 (Perry and Green, 1997). Thus, the lost waster is computed by Eq. (9.65)

$$F_M = F_E + F_B \tag{9.65}$$

While the consumption of water can be computed using the models presented above, the fact that it depends on the weather conditions results in the need to run the model any time. Guerras and Martín (2020) presented simplified models to estimate the water consumption for Rankine, Eq. (9.66):

$$Water_{consumption}(L/kWh) = -2.297 \cdot 10^{-4} \cdot T^2 + 0.798 \cdot H^2 + 7.090 \cdot p^2 + 2.200 \cdot 10^{-2} \cdot T \cdot H$$
$$+ 2.993 \cdot 10^{-2} \cdot T \cdot p - 0.515 \cdot H \cdot p - 1.533 \cdot 10^{-2} T - 1.417 \cdot H - 12.574 \cdot p + 7.6256 \tag{9.66}$$

and Combined cycles, Eq. (9.67):

$$Water_{consumption}(L/kWh) = -4.75 \cdot 10^{-4} T^2 + 1.255 \cdot H^2 + 8.083 \cdot p^2 + 3.453 \cdot 10^{-3} \cdot T \cdot H$$
$$+ 5.833 \cdot 10^{-2} \cdot T \cdot p + 1.292 \cdot H \cdot p - 3.447 \cdot 10^{-2} T - 3.255 \cdot H - 16.555p + 9.690 \tag{9.67}$$

where T in (°C), p in (atm) and H in (fraction):
Example 9.3 shows the design of a wet cooling tower.

EXAMPLE 9.3 Design a wet cooling tower to cooldown the energy from Example 9.2. Assume $V = 1.8 \, V_{min}$ and $(-h_L/k_y)$ equal to -10. The atmospheric air is at 20 °C and 60% humidity, and leaves the column with 95% humidity. The cooling water enters at 20°C and leaves at 28 °C Assume $CT/D_{base} = 1.5$ The contraction of the tower is computed considering $h_c = 0.12 \cdot D_{base}$. Determine the packing height as well as the base diameter and the tower height required.

Solution

We first plot the Hi curve for the operating conditions, see rhomboids in Fig. 9E3.1. For convenience we fit it to a polyonomy, Eq. (9E3.1)

$$H_i(kcal/kg) = -5.99 \cdot 10^{-6} T(°C)^4 + 0.0005157 \cdot T(°C)^3 + 0.0017392 \cdot T(°C)^2 + 0.40925 \cdot T(°C) + 2.28 \tag{9E3.1}$$

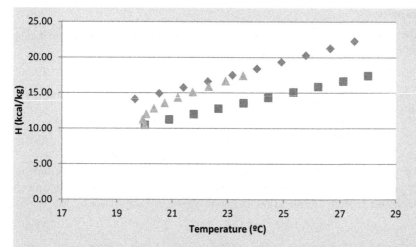

FIGURE 9E3.1 Profile of air temperature across the packing.

For $t_{L2} = 28°C$, we compute H_i to determine the L/V_{min} Eq. (9E3.2)
$H_i = 22.8$ kcal/kg

$$\frac{H_2 - H_1}{t_{L,2} - t_{L,1}} = \frac{f_{Wa}}{f_{air}} = \frac{L}{V} \tag{9E3.2}$$

Thus, $L/V_{min} = 1.55$. Since we will work with 1.8 times the V_{min}

$$L/V = 0.86$$

The energy balance from the Example 9.2 shows

$$Q = 14922 \text{ kW} = L \cdot cp(t_{L,2} - t_{L,1}) = > L = 446\text{kg/s and with } L/V \text{ we have } V = 515 \text{ kg/s} \tag{9E3.3}$$

We can now draw the operating line, squares in Fig. 9E3.1
To compute the NTU we solve Eq. (9E3.4):

$$NTU = \int_{H_1}^{H_2} \frac{dH}{H_i - H} \tag{9E3.4}$$

For each H in the operating line, with the slope of -10 as provided in the information, we compute H_i. Roboids in Fig. 9E3.1 and the profile of the gas temperature. For each as described in Fig. 9.10, see triangles in Fig. 9E3.1. We selected 9 discretization points.

The NTU is computed as follows. We plot 1/Hi-H versus H , see Fig. 9E3.2.
We fit that to a simple polynomial, Eq. (9E3.5):

$$1/Hi\text{-}H = -8.311\text{E-}04 \text{ H}^2 + 1.275\text{E-}02\text{H} + 2.345\text{E-}01 \tag{9E3.5}$$

and integrate it from H = 10.47 to H = 17.40 obtaining:

$$NTU = 1.7$$

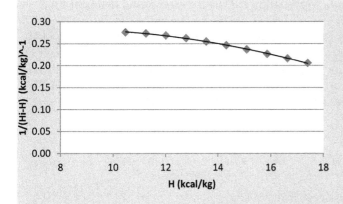

To compute HTU we need the area, Eq. (9E3.6):

$$\text{HTU} = \frac{f_{air}}{k_y A\, S} = \frac{V}{(2.95) \cdot (Gflux)^{0.72} \cdot (Lflux)^{0.26} \cdot S} = \frac{V}{(2.95) \cdot \left(\frac{V}{S}\right)^{0.72} \cdot \left(\frac{L}{S}\right)^{0.26} \cdot S} \qquad (9E3.6)$$

However, the S is related to the height of the tower to secure enough draft.

$$\text{Height} = 1.5 \cdot D \qquad (9E3.7)$$

For this we assume base diameter, since it will provide the lowest velocity and the worst mass transfer rate. Assuming $D_{base} = 24$ m, Height $= 36$ m; $v_{air} = 1.05$ m/s

$$\Delta P_{Loss} = \left(N_{Mist} + N_{Support} + N_{Contraction} + N_{Spray} + N_{Packing}\right) \rho_{air} \frac{v_{air}^2}{2} = 10.04 Pa$$

where:

$N_{Mist} = 2\,\text{m}$

$N_{Support} = 2\,\text{m}$

$$N_{Contraction} = 0.167 \left(\frac{D_{base}}{0.12 D_{base}}\right)^2 = 11.60\,\text{m} \qquad (9E3.8)$$

$$N_{Spray} = 0.16 \cdot (0.12 D_{base}) \cdot \left(\frac{f_{Wa,in}}{f_{air,in}}\right)^{1.32} = 0.38\,\text{m}$$

$$N_{Packing} = Height_{packing} \cdot 2.36 \left(\frac{L}{S \cdot 3.391}\right)^{1.1} \cdot \left(\frac{V}{S \cdot 3.391}\right)^{-0.64} = 0.74\,\text{m}$$

$$\Delta P_{Generated} = CT_{Height} \cdot g \cdot \left(\rho_{g,in} - \rho_{g,out}\right) = 36m \cdot 9.8 \frac{kg}{m \cdot s^2}(1.095 - 1.065)\frac{kg}{m^3} = 10.57 Pa \qquad (9E3.9)$$

Thus, the column is feasible and with this we compute HTU:

HTU $= 0.35$ m and Height$_{packing} = 0.6$ m

Example 9.4 shows a multiobjective formulation for the location of the heliostat field.

EXAMPLE 9.4 Select among the 11 allocations whose conditions are provided below considering both the energy production vs. the water consumption. Assume that the yield of the thermal cycle is 40%, 2000 Heliostats of 120 m² each will be allocated in each position, the cos θ is 0, the optical efficiency of the solar field is 0.6 while $U_L = 0.01$ W/m²K. The pressure in all allocations is 1 atm.

Solution

To compute water consumption per month we use Eq. (9.66)

$$Water_{consumption}(L/kWh) = -2.297 \cdot 10^{-4} \cdot T^2 + 0.798 \cdot H^2 + 7.090 \cdot p^2 + 2.200 \cdot 10^{-2} \cdot T \cdot H$$
$$+ 2.993 \cdot 10^{-2} \cdot T \cdot p - 0.515 \cdot H \cdot p - 1.533 \cdot 10^{-2}T - 1.417 \cdot H - 12.574 \cdot p + 7.6256 \tag{9E4.1}$$

We average the water consumption for each allocation.
To compute the power generated in each allocation monthly we use (9E4.2-4).

$$Q_{(collector)} = DNI_i \cdot A_{field} \cdot \eta_{field} \tag{9E4.2}$$

$$A_{field} = n_h \cdot A_h \cdot \eta_h \tag{9E4.3}$$

$$\eta_{field} = \eta_{optical,field} - U_L \left(\frac{T_m - T_a}{DNI \cdot \cos\theta} \right) \tag{9E4.4}$$

It is a multiobjective optimization problem where each objective is different. We develop the following objective function, Eq. (9E4.5).

$$Obj = \frac{Power_i}{Power_{Max}} + \left(1 - \frac{Water_i - Water_{min}}{Water_{min}} \right) \tag{9E4.5}$$

So that each objective is normalized, by using this, we select location 11 as the most promising considering both water consumption and power production, see Tables 9E4.1 and 9E4.2 for the data and the results respectively.

TABLE 9E4.1 Data for the 11 allocations.

°C	m/place	1	2	3	4	5	6	7	8	9	10	11
	1	12.6	12.7	9.3	6.5	11	8.6	12.1	10.9	10.6	7.8	8.6
	2	13.3	13.8	11.1	8.5	12.4	10.3	12.9	12.5	12.2	9.3	10.3
	3	15.1	15.5	14.4	11.4	14.7	13.1	14.7	15.6	14.3	12.2	13.3
	4	17	16.8	16	13.3	16.1	14.5	16.3	17.3	16.5	13.8	15.1
	5	19.7	19.1	20	17.2	19.2	18.2	19.3	20.7	20	17.6	18.7
	6	23.5	22.4	24.7	22.3	22.8	23.7	23	25.1	24.2	22.9	23.4
	7	26.1	24.6	28	25.3	25.8	27.6	25.5	28.2	27.2	26.2	26.1
	8	26.7	25	28	24.8	25.8	26.9	26	27.9	27.6	26	25.9
	9	24.2	23.3	24.2	21.1	23.4	22.8	23.5	25	24.2	22.4	22.9
	10	20.4	20.3	19.1	16	19.5	17.9	19.5	20.2	19.8	17	17.8
	11	16.4	16.5	13.5	10.6	14.9	12.3	15.7	15.1	14.6	11.7	12.7
	12	13.8	13.9	10.4	7.6	12.3	9.5	13.2	11.9	11.5	8.7	9.7

(Continued)

TABLE 9E4.1 (Continued)

H	m/place	1	2	3	4	5	6	7	8	9	10	11
	1	67	75	76	72	77	70	69	71	65	79	79
	2	67	74	71	67	74	65	68	67	63	73	74
	3	65	71	64	59	68	59	67	59	59	63	65
	4	62	69	60	57	65	58	63	57	53	60	64
	5	63	70	55	51	62	55	59	53	52	55	58
	6	61	69	48	44	57	55	58	48	49	44	52
	7	60	68	41	38	51	40	58	44	50	37	48
	8	63	70	43	42	55	45	61	48	54	39	49
	9	65	71	52	52	61	54	65	54	59	49	56
	10	68	74	66	64	69	64	70	62	64	65	68
	11	67	74	73	72	73	70	71	70	65	76	76
	12	67	76	79	76	78	72	72	74	68	80	82
kW/m^2	m/place	1	2	3	4	5	6	7	8	9	10	11
	1	0.119773	0.105361	0.100209	0.116786	0.099425	0.103495	0.108684	0.104241	0.114583	0.09875	0.101667
	2	0.158317	0.147363	0.1417	0.150174	0.142898	0.14141	0.143932	0.145213	0.150417	0.140417	0.139167
	3	0.204674	0.189031	0.189927	0.191532	0.191607	0.189292	0.18582	0.193772	0.20375	0.199167	0.200417
	4	0.257793	0.243403	0.241397	0.241782	0.243326	0.234838	0.234375	0.244174	0.26125	0.244167	0.24625
	5	0.286626	0.286365	0.279906	0.276882	0.283901	0.274978	0.285096	0.286813	0.29125	0.282917	0.2775
	6	0.321065	0.309992	0.319715	0.319946	0.31875	0.318094	0.315664	0.31821	0.328333	0.332917	0.3375
	7	0.316756	0.309214	0.318362	0.318623	0.318847	0.317578	0.314553	0.317242	0.33	0.33875	0.339167
	8	0.282221	0.278786	0.284013	0.285917	0.287186	0.281847	0.278524	0.282519	0.285417	0.296667	0.299167
	9	0.227508	0.221914	0.221373	0.223688	0.219367	0.219059	0.21848	0.22037	0.233333	0.232917	0.235833
	10	0.161664	0.155839	0.153823	0.156437	0.153898	0.150687	0.149791	0.154533	0.166667	0.15375	0.154583
	11	0.118711	0.113889	0.108912	0.116127	0.114043	0.105941	0.109259	0.113542	0.120833	0.108333	0.110833
	12	0.100545	0.088822	0.083595	0.096438	0.084901	0.084565	0.085312	0.087776	0.059583	0.08125	0.079583

TABLE 9E4.2 Summary of the results comparison.

	1	2	3	4	5	6	7	8	9	10	11
Water (L/kWh)	1.685	1.648	1.682	1.643	1.659	1.670	1.671	1.712	1.705	1.649	1.640
Obj1	0.972	0.995	0.974	0.998	0.989	0.982	0.981	0.956	0.960	0.994	1
Power (kW)	10582	10125	10094	10313	10159	10001	10036	10204	10537	10382	10433
Obj2	1	0.957	0.954	0.975	0.960	0.945	0.948	0.964	0.996	0.981	0.986
obj	1.972	1.952	1.928	1.973	1.949	1.927	1.930	1.920	1.956	1.975	**1.986**

4. Solar technologies

9.4.3 Dry cooling: Air cooling

This section describes the operation of A-frames. However, if a natural draft dry cooling tower is to be designed, the heat transfer section can be designed similarly to an A-frame where the A structure of the pipes becomes flat and the air driving force is generated by the structure of the paraboloid instead of the fans.

9.4.3.1 Mathematical model

The flow of air required to cooldown/condense the steam is computed by an energy balance. Note that the density of the air depends on the pressure and temperature and therefore the weather of the location and the season plays an important role. In addition, the air exiting temperature cannot surpass that of the condensing steam, Eqs. (9.68)–(9.69).

$$m_{air} \cdot C_{p,air} \cdot \left(T_{out,month} - T_{amb,month}\right) = Q_{Cooling} \tag{9.68}$$

$$f_{air} \cdot \rho_{air} = f_{air} \cdot \left(\frac{P_{atm} \cdot M_{W,air}}{R_{gases} \cdot T_{amb,month}}\right) = m_{air} \tag{9.69}$$

A fan can handle a certain maximum flow, over that, a number of fans must be installed. The fan must provide the energy to overcome the pressure drop across the structure. For an A-frame structure the total pressure drop includes the static pressure, the pressure drop across the support, the obstacles before and after the fan, and that across the bundle of pipes, Eq. (9.70):

$$\Delta p_e = - \left[\frac{K_{ts}}{2 \cdot \rho_{a56}} \left(\frac{m_{air}}{N_b \cdot A_{fr}}\right)^2 + \frac{K_{up}}{2 \cdot \rho_{a3}} \left(\frac{m_{air}}{A_e}\right)^2 - \Delta p_{Fs} + \frac{K_{do}}{2 \cdot \rho_{a3}} \left(\frac{m_{air}}{A_e}\right)^2 + \frac{K_{\theta t}}{2 \cdot \rho_{a56}} \left(\frac{m_{air}}{N_b \cdot A_{fr}}\right)^2 \right] \tag{9.70}$$

- **The fan static pressure**, Δp_{Fs}, is a function of the flow rate and the blade angle
- **Coefficient** K_{ts} represents the pressure drop across fan's support platform
- **Coefficient** K_{up} corresponds to the pressure drop across the obstacles before the fan, such as a protector screen. The value of K_{up} is presented in the literature in the form of figures (Kröger, 2004).
- **Coefficient** K_{do} corresponds to the pressure drop across the obstacles after the fan, such as the shaft. The value of K_{do} can be read in the figures presented by Kröger (2004)
- **Coefficient** $K_{\theta t}$ represents the total pressure drop across the heat exchanger bundle and includes the kinetic energy losses across the heat exchanger.

Further details can be seen in Kröger (2004) and Luceño and Martín (2018). In a simplified way it can be computed using the results in (Heyns, 2008), Eq. (9.71). We assume an efficiency, η, of 90% based on the same study:

$$\text{Power}_{\text{fan}}(kW) = 0.001 \cdot n_{fans}\left(\rho_{air}/1.2\right) \cdot \left(0.4762 \cdot \left(f_{air,fan}\right)^2 - 59.414 \cdot f_{air,fan} + 186644\right)\frac{1}{\eta} \tag{9.71}$$

Where f_{air} is the air flow (m^3/s). Note that the fans must physically accommodate within the structure, Eq. (9.72):

$$n_{fans} \cdot 0.25 \cdot \pi \cdot (D_{fan})^2 = 0.25 \cdot tube\_L \cdot Width \cdot n_{bundles} \tag{9.72}$$

In order to compute the contact area that the heat exchanger provides, the design parameters as presented in Manassaldi et al. (2014) are used, Eqs. (9.73)−(9.75):

$$A_{of} = 2\frac{1}{4}\pi \cdot fin_N \cdot \left(tube_{Df}^2 - tube_D^2\right) + \pi \cdot tube_{Df} \cdot fin_{tf} \cdot fin_N \tag{9.73}$$

$$A_{ot} = A_{of} + \pi \cdot tube_D \cdot (tube_L - fin_{tf} \cdot fin_N) \tag{9.74}$$

$$A_{heat,exchanger} = tubes_{Nrow} \cdot tubes_N \cdot A_{ot} \cdot streets \cdot units \cdot N_{bundles} \tag{9.75}$$

The global heat transfer coefficient has three terms: Tube side, the wall and the air side. However, the main resistance to the heat transfer is the film coefficient of the air side (Heyns, 2008; Luceño and Martín, 2018). We approximate the global heat transfer coefficient U_{global} by h_{air} using the correlation presented by Pieve and Salvadori (2011), Eq. (9.76)

$$U_{global} \approx \left(\frac{k_{air}}{tube_D}\right) \cdot 0.134 \cdot (Re_{air})^{0.681} \cdot (Pr_{air})^{0.33} \cdot \left(\frac{(fin_{pf} - fin_{tf})}{(0.5 \cdot (tube_{Df} - tube_D))}\right)^{0.2} \cdot \left(\frac{(fin_{pf} - fin_{tf})}{(fin_{tf})}\right)^{0.11} \tag{9.76}$$

where

$$Re_{air} = \frac{f_{air} \cdot tube_{Deq}}{\mu_{air} \cdot Amin}; \tag{9.77}$$

$$tube_{Deq} = \frac{\left(\frac{(tube_{Df}^2 - tube_D^2)}{(2 \cdot fin_{pf})} + \frac{tube_{Df} \cdot fin_{tf}}{fin_{pf}} + tube_D \cdot \left(1 - fin_{tf}/fin_{pf}\right)\right)}{\left(1 + \frac{(tube_{Df} - tube_D)}{fin_{pf}}\right)}; \tag{9.78}$$

$$Amin = N_{bundles} \cdot streets \cdot units \cdot tubes_N \cdot tube_L$$
$$\cdot (Pt - tube_D - 2 \cdot fin_{tf} \cdot 0.5 \cdot (tube_{Df} - tube_D) \cdot fin_N) \cdot Sin\theta \tag{9.79}$$

$$Pr_{air} = \frac{C_{p,air} \cdot \mu_{air}}{k_{air}}; \tag{9.80}$$

$$Q_{cooling} = U_{global} \cdot LMTD \cdot A \tag{9.81}$$

Example 9.5 shows the design of an A-frame.

4. Solar technologies

EXAMPLE 9.5 According to Example 9.1 facility, design an A-Frame cooling system for cover the cooling requirements of 31,840 kW from a water steam at 332 K. Consider the drop pressure and power equations:

$$\Delta p_{Fs}(Pa) = 0.6351 \cdot 10^{-3} \cdot f_{air,fan}^2 - 0.2975 \cdot f_{air,fan} + 320.045$$

$$\Delta p_e(Pa) = - \begin{bmatrix} 5.02127 \cdot 10^{-8}(m_{air})^2 + 5.67627 \cdot 10^{-7}(m_{air})^2 - \Delta p_{Fs} \\ + 6.7749 \cdot 10^{-7}(m_{air})^2 + 6.64579 \cdot 10^{-6}(m_{air})^2 \end{bmatrix}$$

$$P_{fan}(W) = 0.2057 \cdot f_{air,fan}^2 - 25.667 \cdot f_{air,fan} + 80,630.21 \left(\text{obtained at } \rho_{air} = 1.2 \text{ kg/m}^3\right)$$

Additional data:

$\rho_{air} = \rho_{a56} = \rho_{a3} = 1.25 \text{ kg/m}^3$; $\eta = 0.90$; $T_{amb,month} = 295 \text{ K}$; $T_{out,month} = 310 \text{ K}$; $T_v = 332 \text{ K}$; $C_{p,air} = 1.0035 \text{ kJ/kgK}$; $\mu_{air} = 0.000018 \text{ kg/m} \cdot s$; $k_{air} = 0.024 \text{ J/m} \cdot K \cdot s$; $Pt = 0.15 \text{ m}$; $D_{fan} = 9 \text{ m}$; $units = n_{fans}$; $street = 1$; $n_{bundle,fan} = 4$; $n_{bundle} = n_{bundle,fan} \cdot n_{fan}$; $tubes_N = 60$; $tube_{Df} \approx tube_D + 0.01$; Width $= 9.0 \text{ m}$; $tubes_{Nrow} = 3$; $Pr_{air} = 0.72$; $2 \cdot \theta = 60°$; $fin_{pt} = 0.00117 \text{ m}$; $fin_{tf} = 0.000375 \text{ m}$; $fin_N = 6,038$;

Solution

First, we must determine m_{air} and f_{air} with an energy balance:

$$m_{air} \cdot C_{p,air} \cdot (T_{out,month} - T_{amb,month}) = Q_{Cooling} \tag{9E5.1}$$

$$m_{air} = \frac{Q_{Cooling}}{C_{p,air} \cdot (T_{out,month} - T_{amb,month})} = 2,115.263 \text{ kg/s}$$

$$f_{air} \cdot \rho_{air} = m_{air} \tag{9E5.2}$$

$$f_{air} = \frac{m_{air}}{\rho_{air}} = 1,692.211 \text{ m}^3/s$$

Once the mass flow is known, we can compute the power consumption of fans:

$$\Delta p_e \cdot f_{air} = n_{fan} \cdot (\rho_{air}/1.2) \cdot P_{fan} \cdot \frac{1}{\eta}$$

$$\Delta p_e \cdot f_{air,fan} = (\rho_{air}/1.2) \cdot P_{fan} \cdot \frac{1}{\eta} \tag{9E5.3}$$

The variable Δp_e should be calculated considering the Eqs. 9E5.4–9E5.5:

$$\Delta p_e = - \begin{bmatrix} 5.02127 \cdot 10^{-8}(m_{air})^2 + 5.67627 \cdot 10^{-7}(m_{air})^2 - \Delta p_{Fs} \\ + 6.7749 \cdot 10^{-7}(m_{air})^2 + 6.64579 \cdot 10^{-6}(m_{air})^2 \end{bmatrix} \tag{9E5.4}$$

$$\Delta p_{Fs} = 0.6351 \cdot 10^{-3} \cdot f_{air,fan}^2 - 0.2975 \cdot f_{air,fan} + 320.045 \tag{9E5.5}$$

Note that 9.E5.4 is given using air mass flow (kg/s), and 9.E5.5 consider volumetric air flow (m³/s).

On the other hand, the power supplied by a fan is computed by Eq. (9E5.6):

$$P_{fan}(W) = 0.2057 \cdot f_{air,fan}^2 - 25.667 \cdot f_{air,fan} + 80,630.21 \qquad (9E5.6)$$

Substituting 9.E5.4–9.E5.6 into 9.E5.3:

$$-\begin{bmatrix} 5.02127 \cdot 10^{-8}(m_{air})^2 + 5.67627 \cdot 10^{-7}(m_{air})^2 \\ - \left(0.6351 \cdot 10^{-3} \cdot f_{air,fan}^2 - 0.2975 \cdot f_{air,fan} + 320.045\right) \\ + 6.7749 \cdot 10^{-7}(m_{air})^2 + 6.64579 \cdot 10^{-6}(m_{air})^2 \end{bmatrix}$$

$$\cdot f_{air,fan} = \left(\rho_{air}/1.2\right) \cdot \left(0.2057 \cdot f_{air,fan}^2 - 25.667 \cdot f_{air,fan} + 80,630.21\right) \cdot \frac{1}{\eta}$$

At this point, all variables are known except $f_{air,fan}$, so substituting the values in the previous expression we can compute $f_{air,fan}$ and also n_{fan} as:

$$f_{air,fan} = 461.29 \text{ m}^3/\text{s}; n_{air,fan} = \frac{f_{air}}{f_{air,fan}} = 3.67 \approx 4$$

Once the numbers of fans is known, tube_L can be calculated according to the footprint of the equipment (9.E5.7):

$$n_{fans} \cdot 0.25 \cdot \pi \cdot (D_{fan})^2 = 0.25 \cdot \text{tube\_L} \cdot \text{Width} \cdot n_{bundles} \qquad (9E5.7)$$

$$\text{tube\_L} = \frac{n_{fans} \cdot 0.25 \cdot \pi \cdot (D_{fan})^2}{0.25 \cdot \text{Width} \cdot n_{bundles}} = \frac{4 \cdot 0.25 \cdot \pi \cdot (9)^2}{0.25 \cdot 9.0 \cdot 4 \cdot 4} = 7.07 \text{ m}$$

In order to determine the tube diameter, an energy balance is formulated, employing the design area A and the global heat transfer coefficient U_{global}:

$$Q_{cooling} = U_{global} \cdot LMTD \cdot A \qquad (9E5.8)$$

According to the values of $T_{amb,month}$, $T_{out,month}$ and T_v, it is trivial to calculate LMTD:

$$LMTD = 28.85 \text{ K}$$

The global coefficient U_{global} evolves mechanical design variables, as seen in 9.E5.9:

$$U_{global} \approx \left(\frac{k_{air}}{\text{tube}_D}\right) \cdot 0.134 \cdot (Re_{air})^{0.681} \cdot (Pr_{air})^{0.33} \cdot \left(\frac{(fin_{pf} - fin_{tf})}{(0.5 \cdot (tube_{Df} - tube_D))}\right)^{0.2} \cdot \left(\frac{(fin_{pf} - fin_{tf})}{(fin_{tf})}\right)^{0.11}$$

$$(9E5.9)$$

Substituting the values of the example:

$$U_{global} = \left(\frac{0.024}{\text{tube}_D}\right) \cdot 0.134 \cdot (Re_{air})^{0.681} \cdot (0.72)^{0.33} \cdot \left(\frac{(0.00117 - 0.000375)}{(0.5 \cdot (tube_D + 0.01 - tube_D))}\right)^{0.2} \cdot \left(\frac{(0.00117 - 0.000375)}{(0.000375)}\right)^{0.11}$$

$$U_{global} = \left(\frac{0.024}{\text{tube}_D}\right) \cdot 0.134 \cdot (Re_{air})^{0.681} \cdot (0.72)^{0.33} \cdot (0.159)^{0.2} \cdot (2.12)^{0.11}$$

Reynolds number is computed as Eqs. 9E5.10 and 9E5.11:

$$Re_{air} = \frac{f_{air} \cdot tube_{Deq}}{\mu_{air} \cdot Amin} \tag{9E5.10}$$

$$tube_{Deq} = \frac{\left(\frac{(tube_{Df}^2 - tube_D^2)}{(2 \cdot fin_{pf})} + \frac{tube_{Df} \cdot fin_{tf}}{fin_{pf}} + tube_D \cdot \left(1 - fin_{tf}/fin_{pf}\right)\right)}{\left(1 + \frac{(tube_{Df} - tube_D)}{fin_{pf}}\right)} \tag{9E5.11}$$

but $tube_{Deq}$ can be simplified using the values:

$$tube_{Deq} = \frac{\left(\frac{((tube_D + 0.01)^2 - tube_D^2)}{(2 \cdot 0.00117)} + \frac{(tube_D + 0.01) \cdot 0.000375}{0.00117} + tube_D \cdot \left(1 - \frac{0.000375}{0.00117}\right)\right)}{\left(1 + \frac{(0.01 + tube_D - tube_D)}{0.00117}\right)}$$

$$tube_{Deq} = \frac{\left(\frac{0.0001 + 0.02 \cdot tube_D}{0.00234} + (0.32 \cdot tube_D + 0.0032) + 0.6795 \cdot tube_D\right)}{9.547}$$

$$tube_{Deq} = \frac{0.04274 + 8.547 \cdot tube_D + (0.32 \cdot tube_D + 0.0032) + 0.6795 \cdot tube_D}{9.547} = tube_D + 0.00481$$

And Amin can be calculated as (9.E5.12):

$$Amin = N_{bundles} \cdot streets \cdot units \cdot tubes_N \cdot tube_L \cdot (Pt - tube_D - 2 \cdot fin_{tf} \cdot 0.5 \cdot (tube_{Df} - tube_D) \cdot fin_N) \cdot Sin\theta \tag{9E5.12}$$

$$Amin = 4 \cdot 1 \cdot 4 \cdot 60 \cdot 7.07 \cdot (0.15 - tube_D - 2 \cdot 0.000375 \cdot 0.5 \cdot (0.01) \cdot 6,038) \cdot 0.5$$
$$Amin = 3,393.6 \cdot (0.1274 - tube_D) = 432.345 - 3,393.6 \cdot tube_D$$

Thus, the final Re_{air} expression obtained is:

$$Re_{air} = \frac{1,692.211 \cdot (tube_D + 0.00481)}{0.000018 \cdot (432.345 - 3,393.6 \cdot tube_D)}$$

Finally, introducing the obtained expression of each term, the new U_{global} equation becomes:

$$U_{global} = \left(\frac{0.024}{tube_D}\right) \cdot 0.134 \cdot \left(\frac{1,692.211 \cdot (tube_D + 0.00481)}{0.000018 \cdot (432.345 - 3,393.6 \cdot tube_D)}\right)^{0.681} \cdot (0.72)^{0.33} \cdot (0.159)^{0.2} \cdot (2.12)^{0.11}$$

$$U_{global} = \left(\frac{0.024}{tube_D}\right) \cdot \left(\frac{1,692.211 \cdot (tube_D + 0.00481)}{0.000018 \cdot (432.345 - 3,393.6 \cdot tube_D)}\right)^{0.681} \cdot 0.090$$

The area A is computed considering the tubes of all bundles and its fins (9.E5.13–15):

$$A_{of} = 2\frac{1}{4}\pi \cdot fin_N \cdot \left(tube_{Df}^2 - tube_D^2\right) + \pi \cdot tube_{Df} \cdot fin_{tf} \cdot fin_N;$$

$$A_{of} = 2\frac{1}{4}\pi \cdot 6{,}038 \cdot \left(tube_D^2 + 0.0001 + 0.02 \cdot tube_D - tube_D^2\right) + \pi \cdot (tube_D + 0.01) \cdot 0.000375 \cdot 6{,}038;$$

$$A_{of} = \frac{\pi}{2} \cdot 6{,}038 \cdot (0.0001 + 0.02 \cdot tube_D) + 2.26425\pi \cdot (tube_D + 0.01);$$

$$A_{of} = 0.3245\pi + 62.6443\pi \cdot tube_D$$

$$\tag{9E5.13}$$

$$A_{ot} = A_{of} + \pi \cdot tube_D \cdot (tube_L - fin_{tf} \cdot fin_N);$$
$$A_{ot} = 0.3245\pi + 62.6443\pi \cdot tube_D + \pi \cdot tube_D \cdot (7.07 - 0.000375 \cdot 6{,}038)$$
$$A_{ot} = 0.3245\pi + 67.4501\pi \cdot tube_D \tag{9E5.14}$$

$$A_{heat,exchanger} = tubes_{Nrow} \cdot tubes_N \cdot A_{ot} \cdot streets \cdot units \cdot N_{bundles}$$
$$A_{heat,exchanger} = 3 \cdot 60 \cdot (0.3245\pi + 67.4501\pi \cdot tube_D) \cdot 1 \cdot 4 \cdot 4 \tag{9E5.15}$$
$$A_{heat,exchanger} = 934.56\pi + 194{,}256.288\pi \cdot tube_D$$

Coming back to $Q_{cooling}$, the final expression for the energy balance is:

$$Q_{cooling} = \left(\frac{0.024}{tube_D}\right) \cdot \left(\frac{1{,}692.211 \cdot (tube_D + 0.00481)}{0.000018 \cdot (432.345 - 3{,}393.6 \cdot tube_D)}\right)^{0.681} \cdot 0.090 \cdot (28.85) \cdot (934.56\pi + 194{,}256.288\pi \cdot tube_D)$$

As $tube_D$ is the only unknown variable, due to $Q_{cooling} = 31{,}840{,}000\,W$, its value can be obtained:

$$tube_D = 0.0453\,m$$

9.5 Economics of the renewable electricity

In electrical systems, an interesting variable that is used to evaluate the cost of the production of electricity is the Levelized cost of electricity (LCOE). It can be defined as the minimum price of electricity which can generate enough revenues to pay back the CAPEX, cover the OPEX and generates enough cash for the decommission of the facility, the break-even point of the facility, it can be computed as.

$$LCOE = \frac{\displaystyle\sum_{t=1}^{n} \frac{CAPEX_t + OPEX_t}{(1+r)^t}}{\displaystyle\sum_{t=1}^{n} \frac{Electricity_t}{(1+r)^t}} \tag{9.82}$$

Where "r" is the discount rate and "n" the lifetime of the facility.

Exercises

P9.1 For the cycle presented in Example 9.1, evaluate the performance for January when the DNI is 650 W/m^2

P9.2 Evaluate the production of electricity from a solar tower CSP plant allocated in the South of Spain operates a Brayton cycle. It consists of a three stage compression system with intercooling. The yield, compressor and expansion performance, assumed polytropic, are found below

Yield_field: 0.6; Solar radiation kW per m^2 /0.65/; polytropic efficiency 0.80; polytropic coefficient compression: 1.4; expansion polytropic coefficient: 1.3; Heliostat efficiency: 0.9; Area heliostat (m^2) /120; number of heliostats: 2000;

DNI of 500; 550; 600; 700, 800; 900; 900; 825; 725; 650; 550, 500 W/m^2,

P9.3 Compute the cooling that a tower of 100 m high and 67 m of diameter that operates with a with V = 1.8 V min. Assume ($-h_L/k_y$) equal to -10. The air is at 20°C ad 60% humidity and leaves the column with 95% humidity. The cooling water enters at 20°C and leaves at 28°C The contraction of the tower is computed considering $h_c = 0.10 \cdot D_{base}$.

P9.4 Formulate the optimization problem for the selection of the allocation considering water and energy consumption.

P9.5 For the cycle presented in Example 9.1, determine the yield of the solar field if the cooling that is being removed from the system is 30 MW

P9.6 Compare the performance of the cycle in Example 9.1 with a cycle that includes reheating as in Fig. 9P6.1. Consider that the reheating stage heats stream 4 up to 555 °C using a fraction of the flow of HTF generated at the solar field.

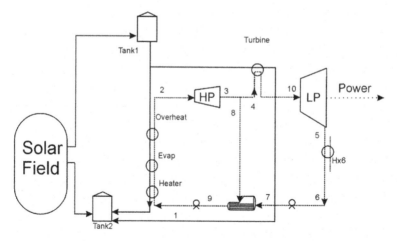

FIGURE 9P6.1 Scheme of the cycle.

References

Almási, J., 1981. Approximate determination of cooling tower dimensions. Period. Polytech. Civ. Eng. 25 (1–2), 95–110.

Brannan, C.R., 2005. Rules of Thumb for the Chemical Engineer, 4th Edition Elsevier, Oxford.

Coulson, J., Richardson, J., 1999. Chemical engineering. Butterworth Heinemann, UK.

Desai, N.B., Kedare, S.B., Bandyopadhyay, S., 2014. Optimization of design radiation for concentrating solar thermal power plants without storage. Sol. Energy 107, 98–112.

El-Gharbi, N., Derbal, H., Bouaichaoui, S., Said, N., 2011. A comparative study between parabolic trough collector and linear Fresnel reflector technologies. Energy Procedia 6, 565–572.

EPA, 1970. A method for predicting the performance of natural draft cooling towers. Pacific Northwestern Water Laboratory.

Erhart, T.; Eicker, U.; Infield, D. (2011). Part-load characteristics of Organic-Rankine-Cycles. In Proceedings of the 2nd European Conference on Polygeneration, Tarragona, Spain, 30 March–1 April 2011; pp. 1–11.

García, S., Martín, M., 2021. Analysis of the performance of CSP facilities using different thermal fluids. Chem. Eng. Res. Des. 168, 46–58.

Geankoplis, C.J., 1993. Transport Processes and Unit Operations, 3rd Edition Prentice Hall, Upper Saddle River, New Jersey. U.S.A..

General Electric, 2001. Next Generation Gas Turbine (NGGT) Systems Study. DOE Contract No. DE-AC26-00NT40846 .

Gould, P.L., Kratzig, W.B., 1999. Cooling tower structures. In: Wai-Fah, C. (Ed.), Structural engineering handbook. CRC Press, Boca Raton.

Guerras, L.S., Martín, M., 2020. On the water footprint in power production: Sustainable design of wet cooling towers. Appl. Energy 263, 114620.

Günther, M., Joemann, M., Csambor, S., 2011. Advanced CSP Teaching Materials Chapter 5 Parabolic Trough Technology. EnerMENA. Available from: http://edge.rit.edu/edge/P15484/public/Detailed%20Design%20 Documents/Solar%20Trough%20Preliminary%20analysis%20references/Parabolic%20Trough%20Technology.pdf.

Heyns, J.A. (2008) Performance characteristics of an air-cooled steam condenser incorporating a hybrid (dry/wet) dephlegmator MEng Thesis. Department of Mechanical Engineering University of Stellenbosch.

International Energy Agency, IEA (2010) Technology Roadmap Concentrating Solar Power France. Paris.

Irena (2012) Concentrating solar power. RENEWABLE ENERGY TECHNOLOGIES: COST ANALYSIS SERIES Volume 1: Power Sector Issue 2/5.

Jebamalai, J.S.M. (2016) Receiver Design Methodology for Solar Tower Power Plants. M.Sc. Thesis, KTH School of Industrial Engineering and Management.

Jüdes, M., Vigerske, S., Tsatsaronis, G., 2009. Optimization of the Design and Partial-Load Operation of Power Plants Using Mixed-Integer Nonlinear Programming. In: Kallrath, J., Pardalos, P.M., Rebennack, S., Scheidt, M. (Eds.), Optimization in the Energy Industry. Energy Systems. Springer, Berlin, Heidelberg, pp. 193–220.

Kelly B. (2006a) Nexant Parabolic Trough Solar Power Plant Systems Analysis Task 1: Preferred Plant Size January 20, 2005—December 31, 2005 Subcontract Report NREL/SR-550-40162 July 2006.

Kelly B. (2006b) Nexant Parabolic Trough Solar Power Plant Systems Analysis Task 2: Comparison of Wet and Dry Rankine Cycle Heat Rejection January 20, 2005—December 31, 2005 Subcontract Report NREL/SR-550-40163 July 2006.

Kelly B. (2006c) Nexant Parabolic Trough Solar Power Plant Systems Analysis Task 3: Multiple Plants at a Common Location January 20, 2005—December 31, 2005 Subcontract Report NREL/SR-550-40164 July 2006.

Kreith, F., Goswami, D.Y., 2005. The CRC Handbook of Mechanical Engineering, 2nd Edition CRC Press, Boca Ratón.

Kröger, D.G., 2004. Air-cooled heat exchangers and cooling towers: thermal-flow performance evaluation and design, Volume II. Pennwell, Oklahoma, Tulsa.

Kuravi, S., Trahan, J., Goswami, D.Y., Rahman, M.M., Stefanakos, E.K., 2013. Thermal energy storage technologies and systems for concentrating solar power plants. Prog. Energ. Comb. Sci. 39 (4), 285–319.

Lang, C., Straub, J. (2011) Natural Draft Cooling Tower Design and Construction in Germany - Past (since 1965), Present and Future http://www.coolingtower-analysis.com/mediapool/135/1355294/data/Publikationen/ Lang_SEWC2011.pdf.

Lipps, F.W., Vant-Hull, L.L. (1980). Programmer's manual for the University of Huston computer code RCELL: cellwise optimization of the solar central receiver project, p. 140.

Luceño, J.A., Martín, M., 2018. Two-step optimization procedure for the conceptual design of A-frame systems for solar power plants. Energy 165, 483–500.

Manassaldi, J., Scenna, N.J., Mussati, S.F., 2014. Optimization mathematical model for the detailed design of air cooled heat exchangers. Energy 64, 734–746.

Martin, M., 2015. Optimal annual operation of the dry cooling system of a Concentrated Solar Energy Plant in the South of Spain. Energy 84, 774–782.

Martín, L., Martín, M., 2013. Optimal year-round operation of a Concentrated Solar Energy Plant in the South of Europe App. Therm. Eng. 59, 627–633.

Martin, M., Martín, M., 2017. Cooling limitations in power plants: Optimal multiperiod design of natural draft cooling towers. Energy 135, 625–636.

Mohamed, M.H., El-Sayed, A.Z., Megalla, K.F., Elattar, H.F., 2019. Modelling and performance study of a -Modeling and performance study of a parabolic trough solar power plant using molten salt storage tank in Egypt: effects of plant site location. Energy Syst. 10, 1043–1070.

Montes, M.J., Abánades, A., Martínez-Val, J.M., Valdés, M., 2009. Solar multiple optimization for a solar-only thermal power plant, using oil as heat transfer fluid in the parabolic trough collectors. Sol. Energy 83 (12), 2165–2176.

Moore R.,Vernon M., Ho C.K., Siegel N.P., Kolb G.J. Design Considerations for Concentrating Solar Power Tower Systems Employing Molten Salt. (2010) SANDIA REPORT SAND2010-6978. Albuquerque, U.S.A.

Moran, M.J., Shapiro, H.N., Boettner, D.D., Bailey, M.B., 2014. Fundamentals of engineering thermodynamics, 8th Edition Wiley, New York.

Nezammahalleh, H., Farhadi, F., Tanhaemami, M., 2010. Conceptual design and techno-economic assessment of integrated solar combined cycle system with DSG technology. Solar Energy 84, 1696–1705.

Noone, C.J., Torrilhon, M., Mitsos, A., 2012. Heliostat field optimization: a new computationally efficient model and biomimetic layout. Sol. Energy 86 (2), 792–803.

NREL (2010) Concentrating Solar Power Projects www.nrel.gov/csp/solarpaces/project_detail.cfm/projectID = 40 (last accessed January 2013).

NREL, 2019. System advisor model. SAM. https://sam.nrel.gov/node/68897. NREL.

Ong, S., Campbell, C., Denholm, P., Margolis, R., Heath, G. (2013) Land-Use Requirements for Solar Power Plants in the United States. Technical Report NREL/TP-6A20-56290 June 2013.

Patnode, A.M. (2006). Simulation and Performance Evaluation of Parabolic Trough Solar Power Plants. Ph.D. Thesis, University of Wisconsin-Madison, USA.

Perry, R.H., Green, D.W., 1997. Perry's chemical Engineer's handbook. McGraw-Hill, New York. U.S.A.

Pieve, M., Salvadori, G., 2011. Performance of an air-cooled steam condenser for a waste-to-energy plant over its whole operating range. Energy Conver. Manage 52, 1908–1913.

Sattler, J., Hoffchmidt, B., Günther, M., Joemann, M., 2011. Chapter 9 Thermal Energy storage. Advanced CSP Teaching Materials. ENERMENA.

Siala, F., Elayeb, M., 2001. Mathematical formulation of a graphical method for a noblocking heliostat field layout. Renew. Energy 23 (1), 77–92.

Walas, S.M., 1990. Chemical process equipment. Selection and Design. Butherworth Heinemann, Newton MA.

Wei, X., Lu, Z., Wang, Z., Yu, W., Zhagn, H., Yao, Z., 2010. A new method for the design of the heliostat field layout for solar tower power plant. Renew. Energy 35 (9), 1970–1975.

Zhou E., Xu, K., Wang, C. (2019) Analysis of the Cost and Value of Concentrating Solar Power in China Technical Report NREL/TP-6A20-74303 October 2019.

Zhuang, X., Xu, X., Liu, W., Xu, W., 2019. LCOE Analysis of Tower Concentrating Solar Power Plants Using Different Molten-Salts for Thermal Energy Storage in China. Energies 9 (12), 1394. Available from: https://doi.org/10.3390/en12071394.

10

Photovoltaic solar energy

César Ramírez-Márquez[1] and Mariano Martín[2]

[1]Universidad de Guanajuanto, División de Ciencias Exactas y Naturales, Guanajuato, Mexico
[2]Department of Chemical Engineering, University of Salamanca, Salamanca, Spain

10.1 Silicon photovoltaic

10.1.1 Introduction

Photovoltaic silicon offers an unmatched potential as a substrate and/or component to a number of devices that currently epitomize modern lifestyle. Photovoltaic silicon has also become one of the leading utilities that aid in the development of a cleaner and more sustainable society.

Photochemical reaction, solar thermal energy, and solar photovoltaic energy are several of the different routes where solar energy can be operated (Yang, 2019). Photovoltaic solar energy transforms solar energy into electrical energy through the photovoltaic (PV) effect. This principle, the photovoltaic effect, dates back to around 1840, when French physicist Becquerel made the discovery that under the illumination of sunlight it was possible to detect an electrical voltage in two pieces of metal in diluted hydrochloric acid. A century later, in the Bell Laboratories, the first silicon solar cell was created, which triggered its research and application, still in use today (Green, 2005).

Nowadays, solar cells are a worldwide commodity and used in various sectors. One of these sectors is the space industry, where solar cells are being used to supply electricity to satellites and space stations. Remote meadows have also been created all throughout deserts, mountains, and islands in order to provide electricity to certain regions of the world, and it has become more and more common to see solar cells installed on roof-tops, apartments, and public buildings to generate electricity on the grid. Due to the amount of energy required to build large photovoltaic power stations which produce megawatts (MW) or even gigawatts (GW) of electricity, sets of solar cells that group together with these constructions have become a regular sight (Yang, 2019).

Since the introduction of the first solar cell, with an efficiency rate of around 6%, the photovoltaic industry has been in continuous development due in large part to the support from the United States, Japan, Europe, China, India, etc. The slow growth of this

Sustainable Design for Renewable Processes
DOI: https://doi.org/10.1016/B978-0-12-824324-4.00029-9

industry together with the use of conventional electrical energy caused the price of solar cells to increase. As the decades passed, the popularity and efficiency of solar cells improved which in turn caused their cost to decrease. All of this was achieved thanks to the support of different government owned projects such as "Project Sunshine" and the "100,000 Roof Project" from the United States, Germany, and Japan (Green, 2016). The rise of the photovoltaic industry has also led to the legislation of laws by different countries for the improvement and use of solar cells both in the domestic and in the industrial sectors. In the last 5 years, the use of solar cells has grown exponentially. Data from 2017 show that around 99 GW of solar energy capacity were installed worldwide. This is a 60 times increase from 2006 when only 1.5 GW of solar energy capacity were installed (Margolis et al., 2017). During the last decade, the photovoltaic industry has shown a growth rate of more than 40% per year, which is more than that of the microelectronic industry. This growth has caused the price of solar cells to decrease, and from 2006 to 2017 a reduction of 15% was projected (Feldman et al., 2018). Many more photovoltaic panels are expected to be installed in the immediate future; and by 2050, worldwide photovoltaic energy is forecasted to be more than 1 terawatt (Razykov et al., 2011). From all this collected data it is obvious that the photovoltaic industry is one of the most promising industries in the world.

The main material within the photovoltaic industry is silicon. More than 90% of solar cells sold in the last decades have a polycrystalline silicon base (Tyagi et al., 2013). In its beginnings, remains of silicon materials or fragments of silicon wafers from the microelectronic industry were used as raw material. At the beginning of the century, however, there was a marked shortage of polycrystalline silicon due to the growth of the photovoltaic industry. Nowadays, most of the polycrystalline silicon used in the industry derives from silica sand (SiO_2). A detailed process of the origins of polycrystalline silicon will be discussed in later sections of the chapter. However, it is important to note that after the refining of silicon by means of various technologies, a high purity silicon is required. It is for this reason that the purity required for its used must be specified. This addresses the main concern of the next section of the chapter.

10.1.2 Purity requirements for solar grade silicon in photovoltaic uses

Polycrystalline silicon is the main material used in the production of solar cells, and it requires a purity above 99.99999%, often called "7-nines" or "7Ns" (Yang, 2019). The quality of solar cells produced depends on the purity of the silicon material. And, it is worth considering that the higher the quality of silicon, the higher the cost of the cell.

Metallurgical silicon (Si_{MG}), which has a purity of around 98.5%, is used to attain silicon of high purity. Si_{MG} is used regularly in the metallurgical, chemical and electrical industries (Ciftja, 2008). Only a small fraction, less than 10% of the total Si_{MG}, is consumed for photovoltaic applications. For solar applications, ultra-pure silicon is separated into electronic grade silicon (Si_{EG}) and into high purity polycrystalline silicon. Along with the use of the term *high purity polycrystalline silicon*, the term *solar grade silicon* (Si_{SG}) was also coined to describe silicon feedstock that can be used to produce solar cell material (Ahmed and Manassah, 2014).

In the past two decades, the demand for Si_{SG} has grown in an exorbitant manner, and due to currently existing processes in the production of solar cells, the purity requirement for Si_{SG} has lowered compared to Si_{EG}. Therefore there is a diversification of Si_{SG} sources, which include virgin silicon (produced directly from Siemens reactors or fluidized bed reactors (FBR), pot waste, and recycled silicon). Si_{SG} also includes metallurgical silicon purified by physical processes or improved Si_{MG} as it is sometimes named (Goetzberger et al., 1998).

The quality of the Si_{SG} is determined by its degree of purity as it commonly contains impurities to a certain degree such as concentration of donors, acceptors, metals, oxygen, carbon, etc. Because of this, standards to provide the Si_{SG} specifications have been implemented. Among the most widely known are: the Chinese national standard for Si_{SG} (with code GB/T25074−2010); and the SEMI standard (Semiconductor Equipment and Materials International) (Yang, 2019). Each of these standards provide a classification of purity requirements (as seen in Table 10.1, referring to the Chinese national standard). Si_{SG} purity is a difficult subject as impurity concentrations may change during crystallization or other processing steps (see Fig. 10.1). Since the efficiency of the solar cell increases as impurities are reduced, with the improvement of the operational efficiency of solar cells, new and higher requirements will be imposed on Si_{SG} purity. This will also cause the renewal of the standards for Si_{SG}, and the different specifications from various organizations will most likely coincide with one another. Along with Si_{SG} purity, the requirements for particle size, size distribution, and surface impurities will be adjusted (Delannoy, 2012). These new guidelines will apply not only to the parts of polysilicon produced by conventional processes, but also to polysilicon produced by emerging processes.

10.1.3 Silicon photovoltaic industry through the time

The first known technology in the Si_{SG} generation was described in patents first filed in the 1950s by the Siemens Corporation, where the decomposition of a silicon-containing gas (commonly trichlorosilane) is produced in thin heated silicon filaments in a chemical

TABLE 10.1 Specifications for the solar grade silicon (Si_{SG}) (Yang, 2019).

Factors	Specifications		
	Grade I	Grade II	Grade III
Resistivity (P, As, Sb): (Ω.cm)	≥ 100	≥ 40	≥ 20
Resistivity (B, Al): (Ω.cm)	≥ 500	≥ 200	≥ 100
Donors (P, As, Sb): ppba	≤ 1.5	≤ 3.76	≤ 7.74
Acceptors (B, Al): ppba	≤ 0.5	≤ 1.3	≤ 2.7
Minority lifetime: μs	≥ 100	≥ 50	≥ 30
Oxygen: (atoms/cm³)	$\leq 1.0 \times 10^{17}$	$\leq 1.0 \times 10^{17}$	$\leq 1.5 \times 10^{17}$
Carbon: (atoms/cm³)	$\leq 2.5 \times 10^{16}$	$\leq 4.0 \times 10^{16}$	$\leq 4.5 \times 10^{16}$
Total metal impurities (Fe, Cr, Ni, Cu, Zn, etc.)	≤ 0.05	≤ 0.1	≤ 0.2

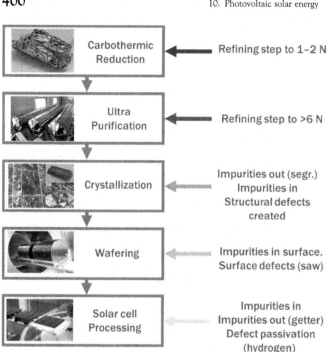

FIGURE 10.1 The solar grade silicon (Si_{SG}) value chain which indicates the main effects in terms of segregation of impurities and incorporation of defects in each step.

vapor deposition reactor. A more modern technology was developed in the late 1970s where deposition of the FBR to produce polysilicon granules is presented (Siffert et al., 2005).

Si_{SG}'s production processes are well known and have been used by the major polysilicon suppliers in the United States, Japan, and Europe since the 1950s. Nowadays, these plants continue to expand in their capacity, and new plants have been built in Japan, China, Taiwan, Korea, and Europe.

Since 1950 the quantity of silicon supply has grown. The increasing demand (between 1997 and 2009) for Si_{SG} for solar cells caused an increase in supplies of polycrystalline silicon. The supply increased from 16,050 tons in 1997 to 92,100 tons in 2009 (Takiguchi and Morita, 2011). The level of wafer production decreased in 2009 with respect to 2008, most probably because of the financial crisis and the global economic recession. However, it is worth noting that the supply of polycrystalline silicon did not decrease despite downward trends in the global economy, and since 2009 there has been an increasing demand for Si_{SG} (Takiguchi and Morita, 2011).

Fig. 10.2 broadly depicts the evolution of solar photovoltaic energy (mainly silicon-based energy) which represents one of the fastest growing, most established, and profitable renewable energy technologies out there. With this in mind, the silicon photovoltaic industry marks a rapid and mature growth within technological energy; therefore the future projections of the industry (Europe, 2019) are to be discussed in the next section.

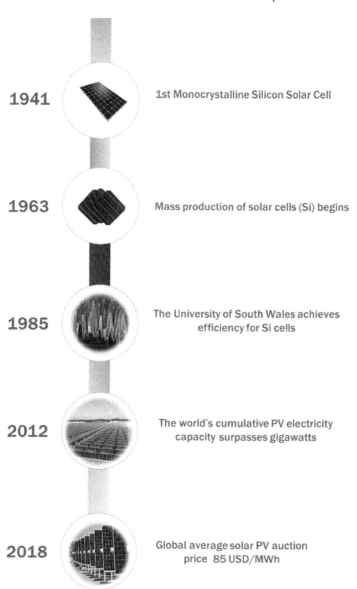

FIGURE 10.2 Silicon photovoltaic industry: an established and rapidly growing renewable energy technology (Europe, 2019).

10.1.4 Trends and the future of silicon photovoltaic energy

The growth of the silicon-based photovoltaic solar energy industry, from its start up until today, has been considerable. It has achieved several milestones in the last decade in terms of facilities, cost reductions, and technological advances as well as established key partnerships for solar energy. The silicon industry is projected to remain an essential renewable option within the solar energy industry for years to come. Overall, the renewable energy market has been thriving at a fast rate in recent years, reaching record levels and often exceeding annual levels of conventional energy capacity in many regions. However, within the group of renewable technologies, silicon-based photovoltaic solar energy installations have dominated the renewable energy industry for several years, reaching in 2018 a worldwide capacity of 480 GW photovoltaic solar energy. This symbolized a growth of around 20% in a single year (Bódis et al., 2019). It is estimated that for a high market potential, reduced costs, and the elemental resource (Si_{SG}) was widely available then photovoltaic solar energy would be expected to continue to drive an overall growth in renewables in various regions for decades to come. A representative scheme of this scenario can be seen in Fig. 10.3. Fig. 10.3 shows the forecast of the globally installed capacity of solar photovoltaic energy, which would increase six times by 2030 (2840 GW) and would reach 8519 GW by 2050. These values show a reference in comparison to the installations available in 2018 (480 GW) (IRENA, 2019).

The rapid worldwide growth in photovoltaic capacity cannot be denied. There are numerous research projects along with the creation of new prototypes in progress; all of which are leading to an encouraging growth in the market for solar technologies at the application level. All this can be exemplified with the integration of photovoltaic solar panels into smart buildings. These panels provide a variety of advantages such as an adaptability to exposed surfaces, cost effectiveness (savings in roofing material), and design flexibility (size and shape). The efficiency and power of solar panels has improved considerably in recent years and future improvements are anticipated. In other words, since the reported efficiency of polycrystalline PV panels has reached 17%, this positive trend is expected to continue until 2030 (IRENA, 2012). Despite all of this, as the global PV market increases, so will the need to avoid panel degradation and the management of the volume of retired PV panels leading to circular economy practices will be necessary. This includes innovative and alternative methodologies to reduce material use and module degradation as well as opportunities to reuse and recycle photovoltaic panels at the end of their useful life, hence the importance in the discussion of environmental impact of solar panels in the next section.

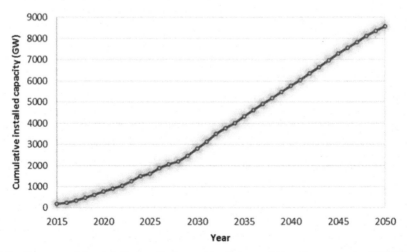

FIGURE 10.3   Forecasts of the globally installed capacity of photovoltaic solar energy for 2050 (IRENA, 2019).

10.1.5 Environmental impacts of silicon photovoltaic panels

The use of silicon-based solar energy technology is growing, and a bright future is expected. However, as the market grows, so will the need to manage both the degradation and the volume of retired solar panels. It is expected that the environmental impact in the area of silicon PV will decrease due to additional improvements such as a greater cell efficiency, a decrease in energy consumption during module production, panel recycling, etc.

The study of the environmental impact of silicon photovoltaic panels must take into consideration a complete life cycle analysis. Numerous authors have enacted these life cycle studies and among their most important results have been that the most critical phases are both the transformation of SiO_2 into Si_{SG} (1190.1 MJ/panel), and the assembly of the panel (272.7 MJ/panel) (IRENA, 2019). Undoubtedly, the first process is identified by a high electricity consumption, even if the most efficient conversion technology has been considered. The second process is identified by the use of aluminum frames and glass ceilings, both of which are high energy intensive materials. In this last phase, the consumption of materials has a greater impact than the consumption of electricity. In reality, the aluminum frame contributes with 22.4%, the glass roof with 23%, ethyl vinyl acetate with 18.9%, and polyester with the electrical box, with 14.6%; while the contribution of electricity use is just 15.6% (IRENA, 2019), the transformation of SiO_2 to solar energy is responsible for the greatest contribution to the greenhouse effect due to its high electricity consumption. Assuming, of course, that the use of coal and wood chips during the transformation of silica into Si_{SG} contributes negatively to the production of CO_2. However, most of the life cycle analysis results support the idea that photovoltaic power generation is advantageous to the environment. The panel production process must also be taken into consideration in comparison to conventional electricity production technologies.

Innovative and alternative ways to reduce material use and module degradation as well as opportunities to reuse and recycle photovoltaic panels at the end of their life cycle must be explored. The assumption of a circular economy framework and the conventional principles of waste reduction (reduce, reuse, and recycle) should also apply to photovoltaic panels.

- *Reduce: Material savings in silicon photovoltaic panels*
  In this area, it should be considered that the main option is to increase the efficiency of the panels by reducing the amount of material used. To date, there has not been any substantial change in the mixture of materials for the production of silicon photovoltaic panels. However, efficient mass production, material renewal, and more efficient technologies are taking place thanks to a strong market growth, a shortage of raw materials, and the decrease in prices for photovoltaic panels. While research is currently underway to reduce the amount of hazardous materials as well as decrease the amount of materials used per panel in order to save costs; Si_{SG} availability is not a major short-term concern, although critical materials may impose long-term constraints. At the same time, higher prices improve the economics of recycling activities and encourage investment in more efficient mining processes, such as in the extraction of metals used in the manufacture of photovoltaic panels, that is, silver, aluminum, copper, and tin.

Some producers of silicon photovoltaic panels are focusing in the reduction or replacement of different components used (Corcelli et al., 2018).

- *Reuse: Repairing photovoltaic panels*

    Most photovoltaic systems have been installed in the last decade. The average useful life of a panel is around 30 years. A panel ages 33% in a decade. If a flaw were to be found in the installed panel within this first decade, reparation of the panel or a replacement warranty can be claimed, and the insurance companies would compensate some or all of the reparation or replacement costs. By reselling the product, the retailer or user may recover some of its value of the returned panel. This is done by running quality tests to verify electrical safety and output power, such as the flash test characterization and a wet leak test. Repaired PV panels can be resold as replacements or as used panels at a reduced market price of approximately 65% of the original retail price, and partially repaired panels or components can be sold on second-hand markets (Cross and Murray, 2018).

- *Recycling: Dismantling and treatment of photovoltaic panels*

    The future disposal of waste from installed photovoltaic systems will mostly be a result of its type and size. That is, while rooftop photovoltaic systems, due to their small and highly dispersed nature can add high costs to the dismantling and to the collection and transportation of expired photovoltaic panels, waste management of large-scale photovoltaic systems is logistically more convenient. Today, the amounts of photovoltaic waste are almost nonexistent which reduces the economic incentive to create dedicated photovoltaic panel recycling plants. This causes photovoltaic panels, at the end of their useful life, to be processed in nonspecialized recycling plants. However, the construction of dedicated photovoltaic panel recycling plants are projected in the short run (Weckend et al., 2016).

10.1.6 The silicon photovoltaic routes

Before 1997, waste from the electronic industry provided the production of solar cells with silicon (Xakalashe and Tangstad, 2012). With the growing demand for silicon in the photovoltaic industry, this practice became unsustainable. Therefore new technologies had to be developed to supply the Si_{SG} photovoltaic industry. Two main routes were developed, the chemical route and the metallurgical route (see Fig. 10.4).

10.1.6.1 Conventional processes of solar grade silicon

Although there are a large number of patents and Si_{SG} production routes, there are only two conventional technologies (within the chemical route) which currently hold a very important market share: (1) The Siemens process, with a market share of more than 90%; and (2) The "fluidized bed reactor" process, which holds a market share of 3%−5% (Woodhouse et al., 2019).

10.1.6.1.1 Siemens: process description

The Siemens process (Fig. 10.5) uses as a precursor to Si_{SG}, trichlorosilane ($SiHCl_3$). $SiHCl_3$ originates from a process that begins with quartz reduction in an electric arc furnace in order

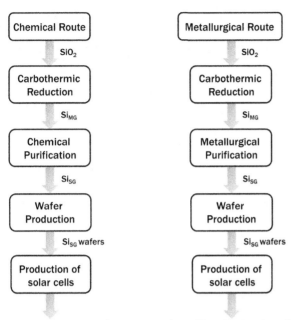

FIGURE 10.4   Phase diagram for the production of solar grade silicon (Si_{SG}) solar cells made from quartz using the chemical route and the metallurgical route (Xakalashe and Tangstad, 2012).

to obtain metallurgical grade silicon (Si_{MG}). The Si_{MG} produced reacts with hydrogen chloride (HCl) in a hydrochlorination synthesis reactor to produce a gas stream composed of a mixture of chlorosilanes. Among them, the most important is trichlorosilane, which will be used as a precursor in the polysilicon production stage (Islam et al., 2011). In the next stage, the purification process runs through a sequence of distillation columns used to obtain ultrapure trichlorosilane. In the last stage, the ultrapure trichlorosilane is disintegrated in a chemical vapor deposition reactor known as the Siemens reactor (Ramos et al., 2015). A relevant drawback in the Siemens process is the need to use extremely pure HCl for the synthesis of chlorosilanes as this compound drags along high environmental and safety risks.

10.1.6.1.2 Union Carbide: process description

The process developed by Union Carbide Co., shown in Fig. 10.6, uses silane (SiH_4) as a source of polycrystalline silicon. As in the Siemens process, Si_{MG} is produced through the metallurgical reduction of SiO_2 and C. Both Si_{MG} and silicon tetrachloride (which is recirculated from the following step) are hydrogenated in a FBR to produce chlorosilanes: $SiCl_4$, SiH_2Cl_2, and $SiHCl_3$ (Zadde et al., 2002). After this, the separation and purification of chlorosilanes is carried out. Trichlorosilane can be transformed into silane through successive redistribution reactions or by using reactive distillation columns (RDC) (Schmid et al., 2019). Afterwards, the high purity silane that is obtained is introduced into a vapor deposition reactor where it disintegrates to produce the Si_{SG}. The Union Carbide Co. process achieves higher efficiencies since the conversion of silane in silicon is greater than the trichlorosilane to silicon transformation used in the Siemens process. However, this process operates under more extreme conditions than to those used in the Siemens process.

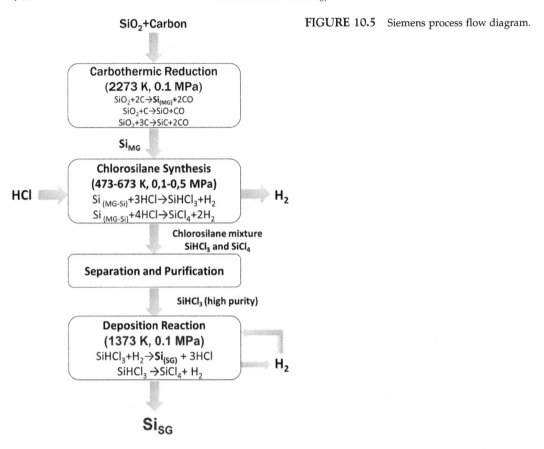

FIGURE 10.5   Siemens process flow diagram.

10.1.6.2 *Units analysis*

For each of the processes described, a series of processing units allow the transformation of one precursor to another to obtain the Si_{SG}. In this section, the units involved in each of the processes will be analyzed based on the mechanisms that govern them.

10.1.6.2.1 Carbothermic reduction of silicon dioxide

Si_{MG} is produced industrially by a carbothermal reduction of silicon dioxide (SiO_2) in electric submerged arc furnaces. Feeding materials include a silicon source (such as quartz) and a standard reducing mixture comprising coal and wood chips. The feedstock is impure, and the impurities are brought into the stream of silicon products to some extent. The carboreduction unit can be seen in Fig. 10.7. This processing unit is involved in both the Siemens process and the Union Carbide process. The model for the carboreduction reactor is based on the work by Wai and Hutchison (1989). The work shows that the reaction between SiO_2 and C, Eq. (10.1) consists of a series of stages.

$$SiO_{2(S)} + 2C_{(s)}\Delta \rightarrow Si_{(l)} + 2CO_{(g)} \tag{10.1}$$

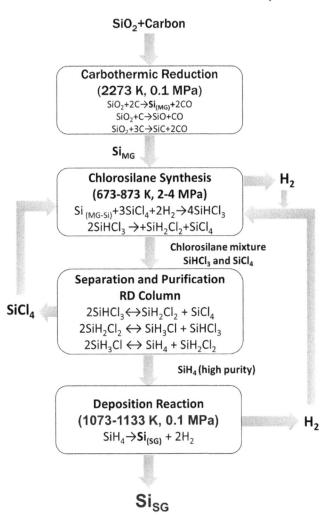

FIGURE 10.6 Process flow diagram of Union Carbide Co.

However, the reaction progress as a series of reactions of SiO_2 and C at high temperatures forming multiple products. Eqs. (10.2)–(10.8) show the possible reactions carried out during the carboreduction process.

$$SiO_{2(l)} + C_{(s)} \rightarrow SiO_{(g)} + CO_{(g)} \tag{10.2}$$

$$SiO_{2(l)} + C_{(s)} \rightarrow Si_{(l)} + CO_{2(g)} \tag{10.3}$$

$$SiO_{2(l)} + 2C_{(s)} \rightarrow Si_{(g)} + 2CO_{(g)} \tag{10.4}$$

$$SiO_{2(l)} + 2C_{(s)} \rightarrow Si_{(l)} + 2CO_{(g)} \tag{10.5}$$

$$SiO_{2(l)} + 2.5C_{(s)} \rightarrow 0.5Si_2C_{(g)} + 2CO_{(g)} \tag{10.6}$$

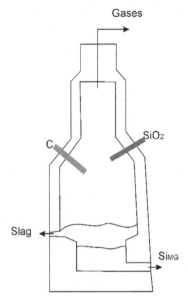

Gases

C

SiO₂

Slag

S$_{iMG}$

$$SiO_{2(l)} + 3C_{(s)} \rightarrow SiC_{(s)} + 2CO_{(g)} \tag{10.7}$$

$$SiO_{2(l)} + 4C_{(s)} \rightarrow SiC_{2(l)} + 2CO_{(g)} \tag{10.8}$$

The distribution of the products obtained by Wai and Hutchison (1989) is for a feeding molar ratio 2:1 for C and SiO₂, with temperatures ranging from 2500 to 3500 K and at atmospheric pressure. It is possible to use that work to develop correlations that estimate the distribution of products as a function of the reaction temperature (K), considering a temperature range from 2600 to 3100 K. To produce such correlations, the information from a figure must be read. It is possible to use software that transforms the figures in data points such as *Digitalizer plot*. The next step is just to find the form of the equation to fit the data. Ramírez-Márquez et al. (2020) developed a surrogate model of each of the species (mole fraction) produced as a function of operating temperature in a range from 2600 to 3100 K, see Eqs. (10.9)–(10.16), where x_i is the mole fraction of each species i, and T in the temperature range between 2600 and 3100 K. Note that not all correlations show the same mathematical form; this is due to the complex shape of the distribution profiles.

$$x_{Si(l)} = -2.48131 \times 10^{-9}T^3 + 1.90239 \times 10^{-5}T^2 - 4.79395 \times 10^{-2}T + 39.71359 \tag{10.9}$$

$$x_{CO(g)} = 9.82689 \times 10^{-5}T + 3.74066 \times 10^{-1} \tag{10.10}$$

$$x_{Si(g)} = 5.93093 \times 10^{-10}e^{6.31510 \times 10^{-3}T} \tag{10.11}$$

$$x_{SiC(s)} = 7.14539 \times 10^{-7}T^2 - 4.50044 \times 10^{-3}T + 7.08465 \tag{10.12}$$

$$x_{Si2C(g)} = 1.72881 \times 10^{-7}T^2 - 9.13915 \times 10^{-4}T + 1.20759 \tag{10.13}$$

$$x_{SiC2(g)} = -1.19611 \times 10^{-14}T^5 + 1.65491 \times 10^{-10}T^4 - 9.14807 \times 10^{-7}T^3$$

$$+ 2.52572 \times 10^{-3}T^2 - 3.48320T + 1919.64937 \tag{10.14}$$

$$x_{SiO(g)} = 7.58739 \times 10^{-7}T^2 - 4.47932 \times 10^{-3}T + 6.69671 \tag{10.15}$$

$$x_{Si2(g)} = 1 - x_{Si(l)} - x_{CO(g)} - x_{Si(g)} - x_{SiC(s)} - x_{Si2C(g)} - x_{SiC2(g)} - x_{SiO(g)} \tag{10.16}$$

Once the silicon is obtained, it is extracted from the furnace through an opening at the bottom and refined by means of a waste treatment or gas purge. In refining, inclusions are removed and the composition is adjusted to the specified value. After refining is completed, the molten alloy is allowed to cool in a mold and is then ground to a specific size. Si_{MG} has a typical purity specification of 98.5%–99.5%. Typical impurities in Si_{MG} include carbon, alkali metals, and transition metals, as well as boron and phosphorus (Xakalashe and Tangstad, 2012). A large consumption of power is required to melt the silica, around 10–11 kWh to produce a kilogram of silicon (Brage, 2003). Example 10.1 shows the use of a particular software to read data from figures.

10.1.6.2.2 Chlorosilane synthesis reactor

Once a Si_{MG} stream is obtained, it is transferred to a chlorosilane synthesis reactor. Depending on the chosen process, it can either be in the presence of a different precursor and at different operating conditions.

10.1.6.2.2.1 Chlorosilane synthesis reactor for Siemens process For the Siemens process, the hydrochlorination of Si_{MG} takes place in the presence of HCl at 1.5 bar aiming at the production of $SiHCl_3$. In the reactor, the following reactions take place, Eqs. (10.17) and (10.18):

$$Si_{(MG)} + 3HCl \leftrightarrow SiHCl_3 + H_2 \tag{10.17}$$

$$Si_{(MG)} + 4HCl \leftrightarrow SiCl_4 + 2H_2 \tag{10.18}$$

Both reactions are quick and exothermic so that there is no need for catalyst and the use of a fluidized bed is justified. To maximize the production of trichlorosilane ($SiHCl_3$), a temperature of 533 K and an excess of HCl, 10% with respect to the stoichiometric one (Jain et al., 2011), is

---

EXAMPLE 10.1 Extract the necessary information from the diagram proposed by Wai and Hutchison (1989), with the PlotDigitizer software.

Solution

In some cases, it is necessary to extract values from figures and graphics since some scientific publications only show the graphs, but the data values are not published. The PlotDigitizer software was used to data extraction. The data extraction process by this software is straightforward: (1) Import the graphic from a file, (2) the axes system is defined, (3) it is digitized either automatically or manually, and (4) the data values are copied in Excel for their manipulation. Once numerical data are obtained from the plot, to check the accuracy of the extraction process, the data are plot and compared to the original diagram (Fig. 10E1.1).

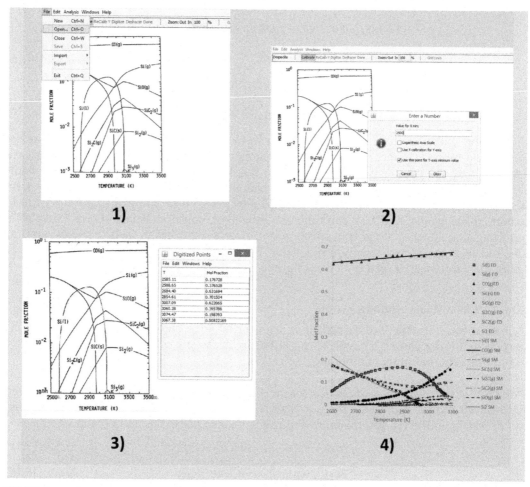

FIGURE 10E1.1   Data extraction using PlotDigitizer software.

recommended, achieving 90% selectivity to SiHCl₃ while the rest follows the second reaction to tetrachlorosilane (SiCl₄) (Kotzsch et al., 1977; Jain et al., 2011).

10.1.6.2.2.2 Chlorosilane synthesis reactor for Union Carbide process   For the Union Carbide process, the hydrochlorination reactor requires $SiCl_4$ (see Fig. 10.8), which is hydrogenated in the presence of Si_{MG}. For this process, Ding et al. (2014) show a thermodynamic analysis of the $SiCl_4 - H_2 - Si_{MG}$ system. Reactions of the $SiCl_4 - H_2 - Si_{MG}$ system involve the hydrogenation of $SiCl_4$ in the gas phase, Eq. (10.19), and the hydrochlorination of Si_{MG} with HCl, Eq. (10.20).

$$SiCl_{4(g)} + H_{2(g)} \leftrightarrow SiHCl_{3(g)} + HCl_g \tag{10.19}$$

$$HCl_{(g)} + \frac{1}{3}Si_{MG(s)} \rightarrow \frac{1}{3}SiHCl_{3(g)} + \frac{1}{3}H_{2(g)} \tag{10.20}$$

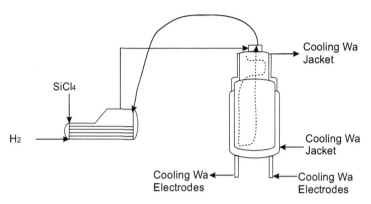

FIGURE 10.8 Hydrochlorhination reactor.

The feed gas is heated using hot process gas leaving the converter. Afterwards it is heated to 1100°C by electrical graphite heaters. The typical consumption is from 0.7 to 3 kWh/kg of SiCl$_4$ (Silicon Products, 2020).

By combining Eqs. (10.19) and (10.20), the reactants SiCl$_4$ − H$_2$ − Si$_{MG}$, produce SiHCl$_3$, Eq. (10.21)

$$\text{SiCl}_{4(g)} + \frac{2}{3}\text{H}_{2(s)} + \frac{1}{3}\text{Si}_{\text{MG(s)}} \leftrightarrow \frac{4}{3}\text{SiHCl}_{3(g)} \tag{10.21}$$

To compute the portfolio of products, it is possible to study the system as an equilibrium using the total Gibbs free energy minimization approach, assuming that the reaction system behaves ideally, given by Eq. (10.22)

$$G^T = \sum_{i=1}^{N} n_i \mu_i = \sum_{i=1}^{N} n_i \left(G_i^o + RT\ln\frac{\hat{f}}{f_i^o} \right) \tag{10.22}$$

The restriction includes the molar balances to each species,

$$\sum_i n_i a_{ik} = A_K \quad (k = 1, 2, \ldots, w) \tag{10.23}$$

For the gaseous phase,

$$\hat{f} = y_i \hat{\varphi} P, \quad f_i^o = P^o, \quad y_i = \frac{n_i}{\sum_i^N n_i} \tag{10.24}$$

"Defining fugacity $\hat{f}$, standard fugacity, f_i^o, and molar fraction, y_i."
In the case of solid silicon,

$$RT\ln\frac{\hat{f}}{f_i^o} = \int_{P_o}^{P} V_{\text{Si}_{MG}}(P - P^o)dP \tag{10.25}$$

"V_{SiMG} is almost independent of the pressure, it can be approximated by,"

$$RT\ln\frac{\hat{f}}{f_i^o} = V_{\text{Si}_{MG}}(P - P^o) \tag{10.26}$$

4. Solar technologies

Consequently, by combining Eqs. (10.22)–(10.26) the following Eq. (10.27) is obtained, which describes the total free Gibbs energy of the system:

$$G^T = \sum_{i=1}^{N} n_i \left(G_i^o + RT\ln\frac{y_i P}{P^o} \right) + n_{Si}G_{Si}^o + V_{Si_{MG}}(P - P^o) \tag{10.27}$$

In which, the Gibbs–Helmholtz relationship is defined by Eq. (10.28).

$$G_i^o = H_i^o + TS_i^o \tag{10.28}$$

Eqs. (10.29) and (10.30) define the standard enthalpy and standard entropy, respectively.

$$H_i^o = H_{i,298}^o + \int_{298}^{T} C_{p,i}\, dt \tag{10.29}$$

$$S_i^o = S_{i,298}^o + \int_{298}^{T} \frac{C_{p,i}}{T} dt \tag{10.30}$$

"G^T is the total Gibbs free energy; N is the number of species in the reaction system; n_i is the number of moles; μ_i is the chemical potential; G_i^o is the standard Gibbs free energy; $\hat{f}$ is the fugacity; f_i^o is the standard fugacity of species i; R is the molar gas constant; T is the temperature; a_{ik} is the number of k_{th} atoms in each molecule of species i; A_K is the total atomic mass of the k_{th} element in the system; w is the total number of elements in the system; P is the total pressure. $\hat{\varphi}$ is the fugacity coefficient; y_i is the molar fraction of species i; P^o is the standard-state pressure (100 kPa); $V_{Si_{MG}}$ is the molar volume of silicon; and H_i^o, S_i^o, and $C_{p,i}$ are the standard enthalpy, standard entropy, and heat capacity, respectively, of species i." Ding et al. (2014) show the thermodynamic data for the chemical species involved.

In either case, the trichlorosilane output stream from the chlorosilane synthesis reactor will be subjected to the separation stages. Example 10.2 shows a the model for the reactor.

10.1.6.2.3 Separation and purification process

Since there is an adequate difference in the boiling points between the chlorosilanes (281 K for dichlorosilane, 305 K for trichlorosilane, and 330.8 K for tetrachlorosilane), the liquid stream that is obtained is subjected to a distillation to separate the species (Paetzold et al., 2015). The distillation is carried out using conventional columns, where at the bottom, a stream of the heaviest, the chlorosilane, is extracted with high purity, while through the dome, a stream of the lighter chlorosilanes with a purity of at least 99.99% molar is obtained. This purity is enough to be used in the production of Si_{SG} in a Siemens type deposition reactor, or to be used as a precursor for silane (SiH_4) in a vapor phase deposition reactor where it decomposes to produce hydrogen and a highly pure silane (Filtvedt et al., 2012). Example 10.3 shows the model of the column using ASPEN plus.

10.1.6.2.4 Silane production for the Union Carbide process

The high purity $SiHCl_3$ stream is introduced into a series of RDC, where the reaction takes place in three stages, having dichlorosilane (SiH_2Cl_2) and monochlorosilane (SiH_3Cl)

EXAMPLE 10.2 Model the hydrochlorination reactor using a total Gibbs free energy minimization approach.

Solution

```
Set
J 'Components' /H2,SiCl4,SiHCl3,SiH2Cl2,HCl,Si/
gas(J) /H2,SiCl4,SiHCl3,SiH2Cl2,HCl/
cop /a, b, c, d, e/
;


Table
*Heat Capacity
*Cp = a + bT + cT2 + dT3 + eT − 2
        cp_v(J,cop)
```

	a	b	c	d	e
H2	33.06	− 11.36e − 03	11.43e − 06	− 2.77e − 09	− 0.16e06
SiCl4	105.57	2.12e − 03	− 0.57e − 06	0.05e − 09	− 1.45e06
SiHCl3	64.28	85.31e − 03	− 72.09e − 06	23.43e − 09	− 0.75e06
SiH2Cl2	50.10	84.35e − 03	− 50.00e − 06	11.04e − 09	− 0.80e06
HCl	32.12	− 13.46e − 03	19.87e − 06	− 6.85e − 09	− 0.05e06
Si	22.82	3.90e − 03	− 0.08e − 06	0.04e − 09	− 0.35e06

```
;


Parameter
*H298 ° (kJ/mol)
Ho(J)
/
H2        0.00
SiCl4     −662.75
SiHCl3    −496.22
SiH2Cl2   −320.49
HCl       − 92.31
Si        0.00/
;


Parameter
*S298 ° (J·mol − 1·K − 1)
So(J)
/H2       130.70
SiCl4     330.86
SiHCl3    313.71
SiH2Cl2   286.72
HCl       186.90
Si        18.82/
```

```
;
Scalar
R         'cont gases (J/molK)' /8.314472/
V         'volume of silicon (m3/mol)' /12.06e-06/
Po        'Standar pressure (kPa)' /100/
To        'Standar temperature (K)' /298/
P         'Pressure (kpa)' /101.325/
nH2in     'no mol H2'/1/
nSiin     'no mol Si' /0.2/
nSiCl4in  'no mol SiCl4' /6.5/
;

Variables
*standard enthalpy
H(J)
*standard entropy
S(J)
*Gibbs-Helmholtz relation
Go(J)
*Temperature ref
;
Positive variables
T
*Mol Fraction
y(J)
*Mol
n(J)
;

Variables
Z
;
T.fx = 373;
n.fx('Si') = 0;

Equations
Eq1, Eq2, Eq3, Eq4, Eq5, Eq6, Eq7, Eq8;
Eq1(J).. H(J) = e= Ho(J) + ((T**2-To**2)*cp_v(J,'b')/2) + ((T**3-To**3)*cp_v
(J,'c')/3) + ((T**4-To**4)*cp_v(J,'d')/4) + ((T-To)*cp_v(J,'a')) - (cp_v(J,'e')*
(1/(T+1e-05)-1/(To)));
Eq2(J).. S(J) = e= So(J) + ((T**2-To**2)*cp_v(J,'c')/2) + ((T**3-To**3)*cp_v
(J,'d')/3) + (cp_v(J,'a')*(log(T+1e-05)-log(To))) + (cp_v(J,'b')*(T-To)) -
(cp_v(J,'e')*(1/(2*T**2+1e-05)-1/(2*To**2)));
Eq3(J).. Go(J) = e= H(J) - (T*S(J));
```

```
Eq4(J)$(gas(J))..y(J)*((n('H2'))+(n('SiCl4'))+(n('SiHCl3'))+(n('SiH2Cl2'))+
(n('HCl')))=e=n(J);
Eq5..sum(J$gas(J),y(J))=e=1;
Eq6..2*nH2in=e=n('SiHCl3')+2*n('SiH2Cl2')+n('HCl')+2*n('H2');
Eq7..nSiin+nSiCl4in=e=n('SiCl4')+n('SiHCl3')+n('SiH2Cl2');
Eq8..4*nSiCl4in=e=4*n('SiCl4')+3*n('SiHCl3')+2*n('SiH2Cl2')+n('HCl');
Equations
obj;
obj..Z=e=(sum(J$gas(J),n(J)*(Go(J)+R*T*log(1e-05+y(J)*P/Po))))+n('Si')*Go
('Si')+n('Si')*V*(P-Po);

model min/all/;
option NLP=snopt;
```

EXAMPLE 10.3 Perform the separation of SiCl$_4$ and SiHCl$_3$ with the help of a conventional distillation column in Aspen Plus V8.8.

Solution

1) Load the components as seen in Fig. 10.11A, they are trichlorosilane (SiHCl$_3$) and tetrachlorosilane (SiCl$_4$).
2) The "Peng-Robinson" thermodynamic method was selected.
3) Go to the "Simulation" section and locate the "Columns-RadFrac" part to add the column.
4) Add the stream lines in the "Material" module, the feed stream (F), distillate stream (D), and the bottoms stream (B).
5) Feed stream data (F) is introduced. Data such as temperature (20°C), pressure (5.0 atm), feed flow (125 kg/h of trichlorosilane and 900 kg/h of tetrachlorosilane).
6) In the "Configuration" tab, the operation specifications are introduced. The "Reflux ratio" and the "Distillate rate" (30 and 120 correspondingly) are chosen as variables, as shown in Fig. 10.11B.
7) Complete the "Streams" and "Pressure" tabs. In "Streams" the feeding stage is set to 14 "On Stage." In the "Pressure" tab, an operating pressure of 0.4 MPa and a pressure drop of 10 psia are set.
8) Define the design specification to meet the purity requirement of 0.999 mole fraction. In the same folder of "Specifications" of block B1 the option of "Design Specifications" is chosen, add a new one and in the tab indicated as "Specifications" on top, enter "Mole Purity" and a target of 0.995, see Fig. 10.13. Choose the component of interest in the "Components" tab by selecting it and passing it on the right side of the white boxes with the arrow of the corresponding direction. Select the product stream of interest in "Feed/Product Stream." Finally, in the "Vary" option, choose the "Reflux ratio" variable (Fig. 10E3.1).

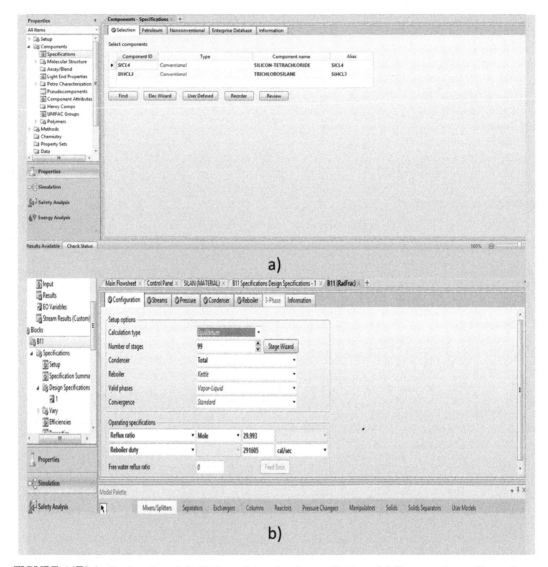

FIGURE 10E3.1 Explanation of distillation column for the purification of $SiCl_4$ using Aspen Plus software with a purity of 99.9% purity.

as intermediates, and as a by-product to tetrachlorosilane ($SiCl_4$) (Ramírez-Márquez et al., 2016), according to the following equations, Eqs. (10.31), (10.32) and (10.33):

$$2SiHCl_3 \overset{cat}{\leftrightarrow} SiH_2Cl_2 + SiCl_4 \tag{10.31}$$

$$2SiH_2Cl_2 \overset{cat}{\leftrightarrow} SiH_3Cl + SiHCl_3 \tag{10.32}$$

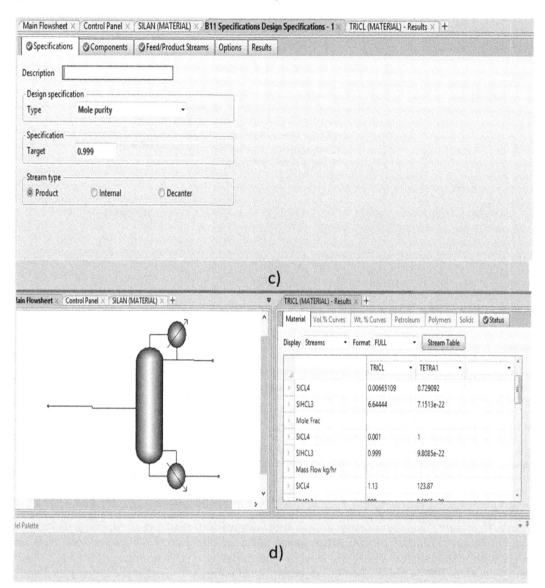

c)

d)

FIGURE 10E3.1 Continued

$$2SiH_3Cl \overset{\text{cat}}{\leftrightarrow} SiH_4 + SiH_2Cl_2 \qquad (10.33)$$

The disproportion of $SiHCl_3$ can be catalyzed by various types of catalysts (Huang et al., 2013). However, most of the studies use "Amberlyst" (A-21). Catalyst A-21 exhibits reasonable reaction rates between 30°C and 80°C, with a suggested operating thermal resistance of 100°C (Huang et al., 2013). The RDC, being an intensified equipment (where

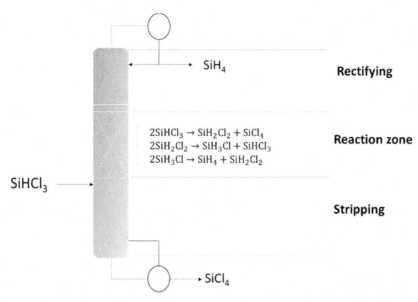

FIGURE 10.9   Reactive distillation column for the production and purification of SiH_4.

at the same time, separation reaction can be carried out (see Fig. 10.9)) is presented as the best option, with a good reduction in energy consumption, achieving purities of at least 99.99 mol% of SiH_4 (Ramírez-Márquez et al., 2016). Example 10.4 shows the model of a reactive distillation column using ASPEN plus.

10.1.6.2.5 Process for obtaining solar grade silicon

Depending on the process, either the Siemens process or the Union Carbide process, the reactor in the acquisition of Si_{SG} differs. Each of the equipment used in both processes will be shown next.

10.1.6.2.5.1 Siemens reactor   Once pure trichlorosilane is obtained, it is used as a silicon source to obtain Si_{SG} by means of a vapor deposition reaction in a Siemens reactor Eqs. (10.34) and (10.35). The deposition takes place on ultrapure U-shaped silicon rods situated within the reactor. These are heated internally by electrical energy (see Fig. 10.10). Del Coso et al. (2008) show the Si_{SG} deposition. The study provides the operating conditions required for the deposition of polycrystalline silicon in the traditional Siemens reactor (temperature range from 1372 to 1500 K and atmospheric pressure). The growth rate of Si_{SG}, the deposition efficiency, and the temperature of the system were the variables analyzed. These are studied in the reaction system formed by the reactions shown in the following equations:

$$SiHCl_3 + H_2 \leftrightarrow Si + 3HCl \tag{10.34}$$

$$SiHCl_3 + HCl \leftrightarrow SiCl_4 + H_2 \tag{10.35}$$

EXAMPLE 10.4 Silane (SiH₄) production by means of a reactive distillation column with Aspen Plus V8.8 with a purity of 99.5% purity.

Solution

1) Load the components as seen in Fig. 10E4.1A, they are trichlorosilane ($SiHCl_3$), tetrachlorosilane ($SiCl_4$), dichlorosilane (SiH_2Cl_2), silane (SiH_4), and monochlorosilane (ClH_3Si).

2) Select the appropriate thermodynamic method for the components, the "Peng-Robinson" method was selected.

3) Go to the "Simulation" section and locate the "Columns-RadFrac" part to add the column.

4) Add the stream lines in the "Material" module as shown in Fig. 10E4.1C, the feed stream (F), distillate stream (D), and the bottoms stream (B).

5) Feed stream data (F) is introduced. Data such as temperature (50°C), pressure (5.5 atm), feed flow (10 kmol/h of trichlorosilane).

6) In the "Configuration" tab, the operation specifications are introduced. The "Reflux ratio" and the "Distillate to feed ratio" (63 and 0.25 correspondingly) are chosen as variables, as shown in Fig. 10E4.1E.

7) Complete the "Streams" and "Pressure" tabs. In "Streams" the feeding stage is set to 46 "On Stage." In the "Pressure" tab, an operating pressure of 5 atm and a pressure drop of 0.5 kPa are set.

8) Enter reactions in the "Stoichiometry" tab with the "New" option. In the "Edit Reaction" window in the upper right "Reaction type" modify from "Equilibrium" to "Kinetic" for each reaction.

9) Complete the kinetic parameters of each reaction in the "Kinetic" tab: the preexponential (k), the activation energy (E), and set the base to "Mole Fraction," as seen in Fig. 10E4.1G and H.

10) Define reactive stages in the folder located on the left side "Specification" of the block (B12) the option "Reactions" is chosen to enter the initial stage ("Starting stage") and the final stage "Ending stage" 16 and 45, respectively, as well as select the reaction in "Reaction ID," as shown in Fig. 10E4.1I

11) Fix the residence time required to carry out the reaction, a time of 2.5 seconds was fixed in the liquid phase for each stage, as observed in Fig. 10E4.1I

12) Define the design specification to meet the purity requirement of 0.995 mole fraction. In the same folder of "Specifications" of block B1 the option of "Design Specifications" is chosen, add a new one and in the tab indicated as "Specifications" on top, enter "Mole Purity" and a target of 0.995, see Fig. 10E4.1J. Choose the component of interest in the "Components" tab by selecting it and passing it on the right side of the white boxes with the arrow of the corresponding direction. Select the product stream of interest in "Feed / Product Stream." Finally, in the "Vary" option, choose the "Reflux ratio" variable. Fig. 10E4.1K shows the composition profile quantitatively and the temperature profile graphically.

4. Solar technologies

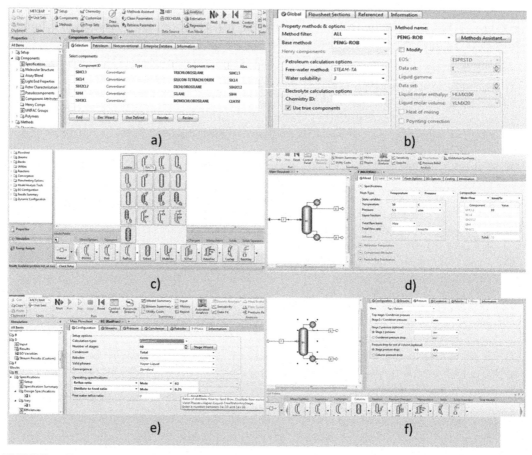

FIGURE 10E4.1 Modeling of a reactive distillation column for the production of silane (SiH_4) using Aspen Plus software with a purity of 99.5% purity.

Silicon deposition is presumed to follow second-order kinetics, where the consumption or generation mass rate of species i on the surface of the bars can be expressed as in Eq. (10.36).

$$R_i = v_i \mu_i k [SiHCl_3][H_2] \tag{10.36}$$

R_i is the mass rate of change in species i by chemical reaction ($kg/m^2 s$); μ_i is the viscosity of the species i ($kg/m\ s$); v_i corresponds the stoichiometry coefficients of the compounds involved in the reactions (-1 for $SiHCl_3$ and H_2 and 3 for HCl); k is the overall reaction coefficient; and (i) is the mole concentration of species i on the surface (Del Coso et al., 2008). The global deposition reaction coefficient is a function of the

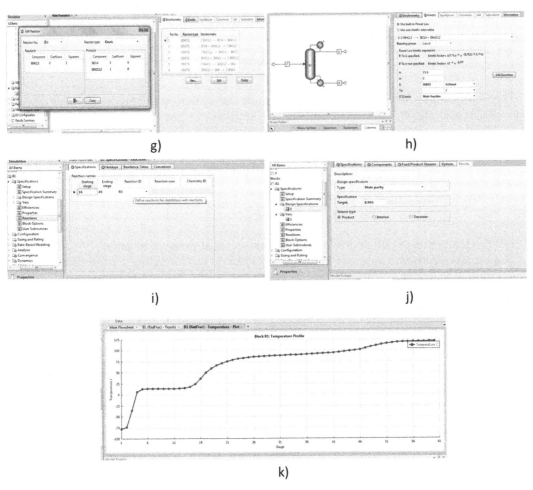

g)

h)

i)

j)

k)

FIGURE 10E4.1   Continued.

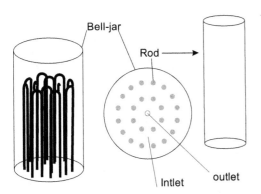

FIGURE 10.10   Scheme of the Siemens reactor.

adsorption of $SiHCl_3$ on the surface of the rod (k_{ad}) and the its decomposition (k_r) as follows, Eq. (10.37):

$$\frac{1}{k} = \frac{[SiHCl_3]}{k_r} + \frac{[H_2]}{k_{ad}}$$

(10.37)

There is a dependency of the temperature on the kinetic coefficients. For this reason, Eqs. (10.38) and (10.39) show Arrhenius's law applied at atmospheric pressure (Del Coso et al., 2008).

$$k_{ad}(T) = 2.72 \times 10^6 \exp\left(\frac{-1.72 \times 10^5}{RT}\right)$$

(10.38)

$$k_r(T) = 5.63 \times 10^3 \exp\left(\frac{-1.80 \times 10^5}{RT}\right)$$

(10.39)

T is the temperature (K); and R is the constant of ideal gases in SI units.

The typical deposition rate can be of the order of 30 kg/h feeding a ratio of H_2 to $SiCl_4$ of 3.5 mol/mol (https://silicon-products-gmbh.com). The polycrystalline silicon deposition itself, is the largest contributor to the energy consumption of the overall process and is assumed to be 45−60 kWh/kg (Ramos et al., 2015). The silicon rods grow continuously to a thickness of 80−150 mm per rod (Ramos et al., 2015). Electrical power is used to heat the rods. The deposition process takes about 3−5 days (Ramos et al., 2015). Therefore it is necessary to use several deposition reactors for the required production.

Finally, in both cases, the gas streams from the reactor in turn are introduced to a series of heat exchangers in order to be cooled down, separated, and purified to be used as recirculation in the process.

10.1.6.2.5.2 Decomposition in the fluidized bed reactor for Union Carbide process For the Union Carbide process, SiH_4 must be heated above its deposition temperature to obtain Si_{SG}. The production of polysilicon in the Union carbide process follows the decomposition of silane as follows, Eq. (10.40):

$$SiH_{4(s)} \rightarrow SiH_2 + H_2$$

(10.40)

The growth rate is controlled by the adsorption of the reactants at high temperature and the desorption of the H_2 absorbed at low temperatures. The competition for the adsorption sites reduces the rate of reactions. From silane the reaction mechanisms is presented as composed of the following stages, where (*) represents the adsorption sites, Eqs. (10.41), (10.42), (10.43), (10.44), (10.45), (10.46), (10.47), (10.48) and (10.49):

$$SiH_{4(s)} \xrightarrow{ka} SiH_{4(g)} \quad \text{(Diffusion across the stagnant layer)}$$

(10.41)

$$SiH_{4(g)} + * \underset{k_{-1},k_1}{\longleftrightarrow} SiH_4^* \quad \text{(Silane adsorption)}$$

(10.42)

$$SiH_4^* \underset{k_{-2},k_2}{\longleftrightarrow} Si(ad) + H_2 \quad \text{(Reaction)}$$

(10.43)

$$\frac{1}{2}H_2 + * \overset{k_{-3}, k_3}{\longleftrightarrow} H* \quad \text{(Hydrogen adsorption)} \tag{10.44}$$

$$Si(ad) \overset{k_4}{\to} Si_{(g)} + * \quad \text{(Silicon evaporation)} \tag{10.45}$$

$$Si(ad) \overset{k_5}{\to} Si_{(s)} \quad \text{(Silicon growing)} \tag{10.46}$$

$$Si(ad) + 2HCl \overset{k_6}{\to} SiCl_2 + H_2 + * \quad \text{(Graven)} \tag{10.47}$$

$$HCl + * \overset{k_{-7}, k_7}{\longleftrightarrow} Cl* + \frac{1}{2}H_2 \quad \text{(HCl adsorption)} \tag{10.48}$$

$$SiH_{4(g)} \overset{k_8}{\to} Si_{(g)} + 2H_2 \quad \text{(Decomposition in gas phase)} \tag{10.49}$$

Thus the reaction rate becomes, Eq. (10.50):

$$R = \frac{k_1 \cdot p_{SiH_4}}{1 + k_2 \cdot p_{H_2}^{0.5} + k_3 \cdot p_{SiH_4}} \tag{10.50}$$

R is the reaction rate, p_i is the partial pressure of each species i, and k_1, k_2, and k_3 are the rate constants as follows, Eqs. (10.51), (10.52) and (10.53):

$$k_1 = 1.6 \times 10^9 e^{\left(\frac{-18500}{T}\right)} [\,=\,]\frac{\text{mol Si}}{\text{m}^2 \cdot \text{s} \cdot \text{atm}} \tag{10.51}$$

$$k_2 = 0.6 \times 10^2 [\,=\,]\text{atm}^{-(1/2)} \tag{10.52}$$

$$k_3 = 0.7 \times 10^5 [\,=\,]\frac{1}{\text{atm}} \tag{10.53}$$

The reaction rate is based on the equilibrium at the substrate surface taking into account the inhibition as a result of the gaseous hydrogen (Roenigk and Jensen, 1985). The denominator is almost 1 at high temperatures. Three different configurations can be used, horizontal/vertical, low/high pressure, and cold/hot wall. The most common are the low pressure systems, from 10 mTorr to 1 Torr (Low pressure Chemical Vapor Deposition, LPCVD). Hot walled furnaces are appropriate for batch processing providing thermal uniformity and high throughput. However, the silicon may deposit on the walls. Cold wall units require lower maintenance since no deposition on the wall occurs. Vertical units use less footprint than horizontal ones. There are three regions in the furnace. Gases enter and go through the pipe. The heating mechanisms is by means of electric resistances. The mass transfer due to the forced convection is limited by providing room between the silicon wafers determining the size of the reactors, see Fig. 10.11.

To model the deposition reactor (Jensen and Graves, 1983), the following assumptions are used (Setalvad et al., 1985):

1. There are no radial temperature gradient. The reactor walls and substrate are heated and the slow reaction rates imply small heats of reaction.

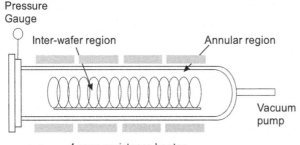

FIGURE 10.11   Wafer reactors.

2. The axial temperature profile is controlled by the furnace design and settings. The heat-up lengths are small and most heat transfer occurs by radiation at LPCVD conditions.
3. There is no axial variation of the gas-phase composition in the interwafer region between any two consecutive wafers since the interwafer spacing is small.
4. There is no radial variation of the gas-phase composition in the annular region. The annular region is small and there is rapid diffusion at LPCVD conditions.
5. The gas phase is at steady state. Chemical Vapor Deposition (CVD) growth processes are slow compared to gas-phase dynamics.

Developing an ordinary set of differential equations to represent the operation of the units assuming that axial transport only takes place in the annual region while radial transport occurs in the interwafer region. The model becomes:

• Interwafer region, Eq. (10.54):

$$\frac{\Delta}{r}\frac{d}{dr}(r \cdot N_{r1}) = -2R \tag{10.54}$$

where N_{r1} corresponds to the flux in the r direction and Δ is the distance between wafers, with the boundary condition of, Eq. (10.55)

$$\frac{dx_1}{dr}\Big|_{r=0} = 0; x_1\left(r_w^-\right) = \left(r_w^+\right) \tag{10.55}$$

where r_w is the wafer radius, and $+$, $-$ refer to the infinitesimal distance in the positive or negative distance in the r direction
• Annulus region, Eq. (10.56)

$$\frac{dN_{z1}}{dz} = -\frac{2R}{(r_t^2 - r_w^2)}\left[r_t \cdot (1 + a) + \frac{r_w^2}{\Delta} \cdot \eta\right] \tag{10.56}$$

where r_t is the pipe radius and a is the area of the wafer plus that of the support related to that of the reactor. The flows can be computed following Fick's law where c is the gas concentration and D is the silane diffusivity, Eqs. (10.57) and (10.58):

$$N_{r1} = -cD\frac{dx_1}{dr} \tag{10.57}$$

$$N_{z1} = -cD\frac{dx_1}{dz} \tag{10.58}$$

The boundary conditions for this region are given at the entrance ($z = 0$) and at the exit point ($z = L$), Eq. (10.59):

$$N_{z1}\big|_{z=0} = v_0 \cdot c_0 \cdot x_{10} \quad \text{and} \quad \frac{dx_1}{dz}\bigg|_{z=L} = 0 \tag{10.59}$$

where the gas velocity, v_o, the total concentration, c_o, and the fraction of silane at the entrance x_{10} provide the flux at the entrance. The efficiency factor, η, is given by Eq. (10.60)

$$\eta = \frac{2 \cdot \int_0^r r \cdot R(r)dr}{r_w^2 \cdot R\big|_{r_w}} \tag{10.60}$$

η is the ratio of the average deposition velocity over a wafer versus the velocity at the extreme presenting a metric to evaluate the uniformity of the deposition. The reaction rate at r_w varies with z and the efficiency factor is a function of the axial position. If the superficial velocity is controlled by the reaction rate, η is equal to 1, while if it is smaller, diffusional resistances control. The optimization of the reactor can be seen in Setalvad et al. (1985) using an orthogonal collocation approach for the discretization of the differential equation optimizing an NLP model.

10.2 Solar cells

10.2.1 Introduction

In 1939 Becquerel observed the photovoltaic effect for the first time. In 1946 Russel created the first silicon-based photovoltaic cell (Kurinec, 2018). These early solar cells consisted of small silicon wafers which were capable of transforming solar energy into electrical energy, although they had a rather low efficiency. With time, the principle of electron hole creation was discovered. With this principle, cells composed of two or more layers of different materials (type p and type n materials) could be implemented. The benefit of this structure is that when a photon collides with a p-type or n-type material, an electron is ejected moving it from layer to layer creating an electron and a hole in the process, this achieves the creation of electrical energy (Gray, 2011). The diverse photovoltaic cells developed till date can be categorized into four groups called generations (Luceño-Sánchez et al., 2019):

- The first generation: mono or polycrystalline silicon cells and gallium arsenide;
- The second generation: thin-film technologies;
- The third generation: offering the use of novel materials, as well as a great variability of designs, and includes expensive but very efficient cells;
- The fourth generation: inorganics-in-organics.

The aim of this review is to show the current state of art on photovoltaic cell technology in terms of the materials used for the manufacture, efficiency, and production costs. Therefore there are numerous types of materials applied in the making of photovoltaic solar cells. PV cells can be classified according to the material that is used (Parida et al., 2011):

10.2.1.1 Silicon

It is the most used material, and it could be classified in two main groups:

- Crystallized silicon
  - Monocrystalline
  - Poly-crystalline
- Amorphous silicon

Other materials. These ones can be divided into:

- Materials used in actual commercial cells such as cells of cuprum-indium/gallium-diselenide, cadmium tellurium, gallium arsenide, or titanium dioxide (O'regan and Grätzel, 1991).
- New materials in development such as graphene (Liu et al., 2008), silicene (Ye et al., 2014), di-sufide spices, or perovskite (Ning et al., 2015).

On the other hand, they can be classified according to the use required:

- PV cells in panels which are the most common ones to capture energy in houses or lands;
- "Thin-film" cells for applications where the weight is important;
- PV concentration cells using Fresnel lenses.

10.2.2 Progress of efficiency silicon solar cells

The energy conversion efficiency of a solar cell can be defined as the ratio of energy from light, and the resulting electrical energy (Mundus, 2016). Depending on the material with which the cells were made, efficiency may vary. Fig. 10.12 shows the progress in the efficiency of various types of solar cells. Dye-sensitized solar cells, organic photovoltaic energy, and quantum dot cells started with a very low efficiency percentage and have slowly increased over the years to around 12%, where they have stabilized. Currently, the value of silicon solar cells is much higher. Their range is about 20%–25%. However, as Fig. 10.12 illustrates, at least 45 years were required for the technology to develop to obtain a purer material and hence increase their efficiency.

10.2.3 Lifetime characterization

Typically, the conversion efficiency of the solar cell increases as the concentration of impurities is reduced (mostly metallic impurities), until the useful life of the wafer does not show a significant effect on the efficiency (Mitchell et al., 2013). However, the controlled introduction of donor and acceptor impurities in a semiconductor allows the

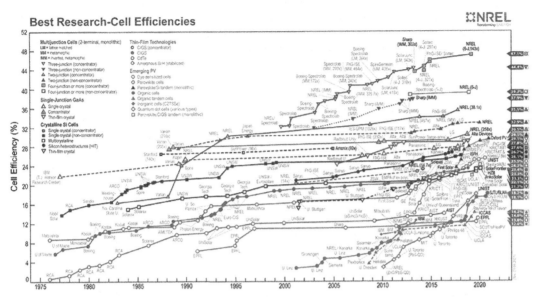

FIGURE 10.12   Best research-cell efficiencies (NREL, 2021). *The chart "Best Research-Cell Efficiencies" is reprinted with permission from the National Renewable Energy Laboratory, https://www.nrel.gov/pv/assets/pdfs/best-research-cell-efficiencies.20200104.pdf, Accessed January 25, 2021.*

creation of n-type (electrons are the primary source of electrical conduction) and p-type (holes are the majority carriers), respectively. In an n-type semiconductor, electrons are known as majority carriers, and their density is equal to the density of N_D donors, while holes are called minority carriers. This terminology is inverted in the p-type region. This is the basis for the construction of all semiconductor devices, including solar cells (Paranthaman et al., 2016). Therefore by improving the solar cell manufacturing process, the minority life requirement of wafers will be increased in order to improve solar efficiency. The need to improve the operational efficiency of solar cells will bestow novel requirements over the manufacturing of cell precursor materials.

Nowadays, efforts are underway to predict the efficiency of solar cells before they are manufactured. That is, which materials, purity, geometry, and thickness should be used to guarantee a longer life. In the case of silicon cells, it allows manufacturers to waive the inadequate wafers in the production of the cells with the purpose of saving costs. Lifetime is an appropriate parameter that will strongly influence the efficiency of the cells. It is well known that the lifetime can change during each processing stage. This can happen with silicon cells due to the absorption, contamination, and hydrogenation. In other words, it is a somewhat complex task to use the initial half-live method to predict cell efficiency. However, once on the market, solar cells have a typical slow rate of degradation, often down to less than 1% per year, and in many cases, it is even undetectable during the first years of operation (Jordan and Kurtz, 2013). There are methods to predict their lifetime fairly accurately. Optical and electrical transient measurements are often utilized in order to address this. Noncontact optical methods such as transient absorption, transient photoluminescence, and transient photo-conductance measurements provide information related to the overall disappearance rate of excited states or charge carriers in the active material. Direct electrical measurements of the photovoltaic materials in contact test the rate of disappearance of the electrochemical potential of the excess carrier densities (e.g., the photoconductivity decay method and the surface photovoltage measurement). For semiconductors, this parameter is known as the division of quasi-Fermi levels for electrons and holes. It is also equivalent to the term that is often found in impedance analysis as the diffusion capacity or chemistry (Mitchell et al., 2013).

10.2.3.1 Photoconductivity decay method

Photodecay methods are employed for research of the photoconductive materials and the photovoltaic structures. The materials suffer the photocurrent (photoconductivity) decay. For the solar cells, one can measure the photovoltage or the photocurrent decay (Pisarkiewicz, 2004). Observation of the decay current after a sudden interruption of light belongs to classical methods of semiconductor investigation. This method for measuring minority carrier lifetime was proposed by Stevenson and Keyes already in 1955 (Cuevas et al., 1998). The method is shown schematically in Fig. 10.13.

The photoconductor is homogeneously illuminated by the light pulses that causes adequate generation of the excess carriers and photocurrent variation. The voltage will increase to a steady-state value after which it will remain constant until light falls on it, since the rate at which the carriers are created and recombined is the same. However, by turning off the light source, the recombination rate of minority carriers is greater than the generation rate. Originally, since the number of carriers is greater, the decomposition rate is swift, and the nature of this decay is somewhat exponential. We obtain the exponential decay curve of the open circuit voltage through the junction of the solar cell. This is allowed only for low intensity lighting, since the number of photo-induced carriers is small. In the case of high intensity lighting where a large number of carriers are produced, this exponential curve will tend toward a straight line with a high slope. With an intermediate intensity, a straight-line curve is also obtained yet it is obvious that the slope will be smaller (Mahan et al., 1979).

10.2.3.2 Surface photovoltage measurement

Surface photovoltage semiconductor characterization techniques have become great method to define the generation lifetime and recombination lifetime of the semiconductors. It is a method that is convenient because it is typically contactless with the potential measured with a Kelvin probe. Kelvin probes have conventionally been used to measure the sample work function as a function of surface treatment (Schroder, 2001). Over time, commercial equipment has been launched on the market to carry out this method in the semiconductor industry.

A surface photovoltage is usually the consequence of a surface or insulator charge or work function difference and it is most commonly perceived with a noncontacting probe. The types of probes are shown in Fig. 10.14. For the Kelvin probe, the electrode vibrates vertically, this causes a change in capacitance between the sample and the probe. On the other hand, for the Monroe Probe, it has a shutter that connects to ground, and a fixed electrode. In this case there is a horizontal vibration. The main difference is that Monroe

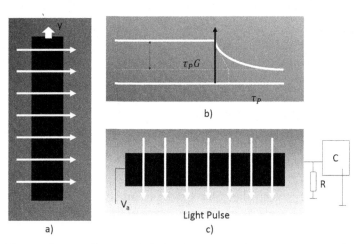

FIGURE 10.13 Photoconductivity decay method after Stevenson and Keyes; n-type sample under constant illumination (A), decay of photoexcited carriers with time (B), and the experimental setup (C) (Pisarkiewicz, 2004).

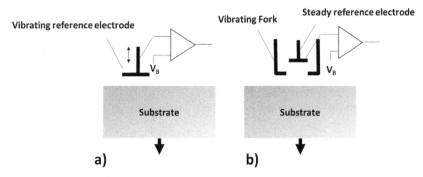

FIGURE 10.14 Kelvin probe (A) and Monroe probe (B) for contact potential difference measurements.

probes are less susceptible to external vibrations. In both cases the conventional frequencies are 500–600 Hz (Schroder, 2001).

The contact potential difference V_{cpd} is determined by measuring the ac current, Eq. (10.61):

$$I = V_{cdp}\frac{dC}{dt} \tag{10.61}$$

where C is the capacitance between the probe and the sample. The V_{cpd} is determined by either calibrating the current versus the bias voltage V_B, or by adjusting V_B until $I = 0$ in which case $V_B = V_{cpd}$.

10.2.4 Efficiency losses

Loss processes carried out in a solar cell that is under solar irradiation fall into two categories (Hirst and Ekins-Daukes, 2011):

- Extrinsic losses, such as series resistance, can limit the efficiency of the device. However, they can theoretically be avoided, allowing the calculation of the fundamental limiting efficiency by excluding them.
- Intrinsic losses are inevitable in the design of this device and will continue to be present in an idealized solar cell.

Usually for intrinsic loss processes (Henry, 1980) (Fig. 10.15):

1. Below E_g loss: A lag between the solar spectrum and the mono-energetic absorption of a single bandgap (E_g), leads to the nonabsorption of photons with energy below the bandgap.
2. Thermalization loss: The above, together with the strong interaction between the excited carriers and the lattice photons, result in a loss of thermalization.
3. Emission loss: It all comes down to the fact that the absorbers turn out to be emitters (Kirchoff's law), and therefore the emission from the solar cell tends to reduce efficiency.
4. Carnot loss: Equating a solar cell as a heat engine, where the sun is the hot deposit and the solar cell as the cold deposit, the conversion of thermal energy to electricity demands an energy penalty. In the solar cell, the loss take place in a voltage drop, called the Carnot factor.
5. Boltzman loss: In conclusion, the discrepancy in the absorption and emission angles results in an entropy generation process (Hirst and Ekins-Daukes, 2011).

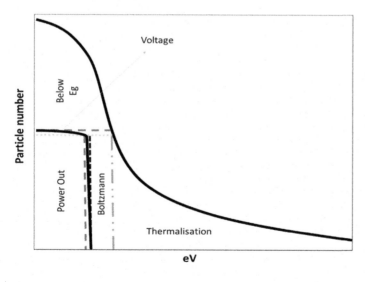

FIGURE 10.15 Intrinsic losses taking place in a device with optimal E_g under one sun radiation.

10.2.5 Silicon solar cell efficiency limits

Until 1999 the efficiency limit of a laboratory silicon solar cell reached a value of 25%. In 2014 the efficiency limit increased to 25.6%, and in 2017 it reached 26.7% (thin layer heterojunction technology based on thin a-Si passive layers and interdigitated posterior contacts in n-type silicon wafers). In wafer-based silicon modules, the limit has reached 24.7%, and continues to increase constantly in both commercialized and laboratory-produced modules (Andreani et al., 2019). Shockley and Queisser (1961), for the first time, calculated the theoretical limiting efficiency of a single junction solar cell, see Eq. (10.16).

In Fig. 10.16, the curve (f) is the detailed balance limit of efficiency, assuming the cell is a blackbody. Curve (j) is the semiempirical limit, or limit conversion efficiency of Prince. Curves (g), (h), and (i) are modified to correspond to 90% absorption of radiation and 100-mW incident solar energy (Shockley and Queisser, 1961). The limits of current solar cell production efficiencies vary depending on the band gap of the semiconductor material.

Shockley and Queisser (1961) showed the detailed balance limit of efficiency, assuming the cell is a blackbody. The limit of efficiency was calculated for an ideal case in which the only recombination mechanism of hole-electron pairs is radioactive as required by the principle of detailed balance. The limits of current solar cell production efficiencies vary depending on the band gap of the semiconductor material.

10.3 Electricity production

10.3.1 Introduction

PV solar is a type of solar energy based on the photovoltaic effect. When electrons are excited by the incident energy of photons some acquire the necessary energy potential to

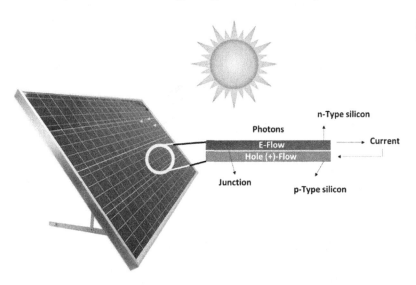

FIGURE 10.16 Scheme for the photovoltaic effect.

reach the conduction band, according to the valence-conduction band model. Commonly, a p-n connection is used in order to help the creation of an electric field in the cell, see Fig. 10.16.

10.3.2 Panel generation of electricity

For electrical applications, PV cells are commonly represented as it is showed in Fig. 10.17, where I_{ph} is the photo-current of the cell and proportional equivalent current in the period of sunshine received by the cell, I_d is reverse saturation current of diode and R_{sh} and R_s are the intrinsic shunt and series resistance of the cell, respectively. Usually the value of R_{sh} is very large if we compare it with R_s. Finally, V is the voltage produced by the cell which can give energy to feed a determinate resistance, R_d. Based on this circuit representation of the PV cell it is possible to model the system (Vinod et al., 2018), Eqs. (10.62), (10.63), (10.64) and (10.65).

$$I = I_{\text{ph}} - I_d - I_{\text{sh}} = I_{\text{ph}} - I_o \qquad (10.62)$$

$$I = I_{\text{ph}} - I_s \left(e^{\frac{q(V+R_sI)}{N_sKAT_o}} - 1 \right) - \frac{V + R_sI}{R_{\text{sh}}} \qquad (10.63)$$

$$(V + R_sI) = V_{\text{oc}} \qquad (10.64)$$

N_s is the number of cells connected in series. R_{sh} is many times considered to be infinite.

$$I_{\text{ph}} = \left(I_{\text{ph,ref}} + \alpha(T - T_{\text{ref}}) \right) \frac{G}{G_{\text{ref}}} \qquad (10.65)$$

G: Solar irradiance received by the module area (W/m^2). α: Temperature coefficient of the short circuit current, 0.005253 (A/K). G_{ref}: Solar irradiance at the Standard Test Condition (1000 W/m^2). T_{ref}: Temperature at the Standard Test Condition (25°C). $I_{\text{ph, ref}}$: Light generated at the reference condition, is practically equal of short circuit current at the reference (Jahangiri et al., 2018), Eqs. (10.66) and (10.67).

$$I_{\text{rs}} = \frac{I_{\text{sc}}}{\left(e^{\frac{q(V+R_sI)}{N_sKAT}} - 1 \right)} \qquad (10.66)$$

$$I_s = I_{\text{rs}} \left(\frac{T}{T_{\text{ref}}} \right)^3 e^{\left[\left(\frac{q \cdot E_g}{AK} \right) \right] \left(\frac{1}{T_{\text{ref}}} - \frac{1}{T} \right)} \qquad (10.67)$$

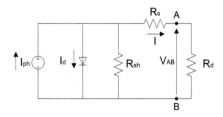

FIGURE 10.17   Equivalent circuit of a cell.

The ideality factor A and energy band gap E_g are taken from the manufacturer of the selected module. q is the charge of an electron $= 1.602 \times 10^{-19}$ C, K Boltzmann's constant $= 1.380 \times 10^{-23}$ J/K Eg Forbidden Energy band gap, for silicon $= 1.1$ eV.

10.3.3 Power produced

Thus the fundamental equation to compute the power produced by a cell is given by Eq. (10.68),

$$P_{\text{Max}} = I_{\text{SC}} \cdot V_{\text{OC}} \cdot F_{\text{F}} \tag{10.68}$$

where I_{SC} is the short-circuit current, it is the maximum photoelectric current generated by the panel, V_{OC} corresponds to the open circuit potential, and F_{F} is the fill factor. Fill factor is a constant, with a maximum value of 1 (Hegedus and Luque, 2003). Fig. 10.18 shows the power curve diagram (Ramos-Hernanz et al., 2013).

The efficiency, η, is defined as the ratio between the power produced (P_{produced}) and that received (P_{received}), where A is the area and P_{incident} is the power per area unit, Eq. (10.69).

$$\eta = \frac{P_{\text{Produced}}}{P_{\text{Received}}} = \frac{I_{\text{SC}} \cdot V_{\text{OC}} \cdot F_{\text{F}}}{A \cdot P_{\text{Incident}}} \tag{10.69}$$

The open circuit voltage and the fill factor decrease substantially with temperature. The net effect of the temperature on the efficiency can be computed as Eq. (10.70)

$$\eta = \eta_{\text{Tref}}\left(1 - \beta_{\text{ref}}(T_{\text{c}} - T_{\text{ref}}) + \gamma \log_{10} I\right) \tag{10.70}$$

where $\eta_{T_{\text{ref}}}$ is the module electrical efficiency at the reference temperature, T_{ref}, and at solar irradiance of 1000 W/m^2. The temperature coefficient, β_{ref}, and the solar radiation coefficient, γ, depend on the material and typically take values of 0.004 K^{-1} and 0.12 for crystalline silicon. γ usually takes to be 0 lately (Dubey et al., 2013). The panel temperature, T_c can be computed as (Luque and Hegedus, 2011), Eq. (10.71).

$$T_{\text{C}} = T_{\text{Env}} + \frac{I\left(\frac{W}{m^2}\right)}{800}(T_{\text{o}} - 20^{\circ}\text{C}) \tag{10.71}$$

where T_o is the nominal operating cell temperature at ideal conditions ($^{\circ}$C), assume 25°C and I is the solar incident radiation.

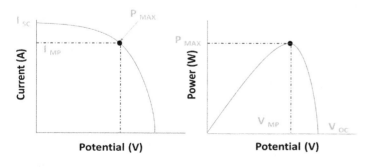

FIGURE 10.18 Power and current curves of a regular photovoltaic cell.

10.3.4 Modules, inverters, and silicon solar photovoltaic systems

The typical installation for the production of power at industrial scale consists of the following units, see Fig. 10.19:

- PV cells, that we have described above.
- Controller. It helps to protect the battery in case of overload due to the different voltage emitted by the cell, when the charge required by the consumer is more than the system is able to provide or due to fails caused in the connection/disconnection of batteries.
- Batteries are used to save energy and used during a period when the PV cells do not work.
- DC/AC inverter. It maximizes the production of energy and transforms the produced energy from DC to AC.

Other optional equipment in PV installations is the transformers to adapt the AC to the requirements of the system or the electrical grid, and an electric meter that could measure the current in both senses, from the cells to the grid and reverse (Kouro et al., 2015).

10.3.5 Global solar power generating capacity

Due to the reduced cost of solar cells, solar energy has become recognized as a reliable and clean source of energy. By 2025 the installed capacity of solar energy is forecast to exceed that of wind energy. The solar revolution started in earnest in 2008 when the new facilities shot up to 6.7 GW from 2.5 GW the year before. For 2010 the new facilities doubled to 17.2 GW, bringing the global cumulative capacity to 40.3 GW. However, for 2012 there was a decrease of 0.1 GW. Due to projections from previous years, the industry was left with overcapacity and the prices of photovoltaic products, such as panels, modules, and cells, decreased and manufacturers with high levels of debt quickly found themselves in a fight to survive. The industry recovered in 2013 and 2014. Solar energy capacity opened up a wide gap with 102.4 GW installed in 2018. From 2020 a decade of strong growth in solar energy is forecast. SolarPower Europe forecasts a level of new installations between 84.5 GW (low) and 165.4 GW (high) (Forecast International's Energy Portal, 2020). Example 10.5 shows the formulation for the optimal allocation of PV panels.

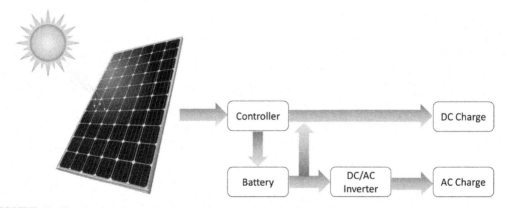

FIGURE 10.19   Configuration of a photovoltaic electricity production facility.

EXAMPLE 10.5 Select among the 11 locations whose conditions are provided in Example 9.4, considering the energy production. Assume that the yield of the photovoltaic panel to be 0.175 at reference condition, and the solar filed consists of 2000 panels of 8 m² each. b_{ref} is 0.004 K^{-1} and g is assumed to be 0.

Solution

We formulate the problem as an MILP. The Gams code is as follows, Eq. (10E5.1):

$$Max \quad y_l \cdot P_l$$
$$st.$$

$$T_{c(m,l)} = T_{amb(m,l)} + \frac{I_{(m,l)}\left(\dfrac{W}{m^2}\right)}{800}(T_o - 20°C)$$

$$\eta_{(m,l)} = \eta_{T_{ref}}\left(1 - \beta_{ref}(T_{c(m,l)} - T_{ref})\right) \quad\quad (10E5.1)$$

$$P_{Produced(m,l)} = \eta_{(m,l)}P_{Received} = \eta_{(m,l)}P_{incident} = \eta_{(m,l)} \cdot A \cdot I_{(m,l)}\left(\frac{kW}{m^2}\right)$$

$$P_l = \frac{1}{12}\sum_m P_{Produced(m,l)}$$

$$\sum^l y_l = 1$$

The code in Gams is as follows:

```
Tempc(month,loc) = Temperature(month,loc) - (T0-20)*(DNI(month,loc)/0.8);
Fyield(month,loc) = Ropt*(1 - beta*(Temperature(month,loc) - Tref));
Power(month,loc) = Fyield(month,loc)*Panels*area*DNI(month,loc)*1000;
Powerloc(loc) = sum(month, Power(month,loc))/12;
eq1.. Z = E = sum(loc, Region(loc)* Powerloc(loc));
eq2.. sum(loc, Region(loc)) = E = 1;
```

We select region 1. Note that in Chapter 9, Example 9.4, with the same locations, we select the 11 due to the combination between water consumption and power production. This is an important result since while using solar, PV and CSP will be located in different places.

10.3.6 Policies and incentives on silicon solar photovoltaic systems

Solar power is a palpable choice for a low carbon future with reliable and long-lasting energy. Greater adoption of this underexploited energy resource to date will help mitigate climate change while stimulating economies, creating jobs, and improving the integrity and security of the grid. However, without strong support from national and international policies in favor of solar energy sources and other renewable energies, society will continue on the path of overdependence on unsafe energy sources, with highly volatile prices and large CO_2 emissions. Incentives must be established for those who adopt these energies in advance, as well as regulatory frameworks and educational initiatives that promote the mass adoption of solar energy by the market. With clear indications to the market, the sector can produce low-carbon solar energy with enough capacity to help meet global energy challenges.

Globally, a series of policies and incentives have been implemented to increase the use of renewable energy, using mostly preferential rate systems, incentives, and subsidies. In general, the policies can be divided as follows (Hosenuzzaman et al., 2015):

- Market incentives: They raise people's awareness as the resulting market expands rapidly. Market incentives offer the following advantages: Feed-in-Tariffs, investment subsidies, loans, and tradable green certifications.
- Technological, research, and development incentives: They provide better opportunities for researchers to increase efficiency, lifespan, and develop new systems that increase their use and reliability.

Exercises

P10.1. Determine the optimal operating temperature that maximizes the liquid silicon produced and the distribution of products.

P10.2. Gibbs minimization optimization. Evaluate the effect of temperature on the production of $SiHCl_3$ in the reactor using the Gibbs minimization model for a 1:5 ratio of H_2 and $SiCl_4$.

P10.3. Model, using a process simulator, a sequence of distillation columns to separate a mixture of 1 kg/s SiH_2Cl_2, 1 kg/s $SiHCl_3$, and 9 kg/s $SiCl_4$ producing streams with 99% molar fraction of each of the three.

P10.4. Select among the 11 allocations whose conditions are provided in Example 9.4, considering the energy production. Assume that the yield of the PV panel to be 0.175 at reference condition, and the solar filed consists of 2000 panels of 8 m^2 each. b_{ref} is 0.004 K^{-1} and g is assumed to be 0. Formulate the problem to obtain the 2 and 3 best options.

P10.5. Which panel efficiency is to be installed in region 2 to be the one selected in Example 10.5.

References

Ahmed, S.A., Manassah, J.T., 2014. Prospects for photovoltaic conversion of solar energy. Alternative Energy Sources 355.

Andreani, L.C., Bozzola, A., Kowalczewski, P., Liscidini, M., Redorici, L., 2019. Silicon solar cells: toward the efficiency limits. Adv. Physics: X 4 (1), 1548305.

Bódis, K., Kougias, I., Taylor, N., Jäger-Waldau, A., 2019. Solar photovoltaic electricity generation: a lifeline for the european coal regions in transition. Sustainability 11 (13), 3703.

Brage, F.P., 2003. Contribución al modelado matemático de algunos problemas en la metalurgia del silicio. Doctoral dissertation, Universidade de Santiago de Compostela.

Ciftja, A., 2008. Refining and Recycling of Silicon: A Review.

Corcelli, F., Ripa, M., Leccisi, E., Cigolotti, V., Fiandra, V., Graditi, G., et al., 2018. Sustainable urban electricity supply chain—indicators of material recovery and energy savings from crystalline silicon photovoltaic panels end-of-life. Ecol. Indic. 94, 37–51.

Cross, J., Murray, D., 2018. The afterlives of solar power: waste and repair off the grid in Kenya. Energy Res. Soc. Sci. 44, 100−109.

Cuevas, A., Stocks, M., Macdonald, D., Sinton, R., 1998. Applications of the Quasi-Steady-State Photoconductance Technique.

Del Coso, G., Del Canizo, C., Luque, A., 2008. Chemical vapor deposition model of polysilicon in a trichlorosilane and hydrogen system. J. Electrochem. Soc. 155 (6), D485−D491.

Delannoy, Y., 2012. Purification of silicon for photovoltaic applications. J. Cryst. Growth 360, 61−67.

Ding, W.J., Yan, J.M., Xiao, W.D., 2014. Hydrogenation of silicon tetrachloride in the presence of silicon: thermodynamic and experimental investigation. Ind. Eng. Chem. Res. 53 (27), 10943−10953.

Dubey, S., Sarvaiya, J.N., Seshadri, B., 2013. Temperature dependent photovoltaic (PV) efficiency and its effect on PV production in the world: a review. Energy Procedia 33 (2013), 311−321.

Europe, S.P., 2019. Global Market Outlook for Solar Power. SolarPower Europe, Brussels, Belgium.

Feldman, D.J., Margolis, R.M., Hoskins, J., 2018. Q4 2017/Q1 2018 Solar Industry Update (No. NREL/PR-6A20−71493). National Renewable Energy Lab. (NREL), Golden, CO, United States.

Filtvedt, W.O., Holt, A., Ramachandran, P.A., Melaaen, M.C., 2012. Chemical vapor deposition of silicon from silane: review of growth mechanisms and modeling/scaleup of fluidized bed reactors. Sol. Energy Mater. Sol. Cell 107, 188−200.

Forecast International's Energy Portal, 2020. A Leading Provider of Power Systems Market Intelligence and Consulting. Online http://www.fi-powerweb.com/Renewable-Energy.html.

Goetzberger, A., Knobloch, J., Voss, B., 1998. Crystalline silicon solar cells. N. Y. 114−118.

Gray, J.L., 2011. The physics of the solar cell, Handbook of Photovoltaic Science and Engineering, 2. Wiley Online Library, pp. 82−128.

Green, M.A., 2005. Silicon photovoltaic modules: a brief history of the first 50 years. Prog. Photovoltaics: Res. Appl. 13 (5), 447−455.

Green, M.A., 2016. Commercial progress and challenges for photovoltaics. Nat. Energy 1 (1), 1−4.

Hegedus, S.S., Luque, A., 2003. Status, trends, challenges and the bright future of solar electricity from photovoltaics. Handbook of photovoltaic science and engineering, 1−43.

Henry, C.H., 1980. Limiting efficiencies of ideal single and multiple energy gap terrestrial solar cells. J. Appl. Phys. 51 (8), 4494−4500.

Hirst, L.C., Ekins-Daukes, N.J., 2011. Fundamental losses in solar cells. Prog. Photovolt.: Res. Appl. Prog. Photovoltaics 19 (3), 286−293.

Hosenuzzaman, M., Rahim, N.A., Selvaraj, J., Hasanuzzaman, M., Malek, A.A., Nahar, A., 2015. Global prospects, progress, policies, and environmental impact of solar photovoltaic power generation. Renew. Sustain. Energy Rev. 41, 284−297.

Huang, X., Ding, W.J., Yan, J.M., Xiao, W.D., 2013. Reactive distillation column for disproportionation of trichlorosilane to silane: reducing refrigeration load with intermediate condensers. Ind. Eng. Chem. Res. 52 (18), 6211−6220.

IRENA, 2012. Renewable Energy Cost Analysis: Solar Photovoltaics.

IRENA, 2019. Future of Solar Photovoltaic: deployment, investment, technology, grid integration and socio-economic aspects (A Global Energy Transformation: paper), International Renewable Energy Agency, Abu Dhabi.

Islam, M.S., Rhamdhani, M.A., Brooks, G.A., 2011. Solar-grade silicon: current and alternative production routes. In: Proceedings of 'Engineering a Better World', the Chemeca Conference, Sydney, Australia.

Jahangiri, M., Rizi, R.A., Shamsabadi, A.A., 2018. Feasibility study on simultaneous generation of electricity and heat using renewable energies in Zarrin Shahr, Iran. Sustain. Cities Soc. 38, 647−661.

Jain, M.P., Sathiyamoorthy, D., Rao, V.G., 2011. Studies on hydrochlorination of silicon in a fixed bed reactor. Indian. Chem. Eng. 53 (2), 61−67.

Jensen, K.F., Graves, D.B., 1983. Modeling and analysis of low pressure CVD reactors. J. Electrochem. Soc. 130, 1950−1957.

Jordan, D.C., Kurtz, S.R., 2013. Photovoltaic degradation rates—an analytical review. Prog. Photovoltaics: Res. Appl. 21 (1), 12−29.

Kotzsch, H.J., Vahlensieck, H.J., Josten, W., 1977. U.S. Patent No. 4,044,109. U.S. Patent and Trademark Office, Washington, DC.

Kouro, S., Leon, J.I., Vinnikov, D., Franquelo, L.G., 2015. Grid-connected photovoltaic systems: an overview of recent research and emerging PV converter technology. IEEE Ind. Electron. Mag. 9 (1), 47–61.

Kurinec, S.K. (Ed.), 2018. Emerging Photovoltaic Materials: Silicon & Beyond. John Wiley & Sons.

Liu, Z., Liu, Q., Huang, Y., Ma, Y., Yin, S., Zhang, X., et al., 2008. Organic photovoltaic devices based on a novel acceptor material: graphene. Adv. Mater. 20 (20), 3924–3930.

Luceño-Sánchez, J.A., Díez-Pascual, A.M., Peña Capilla, R., 2019. Materials for photovoltaics: state of art and recent developments. Int. J. Mol. Sci. 20 (4), 976.

Luque, A., Hegedus, S., 2011. Handbook of Photovoltaic Science and Engineering, second ed John Wiley & Sons, Ltd, Chichester, West Sussex, U.K.

Mahan, J.E., Ekstedt, T.W., Frank, R.I., Kaplow, R., 1979. Measurement of minority carrier lifetime in solar cells from photo-induced open-circuit voltage decay. IEEE Trans. Electron. Devices 26 (5), 733–739.

Margolis, R., Feldman, D., Boff, D., 2017. Q4 2016/Q1 2017 Solar Industry Update (No. NREL/PR-6A20–68425). National Renewable Energy Lab. (NREL), Golden, CO (United States).

Mitchell, B., Wagner, H., Altermatt, P.P., Trupke, T., 2013. Predicting solar cell efficiencies from bulk lifetime images of multicrystalline silicon bricks. Energy Procedia 38, 147–152.

Mundus, M., 2016. Ultrashort Laser Pulses for Electrical Characterization of Solar Cells. BoD–Books on Demand.

Ning, Z., Gong, X., Comin, R., Walters, G., Fan, F., Voznyy, O., et al., 2015. Quantum-dot-in-perovskite solids. Nature 523 (7560), 324–328.

NREL, 2020. http://www.nrel.gov/ncpv/, accessed June 2020.

O'regan, B., Grätzel, M., 1991. A low-cost, high-efficiency solar cell based on dye-sensitized colloidal TiO_2 films. Nature 353 (6346), 737–740.

Paetzold, U., Haeckl, W., Prochaska, J., 2015. United States Patent No. 9,089,788. Washington, DC: United States Patent and Trademark Office.

Paranthaman, M.P., Wong-Ng, W., Bhattacharya, R.N. (Eds.), 2016. Semiconductor Materials for Solar Photovoltaic Cells, Vol. 218. Springer International Publishing, Switzerland.

Parida, B., Iniyan, S., Goic, R., 2011. A review of solar photovoltaic technologies. Renew. Sustain. Energy Rev. 15 (3), 1625–1636.

Pisarkiewicz, T., 2004. Photodecay method in investigation of materials and photovoltaic structures. Opto-Electron Rev. 12, 33–40.

Ramírez-Márquez, C., Sánchez-Ramírez, E., Quiroz-Ramírez, J.J., Gómez-Castro, F.I., Ramírez-Corona, N., Cervantes-Jauregui, J.A., et al., 2016. Dynamic behavior of a multi-tasking reactive distillation column for production of silane, dichlorosilane and monochlorosilane. Chem. Eng. Processing: Process. Intensif. 108, 125–138.

Ramírez-Márquez, C., Martín-Hernández, E., Martín, M., Segovia-Hernández, J.G., 2020. Surrogate based optimization of a process of polycrystalline silicon production. Comput. Chem. Eng. 106870.

Ramos, A., Filtvedt, W.O., Lindholm, D., Ramachandran, P.A., Rodríguez, A., Del Cañizo, C., 2015. Deposition reactors for solar grade silicon: a comparative thermal analysis of a Siemens reactor and a fluidized bed reactor. J. Cryst. Growth 431, 1–9.

Ramos-Hernanz, J.A., Campayo, J.J., Zulueta, E., Barambones, O., Eguía, P., Zamora, I., 2013. Obtaining the characteristics curves of a photocell by different methods. In: International Conference on Renewable Energies and Power Quality, vol. 11, pp. 1–6.

Razykov, T.M., Ferekides, C.S., Morel, D., Stefanakos, E., Ullal, H.S., Upadhyaya, H.M., 2011. Solar photovoltaic electricity: current status and future prospects. Sol. Energy 85 (8), 1580–1608.

Roenigk, K.F., Jensen, K.F., 1985. Analysis of multicomponent LPCVD processes. J. Electrochem. Soc. 132, 448–453.

Schmid, C., Hahn, J., Fuhrmann, C.A., 2019. United States Patent No. 10,384,182. Washington, DC: United States Patent and Trademark Office.

Schroder, D.K., 2001. Surface voltage and surface photovoltage: history, theory and applications. Meas. Sci. Technol. 12 (3), R16.

Setalvad, T., Trachtenberg, I., Bequette, B.W., Edgar, T.F., 1985. Optimization of a low-pressure chemical vapor deposition reactor for the deposition of thin films. Ind. Eng. Chem. Res. 28 (8), 1162–1170.

Shockley, W., Queisser, H.J., 1961. Detailed balance limit of efficiency of p-n junction solar cells. J. Appl. Phys. 32 (3), 510–519.

Siffert, P., Krimmel, E., Singh, D.V., 2005. Library-silicon: evolution and future of a technology. MRS Bull. Mater. Res. Soc. 30 (7), 561.

Silicon Products, Research Engineering Production, 2020. Online https://silicon-products-gmbh.com/hydrogenation-reactor-converter/.

Takiguchi, H., Morita, K., 2011. Global flow analysis of crystalline silicon. Crystalline Silicon-Properties and Uses. IntechOpen.

Tyagi, V.V., Rahim, N.A., Rahim, N.A., Jeyraj, A., Selvaraj, L., 2013. Progress in solar PV technology: research and achievement. Renew. Sustain. Energy Rev. 20, 443–461.

Vinod, Kumar, R., Singh, S.K., 2018. Solar photovoltaic modeling and simulation: as a renewable energy solution. Energy Rep. 4 (2018), 701–712.

Wai, C.M., Hutchison, S.G., 1989. Free energy minimization calculation of complex chemical equilibria: reduction of silicon dioxide with carbon at high temperature. J. Chem. Educ. 66 (7), 546.

Weckend, S., Wade, A., Heath, G.A., 2016. End of Life Management: Solar Photovoltaic Panels (No. NREL/TP-6A20–73852). National Renewable Energy Lab. (NREL), Golden, CO, United States.

Woodhouse, M.A., Smith, B., Ramdas, A., Margolis, R.M., 2019. Crystalline Silicon Photovoltaic Module Manufacturing Costs and Sustainable Pricing: 1H 2018 Benchmark and Cost Reduction Road Map (No. NREL/TP-6A20–72134). National Renewable Energy Lab. (NREL), Golden, CO, United States.

Xakalashe, B.S., Tangstad, M., 2012. Silicon processing: from quartz to crystalline silicon solar cells. Chem. Technol. (March), 6–9.

Yang, D. (Ed.), 2019. Handbook of Photovoltaic Silicon. Springer, Berlin.

Ye, X.S., Shao, Z.G., Zhao, H., Yang, L., Wang, C.L., 2014. Electronic and optical properties of silicene nanomeshes. RSC Adv. 4 (72), 37988–38003.

Zadde, V.V., Pinov, A.B., Strebkov, D.S., Belov, E.P., Efimov, N.K., Lebedev, E.N., et al., 2002). New method of solar grade silicon production. In: 12th Workshop on Crystalline Silicon Solar Cell Materials and Processes: Extended Abstracts and Papers from the workshop held 11–14 August 2002, Breckenridge, Colorado (No. NREL/CP-520–35650). National Renewable Energy Lab., Golden, CO, United States.

Wind based processes

Wind energy: collection and transformation

Mariano Martín

Department of Chemical Engineering, University of Salamanca, Salamanca, Spain

11.1 Wind turbines analysis

11.1.1 Physics of wind power

The energy that can be extracted from the wind flow is evaluated using the continuity equation and an energy balance. When the wind approaches the disk, the streamlines are diverted as it can be seen in Fig. 11.1. As a result, its velocity is decelerated by the resistance originated by the disk. Fluid mechanics of the flow say that the decrease in the velocity generates an increase in pressure. On the contrary, the flow that passes through the disk suffers a sudden pressure drop below atmospheric. Pressure recovers gradually to reach atmospheric once again downstream of the disk.

The mass flow reaching the disk can be written as Eq. (11.1):

$$fc_2 = \rho \cdot A_2 \cdot v_2 \tag{11.1}$$

where ρ is the air density and A_2 is area of turbine disk. The axial force acting on the disk can be determined using Eq. (11.2):

$$F = fc_2(v_3 - v_1) \tag{11.2}$$

Thus the power extracted by the turbine is given by Eq. (11.3):

$$P = Fv_2 = fc_2(v_3 - v_1)v_2 \tag{11.3}$$

Using the continuity equation, we have:

$$fc_1 = fc_2 = fc_3 \tag{11.4}$$

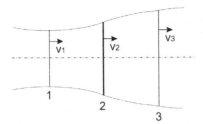

FIGURE 11.1 Streamlines of the air passing across a wind turbine.

The power output from the upstream and downstream winds is computed as in Eq. (11.5):

$$P_w = \frac{1}{2}fc(v_3^2 - v_1^2)$$

$$P_w = \frac{1}{2}fc(v_3 - v_1)(v_3 + v_1)$$

(11.5)

If the velocities up and downstream are the same, no power is produced. The velocity downstream cannot be 0; otherwise, the mass flow rate would also be 0. Therefore to achieve maximum power, v_3 is required to be within 0 and the incoming velocity v_1. For that, we need to know v_2. Assuming that there are no losses, the power that the wind turbine absorbs is the one that the wind loses, Eq. (11.6):

$$P = P_w$$

(11.6)

Comparing Eq. (11.3) with Eq. (11.5), the velocity at the disk can be obtained as follows:

$$v_2 = \frac{1}{2}(v_3 + v_1)$$

$$P_w = \frac{1}{2}\rho A_2 v_1^3 \left[\frac{1}{2}\left(1 + \frac{v_3}{v_1}\right)\left(1 - \left(\frac{v_3}{v_1}\right)^2\right) \right]$$

$$P_w = \frac{1}{2}\rho A_2 v_1^3 C_p$$

(11.7)

$$C_p = \frac{1}{2}\left(1 + \frac{v_3}{v_1}\right)\left(1 - \left(\frac{v_3}{v_1}\right)^2\right)$$

$$C_p = \frac{P_{out}}{P_{available}}$$

After this development, the power available in the wind is given as follows (Kalmikov and Dykes, 2011):

$$P_{available} = \frac{1}{2}\rho A v^3$$

(11.8)

However, the maximum power available is computed using Betz limit as given by Eq. (11.9):

$$P_{Max} = \frac{16}{27}\frac{1}{2}\rho A v^3 \qquad (11.9)$$

It is possible to obtain values of C_P from 0.3 to 0.35 for most machines of good design. Thus the efficiency of each of the turbines is defined as the parameter ν given by Eq. (11.10):

$$P = \nu P_{Max} \qquad (11.10)$$

The measurements of the wind velocity are provided by meteorological services. Sometimes, the velocities are not provided at the proper altitude or have not been previously corrected. Due to the closer presence to the ground, the wind velocity increases with height as the shear diminishes. It is possible to correct the velocity to the one to be used by the turbine at the turbines' hub by the power law method, Eq. (11.1), where the friction coefficient α has been reported in the literature for different terrain types (Patel, 1999), see Table 11.1.

$$v_{H_2} = v_{H_1}\left(\frac{H_2}{H_1}\right)^a \qquad (11.11)$$

11.1.2 Types of turbines

Two major turbines designs can be identified, horizontal axis wind turbine (HAWT) and vertical axis wind turbine (VAWT) (Eriksson et al., 2008) (Fig. 11.2).

✓ HAWT turbines are the most extended for large wind farms due to the possibilities of accessing to higher wind speed, lower ground usage, control of

TABLE 11.1 Ground friction coefficient.

Terrain type	α
Lake, oceans, and smooth hard ground	0.1
Foot high grass on level ground	0.15
Tall crops, hedges, and shrubs	0.2
Wooded country with many trees	0.25
Small town with some trees and shrubs	0.3
City area with tall buildings	0.4

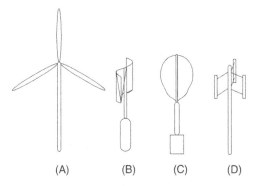

FIGURE 11.2 Wind turbine designs: (A) HAWT; (B) Savonius; (C) Darrieus; (D) H-Rotor. (*HAWT*, Horizontal axis wind turbine).

(A)   (B)   (C)   (D)

the angle of attack, and high efficiency because the blades face the wind perpendicularly.

It has been shown that a three-bladed design is the most efficient configuration since it is possible to achieve better balance among efficiency, blade stiffness, vibration, and noise in spite of the lower weight and cost that two-blade turbines can provide. Sound is one of the disadvantages of horizontal axis turbines together with the difficult transportation and installation as a result of the size of the units. In addition, for better efficiency, there is a need to turn the blades to the wind.

The geometry of the blade and, in particular, the blade pitch angle, determines the efficiency of the turbine (Khalfallaha and Koliubb, 2007). The lift-driven turbines force balance is described in Fig. 11.3.

The lift and drag forces generated by the air flow as it passes by the airfoil are given by Eq. (11.12) and Eq. (11.13), respectively:

$$F_L = \frac{1}{2}\rho v^2 c C_l \tag{11.12}$$

$$F_D = \frac{1}{2}\rho v^2 c C_d \tag{11.13}$$

The C_p of the turbine blade can be determined from them as in Eq. (11.14):

$$C_p = \frac{P}{P_{wind}} = \frac{8}{\lambda^2}\int_{\lambda_r}^{\lambda} Q\lambda_r^3 a'(1-a)\left(1 - \frac{C_d}{C_l}\tan\beta\right)d\lambda_r \tag{11.14}$$

Collective angle β is the sum of the mounting pitch angle θ and the design twist angle α. The tip velocity, λ_r, is calculated as Eq. (11.15):

$$\lambda_r = \frac{\varpi r}{V_1} \tag{11.15}$$

While a and a' are given by Eq. (11.16):

$$\frac{a}{1-a} = \frac{\sigma(C_l\cos\beta + C_d\sin\beta)}{4Q\sin^2\beta}$$

$$\frac{a'}{1-a} = \frac{\sigma(C_l\sin\beta - C_d\cos\beta)}{4Q\lambda_r\cos^2\beta} \tag{11.16}$$

where

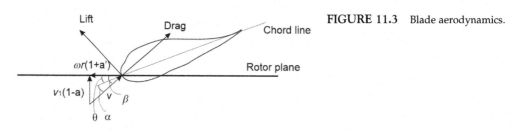

FIGURE 11.3   Blade aerodynamics.

$$\beta = \arctan\left(\frac{\lambda_r(1+a')}{(1-a)}\right) = \frac{2}{3}\arctan\left(\frac{1}{\lambda_r}\right) \tag{11.17}$$

$$\sigma = \frac{Bc}{2\pi r} \tag{11.18}$$

where B *corresponds to* the number of blades, c—the local cross-section chord, r—the radial distance of the cross-section from the hub center:

$$Q = Q_1 Q_2 = \frac{2}{\pi}arcos\left(e^{\frac{3(R-r)}{2r\sin\beta}}\right)\frac{2}{\pi}arcos\left(e^{\frac{3(r-R_{hub})}{2R_{hub}\sin\beta}}\right) \tag{11.19}$$

In addition, the C_p has been correlated as given by Eq. (11.20):

$$C_p = 0.73\left(\frac{151}{\lambda_i} - 0.58\theta - 0.002\theta^{2.14} - 13.2\right)e^{\frac{-18.4}{\lambda_i}}$$

$$\lambda_i = \frac{1}{\frac{1}{\lambda_r - 0.02\theta} - \frac{0.003}{(\theta^3 + 1)}} \tag{11.20}$$

where θ is the pitch angle of rotor blades and the tip speed ratio, λ_r, is defined above.

The geometry of the blade can be designed using two design methods, the direct and the indirect one:

✓ Direct design method: It also receives the name of design by analysis approach. The curvature of the blade determines the flow separation that occurs at the point of discontinuity of the blade surface. To achieve smooth surface curvature, it is characterized by a series of parameters to describe its geometry. The parameters are optimized to achieve the proper pressure distribution through successive iterations in computational fluid dynamics (CFD) simulation codes where the performance of the shape is analyzed.
✓ The inverse method is defined as the blade geometry found in a prescribed pressure distribution at specified operational conditions.
✓ The VAWT turbines present several designs, see Fig. 11.2 including the Darrieus/Egg-Beater characterized by the fact that it is lift force driven as well as the H-rotor one while the Savonius is drag force driven. While these types of turbines are easy to maintain and show low-investment cost and low noise, they have low efficiency.

11.1.3 The power curve

The design of the blade determines the power that the turbine can extract from the wind. While the theoretical power is a cubic polynomial, the nonidealities curve down the production of power out of the wind velocity. It is possible to find the power curves of different industrial turbines in software packages such as System Advisor Model (SAM). Fig. 11.4 shows an example for a particular turbine. The power curve is characterized by different velocities as the curvature changes such as the cut-in speed (V_c), where the

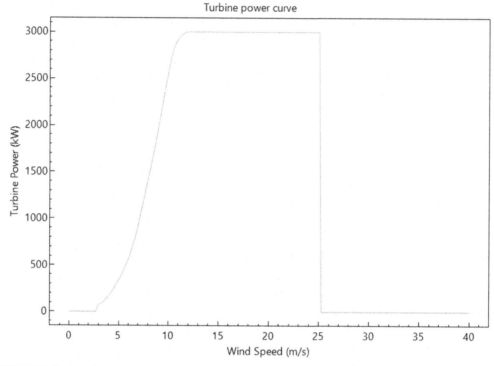

FIGURE 11.4 Commercial turbine power curve (NREL, 2020).

turbine starts producing power, and the velocity at nominal speed (V_r). Different models have been used to represent the curve (Jangamshetti and Rau, 1999; Salameh and Safari, 1992; Dialynas and Machias, 1989; Justus et al., 1976; Giorsetto and Utsurogi, 1983; Pallabazzer, 1995; Albady and El-Saadany, 2010; De la Cruz and Martin, 2016).

1. The linear model assumes a linear increase in the turbine output between the cut-in and the nominal speeds. This model, generally, overestimates wind potential. The linear model is given by Eq. (11.21):

$$P(v) = \frac{v - V_c}{V_r - V_c} \tag{11.21}$$

2. Cubic model 1, Eq. (11.22), assumes a constant overall efficiency of the turbine throughout the ascending segment of the power curve:

$$P(v) = \frac{(v - V_c)^3}{(V_r - V_c)^3} \tag{11.22}$$

3. Cubic model 2, Eq. (11.23), does not include V_c in the model:

$$P(v) = \frac{(v)^3}{(V_r)^3} \tag{11.23}$$

4. Quadratic model, Eq. (11.24), assumes that between $(V_c + V_r)/2$ and V_r that the output of the turbine increases according to the cubic model:

$$P(v) = a_0 + a_1 v + a_2 v^2 \qquad (11.24)$$

5. Quadratic model 2, Eq. (11.25), does not include the linear term $(a_1 v)$ of the previous model:

$$P(v) = \frac{(v^2 - V_c^2)}{(V_r^2 - V_c^2)} \qquad (11.25)$$

6. Based on the particular shape of the power output curves, instead of using piecewise modeling, like most of the models presented above, which are only valid for the curve, we use the following correlation given by Eq. (11.26) to fit the power curve of the wind turbine:

$$P(v) = \frac{P_{nom}}{1 + e^{(-(v-a)m)}} \qquad (11.26)$$

where v is the wind velocity (m/s), and a and m are characteristic adjustable parameters for each turbine, see Table 11.2 and Table 11.3. The utility-scale turbines that dominate the

TABLE 11.2 Characterization of onshore wind turbines (De la Cruz and Martin, 2016).

Turbine	P_{nom} (kW)	a	m	Diameter (m)
Liberty C99	2500	8.99612238	0.75858984	100
Liberty C96	2500	9.11954338	0.74777261	96
Liberty C93	2500	9.31573362	0.72649051	93
Liberty C89	2500	9.50860137	0.71727851	89
Gamesa G90 2.0 MW	2000	8.12256200	0.77318347	90
GE 1.5sle	1500	8.20833453	0.76436618	77
GE 2.5xl	2500	8.46782489	0.77313734	100
Mitsubishi MWT 92/2.4	2400	8.1762053	0.75682497	92
Nordex N100−2500	2500	8.22620202	0.80526163	100
Nordex N90−2500 HS	2500	8.86235320	0.73650534	90
Nordex N90−2500 LS	2500	8.95156390	0.69633716	90
Nordex N80−2500	2500	9.71315552	0.63333904	80
RePower MM92	2050	8.78404079	0.75732222	92
RePower MM82	2050	9.64379092	0.62035901	82
Sulzon S82 1.5	1500	8.08403279	0.77725327	82
Vestas V90−2.0	2000	8.37466968	0.77297558	90
Vestas V112−3.0	3000	8.05709522	0.77613855	112
Vestas V90−3.0	3000	9.78894949	0.60363792	90

total wind energy production range in size from 500 kW to 5 MW, depending on application and location. Onshore turbines are typically smaller, up to 3 MW while offshore ones reach 5 MW (NREL, 2020).

For wind turbines, the **capacity factor** is defined as the fraction of the year the turbine generator is operating at rated (peak) power.

11.1.4 Cost of the wind turbines

In Chapter 1, we presented the typical cost for the installed turbine as an average value per kilowatt installed. However, a more detailed analysis can be performed. Fig. 11.5 presents the breakdown of the costs for the different major components. In particular, the tower constitutes 26% of the cost, the blades 22%, and the gearbox around 13%. Apart from the major contributions to the costs, there are a number of mechanical and electrical elements that transform the movement of the blades into electricity such as the rotor hub, the bearings, the Yaw system and the pitch systems, the generator, the power converter, the transformer, or just the main shaft (Blanco, 2009).

TABLE 11.3 Characterization of offshore wind turbines (De la Cruz and Martin, 2016).

Turbine	P_{nom} (kW)	a	m	Diameter (m)
BARD 5.0	5000	9.189476878	0.77198127	122
Areva Multibrid m5000	5000	9.48633171	0.73235885	116
RePower 5 M	5000	9.34351326	0.72201545	126
NREL 5 MW offshore ref	5000	8.71181297	0.75573577	112

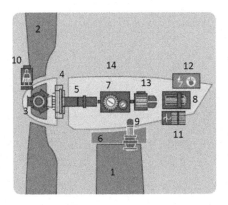

Cost distribution
1. Tower: 26.3%
2. Rotor Blades: 22.2%
3. Rotor hub: 1.37%
4. Rotor bearings: 1.22%
5. Main shaft: 1.91%
6. Main frame: 2.8%
7. Gearbox: 12.91%
8. Generator: 3.44%
9. Yaw system: 1.25%
10. Pitch system: 2.66%
11. Power converter: 5.01%
12. Transformer: 3.59%
13. Brake system: 1.32%
14. Nacelle housing: 1.35%
Cables 0.96% ; Screws: 1.04%

FIGURE 11.5 Wind turbine costs (EWEA, 2020). Source: *Data from EWEA, 2020. European Wind Energy Association and International Economic Development Council, The Economics of Wind Energy A report by the European Wind Energy Association. <https://www.ewea.org/>.*

11.1.5 Production of electricity

In the previous section, we have described the different components of a wind turbine so that the kinetic energy from the wind is transformed into electricity. It is possible to classify the utility-scale wind turbines into four types with regard to the conversion of energy to electricity (Ellis et al., 2011).

Type 1 (fixed speed): Most of the original utility-scale turbines use fixed speed rotors. They employed squirrel-cage AC induction machines with a direct grid connection, and many did not have the ability to feather their blades. The advantage of this design was that they are relatively reliable; however, the fixed speed rotor limits the energy capture from the wind and requires reactive power compensation, often in the form of capacitor banks. See Fig. 11.6A.

Due to the disadvantage of the use of fixed rotor, the rest of the turbine types are designed with variable speed rotors to allow for improved wind energy capture and utilize blade pitching for this purpose.

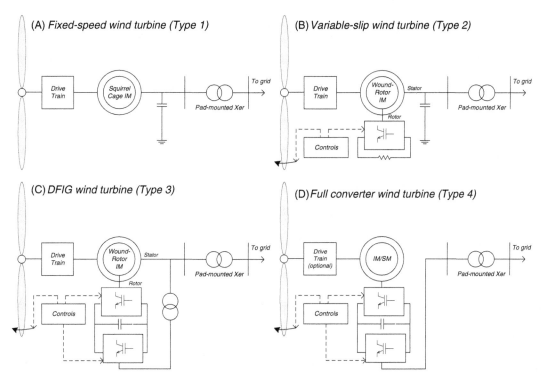

FIGURE 11.6 Wind turbine technologies (Singh et al., 2014). Source: *Reprinted from Singh, M., Muljadi, E., Jonkman, J., Gevorgian, V., Girsang, I., Dhupia, J., 2014. Simulation for Wind Turbine Generators — With FAST and MATLAB-Simulink Modules. National Renewable Energy Laboratory, Golden, CO, with permission from the National Renewable Energy Laboratory, https://www.nrel.gov/docs/fy14osti/59195.pdf, Accessed June 18, 2020. Also please note that the NREL developed figures are not to be used to imply an endorsement by NREL, the Alliance for Sustainable Energy, LLC, the operator of NREL, or the United States Department of Energy.*

Type 2 (variable-slip) turbines, see Fig. 11.6B, offer a wider range of operation speed variations, but the rotor resistance required to do so results in power losses.

Type 3 (doubly fed induction generator) turbines, see Fig. 11.6C, can recover some of this lost power using an AC/DC/AC converter. This also allows for the provision of both real and reactive power output.

Type 4 (full converter) turbines, see Fig. 11.6D, utilize an AC/DC/AC converter as the sole connection to the grid.

11.2 Turbines layout and selection

The wind farm design problem has been extensively studied in the past years. We can distinguish between two problems. The first one deals with the *layout of the turbines* to make the most of the wind resource. The first of these studies dates back to 1994 where the wind farm area was grid-discretized and the optimizer was set to obtain layouts that would increase the wind farm efficiency (Mosetti et al., 1994). Instead of a discretized area, the next step was to consider the wind farm as a continuous space (Lackner and Elkinton, 2007). The allocation of the turbines is based on the wake model, the generation of wake due to the presence of a wind turbine upstream that modifies the wind flow pattern reducing the energy that can be obtained from a turbine downstream. Tools such as SAM model by the NREL allow developing such layouts. Another problem related to the wind farm corresponds to the *selection of the turbine* (Jangamshetti and Rau, 1999; Salameh and Safari, 1992). The particular power curve of each turbine adjusts better to the wind profile of a region (De la Cruz and Martin, 2016).

11.2.1 Wind farm

The studies on turbine layout result in a number of guidelines for the design of the wind farm. It is based on the analysis of the wake. While CFD studies are required for each particular location, simpler analyses are also very useful for initial stages. To compute the velocity downstream we can use the wake model given by Eq. (11.27):

$$Deficit = \left(1 - \frac{v}{v_o}\right) = \frac{2a}{\left(1 + k\frac{x}{R}\right)^2} \tag{11.27}$$

where v is the velocity downstream, v_o is the unaffected wind speed, a is the axial induction factor, defined by the wind speed far away from the rotor, v_∞, and that at the rotor, v_{rotor}

$$a = \frac{v_\infty - v_{rotor}}{v_\infty} \tag{11.28}$$

k is the entrainment constant, Eq. (11.29), x the distance downstream, and R is the downstream rotor radius.

$$k = \frac{0.5}{\ln\left(\frac{z}{z_o}\right)} \tag{11.29}$$

where z_o is the surface roughness and z is the hub height.

$$C_P = 4a(1-a)^2 \qquad (11.30)$$

$$R = \left(\frac{D}{2}\right)\sqrt{\frac{1-a}{1-2a}} \qquad (11.31)$$

where C_p is the thrust coefficient that has the Betz limit of 16/27 as presented in the first section of this chapter and D the diameter.

To extract the most power from the wind, typical turbine density is one turbine per 4D abreast 7D in line with the prevailing wind resulting in an area of influence of each turbine of 28 D^2 (Archer, 2004), to avoid interference among turbines that hinders the efficiency. Typically, shorter spacing is recommended in the case of onshore turbines due to the wake generated, requiring 3–5D, compared to offshore ones that require larger spacing, 5–8 D. Thus the manufacturers recommend separation distances of seven to eight rotor diameters for turbines in line and five rotor diameters for turbines abreast (see Example 11.1).

EXAMPLE 11.1 Optimize the downstream distance between two turbines when the incoming free velocity is 8 m/s and the turbine power curve is the one given by the Vestas 112 model with a turbine diameter of 100 m. Assume $C_p = 0.5$, a hub height of 80 m and a roughness of 0.3 m.

Solution

We formulate the problem as a multiobjective one due to the conflicting variables distance between turbines that reduce the velocity for the second turbine and the power that can be extracted, and the cost or area availability. Thus we produce a Pareto curve of the power as a function of the distance.

```
positive variable

a,Distance,Power2, Power1, k, Radius,vel, area;

variable
obj, Powerfarm;

Scalar

cp              /0.5/
zo        m     /0.3/
zhub      m     /80/
vo        m per s      /8/
Diameter m    /100/
atur          /8.0571/
mtur          /0.77614/
Carea      € per m2    /10/;
equations

eq1, eq2, eq3, eq4, eq5, eq6, eq7, eq8, eq9;
```

```
eq1..       cp = E = 4*a*(1-a)*(1-a);
eq2..        k*log(zhub/zo) = E = 0.5;
eq3..     Radius  = E = (Diameter/2)*((1-a)/(1-2*a))**0.5;
eq4..     Power1  = E = 3000/(1+EXP(-(vo-atur)*mtur));
eq5..       Power2 = E = 3000/(1+EXP(-(vel-atur)*mtur));
eq6..         (1-vel/vo)*(1+k*distance*Diameter/(Radius+0.001))**2 = E = 2*a;
eq7.. Powerfarm = E = Power1 + Power2;
eq8.. area = E = 4*Diameter*Diameter *distance;
eq9.. obj  = E = area*Carea;

Distance.UP = 12;
Option MINLP = Baron;
Model separation /ALL/;
Solve separation Using MINLP Maximizing Powerfarm;
```

See Fig. 11E1.1.

FIGURE 11E1.1   Pareto curve of power versus separation.

11.2.2 Turbine design location problems

The selection of turbine and site location depends on the characteristic power curve of each turbine and the wind profile of the site over time (De la Cruz and Martin, 2016). Thus, there exists an optimal match between site and turbine design. To find that match, a formulation is developed, Eq. (11.32), aiming at the highest power production. The number of wind turbines is considered as an integer variable. We define $y_{(i,j)}$ as the binary variable that allows the selection of the best combination of turbine (i) and site location (j). Since we will only allow the selection of one turbine and site location, if a turbine is not selected within a site location, it will not produce any power and will not contribute to the cost. Thus variable $Pr_{(i,j)}$ is defined using a big-M type of constraint. The maximum area available is to be defined. Note that the turbines do not actually use this area since it can still be used for other purposes. The wind velocity is taken from the site location.

$$Max \quad TotalPower = \sum_i \sum_j Pr_{(i,j)} \cdot n_{turbine(i,j)}$$

s.t.

$$P_{(t,i,j)} = \frac{Pr_{(i,j)}}{1 + e^{(-(v_{(t,j)} - a_{(i)}) \cdot m_{(i)})}} \quad \forall t, i, j$$

$$Pr_{(i,j)} = P_{nom(i)} \cdot y_{(i,j)} \quad \forall i, j$$

$$n_{turbine(i,j)} \leq N \cdot y_{(i,j)} \quad \forall i, j$$

$$n_{turbine(i,j)} \cdot 28 \cdot D^2 \leq A_{Max}$$

$$\sum_i \sum_j y_{(i,j)} = 1$$

(11.32)

However, wind velocity presents uncertainty. To characterize the uncertainty in the wind velocity, a Weibull distribution is typically used (Albadi and El-Saadany, 2010). The probability of occurrence of a particular wind velocity is given by Eq. (11.33):

$$f(v) = \frac{k}{c} \left(\frac{v}{c}\right)^{k-1} e^{-\left(\frac{v}{c}\right)k}$$

(11.33)

where c and k are the scale and shape parameters, respectively. The distribution function is as follows:

$$F(v) = 1 - e^{-\left(\frac{v}{c}\right)k}$$

(11.34)

To add the variability in the wind velocity to the previous formulation for the selection of the turbine design, it is possible to formulate a design model in for the maximum power produced. $\theta_{(i,sc)}$ is the probability that a certain scenario, sc, characterized by a wind velocity, occurs. Example 11.2 shows an example of turbine design selection

---

EXAMPLE 11.2 For 11 allocations, with the monthly wind profile in Table 11E2.1 select the best turbine among Liberty C99, Nordex N100–2500, Vestas V112–3.0, and the best location. Consider a maximum of 5,000,000 m².

Solution

We formulate the problem using Eq. (11.23) considering a set of j consisting of 11 allocations and a set of I of three turbines; 12 monthly velocities are available (Table 11E2.1).

TABLE 11E2.1 Wind monthly (m) velocities for 11 locations (p).

m/p	1	2	3	4	5	6	7	8	9	10	11
1	8	5	5	7	5	4	7	4	6.5	8	6
2	7	9.5	5	7	6	6	7.5	5.5	7.5	14	10
3	7.5	6.5	5.5	5.5	6	5	5	5	6	8.5	6
4	9.5	9.5	6	7	6	6	7	6	8.5	9	7.5

(Continued)

TABLE 11E2.1   (Continued)

m/p	1	2	3	4	5	6	7	8	9	10	11
5	7.5	6	4.5	5	6	4	5.5	4.5	5.5	7	5.5
6	10	11	4.5	6.5	4	5	7	4.5	5.5	7	6.5
7	8	8	4.5	5	5.5	4.5	6	4.5	6.5	8	5
8	7.5	8	4.5	5	5	4.5	6	4.5	7	7	5
9	8.5	9.5	4	4.5	4	4	6	4	6	6.5	4.5
10	6.5	8	5	7	5	4.5	6.5	5	5.5	8	5
11	7	8.5	5	4.5	7	4	7	5.5	7.5	8	5
12	9.5	7.5	7	7.5	7	6.5	9	6	8	9	8.5

We model the problem in GAMS as follows. Note that the sets, parameters, and tables definitions have not been included for the sake of the space available and can be found on-line.

```
Positive variables
P(i,j,city), sumaPo(i,city), Pot(i,city);

Integer variable
ntur(i,city);
Variable
z;
binary variables
y(i,city);

Equations
Power, Obj, totalturb, potvend, bin, area, numero_t;

Power(i,j,city)..    P(i,j,city)*(1+exp((a(i)-V(j,city))*m(i)))=e=Pot(i,city);
totalturb(i,city).. sumaPo(i,city) =E= ntur(i,city)*Sum(j,P(i,j,city));
potvend(i,city)..    Pot(i,city) =E= Pnom(i)*(y(i,city));
area(i,city)..    ntur(i,city)*28*Diam(i)*Diam(i) =L= 5000000;
numero_t(i,city)..            ntur(i,city) =L= 100*y(i,city);
Obj..z=e=    sum((i,city),sumaPo(i,city));
bin..Sum((i,city),y(i,city))=e=1;
Option MINLP=Baron;
Model WTselection /ALL/;
Solve WTselection    Using MINLP Maximizing z;
```

Location #2 and turbine Vestas V112−3.0 is the one selected, installing 14 turbines.

11.3 Electrolysis

The production of electricity from wind is intermittent as the resource itself. That variability has encouraged the development of technology for the use and storage of wind energy in the form of more handy energy carriers. It is here where water electrolysis for the production of hydrogen has taken the lead. While energy storage will be covered in more detail in Chapter 14, here we focus on the production of hydrogen toward its use for the production of bulk chemicals including methane, methanol, and dimethylether (DME).

11.3.1 Electrolyzer types

The most typical electrolyzer technologies are alkaline electrolysis cells, proton exchange membrane electrolysis cells (PEMEC), and solid oxide electrolysis cells (SOEC), Fig. 11.7 shows the three cases:

Alkaline electrolyte solutions: The semireactions taking place are as follows:

$$H_2O + 2e^- \rightarrow H_2 + 2OH^- \quad \text{(Cathode)}$$
$$2OH^- \rightarrow \frac{1}{2}O_2 + H_2O + 2e^- \text{(Anode)} \tag{11.35}$$

There are two typical designs, Tank-type electrolyzers/unipolar and filter press cells/bipolar electrolyzers, that date back to the 1920s, but they are losing interest over time due to limitations in coupling with variable energy sources.

1. Tank-type electrolyzers: The electrodes are connected in parallel, and the number of cells needed to reach the production capacity represents the electrolytic system. The potential difference is the same for each cell. The H_2 and the O_2 are collected through pipes connected to the cathodic and anodic places, respectively, see Fig. 11.7A. The low current density and operating pressure increase the hydrogen production cost. In addition, they do not respond quickly to the variable power input of renewable energy.

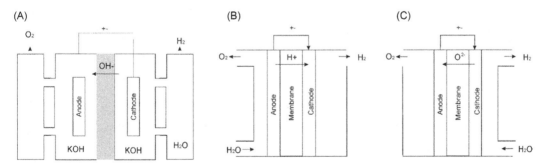

FIGURE 11.7   Scheme of the (A) AEC; (B) PEMEC; (C) SOEC electrolyzers. *AEC,* Alkaline electrolysis cell; *PEMEC,* proton exchange membrane electrolysis cell; *SOEC,* solid oxide electrolysis cell.

2. Filter press cells, bipolar electrolyzers, have their electrodes in series. One electrode acts as both cathode and anode, one surface each so that the configuration is compact. Between electrodes, there is a diaphragm. The unit is connected to the grid through the terminal electrodes. This configuration allows operating at higher potential difference that is the summation of that corresponding to each element and low current intensity.

In both cases, the electrolyte solution flows to be cooled down continuously and is recycled back to the electrolyzer. This stage represents the major water consumption of the plant.

Solid polymer electrolytes: The structure can be seen in Fig. 11.7B. A solid electrolyte is located between two electrodes. The most typical membrane is perfluorosulfonate polymer membrane and operates from 20°C to 100°C. Water is in contact with the anode while hydrogen diffuses across the polymer. H_2 is produced at the cathode. The advantages compared to alkaline technology are that the electrolyte is not corrosive, the design is 4 or 5 times more compact, it shows high power density and cell efficiency as well as a quick response to the current signal, making it convenient for its use with solar and wind power (Schmidt et al., 2017). Hydrogen is produced at moderate or high pressure while oxygen is produced at atmospheric pressure (differential pressure technology). Currently, the production costs are below $10/kg, but price reductions are expected with technical development (USDOE, 2020):

$$2H^+ + 2e^- \rightarrow H_2 \quad \text{(Cathode)}$$
$$H_2O \rightarrow 2H^+ + \frac{1}{2}O_2 + 2e^- \text{(Anode)} \tag{11.36}$$

SOEC: SOECs use solid ion-conducting ceramics as the electrolyte, enabling operation at significantly higher temperatures, see Fig. 11.7C. Potential advantages include high electrical efficiency, low material cost, and the option to operate in reverse mode as a fuel cell or in coelectrolysis mode producing syngas ($CO + H_2$) from water steam (H_2O) and carbon dioxide (CO_2):

$$O^{2-} \rightarrow \frac{1}{2}O_2 + 2e^- \quad \text{(Anode)}$$
$$H_2O + 2e^- \rightarrow H_2 + O^{2-} \text{(Cathode)} \tag{11.37}$$

Table 11.4 summarizes the main characteristics of these three systems (Schmidt et al., 2017) including the consumption of energy and the cost.

11.3.2 Analysis of polymer electrolyte membrane electrolyzers

Water decomposition into its components requires providing its formation energy, ΔH_f, is 68.3 kcal/mol at 25°C. However, only the fraction corresponding to the Gibbs free energy can be provided as work, $\Delta G = 56.7$ kcal/mol at 25°C, and the rest, $T \Delta S$, needs to be provided as heat.

$$\Delta G_f = \Delta H_f - T \Delta S \tag{11.38}$$

TABLE 11.4 Summary of operating conditions for the AEC, PEMEC, and SOEC electrolyzers.

	AEC	PEMEC	SOEC
Electrolyte	KOH (20%–40%)	Polymer membrane	YSZ
Cathode	Ni, Ni–Mo Alloys	Pt, Pt–Pd	Ni/YSZ
Anode	Ni, Ni–Co Alloys	RuO_2, IrO_2	LSM/YSZ
T(°C)	60–80	50–80	650–1000
P(bar)	<30	<200	<25
Voltage efficiency (%$_{HHV}$)	62–82	67–82	<110
Production rate (m^3 H_2/h)	<760	<40	<40
Energy (kWhe/m^3 H_2)	4.5–6.6	4.2–6.6	>3.7
System response	s	ms	s
Cold start time (min)	<60	<20	<60
Stack lifetime (h)	60,000–90,000	20,000–60,000	<10,000
Cost (€/kW)	1000–1200	1660–2320	>2000

AEC, Alkaline electrolysis cell; *PEMEC*, proton exchange membrane electrolysis cells; *SOEC*, solid oxide electrolysis cell; *YSZ*, yttria-stabilized zirconia; *HHV*, high heating value; *LMS*, $La_{0.8}Sr_{0.2}MnO_3$.
Data from Schmidt, P., Gambhir, A., Staffell, I., Hawkes, A., Nelson, J., Few, S., 2017. Future cost and performance of water electrolysis: an expert elicitation study. Int. J. Hydrog. Energy 43, 30470–30492.

The theoretical minimum for the reversible water decomposition can be computed using the Nerst equation as follows:

$$\Delta G = nFE = 2 \times 96,540 \times E$$
$$\Delta G \rightarrow E = 1.23 \text{ V at } 25°C$$
$$\Delta H \rightarrow E_{eq} = 1.48 \text{ V at } 25°C \tag{11.39}$$

The potential difference between both has to be provided by heat. In practice, the actual potential to be applied needs to be from 1.8 to 2.2 V due to the Ohmic loss in the electrolyte, those of the electrodes, and the polarization overpotential (Falcao and Pinto, 2020).

The voltage that must be applied include all the terms related to the losses as in Eq. (11.40):

$$V = E_{Cell} + V_{act} + iR_{Cell} \tag{11.40}$$

Open circuit voltage: is usually determined using the Nernst equation, Eq. (11.41):

$$E_{Cell} = E_{rev}^0 + \frac{RT}{2F}\left[\ln\left(\frac{p_{H_2} \cdot p_{O_2}^{0.5}}{p_{H_2O}}\right)\right] \tag{11.41}$$

where p is the partial pressure of species involved, T is the temperature, F is the Faraday constant, and E_{rev}^0 is the reversible cell potential at standard temperature and pressure. To compute it, there are some empirical correlations from the literature as a function of the pressure and temperature:

$$E_{rev}^0 = 1.5241 - 1.2261 \cdot 10^{-3}T + 1.1858 \cdot 10^{-5}T\ln(T) + 5.6692 \cdot 10^{-7} \cdot T^2$$
$$E_{rev}^0 = 1.229 - 0.9 \cdot 10^{-3}(T - 298)$$
$$E_{rev}^0 = 1.229 - 0.9 \cdot 10^{-3}(T - 298) + 2.3\frac{RT}{4F}\log\left(p_{H_2}^2 p_{O_2}\right) \tag{11.42}$$

Alternatively the open circuit voltage can also be estimated using empirical correlations for atmospheric pressure, that is, Eq. (11.43):

$$E_{cell} = 1.5184 - 1.5421 \cdot 10^{-3}T + 9.523 \cdot 10^{-5}T\ln(T) + 9.84 \cdot 10^{-8}T^2 \tag{11.43}$$

Activation overpotential: Activation losses occur because a fraction of the potential is needed to activate the electrochemical reactions taking place at anode and cathode sides. Although activations at both electrodes can be computed, alternatively it can be simplified by Eq. (11.44) that can be applied to each electrode, Tafel's equation:

$$V_{Act} = \frac{RT}{n\alpha F}\ln\left(\frac{i}{i_o}\right) \tag{11.44}$$

where i_o is the exchange current density (at anode and cathode) and α the charge transfer coefficient (one for the anode and another for the cathode, respectively).

The mass transport overpotential occurs when the current density is high enough to impede the access of reactants to active sites by the overpopulation of reacting molecules and for this reason slowing down the reaction rate. It can also be called diffusion overpotential (V_{Diff}) and can be estimated using the Nernst equation:

$$V_{Diff} = \frac{RT}{nF}\ln\left(\frac{C}{C_o}\right) = -\frac{RT}{nF}\ln\left(1 - \frac{i}{i_{lim}}\right) \tag{11.45}$$

Ohmic losses: The ohmic overpotential is related with the materials resistance to the protons flux.

$$V_{Ohm} = RI = \frac{\delta}{\sigma}I \tag{11.46}$$

where δ is the material thickness and σ the material conductivity. Example 11.3 shows the energy consumption in the production of hydrogen via electrolysis.

EXAMPLE 11.3 We would like to produce 0.5 kg/s of hydrogen at 8 bar via water electrolysis at 1 bar in region no. 2 in Example 11.2. The oxygen produced has to be compressed up to 100 bar for its storage. Assume complete water breakup into hydrogen and oxygen, and that out of the cathode we produce pure hydrogen, and we collect oxygen from the anode. The compressors efficiency is 85% and the electrolysis occurs at 80°C. The electrolytic cell has an energy ratio defined as $\eta = \frac{n_{electrons} \cdot F \cdot V}{E_e}$ is equal to 65%. Assume that the polytropic coefficient is $k = 1.4$.

    a. Compute the energy required for the process.

b. Compare the cost of the wind farm if a Vestas 112–3.0 or a GE 2.5xl are used at 1600 €/kW.
Solution

a. The energy required is that needed to split the water and for gases compression.

$$\text{Energy} = E + W(\text{Oxygen line}) + W(\text{Hydrogen line})$$

For the electrolyzer we have the following reaction:

$$H_2O \rightarrow \frac{1}{2}O_2 + 2H^+ + 2e^- \quad E = -1.23 \text{ V}$$

$$2H^+ + 2e^- \rightarrow H_2 \qquad\qquad E = 0 \text{ V}$$
$$E = 0 + (-1.23) = -1.23 \text{ V}(\text{Not spontaneous})$$

$$\eta = 0.65 = \frac{n_{H_2} \cdot e \cdot F \cdot V}{E_e} = \frac{\frac{500 \text{ g/s}}{2} \cdot 2 \cdot 96500 \cdot 1.23}{E_e} \Rightarrow E_e = 91304 \text{ kW} \tag{11E3.1}$$

The energy consumption of the compressors is computed using the following equation. Considering that the compression cannot be made in one stage, due to the pressure difference, but in two or three stages for hydrogen and oxygen, respectively, due to the pressure ratio 8 and 100, the work used is computed as follows:
The consumption of the H_2 line, Eq. (11E3.2), is equal to 2090 kW:

$$W(Comp) = 2(F) \cdot \frac{8.314 \cdot k \cdot (T_{in} + 273.15)}{((MW) \cdot (k-1))} \frac{1}{\eta_s} \left(\left(\frac{P_{out}}{P_{in}} \right)^{\frac{1k-1}{2\,k}} - 1 \right) \tag{11E3.2}$$

The consumption of the O_2 line, Eq. (11E3.3), is equal to 2464 kW:

$$W(Comp) = 3(F) \cdot \frac{8.314 \cdot k \cdot (T_{in} + 273.15)}{((MW) \cdot (k-1))} \frac{1}{\eta_s} \left(\left(\frac{P_{out}}{P_{in}} \right)^{\frac{1k-1}{3\,k}} - 1 \right) \tag{11E3.3}$$

The number of turbines of each of the two types is computed in the following table as a function of the wind velocity and the power curve. The Vestas requires a smaller number of units, 63, but since the nominal power is larger, the cost at 1600 €/kW is similar.

		Vestas	GE
	Pnom	3000	2500
	atur	8.0571	8.4678
	mtur	0.77614	0.7731
Power average		1524.64453	1120.129678
n turbines		62.872315	85.57753036
		63	86
€		302,400,000	344,000,000

11.3.3 CO_2 hydrogenation

Once the water is split, both gas streams must be purified. If the gas stream contains water, it is first condensed, and later it is compressed and dehydrated using silica gel or similar adsorbent. For the hydrogen-rich stream, we do not only need to remove the moisture but also the traces of oxygen. Therefore a deoxo reactor uses a small amount of the hydrogen produced to convert the traces of oxygen into water. It typically operates at 90°C and 8 bar. The reaction is exothermic and follows the stoichiometry presented as follows:

$$2\,H_2 + O_2 \rightarrow 2\,H_2O \tag{11.47}$$

Next, a final dehydration step using adsorbent beds is used before final compression.

Renewable hydrogen can be used to hydrogenate CO_2, either from chemical and power production facilities (Davis and Martín, 2014; Martín, 2015, 2016) generating methane, methanol, or DME, but also the CO_2 within the biogas to enhance the production of biomethane (Curto and Martín, 2019). Table 11.5 shows the major results on the processes for the production of chemicals from CO_2 hydrogenation.

The production of methanol and DME has been commented in a previous chapter, we focus here on the analysis of the methanation. It consists of a multibed reactor. Each bed operates adiabatically. In Example 11.4 we model the operation of one of those beds.

TABLE 11.5　Basic mass balances to CO_2-H_2-based processes.

Main product/other products and raw materials	CH_4	CH_3OH	DME
CO_2 captured (kg/kg)	3.0	1.4	1.9
H_2O consumption (kg/kg)	2.8	1.2	1.3
O_2 production(kg/kg)	3.3	1.5	2.1

EXAMPLE 11.4 **Model the operation of the first bed of a methanation reactor. Assume a 4:1 CO_2–to-hydrogen ratio and traces of CO and H_2O. The diameter of the reactor bed is 5 m. The catalysis properties are presented as follows.**

Solution

The methanation is governed by the set of reactions below carried out in a multibed reactor as the one in Fig. 11E4.1. The reactions are equilibrium ones whose yield depends on the temperature and feed ratio.

$$\begin{aligned} CO + 3H_2 &\leftrightarrow CH_4 + H_2O\,(I) \\ CO_2 + H_2 &\leftrightarrow CO + H_2O\,(II) \\ CO_2 + 4H_2 &\leftrightarrow CH_4 + 2H_2O\,(III) \end{aligned} \tag{11E4.1}$$

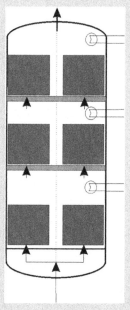

FIGURE 11E4.1 Scheme of the methanation reactor.

The kinetics of each reaction is given as follows:

$$r_I = \frac{k_1}{P_{H_2}^{2.5}} \left(P_{CH_4} \cdot P_{H_2O} - \frac{P_{H_2}^3 \cdot P_{CO}}{K_1} \right) \frac{1}{DEN^2}$$

$$r_{II} = \frac{k_2}{P_{H_2}} \left(P_{CO} \cdot P_{H_2O} - \frac{P_{H_2} \cdot P_{CO_2}}{K_2} \right) \frac{1}{DEN^2}$$

$$r_{III} = \frac{k_3}{P_{H_2}^{3.5}} \left(P_{CH_4} \cdot P_{H_2O}^2 - \frac{P_{H_2}^4 \cdot P_{CO_2}}{K_1 \cdot K_2} \right) \frac{1}{DEN^2} \quad r[=]\frac{kmol}{kg_{catalizador} \, h}$$

$$DEN = 1 + K_{CO} \cdot P_{CO} + K_{H_2} \cdot P_{H_2} + K_{CH_4} \cdot P_{CH_4} + \frac{K_{H_2O} \cdot P_{H_2O}}{P_{H_2}}$$

(11E4.2)

where the kinetic constants are given by

$$k_1 = 9.490 \cdot 10^{16} e^{\left(-\frac{28879}{T} \right)} [=] \frac{kmol \cdot kPa^{0.5}}{kg \cdot h}$$

$$k_2 = 4.390 \cdot 10^4 e^{\left(-\frac{8074.3}{T} \right)} [=] \frac{kmol \cdot kPa^{-1}}{kg \cdot h}$$

(11E4.3)

$$k_3 = 2.290 \cdot 10^{16} e^{\left(-\frac{29336}{T} \right)} [=] \frac{kmol \cdot kPa^{0.5}}{kg \cdot h}$$

and the adsorption constants as follows:

$$K_{CH_4} = 6.65 \cdot 10^{-6} e^{\left(\frac{4604.28}{T}\right)} \quad [=]kPa^{-1}$$

$$K_{H_2O} = 1.77 \cdot 10^{3} e^{\left(-\frac{10666.35}{T}\right)} \quad [=]kPa^{-1}$$

$$K_{H_2} = 6.12 \cdot 10^{-11} e^{\left(\frac{9971.13}{T}\right)} \quad [=]kPa^{-1} \tag{11E4.4}$$

$$K_{CO} = 8.23 \cdot 10^{-7} e^{\left(\frac{8497.71}{T}\right)} \quad [=]kPa^{-1}$$

Finally, the equilibrium constants can be compute using the following expressions:

$$K_1 = 10266.76 e^{\left(-\frac{26830}{T} + 30.11\right)} \quad [=]kPa^2$$

$$K_2 = e^{\left(\frac{4400}{T} - 4.063\right)} \tag{11E4.5}$$

$$K_3 = K_1 \cdot K_2 [=]kPa^2$$

$$P_i = \frac{F_i}{\sum_i F_i \cdot P_T}$$

Effectivity factors:

$$\eta_i = \frac{v_{actual}}{v_{atsurface}}$$

$$\eta_{CH_4} = a_1 + b_1 Z + c_1 Z^2 + d_1 Z^3 + e_1 Z^4 + f_1 Z^5$$
$$\eta_{CO_2} = a_2 + b_2 Z + c_2 Z^2 + d_2 Z^3 + e_2 Z^4 + f_2 Z^5 \tag{11E4.6}$$

See Table 11E4.1.

TABLE 11E4.1 Coefficient for the effectivity factors.

Z (m)	a1	b1	c1	d1	e1	f1
$\eta_{CH_4} Z$						
0–0.2	$3.40271 \cdot 10^{-2}$	$1.50706 \cdot 10^{-1}$	$-1.43056 \cdot 10^{-1}$	$8.95366 \cdot 10^{-1}$	-3.91470	6.22014
0.2–2	$3.46465 \cdot 10^{-2}$	$2.78045 \cdot 10^{-3}$	$-5.60737 \cdot 10^{-3}$	$4.59855 \cdot 10^{-3}$	$-1.80038 \cdot 10^{-3}$	$2.73842 \cdot 10^{-4}$
2–12	$3.53026 \cdot 10^{-2}$	$-5.63342 \cdot 10^{-4}$	$1.04288 \cdot 10^{-4}$	$-1.07611 \cdot 10^{-5}$	$5.56779 \cdot 10^{-7}$	$-1.10707 \cdot 10^{-8}$
$\eta_{CO_2} Z$						
0–0.2	$3.41762 \cdot 10^{-2}$	$1.91920 \cdot 10^{-2}$	$-2.71999 \cdot 10^{-1}$	2.16797	-8.65165	1.35279
0.2–2	$3.46135 \cdot 10^{-2}$	$2.58231 \cdot 10^{-3}$	$4.80816 \cdot 10^{-3}$	$3.77979 \cdot 10^{-3}$	$-1.43009 \cdot 10^{-3}$	$2.11362 \cdot 10^{-4}$
2–11	$3.53703 \cdot 10^{-2}$	$-5.80316 \cdot 10^{-4}$	$1.53452 \cdot 10^{-4}$	$-2.36097 \cdot 10^{-5}$	$1.85678 \cdot 10^{-6}$	$-5.62914 \cdot 10^{-8}$

It typically runs at 580 K and a pressure of 1500 kPa. The catalyst bed is made of Haldor TopsØe Ni/Mg Al_2O_4 Spinel with a particle diameter (D_p) of 0.025, a particle density of 2355.2 kg/m^3, a bed density of 1507.3 kg/m^3, and a porosity (ϕ) of 0.368.

Thus the kinetics is as follows:

$$\frac{dF_{CO_2}}{dz} = r_{CO_2} = \rho_{bed} \cdot A_{cross} \cdot (r_{II} + r_{III}) \cdot \eta_{CO_2}$$

$$\frac{dF_{H_2}}{dz} = r_{H_2} = \rho_{bed} \cdot A_{cross} \cdot (3 \cdot r_I + r_{II} + 4r_{III}) \cdot \eta_{CO_2}$$

$$\frac{dF_{CO}}{dz} = r_{CO} = \rho_{bed} \cdot A_{cross} \cdot (r_I - r_{III}) \cdot \eta_{CO_2} \tag{11E4.7}$$

$$\frac{dF_{H_2O}}{dz} = r_{H_2O} = -\rho_{bed} \cdot A_{cross} \cdot (r_I + r_{II} + 2r_{III}) \cdot \eta_{CH_4}$$

$$\frac{dF_{CH_4}}{dz} = r_{CH_4} = -\rho_{bed} \cdot A_{cross} \cdot (r_I + r_{III}) \cdot \eta_{CH_4}$$

The energy balance is formulated as follows:

$$\frac{dT}{dz} = \frac{\sum_i(-r_i)\left[\Delta H_{R,i}(T_R) + \int_{T_R}^{T} \Delta Cp dT\right]}{F_{Ao}\left(\sum_{i=1}^{n} \Theta_i Cp_i + x\Delta Cp\right)} \tag{11E4.8}$$

Further details of the calculation of the coefficients of the energy balance can be found in (Martín (2019)).

The pressure drop is computed as follows:

$$\frac{dP}{dz} = -\frac{G(1-\phi)}{\rho_o D_p \phi^3}\left[\frac{150(1-\phi)\mu}{D_p} + 1.752G\right]\frac{P_o}{P}\frac{T}{T_o}\frac{F_T}{F_{To}}$$

$$G = \frac{\sum_i F_{io} M_i}{A_c} \tag{11E4.9}$$

$$A_c = \pi D^2 4$$
$$W = \rho_b A_c z$$
$$\mu \approx cte$$

We implement the model in Python

```
import numpy as np
import matplotlib.pyplot as plt
from scipy import integrate

#[kJ/mol]
#300°C        k en min-1

Temperature = 580;
Pressure = 1500;
```

```python
#%FACTORES DE EFECTIVIDAD
robed = 1507.3;
#area in m2, D in m;
Dreactor = 5;
At = 3.14*(Dreactor/2)**2;
Fini = [100,400,5,10,0,580,1500];

Fao = Fini[0];
Fbo = Fini[1];
Fco = Fini[2];
Fdo = Fini[3];
Feo = Fini[4];
F = [(),(),(),(),(),(),()];
#We create an empty list to store the concentrations data

wei = np.linspace(0,0.4,100)
#Equations: We define a function which collect the equations to solve

def    kinetics(F,wei):

        Fa = F[0];
        Fb = F[1];
        Fc = F[2];
        Fd = F[3];
        Fe = F[4];
        T = F[5];
        P = F[6];

#T = K
        To = Temperature;
        Po = Pressure;

        Fto = Fao + Fbo + Fco + Fdo + Feo;

#constantes de velocidad K1 en (kmol·kPa^0.5/kg·h)K2 en (kmol·kPa^-1/kg·h)
        k1 = ((9.49E + 16)*np.exp(-28879.0/T));
        k2 = ((4.39E + 4)*np.exp(-8074.3/T));
        k3 = ((2.29E + 16)*np.exp(-29336/T));
#Presiones parciales
        Pe = ((Fe)/(Fa + Fb + Fc + Fd + Fe))*P;
        Pa = ((Fa)/(Fa + Fb + Fc + Fd + Fe))*P;
        Pb = ((Fb)/(Fa + Fb + Fc + Fd + Fe))*P;
        Pc = ((Fc)/(Fa + Fb + Fc + Fd + Fe))*P;
        Pd = ((Fd)/(Fa + Fb + Fc + Fd + Fe))*P;
#%Constantes de equilibrio Kpa
        Keq1 = 10266.76*np.exp(-(26830/T) + 30.11);
```

```
        Keq2 = np.exp((4400/T)-4.063);
        Keq3 = Keq1*Keq2;
#%Constantes de adsorción en el equilibrio 1/Kpa
        Kc = 8.23E-7*np.exp(8497.71/T);
        Kb = 6.12E-11*np.exp(9971.13/T);
        Ke = 6.65E-6*np.exp(4604.28/T);
        Kd = 1.77E+3*np.exp(-10666.35/T);

        DEN = 1 + Kc*Pc + Kb*Pb + Ke*Pe + (Kd*Pd/Pb);

#%ecuaciones de velocidad
        r1 = (k1/(Pb)**2.5)*(Pe*Pd-(((Pb)**3*Pc))/Keq1)*(1/DEN**2);
        r2 = (k2/(Pb))*(Pc*Pd-((Pb*Pa)/Keq2))*(1/DEN**2);
        r3 = (k3/(Pb)**3.5)*(Pe*(Pd**2)-(((Pb)**4*Pa))/Keq3)*(1/DEN**2);

#%Balance de materia
#%Flujos en kmol/(h*m)
#%MULTIPLICAR POR RO AREA TRANSVERSAL Y FACTOR DE EFICACIA??
#%Factor de eficacia metano

        if wei > 2:
                a1 = 3.53026E-2
                b1 = -5.63342E-4
                c1 = 1.04288E-4
                d1 = -1.07611E-5
                e1 = 5.56779E-7
                f1 = -1.10707E-8
        else:
                if wei < 0.2:
                        a1 = 3.40271E-2
                        b1 = 1.50706E-1
                        c1 = -1.43056E-1
                        d1 = 8.95366E-1
                        e1 = -3.91470
                        f1 = 6.22014
                else:
                        a1 = 3.46465E-2
                        b1 = 2.78045E-3
                        c1 = -5.60737E-3
                        d1 = 4.59855E-3
                        e1 = -1.80038E-3
                        f1 = 2.73842E-4

        nch4 = a1 + b1*wei + c1*wei**2 + d1*wei**3 + e1*wei**4 + f1*wei**5;

        if wei > 2:
```

5. Wind based processes

```
                    a2 = 3.53703E-2
                    b2 = -5.80316E-4
                    c2 = 1.53452E-4
                    d2 = -2.36097E-5
                    e2 = 1.85678E-6
                    f2 = -5.62914E-8

            else:
                    if wei < 0.2:
                            a2 = 3.41762E-2
                            b2 = 1.91920E-2
                            c2 = -2.71999E-1
                            d2 = 2.16797
                            e2 = -8.65165
                            f2 = 1.35279

                    else:
                            a2 = 3.46135E-2
                            b2 = 2.58231E-3
                            c2 = -4.80816E-3
                            d2 = 3.77979E-3
                            e2 = -1.43009E-3
                            f2 = 2.11362E-4

            nco2 = a2 + b2*wei + c2*wei**2 + d2*wei**3 + e2*wei**4 + f2*wei**5

            Ra = robed*At*((r2 + r3))*nco2;
            Rb = robed*At*((3*r1 + r2 + 4*r3))*nco2;
            Rc = robed*At*(r1-r2)*nco2;
            Rd = -robed*At*(r1 + r2 + (2*r3))*nch4;
            Re = -robed*At*(r1 + r3)*nch4;
            Ft = Fa + Fb + Fc + Fd + Fe;

#%BALANCE DE ENERGÍA
#Capacidades caloríficas en kJ/(Kmol·K)

            Cpa = 19.795 + (7.34E-02*T) + (-5.60E-05*T**2) + (1.72E-08*T**3);
            Cpb = 27.143 + (92.738E-04*T) + (-1.381E-05*T**2) + (76.451E-10*T**3);
            Cpc = 30.869 + (-1.29E-02*T) + (2.79E-05*T**2) + (-1.27E-08*T**3);
            Cpd = 32.243 + 1.92E-03*T + (1.06E-05*T**2) + (-3.60E-09*T**3);
            Cpe = 19.251 + 5.21E-02*T + (1.20E-05*T**2) + (-1.13E-08*T**3);

#%Deltah1 en kJ/kmol

            deltaalfa1 = 30.869 + 32.243 - (19.795 + 27.143);
            deltabeta1 = -1.29E-02 + 1.92E-03 - (7.34E-02 + 92.738E-04);
            deltafi1 = 2.79E-05 + 1.06E-05 - (-5.60E-05 + (-1.381E-05));
```

5. Wind based processes

```
        deltaepsilon1 = -1.27E-08 + (-3.60E-09) - (1.72E-08 + 76.451E-10);

        deltaalfa2 = 19.251 + 32.243 - (30.869 + 3*27.143);
        deltabeta2 = 5.21E-02 + 1.92E-03 - (-1.29E-02 + 3*92.738E-04);
        deltafi2 = 1.20E-05 + 1.06E-05 - (2.79E-05 + 3*(-1.381E-05));
        deltaepsilon2 = -1.13E-08 + (-3.60E-09) + (-1.27E-08 + 3*76.451E-10);

        deltah1 =  -41000 + deltaalfa1*(T-298) + (deltabeta1/2)*(T**2-298**2) +
(deltafi1/3)*(T**3-298**3) + (deltaepsilon1/4)*(T**4-298**4);
        deltah2 = 206000 + deltaalfa2*(T-298) + (deltabeta2/2)*(T**2-298**2) +
(deltafi2/3)*(T**3-298**3) + (deltaepsilon2/4)*(T**4-298**4);
#%Deltas a t media entre 307 y 400°C
#%deltah1 =  4.1379e+04;
#%deltah2 = -2.2987e+05;
#%Ra en kmol/(h·m)
        num = (Ra*deltah1 + Re*deltah2);
        den = (Rd-Fdo)*Cpd + ((Re)-Feo)*Cpe + (Rc-Fco)*Cpc + (Fbo-Rb)*Cpb + (Fao-Ra)
*Cpa;

#%Velocidad superficial en kg/(m^2·s)
        G = ((Fa*44) + (Fb*2) + (Fc*28) + (Fd*18) + (Fe*16))*(1/(At*3600));

#%porosidad del lecho
        porosity = 0.368;
#%densidad en kg/m3
        rogas = 2.6935;
#%Viscosidad del gas en Pa·s
        viscgas = 2.088E-05;
        gc = 9.8;
        Dpext = 0.025;
#%Dpext = 0.0173;

        dxdwei = np.zeros(7)
        dxdwei[0] = Ra
        dxdwei[1] = Rb
        dxdwei[2] = Rc
        dxdwei[3] = Rd
        dxdwei[4] = Re
        dxdwei[5] = num/den
        dxdwei[6] = ((-G*(1-porosity))/(rogas*Dpext*porosity**3))*(((((150*(1-
porosity)*viscgas)/Dpext) + 1.752*G))*((Po/P)*(T/To)*(Ft/Fto))
        return dxdwei

#We collect the results in a list called Res
res_f = integrate.odeint(kinetics,Fini,wei)
```

5. Wind based processes

```
fig,axes = plt.subplots()
axes.plot(wei,res_f[:,0],'k-',label = 'CO2')
axes.plot(wei,res_f[:,1],'k--',label = 'H2')
axes.plot(wei,res_f[:,2],'k.',label = 'CO')
axes.plot(wei,res_f[:,3],'k-.',label = 'H2O')
axes.plot(wei,res_f[:,4],'k-o',label = 'CH4')
axes.set_xlabel('z(m)')
axes.set_ylabel('Concentration(kmol/h)')
legend = axes.legend(loc = 'best', fontsize = 'medium')

fig,axes = plt.subplots()
axes.plot(wei,res_f[:,5],'k-',label = 'Temperature (K)')
axes.plot(wei,res_f[:,6],'k-',label = 'Pressure (kPa)')
axes.set_xlabel('z(m)')
axes.set_ylabel('P & T')
legend = axes.legend(loc = 'best', fontsize = 'medium')
#plt.legend('TP')
```

See Fig. 11E4.2.

FIGURE 11E4.2   Performance of the catalytic reactor bed: (A) species profile; (B) temperature and pressure profiles.

11.4 Fuel cells

11.4.1 Description of the operation

Fuel cells perform the opposite chemical reaction compared to electrolyzer. Hydrogen and oxygen, air, are used to produce power, heat, and water as by product limiting the emissions to it. The advantage of this power system is the absence of mobile parts.

$$H_2 + \frac{1}{2}(O_2 + 3.76N_2) \rightarrow H_2O + Q + Power \qquad (11.48)$$

The fuel cells show 40%–60% power efficiency and 20%–30% thermal efficiency (USDOE, 2006). Therefore from 141 MJ/kg of H_2 as the high heating value (HHV), the power production can be computed and the cooling needs or the thermal energy can be obtained:

$$H_2 + \frac{1}{2}O_2 \rightarrow H_2O \qquad (11.49)$$

Different alternatives can be found including alkaline fuel cells (AFCs), polymer electrolyte membrane (PEM), molten carbonate fuel cells (MCFCs), and solid oxide fuel cells (SOFCs) (http://www.energy.gov). Among them:

AFCs: As in the case of the electrolyzers, AFCs were the first ones to be developed and were used in the United States spacecraft to produce electricity. They operate at 120°C–150°C using a solution of potassium hydroxide in water as electrolyte. The electrodes are made of nonprecious materials, see Fig. 11.8A. Lately, they have evolved as a modification of PEM fuel cells where they use a polymer membrane as the electrolyte, but it is an alkaline one. The reactions

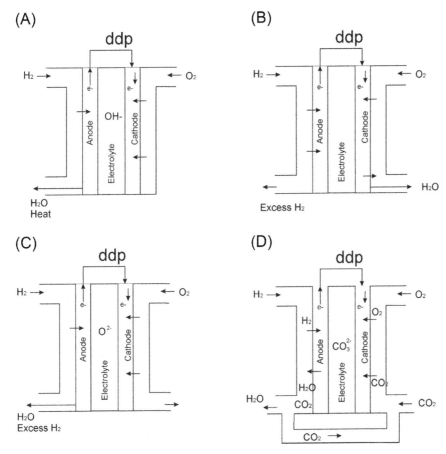

FIGURE 11.8 Scheme of fuel cells: (A) AFC, (B) PEMEC, (C) SOEC, (D) MCFC. *AFC*, Alkaline fuel cell; *MCFC*, molten carbonate fuel cell; *PEMEC*, proton exchange membrane electrolysis cell; *SOEC*, solid oxide electrolysis cell.

taking place are presented as follows. AFCs present high performance due to the rate of the electro-chemical reactions taking place and high efficiency, above 60%:

$$\frac{1}{2}O_2 + H_2O + 2e^- \rightarrow 2OH^- \quad \text{(Cathode)}$$

$$H_2 \rightarrow 2H^+ + 2e^- \qquad\qquad \text{(Anode)}$$

$$2H^+ + 2OH^- \rightarrow H_2O$$

(11.50)

The major drawback of this cell type is the fact that they are susceptible to poisoning by CO_2 due to the formation of carbonates. To reduce that effect, alkaline cells that use liquid electrolyte can run in recirculating mode so that the electrolyte can be regenerated. In addition, liquid electrolyte systems present corrosion issues, wettability, and difficulty in handling differential pressures. That is the main reason for the development of alkaline membrane fuel cells, but they still present challenges such as the tolerance to CO_2, membrane conductivity and durability, higher temperature operation, water management, power density, and anode electrocatalysis.

Polymer electrolyte membrane fuel cells (PEMFC): They also receive the name of proton exchange membrane fuel cells. PEMFCs use a solid polymer as electrolyte while the electrodes are made of porous carbon containing platinum or a platinum alloy catalyst. The reactions that take place are as follows:

$$\frac{1}{2}O_2 + 2H^+ + 2e^- \rightarrow H_2O \quad \text{(Cathode)}$$

$$H_2 \rightarrow 2H^+ + 2e^- \text{(Anode)}$$

(11.51)

Hydrogen and oxygen typically from air and water are needed for the operation. For the acid kind of cell, the hydrogen is fed to the anode region, see Fig. 11.8B, it crosses the anode and reacts with the oxygen at the membrane producing water that exits with the excess of air used. They operate at low temperature, around 80°C, and deliver high-power density. In addition, they are lighter and more compact than other cells. However, to separate the hydrogen's electrons and protons a catalyst made of a noble metal, typically platinum, is used. The platinum is sensitive to carbon monoxide so that it has to be removed first.

SOFCs: The electrolyte is a hard nonporous ceramic compound. They operate the opposite compared to the electrolysis cells, see Fig. 11.8C:

$$\frac{1}{2}O_2 + 2e^- \rightarrow O^{2-} \quad \text{(Cathode)}$$

$$H_2 + O^{2-} \rightarrow H_2O + 2e^- \text{(Anode)}$$

(11.52)

These cells operate at high temperatures, 800°C−1000°C, which avoids the need for catalysts and allows reforming inside. Their efficiency is around 60% in conversing fuel to electricity, but with cogeneration 85% can be achieved. They are not sensitive to sulfur compounds nor poisoned by CO. However, the high operating temperatures slow down the start-up, reduce the durability, and require the use of thermal shields.

MCFCs: They use as electrolyte a molten carbonate salt suspended in a porous chemically inert ceramic lithium aluminum oxide, see Fig. 11.8D. The reactions that govern their operation can be seen as follows:

$$
\begin{aligned}
O_2 + 2CO_2 + 4e^- &\rightarrow 2CO_3^{2-} \quad \text{(Cathode)} \\
2H_2 &\rightarrow 4H^+ + 4e^- \quad\quad\quad \text{(Anode)} \\
4H^+ + 2CO_3^{2-} &\rightarrow 2H_2O + 2CO_2
\end{aligned}
\tag{11.53}
$$

Their high operating temperature, around 650°C, avoids the need to use precious metals as catalysis at the anode and cathode. When coupled with a turbine, it is possible to reach 60%−65% electric efficiency, and if cogeneration is implemented 85%, fuel efficiency can be achieved. However, the major drawbacks include their durability due to corrosion issues related to the electrolyte. They are currently being developed for natural gas and coal-based power plants for electrical utility, industrial, and military applications.

11.4.2 Analysis of a PEMFC

In a PEMFC cell, the electrochemical reaction occurs in which hydrogen and oxygen are combined to produce water. By applying the thermodynamic laws to this reaction, and based on the Nernst equation, the reversible voltage of a cell is obtained, Eq. (11.54):

$$
\begin{aligned}
W &= V \cdot J \cdot A \\
V &= E_{Cell} - losses \\
V &= E_{Cell} - V_{act} - V_{Ohmic} - V_{con}
\end{aligned}
\tag{11.54}
$$

The E_{Cell} can be computed from the Nerst equation, Eq. (11.55):

$$
E_{Cell} = E_0 + \frac{RT}{2F}\left[\ln\left(p_{H_2} \cdot p_{O_2}^{0.5}\right)\right] = \frac{\Delta G}{2F} + \frac{\Delta S}{2F}(T - T_{ref}) + \frac{RT}{2F}\left[\ln\left(p_{H_2} \cdot p_{O_2}^{0.5}\right)\right]
\tag{11.55}
$$

As well as using empirical correlations such as Eq. (11.56):

$$
E_{Cell} = 1.229 - 0.85 \cdot 10^{-3}(T - 298) + 4.3083 \cdot 10^{-5}T\left(\ln p_{H_2} + 0.5 \ln p_{O_2}\right)
\tag{11.56}
$$

The activation efficiency can be computed from

$$
\begin{aligned}
V_{Act} &= \xi_1 + \xi_2 T + \xi_3 \cdot T \ln(C_{O_2}) + \xi_4 T \ln(I) \\
C_{O_2} &= \frac{p_{O_2}}{5.08 \cdot 10^6 e^{(-498/T)}}
\end{aligned}
\tag{11.57}
$$

The ohmic voltage drop results from the resistance to the electrons transfer through the collecting plates and carbon electrodes (R_m), and the resistance to the protons transfers through the solid membrane (R_c), Eq. (11.58). R_c is usually considered constant.

$$
V_{Ohm} = I(R_m + R_c)
\tag{11.58}
$$

$$
R_m = \frac{\rho_M \cdot l}{A}
\tag{11.59}
$$

5. Wind based processes

where ρ_M is the specific resistance of the membrane (Ω cm), computed as in Eq. (11.60). A is the membrane active area (cm^2), and l is the thickness of the membrane (cm).

$$\rho_M = \frac{181.61 + 0.03 \cdot (T/A) + 0.062(T/303)^2 \cdot (I/A)^{2.5}}{\left(\Psi - 0.634 - 3(I/A)\right)e^{\left(4.18\frac{T-303}{T}\right)}} \tag{11.60}$$

The concentration voltage drop due to the mass transport can be determined by Eq. (11.61):

$$V_{con} = -b \ln\left(1 - \frac{J}{J_{\max}}\right) \tag{11.61}$$

where b is a parametric coefficient, which depends on the cell and its operation state, and J represents the actual current density of the cell (A/cm^2).

The efficiency of the fuel cell is computed as Eq. (11.62):

$$\eta = \eta_{thermo} \times \eta_{voltage} \times \eta_{fuel} \tag{11.62}$$

where the thermal efficiency, η_{thermo}, is given by the ratio between the Gibbs free energy and the enthalpy of the reaction, Eq. (11.63):

$$\eta_{thermo} = \frac{\Delta G_r}{\Delta H_r} \tag{11.63}$$

The efficiency of the voltage, $\eta_{voltage}$, is due to the losses as presented previously between the thermodynamic reversible one and the actual used, Eq. (11.64):

$$\eta_{voltage} = \frac{V}{E_{cell}} \tag{11.64}$$

Finally, the fuel efficiency, η_{fuel}, is due to the crossover of fuel across the membrane, Eq. (11.65), so that a fraction is not used:

$$\eta_{fuel} = \frac{F_{used}}{F_{used} + F_{crossover}} \tag{11.65}$$

See Example 11.5.

EXAMPLE 11.5 In a particular region, a company is evaluating the installation of a wind farm consisting of 50 turbines type GE 2.5xl. in five possible locations with different wind profiles over a year, measured at 10 m and characterized by a Weibull distribution and a ground coefficient that is presented in Table 11E5.1. The electricity produced by the wind farm is expected to produce hydrogen in an electrolyzer that consumes 175 MJ/kg of H$_2$ produced. The hydrogen is stored and used to produce power in a fuel cell with a yield of 60%$_{HHV}$. Select the allocation.

Solution

The Weibull distribution is used to compute the probability of a particular velocity to occur. Next, the velocity is corrected to be at 80 m and with the ground coefficient.

For each corrected velocity, using the power curve of the turbine, see Table 11.2, we compute the power that can be obtained by turbine. Table 11E5.2 shows the results for zone 3.

TABLE 11E5.1 Wind data.

Zone	K	C	α
1	2	9	0.2
2	1.5	8.5	0.15
3	3	10	0.1
4	1.5	10	0.1
5	3	9	0.12

TABLE 11E5.2 Results for the operation of the turbine in zone 3.

v (m/s)	p	vco (m/s)	P (kW)	$p \cdot v$ (m/s)	$p \cdot P$ (kW)
0.5	0.00074991	0.61557221	5.75757178	0.00046162	0.00431764
1	0.002997	1.23114441	9.25406557	0.00368974	0.02773445
2	0.01190438	2.46228883	23.8335027	0.02931203	0.28372314
3	0.02628075	3.69343324	60.8214144	0.09706621	1.5984326
4	0.04502424	4.92457765	151.695342	0.22172537	6.82996746
5	0.06618727	6.15572207	358.395086	0.40743042	23.7211915
6	0.08701941	7.38686648	756.063469	0.64280078	65.7921989
7	0.10431682	8.61801089	1322.4967	0.89900347	137.958647
8	0.11506479	9.84915531	1860.55285	1.133291	214.084125
9	0.11722105	11.0802997	2207.18311	1.29884433	258.728315
10	0.11036383	12.3114441	2378.21228	1.35873816	262.468622
11	0.09590929	13.5425885	2451.53961	1.29886003	235.12542
12	0.07674019	14.773733	2481.06864	1.13373911	190.397684
13	0.05634599	16.0048774	2492.65823	0.9018106	140.451285
14	0.0378158	17.2360218	2497.16091	0.65179392	94.432132
15	0.02309723	18.4671662	2498.90333	0.42654038	57.7177445
16	0.01277883	19.6983106	2499.57656	0.25172132	31.9416588
17	0.0063728	20.929455	2499.83653	0.13337924	15.9309589
18	0.0028501	22.1605994	2499.9369	0.06316003	7.12508243
19	0.00113711	23.3917439	2499.97564	0.02659899	2.84274813
20	0.00040256	24.6228883	2499.9906	0.00991207	1.0063841
21	0.00012576	25.8540327	2499.99637	0.00325152	0.31441121

Subsequently, we compute the power of the wind farm of 50 turbines.

With the power of the wind farm, and the consumption of the electrolyzed, the H_2 is computed.

Finally, with the hydrogen and its HHV, we compute the power of the entire system, see Table 11E5.3.

TABLE 11E5.3 Summary of the results.

Zone	1	2	3	4	5
V(ave) (m/s)	12.0	9.9	**11.0**	9.8	10.3
P_{WT}(kW)	1680	1314	**1749**	1320	1636
$P_{Windfarm}$ (MW)	84	65.7	**87.4**	66.0	81.8
Power (MW)	40.6	31.8	**42.3**	31.9	39.5

11.5 Problems

P11.1 Optimize power per unit of area of a system of two turbines when the incoming free velocity is 8 m/s and the turbine power curve is the one given by the Vestas 112 model with a turbine diameter of 100 m. Assume $C_p = 0.5$, a hub height of 80 m and a roughness of 0.3 m.

P11.2 For 11 allocations, with the monthly wind profile in Table 11E2.1, select the best turbine among Liberty C99, Nordex N100−2500, Vestas V112−3.0. Consider the subsidy required to select allocation 1 for the production of at least 200 MW.

P11.3 Formulate the allocation problem for a power curve given by $a + bv + cv^3$

P11.4 Simulate a three-bed methanation reactor with intercooling.

P11.5 Model the operation of the hydrogenation of CO_2 to methanol.

P11.6 In a particular region, a company is evaluating the installation of a wind farm consisting of 50 turbines type Nordex N100−2500, 1600 €/kW$_{Nominal}$, in five possible locations with different wind profiles over a year, measured at 10 m and characterized by a Weibull distribution and a ground coefficient that is presented in Table 11P6.1. The installation of each turbine costs 1600 €/kW, but each region provides a subsidy.

TABLE 11P6.1 Wind and zone data.

Zone	K	C	α	Discount (€/kW)
1	2	9	0.2	100
2	1.5	8.5	0.15	350
3	3	10	0.1	75
4	1.5	10	0.1	400
5	3	9	0.12	200

The electricity produced by the wind farm is expected to produce hydrogen in an electrolyzer that consumes 175 MJ/kg of H_2 produced. The cost of the electrolyzer is 1000 €/kW. The hydrogen is stored and used to produce power in a fuel cell with a yield of $60\%_{HHV}$. The cost of the fuel cell is 2000 €/kW. Select the allocation.

References

Archer, C.L., 2004. The Santa Cruz Eddy and United States Wind Power (Ph.D. thesis). Stanford University, Stanford, 190 pp.

Albadi, M.H., El-Saadany, E.F., 2010. Overview of wind power intermittency impacts on power systems Author links open overlay panel. Elect. Pow. Syst. Res. 80 (6), 627−632.

Blanco, M.I., 2009. The economics of wind energy. Renew. Sustain. Energy Rev. 13 (6−7), 1372−1382.

Curto, D., Martín, M., 2019. Renewable based biogas upgrading. J. Clean. Prod. 224, 50−59.

Davis, W., Martín, M., 2014. Optimal year-round operation for methane production from CO_2 and water using wind energy. Energy 69, 497−505.

De la Cruz, V., Martin, M., 2016. Characterization and optimal site matching of wind turbines: effects on the economics of synthetic methane production. J. Clean. Prod. 133, 1302−1311.

Dialynas, E.N., Machias, A.V., 1989. Reliability modeling interactive techniques of power systems including wind generating units. Arch. Elektrotech. 72, 33−41.

Ellis, A., Kazachkov, Y., Muljadi, E., Pourbeik, P., Sanchez-Gasca, J.J., 2011. Description and technical specifications for generic WTG models − a status report. In: Power Systems Conference and Exposition.

Eriksson, S., Bernhoff, H., Leijon, M., 2008. Evaluation of different turbineconcepts for wind power. Renew. Sustain. Energy Rev. 12 (5), 1419−1434.

Falcao, D.S., Pinto, A.M.F.R., 2020. A review on PEM electrolyzer modelling: guidelines for beginners. J. Clean. Prod. 261, 121184.

Giorsetto, P., Utsurogi, K.F., 1983. Development of a new procedure for reliability modelling of wind turbine generators. IEEE Trans. Power Appar. Syst. 102, 134−143.

Jangamshetti, S.H., Rau, V.G., 1999. Site matching of wind turbine generators: a case study. IEEE Trans. Energy Convers. 14, 1537−1543.

Justus, C.G., Hargraves, W.R., Yalcin, A., 1976. Nationwide assessment of potential output from wind-powered generators. J. Appl. Meteorol. 15, 673−678.

EWEA, 2020. European Wind Energy Association and International Economic Development Council, The Economics of Wind Energy A report by the European Wind Energy Association. < https://www.ewea.org/ > .

Kalmikov, A., Dykes, K., 2011. Wind power fundamentals. <http://web.mit.edu/windenergy/windweek/Presentations/Wind%20Energy%20101.pdf>.

Khalfallaha, M.G., Koliubb, A.M., 2007. Suggestions for improving wind turbines power curves. Desalination 209, 221−229.

Lackner, M.A., Elkinton, C.N., 2007. An analytical framework for offshore wind farm layout optimization. Wind. Eng. 31, 17−31.

Martín, M., 2015. Optimal year-round production of DME from CO_2 and water using renewable energy. J. CO_2 Utilization 13, 105−113.

Martin, M., 2016. Methodology for solar and wind based process design under uncertainty: methanol production from CO_2 and hydrogen. Comp. Chem. Eng. 92, 43−54.

Martin, M., 2019. Introduction to software for chemical engineers, 2nd CRC Press.

Mosetti, G., Poloni, C., Diviacco, B., 1994. Optimization of wind turbine positioning in large windfarms by means of a genetic algorithm. J. Wind. Eng. Ind. Aerodyn. 51 (1), 105−116.

NREL, 2020. https://sam.nrel.gov/. System advisor Model. NREL.

Pallabazzer, R., 1995. Evaluation of wind-generator potentiality. Sol. Energy 55, 49−59.

Patel, M.R., 1999. Wind and Solar Power Systems. CRC Press, New York.

Salameh, Z.M., Safari, I., 1992. Optimum windmill-site matching. IEEE Trans. Energy Convers. 7, 669−676.

Schmidt, P., Gambhir, A., Staffell, I., Hawkes, A., Nelson, J., Few, S., 2017. Future cost and performance of water electrolysis: an expert elicitation study. Int. J. Hydrog. Energy 43, 30470−30492.

Singh, M., Muljadi, E., Jonkman, J., Gevorgian, V., Girsang, I., Dhupia, J., 2014. Simulation for Wind Turbine
    Generators – With FAST and MATLAB-Simulink Modules. National Renewable Energy Laboratory,
    Golden, CO.
USDOE, 2006. Hydrogen fuel cells. <https://www.californiahydrogen.org/wp-content/uploads/files/doe_fuel-
    cell_factsheet.pdf>.
USDOE, 2020. Types of fuel cells. <https://www.energy.gov/eere/fuelcells/types-fuel-cells>.

Geothermal processes

12

Geothermal energy

Mariano Martín

Department of Chemical Engineering, University of Salamanca, Salamanca, Spain

In this chapter, we cover geothermal processes. Within this topic, we evaluate organic Rankine cycles (ORC) that can also be used for waste heat valorization into energy. In addition, the use of refrigerants and hydrocarbons present a number of challenges related to their flammability and toxicity. As a second section of this chapter, safety issues and the evaluation of the risk involved in the operation of ORC with this type of fluids are presented.

12.1 Hot brine

The source of energy for geothermal power facilities is a hot brine extracted from the depths of the Earth (Moon and Zarrouk, 2012). Typically, two wells are drilled, a production well that brings up the hot brine; after it is used to produce energy in the ORC, it is reinjected into the ground through the injection well at a lower temperature. Fig. 12.1 shows a scheme of the well. Typically, geothermal wells reach from 400 to 2000 m deep for medium-temperature brines and from 700 to 3000 m for high-temperature ones. Two types of wells are drilled, vertical and directional. The diameter of the well decreases with the depth along the different drill stages so that any next stage presents a smaller diameter. The well is secured by steel casings that are cemented before drilling the subsequent stage. The casing diameters of the three stages are 20 in., 13−3/8 in., 9−5/8 in., and slotted liners of 7 in. by the end of the well. The final section of the well uses a perforated uncemented liner that allows the geothermal fluids to pass into the pipe (Kipsang, 2015).

The energy that the ORC will produce depends on the brine temperature at the wells head. The fluid can be in liquid phase or biphasic. Biphasic requires far detailed analysis, and no analytic expressions can be easily presented to predict the operation. Thus, in this analysis we assume monophasic fluid. For this case, the temperature of the fluid as a

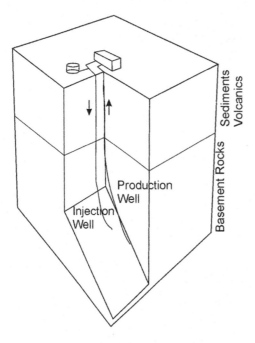

FIGURE 12.1   Scheme of the structure of a geothermal plant.

function of the depth can be estimated using Eq. (12.1), considering the temperature gradient inside the Earth and the heat transfer rate (DiPippo, 2016):

$$T = \left(T_{bh} - a \cdot y\right) + a \cdot A \cdot \left(1 - e^{-y/A}\right) + (T_0 - T_{bh}) \cdot e^{-y/A} \tag{12.1}$$

where T_{bh} is the downhole reservoir temperature that it is to be measured or provided by charts, and T_o is the inflowing fluid temperature, which corresponds to the one if there is no extraction from the geothermal production well as if the brine is lying at the bottom of the reservoir without ascending by the well. Usually both take the same value; a is the temperature gradient ($43.49 \cdot 10^{-3}$ °C/m), y is the vertical distance from the bottom of the well to the surface considering the feed to be at the bottom. A corresponds to the diffusion depth parameter that is a function of time as follows, Eq. (12.2):

$$A(t) = \frac{fc_{Brine} \cdot c_p \cdot f(t)}{2\pi \cdot K} \tag{12.2}$$

where fc_{Brine} is the brine flow rate, c_p is the heat capacity of the brine, K is the thermal conductivity, and $f(t)$ is a dimensionless time function that represents the transient heat transfer to the formation. For stable systems, typically those with flowing times larger than 30 days, $f(t)$ can be approximated as follows, Eq. (12.3):

$$f(t) = -\ln\left(\frac{r}{2\sqrt{\propto \cdot t}}\right) - 0.290 \tag{12.3}$$

where r is the internal radius of the well, $\propto$ is the thermal diffusivity of the formation, and t is assumed to be 30 days.

For biphasic flows, the temperature corresponds to the saturated state at the pressure of the production wellhead. To determine the type of flow, either mono- or biphasic, DiPippo proposed an algorithm that computes the depth at which the brine turns biphasic and estimates the pressure needed there so that there is no such a transition when the flow ascends. Alternatively, based on fluid mechanics, using the pressure at the bottom of the well and Eq. (12.4), if the pressure computed at the head is above the saturation point of the brine, there will not be a biphasic flow, and therefore there is no need for an inner pump.

The pressure drop between the bottom of the reservoir and the entrance of the production well is computed as given by Eq. (12.4):

$$P_{yac} - P_{In} = \frac{\mu \cdot fc_{Brine,well} \cdot \ln(r_{yac.}/r_{well})}{2\pi \cdot K \cdot L_{yac.} \cdot \rho_{yac.}} \tag{12.4}$$

where P_{yac} is the pressure down the well, P_{In} is the pressure at the shaft's entrance, μ is brine viscosity, fc_{Brine} is the flow rate of brine, r_{yac} is the influence radius of the reservoir, r_{well} is the external radius of the well, K the permeability, $L_{yac,}$ is the depth of the source, and ρ_{yac} is the density of the brine.

Next, the pressure drop along the shaft is computed as given by Eq. (12.5), where P_{head} is the pressure at the well's head, u is the brine velocity, y is the well depth, and f is the friction factor of the brine along the shaft.

$$P_{In} - P_{Head} = \frac{f\rho_{yac}u^2y}{D \cdot 2} + \rho_{yac}gy \tag{12.5}$$

The temperature in a well, T_1, into which single-phase liquid or gas is injected, is given by Eq. (12.6):

$$T_1 = (T_{surf} + a \cdot z) - a \cdot A + (T_{inj} - T_{surf} + a \cdot A) \cdot e^{-z/A} \tag{12.6}$$

where z is the distance downward from the top of the well, T_{surf} is the surface temperature of the earth, and T_{inj} is the temperature of the injected fluid, Example 12.1 shows an example for calculus.

EXAMPLE 12.1 Compute the temperature of the brine to be used in a geothermal plant where the reservoir shows the following characteristics.

$r = 0.15$ m; $\alpha = 10^{-6}$ m/s^2; $c_p = 4500$ J/kg/K; $K = 2.5$ W/m °C; $T_{Well} = 150$ °C; $a = 4.39 \cdot 10^{-2}$ °C/m; $y = 3000$ m for the production of 75 kg/s of brine

Solution

Assuming stable operation, t is equal to 30 days, Eq. (12E1.1) and (12E1.2):

$$f(t) = -\ln\left(\frac{r}{2\sqrt{\alpha \cdot t}}\right) - 0.290 = -\ln\left(\frac{0.15 \text{ m}}{2\sqrt{10^{-6} \text{ m/s}^2 \cdot t}}\right) - 0.290 = 2.776 \tag{12E1.1}$$

$$A(t) = \frac{fc_{Brine} \cdot c_p \cdot f(t)}{2\pi \cdot K} = \frac{75\frac{kg}{s} \cdot 4500\frac{J}{kgK} \cdot 2.776}{2 \cdot \pi \cdot 2.5\frac{J}{smK}} = 59{,}655 \text{ m} \tag{12E1.2}$$

Assuming that T_o is T_{bh}, Eq. (12E1.3)

$$T = (T_{bh} - a \cdot y) + a \cdot A \cdot \left(1 - e^{-y/A}\right) + (T_0 - T_{bh}) \cdot e^{-y/A} = (150 - 4.39 \cdot 10^{-2} \cdot 3000)$$

$$+ 4.39 \cdot 10^{-2} \cdot 59655 \cdot \left(1 - e^{-\frac{3000}{59655}}\right) = 147°C \tag{12E1.3}$$

See Example 12.2 that evaluates if a flow of brine is monophasic.

EXAMPLE 12.2 For Example 12.1 assuming a brine viscosity of $1.6 \cdot 10^{-4}$ Pa s and a density of 950 kg/m^3, $r_{yac} = 1000$ m, and $L_{yac} = 150$ m, considering a friction factor f of 0.012, and a permeability of $9.861 \cdot 10^{-14}$ m^2, verify that monophasic flow is reaching the end of the well.

Solution

At 147°C, the saturation pressure is $4.4 \cdot 10^5$ Pa. Therefore for the brine to remain liquid, the pressure must be above that, Eq. (12E2.1) and (12E2.2):

$$P_{head} > 4.4 \cdot 10^5 \text{ Pa}$$

$$P_{yac} - P_{In} = \frac{\mu \cdot fc_{brine,well} \cdot \ln(r_{yac.}/r_{well})}{2\pi \cdot K \cdot L_{yac.} \cdot \rho_{yac.}}$$

$$P_{In} = P_{yac} - \frac{\mu \cdot fc_{brine,well} \cdot \ln(r_{yac.}/r_{well})}{2\pi \cdot K \cdot L_{yac.} \cdot \rho_{yac.}} = 35 \cdot 10^6 \text{Pa} - \frac{1.6 \cdot 10^4 \text{ Pa s} \cdot 75 \text{ kg/s} \cdot \ln(1000/0.15)}{2\pi \cdot 9.861 \cdot 10^{-14} \text{ m}^2 \cdot 150 \text{ m} \cdot 950 \text{ kg/m}^3}$$

$$= 3.38 \cdot 10^7 \text{ Pa} \tag{12E2.1}$$

$$P_{In} - P_{Head} = \frac{f\rho_{yac}u^2 y}{D \cdot 2} + \rho_{yac}gy$$

$$P_{Head} = P_{In} - \left(\frac{f\rho_{yac}u^2 y}{D \cdot 2} + \rho_{yac}gy\right) = 3.38 \cdot 10^7 \text{ Pa}$$

$$- \left(\frac{0.012 \cdot 950 \text{ kg/m}^3 \left(\frac{75}{950 \cdot \pi \cdot 0.15^2}\right)^2 3000}{(2 \cdot 0.15) \cdot 2} + 950 \text{ kg/m}^3 \cdot 9.8 \text{ m/s}^2 3000\right) = 5.8 \cdot 10^6 \text{ Pa} \tag{12E2.2}$$

Therefore the fluid is monophasic.

12.2 Power cycle

For the use of the hot brine toward the production of energy, different cycles can be used. The brine can be used directly as thermal fluid. In this case, flash types of cycles are designed, where a vapor phase is generated by expansion of the brine that is used in a turbine for power production. Alternatively, the brine can be used in Rankine cycles as a heat transfer fluid within a binary cycle to heat up another fluid, the organic one, that is later fed to the turbine system. Thus, the binary cycles use an ORC. It is a particular case of a Rankine cycle that uses organic fluids, with boiling temperatures lower than those of water for the production of energy. It is also possible to use a binary flash cycle, where the organic fluid after being heated up with the brine is flashed and the vapor phase fed to a turbine. In this section, we present the two basic cases. The section is somehow an extension to the one presented in Chapter 9, for the concentrated solar power. The more general section in terms of cycles divides it into the definitions of the types of fluids and the description of the cycles (Lee et al., 2019; Chen et al., 2010; Wang et al., 2013). In addition, binary cycles can be used to recover waste energy from industrial processes (Sun and Li, 2011; Yu et al., 2016).

12.2.1 Types of fluids

Three types of fluids can be distinguished depending on their behavior in the expansion within the turbine: Wet fluids, such as steam, may condense due to the cooling in the isentropic expansion. They are characterized by a $ds/dT < 0$ and a symmetric biphasic bell, see Fig. 12.2. Because of the condensation, the turbine must operate under certain conditions that allow smaller liquid formation, typically below 8%. Isentropic fluids are characterized by a vertical right side of the biphasic bell, $ds/dT = 0$. Examples of them are refrigerants such as R11, see Fig. 12.2. Finally, dry fluids are particularly interesting since there is no risk

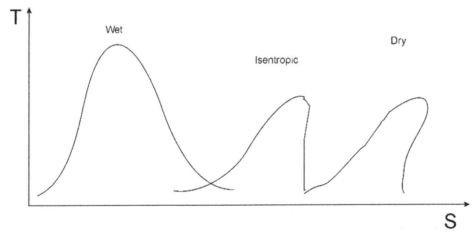

FIGURE 12.2   Types of fluids in Rankine cycles.

of formation of a liquid phase in the expansion at the turbine. They are characterized by $ds/dt > 0$. Organic hydrocarbons such as *n*-hexane or pentane are of this type of fluids, see Fig. 12.2.

12.2.2 Types of cycles

Two types of cycles are presented, binary cycles, where the hot source is used to evaporate a fluid that will be used in the turbine system, and the flash cycles. The combination of both generates the hybrid flash/binary cycles. Within each type, some variants can be presented (Lee et al., 2019).

12.2.2.1 Flash cycles

Description: A flash cycle consists of using the high pressure (HP) and high-temperature brine as operating fluid for the cycle. Fig. 12.3A shows the flowsheet described previously and Fig. 12.3B the corresponding T-S diagram. The brine (1) is expanded in a valve (2) generating a biphasic flow that is separated in a flash so that only the vapor phase (3) is used in the turbine to produce power. Once condensed (6), it is mixed with the liquid phase and pumped back to the well (7).

To improve the efficiency of the system, double-flash systems can be used. Fig. 12.4A shows the flowsheet described in the previous lines and Fig. 12.4B presents the T-S diagram of the cycle. In this case, the liquid from the first expansion (4) is further expanded generating a second vapor phase (8). The second vapor stream is mixed with the initial vapor phase after its expansion in the first turbine (6) and both are fed to the second lower pressure turbine. The expanded fluid is condensed (11), mixed with the liquid and pumped back to the well. Alternatively, both streams can be pumped separately and reinjected in the well.

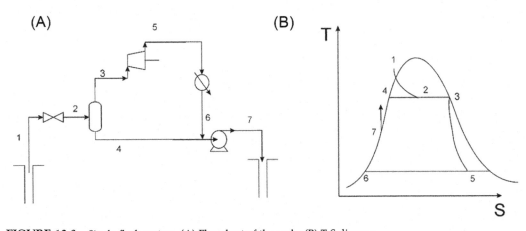

FIGURE 12.3   Single-flash system. (A) Flowsheet of the cycle; (B) T-S diagram.

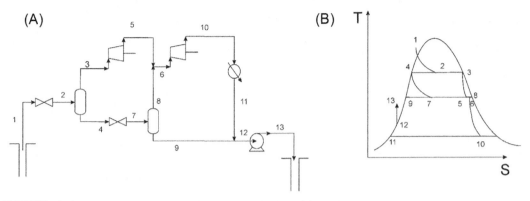

FIGURE 12.4 Double-flash geothermal cycle. (A) Flowsheet; (B) T-S diagram.

Analysis:

1. The hot brine temperature and pressure are computed from the well characteristics and used to determine the enthalpy, Eq. (12.7):

$$h_{hotbrine} = f(T_{brine}, p_{brine}) \tag{12.7}$$

2. An isenthalpic expansion to the steam is performed generating two phases, Eq. (12.8):

$$h_{hotbrine} = h_f \tag{12.8}$$

3. The mass and energy balances to the separator are shown in Eq. (12.9):

$$F = V + L$$
$$h_f\, F = H_V \cdot V + h_L L$$
$$h_f = f(T_2, p_2) \tag{12.9}$$
$$h_L = f(T_4, p_4)$$
$$H_V = f(T_3, p_3)$$

where $T_2 = T_3 = T_4$, and the pressure corresponds to the saturation one at those temperatures and can be computed using Antoine's correlation, Eq. (12.10):

$$p_2(\text{mmHg}) = e^{\left(18.3036 - \frac{3816.44}{227.02 + T_2(^\circ C)}\right)} \tag{12.10}$$

The correlations to compute the steam enthalpies and entropies can be seen in the Appendix B.

4. The turbine expansion is isentropic with an isentropic efficiency, Eq. (12.11). The fact that the fluid is wet and the feed is saturated steam results in the fact that the exhaust stream is a liquid–vapor mixture, see Fig. 12.3B.

$$\eta = \frac{H_5 - H_V}{H_{5,iso} - H_V} \tag{12.11}$$

where the difference between the isentropic expansion and the actual one is the faction of liquid, $f_{L,iso}$, produced if the expansion is ideal compared to the one in case there is an efficiency, f_L, Eq. (12.12):

$$H_{5,iso} = f_{L,iso}h_6 + (1 - f_{L,iso})H_V$$
$$H_5 = f_L h_6 + (1 - f_L)H_V \tag{12.12}$$

The efficiency determines the operating temperature computed as in Eq. (12.13):

$$s_{(3)} = f(T_{steam}, p_{steam}) = s_{(4,s)} \tag{12.13}$$

If multiple expansions are considered, the calculation of the power obtained at each stage is similar and it can be computed as in Eq. (12.14):

$$W_{(Turb1)} = fc_{steam} \cdot (H_3 - H_5) \tag{12.14}$$

5. The condensation is carried out using water that is later cooled again using either an air cooler or a cooling tower. For the analysis of these systems we refer the reader to Chapter 9, where they are evaluated.
6. The liquid is pumped to be reinjected. To compute the energy a simple energy balance is formulated, Eq. (12.15):

$$W_{(Pump)} = \Delta P \cdot Q = \Delta P \cdot \frac{fc_{(water)}}{\rho} \tag{12.15}$$

See Example 12.3.

EXAMPLE 12.3 Compute the power that can be produced using a flash cycle that processes the brine extracted from the well obtained in Examples 12.1 and 12.2. Assume that the first flash expands the brine at 1 bar, the turbine expands the vapor to 0.08 bar, and the injection of the brine is at the same pressure as that at the wellhead.

Solution

We use a process simulator to model the flowsheet of Fig. 12E3.1.

Simulating the flowsheet presented in Fig. 12E3.1, the flow of 75 kg/s of brine at 147°C and 58 bar produces 2.5 MW, and the two pumps consume 0.0006 MW and 0.44 MW respectively, leading to a net production of around 2 MW.

The flash generates 9% of vapor stream and the rest remains as liquid. The liquid brine is reinjected at 95°C.

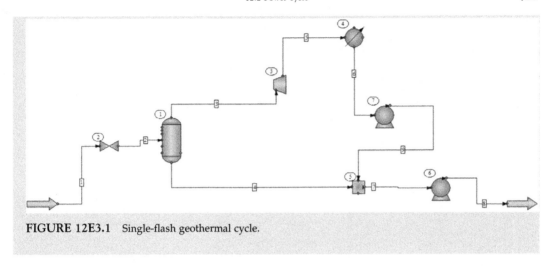

FIGURE 12E3.1 Single-flash geothermal cycle.

12.2.2.2 Binary cycles

Description: Binary cycles do not use the hot brine as the fluid in the turbine but an auxiliary fluid with a lower boiling point such as refrigerants and hydrocarbons. They are appropriate for medium- and low-temperature hot sources. Two alternative structures are presented as follows (Peña et al., 2018).

1. The basic cycle consists of using the hot brine to heat up and evaporate an organic fluid (3) that is later fed to a turbine where it is expanded (4). The working fluid is condensed (5), compressed (1), and sent back to the train of heat exchangers for reevaporation. Fig. 12.5A shows the flowsheet just described and Fig. 12.5B shows the T-S diagram corresponding to such a cycle. Note that a dry working fluid is considered.
2. The two-stage ORC uses two different pressure levels and can be configured in series or in parallel. For the series configuration, the working fluids are pressurized at a low pressure

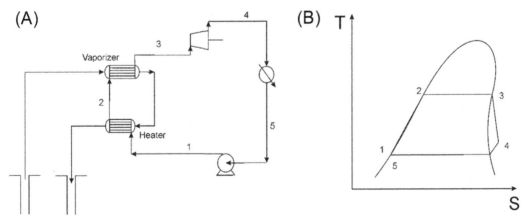

FIGURE 12.5 Basic binary cycle. (A) Flowsheet; (B) T-S diagram.

(A)

(B)

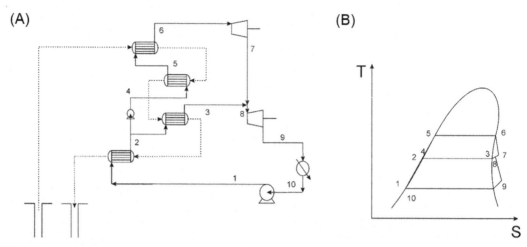

FIGURE 12.6 Two-stage ORC. *ORC,* Organic Rankine cycle. (A) Flowsheet; (B) T-S diagram.

(LP) (1), preheated with the hot brine (2), and divided into two streams, one is pressurized at HP (4), heated up (5), and evaporated (6) before being fed to the HP turbine, while the other (3) is mixed with the exhaust of the HP and fed to the LP turbine (8). The exhaust of the LP turbine (9) is condensed (10) and fed to the preheater (1). The parallel structure consists of splitting the cold working fluid. A fraction is pressurized and evaporated to HP to be fed to the HP turbine, while the second fraction is only pressurized at lower pressure and heated up to be fed to the LP turbine together with the exhaust from the HP turbine. The exhaust of the LP turbine is condensed. Fig. 12.6A shows the flowsheet and Fig. 12.6B the T-S diagram.

Analysis:

1. The brine is extracted from the well at the wellhead conditions as presented in the previous section, and it is used to heat up and evaporate the working fluids.
2. The heating process is performed under counter-current conditions, to make the most of the operating conditions, Eq. (12.16):

$$Q_{(HXi)} = - fc_{(Brine)}c_p \cdot \left(T_{(Brine,Out)} - T_{(Brine,in)}\right)$$
$$Q_{(HXi)} = fc_{(Fluid)} \cdot \left(H_{(Fluid,Out)} - H_{(Fluid,in)}\right)$$
$$T_{(Fluid,Out)} + \Delta T_{min} \leq T_{(Brine,in)}$$
$$T_{(Fluid,in)} + \Delta T_{min} \leq T_{(Brine,Out)}$$

(12.16)

The heat transferred from the brine can be computed using heat capacities. However, to capture the thermodynamics of the working fluid, the enthalpy, H, Eq. (12.17), and entropy, s, Eq. (12.18), of the organic fluids are used to perform the energy balances. Examples for several fluids can be seen in the Appendix.

$$H_{(Fluid,i)} = f(T_{Fluid}, P_{Fluid})$$

(12.17)

$$s_{Fluid(i)} = g(T_{Fluid}, P_{Fluid})$$

(12.18)

3. The expansions at the turbine are assumed to have a certain isentropic efficiency. Therefore the stream exiting the first body can be calculated using Eq. (12.19):

$$\eta = \frac{H_{Fluid,out} - H_{Fluid,in}}{H_{Fluid,out(iso)} - H_{Fluid,in}} \tag{12.19}$$

The efficiency determines the operating temperature computed as Eq. (12.20):

$$s_{Fluid(6)} = f(T_{Fluid}, P_{Fluid}) = s_{Fluid(7,s)} \tag{12.20}$$

The Antoine correlation is used as a bound for the points of saturated vapor, Eq. (12.18), so that we ensure that a particular stream is not in a two-phase flow. Saturated temperatures are also computed using Eq. (12.21):

$$P_{turb1} \cdot 760 = e^{\left(A(OF) - \frac{B(OF)}{(C(OF) + T_{Sat})}\right)} \tag{12.21}$$

Each expansion at the turbine is modeled similarly.
4. The energy obtained from the expansions is given by Eq. (12.22):

$$W_{(Turb)} = \sum_{i \in \{turbines\}} fc_{(fluid,turbine)} \cdot (H_{In,turbine} - H_{Out,turbine}) \tag{12.22}$$

where fc is the flow of fluid.
5. The condensation is carried out using water that is later cooled again using either an air cooler or a cooling tower. Again, Chapter 9, presents the analysis of cooling systems.
6. The exhaust steam is condensed and compressed before restarting the cycle again. The power consumed by each of the pumps can be computed as follows, Eq. (12.23):

$$W_{(Pump)} = \Delta P \cdot Q = \Delta P \cdot \frac{fc_{(Fluid,unit,Pump)}}{\rho} \tag{12.23}$$

See Example 12.4.

EXAMPLE 12.4 Evaluate the performance of a binary cycle that uses toluene as organic fluid. The cold extreme of the cycle is 25°C so that water can be used to refrigerate the heat exchanger and assume pressurization up to 0.5 bar to feed the turbine. The brine is at 177°C, similar to previous examples and a flow of 50 kg/s can be available. The isentropic efficiency of the turbine is 0.85. Assume a heat recovery approach temperature (HRAT) of 5°C between the hot brine and the organic fluid.

Solution

Following the analysis of the cycle presented previously we model it in GAMS using the correlations for the enthalpy presented in Appendix B (Peña et al., 2018). We follow the cycle shown in Fig. 12.5 for reference related to the stream names. First, the sets, scalar, variables, and parameters are defined

```
Set
*                 running variable for vapor pressure correlation
l /1*3/
* streams of
s /1*10/
* streams brine
h /1*3/;
Parameter
*                 vapor pressure coefficients: mmHg, T = C
*                 750 mmHg = 1 bar, 760 mmHg = 1 atm
*                 ln(p_k) = coef_p(h,1) − coef_p(h,2)/(T + coef_p(h,3))
*                 Data from Biegler's database and Coulson
                  coef_p(l)
                  /1     16.0137
                  2      3096.52
                  3      219.48 /;
Positive Variables
foc(s)        individual components streams       kg per s
fbrine(h)        individual components streams      kg per s
Toc(s)      temperature in °C
Tbrine(h)     temperature in °C
Soc(s)      entropy of octer stream kJ per kg K
Poc(s);
Variable
Hoc(s)      enthalpy of octer stream kJ per mol
W_turbine
W_pump1
Q_HX1
Q_HX2
Q_HX3;
```

We start by the well; however, it is possible to use another starting point. We present the energy balances by which the hot brine, a flow of 50 kg/s that is fixed, provides the energy to evaporate toluene at a pressure of 0.5 bar in HX1. To compute the temperature, Antoine correlation using toluene coefficients is used. Note that to avoid temperature crossing we impose bounds to the temperature of the brine (Tbrine) and of the organic fluid (Toc). In addition Hoc(3) corresponds to saturated vapor, while Hoc(2) is that of a saturated liquid. The second energy balance is used to heat up the compressed liquid, Hoc(1), to saturated liquid. Again, temperature bounds are to be included to avoid infeasible operation:

```
*** Well
Scalar
c_p_ind              /4.18/
```

```
TMax_brine                    /177/
Tmin_brine              /120/
T_ref       en K                /298/
h_ref                       Entalpía de referencia para balances (h_25ºC) (kJ·mol^−1)
              /−14.554248/
MWoc             g per mol       /92/;
Equations
we1, we2, we3, we4, we5, we6, we7, we8, we9, we10,
we11, we12, we13, we14, we15, we16,we17, we18, we19, we20;
*,well13, well14, well15, well16,well17;
we1..      Q_HX1 = E = fbrine('1')*c_p_ind*(Tbrine('1') − Tbrine('2') );
fbrine.fx('1') = 50;
Tbrine.fx('1') = Tmax_brine;
Poc.fx('1') = 0.5;
we2.. Poc('1') = E = Poc('2');
we3..     Poc('2')*760 = E = exp(coef_p('1')-coef_p('2')/(coef_p('3') + Toc('2')));
we4..     Tbrine('1') = G = Toc('3') + 5;
we5..     Tbrine('2') = G = Toc('2') + 5;
we6..     Toc('3') = E = Toc('2');
we7..     Hoc('3') = E = − 3.15523*10**(−7)*(Toc('3') + 273.15)**3
               + 4.51006*10**(−4)*(Toc('3') + 273.15)**2−8.2643*10**(−2)*
               (Toc('3') + 273.15) + 16.3670;
we8..     Hoc('2') = E = 2.413*10**(−13)*(Toc('2') + 273.15)**6
               − 5.986*10**(−10)*(Toc('2') + 273.15)**5 + 6.123*10**(−7)*
               (Toc('2') + 273.15)**4−0.0003303*(Toc('2') + 273.15)**3
               + 0.09922*(Toc('2') + 273.15)**2−15.6*(Toc('2') + 273.15) + 971.6;
we9..      Q_HX1 = E = (foc('2')/MWoc)*1000*(Hoc('3') − Hoc('2'));
we10.. Poc('3') = E = Poc('2');
we11.. Tbrine('2') = G = Tbrine('3') + 1;
we12..    foc('2') = E = foc('3');
we13..    foc('2') = E = foc('1');
we14.. fbrine('1') = E = fbrine('2');
we15.. fbrine('2') = E = fbrine('3');
we16..     Q_HX2 = E = fbrine('2')*c_p_ind*(Tbrine('2') − Tbrine('3'));
we17..     Q_HX2 = E = (foc('1')/MWoc)*1000*(Hoc('2') − Hoc('1'));
we18..    Hoc('1') = E = h_ref + (140140−152.3*((Toc('1') + 273.15 + T_ref)/2) +
               0.695*((Toc('1') + 273.15 + T_ref)/2)**2)*10**(−6)*(Toc('1') +
273.15-T_ref);
we19..    Hoc('2') = E = 2.413*10**(−13)*(Toc('2') + 273.15)**6
               − 5.986*10**(−10)*(Toc('2') + 273.15)**5 + 6.123*10**(−7)*
               (Toc('2') + 273.15)**4−0.0003303*(Toc('2') + 273.15)**3
               + 0.09922*(Toc('2') + 273.15)**2−15.6*(Toc('2') + 273.15) + 971.6;
we20..     Tbrine('3') = G = Toc('1') + 5;
Toc.fx('1') = 25;
```

Toluene is a dry fluid; therefore the expansion will not generate a liquid phase. In the turbine, the fluid is expanded from saturated vapor, 3, to overheated vapor at lower pressure. We consider the exhaust pressure that corresponds to a temperature of 25°C so that water is used as cooling agent. To compute that, Antoine correlation is used. The expansion is not ideal; therefore an isentropic yield is defined. We assume it to be 0.85. Eq. (12.20) allows computing the actual temperature:

```
*** Turbine
Variable
H4iso;
Positive variable
Tiso4
Pmin;
Scalar
rendiso /0.85/;
Equations
turb1, turb2, turb3, turb4, turb5, turb6, turb7, turb8, turb9, turb10;
turb1.. Soc('3') = E = -8.01*10**(-11)*(Toc('3') + 273.15)**5
            +1.683*10**(-7)*(Toc('3') + 273.15)**4-0.0001422*
            (Toc('3') + 273.15)**3 + 0.06072*(Toc('3') + 273.15)**2
            -13.06*(Toc('3') + 273.15) + 1208;
turb2..    Soc('3') = E = Soc('4');
turb3..    foc('3') = E = foc('4');
turb4.. Soc('4') = E = -4.367 + 0.3199*(Tiso4 + 273.15) - 323.9*Poc('4')
+4.004*10**(-5)*
            (Tiso4 + 273.15)**2 + 0.2449*Poc('4')*(Tiso4 + 273.15) + 1156*Poc
('4')**2
            -0.0003046*(Tiso4 + 273.15)**2*Poc('4') - 0.03606*(Tiso4 + 273.15)
*Poc('4')**2
            -1774*Poc('4')**3;
turb5..    H4iso = E = 7.797 + 0.0007286*(Tiso4 + 273.15) - 1.743*Poc('4') + 0.000174*
            (Tiso4 + 273.15)**2 + 0.003393*Poc('4')*(Tiso4 + 273.15) -
0.01305*Poc('4')**2;
turb6..    Hoc('4') = E = 7.797 + 0.0007286*(Toc('4') + 273.15) - 1.743*Poc('4') +
0.000174*
            (Toc('4') + 273.15)**2 + 0.003393*Poc('4')*(Toc('4') + 273.15) -
0.01305*Poc('4')**2;
turb7..    rendiso*(Hoc('3')-H4iso) = E = (Hoc('3')-Hoc('4'));
turb8.. W_turbine = E = (foc('3')/MWoc)*1000*(Hoc('4') - Hoc('3'));
turb9.. Pmin*760 = E = exp(coef_p('1') - coef_p('2')/(coef_p('3') + 25));
turb10..    Poc('4') = E = Pmin;
```

The cooling stage condenses the exhaust vapor from the turbine to 25°C and pressurizes the liquid to 0.5 bar to feed HX2:

```
**Cooling
```

```
Equations
cool1, cool2,cool3, cool4, cool5;
cool1.. foc('4') = E = foc('5');
Toc.LO('5') = 25;
cool2.. Hoc('5') = E = 2.413*10**( − 13)*(Toc('5') + 273.15)**6
            − 5.986*10**( − 10)*(Toc('5') + 273.15)**5 + 6.123*10**( − 7)*
            (Toc('5') + 273.15)**4−0.0003303*(Toc('5') + 273.15)**3
            + 0.09922*(Toc('5') + 273.15)**2−15.6*(Toc('5') + 273.15) + 971.6;
cool3.. Poc('5') = E = Poc('4');
cool4..   Q_HX3 = E = (foc('4')/MWoc)*1000*(Hoc('5') − Hoc('4'));
cool5..   W_pump1 = E = foc('5')*(Hoc('1') − Hoc('5'));
```

We optimized the power produced form the cycle so that the temperatures of the brine and those of the toluene come close:

```
Equations
                Obj;
Variables
                Z;
Obj..      Z = E = 2 − W_turbine;
Model RankinePT /ALL/;
option NLP = CONOPT;
Solve RankinePT Using NLP Maximizing Z;
```

See Table 12E4.1 for the results.

TABLE 12E4.1   Stream characteristics.

	Hoc (kJ/mol)	Toc (°C)	Poc (bar)	Flow (kg/s)	Tbrine (°C)	Flow (kg/s)
1	− 14.5	25	0.5	44.6	177	50
2	− 5.7	87.9	0.5	44.6	92.9	50
3	30.4	87.9	0.5	44.6	72.5	50
4	24.0	30.3	0.037	44.6		
5	− 16.0	25	0.037	44.6		

The turbine produces 3.1 MW and the pump consumes 68 kW.

12.2.2.3 Flash/binary cycles

It is possible to combine flash and binary cycles, for instance, it is possible to use the brine to produce stream of organic fluid at high temperature and pressure that is flashed to obtain power in a turbine. Single- and double-flash systems are possible. Fig. 12.7 shows an example of a hybrid cycle.

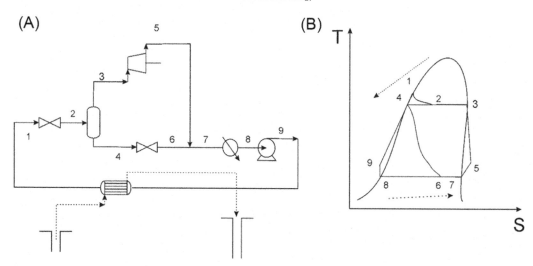

FIGURE 12.7   Basic binary cycle. (A) Flowsheet; (B) T-S diagram.

12.3 Safety issues

Binary cycles with the use of organic species, refrigerants, and hydrocarbons present a number of safety challenges that must be addressed related to the selection of the fluid and the operating conditions of the cycle. Quantitative risk analysis methods allow identifying whether operations, engineering, or management systems can be modified to reduce risk. The best results can be obtained at conceptual design and must be maintained throughout the facility's life cycle. It provides a quantitative method to help managers make decisions and evaluate the overall risk of the process (Crowl and Louvar, 2011). This method includes four major steps, and Fig. 12.8 shows the scheme to estimate the risk.

1. Identification of incidents and the potential event sequences.
2. Evaluating the incident consequences (the typical tools for this step include dispersion modeling and fire and explosion modeling).
3. Estimating the potential incident frequencies using event trees and fault trees.
4. Estimating the incident impacts on people, environment, property, and the risk by combining the impacts and frequencies.

12.3.1 Incident identification

The identification of the incidents can be performed using qualitative risk analysis methods such as HAZOP (hazard and operability)(Kletz, 1984). The most common scenario in the chemical process industry is the loss of containment of hazardous material that, in this case, corresponds with the organic fluids used in the cycle. This can occur through a variety of incidents, such as a leak from a vessel, a ruptured pipeline, a gasket failure, or release from a relief valve. We can distinguish between continuous and instantaneous release.

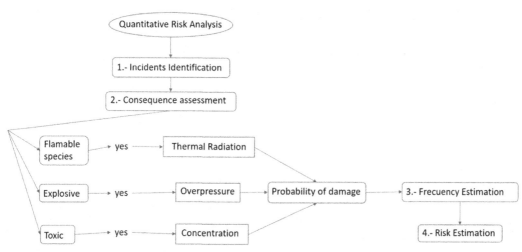

FIGURE 12.8 Scheme of the QRA. *QRA, Quantitative risk analysis.*

12.3.1.1 Continuous release

Due to a puncture or as a result of corrosion a hole can be generated in a pipe or vessel resulting in the leak of the chemical species. The flow that will result is responsible for the incident. This type of release can generate jet fire, flash fire, and toxic release.

To compute the flow of chemicals, fluid mechanics is applied:

- Flow of Liquid through a Hole, where A is the area of the hole, C_o is the discharge coefficient, ρ is density, and P_g is the relative pressure, Eq. (12.24):

$$fc = \rho \cdot Q = A \cdot C_o \cdot \sqrt{2\rho g_c P_g} \tag{12.24}$$

- Flow of Liquid through a Hole in a Tank, where h_L is the liquid height in the tank, Eq. (12.25):

$$fc = \rho \cdot Q = A \cdot C_o \cdot \sqrt{2\left(\frac{g_c P_g}{\rho} + gh_L\right)} \tag{12.25}$$

- Flow of Vapor through Holes, where M is the molecular weight, P is the pressure, k is the ratio of heat capacities, T is temperature, Eq. (12.26):

$$fc = A \cdot C_o \cdot P_o \sqrt{\frac{2g_c M}{R_g T_o} \frac{k}{k-1} \left(\left(\frac{P}{P_o}\right)^{2/k} - \left(\frac{P}{P_o}\right)^{(k+1)/k}\right)} \tag{12.26}$$

- Flashing liquids for liquids stored under pressure that exposed to the atmosphere evaporate. The process is assumed to be adiabatic, Eq. (12.27):

$$fc = \rho \cdot Q = A \cdot C_o \cdot \sqrt{2\rho g_c(P - P^{sat})} \tag{12.27}$$

- Flow of evaporation from pools, where K is the mass transfer coefficient for an A, T is the liquid temperature, and P^{sat} is the saturation pressure, Eq. (12.28):

$$fc = \frac{MKAP^{sat}}{RT_L} \tag{12.28}$$

12.3.1.2 *Instantaneous release*

The mass inside the unit is released generating boiling liquid expanding vapor explosion (BLEVE), unconfined vapor cloud explosion (UVCE), flash fire, and toxic releases.

12.3.2 The consequence

The releases result in fires, explosions, or toxicity. The idea is to quantify, using dispersion modeling and the analysis of the consequences, the effect that they have downwind. To assess that effect, the different consequences are characterized by the physical variables that cause the accidents. It depends on the operating conditions of a process and the characteristics of the substances (toxicity, flammability, and explosiveness). Thus the consequence is expressed as a function of the operating variables of each component of the cycle and of the properties of each working fluid. For flammable species, the *radiation dose* is the variable that characterizes the accidents such as jet, flash and pool fires, and BLEVEs, the *concentration* determines the toxicity, and for explosive species, the *overpressure* generated characterizes the consequences of the release. In the following, we discuss on the estimation of the different variables for the accidents.

12.3.2.1 *Physical variables to characterize the consequence*

12.3.2.1.1 Radiation dose (flammable species)

12.3.2.1.1.1 Jet fire   The jet fire is produced by flammable material leakage, which suffers an immediate ignition. Although this incident generates less damage than other incidents, its frequency of occurrence is higher. This accident occurs *in gas phase release* and its reach depends significantly on the pressure. To compute the radiation dose, q (kW/m^2), as a result of the jet fire, the procedure is shown in Eq. (12.29):

$$\begin{aligned}
\frac{L_{Flame}}{d} &= \frac{15}{C_T}\sqrt{\frac{M_a}{M_f}} \\
X &= \sqrt{L^2 + L_{Flame}^2} \\
P_w(\text{Pa}) &= 101335(RH)e^{\left(14.4114 - \frac{5328}{T_a}\right)} \\
\tau_a &= 2.02(P_w(\text{Pa})X(\text{m}))^{-0.09} \\
F_p &= \frac{1}{4\pi L^2} \\
R &= \tau_a \cdot \eta \cdot m_r \cdot \Delta H_{comb} \cdot F_p \\
d &= d_{nozzle}\sqrt{\frac{\rho}{\rho_{air}}} \\
q &= RF_{12}
\end{aligned} \tag{12.29}$$

where L_{Flame} is the length of the visible turbulent flame measured from the break point, d is the jet diameter, physical diameter of the nozzle, C_T is the fuel mole fraction in the mixture air–fuel, M_a is the air molecular weight, M_f is the molecular weight of the n-octane, η is the fraction of the energy converted to radiation, M_r is the mass released due to the release, L is the distance from the incident to the receptor, typically 50 m is assumed, RH is the relative humidity as a fraction, and T_a is the ambient temperature in K.

The view factor F_{12} (Assael and Kakosimos, 2010) is the fraction of energy exiting an isothermal, opaque, and diffuse surface 1 (by emission or reflection), that directly impinges on surface 2 (to be absorbed, reflected, or transmitted), Eq. (12.30):

$$
\begin{aligned}
\pi F_{12H} = {} & \tan^{-1}\left(\frac{1}{D}\right) + \frac{\sin\Theta}{C}\left[\tan^{-1}\left(\frac{\alpha\beta - F^2\sin\Theta}{FC}\right) + \tan^{-1}\left(\frac{F\sin\Theta}{C}\right)\right] \\
& - \left[\frac{\alpha^2 + (\beta+1)^2 - 2(\beta + 1 + \alpha\beta\sin\Theta)}{AB}\right]\tan^{-1}\left(\frac{AD}{B}\right) \\
\pi F_{12V} = {} & -E\tan^{-1}D + E\left[\frac{\alpha^2 + (\beta+1)^2 - 2\beta(1 + \alpha\sin\Theta)}{AB}\right]\tan^{-1}\left(\frac{AD}{B}\right) \\
& + \frac{\cos\Theta}{C}\left[\left[\tan^{-1}\left(\frac{\alpha\beta - F^2\sin\Theta}{FC}\right) + \tan^{-1}\left(\frac{F\sin\Theta}{C}\right)\right]\right] \\
F_{12} = {} & \sqrt{F_{12V} + F_{12H}}
\end{aligned}
\tag{12.30}
$$

where $\alpha, \beta, A, B, C, D, E, F,$ and Θ are given as defined in Eq. (12.31):

$$
\begin{aligned}
\alpha &= \frac{L_{Flame}}{R}; \beta = \frac{L}{R} \\
A &= \sqrt{\alpha^2 + (\beta+1)^2 - 2\alpha(\beta + 1)\sin\Theta} \\
B &= \sqrt{\alpha^2 + (\beta-1)^2 - 2\alpha(\beta - 1)\sin\Theta} \\
C &= \sqrt{1 + (\beta^2 - 1)\cos^2\Theta} \\
D &= \sqrt{(\beta - 1)(\beta + 1)} \\
E &= \alpha\cos\Theta/(\beta - \alpha\sin\Theta) \\
F &= \sqrt{(\beta^2 - 1)} \\
\tan\Theta/\cos\Theta &= 0.666 Fr^{0.333} Re^{0.117} \\
Fr &= \frac{u_w^2}{gD} \\
Re &= \frac{u_w \rho_{air} D}{\mu_{air}}
\end{aligned}
\tag{12.31}
$$

The exposure time, t_{eff} (s), used to compute the thermal radiation dose, is usually considered to be the sum of an initial reaction time (5 seconds) plus the time required to reach a safe point moving at 4 m/s.

12.3.2.1.1.2 Pool fire When a pool of flammable liquid ignites, a pool fire is generated. The physical variable calculated is the radiation dose, q. To estimate it, Mudan models (Assael and Kakosimos, 2010) are among the most used. The burning rate is computed depending on where the boiling temperature of the liquid is above or below the atmospheric one. Thus

1. For liquids with $T_{eb} > T_{atm}$, Eq. (12.32):

$$m' = \frac{0.001\,(\mathrm{kg/m^2}) \cdot \Delta H_{comb}}{c_p(T_{eb} - T_{atm}) + \Delta H_{vap}} \tag{12.32}$$

2. For liquids with $T_{eb} < T_{atm}$, Eq. (12.33):

$$m' = \frac{0.001 \cdot \Delta H_{comb}}{\Delta H_{vap}} \tag{12.33}$$

The radiation dose, q (kW/m^2), is computed as in Eq. (12.34), with a time endurance, t, of the fire:

$$q = E \cdot F_{12} \cdot \tau$$
$$t = \frac{m}{\pi (D_{pool}/2)^2 m'} \tag{12.34}$$

where the emissive power can be calculated as in Eq. (12.35) in (kW/m^2):

$$E = 140 \cdot e^{-0.12D} + 20(1 - e^{-0.12D}) \tag{12.35}$$

The flame height is given by Eq. (12.36):

$$H_f = 42 \cdot D\left(\frac{m'}{\rho_{air}\sqrt{gD}}\right)^{0.61} \tag{12.36}$$

And the transmissivity can be computed by Eq. (12.37):

$$\tau = 2.02(P_w \cdot X)^{-0.09} \tag{12.37}$$

where D is the pool diameter and X is the distance to the receptor, Eq. (12.38):

$$X = L - D/2 \tag{12.38}$$

The view factor F_{12} is the fraction of energy exiting an isothermal, opaque, and diffuse surface 1 (by emission or reflection) that directly impinges on surface 2 (to be absorbed, reflected, or transmitted) given by Eq. (12.39):

$$F_{12H} = \frac{B - \frac{1}{S}}{\pi\sqrt{B^2 - 1}} \tan^{-1}\sqrt{\frac{(B+1)(S-1)}{(B-1)(S+1)}} - \frac{A - \frac{1}{S}}{\pi\sqrt{A^2 - 1}} \tan^{-1}\sqrt{\frac{(A+1)(S-1)}{(A-1)(S+1)}}$$

$$F_{12V} = \frac{1}{\pi S} \tan^{-1}\left(\frac{h}{\sqrt{S^2 - 1}}\right) - \frac{h}{\pi S} \tan^{-1}\left(\sqrt{\frac{S-1}{S+1}}\right) + \frac{Ah}{\pi S\sqrt{A^2 - 1}}\sqrt{\frac{(A+1)(S-1)}{(A-1)(S+1)}}$$

$$F_{12} = \sqrt{F_{12V} + F_{12H}}$$
(12.39)

where

$$A = \frac{h^2 + S^2 + 1}{2S}; B = \frac{1 + S^2}{2S}; S = \frac{2L}{D}; h = \frac{2H_f}{D}$$

12.3.2.1.1.3 Boiling liquid expanding vapor explosions (BLEVE) The BLEVE consists of the abrupt release of the total mass contained in a unit when the tank containing a liquid held above its atmospheric pressure boiling point breaks, resulting in the explosive vaporization of the liquid. If the liquid is flammable, a VCE might be generated; if it is toxic, the surrounding area will be exposed. In both cases, the energy released by the BLEVE process itself can result in considerable damage.

For a BLEVE to occur, a sudden failure of the container is to happen. The most common type of BLEVE is caused by fire as a result of the following steps:

1. A fire develops adjacent to a tank containing a liquid.
2. The fire heats the walls of the tank.
3. The tank walls below liquid level are cooled by the liquid, increasing the liquid temperature and the pressure in the tank.
4. If the flames reach the tank walls or roof where there is only vapor and no liquid is present to remove the heat, the tank metal temperature rises until the tank loses it structural strength.
5. The tank ruptures, explosively vaporizing its contents.

If the liquid is flammable and a fire is the cause of the BLEVE, the liquid may ignite as the tank ruptures. Often, the boiling and burning liquid behaves as a rocket fuel, propelling vessel parts for great distances. If the BLEVE is not caused by a fire, a vapor cloud might form, resulting in a VCE. The vapors might also be hazardous to personnel by means of skin burns or toxic effects.

The effects of a BLEVE can be estimated by the radiative dose that can be computed as given by Eq. (12.40) (CPS, 1994):

$$
\begin{aligned}
D_{Max} &= 5.8 \cdot M^{1/3} \\
t_{BLEVE} &= \begin{cases} 0.45 \cdot M^{1/3} \text{ for } M < 30,000 \text{ kg} \\ 2.6 \cdot M^{1/6} \text{ for } M > 30,000 \text{ kg} \end{cases} \\
H_{BLEVE} &= 0.75 D_{Max} \\
X &= \sqrt{L^2 + H_{BLEVE}^2} - \frac{D_{Max}}{2} \\
P_W &= 101335(RH)\exp^{\left(14.4114 - \frac{5328}{T_{atm}}\right)} \\
\tau_a &= 2.02(P_W X)^{-0.09} \\
F_{12} &= \frac{H_{BLEVE}\left(\frac{D_{Max}}{2}\right)^2}{\left(L^2 + H_{BLEVE}^2\right)^{1.5}} \\
E &= \frac{RM\Delta H_{comb}}{\pi D_{Max}^2 t_{BLEVE}} \\
q &= E\tau_a F_{12}
\end{aligned}
\tag{12.40}
$$

where, D_{Max} is the maximum diameter of the fireball (m), X is the distance from the flame surface to an individual, t_{BLEVE} is the duration of the fire ball (s), L is the distance from the incident to the receptor, typically 50 m, H_{BLEVE} is the vertical distance from the center of the fireball to the ground, P_w is the water partial pressure (Pa), t_a is the atmospheric transmissivity (-), E is the radiative emissive flux (kW/m²), R is the radiative fraction of the heat of combustion (0.4), M is the mass due to an instantaneous release, ΔH_{comb} is the combustion energy (kJ/kg), F_{12} is the view factor (-), q is the radiant flux at the receptor (kW/m²), and RH is the relative humidity.

12.3.2.1.1.4 Flash fire   The flash fire is the combustion without explosion of flammable material. In this section, Eq. (12.41), we compute the thermal radiation generated by such an event Moran (2017). However, in Section 12.2 we show the calculus of the concentration, the computation of the lower flammability limits (LFL) as a function of the distance. If the concentration (Pasquill−Gifford model) is greater than or equal to LFL, the flash fire thermal radiation generates one death.

$$
\begin{aligned}
D_{Max} &= 6.48 \cdot M^{0.325} \\
M &= \text{flash fire mass (kg)} \\
t_{FF}(s) &= 0.825 M^{0.26} \\
H_{FF}(s) &= 0.75 D_{Max} \\
E &= \frac{RM\Delta H_{comb}(\text{J/kg})}{\pi D_{Max}^2 t_{FF}} \\
R &= 0.27 \cdot P(\text{MPa})^{0.32} \\
\tau_a &= 1.30(P_W RH)^{0.69} \\
P_W &= 101335(RH)\exp^{\left(14.4114 - \frac{5328}{T_{atm}}\right)} \\
F_{12} &= \frac{L\left(\frac{D_{Max}}{2}\right)^2}{\left(L^2 + \left(\frac{D_{Max}}{2}\right)^2\right)^{1.5}} \\
q(\text{kW/m}^2) &= E\tau_a F_{12}
\end{aligned}
\tag{12.41}
$$

The definitions of the variables are similar to previous cases see Example 12.5 for a case study.

EXAMPLE 12.5 Compute the radiation dose of a BLEVE generated at a geothermal plant that uses toluene as heat transfer fluid (HTF). Assume that the storage tank of toluene contains 2500 kg and the exposure time equal to that of the duration of the fire is lower than 10 seconds. The relative humidity is 68%, $P = 1$ atm, and the ambient temperature 20°C.

Solution

Following the procedure presented in Eq. (12.37), we compute the variables involved:

T_{atm}	293	K
MW	92	kg/kmol
ΔH_{comb}	−3910.9	kJ/mol
M	2500	kg
D_{Max}	78.72	m
t_{BLEVE}	6.11	s
H_{BLEVE}	59.04	m
L	50	m
X	38.01	m
RH	0.68	
P_w	1584	Pa
F_{12}	0.197	
R	0.4	
E	357.54	kW/m^2
t_a	0.750	
q	52.98	kW/m^2

12.3.2.1.2 Concentration (toxic)

The consequences of toxic releases are based on computing the concentration of the species over the area affected. The dispersion mode of Pasquill–Giffrod model is used based on solving the continuity Eq. (12.42):

$$\frac{\partial C}{\partial t} + \frac{\partial}{\partial x_j}\left(u_j C\right) = 0 \tag{12.42}$$

There are a number of cases where the release is an instantaneous puff or a continuous plume, with or without wind. The full set of scenarios can be seen in Crowl and Louvar (2011), here we just present two typical cases for reference:

- Puff with instantaneous point source at ground level, coordinates fixed at release point, constant wind only in x direction with constant velocity u, Eq. (12.43):

$$C(x,y,z,t) = \frac{fc}{\sqrt{2}\pi^{3/2}\sigma_x\sigma_y\sigma_z} e^{\left\{-\frac{1}{2}\left[\left(\frac{x-ut}{\sigma_x}\right)^2 + \frac{y^2}{\sigma_y^2} + \frac{z^2}{\sigma_z^2}\right]\right\}} \tag{12.43}$$

- Plume with continuous steady-state source at ground level and wind moving in x direction at constant velocity u, Eq. (12.44):

$$C(x,y,z) = \frac{fc}{\pi\sigma_y\sigma_z} e^{\left\{-\frac{1}{2}\left[\frac{y^2}{\sigma_y^2}+\frac{z^2}{\sigma_z^2}\right]\right\}}$$

(12.44)

Because the dispersion of the species depends on the atmospheric conditions and the wind speed, the values for the σ coefficients are tabulated as a function of the day and night period, the insolation during the day, and the presence of clouds during the night

TABLE 12.1 Characterization of the atmospheric conditions for the dispersion model.

	Relative cloud coverage				
	Daytime			Night time	
Wind speed (m/s)	Strong insolation	Moderate insolation	Slight insolation	>4/8 low clouds	<3/8 clouds
<2	A	A−B	B	F	F
2−3	A−B	B	C	E	F
3−4	B	B−C	C	D	E
4−6	C	C−D	D	D	D
>6	C	D	D	D	D

A, Extremely unstable; B, moderately unstable; C, slightly stable; D, neutrally stable; E, slightly stable; F, moderately stable.

TABLE 12.2 Equations for the dispersion coefficients $(x[=]m)$. Plume dispersion.

Stability	σ_x o σ_y	σ_z
Rural conditions		
A	$0.22x(1+0.0001x)^{-1/2}$	$0.20x$
B	$0.16x(1+0.0001x)^{-1/2}$	$0.12x$
C	$0.11x(1+0.0001x)^{-1/2}$	$0.08x(1+0.0002x)^{-1/2}$
D	$0.08x(1+0.0001x)^{-1/2}$	$0.06x(1+0.0015x)^{-1/2}$
E	$0.06x(1+0.0001x)^{-1/2}$	$0.03x(1+0.0003x)^{1}$
F	$0.04x(1+0.0001x)^{-1/2}$	$0.016x(1+0.0003x)^{-1}$
Urban conditions		
A−B	$0.32x(1+0.0004x)^{-1/2}$	$0.24x(1+0.0001x)^{+1/2}$
D	$0.22x(1+0.0004x)^{-1/2}$	$0.20x$
D	$0.16x(1+0.0004x)^{-1/2}$	$0.14x(1+0.0003x)^{-1/2}$
E−F	$0.11x(1+0.0004x)^{-1/2}$	$0.08x(1+0.0015x)^{-1/2}$

A, Extremely unstable; B, moderately unstable; C, slightly stable; D, neutrally stable; E, slightly stable; F, moderately stable.

TABLE 12.3 Equations for the dispersion coefficients $(x[=]m)$. Puff dispersion.

Stability	$\sigma_x \; 0 \; \sigma_y$	σ_z
A	$0.18x^{0.92}$	$0.60x^{0.75}$
B	$0.14x^{0.92}$	$0.53x^{0.73}$
C	$0.10x^{0.92}$	$0.34x^{0.71}$
D	$0.06x^{0.92}$	$0.15x^{0.70}$
E	$0.04x^{0.92}$	$0.10x^{0.65}$
F	$0.02x^{0.89}$	$0.05x^{0.61}$

A, Extremely unstable; B, moderately unstable; C, slightly stable; D, neutrally stable; E, slightly stable; F, moderately stable.

and the wind speed, see Tables 12.1–12.3. The wind speed, as presented in Chapter 11, is typically measured in a certain height and may need to be corrected.

12.3.2.1.2.1 Toxicity The variable used to represent the toxicity of a species is the LC_{50}. It refers to the concentration in ppm that can kill 50% of a population. Thus, the concentration at a distance is the key variable for assessing the toxic effects of a substance, thus establishing a correlation between emission flow, distance, and concentration. The degree of toxicity is divided into six categories, dangerous toxic, <0.1 mg/$kg_{bodyweight}$, seriously toxic, 0.1–50 mg/$kg_{bodyweight}$, highly toxic, 50–500 mg/$kg_{bodyweight}$, moderately toxic, 0.5–5 g/$kg_{bodyweight}$, slightly toxic, 5–15 g/$kg_{bodyweight}$, extremely low toxicity > 12 g/$kg_{bodyweight}$. The magnitude of the effect on a person can be reduced and controlled by *hygiene* methods; however, toxicity cannot be controlled.

12.3.2.1.2.2 Flash fire The accident of a flash fire has been described from the radiation dose point of view in the previous section. However, it is also controlled by the concentration, in particular the LFL. If the concentration (Pasquill–Gifford model) is greater than or equal to the LFL, the flash fire thermal radiation generates one death. Thus the dispersion model computes the concentration with the distance while the radiation is computed in the previous section.

To compute the LFL, we take a look at the stoichiometry of the combustion of a particular species, Eq. (12.45):

$$C_mH_xO_y + zO_2 \rightarrow mCO_2 + \frac{x}{2}H_2O \tag{12.45}$$

Thus, Eq. (12.46)

$$z = m + \frac{x}{4} - \frac{y}{2} \tag{12.46}$$

We define the fuel mole ratio, C_{st}, as follows, Eq. (12.47):

$$C_{st} = \frac{\text{moles fuel}}{\text{moles fuel} + \text{moles air}} \times 100 = \frac{100}{1 + \left(\frac{1}{0.21}\right)\left(\frac{\text{moles } O_2}{\text{moles fuel}}\right)} = \frac{100}{1 + \left(\frac{z}{0.21}\right)} \tag{12.47}$$

The low and upper flammability indexes are computed as given by Eqs. (12.48) and (12.49):

$$LFL = 0.55C_{st} = \frac{0.55(100)}{4.76m + 1.19x - 2.38y + 1} \tag{12.48}$$

$$UFL = 3.50C_{st} = \frac{3.50(100)}{4.76m + 1.19x - 2.38y + 1} \tag{12.49}$$

The effect of the temperature on the limits can be estimated as given by Eqs. (12.50) and (12.51):

$$LFL_T = LFL_{25} - \frac{0.75}{\Delta H_{comb}}(T - 25) \tag{12.50}$$

$$UFL_T = UFL_{25} + \frac{0.75}{\Delta H_{comb}}(T - 25) \tag{12.51}$$

The pressure has little effect on the LFL except at very LPs, but for the upper flammability limit (UFL) it can be computed using Eq. (12.52):

$$UFL_P = UFL_{atm} + 20.6(\log P(MPa) + 1) \tag{12.52}$$

To compute the LFL and UFL for vapor mixtures Eq. (12.53) can be used:

$$L(U)FL_{mix} = \frac{1}{\sum_i \frac{y_i}{L(U)FL_i}} \tag{12.53}$$

Below the limiting oxygen concentration (LOC), the reaction cannot generate enough energy to heat the entire mixture of gases (including the inert gases) to the extent required for the self-propagation of the flame, Eq. (12.54):

$$LOC = zLFL \tag{12.54}$$

See Fig. 12.9 for the graphic representation of the flamability limits.

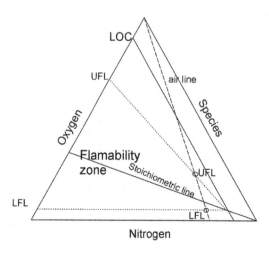

FIGURE 12.9   Flammability limits.

See Example 12.6.

EXAMPLE 12.6 A puff of toluene from our geothermal plant of 1 kg, considered to be instantaneous at ground level, is released in a region with a wind velocity of 4 m/s with clear sky in a rural area in direction toward the closest city located at 2 km. Determine if the population will be affected below the ERPG1, 50 ppm.

Solution
At the center of the cloud, the concentration is maximum, and Eq. (12.43) becomes Eq. (12E6.1)

$$C(x,0,0,t) = \frac{fc}{\sqrt{2}\pi^{3/2}\sigma_x\sigma_y\sigma_z} \tag{12E6.1}$$

With the location and weather conditions, Eqs. (12E6.2) and (12E6.3):

$$\sigma_x = \sigma_y = 0.14x^{0.92} \tag{12E6.2}$$

$$\sigma_z = 0.53x^{0.73} \tag{12E6.3}$$

For the ERPG1, 50 ppm correspond to $188 \text{ mg/m}^3 = 188 \cdot 10^{-6} \text{ kg/m}^3$. Thus at 2 km from the source of the puff, the concentration is $4 \cdot 10^{-8} \text{ kg/m}^3$ so the population is safe.

12.3.2.1.3 Overpressure (explosive species)

12.3.2.1.3.1 Vapor cloud explosions The most dangerous and destructive explosions in the chemical industry are the vapor cloud explosions (VCEs). They result from the ignition of a flammable mixture of vapor, gas, aerosol, or mist, in which the flame speed accelerates to sufficiently high velocities to produce significant overpressure. These explosions occur in a sequence of steps:

1. Sudden release of a large quantity of flammable vapor, typically as a result of the rupture of a vessel containing a superheated and pressurized liquid.
2. Dispersion of the vapor throughout the plant site while mixing with air.
3. Ignition of the resulting vapor cloud.

If the substance released is flammable and explosive, this accident can occur in an instantaneous and/or a continuous release. The unconfined vapor cloud explosion (UVCE), is an explosion that is given by a fast release of a large liquid/gas amount flammable, which forms a cloud shape and disperses in the air. For a UVCE, the variable that characterizes its effect is the overpressure profile for a cloud mass. To estimate the overpressure, the scaled distance, z_e, is computed from the distance, L, as given by Eq. (12.55):

$$z_e = \frac{L}{m_{TNT}^{1/3}} \tag{12.55}$$

where the equivalent mass of TNT is computed from the energy produced by the combustion of the species compared to that of TNT (4760 kJ/kg), Eq. (12.56):

$$m_{TNT} = \frac{\eta \cdot m \cdot \Delta H_{comb}}{E_{TNT}} \quad (12.56)$$

where η is the empirical explosion efficiency. Thus the overpressure can be computed using Eq. (12.57) (Assael and Kakosimos, 2010):

$$p(kPa) = \frac{80,800\left[1 + \left(\frac{z_e}{4.5}\right)^2\right]}{\sqrt{1 + \left(\frac{z_e}{0.048}\right)^2}\sqrt{1 + \left(\frac{z_e}{0.32}\right)^2}\sqrt{1 + \left(\frac{z_e}{1.35}\right)^2}} \quad (12.57)$$

See Example 12.7.

EXAMPLE 12.7 Compute the overpressure generated by the UVCE of 2500 kg of toluene stored in a facility at 50, 100, and 250 m from the source.

Solution

The toluene has a combustion enthalpy of 3910.9 kJ/mol and a molecular weight of 93 kg/kmol. Using Eqs. (12.52)–(12.54) the results are shown in Table 12E7.1.

TABLE 12E7.1 Overpressure as a result of UVCE of toluene.

L	50	100	2000
MTNT (kg)	18,978	18,978	18,978
Ze (m)	1.87	3.75	9.37
P (kPa)	238	50.5	10.75

UVCE, Unconfined vapor cloud explosion.

12.3.2.2 Probability of damage

The probability of damage as a result of the exposure to radiation, overpressure, and toxicity is determined based on single exposure dose–response. The levels of response to toxicants with numbers or percentage affected at each dose-level responses of a large number of people follows a normal (Gaussian) distribution as in Eq. (12.58):

$$f(x) = \frac{1}{\sigma\sqrt{2\pi}} e^{-\frac{1}{2}\left(\frac{x-\mu}{\sigma}\right)^2} \quad (12.58)$$

where x is the response of the individual to the dose, $f(x)$ is the fraction of individuals with a certain response level, μ is the mean of the response, and σ is the standard deviation of the response. Using the distribution, the number of individuals affected, $N_{affected}$, of the total population N is given by Eq. (12.59):

$$N_{affected} = f(x)N \quad (12.59)$$

Representing the response versus the log(dose) for each cause yields a similar sigmoid curve. The idea is to develop a convenient equation to predict the consequence severity of a causative variable, for example, concentration and time, pressure, impulse, radiation intensity, and time.

To predict the percentage affected by a cause, for example, chemical dose–response curve is transformed into a linear equation as follows, Eq. (12.60). Using the normal distribution function, $f(x)$, to represent the dose–response data, let $u = (x-\mu)/\sigma$:

$$f(x) = \frac{1}{\sigma\sqrt{2\pi}}e^{-\frac{1}{2}\left(\frac{x-\mu}{\sigma}\right)^2} = \frac{1}{\sigma\sqrt{2\pi}}e^{-\frac{1}{2}(u)^2} \tag{12.60}$$

Defining Y as the probit variable to estimate probability or percentage of individuals affected, Eq. (12.60) becomes Eq. (12.61):

$$\text{prob} = \frac{1}{\sqrt{2\pi}}\int_{-\infty}^{Y-5}e^{-\frac{1}{2}(u)^2}du \tag{12.61}$$

On a linear probit scale, the sigmoidal dose–response curve is converted into a straight line. The probit variable Y ranges from 2 to 8 resulting in the range from 0.2 to 99.7 probability with $Y = 5$ as the mean, probability 50%. Thus the equation to represent the dose–response data for all agents is given by Eq. (12.62):

$$Y = k1 + k2\ \ln V \tag{12.62}$$

where $k1$ and $k2$ are probit parameters, and V is the *causative variable. For radiation*, we considered the Probit function referred to death by third-degree burns, which applies to radiation by jet fire and pool fire. With regard to *pressure*, it is associated with deaths from pulmonary hemorrhage due to exposure to shockwave. The effects of *toxicity* were measured through the LC_{50} (Crowl and Louvar, 2011). Table 12.4 shows the probit variables.

The Probit values are converted to probability of damage, which allows calculating the probability value "prob" of any physical variable, from the Probit value of this, at a distance d (Martínez et al., 2017), Eq. (12.63):

$$\text{prob}\left(1 + e^{\left(\frac{5.004-(Y)}{0.612}\right)}\right) = 1.005 \tag{12.63}$$

TABLE 12.4 Probit parameters for different accidents.

Concentration	Variable	Probit variables	
		$k1$	$k2$
Toluene	$C\ (\text{ppm})^{2.5}t\ (\text{min})$	-6.79	0.41
Radiation	$\left(\frac{t_{eff}(s)\left(I\left(\frac{W}{m^2}\right)\right)^{4/3}}{10^4}\right)$	-14.9	2.56
Overpressure	$P\ (\text{Pa})$	-77.1	6.91

In the case of toxicity, a linear relationship can be established based on the value of its LC_{50}, thus for a given value of concentration is possible to know directly the value of the probability of affectation, Eq. (12.64):

$$\frac{prob \cdot LC_{50}}{C} = 0.5 \tag{12.64}$$

See Example 12.8.

EXAMPLE 12.8 Compute the probability of damage for the accidents of Examples 12.5 (BLEVE) and 12.7 (UVCE) at a distance of 50 m. Assume that the exposure time is that of the duration of the fire if lower than 10 seconds.

Solution

	Variable	Probit variables			
		$k1$	$k2$	Y	Prob
Radiation	$\left(\frac{t(s)\left(I\left(\frac{W}{m^2}\right)\right)^{4/3}}{10^4}\right)$	− 14.9	2.56		
E12.5 (BLEVE)	2565			5.2	0.058
Overpressure	P (Pa)	− 77.1	6.91		
E12.7 (UVCE)	238,697			8.47	0.99

12.3.3 Frequency

The frequency of occurrence of an event is tabulated based on the analysis of previous accidents. It is given in events per year. Table 12.5 shows a few representative cases (Haag and Ale, 2005).

TABLE 12.5 Frequency of occurrence of specific accidents.

Event	f (events per year)
Pressure vessel failure	10^{-5} to 10^{-7}
Piping failure	10^{-5} to 10^{-6}
Piping leakage	10^{-3} to 10^{-4}
Atmospheric tank failure	10^{-3} to 10^{-5}
Cooling water failure	1 to 10^{-2}

For each event, a tree event is developed evaluating the different incidents and accidents that it may cause. The frequency of a consequence of a specific scenario endpoint is computed using Eq. (12.65):

$$f = f_i \cdot \prod_{j=1}^{i} \text{PFD} \tag{12.65}$$

where PFD is the probability of the failure to produce a particular accident.

12.3.4 Risk estimation

The total risk is the summation of the individual risks of the different elements and incidents for continuous and instantaneous releases. Each incident may evolve into a set of accidents, which vary according to the chemical nature of each working fluid. The individual risk is obtained as the product of the probability of affectation (prob), for an accident from a distance and the frequency (f) of occurrence of the accident, Eq. (12.66).

$$\text{Risk} = \text{Frequency} \cdot \text{Probability of damage} \tag{12.66}$$

Problems

P12.1 The flow of 100 kg/s of brine from a geothermal reservoir exits saturated. Compute pressure at the well. The characteristics of the well can be seen as follows:
$r = 0.15$ m; $\propto\, = 10^{-6}$ m/s^2; $c_p = 4500$ J/kg/K; $K = 2.5$ W/m °C; $T_{Well} = 175$°C; $a = 4.39 \cdot 10^{-2}$; $y = 2000$ m; brine viscosity of $1.6 \cdot 10^{-4}$ Pa $\cdot$ s and density of 950 kg/m^3, $r_{yac} = 1000$ m and $L_{Yac} = 150$ m, considering a friction factor of 0.012, and a permeability of $9.861 \cdot 10^{-14}$m^2, so that monophasic flow is reaching the end of the well.

P12.2 For the reservoirs in the following map, assuming a flow of 100 kg/s, compute the temperature of the brine. Assume $r = 0.15$ m; $\propto\, = 10^{-6}$ m/s^2; $c_p = 4500$ J/kg/K; $K = 2.5$ W/m °C and; $a = 4.39 \cdot 10^{-2}$

P12.3 Evaluate the effect of the expansion pressure on the power production in a single-flash Rankine cycle that uses a flow of 100 kg/s of brine at 172°C and 133 bar. Assume turbine exhaust pressures of 0.08 bar (Fig. 12P3.1).

P12.4 For the reservoirs of Aragon, Salamanca, Sevilla, and Canarias, compute the power that can be produced using a single-stage binary cycle that uses toluene as organic fluid. The cold extreme of the cycle is 25°C so that water can be used to refrigerate the heat exchanger and assume pressurization up to 0.5 bar to feed the isentropic efficiency of the turbine is 0.85. Assume a HRAT of 5°C between the hot brine and the organic fluid.

P12.5 Evaluate the performance of a flash binary cycle shown in Fig. 12P5.1 that uses 50 kg/s of brine at 147°C to evaporate a flow of 125 kg/s of toluene at 5 bar and 130°C. Assume an exhaust pressure for the turbine of 0.037 bar and an isentropic yield of 0.85 at the turbine.

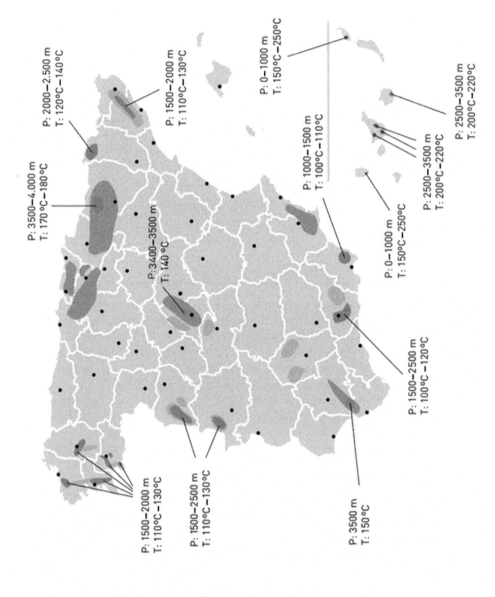

FIGURE 12P3.1 Location of geothermal sources in Spain (Geoplat.org with permission).

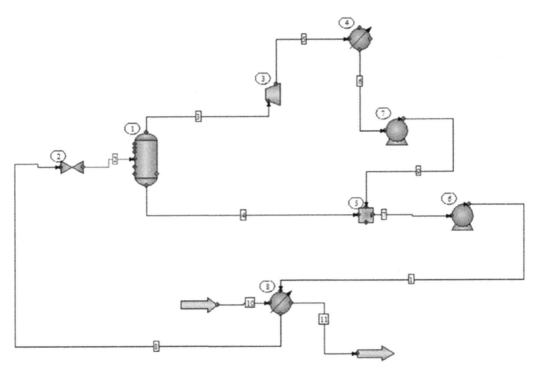

FIGURE 12P5.1   Flowsheet of the flash/binary cycle.

P12.6 Evaluate the performance of a binary cycle that uses cyclohexane as organic fluid. The cold extreme of the cycle is 25°C so that water can be used to refrigerate the heat exchanger and assume pressurization up to 0.5 bar to feed the turbine. The brine is at 177°C, similar to previous examples and a flow of 50 kg/s can be available. The isentropic efficiency of the turbine is 0.85. Assume a HRAT of 5°C between the hot brine and the organic fluid.

P12.7 Compute the radiation dose of a Flash generated at a geothermal plant that uses toluene as HTF. Assume that the storage tank of toluene contains 1500 kg. The relative humidity is 68%, $P = 1$ atm, and the ambient temperature 20°C.

P12.8 The new geothermal plant is to be allocated in a shire. The predominant winds are in the x direction with an average velocity of 1.5 m/s. It is a region with an average moderate cloudiness over the year. Assuming by design that a maximum puff of toluene from our geothermal plant of 10 kg can be released instantaneously at ground level. Determine the maximum distance of the facility to avoid health issues to the population for an ERPG1, 50 ppm.

P12.9 Compute the storage among of toluene for a BLEVE or a UVCE show a probability of damage below 0.5 at 50 m.

References

Assael, M.J., Kakosimos, K.E., 2010. Fires, Explosions, and Toxic Gas Dispersions Effects Calculation and Risk Analysis. CRC Press, Boca Raton, FL.

Chen, H., Goswami, D.Y., Stefanakos, E.K., 2010. A review of thermodynamic cycles and working fluids for the conversion of low grade heat. Renew. Sustain. Energy Rev. 14, 2059–3067.

CPS, 1994. Guidelines for Evaluating the Characteristics of Vapor Cloud Explosions, Flash Fires, and BLEVEs. AIChE.

Crowl, D.A., Louvar, J.F., 2011. Chemical Process Safety. Fundamentals with Applications, third ed. Prentice Hall, New York.

DiPippo, R., 2016. Geothermal Power Generation. Developments and Innovation. Woodhead Publishing, Sawaton, 2016. ISBN: 978-00-8100-337-4.

Haag, U.P.A.M., Ale, B.J.M., 2005. Guidelines for Quantitative Risk Assessment, Purple Book, Part One: Establishments, third ed. Ministerie van Verkeer en Waterstaat, Amsterdam.

Kipsang, C., 2015. Cost Model for Geothermal Wells Proceedings World Geothermal Congress 2015, 19–25 April 2015, Melbourne, Australia.

Kletz, T., 1984. Cheaper, Safer Plants, or Wealth and Safety at Work: Notes on Inherently Safer and Simpler Plants. IchemE.

Lee, I., Tester, J.W., You, F., 2019. Systems analysis, design, and optimization of geothermal energy systems for power production and polygeneration: State-of-the-art and future challenges. Renew. Sust. Energ. Revs. 109, 551–557.

Martínez, J., Peña, J., Ponce-Ortega, J.M., Martín, M., 2017. A multi-objective optimization approach for the selection of working fluids of geothermal facilities: economic, environmental and social aspects. J. Environ. Manage. 203, 962–972.

Moon H., Zarrouk, S.J., 2012. Efficiency of geothermal power plants: a worldwide review. In: New Zealand Geothermal Workshop 2012 Proceedings, 19–21 November 2012 Auckland, New Zealand.

Moran, S., 2017. Process Plant Layout. Elsevier, Oxford.

Peña, J., Martínez, J., Martín, M., Ponce-Ortega, J.M., 2018. Optimal production of power from mid-temperature geothermal sources: scale and safety issues. Energy Conv. Manag. 165, 172–182.

Sun, J., Li, W., 2011. Operation optimization of an organic Rankine cycle (ORC) heat recovery plant. Appl. Therm. Eng. 31, 2032–2041.

Wang, J., Yan, Z., Wang, M., Ma, S., Dai, Y., 2013. Thermodynamic analysis and optimization of an organic Rankine Cycle (Orc) using low grade heat source. Energy 49, 356–365.

Yu, H., Feng, X., Wang, Y., Biegler, L.T., Eason, J.A., 2016. Systematic method to customize an efficient organic Rankine cycle (ORC) to recover waste heat in refineries. Appl. Energy. 179, 302–315.

Water as energy resource

13

Water as a resource: renewable energies and technologies for brine revalorization

Borja Hernández and Mariano Martín

Department of Chemical Engineering, University of Salamanca, Salamanca, Spain

13.1 Overview of resources and applications

Most of the research studies regarding the extraction, supply chain, and production of energy and chemicals mainly focus on the use of land resources. Even though human population is based on land locations and most of the materials used every day come from it, it is important not to forget that water represents 71% of the Earth's surface. Within that 71% of water, 96.5% corresponds to oceans and seas and the remaining 3.5% corresponds to the ice and freshwater resources. Due to the scarcity of safe freshwater, only 0.02% of the overall water on Earth, most of the work has focused on the supply of water from the sea. Different technologies have been used such as: reverse osmosis, evaporation, or multi-effect distillation. The use of seawater for the production of freshwater have been widely covered by a lot of books and it is not the aim of this chapter. For further details about freshwater production see Martín (2016), Kucera (2014), Voutchkov (2013), and Gnaneswar (2018).

Alternative to the production of freshwater, the sea can also be used to produce renewable energy, chemicals, or algae (United Nations, 2005; Giwa et al., 2017). A summary of direct and indirect products is presented in Fig. 13.1. However, we should not forget hydropower that nowadays represents a major share in the contribution of renewable resources to power production. The first section of the chapter, Section 13.2, focuses on renewable energy production from sea water and rivers and dams. The second part, Section 13.3, focuses on the use of the brine obtained from desalination plants to produce chemicals in a more sustainable manner.

517

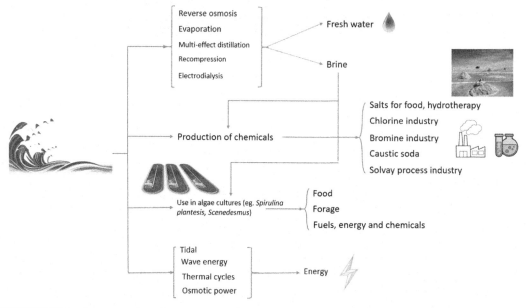

FIGURE 13.1   Summary of products and technologies for seawater revalorization.

13.2 Energy production from water

The reduction of CO_2 emissions pledged by most of the countries in different intergovernmental agreements has continuously promoted the use of renewable energies. Some of the renewable energies such as biomass-based processes, solar, wind, or hydro are currently used at commercial scale. However, the number of commercial facilities implemented of ocean energy is much smaller. Ocean energy can be extracted from different methods as explained in the following subsections for tidal, waves, temperature gradient, or salinity gradient energy sources. In this section we consider them all as well as hydro.

13.2.1 Tidal energy

13.2.1.1 Introduction

Tidal energy is produced by the differences in the elevation of the sea generated by the tides. The tides are generated by the gravitational force of the Moon on the Earth. Since the density of the Earth and the water availability are highly heterogeneous, the distribution is irregular in our planet. Neill et al. (2018) presented the regions with highest potential for the production tidal energy. Hudson Bay of Canada, the Atlantic coast of Western Europe, and the south Atlantic coast of Argentina were determined as the regions with highest potential. This tidal energy is presented in two forms: *Tidal potential energy*, which is generally extracted by constructing a dam or barrage, and *tidal current energy*, which extracts the kinetic energy. An overview of current tidal commercial facilities is provided in Neill et al. (2018).

13.2.1.2 *Physics of tidal power*

1. *Tidal current turbines* extract the kinetic energy from the moving unconstrained tidal streams to generate electricity, see Fig. 13.2. Currents have the same periodicity as vertical oscillations, being thus somehow predictable, although they tend to follow an elliptical path. The maximum kinetic energy to be extracted is calculated as given by Eq. (13.1):

$$E = mv = \rho v^2 \tag{13.1}$$

Where m is the mass of water, ρ is the seawater density (approximately $1022 \, kg/m^3$), and v is the water velocity.

The maximum power for a mass of water passing through the rotor with a cross sectional area, A, can be expressed as follows, Eq. (13.2), see Chapter 11 for the development of this equation since it is also common to wind turbines:

$$P_{Avail} = Cp \frac{1}{2} \rho A v^3 \tag{13.2}$$

where P_{Avail} is the power developed by the rotor (W), A is the area swept out by the turbine rotor (m^2), v is the stream velocity (m/s), and Cp is the power coefficient of the turbine, which is the percentage of power that the turbine can extract from the water flowing through it. According to the studies carried out by Betz, the theoretical maximum amount of power that can be extracted from a fluid flow (water or air) is about 59%, which is referred to Betz limit (Garrett and Cummins, 2004; Hagerman et al., 2006).

The effect of the seabed on the fluid flow velocity can be computed using a power law relationship, Eq. (13.3):

$$v_{Z_2} = v_{Z_1} \left(\frac{Z_2}{Z_1} \right)^{1/10} \tag{13.3}$$

The reference is the seabed where $z = 0$. Thus for a particular depth and velocity, the depth average velocity is computed as given by Eq. (13.4):

$$\overline{v} = \frac{\int_{h_1}^{h_2} v \, dz}{\int_{h_1}^{h_2} dz} = \frac{\int_{h_1}^{h_2} v_o \left(\frac{Z}{Z_o} \right)^{1/10} dz}{\int_{h_1}^{h_2} dz} = \left(\frac{10}{11} \right) \frac{v_o \left(\frac{1}{Z_o} \right)^{1/10}}{h_2 - h_1} \left(h_2^{11/10} - h_1^{11/10} \right) \tag{13.4}$$

FIGURE 13.2 Flow across an area.

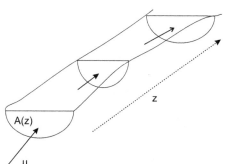

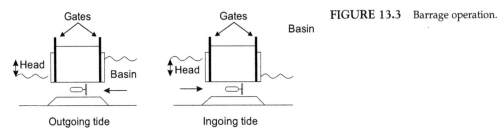

FIGURE 13.3   Barrage operation.

The power obtained from a turbine suffers from losses of efficiency due to the drive-train (0.96), the generator (0.95), and the power conditioning (0.98).

2. *Tidal barrage* extracts the energy from the differences of height between two regions divided by a barrage, see Fig. 13.3. The gross energy potential of tidal basin is computed as given by Eq. (13.5):

$$E = \int \rho g A z\, dz = \frac{1}{2} \rho g A h^2 \tag{13.5}$$

Where E is the energy (J), g gravity (9.8 m/s^2), ρ the seawater density (approximately 1022 kg/m^3), A the sea area (m^2), z the vertical coordinate of the ocean surface (m), and h the tide amplitude (m). The factor of 1/2 arises due to the assumption of a linear reduction in hydraulic head as the basin empties (Roberts et al., 2016).

The power depends on the time of the tide as a function of the number of tides per month (N_{tides}), Eq. (13.6)

$$t_{tide} = \frac{t_{month}}{N_{tides}} \tag{13.6}$$

The gross power potential per tide is computed as Eq. (13.7)

$$P_{tide} = \frac{E}{t_{tide}} \tag{13.7}$$

The total power in a month can be calculated by the number of tides per month and the efficiency of the barrage, $\eta_{narrage}$, typically in the range of 20%–40%, Eq. (13.8)

$$P_{montly} = P_{tide} \cdot N_{tide} \cdot \eta_{narrage} \tag{13.8}$$

13.2.1.3 *Types of turbines*

The design of the turbines depends on the source of energy. We divide this section into the turbine designs used for tidal current and those aimed at barrage operation (Coiro, 2007).

1. *Tidal current turbines.*

Horizontal axis tidal current turbines: The turbine blades rotate about a horizontal axis which is parallel to the direction of the water flow, see Fig. 13.4A. They are arrayed under-water in rows.

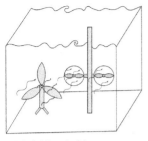

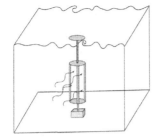

(A) Axial flow turbines

(B) Cross-flow turbines

FIGURE 13.4 Major types of current tidal turbines. (A) Axial flow turbine. (B) Cross flow turbine.

TABLE 13.1 Models for turbine power curves.

Linear models	$P(v) = \frac{v - V_c}{V_r - V_c}$
Cubic model 1	$P(v) = \frac{(v - V_c)^3}{(V_r - V_c)^3}$
Cubic model 2	$P(v) = \frac{(v)^3}{(V_r)^3}$
Quadratic model 1	$P(v) = a_0 + a_1 v + a_2 v^2$
Quadratic model 2	$P(v) = \frac{(v^2 - V_c^2)}{(V_r^2 - V_c^2)}$
Logistic model (Hardisty, 2011)	$P(v) = \frac{P_{nom}}{1 + e^{(-(v-a)m)}}$

Vertical axis tidal current turbines: Vertical axis turbines are cross-flow turbines, with the axis positioned perpendicular to the direction of the water flow, see Fig. 13.4B. Cross flow turbines allow the use of a vertically oriented rotor which can transmit the torque directly to the water surface without the need of complex transmissions systems or an underwater nacelle. The vertical axis design permits the harnessing of tidal flow from any direction, making use of a full tidal ellipse of the flow. However, the implementation of this type of turbines is not widely extended since they experiment a lot of vibrations, as the forces exerted on the bladders are very different.

The design of the turbine determines the power that a turbine can extract. While the theoretical power is a cubic polynomial, the nonidealities curve down the production of power out of the water flow similar to the wind turbines (Hagerman et al., 2006) as presented in Chapter 11 see Table 13.1. Thus the same models as for wind turbines can be used. For further explanation, we refer to Chapter 11. Table 13.2 shows the characteristics of the power curve of some commercial turbines (Roberts et al., 2016). Where V_c is the cut-in and V_r the nominal speeds, v is the water velocity (m/s), and a and m are characteristic adjustable parameters for each turbine.

2. *Tidal barrage turbines*

Bulb turbines: They receive this name due to the shape of the upstream watertight which contains the generator located in the horizontal axis and is mounted inside the water passageway. This structure reduces the size and costs of the installation (Fig. 13.5A).

TABLE 13.2   The characteristics of some commercial turbines (Roberts et al., 2016).

	Atlantis AR1000	Bourne RiverStar	MCT SeaGen S	Voith 1 MW
Power (kW)	1000	50	2000	1000
Rated flow speed (m/s)	2.65	2.05	2.4	2.9
Nạ Rotors	1	1	2	1
Diameter rotor (m)	18	6,09	2	16
Cp	0.41	0.39	0.45	0.4

(A)

(B)

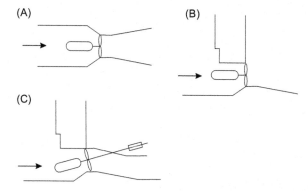

(C)

FIGURE 13.5 Tidal barrage turbine designs. (A) Bulbe turbines. (B) Rim Turbine; (C) Tubular turbine.

Rim turbine: In this design, the generator is separated from the turbine so that only the turbine is in the water while the generator is mounted on the barrage and connected though a shaft that moves with the turbine. In this way the generator is protected from sea water (Fig. 13.5B).

Tubular turbines: In tubular turbines the generator is mounted on the top of the barrage at a 45 degree angle with the turbine, and the blades are connected to a long shaft (Fig. 13.5C).

13.2.1.4 Tidal harmonic analysis

Tidal range and/or currents at a site should be predicted by considering a finite number of harmonic constituents particular to a site, whose phases and periods can be obtained from published tide tables, determined from navigational chart data, or by mathematical analysis of directly measured tidal range and/or current speeds. The basis of harmonic analysis is the assumption that the tidal variations can be represented by a finite number (N) of harmonic terms of the form, Eq. (13.9):

$$v_n \cdot \cos(\sigma_n - g_n) \tag{13.9}$$

where v_n is amplitude, g_n is phase lag on the equilibrium tide at Greenwich, and σ_n is an angular speed.

Depending on the aim of the study, harmonic analysis may be simplified by considering only the principal tidal harmonics or may be made more detailed by incorporating further tidal constituents, which should lead to greater accuracy in predictions. The Kelvin equation when rewritten as the harmonic current equation becomes Eq. (13.10). It may be used to predict the tidal range and/or currents at the site, where A are the amplitudes, τ the periods, and ρ the harmonic constituents (Hardisty, 2011).

$$v(t) = v_o + A_{MU2}\cos\left[\frac{2\pi t}{\tau_{MU2}} + \rho_{MU2}\right] + A_{SU2}\cos\left[\frac{2\pi t}{\tau_{SU2}} + \rho_{SU2}\right] + A_{MU4}\cos\left[\frac{2\pi t}{\tau_{SU4}} + \rho_{SU4}\right] +$$

$$A_{OU1}\cos\left[\frac{2\pi t}{\tau_{OU1}} + \rho_{OU1}\right] + A_{KU1}\cos\left[\frac{2\pi t}{\tau_{KU1}} + \rho_{KU1}\right] + A_{KU2}\cos\left[\frac{2\pi t}{\tau_{KU2}} + \rho_{KU2}\right]$$

$$(13.10)$$

where A_{MU2}, A_{SU2}, A_{MU4}, A_{KU2}, A_{KU1}, and A_{OU1} are the amplitudes of the lunar semidiurnal, solar semidiurnal, lunar quarter diurnal, luni-solar semidiurnal, luni-solar diurnal, and lunar diurnal current velocities. τ_{MU2}, etc. are the corresponding periods and ρ is the phase difference. The values can be seen in Table 13.3. By integrating the power generated by these oscillations, the energy can be computed over time as in Eq. (13.11). Thus the annual energy is to be computed over time as follows:

$$E = \int_{t=0}^{t} P(v) \qquad (13.11)$$

13.2.2 Ocean wave power energy

13.2.2.1 Introduction

Ocean wave power energy is one of the most abundant renewable energy resources around the globe and with the highest power density ($2-3\,kW/m^2$) in comparison to other resources (wind $0.5\,kW/m^2$ or solar $0.15\,kW/m^2$) (López et al., 2013). However, the exploitation cost is too high, ranging between \$800 and \$3300/MW (Astariz and Iglesias, 2015).

TABLE 13.3 Tide generating constituents (Hagerman et al., 2006).

Symbol	Definition	Period (h)	Amplitude
M_{U2}	Principal lunar semidiurnal constituent	12.42	100
S_{U2}	Principal solar semidiurnal constituent	12	46.6
M_{U4}	Lunar quarter diurnal	6.21	0.5/1.4*100
O_{U1}	Lunar diurnal constituent	25.82	41.5
K_{U1}	Luni-solar diurnal constituent	23.93	58.4
K_{U2}	Luni-solar semidiurnal constituent	11.97	12.7

The energy contained in the waves is generated by several factors such as the wind, the gravity interaction between the Moon and the Earth, storms, earthquakes, etc. Among them, wind plays the main role, and it is highly representative to estimate the power. Thus, the estimation of wave energy is typically carried out as for wind resources. The wave energy is experimentally evaluated together with its predominant direction. Several examples of analysis of the power in waves at global and local scales are available in Folley and Whittaker (2009), López et al. (2013), Waters et al. (2009), and Rusu and Onea (2016).

13.2.2.2 Power potential

The waves are typically characterized by a harmonic behavior with characteristic periods (T), length (L), height (H), frequency ($f = 1/T$), or propagation velocity (c), see Fig. 13.6. The characterization of these parameters is necessary for the estimation of the power available in the waves. However, it is important to consider that overlapping of the waves typically takes place. To determine the characteristic wave is then necessary to compute the power spectrum of the frequencies at a given location. From the characteristic parameters, the energy potential per horizontal meter of wave can be estimated as given by Eq. (13.12).

$$P = \frac{g^2 \rho H^2 T}{64\pi} \tag{13.12}$$

13.2.2.3 Technologies for power production from ocean wave energy

The low development of the technologies to produce power from the waves and its high cost represent the main barrier for the exploitation of this energy resource. The technologies, most of them under development, have been classified attending to different approaches. One of the first classifications was given by Hagerman (1995), who classified the systems attending to the mode of energy absorption. Other approach can be followed as defined by Cruz (2008), who distinguish the different types of systems according to the development of the technology.

First generation systems are the devices constructed with conventional technologies, typically in onshore locations. The first system registered was based on a buoy connected by a lever to a mechanical system used onshore. The use of solid mechanical systems was

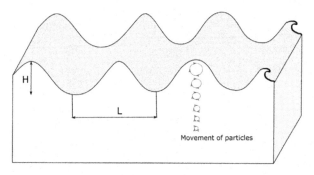

FIGURE 13.6 Characteristic lengths and movement of particles in the waves.

then replaced by oscillating water columns systems to transfer the energy. These systems have covered a wide range of the technologies developed in this area and they are the most explored systems for onshore and nearshore locations (Cruz, 2008; Falcao, 2010). The reader can find a more detailed analysis with models for estimating the power in Cruz (2008). Most of these systems follows the structure presented in Fig. 13.7:

- A compression chamber composed by a partially submerged structure. Inside this chamber the water level is oscillating periodically due to the movement of the waves.
- A valve system that controls the air contained in the compression chamber.
- A turbine dedicated to the production of energy.

Apart from the use of water oscillating columns, other alternatives are the pendulum-based systems and the Tapchan systems. The pendulum systems are based on an oscillating panel that is moved by the waves. The panel is then connected to a hydraulic engine that produces the energy. One example can be found in the "Pendulor" device (Gunawardane et al., 2016). The Tapchan system is based on constructing a channel with certain height. The kinetic energy of the waves is transformed first into potential energy when it accesses the structure, which works as a tank. Then, the water flows from the channel back to the ocean passing through a Kaplan turbine that generates the energy (Mehlum, 1986).

Second generation systems involve the systems based on float pumps that can be found in offshore and nearshore sites with high levels of energy. The working method for extracting the energy in these systems is diverse. Some of the units used are also based on oscillating water columns such as "Masuda" buoy or the "Mighty Whale." Other systems are similar to the Tapchan system where the devices have a concave shape to recover water from the impacting waves, which moves then a small turbine, see "Wave Dragon" system for further information. Another type of floating systems is the one based on oscillating solid bodies with internal hydraulic motors and generators, see "Pelamis" device. An alternative to floating systems also included in this group, is the use of submerged devices. These submerged devices follow similar principles for the extraction of energy: On the one hand, it is possible to find systems based on oscillating water columns such as the "Archimedes Wave Swing." On the other hand, there are also mechanical systems anchored at the bottom of the sea that move by the oscillation of the waves and generate energy, see "WaveRoller" or "Oyster" systems (Falcao, 2010).

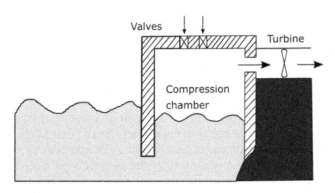

FIGURE 13.7 Schematic diagram of an oscillating water column system for power production.

Third generation systems are composed of the large-scale offshore devices. These systems could suppose the final stage of the development for power extraction before a successful commercialization (Cruz, 2008) but further development of these technologies is still required to make the systems profitable.

13.2.3 Gradient-based methods

The last type of the power production methods is the one based on gradients, either temperature or salt concentrations. They have been by far less exploited than the previous ones and they are only available at experimental scale. We divide the presentation into those based on thermal gradients and the ones based on concentration gradients.

13.2.3.1 Ocean thermal energy conversion

This technology is based on the temperature gradients that exist between the deep and surface of the sea. It has the highest potential when compared to other ocean energy technologies. However, they are limited by the high capital costs, starting to be competitive with prices of $0.30/kWh (Rajagopalan and Nihous, 2013; Irena, 2014). The technology has been determined to be more suitable in remote islands of the tropical seas where the generation can be combined with other functions (e.g., air-conditioning and fresh water). Apart from the economic barriers, there are also environmental problems associated to them like the impact on the marine life generated by the pipes. The technologies for extracting the energy are based on different thermodynamic cycles (Irena, 2014). A summary of them is only provided since they are deeply described in Chapters 9 and 12:

- **Open cycle.** The water of the surface of the sea is evaporated in a vacuum chamber and then sent to a turbine where it is condensed producing power. The condensed water is finally cooled down with the water from the depth of the sea before it is sent back to the ocean.
- **Closed cycle** in which the working fluid follows a Rankine cycle. The water in the surface is used in the evaporator and the water from the depth is used in the cooler.
- **Kalina cycle.** It is an alternative to the closed cycle in which water and ammonia are used as working fluids.
- **Hybrid system.** It combines the open and closed cycles where the steam generated is first used to heat-up the working fluid of the closed cycle and then sent to a vacuum chamber where it is evaporated to produce energy as in an open cycle. An example of this type of systems is provided in Fig. 13.8.

13.2.3.2 Osmotic power

These systems are the least explored ones to produce energy. They are based on the production of energy by the salinity gradients between freshwater and seawater. As a result from the salinity gradient between the freshwater and seawater in an osmotic membrane, part of the freshwater is transferred to the stream of the seawater. At the exit of the membrane, the sea water then has higher flowrate of water that can be split and sent to a turbine for producing power (Skilhagen et al., 2008). A summary of the system is provided in Fig. 13.9.

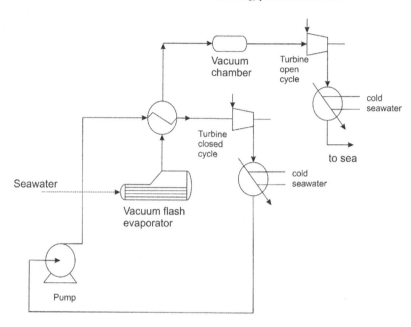

FIGURE 13.8 Hybrid cycle for the production of power with ocean thermal energy conversion.

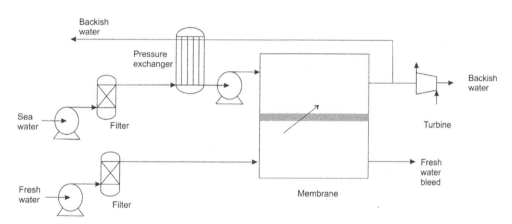

FIGURE 13.9 Schematic osmotic power production system.

13.2.4 The use of hydro as power reservoirs

13.2.4.1 Introduction

The system consists of an upper reservoir and a lower reservoir so that water is allowed to flow downhill to produce energy and when energy is available, the water flow is reversed, see Fig. 13.10. The system depends on the geography of the terrain as well as the need to build dams.

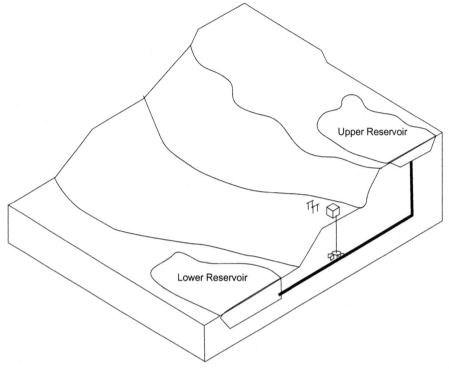

FIGURE 13.10   Scheme of the operation of hydropower.

13.2.4.2 *Energy from the flow*

A common form of Bernoulli's equation, valid at any arbitrary point along a streamline is given by Eq. (13.13):

$$p_1 + \frac{\rho v_1^2}{2} + z_1 \rho h = \text{constant} \qquad (13.13)$$

Hydropower production and the storage of energy are based on two steps. The energy producing stage can be computed, based on Bernouilli's equation as follows, Eq. (13.14):

$$W = Q \cdot \rho \cdot g \cdot H \cdot \eta \qquad (13.14)$$

where W is the mechanical power (W), Q is the flowrate (m^3/s), ρ is the density (kg/m^3), g is the acceleration of gravity (m/s^2), H is the water fall height (m), and the global efficiency ratio, η, is between 0.7 and 0.9.

A more detailed analysis can be performed by applying the energy balance to both points, the reservoir and downstream the turbine, Eq. (13.15):

$$\frac{P_1}{\gamma} + \frac{v_1^2}{2g} + z_1 = \frac{P_2}{\gamma} + \frac{v_2^2}{2g} + hf_2 + z_2 + W$$

$$hf_i = f_i \frac{L_i}{d_i} \frac{v_i^2}{2g} \qquad (13.15)$$

where the friction factor, f_i, can be determined from the Colebrook equation as, Eq. (13.16):

where:

$$\frac{1}{\sqrt{f_i}} = -4.0 \log_{10} \left(\frac{\varepsilon/d_i}{3.7} + \frac{1.256}{Re\sqrt{f_i}} \right)$$

$$Re = \frac{\rho v_i d_i}{\mu}$$

(13.16)

ε is the roughness, d_i the pipe diameter (m), μ the liquid viscosity, v the liquid velocity (m/s), and ρ is the density (kg/m³).

When an excess of energy is available, water can be pumped up in a second stage to use the upper reservoir as storage energy system. Thus by formulating an energy balances between two points, the energy balance becomes Eq. (13.17). See Example 13.1 for the application of these equations:

$$\frac{P_1}{\gamma} + \frac{v_1^2}{2g} + W + z_1 = \frac{P_2}{\gamma} + \frac{v_2^2}{2g} + hf_2 + z_2.$$

(13.17)

EXAMPLE 13.1 A hydropower facility produces 250 MW from a reservoir at 500 m from the turbine. The pipe for water is 6 m of a material with a roughness of 0.00005. The equivalent length of the pipe is 1.5 times the head. The yield of the turbine is 80%. **(A)** Determine the flow of water required to achieve this production capacity. **(B)** Determine the power consumption to pump up 70 m³/s assuming that the pump has an efficiency of 80%.

Solution

a. For the power production mode, we assume that at both ends the pressure is the same and that the pipe has constant diameter. Thus the energy balance becomes where $z_1 = 500$ m and $z_2 = 0$ m, Eqs. (13E1.1) and (13E1.2).

$$z_1 = hf_2 + z_2 + W$$
$$hf_i = f_i \frac{L_i}{d_i} \frac{v_i^2}{2g}$$

(13E1.1)

where:

$$\frac{1}{\sqrt{f_i}} = -4.0 \log_{10} \left(\frac{\varepsilon/d_i}{3.7} + \frac{1.256}{Re\sqrt{f_i}} \right)$$

$$Re = \frac{\rho v_i d_i}{\mu}$$

(13E1.2)

By solving the balance

$$f = 0.00217$$

$$v = 2.25 \text{ m/s}$$

$$Q = 63.8 \text{ m}^3/\text{s}$$

b. For the storage mode, Eq. (13E1.3)

$$z_1 + W = hf_2 + z_2 \qquad (13E1.3)$$

$$hf_i = f_i \frac{L_i \, v_i^2}{d_i \, 2g}$$

By solving the balance we have

$$v = 2.48 \text{ m/s}$$

$$f = 0.00215$$

$$W = 429 \text{ MW}$$

13.3 Production of chemicals from the sea

13.3.1 Introduction

Apart from the production of freshwater and energy, the seawater is also a source of chemicals, in particular salts. A summary of the dry matter that can be found in different seas is presented in Table 13.4. It shows that the composition varies depending on the sea but in all the cases it is dominated by sodium chloride. This component has been then the most popular resource extracted for years not only in the form of salt but also used as basis in other industries. The analysis of the NaCl industry can be found in Chapter 4 in Martín (2016) for further reference. Furthermore, other products such as magnesium, calcium, potassium, bromine, or rubidium are also interesting to be generated from sea resources, see Fig. 13.11.

Apart from the direct production of these chemicals, there are also residual streams with high content of salts in several industries. One example with high impact on the environment is the generation of freshwater via reverse osmosis. This system is the most widely used in the

TABLE 13.4 Salt composition of several oceans and seas (Martín, 2016).

%Salt/Ocean-sea	Atlantic	Mediterranean	Dead Sea
%Salts	3.63	3.87	22.3
NaCl	77.03	77.07	36.55
KCl	3.89	2.48	4.57
$CaCl_2$	–	–	12.38
$MgCl_2$	7.86	8,76	45.20
$NaBr = MgBr_2$	1.30	0.49	0.85
$CaSO_4$	4.63	2.76	0.45
$MgSO_4$	5.29	8.34	–
$CaCO_3 + MgCO_3$	–	0.10	–

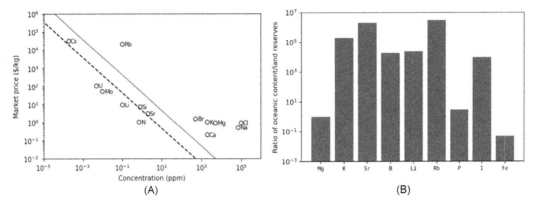

FIGURE 13.11 (A) Interesting products to be obtained from brine (data from Shahmansouri et al., 2015), dashed line is for a flowrate of 50,000 m³/day and continuous line is for 5000 m³/day. (B) Ratio of ocean and land reserves for chemicals summarized from Loganathan et al. (2017). *Source: (A) Adapted from Shahmansouri, A. Min, J. Jin, L. Bellona, C., 2015. Feasibility of extracting valuable minerals from desalination concentrate: a comprehensive literature review. J. Clean. Prod., 100, 4–16; (B) adapted from Loganathan, P. Naidu, G. Vigneswaran, S., 2017. Mining valuable minerals from seawater: a critical review. Environ. Sci. Water Res. Technol. 3, 37.*

production of freshwater from the sea. It requires a high consumption of power and generates brines that damage the environments near the discharge zone (Jones et al., 2019). In order to make the system more sustainable, different renewable energies have been integrated with the reverse osmosis plant (Khan et al., 2018; Mito et al., 2019). However, the management of brine and processes to generate an added value out of this brine have not been widely studied becoming two areas of interest in the upcoming years. In particular, the design of processes, previously aborded for the production of salts, needs to be reformulated and integrated with the brine produced from the reverse osmosis plants. In the following sections, the concentration methods and some of the processes for generating potential chemicals from brine are presented.

13.3.2 Concentration methods and possible chemicals

The first step for generating added value products from the brine is concentration since the brine resultant from reverse osmosis only has concentrations of nearly the double than in the sea. The concentration methods were described in detail in a previous book (Martín, 2016) and are summarized in Table 13.5. As a result from the concentration of brine, several salty products can be obtained. United Nations (2005) provided a description of the production of some of the chemicals presented in Fig. 13.12. Among these chemicals, Martín (2016) mainly focused on $CaCO_3$ and NaCl industries since they are the salts with highest concentration and more widely used. The use of $CaCO_3$ was studied within the Solvay's process for the production of CaO as an intermediate in the production Na_2CO_3 from NaCl. The use of NaCl was also covered to produce Na_2SO_4 and HCl via the Mannheim process. However, more specialised processes were not described. In the following Section 13.3.3 the production of chlorine is presented. Section 13.3.4 focuses on the recovery of bromine, Section 13.3.5 focuses on the magnesia production, Section 13.3.6 provides a brief description for the production of potassium, and Section 13.3.7 is dedicated to other interesting chemicals.

TABLE 13.5 Summary of technologies for the concentration of brine.

Technology	Characteristics
Natural evaporation	– Most used and cheapest method.
	– Used for obtaining dried salts.
Forced evaporation	– It requires a high consumption of energy.
	– Allows the control of the some of the species generated.
	– Evaporator efficiency can be improved by:
	– The use of multieffect evaporators.
	– Counter-current feed of the system.
	– Use of re-compression systems for using the vapor as heating utility.
Freezing	– This method requires less energy consumption than heating.
	– However, it does not ensure a complete separation of salt and water. Salt is trapped within the ice crystals and it also remains in the water.
Reverse osmosis	– This method is based on the diffusion of ions through a membrane.
	– It requires very high pressures and subsequently power to ensure a high removal of salts.
Extraction with chemicals	– This method is based on the extraction of water by combining it with other chemical species (organic or halogenated compounds) that are later separated.
Electro-dialysis	– Similar to reverse osmosis, the method is based on the transport of ions by concentration gradients. However, in this system, the electric potential is used instead of pressure as driving force.
	– It requires high power consumption and does not allow to synthesize salts.
Ion exchange	– In this technology, ionic resins are used to capture the ions with opposite charge.
	– This method does not allow a complete removal of the salts since they need to be recovered later from the resins with other solution.
	– The high quantity of resin required makes this method only reliable for purifying water with low concentration of ions.

13.3.3 Chlorine industry

13.3.3.1 *Production of chlorine*

Chlorine is one of the basic chemicals with highest global production accounting for 65 million of tons per year (Chlorine Institute, 2020). The production of chlorine from brine can be carried out following different alternatives of the Chlor-Alkali process, see Eq. (13.18).

$$2NaCl + 2H_2O \rightarrow Cl_2 + H_2 + 2NaOH \qquad (13.18)$$

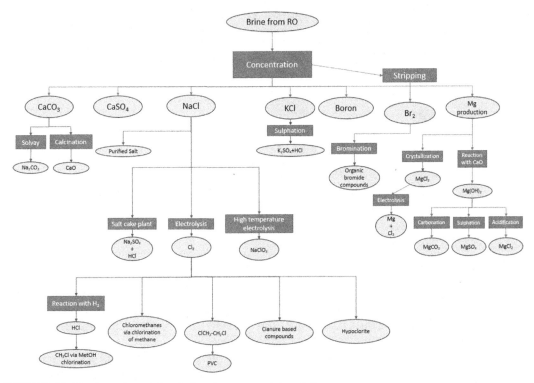

FIGURE 13.12   Summary of chemicals to be generated from brines.

These alternatives depend on the type of electrolytic cell used (Schmittinger et al., 2011):

- **Mercury cell (Castner−Kellner process)**: This type of cell uses a stream of mercury that remains at the bottom (cathode). At the top of the cell, the saturated brine solution floats on the mercury (anode). As a result from the electrolysis in the cell, sodium amalgam is produced in the cathode. The amalgam is decomposed in a vessel where it reacts with water, obtaining mercury (which is recycled to the cell), a caustic solution (which is purified to obtain NaOH), and hydrogen (which is cooled down to remove the traces of Hg). Chlorine is produced in the anode with traces of oxygen. Apart from the amalgam and chlorine, a third stream is also generated in the process containing the unreacted brine. This brine solution recycled has a low concentration and it needs additional salt or a more concentrated solution to maintain the concentration in the cell. These Castner−Kellner process involves the management of mercury requiring measures to prevent environment and healthy issues and it is suggested to be avoided in some of the countries (European Comission, 2017).
- A second process is based on the use of a **diaphragm cell, Grieshem cell process**. The cells used in this process have a permeable diaphragm between the anode and the cathode. The brine is introduced into the anode and flows through the diaphragm to

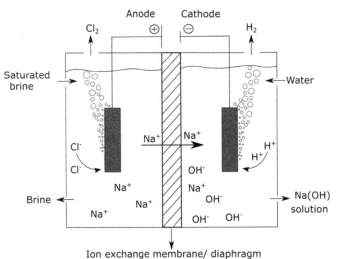

FIGURE 13.13 Example of either diaphragm or membrane cell.

the cathode concentrating there the sodium ions. The Cl$^-$ ions remain in the anode generating the chlorine. On the other side, hydrogen is generated in the cathode from the H$^+$ ions, meanwhile the OH$^-$ ions of the cathode form NaOH. The resulting NaOH liquor is removed, meanwhile the NaCl liquor is recycled as brine.

- **Membrane cell process**. This process is similar to the previous one, but the diaphragm is substituted by a polymer-based ion exchange membrane, see Fig. 13.13. The membrane allows the transfer of sodium ions from the anode to the cathode. As a result, H$_2$ is also generated in the cathode and Cl$_2$ is generated in the anode. The resultant liquid streams are the brine, which can be recycled, and the NaOH. A general description of the process for the use of this technology and the previous diaphragm cell is presented in Fig. 13.13 (Example 13.2).

EXAMPLE 13.2 One of the difficulties for the implementation of chlorine production technologies by means of electrolysis of brine is the high consumption of power. This example provides an estimation of the power consumption in an electrolytic cell. The diaphragm cell is assumed to process 0.2 kg/s of brine with a concentration of 10 g/L. The concentration at the outlet is determined to be 5 g/L. Assume perfect mix in the anodic compartment and that the diffusion of sodium ions through the membrane is equimolar with the reaction. Assume the loses in the ion exchange membrane, the structure and the solution to be 0.2 V each.

Solution.

The power consumption in a cell is given as Eq. (13E2.1):

$$P = I \cdot V \tag{13E2.1}$$

The voltage in the cell is composed of several terms:

- The voltage in the anode can be determined from the electrolytic potential of the reaction in the anode:

$$Cl^- - e^- \rightarrow \frac{1}{2}Cl_{2(g)} \quad E_o = 1.36 \text{ V}$$

- The potential in the cathode is:

$$2H_2O + 2e^- \rightarrow H_{2(g)} + 2OH^- \quad E_o = 1.23 \text{ V}$$

- The voltage difference to be covered in the diaphragm, around 0.4 V. In the same way, there is also a potential for an ion exchange membrane around 0.2 V.
- Other losses in the cell:
  - The one given through the structure: 0.2 V.
  - The one in the anolyte solution: 0.2 V.

The voltage in the cell is then the sum of all the factors, being obtained to be: 3.39 V.

To compute the current, it is necessary to determine the mass balance in the anodic compartment of the cell. Assuming that the salt is completely diluted, the amount of chlorine reacting is computed as in Eq. (13E2.2):

$$0.2\frac{kg}{s} \cdot \frac{10 \text{ g}}{L} - 0.2\frac{kg}{s} \cdot \frac{5 \text{ g}}{L} = 1\frac{g}{s} \tag{13E2.2}$$

By computing the molecular weight, it is determined that for 1 mol of NaCl, 61% of mass is chlorine and the remaining 39% is sodium. Taking into account that the sodium chloride has reacted with an equimolar amount, the chlorine generated is Eq. (13E2.3):

$$1\frac{g}{s} \cdot 0.61 \cdot \frac{1 \text{ mol Cl}^-}{35.5g} \cdot \frac{0.5 \text{ mol Cl}_2}{1 \text{ mol Cl}^-} \cdot \frac{71 \text{ g Cl}_2}{1 \text{ mol Cl}_2} = 0.61 \text{ g Cl}_2 \tag{13E2.3}$$

Since for every mole of chlorine generated, two moles of electrons are required, using the Avogadro number it is possible to compute the flux as follows (13.E2.4):

$$0.61 \text{ g Cl}_2 \cdot \frac{1 \text{ mol Cl}_2}{71 \text{ g Cl}_2} \cdot \frac{2 \text{ mol } e^-}{1 \text{ mol Cl}_2} \cdot \frac{6.02 \times 10^{23} \text{ } e^-}{1 \text{ mol } e^-} = 1.03 \times 10^{22} \frac{e^-}{s} \tag{13E2.4}$$

The current and power are then computed as given by Eq. (13E2.5):

$$I = C \cdot \frac{e^-}{s} = 1.6 \times 10^{-19} \left(\frac{C}{e^-}\right) \cdot 1.03 \times 10^{22}\frac{e^-}{s} = 1.65 \times 10^3 \frac{C}{s} = 1650 \text{ A} \tag{13E2.5}$$

$$P = 1650 \text{ A} \times 3.39 \text{ V} = 5.6 \text{ kW}$$

13.3.3.2 Chlorine derived products

The chlorine generated by these processes can be applied in the production of several derived chemicals as shown in Fig. 13.12. The first, and one of the most widely used chemicals is hydrogen chloride. This product is generated as byproduct in chlorination processes (e.g., production of chloromethanes or vinyl chloride) and in the production of sodium sulfate from NaCl and sulfuric acid by Mannheim process (United Nations, 2005; Von Plessen, 2012). The process takes place in a Mannheim furnace where the endothermic Eqs. (13.19) and (13.20) take place at temperatures between 600°C and 700°C.

$$NaCl + H_2SO_4 \rightarrow HCl + NaHSO_4 \tag{13.19}$$

$$NaCl + NaHSO_4 \rightarrow HCl + Na_2SO_4 \tag{13.20}$$

Furthermore, HCl can be also directly synthesized from chlorine, see Eq. (13.21). The direct synthesis takes place by reacting the chlorine with H_2 at temperatures above 250°C. The process is exothermic and requires an excess of 10% in volume of hydrogen for reduction of the free chlorine. The resultant HCl gas is absorbed later in purified water. This absorption is exothermic and requires intermediate refrigeration to ensure that HCl remains in the liquid phase. Furthermore, the absorption system must be built with noncorrosive materials such as glass. Several technologies can be used such as the use of batch and continuous scrubbers, adiabatic absorbers, and isothermal absorbers (De Dietrich, 2020; Austin and Glowacki, 2000). In this case, the high concentration of HCl in the gas requires the use of refrigerated absorbers instead of a scrubber. In this sense, the use of isothermal absorption is preferred where two technologies are mainly used:

- Falling-film absorbers where the HCl is introduced in cocurrent with weak acid in tubes that are cooled with water (Fig. 13.14A). The mixture introduced in the top of the absorber falls along the tubes and it is recovered at the bottom where a fraction is recycled and solved with more water to capture the HCl gas.
- A packed column with cooling loops. The packed column is divided in a set of sections where part of the resultant product is recycled, cooled down, and used as absorbent again, see Fig. 13.14B.

$$H_2 + Cl_2 \rightarrow 2HCl \tag{13.21}$$

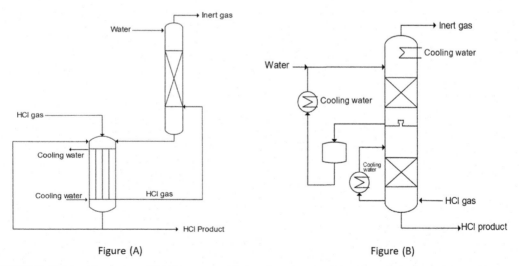

Figure (A)   Figure (B)

FIGURE 13.14   Absorption of HCl by means of falling-film absorber (A) and packed column (B).

An alternative in the case of producing solutions with low HCl concentrations is the use of adiabatic absorption columns with no internal removal of heat. For an adequate operation of this column, a rigorous control of the temperature in the intermediate parts of the column must be carried out (Examples 13.3 and 13.4).

EXAMPLE 13.3 *Modeling of a falling film absorber.* **One of the alternatives to produce HCl is by means of a falling film reactor. This production method allows ensuring a high absorption rate since the heat generated by the dissolution can be easily dissipated. Considering that the concentration of the HCl in the interphase is the maximum allowed by the azeotrope, determine the conversion achieved in a reactor with a film of water (initial mass fraction of HCl of zero) with a thickness of 0.1 mm and in a pipe with a length of 5 m. Note: Assume that the film does not suffer significant changes of thickness in the pipe. The diffusion coefficient of HCl in water is $3.7 \cdot 10^{-9} \ m^2/s$ and the density of water is 1000 kg/m^3.**

Solution:

In this type of absorber, the raw materials can be introduced in cocurrent or counter-current, and they are separated in several internal pipes where absorption takes place. In each of the pipes, it can be assumed that the liquid flows on the wall as shown in Fig. 13E3.1. The water contacts with the gas that is in the core of the pipe. Since the HCl-water system shows an azeotrope, the concentration in the interphase is never higher than 38% of HCl. Thus the gas phase, in either cocurrent or counter-current with the liquid, always ensures to maintain these saturated conditions on the interphase. Apart from considering the saturated conditions in the interphase for the entire pipe, the following assumptions are also taken into account in the following stages of the problem:

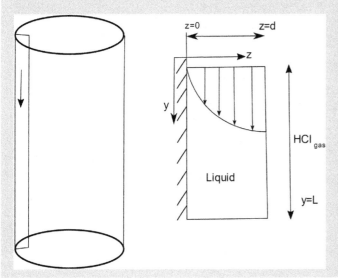

FIGURE 13E3.1 Geometries in a falling film reactor.

- Steady-state operation is considered.
- No counter-diffusion of water in the vapor is assumed.
- The thickness of the film is assumed to remain constant.

Prior to computing the mass transfer, the momentum is computed for the liquid system. The distribution of velocity profiles is determined as follows, Eq. (13E3.1):

$$\mu \frac{d^2 v_y}{dz^2} + \rho g = 0 \tag{13E3.1}$$

This Eq. (13E3.1) is solved with the following boundary conditions of Eqs. (13E3.2) and (13E3.3):

$$\text{At:} z = 0 \rightarrow v_y = 0 \tag{13E3.2}$$

$$\text{At: } z = d \rightarrow v_y = v_{\max} = \frac{\rho g d^2}{2\mu} \tag{13E3.3}$$

Together with the momentum equation, a mass balance can be also presented for the HCl as in Eq. (13E3.4) and simplified to Eq. (13E3.5):

$$D_{HCl-w}\Delta x \Delta y \left(\frac{\partial c_{HCl}}{\partial z} \right)_{z+\Delta z} + u_y c_{HCl}\big|_{y+\Delta y} \Delta x \Delta z = D_{HCl-w}\Delta x \Delta y \left(\frac{\partial c_{HCl}}{\partial z} \right)_{z} + u_y c_{HCl-w}\big|_y \Delta x \Delta z \tag{13E3.4}$$

$$v_y(z)\frac{\partial c_{HCl}}{\partial z} = D_{AB}\frac{\partial c_{Hcl}^2}{\partial z^2} \tag{13E3.5}$$

The mass balance can be solved with the following boundary conditions, Eqs. (13E3.6)–(13E3.8):

$$\text{At: } 0 < z < d; y = 0 \rightarrow c_{HCl} = 0 \tag{13E3.6}$$

$$\text{At: } z = 0; \ 0 < y < L \rightarrow \frac{\partial c_{HCl}}{\partial z} = 0 \tag{13E3.7}$$

$$\text{At: } z = d; 0 < y < L \rightarrow C_{HCl} = C_{interf} \tag{13E3.8}$$

Finally, the average concentration based on the mass, proportional to the velocity, distribution, can be computed as in Eq. (13E3.9).

$$\overline{c_{HCl}} = \frac{\sum\limits_{z=0}^{z=d} c(z)v(z)}{\sum\limits_{z=0}^{z=d} v(z)} \tag{13E3.9}$$

Solving these equations with gProms® (see the code attached), we obtain an average moisture content of 33.7 wt.%.

EXAMPLE 13.4 Model of a scrubber. Determine the percentage of HCl removed by a scrubber. The scrubber is composed of five trays where the contact between the gas phase and the water takes place. The gas is composed of the resultant gases of chloromethane production via methane chlorination where it is assumed that only HCl is absorbed in water. The initial molar fraction of HCl in the gas is 4% and the molar flowrates are 1 kmol/s for the gas and 0.2 kmol/s for the water used for absorption. Assume that the average molecular weight of the gases is 20 kg/kmol and the dissolution enthalpy of the HCl in water is -74.84 kJ/kg. The gas is fed at 320 K and the water is fed at 298 K. The specific heat of water is 4.185 kJ/kgK and the specific heat of the gas is 2 kJ/kgK and the Henry's constant is dependent on the temperature as (Marsh and McElroy, 1985):

$$\log_{10} H = -1.524 + \frac{878.6}{T(K)} \tag{13E4.1}$$

Solution

For determining the percentage of HCl removed, it is necessary to model each trays of the scrubber. In each of them, equilibrium is assumed and the following Eqs. (13E4.2–13E4.5) are solved together Eq. (13E4.1).

$$L_{\text{wat}} \left(\frac{\text{mol}_{\text{wat}}}{s} \right) \cdot x_{\text{in}} \left(\frac{\text{mol}_{\text{HCl}}}{\text{mol}_{\text{wat}}} \right) + G_{\text{gas}} \left(\frac{\text{mol}_{\text{gas}}}{s} \right) \cdot y_{\text{in}} \left(\frac{\text{mol}_{\text{HCl}}}{\text{mol}_{\text{gas}}} \right)$$

$$= L_{\text{wat}} \left(\frac{\text{mol}_{\text{wat}}}{s} \right) \cdot x_{\text{out}} \left(\frac{\text{mol}_{\text{HCl}}}{\text{mol}_{\text{wat}}} \right) + G_{\text{gas}} \left(\frac{\text{mol}_{\text{gas}}}{s} \right) \cdot y_{\text{out}} \left(\frac{\text{mol}_{\text{HCl}}}{\text{mol}_{\text{gas}}} \right) \tag{13E4.2}$$

$$x_{\text{out}} = H y_{\text{out}} \tag{13E4.3}$$

$$L_{\text{wat}} \cdot (1 + x_{\text{in}}) \cdot cp_{\text{water}} \cdot T_{\text{Lin}} + G_{\text{gas}} \cdot (1 + y_{\text{in}}) \cdot cp_{\text{gas}} \cdot T_{\text{Gin}} + \Delta H_{\text{dis}} \cdot L_{\text{wat}} \cdot (x_{\text{in}} - x_{\text{out}})$$

$$= L_{\text{wat}} \cdot (1 + x_{\text{out}}) \cdot cp_{\text{water}} \cdot T_{L\,\text{out}} + G_{\text{gas}} \cdot (1 + y_{\text{out}}) \cdot cp_{\text{gas}} \cdot T_{\text{Gout}} \tag{13E4.4}$$

$$T_{\text{Gout}} = T_{\text{Lout}} \tag{13E4.5}$$

For solving the system, a Python framework is designed, where each of the trays are defined in Pyomo and the previous equations are solved using Ipopt as solver. The computation in each tray is carried out sequentially from the top to the bottom assuming the same initial gas properties and flows in each tray meanwhile the liquid properties are updated. After finishing the computation from the top to the bottom, the gas properties and fluxes are updated in the opposite way (from bottom to top). This two-way coupling scheme can be carried out for an initial large number of iterations defined or until a threshold is achieved in the error between iterations. In the current system, the minimum error that can be obtained from the iteration is only $1 \cdot 10^{-6}$% based on the mass balance of the absorbed component.

The results obtained show an absorption of 0.03997 mol/s, which represents to remove 99.94% of the HCl contained in the gas. The plot for the distribution of the molar fraction in the absorber is obtained as shown in Fig. 13E4.1.

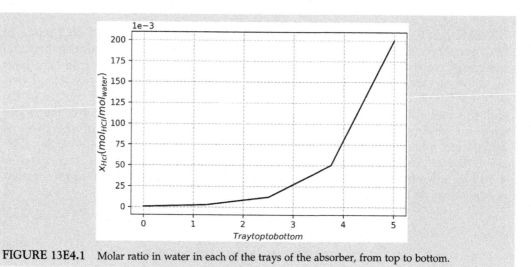

FIGURE 13E4.1   Molar ratio in water in each of the trays of the absorber, from top to bottom.

13.3.3.2.1 Chlorine derived products: chloromethanes

The hydrogen chloride produced by any of the previous methods can be used in the production of mono-chloromethane by a **radical substitution reaction with methanol**, see Eq. (13.22). The reaction occurs between 200°C and 400°C using an alumina and cuprous or zinc chloride catalyst (Schmidt et al., 2013). The reaction shows high yields and the equilibrium to methyl chloride is favored at low temperatures. Values of the equilibrium constants can be found in Thyagarajan et al. (1966) and Schmidt et al. (2013). Together with Eq. (13.22), the following reactions Eqs. (13.23) and (13.24) also take place (Becerra et al., 1992):

$$CH_3OH + HCl \rightarrow CH_3Cl + H_2O \tag{13.22}$$

$$2CH_3OH \leftrightarrow (CH_3)_2O + H_2O \tag{13.23}$$

$$(CH_3)_2O + 2HCl \rightarrow 2CH_3Cl + H_2O \tag{13.24}$$

Alternative to methanol chlorination, the production of mono-chloromethane, can be also carried out by means of the **chlorination of methane**. This process requires temperatures between 300°C and 350°C and it can be carried out following thermal, photochemical, or catalytic methods for the activation (Rossberg et al., 2000). The thermal dissociation leads to a successive substitution of the hydrogen atoms of methane as presented in Eqs. (13.25)–(13.28). The reaction finishes when the system is cooled down and the chlorine ions synthesized to Cl_2.

$$Cl_2 \rightarrow 2Cl^- \text{ Initiation} \tag{13.25}$$

$$CH_4 + Cl^- \rightarrow CH_3^- + HCl \text{ Propagation} \tag{13.26}$$

$$CH_3^- + Cl_2 \rightarrow CH_3Cl + Cl^- \text{ Propagation} \tag{13.27}$$

$$2Cl^- + M \rightarrow Cl_2 + M \text{ Termination} \tag{13.28}$$

13.3.3.2.2 Chlorine derived products: vinyl chloride

Another interesting organic compound is the production of vinyl chloride, which can be then used in the production of one of the most widely use plastics, polyvinylchloride. The product is typically produced by means of the process presented in Fig. 13.15. In the first step ethylene is divided in two fractions (Dreher et al., 2011):

- One fraction is sent to the direct chlorination reactor. Depending on the operating regime, the reaction presented in Eq. (13.29) can take place at low (20°C–70°C) or high (100°C–150°C) temperatures and typically using FeCl$_3$ as catalyst (Icis, 2020; Orejas, 2001).

$$Cl_2 + H_2C = CH_2 \rightarrow ClH_2C - CH_2Cl \tag{13.29}$$

- Another fraction can be sent to the oxychlorination reactor. In this reactor, ethylene reacts with the hydrogen chloride recycled from the pyrolysis at the cracking furnace in the presence of oxygen. The reaction can take place using cuprum as catalyst in a fluidized bed with temperatures between 220°C and 240°C and pressures between 150 and 500 kPa or in fixed bed reactor between 230°C–300°C and 150–1400 kPa (Magistro and Cowfer, 1986), Eq. (13.30).

$$2HCl + H_2C = CH_2 + \frac{1}{2}O_2 \rightarrow ClH_2C - CH_2Cl + H_2O \tag{13.30}$$

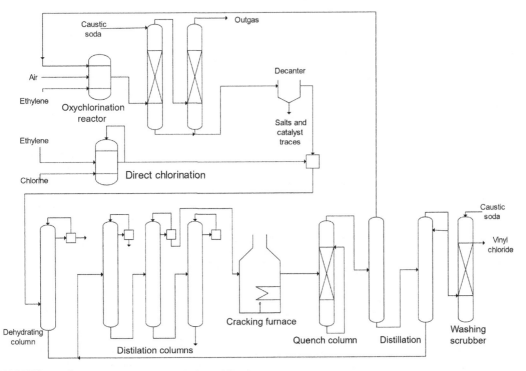

FIGURE 13.15 Process to produce ethylene chloride.

Once ethylene dichloride is produced, the stream is purified to concentrations of ethylene dichloride above 99 wt.%. Initially, the stream is contacted with water removing the water from the reaction, solid impurities from the catalyst, and part of the unreacted HCl. Then, the stream is sent to a set of distillation columns where it is separated from the remaining water and unreacted products (Dreher et al., 2011). After purification, it is sent to a cracking reactor. The reactor works at 500°C and vinyl chloride is produced from ethylene chloride as presented in Eq. (13.31) (Mochida et al., 1996).

$$ClH_2C - CH_2Cl \rightarrow H_2C = CHCl + HCl \tag{13.31}$$

The resulting vinyl chloride is finally separated by means of two distillation columns. The first column separates the HCl in the light fraction and it is recycled to the oxychlorination reactor. The bottom fraction contains vinyl chloride and the unreacted ethylene chloride, which are sent to the second column. The second column separates the ethylene chloride recycling it to the set of distillation columns before the cracking furnace. Finally, the vinyl chloride is sent to a scrubber for the removal of impurities by contact with water.

13.3.3.2.3 Chlorine derived products: isocyanates

Other types of products generated from chlorine are isocyanate compounds. These products are widely used in the purification of water, the most significant being trichloroisocyanuric acid (TCCA) and sodium dichloroisocyanurate (NaDCC).

The production of TCCA ($C_3N_3O_3Cl_3$) is carried out by the reaction of a salt of dichloroisocyanuric acid with chlorine in an aqueous mixture to a pH between 2.5 and 3.5 and atmospheric temperature (Mesiah et al., 1972).

Another compound is the production of NaDCC ($C_3N_3O_3Cl_2Na_2(H_2O)$). This compound is produced by means of the reaction between cyanide, Na(OH), and chlorine. The temperature is required to be between 5°C and 45°C, ratios of Na(OH):CNH must be between 1.9:1 and 2.1:1, and it is necessary to have an excess of chlorine in aqueous medium with pH from 1.5 to 3.5 (Berkowitz, 1972).

13.3.3.2.4 Chlorine derived products: hypochlorite

The chlorine generated via the alkali process can be also directly applied for the production of hypochlorite. In the process the chlorine reacts with a caustic soda aqueous solution following Eq. (13.32). The solution must be alkaline and takes place between 30°C and 35°C into a packed column (Vogt et al., 2010). The resulting product from the column is separated in two streams: one recovered as final product and another one recycled to increase the concentration of hypochlorite in the solution. The final product is stored in a tank with a cooler to avoid temperatures above 35°C and avoid the displacement of the equilibrium.

$$2Na(OH) + Cl_2 \leftrightharpoons NaOCl + NaCl + H_2O \tag{13.32}$$

13.3.3.2.5 Chlorine derived products: sodium chlorate

As an alternative to the chlor-alkali process, chlorine chemicals can also be produced via electrolytic routes. One of the most relevant is sodium chlorate, which is one of the most popular herbicides. Ninety-eight percent of the sodium chlorate produced by this process is generated by means of the electrolytic route as presented in Eq. (13.33). The reaction takes place in a cell between a pH of 5.9 and 6.7 and around 80°C to displace the equilibriums for the generation of the ClO_3^- ion. The resulting solution obtained from the cell is concentrated to approximately 500 g/L and then cooled down to 30°C for generating the $NaClO_3$ crystals. The last part is the recovery of the crystals by means of centrifugation and the drying in case that the chlorate is not desired to be sell as a hydrate (Tilak and Chen, 1999).

$$NaCl + 3H_2O \rightarrow NaClO_3 + 3H_2 \tag{13.33}$$

13.3.4 Recovery of bromine

The recovery of bromine from brines is carried out by following the process patented by the American Potash and Chemical Corp., see Fig. 13.16 (Stewart, 1934; Grinbaum and Freiberg, 2011). The process extracts the ion bromide presented in the brine by oxidation with chlorine and steam in a contact tower as in Eq. (13.34). The reaction requires to be heat up to prevent the bromine to remain in the aqueous solution (Termine Group, 2018).

$$2Br^- (aq.) + Cl_2 \rightarrow Br_2 + 2Cl^- \tag{13.34}$$

Once bromine is captured by the steam, the mixture is separated by means of the following steps:

- First, condensation takes place allowing to separate the chlorine that remains in the gas stream. The chlorine obtained from this separator can then be recycled to the contact tower.

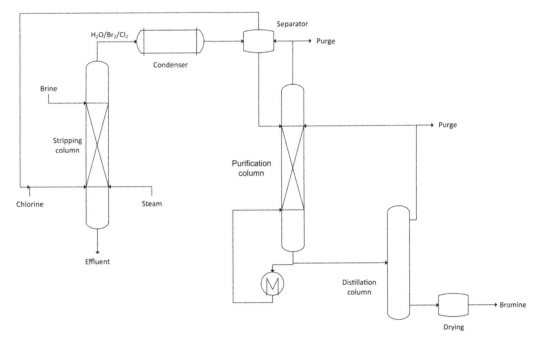

FIGURE 13.16 Bromine production process.

- The liquid mixture is sent to a separator where two layers are formed. The heavy layer is the bromine, and the lighter corresponds to the aqueous phase (mixture of bromine, chlorine, and water) that is recycled to the tower.
- The heavy liquid mixture is composed of bromine and water, which is removed by distillation and recycled to the contact column. The resultant bromine is finally dried, typically with desiccants such as H_2SO_4.

13.3.5 Magnesium industry

An alternative mineral that can be produced from brine is magnesium. This component mainly appears in the form of magnesium chloride and sulfate in brines with different concentrations depending on the sea (Mutaz and Wagialia, 1990). The extraction of magnesium ions from the sea was introduced in 1925 by IG Farben in Germany, developing an electrolyzer for $MgCl_2$, and then extended by Dow Chemical, electrolyzing also crystallization water ($MgCl_2 \cdot nH_2O$). These processes accounted for most of the market until the 1970s, before the Pidgeon process was developed to produce magnesia from dolomite. Instead of Pidgeon process which is the most popular nowadays, in the current chapter we mainly focus on the *IG Farben-Dow Chemical process* due to its possible implementation for the revalorization of brines (Takeda et al., 2014; Thayer and Neelameggham, 2001).

A summary of the IG Farben-Dow Chemical process, which allows to extract the magnesia from brines and dolomites is presented in Fig. 13.17. The first part consists of the

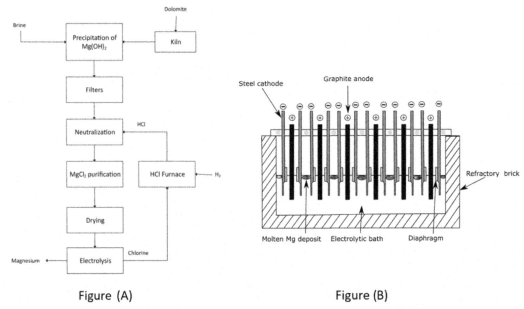

Figure (A)   Figure (B)

FIGURE 13.17   (A) Process to produce magnesium and (B) example of electrolytic cell for the production.

production of oxides by means of calcination (temperatures between 700°C and 900°C) as presented in Eq. (13.35).

$$CaCO_3MgCO_{3(s)} \leftrightarrow CaO_{(s)} + MgO_{(s)} + 2CO_{2(g)} \tag{13.35}$$

The oxides produced are dissolved in brine, which contains additional Na^+, Ca^{2+}, and Mg^{2+} ions. The solution of calcium and magnesium oxides promote the formation of hydroxides as presented from Eqs. (13.36), (13.37) and (13.38), to (13.39) The addition of hydroxides also increases the pH of seawater (typically between 7.5 and 8.4), which must be carefully controlled to be between 8 and 10.5 to favor the generation of $Mg(OH)_2$ at the same time that $Ca(OH)_2$ generation is minimized (Um and Hirato, 2014). Even though this operating region is the one with the highest yield, other hydroxides (mainly calcium, potassium, and sodium) are still produced.

$$CaO_{(s)} + H_2O \rightarrow Ca(OH)_{2(s)} \tag{13.36}$$

$$MgO_{(s)} + H_2O \rightarrow Mg(OH)_{2(s)} \tag{13.37}$$

$$Ca(OH)_{2(s)} \leftrightarrow Ca^{2+}_{(aq)} + 2OH^-_{(aq)} \tag{13.38}$$

$$Mg^{2+}_{(aq)} + 2OH^-_{(aq)} \leftrightarrow Mg(OH)_{2(s)} \tag{13.39}$$

Once magnesium hydroxide precipitates, it is separated together with the remaining hydroxides from water by means of a filter. Then, it is neutralized with a solution of hydrogen chloride (HCl) as presented in Eq. (13.40). Depending on the temperature and the amount of water of the HCl solution, different magnesium chloride hydrates can be generated. The resultant solution is sent to a dryer in order to reduce the moisture content below 27% before the electrolyzer.

$$Mg(OH)_2 + 2HCl \rightarrow MgCl_2 + 2H_2O \tag{13.40}$$

Electrolysis takes place at high temperature, typically around 750°C, with the objective of melting the salt (melting point of $MgCl_2$ is 714°C) and allow the flux of Mg to the cathode. The cathode, composed of steel and a cover of brick, is immersed into the molten salt obtaining the pure magnesium floating in the cathode compartment. The reaction in the cathode is presented in Eq. (13.41). The anode is composed of graphite and chlorine is generated there as in Eq. (13.42) and released through the diaphragm of the cell, see Fig. 13.17. This chlorine is later absorbed in water obtaining HCl, which is recycled for the neutralization of $Mg(OH)_2$.

$$Mg^{2+} + 2e^- \rightarrow Mg(l) \tag{13.41}$$

$$2Cl^- \rightarrow Cl_2(g) + 2e^- \tag{13.42}$$

13.3.6 Potassium-based products

The last product studied here for the revalorization of brines is potassium. Potassium is extracted in the form of potassium chloride. This component is mainly produced from minerals such as sylvite, but it can also be produced from brines. The production of this mineral is only significant in those cases with high concentration and significant differences exist between different seas (United States Geological Surveys, 2020).

The extraction of KCl from the sea is carried out by froth flotation with organoamines (Every et al., 1969), or by selective precipitation with soluble agents that change the solubility such as methanol (Gaska and Mich, 1966).

A more interesting potassium-based component is *potassium sulfate (K₂SO₄)*. This component is widely used as fertilizer and it is produced by means of the Mannheim process. The process is similar than the one used for the production of sodium sulfate where Eq. (13.43) takes place between 500°C and 600°C in an oven.

$$H_2SO_4 + 2KCl \rightarrow K_2SO_4 + 2HCl \tag{13.43}$$

Another interesting potassium component is *potassium nitrate*. This product can be generated by using potassium hydroxide (KOH) or potassium chloride (KCl) as raw materials. The first is mainly produced from the electrolysis of the second so the preferred method for the production mainly uses KCl. It can be combined with ammonium nitrate presented in Eq. (13.44). Alternatively, the production can be also carried out as presented in Eq. (13.45) where KCl reacts with nitric acid (Manor et al., 1983).

$$NH_4NO_3 + KCl \rightarrow NH_4Cl + KNO_3 \tag{13.44}$$

$$NaNO_3 + KCl \rightarrow NaCl + KNO_3 \tag{13.45}$$

13.3.7 Other interesting chemicals

Apart from previous products, other products can be also interesting in the recovery from brines as presented in previous Fig. 13.11:

- Lithium is one of the most demanded products nowadays and it is mainly extracted from inland reservoirs, mainly from shallow brine. However, most of the lithium can be found in the sea, becoming an interesting source in future years if there is a depletion of the reservoirs and the extraction from the sea becomes competitive. For extracting the lithium from brines, first precipitation is carried out in the form of LiCl and then Li metal is obtained by electrolysis. The electrolysis is carried out around 500°C with molten salts in similar way than for magnesium (Averill and Olson, 1977; Yang et al., 2018).
- Boron is another chemical that can be extracted from brines. Its extraction mainly comes from inland shallow brines and internal lakes where the concentration is sufficiently high. This chemical can be produced from brine in the form of boric acid (H_3BO_3) by the addition of HCl (Wang et al., 2018).
- Other chemicals with high interest in the upcoming years and for sufficiently high plants are iodine, rubidium, strontium, and cesium.

Problems

P13.1 Evaluate the pumping up flowrate for capturing the wind energy for a wind farm of 100 turbines. The water velocity in the pipe has to be between 2 and 4 m/s. Determine the piping system, number of pipes assuming a diameter of 2 m each. The power curve of the wind turbine is given by the sigmoidal law with P_{nom} = 3000 kW; a = 8.0581 m/s; and m = 0.77614 s/m; v(i) = 8,9,7,6 m/s.

P13.2 Determine the size of the reservoir situated 300 m above the turbine to produce 200 MW over a month. The pipe for water is 7 m of a material with a rugosity of 0.00005. The equivalent length of the pipe is 1.5 times the head. The yield of the turbine is 80%.

P13.3 Based on Exercise 13.3 (a) determine the percentage of the solution obtained for HCl when the thickness of film falling in the pipe is 0.25 mm. (b) Perform an energy balance and determine the distribution of temperatures in the film. Assume that the gas is in equilibrium with the interphase without heat transfer and that the heat of dissolution of HCl in water remains constant with a value of −74.84 kJ/kg.

P13.4 The production of methyl chloride from methanol takes place very fast under the presence of catalyst, reaching suddenly the equilibrium conditions, see Schmidt et al. (2013) for an example in a microreactor. Assuming equilibrium conditions at the end of the reactor, determine the conversion achieved for the methanol when the reactor works at 2 bar and 550 K with a molar flowrate of F_{CH_3OH} = 1 mol/s; F_{HCl} = 1.2 mol/s; F_{inert} = 1 mol/s. The equilibrium constants are computed as:

$$K_1 = K_2 \cdot K_3$$

$$K_2 = 1.356 * 10^2 - 0.9177 \cdot T(C) + 2.258 \cdot 10^{-3} \cdot T(C)^2 - 1.944 \cdot 10^{-6} \cdot T(C)^3$$

$$K_3 = 122.02 - 0.5367 \cdot T(C) + 0.0011 \cdot T(C)^2 - 8 \cdot 10^{-7} \cdot T(C)^3$$

References

Astariz, S., Iglesias, G., 2015. The economics of wave energy: a review. Renew. Sustain. Energy Rev. 45, 397−408.

Austin, S., Glowacki, A., 2000. Hydrochloric acid. Ullmann's Encyclopedia of Industrial Chemistry. Wiley.

Averill, W.A., Olson, D.L., 1977. A review of extractive processes for lithium from ores and brines. Lithium Needs Resour.

Becerra, A.M., Castro Luna, A.E., Ardissone, D.E., Ponzi, M.I., 1992. Kinetics of the catalytic hydrochlorination of methanol to methyl chloride. Ind. Eng. Chem. Res. 31 (4), 1040−1045.

Berkowitz, S., 1972. Continuous production of sodium dichloroisocyanurate dihydrate.

Chlorine Institute, 2020. Chlorine Manufacture. Available in: https://www.chlorineinstitute.org/stewardship/chlorine/chlorine-manufacture/.

Coiro, D.P., 2007. Horizontal and Vertical Axis Turbines for Wind and Marine Current Energy Exploitation: Design, Developments and Experimental Test. Disponível em: < https://upcommons.upc.edu/bitstream/handle/2099.1/14731/Master%20Thesis_%20TIDAL%20POWER_%20Economic%20and%20Technological%20assessment_%20Tatiana%20Montllonch.pdf> Acesso em: 14 abr. 2019.

Cruz, J., 2008. Ocean Wave Energy: Current Status and Future Perspectives. Springer-Verlang, Berlin Heidelberg, ISBN: 978−3−540−74894-6.

De Dietrich, 2020. Available in: https://www.ddpsinc.com/hcl-treatment#:~:text = Absorption%20of%20HCl%20in%20Water,C%20(104%C2%B0F)%20.

Dreher, E., Toerkelson, T.R., Beutel, K.K., 2011. Chlorethanes and chloroethylenes. Ullmann's Encyclopedia of Industrial Chemistry. Wiley.

European Comission, 2017. Regulation (EU) 2017/852.

Every, R.L., Thieme, J.O., Casad, B.M., 1969. Selective precipitation of potassium chloride from brine using organoamines. United States Patent Office. Available in: https://patentimages.storage.googleapis.com/7e/46/4e/153f47b0486541/United States3437451.pdf.

Falcao, A.F.O., 2010. Wave energy utilization: a review of the technologies. Renew. Sustain. Energy Rev. 14 (3), 899−918.

Folley, M., Whittaker, T.J.T., 2009. Analysis of the nearshore wave energy resource. Renew. Energy 34 (7), 1709−1715.

Garrett, C., Cummins, P., 2004. The power potential of tidal currents in channels. Proc. Roy. Soc. A .

Gaska, R.A., Mich, M., 1966. Recovery of potassium halides from brine. United States Patent Office. Available in: https://patentimages.storage.googleapis.com/e3/9b/d7/7291022a8ed0e8/United States3231340.pdf.

Giwa, A., Dufour, V., Al Marzoooqi, F., Al Kaabi, M., Hasan, S.W., 2017. Brine Management Methods: Recent Innovations and Current Status.

Gnaneswar, V., 2018. Sustainable Desalination Handbook. Butterworth-Heinemann, Oxford, United Kingdom, ISBN: 978−0−12−809240-8.

Grinbaum, B., Freiberg, M., 2011. Bromine. Kirk-Othmer Encyclopedia of Chemical Technology. Wiley.

Gunawardane, S.P., Kankanamge, C.J., Watabe, T., 2016. Study of the performance of the "Pendulor" wave energy converter in an array configuration. Energies 9 (4), 282.

Hagerman, G., 1995. Wave power. Encyclopedia of Energy Technology and the Environment. John Wiley & Sons, pp. 2859−2907.

Hagerman, G., Polagye B., Bedard R., Previsic M., 2006. Methodology for Estimating Tidal Current Energy Resources and Power Production by Tidal in Stream Energy Conversion (TISEC) Devises. EPRO -TP-001 NA Rev 2.

Hardisty, J., 2011. The tidal stream power curve: a case study. Energy Power. Eng. 2012 (4), 132−136.

Icis, 2020. Ethylede Dichloride Production. Available in: https://www.icis.com/explore/resources/news/2007/11/02/9075707/ethylene-dichloride-edc-production-and-manufacturing-process/#:~:text = In%20the%20direct%20chlorination%20EDC,of%20highly%20chlorinated%20by%2Dproducts.&text = The%20catalyst%20can%20be%20used,as%20in%20a%20boiling%20reactor.

Irena, 2014. Ocean Thermal Energy Conversion. Technology Brief. Available in: https://www.irena.org/publications/2014/Jun/Ocean-Thermal-Energy-Conversion.

Jones, E., Qadir, M., van Vliet, M.T.H., Smakhtin, V., Kang, S., 2019. The state of desalination and brine production: a global outlook. Sci. Total. Environ. 657, 1343−1356.

Khan, M.A.M., Rehman, S., Al-Sulaiman, F.A., 2018. A hybrid renewable energy system as a potential energy source for water desalination using reverse osmosis: a review. Renew. Sustain. Energy Rev. 97, 456−477.

Kucera, J., 2014. Desalination: Water from Water.

Loganathan, P., Naidu, G., Vigneswaran, S., 2017. Mining valuable minerals from seawater: a critical review. Environ. Sci. Water Res. Technol. 3, 37.

López, I., Andreu, J., Ceballos, S., Martínez de Alegría, I., Kortabarria, I., 2013. Review of wave energy technologies and the necessary power-equipment. Renew. Sustain. Energy Rev. 27, 413−434.

Magistro, A.J., Cowfer, J.A., 1986. Oxychlorination of ethylene. J. Chem. Educ. 63 (12), 1056−1058.

Manor, S., Bar-Guri, M., Hasidim, K., Alexandron, A., Kreisel, M., 1983. Process for the Manufacture of Potassium Nitrate. Available in: https://patentimages.storage.googleapis.com/28/30/a5/f7c262fae798e6/United States4378342.pdf.

Marsh, A.R.W., McElroy, W.J., 1985. The dissociation constant and Henry's law constant of HCl in aqueous solution. Atmos. Environ. 19 (7), 1075−1080.

Martín, M., 2016. Industrial Chemical Process. Analysis and Design. Elsevier, Oxford, UK.

Mehlum, E., 1986. TAPCHAN. In: Evans, D.V., Falcao, A.F.D. (Eds.), Hydrodynamics of Ocean Wave-Energy Utilization. Springer, Berlin, Heidelberg. Available from: https://doi.org/10.1007/978−3−642−82666-5_3.

Mesiah, R., Chancey, H., Cohen, M., 1972. Trichloroisocyanuric Acid Manufacture.

Mito, M.T., Ma, X., Albuflasa, H., Davies, P.A., 2019. Reverse osmosis (RO) membrane desalination driven by wind and solar photovoltaic (PV) energy: state of the art and challenges for large-scale implementation. Renew. Sustain. Energy Rev. 112, 669−685.

Mochida, I., Tsunawaki, T., Sotowa, C., Korai, Y., Higuchi, K., 1996. Coke produced in the commercial pyrolysis of ethylene dichloride into vinyl chloride. Ind. Eng. Chem. Res. 35 (10), 3803–3807.

Mutaz, I.S., Wagialia, K.M., 1990. Production of magnesium from desalination brines. Resour. Conserv. Recycl. 3 (4), 231–239.

Neill, S.P., Angeloudis, A., Robins, P.E., Walkington, I., Ward, S.L., Masters, I., et al., 2018. Tidal range energy resource and optimization—past perspectives and future challenges. Renew. Energy 127, 763–778.

Orejas, J.A., 2001. Model evaluation for an industrial process of direct chlorination of ethylene in a bubble-column reactor with external recirculation loop. Chem. Eng. Sci. 56 (2), 513–522.

Rajagopalan, K., Nihous, G.C., 2013. Estimates of global Ocean Thermal Energy Conversion (OTEC) resources using an ocean general circulation model. Renew. Energy 50, 532–540.

Roberts, A., Thomas, B., Sewell, P., Khan, Z., Balmain, S., Gillman, J., 2016. Current tidal power technologies and their suitability for applications in coastal and marine áreas. J. Ocean. Eng. Mar. Energy 2, 227–245.

Rossberg, M., Lendle, W., Pfleiderer, G., Tögel, A., Torkelson, T.R., Beutel, K.K., 2000. Chloromethanes. Ullmann's Encyclopedia of Industrial Chemistry. Wiley.

Rusu, E., Onea, F., 2016. Estimation of the wave energy conversion efficiency in the Atlantic Ocean close to the European islands. Renew. Energy 85, 687–703.

Schmidt, S.A., Kumar, N., Reinsdorf, A., Eränen, K., Wärna, J., Murzin, D.Y., et al., 2013. Methyl chloride synthesis over Al_2O_3 catalyst coated microstructured reactor—thermodynamics, kinetics and mass transfer. Chem. Eng. Sci. 95, 232–245.

Schmittinger, P., Flowkiewicz, T., Curlin, L.C., Lüke, B., Scannell, R., Navin, T., et al., 2011. Chlorine. Ullmann's Encyclopedia of Industrial Chemistry. Wiley.

Shahmansouri, A., Min, J., Jin, L., Bellona, C., 2015. Feasibility of extracting valuable minerals from desalination concentrate: a comprehensive literature review. J. Clean. Prod. 100, 4–16.

Skilhagen, S.E., Dugstad, J.E., Aaberg, R.J., 2008. Osmotic power—power production based on the osmotic pressure difference between waters with varying salt gradients. Desalination 1–3, 476–482.

Stewart, L.C., 1934. Commercial extraction of bromine from sea water. Ing. Eng. Chem. Res. 26 (4), 361–369.

Takeda, O., Uda, T., Okabe, T.H., 2014. Chapter 2.9—Rare earth, titanium group metals, and reactive metals production. Treatise on Process Metallurgy. Volume 3: Industrial Processes. Elsevier, pp. 995–1069.

Termine Group, 2018. How Is Bromine Produced Commercially? Available in: https://www.termine.com/archives/542.

Thayer, R.L., Neelameggham, R., 2001. Improving the electrolytic process for magnesium production. JOM 53, 15–17.

Thyagarajan, M.S., Kumar, R., Kuloor, N.R., 1966. Hydrochlorination of methanol to methyl chloride in fixed catalyst beds. Ind. Eng. Chem. Process. Des. Dev. 5 (3), 209–213.

Tilak, B.V., Chen, C., 1999. Electrolytic sodium chlorate technology: current status. In Chlor-Alkali and Chlorate Technology: R.B. Macmullin Memorial Symposium.

Um, N, Hirato, T, 2014. Precipitation behavior of Ca(OH)2, Mg(OH)2, and Mn(OH)2 from CaCl2, MgCl2, and MnCl2 in NaOH-H2O solutions and study of lithium recovery from seawater via two-stage precipitation process. Hydrometallurgy, 146, 142–148.

United Nations. Industrial Development Organization, 2005. Extraction of Chemicals from Seawater, Inland Brines and Rock Salt Deposits. University Press of the Pacific, ISBN: 978–1410223869.

United States Geological Surveys, 2020. Mineral Commodity Summaries.

Vogt, H., Balej, J., Bennett, J.E., Wintzer, P., Sheikh, S.A., Gallone, P., et al., 2010. Chlorine oxides and chlorine oxygen acids. Ullmann's Encyclopedia of Industrial Chemistry. Wiley.

Von Plessen, H., 2012. Sodium sulfates. Ullmann's Encyclopedia of Industrial Chemistry. Wiley, 10.1002/14356007.a24_355.

Voutchkov, N., 2013. Desalination Engineering: Planning and Design.

Wang, H., Zhong, Y., Du, B., Zhao, Y., Wang, M., 2018. Recovery of both magnesium and lithium from high Mg/Li ratio brines using a novel process. Hydrometallurgy 175, 102–108.

Waters, R., Engström, J., Isberg, J., Leijon, M., 2009. Wave climate off the Swedish west coast. Renew. Energy 34 (6), 1600–1606.

Yang, S., Zhang, F., Ding, H., He, P., Zhou, H., 2018. Lithium metal extraction from seawater. Joule 2 (9), 1648–1651.

Integration of resources

Renewable-based process integration

Salvador I. Pérez-Uresti[1], Ricardo M. Lima[2] and Arturo Jiménez-Gutiérrez[1]

[1]Chemical Engineering Department, Tecnológico Nacional de México/Instituto Tecnológico de Celaya, Celaya, GTO, Mexico [2]Computer, Electrical and Mathematical Sciences & Engineering Division, King Abdullah University of Science and Technology (KAUST), Thuwal, Saudi Arabia

14.1 Introduction

Heavy industries such as ceramic, iron, cement, and steel are high-energy-consuming and will require a total reconfiguration to reduce their environmental impact. In this respect, the design of fully renewable-based processes will very likely play a central role. In a renewable-based integrated process, the renewable resources available in a site are included in a single facility and converted into electricity, utilities or chemical compounds through different technologies which have in common several steps (Oh et al., 2014; Kostevšek et al., 2015). However, renewable-based integrated process design and operation remain a challenge because of renewable resources uncertainty.

The Process Systems Engineering (PSE) community has developed several mathematical frameworks and techniques to handle design/synthesis and planning problems under uncertainty. The main motivation to address uncertainty resided on the uncertainty of some parameters, such as raw materials quality, kinetic constants involved in reactors design, transfer coefficients in heat exchanger networks, yields, product demands, and products prices. The uncertainty on these parameters affects process feasibility and profitability, and therefore, it is relevant to consider uncertainty at the design stage.

In the PSE community, Takamatsu et al. (1973) and Grossmann and Sargent (1978) were among the first to address process design with uncertain parameters. Around the same period, Friedman and Reklaitis (1975a,b) focused on the solution of linear programming problems with uncertain parameters, motivated by achieving flexible solutions.

Flexibility and resilience analysis of chemical processes is an area that is related to process design under uncertainty and that has conceptually contributed with methodologies to handle uncertainty (Swaney and Grossmann, 1985; Pistikopoulos and Grossmann, 1989; Grossmann et al., 1983). Flexibility analysis aims to find a flexible and feasible process design that when subject to variations in some operational parameters it remains feasible by adjusting control variables. Zhang et al. (2016a) reviewed the methods applied for flexibility analysis and made the parallel between those methods and robust optimization (Ben-Tal et al., 2009).

Since those initial works, the optimization frameworks developed in PSE have evolved significantly and were applied in many applications, such as the process design/synthesis, scheduling, planning, and supply chain optimization. We refer the reader to Sahinidis (2004), Grossmann et al. (2017), and Li and Grossmann (2021) for reviews on the application of optimization under uncertainty applied to PSE problems. In the last years, the PSE community put efforts to leverage its knowledge on process design/synthesis and optimization under uncertainty to address challenges in renewable processes integration.

In the following sections, we start by reviewing three examples of integrated systems to produce chemical compounds, energy, and utilities, and then we focus on two methodologies for process design under uncertainty: two-stage stochastic programming and clustering methods. Two-stage stochastic programming enables the solution of stochastic optimization problems featuring design and operation decisions in one formulation. Clustering methods are useful to generate scenarios (Baringo and Conejo, 2013) or to approximate long-time horizons with hourly resolutions by a small number of representative periods (Teichgraeber and Brandt, 2019). Two-stage stochastic programming and clustering methods are relevant tools to address design problems under uncertainty involving long- and short-term time scales.

14.2 Examples of renewable-based integrated systems

We discuss three examples of processes to produce utilities, heat, fuel, and chemical compounds using renewable resources.

14.2.1 Design of utility plants

Utility plants are important components of the chemical industry, as they supply the electricity and heat required in different units such as distillation columns, heat exchangers and scrubbers, among others. The structure of a utility plant usually consists of two sections, the steam-generating section, and steam network section. In the first section, steam is generated at a very high pressure (165−100 bar) and is sent to the second section, where it is expanded through steam turbines, which also produce electricity. The traditional utility plants use fossil-based fuels to produce steam, which increases the environmental impact of the chemical industry. Pérez-Uresti et al. (2019) developed a mathematical model aiming at designing renewable-based utility plants. In that work, the authors proposed the integration of biomass, manure, solar radiation, and wind to produce utilities, and different technologies were considered to process them. For instance, biomass can be burnt in a boiler to produce as many as four types of steam, or converted into syngas through indirect gasification or direct gasification, see Fig. 14.1. A

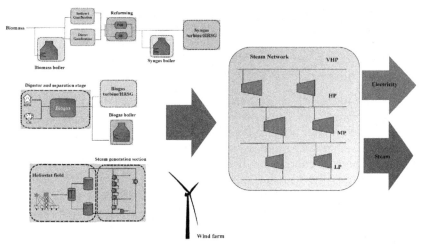

FIGURE 14.1   Superstructure for renewable-based utility plant.

wind farm, a digestor, and a concentrating solar power (CSP) plant are also considered within the superstructure. This system also integrates two types of thermodynamic cycles to produce electricity, a Rankine cycle with regeneration, used to model the steam turbine, and a Brayton cycle related to the gas turbines performance that are linked to heat recovery steam generators. Then, in Pérez Uresti et al. (2020), the authors developed a case study for the south-west region of Mexico and obtained a plant that integrates a biomass boiler, a CSP plant, and wind turbines. However, as they considered a more detailed time discretization, wind farm was no longer selected, and the model relied on a CSP plant integrated with a biomass boiler.

14.2.2   Green-based production of methane

Green-based production of chemical compounds has been recently studied as an alternative option to store renewable energy. For instance, renewable-based ammonia can be obtained from green hydrogen produced from water electrolysis using wind and solar energies (Sánchez and Martín, 2018; Palys et al., 2021). Another example is the thermochemical conversion of biomass into syngas, which can be used in the production of other compounds, such as methanol and methane, among others, which are easier and cheaper to store when compared to hydrogen and batteries.

Martín and Davis (2015) designed an integrated renewable-based plant to produce methane using wind energy, solar radiation, and biomass. Solar panels and wind turbines were considered to carry out the water electrolysis and produce hydrogen and pure oxygen. Then, lignocellulosic biomass is converted into syngas through a gasification process. Two options were considered, indirect gasification and direct gasification. Once leaving the gasifier, syngas is sent to the reformation section. In this way, the hydrogen produced from both sides reacts with carbon dioxide to produce methane. The optimal design for a case study yielded a gasification process combined with solar panels. However, the authors highlighted that the high variability of renewable resources led to large idle units when a constant production of methane was required.

14.2.3 Integrated networks for fuels and energy production

Renewable-based processes have in common several steps, which aid for their integration into multiproduct networks. In these networks the product obtained from one process can be used as an input to another one. In this respect, Martín and Grossmann (2018) developed an integrated network that uses biomass, solar radiation, waste, and wind as raw materials to produce power, thermal energy, methanol, ethanol, and other fuels. The superstructure proposes different processing routes for each renewable resource. For instance, solar radiation is used to produce electricity through PV-technology and a CSP plant. It is also used as an algae growing-up component to produce biodiesel and glycerol. Lignocellulosic biomass is converted into syngas, used to produce electricity, hydrogen, methanol, thermal energy, Fischer—Tropsch liquids, and biodiesel, or converted into ethanol through fermentation. The network also considers the production of hydrogen, which can be obtained from electrolysis and is used to produce methane. The authors studied the scale-up of the integrated network at local, regional, and national levels and described the changes in technology selections. Results showed that the technologies distribution at the regional level changed significantly when the integration is conducted at the national level. The national integrated network allowed to reallocate the production centers to other cities, thus reducing the processing and transportation costs with respect to regional networks.

14.3 Two-stage stochastic programming

The uncertainty and variability of renewable resources add complexity to the operation of steady-state systems, and thus their integration requires proper characterization and handling in process design. The PSE community has a strong tradition on mathematical frameworks to consider uncertainty into optimization models. On this regard, one can distinguish two approaches: robust optimization and stochastic programming (Grossmann et al., 2017). The main difference between both approaches lies on the way the uncertainty is captured. In stochastic programming, the uncertain parameters are characterized by a number of discrete scenarios associated with a probability of occurrence, whereas in robust optimization the uncertain parameters are described by polyhedral sets of uncertainty (Ben-Tal et al., 2009).

Both approaches can handle recourse actions to include corrective actions through control variables that are adjusted on the basis of the random parameters realization. Some relevant examples of robust optimization include short-term time horizon problems for the determination of optimal scheduling for air separation plants (Zhang et al., 2016b), multichiller systems (Saeedi et al., 2019), wind-powered ammonia plants (Allman et al., 2019), and electricity dispatch (Malherio et al., 2015). In this chapter, we provide basic concepts of stochastic programming formulations.

Stochastic programming, usually regarded as a scenario-based approach, optimizes a function, that traditionally relies on the expectation of a random variable calculated over a discrete and finite number of scenarios. Important applications of this approach include multi-year time horizon problems where long-term decisions, such as design or investment decisions, are required (Grossmann et al., 2017; Li and Grossmann, 2021).

In the context of process design, we present the strategy proposed by Halemane and Grossmann (1983) to formulate a stochastic model consisting of two-decision framework stages: a design stage and an operating stage. The main purpose of the first stage is the selection of the design variables, such as equipment sizes, technologies, or even investment decisions that ensure feasibility over the scenarios considered within the formulation. The design variables, also called first-stage decisions, are selected before uncertainty is revealed and remain fixed during the operating stage.

The operating stage involves the control variables, also called second-stage decisions, that are scenario dependent and can be adjusted to ensure operation feasibility once the plant is installed. In this section, we present two formulations: a risk neutral that optimizes the expectation of a random variable, and a risk-averse that includes a risk metric in the objective function.

The general formulation of an optimization problem under uncertainty is given by

$$
\begin{aligned}
&\min E_\theta\big[f(d, z, \theta)\big], \\
&\text{s.t.} \\
&h(d, z(\theta), \theta) = 0 \\
&g(d, z(\theta), \theta) \le 0
\end{aligned}
\qquad z(\theta) \in Z, \quad \theta \in \Theta,
\tag{14.1}
$$

where f is the random variable to minimize, and h and g are equality and inequality constrains. In process design, d denotes design variables, z represents the control variables, and θ represents the random parameters.

In stochastic programming, Equation (14.1) is addressed by approximating the cumulative distribution of the random variables by a finite discrete distribution. This approximation simplifies the calculation of the expectation of f, and generates a finite number of constraints. In this way, Problem (14.1) is written as

$$
\begin{aligned}
&\min f = f^0(d) + \sum_{i \in I}^{I} \pi_i f(z_i(\theta_i), \theta_i), \\
&\text{s.t.} \\
&h^i(d, z_i(\theta_i), \theta_i) = 0 \\
&g^i(d, z_i(\theta_i), \theta_i) \le 0
\end{aligned}
\qquad z_i(\theta_i) \in Z, \quad \theta_i \in \Theta, \quad i \in I
\tag{14.2}
$$

where π_i denotes the probability of occurrence of scenario i, such that $\sum_{i \in I} \pi_i = 1$. Note that the design variables, d, are independent of scenarios realizations, whereas control variables, $z_i(\theta_i)$, are scenario dependent. The beauty of Problem (14.2) is that it seeks for a solution for the design variables d that is valid for all considered scenarios and minimizes the expectation of f. This problem can be solved directly by using a proper optimization solver or a decomposition method that exploit the structure of the problem. For a thorough description of methods to solve these problems the reader is referred to Birge and Louveaux (2011) and Kall and Mayer (2011).

The two-stage stochastic programming formulation can also include a risk measure in the objective function or constraints to drive the optimization to first-stage solutions that minimize a risk measure of the quantity of interest. Risk-averse solutions are relevant for problems with long-term horizons where the uncertainty is significant and worse solutions may have a significant impact. For example, in a cost minimization problem the risk measure helps to drive the solutions to values that avoid the realization of high costs.

A risk measure with coherent properties for optimization problems is the conditional value at risk (CVaR) (Rockafellar and Uryasev, 2000; Rockafellar, 2007). CVaR is defined as the conditional expectation of the random variable to be greater than the value at risk (VaR) for a specified quantile $(1 - \alpha)$:

$$\mathrm{CVaR}_{1-\alpha} := \mathbb{E}_\theta\left[f(d, z, \theta)|f(d, z, \theta) \geq \mathrm{VaR}_{1-\alpha}\right], \qquad (14.3)$$

where θ is a random parameter, $1 - \alpha$ denotes a quantile of the cumulative distribution function of $f(d, z, \theta)$, and VaR is an extreme value of the $(1 - \alpha)$-quantile. Fig. 14.2 shows a schematic representation of the CVaR and the VaR.

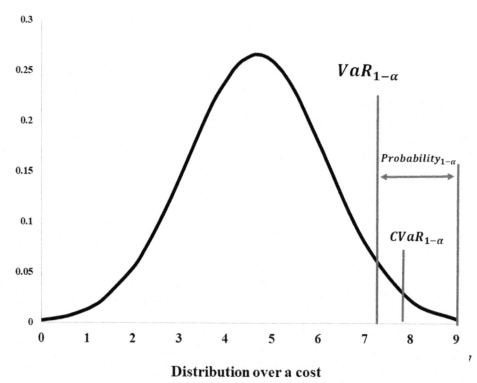

FIGURE 14.2   Value at risk (VaR) and conditional value at risk (CvaR) definitions.

A two-stage risk-averse stochastic programming model considering the minimization of a combination of the expectation of f and the CVaR of f can be defined as

$$\min f = f^0(d) + \beta \sum_{i \in I}^{I} \pi_i f(z_i(\theta_i), \theta_i) + (1 - \beta) \left[\eta + \frac{1}{1 - \alpha} \sum_{i \in I}^{I} \pi_i \left(f(z_i(\theta_i), \theta_i) - \eta \right)^+ \right],$$

$$(14.4)$$

s.t.
$$\begin{aligned} h^i(d, z_i(\theta_i), \theta_i) &= 0 \\ g^i(d, z_i(\theta_i), \theta_i) &\leq 0 \end{aligned}, \quad z_i(\theta_i) \in Z, \quad \theta_i \in \Theta, \quad i \in I,$$

where η is the value at risk, and $\left(f(z_i(\theta_i), \theta_i) - \eta \right)^+$ accounts only for the positive results of the operation inside the parenthesis. Problem (14.2) can be solved directly by using a proper optimization solver or a decomposition method that handles the CVaR. Lima et al. (2018) discuss the derivation of this problem and studied alternative decomposition methods to handle the CVaR (Example 14.1).

EXAMPLE 14.1 Based on the information shown in Table 14E1.1, determine the amount of flour that a baker should purchase to prepare bread so that the risk of losses because of nonpurchased bread are the minimum. The excepted profit of the bakery should be greater than $10/day.

TABLE 14E1.1   Data for problem 14.1.

Scenario	Probability	Demand
1	0.1	100
2	0.13	130
3	0.12	150
4	0.20	170
5	0.15	200
6	0.1	230
7	0.1	240
8	0.1	255
Sale price	$ 1.4	
Flour cost	$ 1	

In this problem, the first-stage decision is the amount of flour (FP) to be purchased. The total profit obtained for the sales are given by the following equation,

$$\text{Profit}_s = \text{Sale}_s * \text{SP} - \text{FP} * \text{PP} \quad \forall s \qquad (14E1.1)$$

where SP is the sale price and PP is the flour cost. Then, we define the variable "delta" to calculate the deviation of profit obtained in each scenario with respect to the threshold profit (PT) set by the company (10 $/day).

$$delta_s \geq PT - (Sale_s * SP - FP * PP) \forall s \tag{14E1.2}$$

The formulation of this problem is given below:

SETS	Positive variable
s / 1*8 /;	FP, sale(s), delta (s), edr;
Parameters	Variable
D(s) Demand in scenario	Z;
/1 100, 2 130, 3 150, 4 170, 5 200, 6 230, 7	Equation
240, 8 255/	sales1, sales2, Obj,
P(s) probability of each scenario	deltadef, edrdef, edrcon;
/1 0.1, 2 0.13, 3 0.12, 4 0.20, 5 0.15, 6 0.1, 7	sales1 (s).. sale (s) = l = FP;
0.1, 8 0.1/	sales2 (s).. sale (s) = l = D(s);
Scalar	deltadef(s).. delta (s) = g = PT − (sale (s) * SP − FP*PP);
PP "flour cost" /1/	edrdef.. edr = e = sum ((s), delta (s) * P(s));
SP "sales price" /1.4/	edrcon.. edr = l = EDRmax;
PT "profit threshold" /10/	Obj.. Z = e = sum((s), sale (s) * P(s))*SP − FP*PP;
EDRmax "max exp downside risk" /1/;	Model bakery /ALL/;
	Option MIP = cplex;

The optimal solution of this problem indicates that the baker should purchase 140 kg of flour, yielding a profit of $48.5.

14.4 Clustering methods

The modeling of renewable-based energy systems usually leads to large problems that are CPU time consuming, as they require integrating long-term decisions, such as investment decisions implying a large-time horizon discretization, with the short-term behavior of renewable resources. These time scales can be effectively included in the same model using representative periods (Tejada-Arango et al., 2019).

The representative period technique makes use of clustering methods to group days with similar characteristics in the same group, known as "time slices," preserving the chronological sequences within each group. In this way, a large group of historical data is represented by one representative period (Nahmmacher et al., 2016). Then, the model finds the optimal solution for the representative sets without a significant increase of the CPU burden.

Representative periods have been used extensively in power generation expansion and transmission planning models. For instance, Heuberger et al. (2017) used representative periods to study possible changes on the Britain electrical grid for the year 2035. Other examples included the Regional energy deployment system (ReEDS), and the integrated planning model that approximate the system operation by using representative periods to capture the average behavior of electricity demand, wind, and solar outputs (Mallapragada et al., 2018).

The calculation methodology proposed by Teichgraeber and Brandt (2019) involves three steps: normalization, assignment, and representation, which are described next.

Normalization. Normalization is required when using multiple data with different orders of magnitude. A common method to conduct the normalization consists on dividing the observations by the largest value. Other normalization methods include the full normalization, the element-based normalization, and the sequence-based normalization.

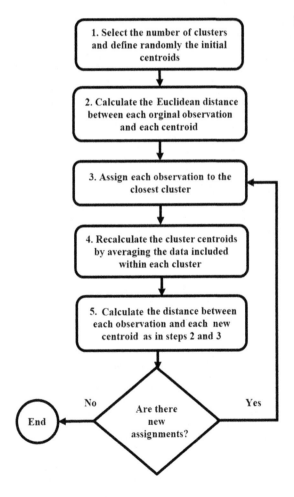

FIGURE 14.3   k-means algorithm.

Assignment. In this step, the data with similar characteristics are assigned to clusters using clustering methods. The clustering methods perform a distance measure of any observation with respect to the centroid of a cluster, assigning the observation to the closest cluster. Among the clustering methods used to calculate representative periods, the k-means and k-medoids algorithms are the most commonly used. The first one uses the average value of data included within each cluster as the centroid, whereas the latter uses its medoid. Fig. 14.3 shows the steps of the k-means algorithm.

Representation. The representation of each cluster is usually conducted by using their centroids. However, they cannot capture the real behavior of renewable resources, as they are calculated as the average value of the data included within each cluster. A more adequate practice is to use the closest day/period of each centroid; in this way, a real day is selected as the most representative one. Then, an associated weight is assigned to each cluster as a function of the total number of days included (Example 14.2).

See Example 14.3.

EXAMPLE 14.2 Let ω, ϕ, φ, ϱ, λ, τ, ∂, μ, ν, π, δ, x, γ, and β be vectors with dimension n, such that: $\omega \in \mathbb{R}^n$, $\phi \in \mathbb{R}^n$, $\varphi \in \mathbb{R}^n$, $\varrho \in \mathbb{R}^n$, $\lambda \in \mathbb{R}^n$, $\tau \in \mathbb{R}^n$, $\partial \in \mathbb{R}^n$, $\mu \in \mathbb{R}^n$, $\nu \in \mathbb{R}^n$, $\delta \in \mathbb{R}^n$, $x \in \mathbb{R}^n$, $\gamma \in \mathbb{R}^n$, $\pi \in \mathbb{R}^n$, $\beta \in \mathbb{R}^n$. Assume the values given below with $n = 5$, and group the vectors in three clusters using the k-means algorithm.

$\omega = [4, 5, 8, 1, 7]$; $\phi = [9, 0, 5, 2, 3]$; $\varphi = [0, 1, 2, 4, 5]$; $\varrho = [3, 5, 5, 2, 7]$; $\lambda = [1, 4, 5, 8, 2]$;
$\tau = [6, 7, 3, 2, 2]$; $\partial = [9, 4, 6, 7, 6]$; $\mu = [5, 9, 3, 10, 8]$, $\nu = [3, 4, 9, 0, 0]$, $\pi = [1, 3, 2, 7, 5]$,
$\delta = [5, 3, 7, 7, 3]$, $x = [6, 7, 7, 8, 8]$, $\gamma = [2, 1, 2, 5, 5]$, and $\beta = [4, 2, 3, 1, 4]$;
This example is solved using the k-means algorithm shown in Fig. 14.3.

1. **Selection of cluster centroids.** In this step, we define the number of clusters used to group the data. We choose randomly three vectors among the observations and set them as the centroids of the clusters. For instance, we take vectors ω, ϱ, and λ. Note that the labels *Var1*, *Var2*, *Var3*, *Var4*, and *Var5* are used to denote the components of the multidimensional vectors (Table 14E2.1).

TABLE 14E2.1 Centroids randomly selected.

		Centroids arguments				
		Var 1	Var 2	Var 3	Var 4	Var 5
Cluster 1	ω	4	5	8	1	7
Cluster 2	λ	1	4	5	8	2
Cluster 3	ϱ	3	5	5	2	7

2. **Distance computation**. Then, we calculate the Euclidean distance between each observation (vector) and each cluster centroid using

$$d(i,j) = \sqrt{\sum_{k=1}^{p} (x_{i,k} - x_{j,k})^2},$$

(14E2.1)

where $x_i = (x_{i,1}, \ldots, x_{i,p})$, $x_j = (x_{j,1}, \ldots, x_{j,p})$; i and j are indexes denoting an observation and the centroid of a cluster, respectively. Results are shown in Table 14E2.2.

TABLE 14E2.2 Euclidean distance calculation.

Observation	Distance			Min	Assignment
	Cluster 1	Cluster 2	Cluster 3		
ω	0.0	9.6	3.3	0.0	1
ϕ	8.7	10.8	8.8	8.7	1
φ	9.0	6.6	6.5	6.5	3
ϱ	3.3	8.1	0.0	0.0	3
λ	9.6	0.0	8.1	0.0	2
τ	7.7	8.6	6.5	6.5	3
∂	8.2	9.1	8.0	8.0	3
μ	11.1	9.2	9.4	9.2	1
ν	7.3	9.4	8.4	7.3	1
π	9.4	4.5	6.8	4.5	2
δ	7.6	4.8	7.3	4.8	2
x	7.7	8.6	7.3	7.3	3
γ	8.7	6.1	6.2	6.1	2
β	6.6	8.4	4.9	4.9	3

3. **Assignment**. The assignment of one observation is conducted on the basis of the minimum Euclidean distance with respect to each centroid. For instance, the distance of vector φ to centroids 1, 2, and 3 are 9, 6.6, and 6.5, respectively. In this case, the minimum distance is 6.5; therefore vector φ is assigned to cluster 3.
4. **Centroids recalculation**. The cluster centroids are recalculated by averaging the data assigned to each cluster (Table 14E2.3).

TABLE 14E2.3 Centroids recalculation.

	Centroids arguments				
	Var 1	Var 2	Var 3	Var 4	Var 5
Cluster 1	5.25	4.50	6.25	3.25	4.50
Cluster 2	2.25	2.75	4.00	6.75	3.75
Cluster 3	4.67	4.33	4.33	4.00	5.33

5. **Calculate reassignments**. The distance of each observation is again calculated as in step 2. Then, we determine if the observations should be reassigned on the basis of the minimum distance criterion (Table 14E2.4).

TABLE 14E2.4 Reassignment of clusters.

	Distance				Previous assignments	New assignment
Observation	Cluster 1	Cluster 2	Cluster 3	Min		
ω	4.0	8.2	5.1	4.02	1	1
ϕ	6.3	8.8	6.9	6.30	1	1
φ	7.7	4.6	6.2	4.61	3	⊡ 2
ϱ	3.8	6.3	3.2	3.23	3	3
λ	7.0	3.0	6.4	2.96	2	2
τ	5.0	7.7	5.1	5.02	3	3
∂	5.5	7.5	5.6	5.54	3	3
μ	9.4	8.7	8.2	8.17	1	⊡ 3
ν	6.6	9.3	8.3	6.61	1	1
π	7.3	2.7	5.5	2.69	2	2
δ	4.4	4.2	4.8	4.15	2	2
x	6.5	7.8	6.3	6.25	3	3
γ	6.6	3.4	5.0	3.43	2	2
β	4.9	6.1	4.3	4.29	3	⊡ 1

As can be observed, vectors φ, μ, and β were reassigned to clusters 2, 3, and 1, respectively. Thus new centroids need to be recalculated as in step 4, and step 5 is repeated (Table 14E2.5).

TABLE 14E2.5 New assignments.

Observation	Distance			Min	Previous assignments	New assignments
	Cluster 1	Cluster 2	Cluster 3			
ω	4.6	8.2	6.3	4.6	1	1
ϕ	5.1	8.8	8.7	5.1	1	1
φ	7.6	3.7	8.7	3.7	2	2
ϱ	4.9	6.1	5.0	4.9	3	▢ 1
λ	8.4	3.5	7.2	3.5	2	2
τ	5.7	7.8	6.0	5.7	3	▢ 1
∂	7.7	8.0	4.3	4.3	3	3
μ	12.3	9.2	5.6	5.6	3	3
ν	5.1	9.4	10.2	5.1	1	1
π	8.5	2.3	6.7	2.3	2	2
δ	6.1	4.9	5.4	4.9	2	2
x	9.4	8.3	3.6	3.6	3	3
γ	7.0	2.6	7.3	2.6	2	2
β	3.5	5.7	7.3	3.5	1	1

In this case two reassignments were observed, so the procedure needs to be repeated until there are no more reassignments.

EXAMPLE 14.3 The energy integrated system shown in Fig. 14E3.1 is being considered to supply electricity to a city located in the south-west region of México. Consider the hourly data of solar irradiance, wind velocity, temperature, and electricity demand for the years 2016−2019 available in the excel document Cap_RP, along with the following assumptions:

- The size of a PV unit is assumed to be 8 m², with the energy converted given by (Malheiro et al., 2015),

$$E_{PV}(t) = PMP_{REF} \frac{IR(t)}{1000 \text{ W/m}^2} \left[1 + \gamma_{PV}(T_{cell}(t)) - 25°C\right]\eta_{DA} \tag{14E3.1}$$

$$T_{cell}(t) = T_{amb} + \frac{NOCT}{800 \text{ W/m}^2} IR(t) \tag{14E3.2}$$

where $PMP_{REF} = 0.25$ kW, $\gamma_{PV} = 0.3\%/°C$, $\eta_{DA} = 93\%$, NOCT = 41.5°C, and IR(t) denotes the solar irradiance in W/m².

- The energy generated by a wind turbine is calculated as follows (Pérez-Uresti et al., 2019)

$$E_{wind_s}(t) = \frac{P_{nom}}{1 + e^{(-(v_s - a)mp)}} \tag{14E3.3}$$

where $v_{sc,t}$ is the wind velocity in m/s, $P_{nom} = 1500$, $a = 8.08$ m/s, and $mp = 0.78$ s/m.

The area used by a wind turbine is calculated as:

$$A_{wind} = 28D_{WT}^2,$$

where $D_{WT} = 82$ m.

- Assume the following economic data (Table 14E3.1).

See Fig. 14E3.1.

TABLE 14E3.1 Data for Example 14.3.

Solar panel cost	1700 $/kW
O&M cost for solar panel	$ 0.025/kWh
Wind turbine cost	1600 $/kW
O&M cost of wind turbines	0.02 $/kW
Batteries cost	$ 230/kWh
O&M cost of batteries	$ 0.015/kWh
Max. PV area	1,500,000 m^2
Max. Wind farm area	36 km^2

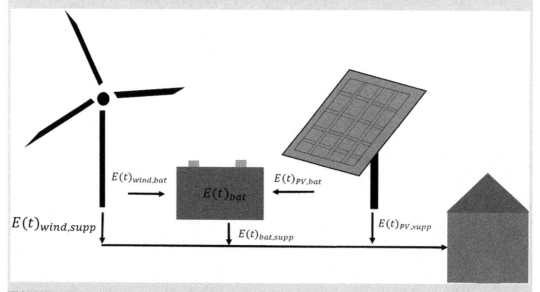

FIGURE 14E3.1 Energy integrated system.

a. Calculate three representative periods to characterize the data.

In this example, we use the k-means function of Matlab to calculate the representative periods. We first normalize the data by taking the largest value of each parameter, and then the following Matlab code is used to calculate the cluster centroids and data assignment,

```
clear
clc
X = xlsread('Cap_RP.xlsx','vectors','B2:CS1461');
%X is a variable used to call the values included in an excel spreadsheet document
K = 3% denotes the number of cluster to be determined
[idx,C,D] = kmeans(X,k,'MaxIter',10000);
%idx: is a vector that denotes the data assignment to each cluster
%C: is a matrix that gives the centroids values of each cluster
%D: is a matrix that indicates the distance between each observation and each cluster
centroid
xlswrite('cluster_idx_K3.xls',idx)
xlswrite('cluster_C_K3.xls',C)
xlswrite('cluster_sum_K3.xls',sumd)
xlswrite('cluster_D_K3.xls',D)
```

The data consist of 24-h series of electricity demand, solar irradiance, wind velocity, and ambient temperature. Thus the matrix, used to feed the previous code, is formulated as follows:

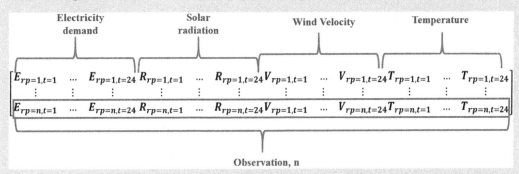

Results showed that clusters 1, 2, and 3 include 611, 278, and 571 days, respectively. Thus the weights assigned to each of them are 41.85%, 19.04%, and 39.11%. Fig. 14E3.2 shows the data series calculated as representative periods for each parameter. In this case, the RP were assumed as the centroids of the cluster given by the matrix C in the Matlab code shown previously.

b. Use the RP calculated in the previous point and formulate an optimization model to determine the optimal number of panels, wind turbines, and batteries that minimize the total annual cost while satisfying the electricity hourly demand of the city. Consider the number of panels, wind turbines, and batteries capacity as the design variables.

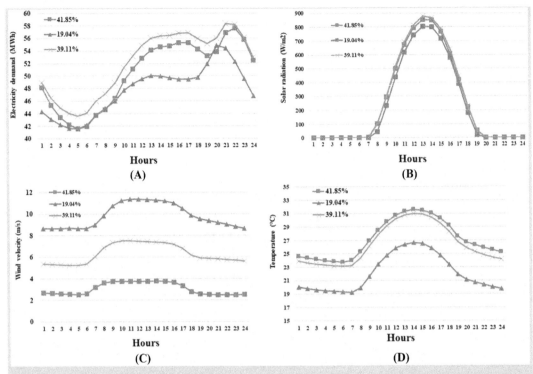

FIGURE 14E3.2   Representative periods for: (A) electricity demand; (B) solar irradiance; (C) wind velocity; (D) ambient temperature.

The objective function for this problem can be written as follows:

$$\min \text{TAC} = \sum_{r \in R} \sum_{t \in T} 365 * \pi_r * \text{Opcost}_{r,t} * K C^{INV}, \qquad (14E3.4)$$

where C^{INV}, is the total investment cost, K is an annualization factor, taken as 1/3 (Douglas, 1988), and π_r is the weight assigned to each representative period, r, considered in the formulation. The problem was solved using the GAMS software environment, as shown below.

Note: parameters and variables sections are defined in the electronic material.

```
***************************solarpanels********************************
solve equation using MIP MINImizing TAC
total_area_PV(sc,t).. A_pv(sc,t) = e = 8*Nt_pv(sc,t);
Area_UB_PV(sc,t).. A_pv(sc,t) = l = 1500000*y_pv;
Area_LB_PV(sc,t).. A_pv(sc,t) = g = 8*y_pv;
number_panels_M(sc,t).. Nt_pv_D_M(t) = g = Nt_pv(sc,t);
number_panels_D(sc,t).. Nt_pv_D = g = Nt_pv_D_M(t);
Inv_cost_PV.. C_inv_pv = e = (1700*Nt_pv_d)/1e6;
op_cost_PV(sc,t).. C_op_pv(sc,t) = e = ((0.025*E_pv(sc,t)))/1e6;
op_cost_pv_prom(sc).. C_op_pv_M(sc) = e = sum((t),C_op_pv(sc,t));
```

op_cost_pv_T.. C_op_pv_T = e = sum((sc),C_op_pv_M(sc)*prob(sc));
*******************balance PV*********
out_pv(sc,t).. E_pv(sc,t) = e = E_pv_bat(sc,t) + E_pv_supp(sc,t);
*********************wind turbines***************************
power_wind(sc,t).. Pt_wind(sc,t) = E = P_turbine(sc,t)*Nt_wind(sc,t);
AREA_wind(sc,t).. A_wind(sc,t) = E = 28*(6724)*Nt_wind(sc,t);
NUMERO_TURB_wind_REST(sc,t).. A_wind(sc,t) = L = A_max_farm_wind*Y_WIND;
NUMERO_TURB_wind_REST2(sc,t).. A_wind(sc,t) = g = 28*(6724)*Y_WIND;
number_turbines_M(sc,t).. Nt_wind_D_M(t) = g = Nt_wind(sc,t);
number_trubines_D(sc,t).. Nt_wind_D = g = Nt_wind_D_M(t);
op_cost_wind(sc,t).. C_op_wind(sc,t) = e = ((0.02*Pt_wind(sc,t)))/1e6;
op_cost_wind_prom(sc).. C_op_wind_M(sc) = e = sum((t),C_op_wind(sc,t));
op_cost_wind_T.. C_op_wind_T = e = sum((sc),C_op_wind_M(sc)*prob(sc));
wind_farm_inv.. C_wind = e = (1600*1500*Nt_wind_D)/1e6;
*******************balance wind*********
out_wind(sc,t).. Pt_wind(sc,t) = e = E_wind_bat(sc,t) + E_wind_supp(sc,t);
**************************Balance batteries**************
In_batterie(sc,t).. E_in_bat(sc,t) = e = (E_wind_bat(sc,t) + E_pv_bat(sc,t))*N_eff_charge;
*****************battery balance for t = 1 for each RP = 1*****************
Balance_bat_to(sc,t)$(ord(t) eq 1 and ord (sc) eq 1).. E_bat(sc,t) = e = E_bat_zero_1(sc) +
E_in_bat(sc,t)-E_bat_out(sc,t);
*****************battery balance for t = 1 for each RP > 1*****************
Balance_bat_to_1(sc,t)$(ord(t) eq 1 and ord (sc) gt 1).. E_bat(sc,t) = e = E_bat_zero_1(sc) +
E_in_bat(sc,t)-E_bat_out(sc,t);
*****************Battery balance for t > 1***************************
Balance_bat(sc,t)$(ord(t) gt 1).. E_bat(sc,t) = e = E_bat(sc,t-1) + E_in_bat(sc,t)-E_bat_out(sc,t)
*******energy accumulated in battery in periods t = 1 RP = 1**************
bat_ac_link_1(sc)$(ord(sc) eq 1).. E_bat_zero_1(sc) = e = 0;
*****energy accumulated in battery in periods t = 1 RP > 1*****************
bat_ac_link_2(sc)$(ord(sc) gt 1).. E_bat_zero_1(sc) = e = E_bat(sc-1,'24');
out_battteries(sc,t).. E_bat_out_supp(sc,t) = e = E_bat_out(sc,t)*N_eff_discharge;
*****************cost of batteries***************************
Batery_capacity(sc,t).. E_bat_d = g = E_bat(sc,t);
inv_bat.. C_bat = e = (230*E_bat_d)/1e6;
op_cost_bat(sc,t).. C_op_bat(sc,t) = e = ((0.015*E_bat(sc,t)))/1e6;
op_cost_bat_prom(sc).. C_op_bat_M(sc) = e = sum((t),C_op_bat(sc,t));
op_cost_bat_T.. C_op_bat_T = e = sum((sc),C_op_bat_M(sc)*prob(sc));
Pot_importada(sc,t).. Pot_disp(sc,t) = e = Pot_imp(sc,t) + Pot_no_imp(sc,t);
pot_importada_prom(t).. Pot_imp_prom(t) = e = sum((sc),Pot_imp(sc,t)*prob(sc));
Pot_import_M(sc).. Pot_imp_M(sc) = e = sum((t),Pot_imp(sc,t));
Pot_import_T.. Pot_imp_T = e = sum((sc),Pot_imp_M(sc)*prob(sc));
costo_imp.. C_pimp = e = sum((sc),10000*Pot_imp_T*365)/1e6;
************************demand supply***************** +

demand_supply(sc,t).. (E_pv_supp(sc,t) + E_wind_supp(sc,t) + E_bat_out_supp(sc,t))/1000 + Pot_imp(sc,t) = g = E_demand(sc,t);

annual_cost..   TAC = e = 0.3333*(C_bat + C_wind + C_inv_pv) + (C_op_bat_T + C_op_wind_T + C_op_pv_T)*365 + C_pimp;

model equation /all/;

option MIP = CPLEX;

solve equation using MIP MINImizing TAC;

The optimal solution for this problem is 145.4 $ MMUSD/y, and consists of a system with 7344 panels, 60 wind turbines, and a battery with a capacity of 758 MWh.

Exercises

P14.1 Calculate 12 representative periods for the data series available in the document Cap_RP. Then, obtain different samples of 12-RP by replicating 10 times the k-means code shown in Example 14.3. Compare the results.

P14.2 Solve Example 14.3b using the 12-RP calculated in the previous problem. Compare the configurations obtained and determine the number of units that remain idle over time.

References

Allman, A., Palys, M.J., Daoutidis, P., 2019. Scheduling-informed optimal design of systems with time-varying operation: a wind-powered ammonia case study. AIChE J. 65 (7), e16434.

Baringo, L., Conejo, A.J., 2013. Correlated wind-power production and electric load scenarios for investment decisions. Appl. Energy 101, 475–482.

Ben-Tal, A., El Ghaoui, L., Nemirovski, A., 2009. Robust Optimization. Princeton Series in Applied Mathematics. Princeton University Press.

Birge, J.R., Louveaux, F., 2011. Introduction to Stochastic Programming. Springer.

Douglas, J.M., 1988. Conceptual Design of Chemical Processes. McGraw-Hill, New York.

Friedman, Y., Reklaitis, G., 1975a. Flexible solutions to linear programs under uncertainty - inequality constraints. AIChE J. 21 (1), 77–83.

Friedman, Y., Reklaitis, G., 1975b. Flexible solutions to linear programs under uncertainty - equality constraints. AIChE J. 21 (1), 83–90.

Grossmann, I., Sargent, R., 1978. Optimum design of chemical-plants with uncertain parameters. AIChE J. 24 (6), 1021–1028.

Grossmann, I.E., Halemane, K.P., Swaney, R.E., 1983. Optimization strategies for flexible chemical processes. Comp. Chem. Eng. 7 (4), 439–462.

Grossmann, I.E., Apap, R.M., Calfa, B.A., Garcia-Herreros, P., Zhang, Q., 2017. Mathematical programming techniques for optimization under uncertainty and their application in process systems engineering. Theor. Found. Chem. Eng. 51 (6), 893–909.

Halemane, K.P., Grossmann, I.E., 1983. Optimal process design under uncertainty. AIChE J. 29 (3), 425–433.

Heuberger, C.F., Staffell, I., Shah, N., Mac Dowell, N., 2017. A systems approach to quantifying the value of power generation and energy storage technologies in future electricity networks. Comp. Chem. Eng. 107, 247–256.

Kall, P., Mayer, J., 2011. Stochastic Linear Programming. Springer, New York.

Kostevšek, A., Petek, J., Čuček, L., Klemeš, J.J., Varbanov, P.S., 2015. Locally Integrated Energy Sectors supported by renewable network management within municipalities. Appl. Therm. Eng. 89, 1014–1022.

Li, C., Grossmann, I.E., 2021. A review of stochastic programming methods for optimization of process systems under uncertainty. Front. Chem. Eng. 2, 34.

Lima, R.M., Conejo, A.J., Langodan, S., Hoteit, I., Knio, O.M., 2018. Risk-averse formulations and methods for a virtual power plant. Comput. Oper. Res. 96, 350–373.

Malheiro, A., Castro, P.M., Lima, R.M., Estanqueiro, A., 2015. Integrated sizing and scheduling of wind/PV/diesel/battery isolated systems. Renew. Energ. 83, 646–657.

Mallapragada, D., Papageorgiou, D., Venkatesh, A., Lara, C.L., Grossmann, I.E., 2018. Impact of model resolution on scenario outcomes for electricity sector system expansion. Energy 163, 1231–1244.

Martín, M., Davis, W., 2015. Integration of wind, solar and biomass over a year for the constant production of CH_4 from CO_2 and water. Comp. Chem. Eng. 84, 314–325.

Martín, M., Grossmann, I.E., 2018. Optimal integration of renewable based processes for fuels and power production: Spain case study. Appl. Energ. 213, 595–610.

Nahmmacher, P., Schmid, E., Hirth, L., Knopf, B., 2016. Carpe diem: a novel approach to select representative days for long-term power system modeling. Energy 112, 430–442.

Oh, S.Y., Binns, M., Yeo, Y.K., Kim, J.K., 2014. Improving energy efficiency for local energy systems. Appl. Energ. 131, 26–39.

Palys, M.J., Wang, H., Zhang, Q., Daoutidis, P., 2021. Renewable ammonia for sustainable energy and agriculture: vision and systems engineering opportunities. Curr. Opin. Chem. Eng. 31, 100667.

Pérez-Uresti, S.I., Martín, M., Jiménez-Gutiérrez, 2019. A. Superstructure approach for the design of renewable-based utility plants. Comp. Chem. Eng. 123, 371–388.

Pérez Uresti, S.I., Martín, M., Jiménez-Gutiérrez, A., 2020. A methodology for the design of flexible renewable-based utility plants. ACS Sust. Chem. Eng. 8, 4580–4597.

Pistikopoulos, S., Grossmann, I., 1989. Optimal retrofit design for improving process flexibility in nonlinear-systems. 1. Fixed degree of flexibility. Computers & Chem. Eng. 13 (9), 1003–1016.

Rockafellar R.T., 2007. Coherent Approaches to Risk in Optimization Under Uncertainty, INFORMS, chap. 4, 38–61.

Rockafellar, R.T., Uryasev, S., 2000. Optimization of conditional value-at-risk. J. Risk. 2, 21–42.

Saeedi, M., Moradi, M., Hosseini, M., Emamifar, A., Ghadimi, N., 2019. Robust optimization based optimal chiller loading under cooling demand uncertainty. Appl. Therm. Eng. 148, 1081–1091.

Sahinidis, N., 2004. Optimization under uncertainty: state-of-the-art and opportunities. Comput. Chem. Eng. 28 (6–7), 971–983.

Sánchez, A., Martín, M., 2018. Optimal renewable production of ammonia from water and air. J. Clean. Prod. 178, 325–342.

Swaney, R., Grossmann, I., 1985. An index for operational flexibility in chemical process design. 1. Formulation and theory. AIChE J. 31 (4), 621–630.

Takamatsu, T., Hashimoto, I., Shioya, S., 1973. On design margin for process system with parameter uncertainty. J. Chem. Eng. Jpn. 6 (5), 453–457.

Teichgraeber, H., Brandt, A.R., 2019. Clustering methods to find representative periods for the optimization of energy systems: an initial framework and comparison. Appl. energy 239, 1283–1293.

Tejada-Arango, D.A., Wogrin, S., Siddiqui, A.S., Centeno, E., 2019. Opportunity cost including short-term energy storage in hydrothermal dispatch models using a linked representative periods approach. Energy 188, 116079.

Zhang, Q., Grossmann, I.E., Lima, R.M., 2016a. On the relation between flexibility analysis and robust optimization for linear systems. AIChE J. 62 (9), 3109–3123.

Zhang, Q., Morari, M.F., Grossmann, I.E., Sundaramoorthy, A., Pinto, J.M., 2016b. An adjustable robust optimization approach to scheduling of continuous industrial processes providing interruptible load. Comp. Chem. Eng. 86, 106–119.

Energy storage

Mariana Corengia and Ana I. Torres

Universidad de la República, Montevideo, Uruguay

15.1 Introduction

The worldwide effort in decarbonization of the economy has largely increased the exploitation of modern renewable energy sources (RES) such as wind and sunlight. Currently, the electricity generation sector has the highest share of renewable use: according to REN21 (2020) 27.3% of the global electricity generated in 2019 came from these renewable sources and 32 countries had at least 10 GW of generation capacity.

Despite its success, integration of these sources into already existing power grids poses challenges that are still not solved. Of interest to this chapter is the production of an excess of energy: in regions in which the power mix is close to 100% renewable, for example, Costa Rica, Uruguay, Norway, Sweden, failures due to the nonprogrammable nature of modern renewable sources have been avoided by installing a large generation capacity. Oversized capacities imply times in which electricity from RES is available in excess of the needs. The simplest way to deal with this surplus is curtailment: "a reduction in the output of a generator from what it could otherwise produce given available resources (e.g., wind or sunlight)" (Bird et al., 2014). Curtailment translates in higher costs for each unit of electric energy that is fed into the grid, and also, each unit of electric energy that is curtailed may be seen as a unit of conventional (fossil) energy whose usage was not avoided (Denholm et al., 2015). Curtailment can be discouraged by policy that favors the use of nonprogrammable RES over other power sources. This has proved to be successful for rather low nonprogrammable RES penetration [e.g., ≤20% in the European Union (Bird et al., 2016)]. However, these actions have a limit: if high nonprogrammable RES penetrations are sought, favoring them over other sources will not be enough to prevent curtailment and other strategies are needed.

The main goal of developing energy storage systems (ESS) is to make better use of RES power generation capacities and thus to prevent curtailment. The term "Energy Storage Systems" was initially conceived as systems in which electric energy is converted into a

Sustainable Design for Renewable Processes
DOI: https://doi.org/10.1016/B978-0-12-824324-4.00028-7

different type of energy and stored in this second type until being, at a later time, converted back into electric energy. This type of systems is referred to as power-to-power (P2P) systems, regardless of how the energy is stored. Storage in batteries is a major example in this sense, and can be centralized by power companies, or decentralized using commercial home-battery storage units (see options in Tesla, 2020; Solar Harmonics, 2020).

A broader definition of energy storage does not require conversion back to electric energy. If the energy is both stored and later used as heat we have power-to-heat (P2H) systems. Energy in excess can also be stored in chemical bonds. The most salient example is the production of hydrogen, an important reactant in the industrial sector, currently produced through reforming of gas or petroleum cuts. Powering water electrolyzers using RES offers both a green path for hydrogen production and a sink to surpluses of nonprogrammable RES. If this hydrogen will not be used in P2P applications, we have power-to-gas (P2G) systems. P2G systems include the production of hydrogen for transportation purposes. Another example is ammonia which can be produced for both its classical use as a fertilizer or as an energy carrier (fuel or hydrogen storage, see, e.g., Palys et al., 2019; Palys and Daoutidis, 2020; Allman et al., 2017; Demirhan et al., 2019). Similarly, the use of RES has been proposed in the production of synthetic fuels such as gasoline and diesel (Zhang et al., 2019a), or chemical compounds that may be used as fuels or as reactants in other processes: methane (Davis and Martín, 2014), methanol (Martín and Grossmann, 2017), dimethyl ether (Martín, 2016) (see Chapter 11: Wind Energy: Collection and Transformation, for hydrogen use in a chemical process). In general, when we talk about Power to X (P2X) systems we refer to systems in which the X may be a fuel, a chemical, or something else, that will not be converted back to electric power, except for transportation purposes. In here we should mention that the boundaries between these definitions are not 100% established, what we have discussed is the most common use rather than a clear-cut definition.

Many times ESS are combined with another generation/demand matching strategy known as demand response programs. Demand response programs, seek a change in electricity consumption patterns (either from industrial, commercial, or residential customers) in response to changes in the price of electricity. There are different ways to stimulate this change, Yan et al. (2018) classified them according to whether there is a price signal over time (price-driven) or if there are incentives for meeting goals at certain moments (event-driven). Each class has its own subcategories. For instance, in the price-driven strategies, the price signal can be quite static with repeated variations through days (and sometimes seasons) or it can vary continuously in response to the wholesale market prices or the mismatch between generation and consumption. The former are usually referred to as time-of-use (TOU) tariffs, whereas the latter are generally called real-time-pricing (RTP) programs.

An ESS specific to industrial processes that consume large amounts of electricity, is production rescheduling to match TOU or RTP programs. These processes may be seen as energy storage entities and the products they produce as the energy "carriers." This type of ESS has been recently analyzed. For instance, Baldea and coworkers have studied the effect of the participation in these pricing programs for different energy-intensive industrial processes. In Seo et al. (2020), possible cost and emission reductions in the glass industry were analyzed by combining energy inputs in glass furnaces from natural gas combustion and day-ahead electric programs. In Otashu and Baldea (2019), the authors take advantage of the fast kinetics of electrochemical processes to match the chlor-alkali process to a 15-minute price step demand response program. In this case, the authors consider a known horizon pricing for 3 consecutive days. Beyond the known prices in the day-ahead markets, a general stochastic approach was developed in Kelley et al. (2020) to account for electric energy cost uncertainties (and also product demand), and applied to industrial air separation units. As pointed out by Zhang and Grossmann (2016), process complexity and safety requirements may be the limiting factors to how much energy consumption can be time-shifted in demand side management at industrial scale.

The goal of this chapter is to introduce the reader on the formulation of optimization problems to select the best energy storage solution in different settings. In doing so, we will take a superstructure-based approach. In Section 15.2 we consider a P2P generic superstructure and divide it in stages. We identify the alternatives that are available at each stage and postulate the equations that link the stages and will act as the constraints of the proposed optimization problem. Although the focus is on P2P, many of what is discussed can be easily extended to analyze generic P2X systems.

For some of the stages, it may be worth considering rigorous models. These rigorous models may be very involved and discussing them in detail may well be worth a chapter for each. Thus in Section 15.3 instead of an in-depth discussion, we comment on what we consider the key points that should be taken into account. In doing so, we provide some of the equations that an undergraduate ChemE student may not be familiar with and leave out what we consider to be usually covered in undergraduate programs. We provide references to several papers that study each system in detail for further study.

Finally, a series of case studies are presented: Section 15.4 discusses problem formulations for finding optimal charge/discharge schedules in self-storage applications based on batteries; Section 15.5 discusses the optimal design of a system that uses energy surpluses for the production of hydrogen.

15.2 Systems for energy storage from the design perspective

Fig. 15.1 shows the general diagram with the options for exploitation of RES in P2P settings: some of the energy is directly fed to the users (via the grid), some of the energy is stored for later use, some of the energy is curtailed.

As electricity cannot be stored as such, it has to be converted to another type of energy. Fig. 15.1 shows the options that are currently used. As seen, conversion to mechanical energy for pumping water is the most popular method and roughly accounts for 98% of the 173 GW worldwide electric power storage capabilities (2018 data from Mongird et al., 2019). In this system, power surpluses from the grid are employed to activate a water pump, that moves water from a lower reservoir to a higher one (see Chapter 13: Water as a Resource: Renewable Energies and Technologies for Brine Revalorization, for details on the technology). The energy is thereby stored as gravitational (potential) energy, and can generate electric power, whenever needed, by flowing through a hydro-turbine. These systems are established and trustworthy technologies, with relatively fast dynamics, so their

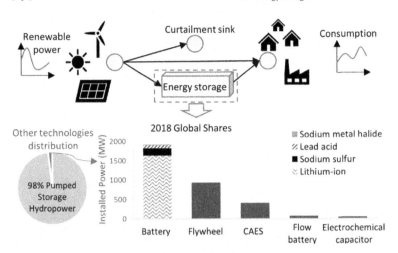

FIGURE 15.1   Top: Use of renewable energy in power-to-power settings schedule. Bottom: Currently installed options for energy storage. Source: *Data from Mongird, K., Fotedar, V., Viswanathan, V., Koritarov, V., Balducci, P., Hadjerioua, B., et al., 2019. Energy storage technology and cost characterization report. Tech. Rep. US Department of Energy.*

high share in already installed ESS should not be a surprise. They represent the energy storage workhorse as can be used for large-scale storage (GWh to TWh) for long periods of time (months) (Aneke and Wang, 2016).

Apart from pumped-hydro, batteries, flywheels, and compressed air energy storage (CAES) (in the order of percentage of share) are important energy storage devices. Batteries make use of reversible reactions to store energy in chemical bonds: electric energy drives a nonspontaneous reaction, and is released when the spontaneous one occurs; flywheels are mechanical devices that store kinetic (rotational) energy: they are accelerated by input power, and slow down when providing power; CAES uses electric power to compress air: the compressed air can be later expanded in a pneumatic air motor coupled to a power generator. Flow batteries and electrochemical capacitors have a much lower participation in current grid storage. Note that we have separated flow batteries from the "battery" group. We will describe these devices in more detail later in the chapter and the rationale of this separation will become clear at that point. Storage time and size scales differ between the energy storage options: batteries store energy in the MWh scale for days, flywheels in the kWh scale for hours, CAES in the GWh scale for days (Aneke and Wang, 2016). This difference in time and size scales can be taken advantage of by designing hybrid ESS: thinking on the variability of the nonprogrammable RES, it is clear that a single system will not be able to capture all the surpluses; ESS that combine devices that guarantee a base storage with devices that capture the generation peaks, may do the job.

Applying a systems thinking, the energy storage options can be organized in a super-structure such as the one shown in Fig. 15.2. In here, we have identified the sources (generation of renewable energy), the sinks (consumers), three main stages (charge, storage, and discharge) and two auxiliary conversion stages.

On the left side of the figure, RES generation is either in alternating current (AC, e.g., wind turbines) or in direct current (DC, e.g., solar panels), whereas charge may also either be AC (e.g., pumps) or DC (e.g., batteries). Then, the first auxiliary conversion stage is required for AC/DC or DC/AC as well as voltage and frequency adjustments.

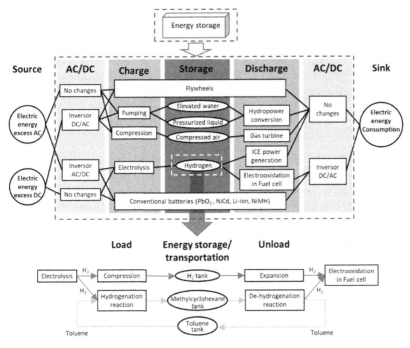

FIGURE 15.2   Top: Example of power-to-power superstructure for energy storage. Bottom: Nested structure with example of hydrogen storage alternatives.

These changes are within the electrical energy domain. Changing from electrical energy to a second type of energy that can be stored is called the charging stage (Aneke and Wang, 2016). The storage stage per se, is the one that keeps the energy at this second type for a certain amount of time, thus allowing for time shifting the energy. At the discharge stage, this second type of energy is changed back to electric energy (in P2P applications). Consumers may uptake AC-based electricity (standard distribution in electric grids, as well as conventional power generation) or DC-based electricity (LED lighting, batteries for small portable electronic devices, computers, and electric vehicles) so auxiliary conversions within the electrical energy domain are also required to couple consumers with the discharge stage.

As in any superstructure, multiple alternatives are available at any stage. In the figure, the blocks within the stages represent some of these alternatives. These individual blocks have to be seen as a family of processes rather than single operations. This is, we could zoom in into each of the blocks to find nested superstructures with also several alternatives. In the figure, this idea is schematized for the case of hydrogen storage, showing two of the many possible options.

Charge/discharge stages are closely related to the storage stage, and in some cases (batteries, flywheels, and super-capacitors) they even belong to the same physical unit. These cases are represented in Fig. 15.2 as a single block that spans charge/storage/discharge stages.

8. Integration of resources

Finding the best ESS can be interpreted in different ways, but in all cases it involves choosing the best(s) pathway(s) to connect given generation and consumption patterns for a chosen objective function. In what follows, we will present the equations that model a generic superstructure such as the one in Fig. 15.2 at the network level. In the derivation of the equations we will look for systematization of the analysis; as will be seen in the examples and exercises many times the models can be greatly simplified.

15.2.1 Modeling the energy storage systems superstructure

We can think of the ESS superstructure as a directed graph in which energy flows through different nodes. We can distinguish nodes of different types and relate them to individual stages: sources (power generators/grid), sinks (consumers/curtailment), nodes whose main goal is to combine and/or distribute energy of the same type (not included in Fig. 15.2, but omnipresent), nodes whose main goal is to change the type of energy (charge/discharge), nodes whose main goal is to delay a signal (storage). Each node has a characteristic set of terms to relate input-to-output flows.

15.2.1.1 Source nodes

The source nodes represent the electric energy generators, that is, wind turbines, water turbines, photovoltaic panels, etc. We might be interested in the energy generated by them, or in the energy that is generated in excess. At this point, it should be mentioned that there is not a unique way to define excess, surplus, or curtailment. Thus some care must be taken to properly account for them. Bird et al. (2016), Zheng et al. (2019), and Cui et al. (2020) can be consulted to check some of the definitions that have been used. To establish a common ground, we need to state a definition which for this chapter will be: "surplus power is the power that could have been generated by a nonprogrammable energy source but was not."

Fig. 15.3 schematizes the idea: both programmable and nonprogrammable generators have a maximum installed capacity IC_G (dotted lines in the figure). At any time the power actually demanded from the generator is represented by the dashed lines in the figure. The difference between installed capacity and demand (shaded area in the right figure) is

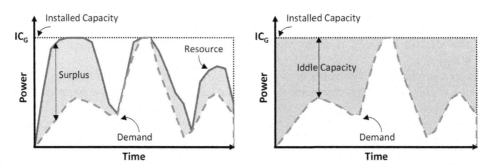

FIGURE 15.3 Definition of surplus: The left plot corresponds to a nonprogrammable generator, the right plot to a programmable one. In both the dashed line is the power demanded by the consumer and the straight line at the top, the capacity of the generator. The solid line on the right plot represents the availability of the resource. Surplus is the difference between the power that would be available from the resource and the power that is actually consumed.

the idle capacity, which in this example is the same for both cases. In a nonprogrammable generator there is also a pattern for the availability of the resource which is represented by the solid line in the left figure. This availability of the resource limits the use of the installed capacity. In nonprogrammable generators the resource that is not used at the time when it is available (shaded area in the left figure), is lost. In programmable sources the resource (which may be a fuel in thermal power stations) that is available, but not used at a certain time, can be saved for later use. Thus programmable sources do not have surpluses, although their capacity may not be fully used. The key point here is to distinguish that not fully using the equipment turns into idle capacity while not fully using the resource turns into a surplus.

Mathematically, the energy flow out of the source node can be represented as a vector whose elements are powers that may vary with time t. Depending on the problem, these powers may be given by the availability of the resources (solid lines) or the surpluses (the difference between solid and dashed lines). Selection of one or the other approach is purely by convenience. For example, if we take the first approach, the equation for the source node of generator G_k is:

$$P_{G_{k,t}}^{max} = \begin{cases} IC_{G_k} & \text{if } k \text{ programmable} \\ IC_{G_k} f_{G_{k,t}} & \text{if } k \text{ nonprogrammable} \end{cases} \tag{15.1}$$

where $P_{G_{k,t}}^{max}$ is the maximum power achievable by generator G_k at time t, IC_{G_k} is the installed capacity of generator G_k and $f_{G_{k,t}}$ is a factor that accounts for the availability of the resource at each time t. $P_{G_{k,t}}^{max}$ caps $P_{G_{k,t}}^{gen}$ which is the power actually produced by generator G_k

$$P_{G_{k,t}}^{gen} \leq P_{G_{k,t}}^{max} \tag{15.2}$$

If generator k is nonprogrammable, the surplus at time t ($P_{G_{k,t}}^{spls}$) is defined as the difference between available and produced power:

$$P_{G_{k,t}}^{spls} = P_{G_{k,t}}^{max} - P_{G_{k,t}}^{gen} \tag{15.3}$$

See Example 15.1.

15.2.1.2 Charge-stage nodes

Charging units, C_l, convert a flow of electric energy into a flow of another type of energy (chemical, mechanical, etc.). These nodes are characterized by a maximum conversion capacity and an efficiency. The maximum conversion capacity is given by the installed capacity of the charging unit IC_{C_l} which is a design variable of the equipment. The efficiency $\eta_{C_{l,t}}$ accounts for energy losses and depends on the specific unit and how it is operated; it may for example vary with the inlet power, thus it may also change over time. Each charging node thus adds the following equations to the model of the superstructure.

$$P_{C_{l,t}}^{in} \leq IC_{C_l} \tag{15.4}$$

$$\eta_{C_{l,t}} = f\left(P_{C_{l,t}}^{in}, \text{others}, \dots\right) \tag{15.5}$$

EXAMPLE 15.1 Generation and consumption curves.

Consider an isolated renewable energy grid composed by three nonprogrammable generators (one wind farm and two solar farms) and one programmable generator (a dam).

The wind farm has installed 24 windmills, 2 MW each. The solar farms have installed capacities of 50 and 20 MW each. The dam has three hydro-turbines, each with a maximum capacity of 30 MW. The grid supplies power to a block of buildings, and the community is considering the installation of an ESS. Where do we start?

The first thing is to graphically represent the system. Figure 15.E1.1 shows that representation, following the rationale of Fig. 15.2 and notation in Section 15.2.1.1 for installed capacity, power generation, and surplus. Note that hydro-power does not have a surplus, this is valid if we assume that the dam is not full (but has enough water to complete the supply the demand).

If the system is going to supply a community, then the power generated by the sources has to match the demand at each time t, in Fig. 15E1.1 this is represented by the linking node. The following equation must hold at the node: $P_{G_{wind,t}}^{gen} + P_{G_{sol1,t}}^{gen} + P_{G_{sol2,t}}^{gen} + P_{G_{hydro,t}}^{gen} = P_{U_{1,t}}^{dem}$.

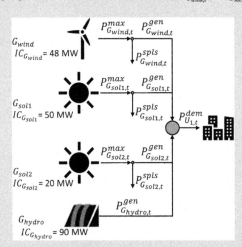

FIGURE 15E1.1 System diagram for the example with different types of generators and no energy storage systems.

To obtain the values for $P_{G_{k,t}}^{max}$ we must gather information on the availability of the nonprogrammable sources. This data is usually collected by government agencies. Fig. 15E1.2 shows typical daily production factors f for wind and solar energy. Then, $P_{G_{k,t}}^{max} = f_{G_{k,t}} IC_{G_k}$. The surplus for each of these generators is found as the difference as stated in Eq. (15.3). Fig. 15E1.3 shows the demand curve (dashed *line* noted as User 1) and a composite of the maximum powers of the three nonprogrammable sources (i.e., $\sum_k P_{G_{k,t}}^{max}$).

If we consider a strategy that maximizes the use of the nonprogrammable resources, then all the energy corresponding to the colored area below the demand line will be actually turn into power. The energy over the gray line cannot be used, thus it will be curtailed. Notice that in this example, solar energy is never curtailed. Also note that, even when the installed capacity of nonprogrammable sources (118 MW) is larger than the consumption peak (106 MW), $\sum_k P_{G_{k,t}}^{max}$ is not, then there exists a mismatch between production and demand that requires the use of the programmable source during many hours.

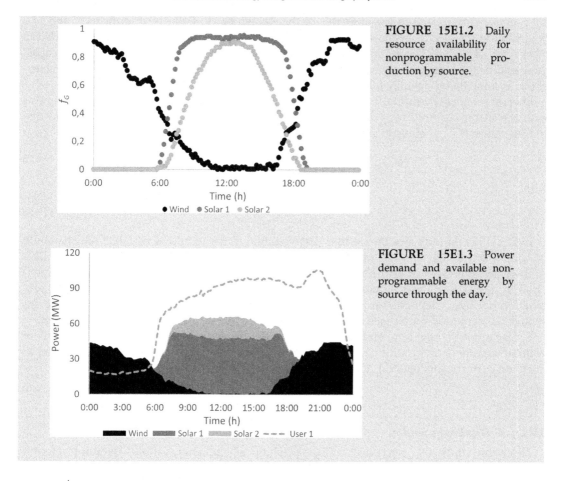

FIGURE 15E1.2 Daily resource availability for nonprogrammable production by source.

FIGURE 15E1.3 Power demand and available nonprogrammable energy by source through the day.

where $P^{in}_{C_l,t}$ is the (electric) power input to node C_l at time t. The energy flow exiting the node is then computed as:

$$\dot{E}^{out}_{C_l,t} = P^{in}_{C_l,t} \eta_{C_l,t} \tag{15.6}$$

Notice that we use $\dot{E}^{out}_{C_l,t}$ to denote this nonelectrical energy flow exiting node C_l as it reinforces the idea that there is a conversion of energy. We will reserve P (power) for electric energy flows.

15.2.1.3 Storage nodes

As mentioned before, storage units S_m time shift the energy but do not involve energy conversion. These units are characterized by the maximum amount of energy, mass or volume that can be stored (i.e., installed capacity IC_{S_m}); the state of charge ($SoC_{S_m,t}$) which is defined as the energy, mass, or volume, that is available in the storage unit at a certain time, generally it is reported as a fraction of the installed capacity ($x^{SoC}_{S_m,t}$, see Eq. 15.8) and the self-discharge processes ($SD_{S_m,t}$) which account for energy, mass, or volume losses as the storage time goes by.

SD processes are related to the efficiency of the storage block. Both these terms (*SoC* and *SD*) originated in the electrochemical field, mostly due to batteries, but they have been extended to other types of storage. For example in storage as elevated water, the installed (maximum) capacity would be the nominal volume of the reservoir tank, and the state of charge (*SoC*) the fraction of this volume that is actually filled with water; *SD* in this example would be a leak.

Energy (or mass/volume) balances link the *SoC* of the storage units at different times t with the incoming flow ($\dot{E}^{in}_{S_{m,t}}$ if energy) from the charge stages, the outgoing flow to discharge stages ($\dot{E}^{out}_{S_{m,t}}$ if energy) and *SD* phenomena if any as written in Eq. (15.7):

$$SoC_{S_{m,t}} = SoC_{S_{m,t-1}} + \Delta t \dot{E}^{in}_{S_{m,t}} - \Delta t \dot{E}^{out}_{S_{m,t}} - SD_{S_{m,t}} \tag{15.7}$$

with

$$SoC_{S_{m,t}} = x^{SoC}_{S_{m,t}} IC_{S_m} \tag{15.8}$$

In Eq. (15.7) Δt indicates the desired discretization in the temporal variable, which should be chosen on the basis of the practicality and solvability of the optimization problem. This is, if we are interested in year-long behaviors suitable Δt's could be in hours, if we are interested in intraday analysis suitable Δt's could be minutes. As in the previous node the efficiency term $SD_{S_{m,t}}$ depends on the specific unit and how it is operated, which phenomena should be accounted for in each specific storage unit will be discussed in the next section.

SoCs cannot be negative, and are limited by the maximum amount of energy, mass, or volume that the storage unit can handle, these limits may sometimes need to be added as inequality constraints (see Eq. 15.9).

$$0 \le SoC_{S_{m,t}} \le IC_{S_m} \tag{15.9}$$

15.2.1.4 Discharge nodes

Discharge units, D_n, convert chemical, mechanical, etc. energy flows back into electric energy flows. Their models are analog but opposite to those described for charging units. Thus the equations are:

$$P^{out}_{D_{n,t}} \le IC_{D_n} \tag{15.10}$$

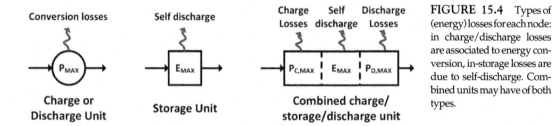

Conversion losses

Charge or Discharge Unit

Self discharge

Storage Unit

Charge   Self   Discharge
Losses discharge Losses

Combined charge/ storage/discharge unit

FIGURE 15.4 Types of (energy) losses for each node: in charge/discharge losses are associated to energy conversion, in-storage losses are due to self-discharge. Combined units may have of both types.

$$\dot{E}_{D_{n,t}}^{in} \eta_{D_{n,t}} = P_{D_{n,t}}^{out} \tag{15.11}$$

$$\eta_{D_{n,t}} = f\left(\dot{E}_{D_{n,t}}^{in}, others, \ldots\right) \tag{15.12}$$

15.2.1.5 Sink nodes

Analogously to sources, sinks are represented as vectors whose elements are powers that may vary with time t. Sink nodes are either the users (of the electric system) or the surplus of energy that is curtailed. Note that the latter were already introduced in Section 15.2.1.1 as possible sources. This is not a mistake, depending on the specific problem it may be convenient to take the surplus as the source, for example, when the objective is to maximize their use, however, even in this case some of the surplus may need to be curtailed, thus a sink node for "leftover surpluses" should always be included in the formulation. Demand (or users) node was also already introduced in Example 15.1: we consider that each user U_h, has a power demand pattern, $P_{U_{h,t}}^{dem}$, that has to be met.

15.2.1.6 Linking nodes

These are not real units but auxiliary nodes that are needed to build the superstructure, and allow connections between the different units in physically meaningful ways. This is, for example the energy flow from a hydrogen storage unit cannot be used to feed a hydro-turbine. These nodes collect/distribute energy flows they do not convert, store or lose energy, thus the equation is:

$$\sum \dot{E}^{in} = \sum \dot{E}^{out} \tag{15.13}$$

or

$$\sum P^{in} = \sum P^{out} \tag{15.14}$$

In the superstructure they should be placed between generators and charging stages (Eq. 15.14 applies), between charging stages and storage (Eq. 15.13 applies), between storage and discharge (Eq. 15.13 applies), and between discharge and users (Eq. 15.14 applies). An example of this node was also introduced in Example 15.1.

15.2.1.7 Modeling with units that combine stages

The previous equations allow for building a generic superstructure. However, some devices combine more than one stage in one equipment. Of particular interest are those in which charge storage and discharge occur in the same unit, and single units whose operation is reversible, that is, can charge and discharge, but do not store energy. Examples of the first were already mentioned, examples of the latter include reversible solid oxide fuel cells (SOFCs), that can also work as electrolyzers, or pumps that can be reversed into turbines.

As illustrated in Fig. 15.4, the easiest way to model the first type of devices is to consider that charge/storage and storage/discharge linking nodes only have one input and one output, in addition to the specific equations for each node discussed in the previous

8. Integration of resources

paragraphs (see Example 15.2). The second type of devices require including equations to prevent simultaneous charge and discharge or the overlap of ramps-up and ramps-down dynamics when switching the operation mode. A usual way to tackle this is by introducing binary variables to separate the operation modes (see, e.g., Jiang et al., 2020; Mehrjerdi and Hemmati, 2019), but as will be seen in the example in the case study in Section 15.4 this is not always needed.

Units that combine stages may require additional equations. For example, in batteries maximum charge/discharge powers and energy storage capacities are related through a parameter known as the c_{rate}. As these additional equations depend on the specific equipment, the discussion is left to Section 15.3.

15.2.2 Formulation of the optimization problem

The equations in Section 15.2.1 represent the core of the constraints of different optimization problems that can formulated to find the best ESS. In P2P applications all stages apply as described, whereas in P2X applications the equations remain the same up to the storage stage and some modifications may be needed in downstream stages. For example, in case of P2G the discharge stage, as an energy conversion step, is bypassed, and the hydrogen requirements are modeled as the user sink. In some cases it could be convenient to couple the superstructure to another one that includes various possible uses for the X.

Binary variables should be assigned to each unit at each stage to select them for the optimal structure or not. As mentioned, logical constraints may need to be added to account for equipment that combine stages, as well as to assign relations between units, force the presence of one or at least one unit for each stage, or eventually limit the number of options selected at each stage. The presence of logical constraints with continuous variables is common in these superstructures; thus additional big-M or convex-hull constraints may also be needed to avoid terms containing products (see Biegler et al., 1997).

Common goals in ESS design problems include minimizing costs, minimizing curtailment, or satisfying a certain demand using renewable sources. Including environmental or safety goals is pertinent as for example many batteries involve hazardous chemicals. These goals can be written as a single objective function choosing appropriate weights for the different terms, or a main objective function may be chosen and the rest of the goals left as constraints. Pareto frontiers may be found by solving the problem multiple times.

Another possible objective for an optimization problem, commonly used when dealing with batteries, is how to optimally operate the system. This is formulated as a scheduling problem as will be discussed in Section 15.4.

The decision variables vary depending on the goal of the problem: in design problems common variables are the presence or not of the different equipment (binaries), their capacities (continuous) and time-dependent operation variables such as $P_{G_{k,t}}^{gen}$, $P_{C_{l,t}}^{in}$, $SoC_{S_{m,t}}$, and $\dot{E}_{D_{n,t}}^{in}$ (continuous); scheduling problems only have the latter.

Nonlinear constraints are common in formulations that have capacity as a decision variable or when detailed equations to relate variables are employed. For example, as mentioned in Section 15.2.1.2, efficiencies ($\eta_{C_{l,t}}$ and $\eta_{D_{n,t}}$) can be considered as constant

EXAMPLE 15.2 Equations to model a superstructure at the network level.

Consider the system schematized in Fig. 15E2.1, it represents a superstructure for an ESS with only one nonprogrammable energy source (G_1), and one consumer (a community U_1). The community wants to install an ESS but also needs to be able to consume directly from the grid: this is represented with the bypass P_t^{direct}. If at any time the maximum possible power generated by G_1 ($P_{G_1,t}^{max}$) is larger than the power demand ($P_{U_1,t}^{dem}$) the surplus is curtailed ($P_{G_1,t}^{spls}$). This is represented with the node *Surplus*.

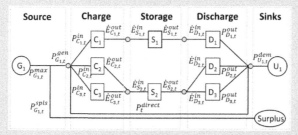

FIGURE 15E2.1   Energy storage systems superstructure for Example 15.2.

Two types of energy storage solutions are considered. The first one is a battery (represented by nodes C_1, S_1, D_1, and two linking nodes between charge/storage and storage/discharge); the second one is a hydrogen-based system which requires one electrolyzer, a system of storage tanks (S_2), and one FC. Two types of electrolyzers (C_2 and C_3), and two types of FCs (D_2 and D_3) are available in the market.

We will formulate the equations that model this superstructure at the network level, following the same order in which they were discussed in Section 15.2.1. The generation node was already discussed in Example 15.1 and here it is assumed that $P_{G_1,t}^{max}$ is known.

As stated in Section 15.2.1.2 each charge node has two characteristic parameters: maximum capacity (IC_C) and efficiency $\eta_{C,t}$. Then the equations for these nodes are:

$$0 \leq P_{C_1,t}^{in} \leq IC_{C_1} \tag{15E2.1}$$

$$\dot{E}_{C_1,t}^{out} = P_{C_1,t}^{in} \eta_{C_1,t} \tag{15E2.2}$$

$$0 \leq P_{C_2,t}^{in} \leq IC_{C_2} \tag{15E2.3}$$

$$\dot{E}_{C_2,t}^{out} = P_{C_2,t}^{in} \eta_{C_2,t} \tag{15E2.4}$$

$$0 \leq P_{C_3,t}^{in} \leq IC_{C_3} \tag{15E2.5}$$

$$\dot{E}_{C_3,t}^{out} = P_{C_3,t}^{in} \eta_{C_3,t} \tag{15E2.6}$$

As stated in Section 15.2.1.3, storage units are characterized by the maximum amount of energy that can be stored, their *SoC*, and losses due to *SD*. An energy balance accounts for the changes in the *SoC* at each storage node:

$$SoC_{S_1,t} = SoC_{S_1,t-1} + \Delta t \dot{E}_{S_1,t}^{in} - \Delta t \dot{E}_{S_1,t}^{out} - SD_{S_1,t} \tag{15E2.7}$$

$$0 \leq SoC_{S_{1,t}} \leq IC_{S_1} \tag{15E2.8}$$

$$SoC_{S_{2,t}} = SoC_{S_{2,t-1}} + \Delta t \dot{E}^{in}_{S_{2,t}} - \Delta t \dot{E}^{out}_{S_{2,t}} - SD_{S_{2,t}} \tag{15E2.9}$$

$$0 \leq SoC_{S_{2,t}} \leq IC_{S_2} \tag{15E2.10}$$

As stated in Section 15.2.1.4 the discharge nodes have two characteristic parameters: maximum capacity and efficiency. Then the equations for these nodes are:

$$0 \leq P^{out}_{D_{1,t}} \leq IC_{D_1} \tag{15E2.11}$$

$$\dot{E}^{in}_{D_{1,t}} \eta_{D_{1,t}} = P^{out}_{D_{1,t}} \tag{15E2.12}$$

$$0 \leq P^{out}_{D_{2,t}} \leq IC_{D_2} \tag{15E2.13}$$

$$\dot{E}^{in}_{D_{2,t}} \eta_{D_{2,t}} = P^{out}_{D_{2,t}} \tag{15E2.14}$$

$$0 \leq P^{out}_{D_{3,t}} \leq IC_{D_3} \tag{15E2.15}$$

$$\dot{E}^{in}_{D_{3,t}} \eta_{D_{3,t}} = P^{out}_{D_{3,t}} \tag{15E2.16}$$

The efficiencies at charge, storage, and discharge nodes can be either assumed as a parameter or related to the energy flows through rigorous models.

U_1 and *Surplus* are the sink nodes; $P^{dem}_{U_{1,t}}$ characterizes U_1 and is usually given as a consumption pattern; $P^{spls}_{G_{1,t}}$ characterizes the *Surplus* node and is computed as the difference:

$$P^{spls}_{G_{1,t}} = P^{max}_{G_{1,t}} - P^{gen}_{G_{1,t}} \tag{15E2.17}$$

There are six linking nodes. The first one connects the source node with the three charge stages and the power directly fed to the user:

$$P^{gen}_{G_{1,t}} = P^{in}_{C_{1,t}} + P^{in}_{C_{2,t}} + P^{in}_{C_{3,t}} + P^{direct}_t \tag{15E2.18}$$

The second and third ones connect charge with storage and storage with discharge nodes in the battery:

$$\dot{E}^{out}_{C_{1,t}} = \dot{E}^{in}_{S_{1,t}} \tag{15E2.19}$$

$$\dot{E}^{out}_{S_{1,t}} = \dot{E}^{in}_{D_{1,t}} \tag{15E2.20}$$

The fourth and fifth ones connect charge with storage and storage with discharge nodes for the second type of storage:

$$\dot{E}^{out}_{C_{2,t}} + \dot{E}^{out}_{C_{3,t}} = \dot{E}^{in}_{S_{2,t}} \tag{15E2.21}$$

$$E^{out}_{S_{2,t}} = E^{in}_{D_{2,t}} + E^{in}_{D_{3,t}} \tag{15E2.22}$$

The sixth node connects the user node with the three discharge stages and the power directly obtained from the energy source (G_1):

$$P_{U_{1,t}}^{dem} = P_{D_{1,t}}^{out} + P_{D_{2,t}}^{out} + P_{D_{3,t}}^{out} + P_t^{direct} \tag{15E2.23}$$

Now we should consider the logical constraints. For this, a binary variable y is defined for each of the charge storage and discharge nodes; $y = 1$ if the node is selected; $y = 0$ if the node is not. In batteries charge, storage, and discharge happen in the same unit. Then:

$$y_{C_1} = y_{S_1} = y_{D_1} \tag{15E2.24}$$

By problem statement, one and only one storage option will be considered, and if storage node S_2 is selected only one charge and only one discharge option will be considered. Then the following equations apply:

$$y_{S_1} + y_{S_2} = 1 \tag{15E2.25}$$

$$y_{S_2} = y_{C_2} + y_{C_3} \tag{15E2.26}$$

$$y_{S_2} = y_{D_2} + y_{D_3} \tag{15E2.27}$$

Equations that relate inlet/outlet energy flows for each charge, storage, or discharge node with its binary should be included. The equations should state that:

$$P_{C_{1,t}}^{in} = 0 \; \forall t \Leftrightarrow y_{C_1} = 0 \tag{15E2.28}$$

$$\dot{E}_{S_{m,t}}^{in} = 0 \; \forall t \Leftrightarrow y_{S_m} = 0 \tag{15E2.29}$$

$$\dot{E}_{D_{n,t}}^{in} = 0 \; \forall t \Leftrightarrow y_{D_n} = 0 \tag{15E2.30}$$

$$\dot{E}_{C_{1,t}}^{out} = 0 \; \forall t \Leftrightarrow y_{C_1} = 0 \tag{15E2.31}$$

$$\dot{E}_{S_{m,t}}^{out} = 0 \; \forall t \Leftrightarrow y_{S_m} = 0 \tag{15E2.32}$$

$$P_{D_{n,t}}^{out} = 0 \; \forall t \Leftrightarrow y_{D_n} = 0 \tag{15E2.33}$$

Each of these equations translates into a set of two inequality constraints. As an example for Eq. (15E2.28) the restrictions are:

$$P_{C_{1,t}}^{in} \leq M y_{C_1} \; \forall t \tag{15E2.34}$$

$$\frac{1}{M} y_{C_1} \leq \sum_t P_{C_{1,t}}^{in} \tag{15E2.35}$$

where M is an arbitrary large parameter. Notice that Eq. (15E2.34) forces all inlet energy flows to be 0 when the equipment is not selected for installation, whereas Eq. (15E2.35) allows for installing the equipment if it is used at some point (i.e., there exists a time t in which $P_{C_{1,t}}^{in} \neq 0$).

Finally, the battery cannot charge and discharge at the same time. Then a binary variable ($z_{S_{1,t}}$) is included: during charge $z_{S_{1,t}} = 1$; during discharge $z_{S_{1,t}} = 0$. Then, Eqs. (15E2.1) and (15E2.11) are replaced by:

$$0 \leq P_{C_{1,t}}^{in} \leq z_{S_{1,t}} IC_{C_1} \tag{15E2.36}$$

$$0 \leq P_{D_{1,t}}^{out} \leq \left(1 - z_{S_{1,t}}\right) IC_{D_1} \tag{15E2.37}$$

parameters or as variables that depend on charge/discharge powers. Many of these more rigorous equations lie within the expertise of chemical engineers, thus including them in the models is one of the main contributions that we can make in the design of optimal ESS. We will review some of these rigorous models in the next section.

15.3 Overview of devices and technologies: considerations for rigorous modeling

15.3.1 Electrochemical devices

In electrochemical devices electric energy is used at the charging stage to drive a non-spontaneous chemical reaction, that is, one in which the change in Gibbs free energy ΔG is positive, and products have more energy than reactants. Energy is released at the discharging stage when the spontaneous reverse reaction occurs. Fig. 15.5 schematizes the general idea. To be able to charge/discharge electrochemical devices, these must have two physically separate compartments: the anode and the cathode. To be able to store the energy in the products (A_{red} and B_{ox} in the figure) these must also be stored in physically separated compartments, otherwise the spontaneous reaction occurs without any work done. Important examples include those that use the hydrogen cycle, conventional batteries, and redox flow batteries (RFBs). Before describing them in detail, we will briefly review some basic concepts of electrochemistry. The interested reader is encouraged to consult Bockris et al. (2002) for a more in-depth discussion.

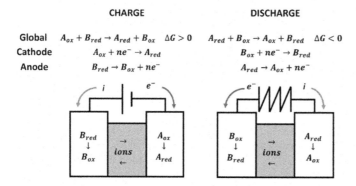

FIGURE 15.5 General diagram for charge and discharge stages in electrochemical devices including redox reactions.

15.3.1.1 General concepts

Recalling thermodynamics, the change in Gibbs free energy (ΔG) is related to the concentration of the species by Eq. (15.15):

$$\Delta G = \Delta G^0 + RT \ln Q \tag{15.15}$$

where ΔG^0 is the change in standard Gibbs free energy, T is the absolute temperature, R is the gas constant and Q is the reaction quotient.

Eq. (15.16) relates ΔG to the open cell voltage $E_{j=0}$, that is, the voltage measured between the positive and negative terminals of the cell, when no current circulates, and Eq. (15.17) relates ΔG^0 to the standard cell potential E^0 (i.e., as measured between the positive and negative terminals of the cell, when no current circulates and the activity and fugacities of the species are all one).

$$\Delta G = -nFE_{j=0} \tag{15.16}$$

$$\Delta G^0 = -nFE^0 \tag{15.17}$$

where F is the Faraday constant, a unit conversion factor: 96,485 Coulomb per mole of electrons, whereas n is the stoichiometric coefficient for the electrons as shown in Fig. 15.5. Substituting Eqs. (15.16) and (15.17) in Eq. (15.15), the open cell potential can be calculated from the standard values:

$$E_{j=0} = E^0 - \frac{RT}{nF} \ln Q \tag{15.18}$$

The reader that has taken Electrochemistry courses recognizes Eq. (15.18) as the Nernst equation. E^0 is usually computed from tabulated values of reduction potentials for each half-cell reaction (i.e., anode or cathode separately, see Zoski, 2007). As the reaction taking place in the anode is the oxidation, then we have to consider the sign opposite to the one in the table. Thus:

$$E^0 = E^0_{cathode} - E^0_{anode} \tag{15.19}$$

As current starts circulating, the potential E differs from the open circuit one by a magnitude η, the overpotential. η is simply defined as the difference in Eq. (15.20). Notice that it may have positive or negative values depending if we are in the charging or discharging process, a simple way to see which sign should it have is to think that it will always be against our interests: need more voltage than open circuit potential for charging and provides less voltage when discharging.

$$\eta = E - E_{j=0} \tag{15.20}$$

It has to be noted that the literature may sometimes be confusing on the sign, as quite frequently overpotentials are reported as positive values (see, e.g., Carmo and Stolten, 2019; Srinivasan, 2006). Notice that in reversible systems (e.g., batteries) the following inequality must always hold: $E_{charge} > E_{j=0} > E_{discharge}$.

The overall overpotential is found by adding contributions of several sources of irreversibility of the operation. Although the phenomena that leads to irreversibility varies for different systems and a system-by-system analysis is recommended, common sources include the charge-transfer resistance in the chemical reaction (η_{act}), mass transport (η_{con}), ohmic losses (η_{ohm}), and loss of active electrode area due to the formation of bubbles (η_{bub}).

All these overpotentials increase with an increase in the reaction rate, that is, the rate of charge/discharge. In an electrochemical reaction, the rate of reaction is proportional to the amount of electricity employed (Faraday's law):

$$\frac{\partial N_{species}}{\partial t} = \frac{i}{nF} \tag{15.21}$$

where i is the electric current, an extensive property. The electric current density $j = i/A$, where A is the area of the electrode, is the parameter employed in the mathematical expressions for estimating the overpotentials. Notice that while the electric current in the anode and the cathode must be equal, the current densities may differ if the areas of the electrodes are different.

15.3.1.1.1 Models for the overpotentials

Activation overpotential, η_{act}. This overpotential is related to the activation energy that must be achieved to break chemical bonds prior to the electrochemical reaction. The Butler–Volmer equation (Eq. 15.22) links the activation overpotential with the electric current density at each electrode, and is derived similarly to the Arrhenius equation (Bockris et al., 2002; Carmo and Stolten, 2019):

$$j = j_0\left(e^{-\frac{\beta F \eta_{act}}{RT}} - e^{-\frac{(1-\beta)F\eta_{act}}{RT}}\right) \tag{15.22}$$

The preexponential factor j_0 (exchange current density) is an intrinsic property of the electrode for a given redox couple under specific working conditions. Therefore, j_0 will vary with temperature, pressure, and electrolyte composition (Bessarabov and Millet, 2018), as well as the condition of the electrode surface. The parameter β is the "symmetry factor" and accounts for the difference in the activation barrier of the direct and reverse reactions. It is defined as a ratio, so its values are always in the 0–1 range.

Eq. (15.22) as such applies only for simple and unimolecular reactions. A generalized form of the Butler–Volmer equation replaces β with another parameter that takes into account stoichiometry factors, number of steps involved, and number of electrons transferred up to the rate-determining step. For more details see Bockris et al. (2002).

Ohmic overpotential, η_{ohm}. This overpotential accounts for the classical resistance to the flow of current between two points of a conductor. It is characterized by an equivalent resistance, R_{int} and related to η_{ohm} with Ohm's law:

$$\eta_{ohm} = jR_{int} \tag{15.23}$$

Notice that in Eq. (15.23), R_{int} has units cm^2 so that η_{ohm} is in V.

Mass transport overpotential, η_{con}. This overpotential is sometimes also referred as the concentration overpotential and accounts for mass transport limitations. As in any heterogeneous catalytic reaction, the reactant must reach the surface of the electrode, and products diffuse out. At high current rates, reaction rates become larger, thus mass transport may become the limiting phenomena. To evaluate this overpotential, the concentration gradient between the bulk electrolyte (C_B) and the electrode surface C_S has to be considered (Bockris et al., 2002):

$$\eta_{con} = -\frac{RT}{nF}\ln\frac{C_S}{C_B} \tag{15.24}$$

Overpotential due to formation of bubbles, η_{bub}. This overpotential accounts for blockage at the surface of the electrode by bubbles that are formed during the reaction. Not all bubbles result in an overpotential, just the ones that prevent reactants reaching the reactive surface. Nouri-Khorasani et al. (2017) derive a model for η_{bub} from the Butler–Volmer equation:

$$\eta_{bub} = -\frac{RT}{\alpha F}\ln(1-\theta) \tag{15.25}$$

where α is the transfer coefficient that is employed instead of β in the generalized version (see Eq. 15.22), and θ is the fraction of the electrode surface blocked by the bubbles. The relation between θ and the current density j is not easy to evaluate; some examples can be found in Li et al. (2018b) and Li et al. (2019b).

Relating the efficiency to the operation conditions in the system. In Sections 15.2.1.2 and 15.2.1.4 we considered Eqs. (15.6) and (15.11) for the efficiencies $\eta_{C_{l,t}}$ and $\eta_{D_{n,t}}$, and mentioned they depended on the operation conditions of the system. For electrochemical systems the link is through the overpotentials. In a system with no side reactions, the conversion efficiency η_{C_l} for the charging stage can be written as:

$$\eta_{C_l} = \frac{E_{j=0}}{E} \tag{15.26}$$

Substituting E, and considering all the discussed contributions to the overpotentials

$$\eta_{C_l} = \frac{E_{j=0}}{E_{j=0} + \eta(j)} \tag{15.27}$$

$$\eta_{C_l} = \frac{E_{j=0}}{E_{j=0} + \eta_{act} + \eta_{ohm} + \eta_{con} + \eta_{bub}} \tag{15.28}$$

For discharging stages efficiency is defined inversely to Eq. (15.26), so that efficiency lies in 0–1:

$$\eta_{D_n} = \frac{E}{E_{j=0}} \tag{15.29}$$

In case of multiple reactions taking place in the electrode, we have to consider that the voltage E is the same for all reactions, but the total current i is split between them.

Finally, electrochemical systems are characterized by their Voltage–Current ($E-I$) curves which are constructed by accounting for all overpotentials at the different current/current density levels. Some examples of $E-I$ curves will be presented in the following sections.

15.3.1.2 Charge-stage devices: electrolyzers

Electrolyzers imply reduction of protons into hydrogen. There are three main technologies: Alkaline Electrolysis Cells (AECs), Proton Exchange Membrane Electrolysis Cells (PEMECs), and Solid Oxide Electrolysis Cells (SOECs). These electrolyzers were schematized in Fig. 11.7 (see Chapter 11: Wind Energy: Collection and Transformation). The three of them have the same global chemical reaction, but different electrolytes and catalysts (at the electrodes). As the materials are different, properties such as R_{int} and j_0 are different. Example values for these parameters can be found in Amores et al. (2014), Sandeep et al. (2017), Falcão and Pinto (2020), and AlZahrani and Dincer (2017).

Activation and ohmic overpotentials are important in all three; in SOEC high temperatures ($\approx \geq 900K$) are required to lessen the effect of the ohmic overpotential and also to increase the value of j_0. Low-temperature (PEMECs, AECs) electrolyzers also have nonnegligible overpotentials due to bubble formation. In PEMECs η_{bub} is important in the anode (O_2 formation). In the cathode, as protons reach through the membrane, formation of H_2 bubbles does not block the reaction. In AECs, the extent of this overpotential depends on the particular configuration of the cell, so it has to be studied case by case. At this point we would like to mention that although important, the effect of η_{bub} is rarely included in the literature as a separate term, and may be accounted for as part of the other overpotentials, in particular as part of the ohmic overpotential (Brauns and Turek, 2020). SOECs do not have this problem as the temperatures required for operation are incompatible with water in the liquid phase. Fig. 15.6 schematizes these contributions to the $E-I$ curve of an electrolyzer.

Important features to consider when coupling these devices to nonprogrammable energy sources are their on−off dynamics and limits for partial load. In Corengia and Torres (2018a) we showed that PEMECs have on−off dynamics in the order of minutes, so they are suitable for matching peaks. That is of course not possible for SOECs as temperatures ramp ups and downs are in the order of hours; according to Robles et al. (2018) AECs cold starts in no less than 15 minutes. These on−off times should be included in the formulation of optimization problems for combined design and schedule of ESS as constraints to switch between idle, operating, start-up, and shut-down modes. Binary variables may be used to organize and prevent this overlapping of the operation modes (Zhang et al., 2016).

With the current technology readiness levels, low-temperature electrolyzers, and PEMECs in particular, are preferred for flexible and variable-rate operation. Still, AECs

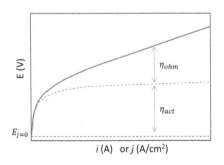

FIGURE 15.6 Typical $E-I$ curve for electrolyzers showing overpotentials η_{act} and η_{ohm}.

are the most mature and cheapest technology, suitable for all-purpose operation. PEMECs are projected to match (or overpass) the installed capacity of AECs in P2X projects by 2020 (Wulf et al., 2020), so they have to be taken into account in any electrolysis-based H_2 production superstructure. When AECs are included, it is important to acknowledge that due to safety issues, they may not be able to operate at any partial load. Then, a restriction accounting for the minimum working load has to be added when formulating an optimization problem that includes them. The reader is directed to Carmo and Stolten (2019) for reference values and to Gahleitner (2013) for real case studies in P2G pilot plants.

Equations to account for investment costs may also need to be added. Regarding this, important considerations are that electrolyzers are modular equipment: according to data in Proost (2019) and Saba et al. (2018), AECs follow the six-tenths rule up to 2.5 MW; electrolyzers larger than 2.5 MW are rare and capacities are build up by stacking multiple units; a linear cost—capacity variation can be assumed for first estimates beyond a certain threshold. PEMECs are not as widely available in the market but some costs are also reported in Proost (2019). However, these may soon be outdated as a large reduction in PEMEC costs is expected due to new materials and improvements in the manufacturing processes of these cells (Schmidt et al., 2017).

15.3.1.3 *Discharge-stage devices: fuel cells*

FCs are discharge devices which basically work opposite to electrolyzers: a fuel (e.g., H_2) is oxidized in a compartment and an oxidant (e.g., air) reduced in a separate one (see Fig. 15.5). Analogously to internal combustion engines (ICEs) in FCs the energy released by a chemical reaction is used to do work. The difference is that in ICEs the energy released by the chemical reaction is used to do mechanical work, whereas in FCs the electron exchange due to the chemical reaction occurs through an outer electrical circuit and is used to do electrical work.

Mirroring technologies for the electrolyzers, proton exchange membrane FCs (PEMFCs), SOFCs, and alkaline FCs (AFCs) have been developed. Other options for the electrolyte are available, popular ones are phosphoric acid FCs (PAFCs) and molten carbonate FCs (MCFCs). Regarding their technology readiness level, all of these systems have been prototyped and used in niche applications (submarines, space, etc.) (Okano, 2016). However, they still struggle with wide adoption due to market barriers: in particular higher costs and lower robustness, reliability and durability when compared to ICEs (Wang, 2015).

FCs are sometimes classified according to their working temperature: SOFCs and MCFCs are high-temperature devices (above 900K); PEMFCs, AFCs, and PAFCs are low-temperature devices (under 400K). Just as happens with the electrolyzers, these different temperature ranges imply different start-up/shut-down times, thus not all options are equally adequate for intermittent operation. As discussed in the previous section binary variables can be used to prevent overlapping of operation modes.

High-temperature FCs can operate with a wide range of fuels and particularly hydrocarbons, without major problem; whereas low-temperature FCs must be adapted (i.e., the catalyst specially designed) when the fuel is not H_2. These alternative fuel options are being investigated mainly to avoid the problems associated with hydrogen storage (Ong et al., 2017).

We will focus on hydrogen FCs, which are the most mature option; in these H_2 is oxidized and O_2 reduced to form H_2O. As before equations linking the chemical reaction rate with voltage and current, are needed to estimate energy conversion efficiencies. Again the link is through the $E-I$ curve for which overpotentials must be estimated. Fig. 15.7 shows a generic $E-I$ curve for FCs. Regarding overpotentials, there is no bubble production in FCs, but in low-temperature FCs the water produced in the cell may condensate and eventually disturb the diffusion of the gaseous reactants. Managing this liquid water may become a relevant operation issue (Dai et al., 2009), which is generally tackled by adding additives to repel water (Wong et al., 2019). The other three overpotentials must be taken into account in FCs. As seen in Fig. 15.7 these overpotentials divide the $E-I$ curve in three zones. In the first zone η_{act} dominates; in the second η_{act} and η_{ohm} are relevant, but changes are due to changes in η_{ohm}; in the third the mass transport overpotential (η_{con}) increases sharply at high currents reaching a limiting current value for operation that cannot be surpassed. Rigorous equations for the $E-I$ curve may be included in the model, but piecewise approximations of the curve provide good estimates and simpler constraints.

15.3.1.4 Devices that combine charge/discharge: reversible electrolyzer/fuel cell

Reversible FCs, regenerative FCs, or unitized regenerative FCs (URFCs) are devices that can work either as an electrolyzer or as a FC. Then, charge and discharge stages take place in a single equipment, which has two operation curves: one similar to that of an electrolyzer and the other one similar to that of an FC.

There are two possible URFC configurations: the first one keeps hydrogen and oxygen in their respective compartments and cathode and anode alternate in the different operation modes; the second one keeps the electrodes working either as anode or as cathode and alternate which gas is present in each compartment (Paul and Andrews, 2017). For modeling purposes, the former is simpler as no idle time is required between charge and discharge, whereas in the latter a purge stage must be included when changing modes. One may think why are we even considering the latter option, the answer is that catalysts at the electrode are usually designed to perform well at reduction or oxidation, not both. Bifunctional catalysts do exist, but are less efficient; this is reflected in parameters that affect the operation curves (such as j_0). Likewise, the rest of the materials inside the cell are also designed to work in one of the operation modes, so parameters such as R_{int} may also change when replacing the materials for ones that can tolerate both modes. Thus

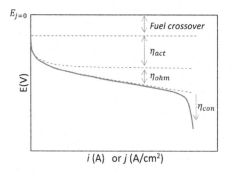

FIGURE 15.7 Typical $E-I$ curve for fuel cells showing overpotentials η_{act}, η_{ohm}, and η_{con}.

there is a trade-off to be considered: frequent (hourly) switches favor the first configuration; for daily switches it is not so clear.

In these devices, the maximum allowed power for the charge and discharge stages cannot be selected individually. As an example, 18 different catalysts combinations in proton exchange membranes URFCs were tested in Doddathimmaiah and Andrews (2009). It was found that in all cases, the ratio of maximum power in electrolyzer mode/maximum power in FC mode lies in the range 7.8−22.

The URFC concept has been explored for the three electrolyzers mentioned in Section 15.3.1.2 (i.e., PEM, AE, and SOEC) and some other options are under study (Paul and Andrews, 2017; Wang et al., 2017). Regarding technology readiness levels, reversible PEM demonstration prototypes have been published over a decade ago (see, e.g., Bents et al., 2006; Sone, 2011).

15.3.1.5 Storage stage devices: hydrogen storage

Storage itself does not imply an electrochemical reaction. However, we include this section here as to be able to displace electric power generation from its consumption using the previous units, another unit (or system) that stores hydrogen is required. H_2 can be stored in tanks as a gas or a liquid, can be adsorbed in a solid and stored as such, or can be used to produce another chemical compound such as a metal hydride, ammonia, or a liquid organic hydrogen carrier (LOHC). Up to date, only compressed or liquid hydrogen storage takes place at a significant scale. As will become clear in the next discussion, all hydrogen storage solutions require energy at some point and release energy at another point. Integration of these energy flows with other processes is an option to consider in order to reduce the net consumption of energy. This reduction is important as a net consumption of energy evidently reduces the efficiency of the overall ESS.

H_2 storage in tanks. To achieve a reasonable energy density (energy stored/storage volume), H_2 needs to be compressed or liquified. Hydrogen-carrying vessels are regulated, thus there are not many design variables to play with. The main variables are the total volume that needs to be stored which defines how many vessels are needed and the storage pressure which is limited by the type of vessel. Type I vessels are metallic vessels employed for industrial applications; they are a mature technology, and can work with pressures up to 50 MPa, although they are generally used in the 20−30 MPa pressure range. The other types of vessels (II, III, and IV) are known as composite overwrapped pressure vessels. Type II vessels have a thick metal liner and are partially covered in fiber resin composites (only the cylindrical surface). In types III and IV the whole vessel is covered with composite, being the difference the composite/metal ratio. The aim of the composite cover is both reaching higher working pressures and higher mass storage efficiencies (defined as hydrogen mass/storage mass). Notice that hydrogen is very light, thus the weight of the vessels themselves becomes an important issue if the stored hydrogen needs to be transported.

An alternative, where geologically possible, is the use of natural caverns instead of vessels. In this sense, salt cavities are preferred, but other options (such as depleted oil fields) have been proposed. Although we acknowledge that caverns are an option, modeling the uptake in them is out of the scope of this chapter. Another option that has been mentioned as a hydrogen storage alternative, is the use of underground pipelines, for example those that are already installed for natural gas. However, recent studies have shown that these may not be adequate for hydrogen (Andersson and Grönkvist, 2019).

Apart from the vessels themselves, the design of the storage of hydrogen needs to account for the compressors, and the energy that they require. Process simulators are very well suited to design and evaluate different configurations for compression systems. Equations for estimation of the energy requirements usually assume polytropic compression as shown in Eq. (15.30) (Martín, 2016):

$$P_{comp,t} = \frac{H_{2,t}RkT}{MW(k-1)\eta_{comp}} \left(\left(\frac{p_{H_{2,t}}}{p_{el}} \right)^{\frac{k-1}{k}} - 1 \right) \tag{15.30}$$

where $P_{comp,t}$ is the power consumed by the compressor, k is the polytropic coefficient, η_{comp} is the compressor efficiency, $p_{H_{2,t}}$ is the pressure inside the tank, p_{el} is the operating pressure of the electrolyzer (assuming the compressor is fed by an electrolyzer), and the molecular weight, $MW = 2\,g/mol$. Notice that the pressure inside the tank varies as its *SoC* varies, then it is a function of time.

Higher final pressures inside the tanks, translate in fewer and smaller storage vessels, but evidently more energy for compression. This energy is basically lost as discharge stages (FCs) operate at rather low pressures. Although the energy during the required decompression could theoretically be recovered, it requires an additional investment. Overall, the point is that there are tradeoffs, that may be worth looking at if designing these systems in detail. For initial estimates Barthelemy et al. (2017) suggest adding an extra 10% to the energy required for hydrogen production (i.e., the energy requirements of the electrolyzer) to account for compression.

For liquid storage, H_2 requires a very low temperature ($\cong 20K$), thus the liquefaction process requires even more energy than compression: up to 35%–40% of energy loss are suggested as initial estimates (Barthelemy et al., 2017; Sundén, 2019). Besides, liquid hydrogen storage is also prone to *SD* as the very low temperatures imply some unavoidable boil-off. Spherical geometries are employed to minimize the heat transfer area in this case, and assuming good isolation, losses can be estimated as 0.1% per day (Andersson and Grönkvist, 2019).

Adsorbed H_2. In this case the idea is that H_2 is adsorbed in a solid that is packed in a vessel. Materials that have been proposed for adsorption of H_2 include: high-surface carbon materials [activated carbon, graphite, nanotubes, nanofibers (Mohan et al., 2019)], clay minerals, zeolites, metal–organic framework materials (Gil and Vicente, 2019) and porous organic polymers (Cousins and Zhang, 2019).

Adsorption is an exothermic process, so heat must be removed during the filling stage. Liquid nitrogen is usually suggested as refrigerant to achieve acceptable storage ratios (kg H_2/kg adsorbent): according to Andersson and Grönkvist (2019) "at least 10 kg of liquid nitrogen need to be evaporated to remove the heat of adsorption of 1 kg of hydrogen." It is quite evident that energy management is a challenge for this storage option as the vessel interior has to be kept very cold. Heat transfer to the unit is only desired for the (endothermic) desorption stage. Estimates for energy efficiency have been reported by Adametz et al. (2014): assuming systems operating at 77K and storage periods up to 500 hours the authors estimate energy losses in the 20%−35% range. Currently, these technologies have only been tested at laboratory scale.

Hydrides. Instead of the weak adsorption interactions involved in adsorbed H_2, hydrides require breaking the molecular bond of H_2 to generate new chemical bonds. Thereby, more energy is involved. Dehydrogenation is the endothermic reaction, and heating the most frequent strategy to release the stored H_2. However, the exothermic reaction (formation of the hydride) may not be spontaneous, so also requiring external energy.

Hydrides are classified in elemental metal hydrides, intermetallic hydrides, complex metal hydrides, and other hydrides. Magnesium hydride (MgH_2) and aluminum hydride (AlH_3) are the most promising alternatives within the elemental metal hydrides, as both are abundant and low-cost metals. They have a relatively high storage capacity: up to 7.6% and 10.1% in mass (Andersson and Grönkvist, 2019). MgH_2 is very stable, and temperatures beyond 550−600K are required for reasonable H_2 release rates (Yartys et al., 2019). AlH_3 releases H_2 at reasonable rates at temperatures below 400K, however they are not easily formed requiring pressures in the order of GPa (Liu et al., 2021). Then, this hydride in particular requires external energy for both hydride formation and decomposition.

Intermetallic hydrides have low enthalpies of dehydrogenation, so better from the energy management side. However, they are more expensive and have lower capacities ($\cong 2\%$ in mass). This is a problem for mobile applications, but may not be a problem for large-scale stationary energy storage (Andersson and Grönkvist, 2019). Some of these hydrides have found other applications, for example, as components of electrodes in batteries as will be seen in Section 15.3.1.6.

LOHCs. LOHCs are chemical compounds that undergo (catalyst-mediated) reversible hydrogenation−dehydrogenation reactions.

$$\text{Hydrogenation:} \quad H_0LOHC + nH_2 \rightarrow H_nLOHC \quad \Delta H < 0$$
$$\text{Dehydrogenation:} \quad H_nLOHC \rightarrow H_0LOHC + nH_2 \quad \Delta H > 0$$

The idea is that in their hydrogen-rich form (H_nLOHC) LOHCs can be easily and safely stored and transported using regular infrastructure. Many chemical pairs have been proposed as LOHC among them: toluene/methyl-cyclohexane, naphthalene/decalin, dibenzyltoluene/perhydro-dibenzyltoluene, or even carbon dioxide/methanol. Research in LOHC is currently very active, Niermann et al. (2019a) provide a pros and cons discussion of many of the currently available options.

The use of LOHC targets large H_2 production capacities, long-term storage (months) and long-distance transportation (including intercontinental). The latter implies that the produced hydrogen may be used (as such or to obtain electricity) in a place geographically far away from its generation, so processes for the hydrogenation and dehydrogenation steps must be designed separately. This has an important consequence for the overall energy efficiency of the ESS: the hydrogenation step is exothermic, so extra energy (in the form of heat) can be recovered in the saturation site (i.e., H_nLOHC producing plant) and possibly integrated to other processes/plants. On the other hand, the dehydrogenation step is endothermic, so the desaturation site needs to account for this extra energy requirement. Overall, the easiest way to include these effects in a model inside the ESS superstructure is through an efficiency factor that discounts from the energy stored as hydrogen, the energy that was required at the desaturation site. Such efficiency factor may be obtained by building an offline simulation of the system; Niermann et al. (2019b) include examples of such designs and provide technoeconomic analysis for different LOHCs pairs. Similarly, transportation of the H_nLOHC to the site of consumption and of the H_0LOHC back to the cite of hydrogenation consumes energy; another efficiency factor dependent on distance and means of transportation should be included in the model.

15.3.1.6 Devices that combine charge, storage and discharge (I): Batteries

A battery is an array of electrochemical cells that can provide power. Batteries are classified according to their ability to be recharged: primary batteries cannot be recharged whereas secondary batteries can. For energy storage applications only secondary batteries are of interest. The working principle of secondary batteries is a reversible reaction such as the one shown in Fig. 15.5, so which is the cathode and which is the anode depends on whether we are at a charging or a discharging stage.

Opposite to FCs and electrolyzers, conventional batteries do not exchange mass with the surroundings. This has important implications in modeling as these devices will never reach a steady state (note that steady state operation is an option for FCs and electrolyzers). During charge/discharge stages, the concentration of active species within the battery changes, then following Nernst equation (Eq. 15.18) $E_{j=0}$ changes. As in any batch process this change is with time, but when modeling batteries it is more convenient to take its SoC as the independent variable instead of the time.

$E-I$ curves can also be used to characterize the operation in a battery; the charge stage follows a curve similar to the electrolyzer (see Fig. 15.6), the discharge stage follows a curve similar to the FC (Fig. 15.7). There is a curve for each SoC (time) which implies that each device will not have one $E-I$ curve for each stage but a family. The overpotentials that play a role are η_{act}, η_{ohm}, and η_{con}. η_{bub} should never be considered as these devices are not designed for chemistries that involve gaseous phases; if formed is due to overcharge conditions that have favored secondary reactions. Operating at these conditions is never desirable: they irreversibly degrade the battery and pose an explosion risk. In formulating an optimization problem in which these issues are important a charge/discharge protocol may be added as a constraint. A simple protocol is charge at constant current up to a certain voltage, then charge at constant voltage. Finding optimal charge/discharge protocols is an active research area (Bharathraj et al., 2020; Wang et al., 2019; Maia et al., 2019).

One important issue to consider in batteries as ESS devices is that they degrade over time. At this point you might be thinking that all equipment degrade with time and we rarely take that into account in design problems. The difference is that battery degradation is noticeable and makes an equipment useless well before reaching zero capacity. Think for example on a cell phone that has to be charged twice instead of once a day, or an electric vehicle that drops its drive range from 240 to 120 km. There are several degradation mechanisms, some depend on the mode of operation some do not. The latter are known as calendar-aging phenomena. All of them depend on the chemistry of the battery. Below, we review some of the available options and briefly discuss degradation mechanisms for each, but before that, we would like to introduce some terminology and parameters that are particular to the operation of batteries:

- c_{rate}: The c_{rate} is related to the current i. Note that in Eq. (15.21) i depends on the mass of active material, thus i is an extensive property; the c_{rate} is the related intensive one:

$$c_{rate} = \frac{i}{IC_S^Q} \tag{15.31}$$

where IC_S^Q is the same IC_S (installed capacity) mentioned in Section 15.2.1.3 but in Ampere-time units. Then c_{rate} has units of $1/time$. The concept is that a charge at 1C (i.e., $c_{rate} = 1$) implies that the battery is fully charged in one hour, whereas a charge at 2C (i.e., $c_{rate} = 2$) would take half an hour.

- **Cycle number:** It is the number of charge/discharge cycles that a battery goes through in a certain period. It is commonly used as a parameter of how much the battery will last, that is, a battery with a cycle number of 5000 lasts longer than one whose cycle number is 2000.
- **Depth of discharge (DoD):** It defines how much of the energy of the battery was used in an operation. It assumes an initially fully charged battery. Notice that $SoC = 1 - DoD$ when both are written as fractions.
- **State of health (SoH):** It indicates the remaining capacity of the battery in comparison to its initial value (i.e., its first cycle).

Lead-acid. These are the batteries inside vehicles that have combustion engines. They have Pb-based electrodes: $Pb/PbSO_4$ in the negative electrode and Pb/PbO_2 in the positive one. Both electrodes are immersed in a sulfuric acid solution that acts as the electrolyte. The discharge half-reactions are:

$$Pb_{(s)} + HSO_{4(aq)}^- \rightarrow PbSO_{4(s)} + H_{(aq)}^+ + 2e^-$$

$$PbO_{2(s)} + HSO_{4(aq)}^- + 3H_{(aq)}^+ + 2e^- \rightarrow PbSO_{4(s)} + 2H_2O_{(l)}$$

The opposite half-reactions take place at each electrode during charge. As seen, there is a net consumption of sulfuric acid during discharge, and a net production during charge. Then, $SoCs$ for these batteries are easy to estimate by measuring the specific gravity of the electrolyte.

These batteries are easy to build and operate, and tolerate high surge currents. Thinking on energy storage applications, they would be good options to deal with peaks, but not very attractive for mobile or user-level applications as their energy/volume or energy/mass ratios are very low compared to other options (Miao et al., 2019). Degradation in these batteries is related the DoD: curves relating DoD to number of cycles (useful for building degradation models) can be found in Gong and Wang (2020).

Li-ion batteries. Li-ion refers to a family of batteries in which one electrode is made of a metallic oxide: for example, $LiCoO_2$ (LCO), $LiMn_2O_4$ (LMO), $LiNi_xCo_yMn_zO_2$ (LNCM), $LiNi_xCo_yAl_zO_2$ (LNCA), or a metallic salt as $LiFePO_4$ (LFP) and the other one of graphite powder. The electrolyte is a liquid organic phase in which a Li-salt is dissolved. These batteries are characterized by their high-energy/volume and energy/mass ratios making them suitable for mobile applications. In particular, LNCM and LFP dominate the energy storage sector (Wang et al., 2021). If we take, for example, a negative electrode of LiFePO$_4$ and a positive electrode of lithiated graphite, the corresponding half-reactions during discharge are:

$$LiFePO_4 \rightarrow FePO_4 + Li^+ + e^-$$

$$C_6 + xLi^+ + xe^- \rightarrow Li_xC_6$$

Note that, as long as no secondary reactions occur, the concentration of Li$^+$ ions remains constant in the electrolyte. Analogous equations can be written for the other Li/metallic oxide combinations.

Degradation mechanisms for Li-ion batteries have been recently reviewed by Thompson (2018): calendar aging and cycling aging appear as the two main components. Calendar aging is associated with degradation whereas the battery is in idle mode (i.e., not while being charged or discharged). Cycling aging is the degradation resulting from battery usage and is dependent on the operation conditions: temperature, *SoC*, c_{rate}, and *DoD*. Models for degradation of these batteries are very complex, and some of the involved phenomena, such as formation of a passivation layer between the electrode and electrolyte, are not yet fully understood (Wang et al., 2018). Then, the way to estimate these effects is through experimentation: a battery is subject to many charge/discharge cycles at different conditions and its *SoH* monitored. Establishing the relationships between *SoH* with c_{rate}, *SoC*, etc. is also an active research area. A compilation of results reported in many experimental papers, together with convex piecewise models for them can be found in Fortenbacher and Andersson (2017).

Sodium−sulfur (Na−S) battery. Na−S batteries are high-temperature batteries in which one electrode is molten sodium and the other one molten sulfur. They are separated by beta-alumina electrolyte, which presents ion conductivity but not electronic conductivity. To keep a good ionic conductivity, operation temperature usually lies in the 300−350°C range. These batteries are characterized by employing inexpensive and abundant materials, having large cycle numbers and high specific energy and power density (Syali et al., 2020; Li et al., 2019a). Then, they have been claimed to be a good alternative for large-scale ESS, for example, centralized in power plants. For this application, commercial units are already available in the market (NGK Insulators, 2020). The half-cell reactions are:

$$2Na \rightarrow 2Na^+ + 2e^-$$

$$2Na^+ + xS + 2e^- \rightarrow Na_2S_x$$

Due to the high operating temperatures, these cells work better if they are used on a regular basis. If they remain in idle mode for long periods of time, heat losses lead to large ohmic resistances and eventually phase changes in the electrodes. An estimation of the amount of time that the battery may remain in idle mode without the need of preheating before the next operation, is found by formulating an energy balance that accounts for heat losses. An example of a detailed energy balance for an Na−S battery can be found in Caprio et al. (2020). When the batteries are employed on a regular basis, this is not a major difficulty, thereby, some examples of grid level electric system design with Na−S batteries available do not consider this phenomena (e.g., Miao and Hossain, 2020; Yamchi et al., 2019; Li et al., 2018a; Chen and Zhao, 2017).

Nickel−metal hydride (Ni−MH). These batteries replaced nickel−cadmium (NiCd) batteries when these were banished in the EU. Their application is mainly for small portable devices, but some vehicles still use them (Toyota, 2021). We include them here as they may have a second life as stationary energy storage devices once they become unsuitable for their original purpose. They also represent an alternative to Li-ion batteries, if Li reserves become scarce.

These batteries have one electrode made of nickel oxide hydroxide (NiOOH) and the other one is an ordered metallic alloy (M) that can form hydrides, the electrodes are connected by an alkaline electrolyte. The half-cell reactions are:

$$Ni(OH)_2 + OH^- \rightarrow NiO(OH) + H_2O + e^-$$

$$H_2O + M + e^- \rightarrow OH^- + MH$$

Regarding degradation, oxygen may be generated as a secondary reaction at the nickel electrode during charge and overcharge. Thus excessive overpotentials, which happen if batteries want to be charged fast by increasing currents and c_{rate}s, must be avoided. This oxygen production lowers the cell performance and life while raising safety concerns (Gu et al., 1999).

15.3.1.7 Devices that combine charge, storage, and discharge (II): flow batteries

Flow batteries or RFBs are devices that have characteristics in between electrolyzer/FCs and batteries. In the most general case, they are composed by three main units: a reversible electrochemical reactor which resembles a reversible electrolyzer/FC; and two reservoir tanks: one for the anolyte and one for the catolyte. One of the more common configurations, the Vanadium redox battery, is shown in Fig. 15.8. Another possible configuration is the undivided RFB: in which there is no separation between cathode and anode and only one electrolyte reservoir is required (Arenas et al., 2017).

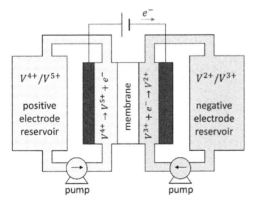

FIGURE 15.8 Diagram of an all-vanadium flow battery in charge mode.

Either with separate or undivided liquids, RFBs are closed systems that do not exchange mass with the surroundings (such as batteries). The main difference with conventional batteries is that their maximum conversion power is not constrained by their energy storage capacity, and vice versa. This turns into more flexible design.

Several electroactive materials (the ones involved in the chemical reaction) have been investigated for use in RFBs being inorganic compounds the more frequently used. Aqueous and nonaqueous electrolytes have been used. Some of the inorganic chemistry systems that reached industry-level demonstration include: iron-chromium, polysulfide-bromide, all-vanadium, and zinc-bromine.

As an example, Fig. 15.8 shows the all-vanadium system: the electrolyte is generally sulfuric acid, and cathode and anode are separated by a proton exchange membrane. Discharge half-reactions are:

$$VO_2^+ + 2H^+ + e^- \rightarrow VO^{2+} + H_2O$$

$$V^{2+} \rightarrow V^{3+} + e^-$$

All the active species should remain dissolved, so solubility limits must be considered to establish the maximum concentration of the active material, in particular $VOSO_4$. For building the E-I curve of these devices, η_{act} and η_{ohm} should be considered. Mass transfer can be improved by increasing the flow to the cell, so lower η_{con} can be obtained at the expense of increased power consumption in the auxiliary pumps.

15.3.2 Nonelectrochemical devices

Looking back at Fig. 15.1, mechanical energy storage currently dominates ESS in P2P applications, being pumped-hydro energy storage (PHES), flywheels and CAES the most common ones. Other mechanical systems have been proposed and are under study (see, e.g., ARES, 2020) and some are even commercially available (see, e.g., Gravity Power, 2020). In general, these systems store energy as mechanical energy and charge/discharge stages imply a change in a subtype of mechanical energy, that is, there is a change in speed, height, or pressure.

For example, in the simplest PHES configuration two water reservoirs at different levels are connected through a pump and a hydro-turbine, such as schematized in Fig. 15.9. In here, water is pumped from the lower to the higher reservoir during the charging stage; the opposite water flow produces electrical power in the turbine during the discharge

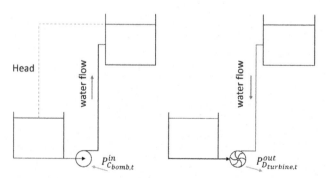

FIGURE 15.9 Pumped-hydro energy storage in closed loop: in charge mode (left) and discharge mode (right).

stage. The reservoirs may operate in open- or closed-loop. Open-loop PHES usually take advantage of an available source of water, such as a river or a preexisting dam, as the lower reservoir. In CAES air is compressed and many times an already existing cavern is taken advantage of to store it at high pressure (Tong et al., 2021). One of the things to consider when looking at CAES is heat management as heat is released during compression and required during expansion. Flywheels are analogous to batteries in the sense that charge, storage, and discharge stages occur in the same unit.

In what follows, we will briefly provide the equations that describe these systems following a similar structure as that of Section 15.3.1. We assume the reader is familiar with these concepts and just include the equations as a refresher.

15.3.2.1 Charge-stage devices: pumps and compressors

Pumps are devices that propel liquids, and compressors devices that propel gases. Although very different in their construction, they are both governed by the mechanical energy balance (Bernoulli's equation) between two conveniently selected points in the system, for example, the tops of the low and high reservoirs in PHES. Bernoulli's equation is generally written as follows:

$$\dot{W} = \Delta \left(\frac{1}{2} \frac{\dot{m} \langle v \rangle^2}{\alpha} \right) + \int_1^2 \frac{\dot{m}}{\rho} dp + \dot{m}g\Delta z + \dot{E}_v \tag{15.32}$$

where $\dot{W}$ is the work done by the pump or compressor to the fluid in units of power; Δ indicates the difference between final (2) and initial (1) points in the system numbered following the direction of the flow; $\dot{m}$ is the mass flowing between the two selected points in units of mass/time; $\langle v \rangle$ is the mean velocity at each of these two points; α is a coefficient that considers the geometry of the conduit, the type of fluid (Newtonian or non-Newtonian) and the flow regime (turbulent or laminar); ρ is the density of the fluid, p is the pressure of the fluid; g is the gravitational acceleration; Δz is the difference in height between the two points and $\dot{E}_v$ is a (positive) term that accounts for friction losses. This last term is estimated taking into account, the length of the pipeline, the accessories (tees, valves, elbows, etc.), the velocity of the fluid through them and the friction factor f_F from the Moody diagram (Tilton, 1997). Notice that the first term on the right-hand side is the kinetic energy term, the second is the static pressure energy term, and the third term is gravitational potential energy.

In the case of liquid fluids, ρ is a constant value then the term in the integral simplifies to $\dot{m}\frac{\Delta p}{\rho}$.

Further, in the context of PHES in which both reservoirs are at atmospheric pressure, and the surface area of the reservoirs is much larger than the flow-through area in the pipes, both the pressure and kinetic energy terms can be neglected, and the overall equation simplifies to:

$$\dot{W} = \dot{m}g\Delta z + \dot{E}_v \tag{15.33}$$

In the case of gases, ρ is not a constant and a proper equation of state has to be included in the integral term. Usually the main simplification for CAES is that the term $\Delta z \cong 0$, as energy is mainly stored as pressure, not height.

The operation point of the charging system (i.e., the pair $\dot{m}$ and $\dot{W}$) is found by intersecting the system curves (i.e., Eq. 15.32 or the simplified versions) and the device curve which is obtained from vendors. For each operation point, vendors also provide efficiency values for converting the electric energy at the input into mechanical energy of the fluid at the output. Combining this efficiency with the efficiency of the system $\left(1 - \dot{E}_v/\dot{W}\right)$ we obtain an expression for the overall efficiency at the charge stage.

15.3.2.2 Discharge-stage devices: turbines

Turbines are the opposite analogs of pumps and compressors: moving fluids transfer energy to the blades of the turbine which are attached to a shaft whose rotation spins an electric generator. Again the construction of turbines highly differs from pumps and compressors, but the Bernoulli's equation must still hold, after properly accounting for the changes in sign as shown in Eq. (15.34):

$$\dot{W}_{turbine} = -\Delta\left(\frac{1}{2}\frac{\dot{m}\langle v\rangle^2}{\alpha}\right) - \int_1^2 \frac{\dot{m}}{\rho}dp - \dot{m}g\Delta z - \dot{E}_v \tag{15.34}$$

where $\dot{W}_{turbine}$ is the work done by the fluid on the blades and we have modified the signs of the other terms to make $\dot{W}_{turbine}$ a positive value. The rest of the discussion is as before; operation curves for the turbines are also obtained from vendors and must be intersected with Bernoulli's equation to obtain the operation point.

15.3.2.3 Devices that combine charge and discharge: reversible pump/turbines

Just as described for URFCs, some equipment can be designed to work both as pump during the charge stage and as turbines during discharge stages. This mode of operation is particularly suitable for small or micro scale applications, as the number of units is reduced. For instance, Morabito and Hendrick (2019) reported a pilot project for a building in Belgium, where a centrifugal device was used both as a pump and as a turbine, in a PHES configuration in which a storm-water basin was employed as the higher reservoir. The overall efficiency obtained in this pilot was lower than specialized equipment (42% vs $\cong 80\%$, Ma et al., 2018). However, it is a good starting point for a very new application of self-storage with little additional investment to the capabilities that buildings already have. Bernoulli's equations as described in the previous sections also apply here, in analogy to what was commented for the URFCs, there will be a set of operation curves for the device acting in charge mode and another set for the device acting in the discharge mode.

15.3.2.4 Storage stage

Water storage in PHES. In here, the important idea is to acknowledge that we have to think of huge volumes of water, thus we need to expand our concept of storage in tanks to storage in for example dams, rivers, or lakes. Artificial lakes are also an option, see, for example, Soha et al. (2017). Notice that in this case energy storage capacity not only depends on the volume of the reservoirs, but also on the difference in height between them. This is a straightforward consequence of Eq. (15.33). PHES as energy storage systems can be very flexible allowing for intra-hour services in grid frequency control applications, to daily, weekly, seasonal and even pluri-annual service when they must match renewable intermittence (Hunt et al., 2020). However, their flexibility may be limited by geology. *SD* processes in PHES include all loses from the upper reservoir, including filtration and evaporation, which may be particularly important in some regions/seasons.

Air storage for CAES. Similar to PHES, storage of compressed air is frequently employed in industrial processes, and the difficulty in CAES is the amount of air that needs to be stored. These large amounts result in that the storage stage ends up being the most expensive component in the ESS (Garvey and Pimm, 2016). Then, when available, the use of natural formations (such as salt caverns) may be attractive for economic feasibility.

For modeling purposes, two limiting operations are considered in the design: isochoric (constant volume) and isobaric (constant pressure). Isochoric conditions represent the case of reservoirs with fixed walls, isobaric conditions are obtained when air reservoirs also have a liquid phase (usually water) connected to a liquid reservoir through a liquid column. Injection of air displaces some of the liquid in the air reservoir, but if the liquid reservoir is large enough the height of the column does not change much, so does not the pressure. This effect is more clearly seen in figures which can be found in Garvey and Pimm (2016) and Olabi et al. (2020). In practice, there are no perfect isochoric or isobaric conditions, but any real situation is between these two limits.

15.3.2.5 Devices that combine charge storage and discharge: flywheels

Flywheels are devices composed by a heavy rotating mass coupled with a motor/generator. In the charge stage, the motor applies a torque that accelerates the rotating mass (rotor). Following the first Newton's law, the mass tends to keep moving until another force acts on it. In the discharge stage, an external force is applied through breaks to deaccelerate the rotor and produce electric power.

In flywheels, electric energy is then stored as angular kinetic energy, E_k, following Eq. (15.35):

$$E_k = \frac{1}{2} I \omega^2 \tag{15.35}$$

where I is the moment of inertia and ω is the angular velocity. Recalling basic physics, the moment of inertia depends on the mass, the radius, and the geometry of the rotating element in the flywheel (e.g., solid vs hollow cylinders). The maximum energy that can be stored in these devices is limited by the strength of the rotor material. The reader is directed to Bender (2016) for equations.

The charge/discharge power depends on the torque that the electric machine can use to accelerate (in charge mode) or deaccelerate (in discharge mode). Overall, the efficiency conversion between mechanical types of energy (ratio of applied torque and increment in kinetic energy) is very high: vacuum-sealed flywheels may reach up to 98%−99%, but the conversion between electrical/mechanical energy is lower to 85%−90%, so this is what must be accounted for charge/discharge efficiencies (Ehsani, 2013).

Besides, the permanent movement of charged flywheels implies friction. This turns into high levels of *SD*: 20% of stored capacity per hour or even higher (Olabi et al., 2021). Still, they are a very good option for short-term storage.

15.4 Case study 1: self-storage strategies to take advantage of time-of-use policies

Energy self-storage in batteries is not an innovative concept. Lead-acid batteries have been employed to supply power in remote areas for decades (Baldsing et al., 1991), and the use of uninterruptible power source (UPS) units in grid-connected systems is a common practice to avoid power interruptions. Still, the combination of changes in the electrical market (evolution of TOU or other demand response programs) and the diminishing costs of simple energy storage devices provide new incentives to install self-storage units. Along these lines, "behind-the-meter" batteries are being proposed to be installed by the users to shift their energy consumption to times when the electricity price is low (Corengia and Torres, 2018b); to reduce the maximum power they consume from the grid (Koskela et al., 2019; Ding and Zhang, 2019), or to combine both effects (Corengia and Torres, 2020a; Zhang et al., 2019b).

In particular, Li-ion batteries present long calendar and cycle lives, high efficiency and reliability, thus they can be employed to take advantage of differential prices without a major disruption in users' daily routines. Although a promising solution, charge and discharge cycles also degrade batteries, thus expected savings in the energy bill may actually be nonexistent, if these savings are counterbalanced by the capacity lost by the battery. As the capacity loss depends on operational parameters such as the c_{rate} (see Section 15.3.1.6), the battery operating schedule affects current-induced degradation. In Corengia and Torres (2018b), we presented a convex optimization problem to obtain the optimal operation schedule for a behind-the-meter battery, considering its degradation. The most salient aspects of the problem formulation are reproduced here.

Fig. 15.10A and C schematizes the problem. Following the structure in Section 15.2.1 the first thing to do is to identify the source node. In this case, instead of the resource

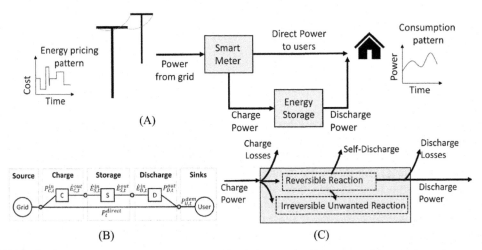

FIGURE 15.10 Graphical representation of the behind-the-meter self-storage problem in time-of-use settings (Case Study 1). (A) A generic (household or industry) uses an energy storage system to avoid consumption from the grid at peak price hours. (B) The superstructure for the system. (C) Phenomena inside the battery.

availability for the generators discussed in Section 15.2.1.1, the time-varying signal is the price of energy. This is we implicitly assume that energy is always available. Some examples of real TOU signals are presented in Fig. 15.11.

Next, we should identify which of the nodes are present in this problem: there is only one charge node, one storage node, and one discharge node, all of them combined in a single unit. The superstructure for this case is in Fig. 15.10B: the charge node can take any power ($P^{in}_{C,t}$) from the grid, and the discharge node can feed the user with power $P^{out}_{D,t}$. At times, it might be more convenient to bypass the battery and feed the load directly from the grid. This is shown in the figure as the direct power flow (P^{direct}_t). Following the same procedure as in Example 15.2 the equation for the battery can be formulated. We will not go through the derivation again as it is exactly as presented before. The only comment is that for this particular case study, daily use of a Li-ion battery, SD losses can be neglected, thus the final equations for the battery are:

$$SoC_t = SoC_{t-1} + \Delta t P^{in}_{C,t} \eta_C - \Delta t \frac{P^{out}_{D,t}}{\eta_D} \tag{15.36}$$

$$0 \leq P^{in}_{C,t} \leq IC_C \tag{15.37}$$

$$0 \leq P^{out}_{D,t} \leq IC_D \tag{15.38}$$

Regarding the limits for the SoC we have considered 20% and 80%, as these are usual safe values for Li-ion batteries:

$$0.2IC_{S,t} = SoC^{MIN} \leq SoC_t \leq SoC^{MAX} = 0.8IC_{S,t} \tag{15.39}$$

where $IC_{S,t}$ denotes the maximum storage capacity, which should be noted that it is not a constant: it varies with time as the battery degrades.

Regarding users and curtailment nodes, in the first work (Corengia and Torres, 2018b) we assumed the user would uptake all the energy stored in the battery (this means the

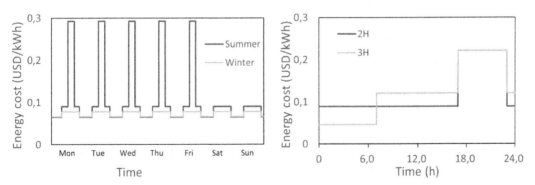

FIGURE 15.11 Examples of residential time-of-use tariffs. Left: Iowa (MidAmerican Energy, 2014). Right: Uruguay two-step time-of-use (2H) and three-step time-of-use (3H) (UTE, 2020). Source: *Data from MidAmerican Energy Company, 2014. MidAmerican Energy Company electric tariff no. 2 filed with Iowa utilities board. <https://www.midamericanenergy.com>; UTE, 2020. Pliego tarifario. <https://www.ute.com.uy>.*

battery is rather small for the user's needs); then in Corengia and Torres (2020a) we assumed the user was a student residence.

Regarding the formulation of the optimization problem, a consequence of having only one equipment is that it cannot operate simultaneously in charge and discharge modes; then a binary variable (y_t) is included in Eqs. (15.37) and (15.38): during charge $y_t = 1$; during discharge $y_t = 0$:

$$P^{in}_{C,t} \leq y_t IC_C \tag{15.40}$$

$$P^{out}_{D,t} \leq (1 - y_t) IC_D \tag{15.41}$$

An alternative to using binary variables could be adding the following constraint: $P^{in}_{C,t} P^{out}_{D,t} = 0$, but is generally not preferred as it introduces nonlinearity.

Regarding the objective function, for a given battery the goal would be to find how to optimally operate it so net savings are maximized. Net savings should discount the cost of degradation from the savings in the electricity bill. Then, with a TOU tariff with varying cost of energy over time ($\$_t$), net savings can be expressed as:

$$\text{Net savings} = \text{cost of energy discharged} - \text{cost of energy charged} \\ - \text{penalty for capacity lost at charge and discharge} \tag{15.42}$$

$$\text{Net savings} = \sum_t \left(\$_t P^{out}_{D,t} \Delta t - \$_t P^{in}_{C,t} \Delta t - Cost^{ES} IC_{S,0} x^{CL}_t \right) \tag{15.43}$$

where $Cost^{ES}$ denotes the cost of the technology in USD/kWh; $IC_{S,0}$ is the initially installed capacity of the storage stage in kWh; and x^{CL}_t is the fraction of capacity lost during operation time t.

Up to now, we have modeled the structure at the network level, but have not included anything that relates this model with the chemistry inside the battery. We will focus on connecting this model with degradation models for Li-ion batteries. We have already mentioned that degradation models were a function of the c_{rate} then: $x^{CL}_t = f(c_{rate,t})$. The functionality f is obtained from experimental data: a battery is cyclically charged and discharged at different values of c_{rate} and its SoH monitored at each cycle. As seen in Fig. 15.12, the x^{CL}_t versus c_{rate} plot is obtained from the slopes of the curves at each c_{rate}.

For the experimental data reported in (Sarker et al., 2017), a second-order polynomial proved to be a reasonable and theoretically expected estimation:

$$x^{CL}_t = \Delta t \left(\alpha_1 \left(c_{rate,t} \right)^2 + \alpha_2 c_{rate,t} \right) \tag{15.44}$$

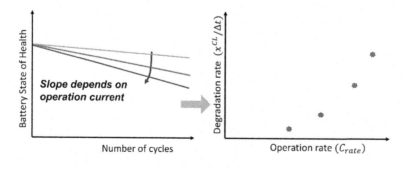

FIGURE 15.12 Example of data for estimating degradation. Left: batteries' state of health is monitored when cyclically charged and discharged at different c_{rate}. Right: the corresponding slope at each c_{rate} provides a point for the degradation rate curve.

Another characteristic of batteries is that the installed powers for charge and discharge stages (IC_C and IC_D) cannot be individually selected, and they are also related to the storage stage capacity. In Corengia and Torres (2018b), this was accounted for by limiting the charge and discharge powers to 3C (c_{rate}). At this point an equation that relates c_{rate} to power is needed. In Li-ion batteries working below the limiting c_{rate} values (i.e., 3C), overpotentials can be neglected and $E_t \cong E_{j=0,t}$, then:

$$c_{rate} = \frac{i_t}{IC_S^Q} \cong \frac{i_t E_t}{IC_S^Q E_{j=0,t}} = \frac{P_{C,t}^{in} + P_{D,t}^{out}}{IC_S} \tag{15.45}$$

The product $i_t E_t$ corresponds to the inlet/outlet power of the battery ($P_{C,t}^{in}/P_{D,t}^{out}$), whereas the product $IC_S^Q E_{j=0,t}$ is the amount of energy that can be stored. Adding $P_{C,t}^{in} + P_{D,t}^{out}$ in the numerator is valid as we have already stated that charge and discharge cannot happen simultaneously.

Relaxations that result in an equivalent convex problem are discussed in Corengia and Torres (2018b). The most relevant are: binary variables for the powers can be omitted as simultaneous charge and discharge result in more losses thus worse values of the objective function; the nonconvex Eq. (15.44) can be relaxed to the convex inequality constraint $x_t^{CL} \geq \Delta t(\alpha_1(c_{rate,t})^2 + \alpha_2 c_{rate,t})$. Note that this restriction is always active at the equality as any other value results in a larger penalization in the degradation term of the objective function.

Taking the previous equations and discussions into account, and after some manipulation of the terms for simplification, the optimization problem is formulated as shown in Eq. (15.46). It can be proven that this problem is strictly convex, thus its solution is unique: the battery charges and discharges as slowly as possible to avoid degradation.

$$\min_{P_{C,t}^{in}, P_{D,t}^{out}, SoC_t, x_t^{CL}, c_{rate,t}, IC_{S,t}} \Delta t \sum_t \$_t \left(P_{C,t}^{in} - P_{D,t}^{out} \right) + Cost^{ES} IC_{S,0} \sum_t x_t^{CL} \tag{15.46}$$

$$\text{s.t.} \quad SoC_t = SoC_{t-1} + \Delta t P_{C,t}^{in} \eta_C - \Delta t P_{D,t}^{out}/\eta_D$$

$$c_{rate,t} = \frac{P_{C,t}^{in} + P_{D,t}^{out}}{IC_{S,0}}$$

$$IC_{S,t+1} = IC_{S,t} - IC_{S,0} x_t^{CL}$$

$$x_t^{CL} \geq \Delta t \left(\alpha_1 \left(c_{rate,t} \right)^2 + \alpha_2 c_{rate,t} \right)$$

$$0.2 IC_{S,t} \leq SoC_t \leq 0.8 IC_{S,t}$$

$$0 \leq P_{C,t}^{in} \leq 3 IC_{S,t}$$

$$0 \leq P_{D,t}^{out} \leq 3 IC_{S,t}$$

We could be interested in seeing if it is worth complicating ourselves by including degradation in the formulation of the problem. This was studied in Corengia and Torres (2019) where everything that was related to degradation was removed from Eq. (15.46) resulting in the following linear problem:

$$\min_{P_{C,t}^{in}, P_{D,t}^{out}, SoC_t} \Delta t \sum_t \$_t \left(P_{C,t}^{in} - P_{D,t}^{out} \right) \tag{15.47}$$

$$\text{s.t.} \quad SoC_t = SoC_{t-1} + \Delta t P_{C,t}^{in} \eta_C - \Delta t P_{D,t}^{out}/\eta_D \quad 0 \leq P_{C,t}^{in} \leq 3 IC_S$$

$$0.2 IC_S \leq SoC_t \leq 0.8 IC_S \qquad\qquad\qquad 0 \leq P_{D,t}^{out} \leq 3 IC_S$$

It should be noted that if degradation is not taken into account IC_S does not depend on time anymore. This problem is convex, but not strictly convex, so multiple operation schedules lead to the same optimal solution. Fig. 15.13A and B shows some of the options.

One of these possible *argmin* coincides with the *argmin* obtained when degradation effects are considered, implying that the same capacity loss over time is achieved. However, most probably any algorithm used to solve the problem will stop at an *argmin* which will not coincide with the one that considers degradation. Thus if the battery degrades, but degradation was not considered when defining the optimal operation schedule, it will happen at a faster rate than if it had been taken into account directly in the optimization problem. How much faster depends on the coefficient of the second-order term in the model for capacity loss (this is α_1 in Eq. 15.44). Simulations of the long-term operation of the battery with different optimal strategies for problem (15.47) are shown in Fig. 15.13C.

Up to this point, we have assumed that the user has a large enough power demand to always employ the power discharged by the battery $P_{D,t}^{out}$. In Corengia and Torres (2020a) we improved this assumption by including a consumption pattern $P_{U,t}^{dem}$ for the user. This implies adding the following restriction:

$$P_{U,t}^{dem} = P_{D,t}^{out} + P_t^{direct} \tag{15.48}$$

Fig. 15.14A shows an example of installing the battery in a student residence. The black line represents the residence consumption profile which has a large peak in the morning (wake up, breakfast time), a medium peak at noon (probably some but not all students come back for lunch), and another large peak at night. The energy cost for this case is 2H TOU in Uruguay (see Fig. 15.11): expensive between 17:00 and 23:00 and cheap the rest of the day. The battery optimal on−off schedule was obtained considering the problem with degradation, and the restriction in Eq. (15.48). The gray lines in the figure represent the power taken from the grid to charge the battery; the dotted gray lines represent the discharge power from the battery to the user. The power taken from the grid at each time (for direct consumption and battery charge) is represented by the dashed black lines. As seen the battery goes through a full charge between 23:00 and 17:00 (cheap time), and a full discharge between 17:00 and 23:00 (most expensive time).

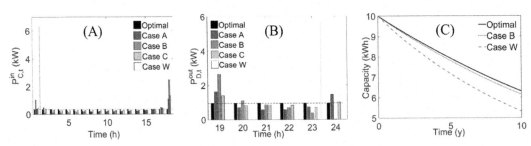

FIGURE 15.13 Different charge (A)/discharge (B) strategies obtained with the nonpenalized problem (Eq. 15.47). The one labeled "optimal" corresponds to the solution of the penalized problem (Eq. 15.46). (C) Change in battery capacity over time for some of the strategies shown in Parts (A) and (B).

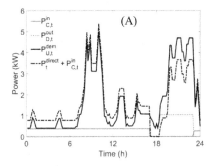
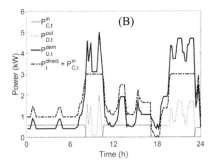

FIGURE 15.14 Solutions for the self-storage problem when the power produced has to match a consumption profile. (A) Electricity bill only charges for the energy consumed. (B) A (fixed) charge for precontracted power is also included in the bill.

In addition, in some countries (e.g., Uruguay), the electricity bill is composed by two terms: one that is related to the energy consumed and one that is related to a maximum power that the user precontracts. This maximum power is tied to a fixed charge, lowering the precontracted power lowers the bill. Fig. 15.14B shows the results for this residence under this type of bill. The optimization problem, was modified to include this extra cost in the objective function and an additional constraint to avoid taking more power from the grid than the precontracted value (P^{grid}):

$$P_t^{direct} + P_{C,t}^{in} \leq P^{grid} \tag{15.49}$$

The figure shows that the battery is not only used to displace energy consumption to avoid the expensive TOU but also to avoid the morning power peak, which allows for keeping the service with lower precontracted powers.

15.5 Case study 2: seizing renewable energy surplus for the production of hydrogen

Besides its possible use as an energy carrier in P2P settings, hydrogen is suggested as an option for decarbonization of the transportation sector, in particular heavy duty applications. In countries with large nonprogrammable RES, power companies are implementing synergies with the fuel sector to use their surpluses for the production of fuel-H_2 (MIEM et al., 2020). In this case study, we analyze the design of H_2 production plants in this setting.

Fig. 15.15 states the problem: the system combines programmable and nonprogrammable energy sources to supply electric power to a community. Following Example 15.1 nonprogrammable RES are used whenever available, and programmable sources are turned on in cases where $P_{U_t}^{dem} > P_{G_{nonprog,t}}^{max}$. Also as discussed in Example 15.1, even when the nonprogrammable sources are prioritized, there may be times at which there exists a surplus $P_{G_{nonprog,t}}^{spls}$. The idea here is to employ part of that surplus to produce H_2. Note that the system is not forced to employ all the surplus for H_2 production, as this situation may

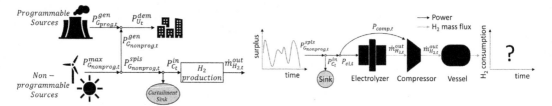

FIGURE 15.15 Graphical representation of the hydrogen production from renewable energy surplus problem (Case Study 15.2). Left: renewable energy sources are integrated to the grid; when generation is greater than demand, the surplus is used to turn on an electrolyzer. Right: diagram of the hydrogen production system.

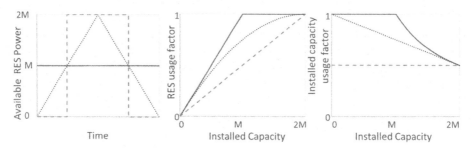

FIGURE 15.16 Three possible power versus time patterns with the same total energy (left). Their corresponding usage factor of energy (middle) and charging unit (right) versus installed capacity of the charging step.

imply designs with large installed capacities that operate at full load less than 1% of the time (Estermann et al., 2016). This of course is not economical, implying that there is an optimal installation capacity that balances resource usage and capacity usage.

To illustrate this, Fig. 15.16 shows three different patterns of power availability over time. The filled curve in the left plot represents the case where power is available at a constant value (M); the dashed curve has two possible values 0 and $2\,M$ (note that this case is a limiting model for solar radiation); the dotted curve ramps up and down. They have the same mean value (M), so the three curves account for the same amount of available energy (i.e., the area below the curves). The middle plot shows how much of the previous power is used for different charge-stage installed capacities (IC_C). For the filled curve below $1\,M$ the usage of the resource increases linearly with IC_C, and above $1\,M$ the resource caps the IC_C that can be used. In the dashed line case, the usage of the resource increases linearly with IC_C, but to be able to use all of it, $IC_C = 2\,M$ is required. In the dotted line case, a $IC_C = 2\,M$ is also required to be able to capture all the resource, but the increment in the usage of the resource is not linear with IC_C (the reader is encouraged to prove this for the triangular shape used in the left plot). The right plot shows the average usage of the charge-stage installed capacity for each of the resource patterns. For the filled curve, below $1M$ the equipment is used at full capacity and any capacity above $1\,M$ is idle capacity. In the dashed line case, the usage of the resource is 0.5 as the resource is always available half of the time. In the dotted line case, the usage factor decreases linearly from 1 to 0.5 as

IC_C varies from 0 to 2 M (again the reader is encouraged to reproduce the right plot based on the left plot).

Real available power curves can be quite complex. If the power comes from RES surplus, they depend both on climatic conditions as well as the consumption patterns, which are usually cyclic. As an example, Fig. 15.17 shows curves for daily nonprogrammable power surplus ($P^{spls}_{G_{nonprog,t}}$) over time in Uruguay (Corengia and Torres, 2020b).

In order to design a system that uses these surpluses for hydrogen, following the rationale of Section 15.2, an H_2 consumption pattern should be known. As a first step in the analysis in Corengia et al. (2020) the focus was on the design of the charge stage assuming different targets of average H_2 consumption needs. This is in Fig. 15.15; this case study focuses on the system up to H_2 flow rate out of the compressor.

Questions pertinent to the design problem include which H_2 production technology to use (i.e., type of electrolyzer as described in Section 15.3.1.2), how many electrolyzers to install and which capacity to select for each. As mentioned in Section 15.3.1.2, low-temperature electrolysis is preferred for flexible and variable-rate operation. And in particular, PEMECs can be reasonably matched to these highly variable surplus curves.

As discussed in Section 15.3.1.5, hydrogen needs to be compressed to be stored, thus always requiring an energy flux which can be assumed as part of the charging stage. This implies that both H_2 production and compression will take place when $P^{spls}_{G_{nonprog,t}}$ is available.

In Corengia et al. (2020), different optimization problems to analyze the trade-off between maximum usage of the resource and maximum usage of installed capacity were proposed and solved. As explained in the reference, the objective function can vary depending on whether the goal of H_2 production is to obtain maximum net present value (*NPV*), or to minimizing energy curtailment.

The constraints of the problem are:

- Source node: It is characterized by $P^{spls}_{G_{nonprog,t}}$ curve that represents Fig. 15.17.
- Linking node 1: The inlet power ($P^{in}_{C_t}$) is capped by the availability of the resource $P^{spls}_{G_{nonprog,t}}$,

$$P^{in}_{C_t} \leq P^{spls}_{G_{nonprog,t}} \tag{15.50}$$

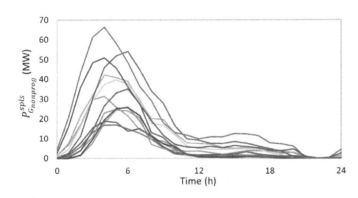

FIGURE 15.17 Surpluses of nonprogrammable renewable energy sources in Uruguay. Each curve corresponds to a representative day for each month, values obtained with SIMSEE software. Source: *Reproduced from Corengia, M., Torres, A.I., 2020b. Exploiting wind/sun energy surplus for production of green-hydrogen: combining PEM and AE electrolyzers for cost efficiency. In: 2020 AIChE Annual Meeting.*

- Linking node 2: This power is employed for both the electrolyzer ($P_{el,t}$) and the compressor ($P_{comp,t}$)

$$P_{C_t}^{in} = P_{el,t} + P_{comp,t} \tag{15.51}$$

- Charge node: The electrolyzer produces hydrogen at a rate $\dot{m}_{H_2,t}$ (in units of mass/time):

$$P_{el,t}\eta_{el,t} = \dot{m}_{H_2,t}\left(-\Delta H_{H_2O}^{form}\right) \tag{15.52}$$

$$P_{el,t} \leq IC_{el} \tag{15.53}$$

where $\eta_{el,t}$ is the efficiency of the electrolyzer and $\Delta H_{H_2O}^{form}$ is the enthalpy of the formation of water in energy/mass units, and IC_{el} is the maximum capacity of this operation in units of power.

- Storage node: As mentioned as a starting point only the compression system was considered. An equation to estimate its power consumption as a function of $\dot{m}_{H_2,t}$, and an equation to limit the maximum operation flow (IC_{comp} in mass/time units) are required.

$$P_{comp,t} = f\left(\dot{m}_{H_2,t}\right) \tag{15.54}$$

$$\dot{m}_{H_2,t} \leq IC_{comp} \tag{15.55}$$

For the case study the functionality f was obtained from simulations in Aspen Plus.

- Relations for economic evaluation. As the interest is in producing H_2 in a profitable manner, constraints to compute economic parameters need to be included in the formulation, these are commonly used equations for *NPV*, capital investment (*IEC*), revenues (*REV*), and operational expenditures (*OPEX*) will not be discussed here but are mentioned in Corengia et al. (2020).

Then, the optimization problem for finding the capacity that maximizes *NPV* can be casted as in Eq. (15.56), where $\$_{H_2}$ is the price of hydrogen and $\$_{energy,t}$ is the price of the electric energy surplus.

$$\max_{P_{el,t},P_{comp,t},\dot{m}_{H_2,t},IC_{el},IC_{comp},NPV,IEC,REV,OPEX} NPV \tag{15.56}$$

$$\text{s.t.} \quad NPV = -IEC + \sum_{y=1}^{N} \frac{REV - OPEX}{(1+r)^y}$$

$$IEC = f_{el}(IC_{el}) + f_{comp}\left(IC_{comp}\right)$$

$$REV = \$_{H_2} \sum_t \Delta t \dot{m}_{H_2,t}$$

$$OPEX = OPEX_{o\&m}IEC + \sum_t \Delta t \$_{energy,t}\left(P_{el,t} + P_{comp,t}\right)$$

$$P_{el,t} + P_{comp,t} \leq P_{G_{nonprog,t}}^{spls}$$

$$P_{el,t}\eta_{el,t} = -\Delta H_{H_2O}^{form}\dot{m}_{H_2,t}$$

$$P_{el,t} \leq IC_{el} \quad P_{comp,t} = f\left(\dot{m}_{H_2,t}\right) \quad \dot{m}_{H_2,t} \leq IC_{comp}$$

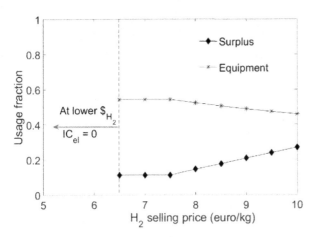

FIGURE 15.18 Optimal usage factors for both resource and installed capacity at several H_2 prices. Surplus cost = 0 USD/kWh. Source: *Adapted from Corengia, M., Estefan, N., Torres, A. I., 2020. Analyzing hydrogen production capacities to seize renewable energy surplus. Computer Aided Chem. Eng. 48, 1549–1554.*

This problem was solved for different assumed H_2 prices. Fig. 15.18 shows the results. For the surplus and technologies discussed here, the optimal solution is not to install the plant if the price is below 6.5 €/kg H_2. Beyond this threshold, the "flat" region corresponds to solutions in which *NPV* increases without any change in IC_C: the operation schedule remains the same, but H_2 price increases. Eventually, as the price increases beyond 7.5 €/kg H_2, it is more profitable to increase the installed capacity to capture more resource. This increment in resource usage factor is done at the expense of a lower usage factor for the equipment.

This example shows how the variability of the resource highly complicates the design problem, the overall conclusion of this case is that although PEMECs have on–off dynamics that can "follow" the resource very well, low usage factors turn into very large production costs.

Problems

P15.1 Consider the Example 15.2 and the ESS superstructure in Fig. 15E2.1. Rewrite the logical constraints for this system if the installation of both a battery and an H_2 is valid in an optimal solution, and the nodes C_3 and D_3 are the charge and discharge stages of a URFC.

P15.2 Due to the availability of raw materials, an industrial plant is projected in a location where there is no grid connection available. The plant is studying a 100% renewable power system that couples solar energy with a battery. The plant is expected to consume a constant power of 250 kW (24 h/day). The expected solar production factor is shown in Table 15P2.1. The battery has a 97% charge and discharge efficiency.

    a. Estimate the energy capacity (MWh) required to operate with only intraday storage and the required capacity of the solar panels (MW).

    b. For the selected capacity, find the energy surplus.

TABLE 15P2.1 Data for problem P15.2.

Hour	Winter	Summer	Autumn /Spring	Hour	Winter	Summer	Autumn /Spring	Hour	Winter	Summer	Autumn /Spring
1–6	0	0	0	11	0.39	0.73	0.51	16	0.37	0.67	0.44
7	0	0.16	0	12	0.41	0.74	0.51	17	0.19	0.62	0.36
8	0	0.43	0.20	13	0.42	0.75	0.52	18	0	0.47	0.14
9	0.22	0.62	0.41	14	0.42	0.73	0.49	19	0	0.10	0
10	0.37	0.71	0.49	15	0.40	0.70	0.47	20–24	0	0	0

    c. Repeat parts (a) and (b) considering the following power consumption pattern for the plant: 10 kW from 22:00 to 6:00, 250 kW from 6:00 to 18:00, and 100 kW from 18:00 to 22:00.

P15.3 Fig. 15P3.1 shows an ESS for intraday operation. It can be assumed that it charges at constant P_C from 11:00 to 15:00, and discharges at constant power P_D from 18:00 to 20:00.

    a. Find the round-trip efficiency in a day cycle if SD is negligible. Round-trip efficiency is defined as the ratio between all the energy that at the outlet of the discharge node over all the energy at the inlet of the charge node.

    b. Assume now that the storage stage has an SD rate of 5% per hour. Find the new round-trip efficiency.

    To simplify (b), assume that both at the beginning and at the end of the day the storage device is completely discharged.

FIGURE 15P3.1 ESS for problem P15.3.

P15.4 Many industries have peaks in power consumption during relatively short periods of time. Electric energy providers discourage these peaks by including in the electricity bill a component that is proportional to the peak of maximum power through the billing period (usually one month). This component is known as a demand charge. According to Oudalov et al. (2007), demand charges can make up as much as one-half of industrial costumer electricity bill. Shaving these consumption peaks to reduce the bill, is one of the reasons that make self-storage attractive to this type of costumers.

    a. Consider a facility with a known consumption profile during the day (P_{demand}). Formulate an LP optimization problem that maximizes the savings that can be obtained by installing a battery of capacity C (kWh), under a pricing policy with a demand charge of $\$_{DC}$ (USD/kW). Assume that the cost for consumed energy (USD/kWh) has a constant value over time.

TABLE 15P4.1 Consumption over time for problem P15.4.

Time	Power (kW)	Time	Power (kW)	Time	Power (kW)	Time	Power (kW)	Time	Power (kW)
0–8:00	50	10:00	380	12:15	260	14:30	620	16:45	650
8:00	80	10:15	510	12:30	140	14:45	800	17:00	250
8:15	80	10:30	430	12:45	140	15:00	880	17:15	140
8:30	80	10:45	230	13:00	140	15:15	1000	17:30	140
8:45	80	11:00	200	13:15	140	15:30	900	17:45	140
9:00	140	11:15	780	13:30	140	15:45	810	18:00	140
9:15	140	11:30	420	13:45	220	16:00	710	18:15	140
9:30	140	11:45	180	14:00	310	16:15	650	18:30	100
9:45	680	12:00	200	14:15	180	16:30	650	18:30–24	50

b. The facility is renewing the battery of its electric forklift and evaluating a second life for it as a peak shaving unit. The battery is an 80 V one with a current capacity of 300 Ah. Additional peripherals and a control system are needed and the plan is to cover their cost with the electric bill savings of one year. Implement the problem formulation in (a) and find the maximum installation cost that can be afforded. Assume a demand charge of 2.1 USD/kW and an energy cost of 0.03 USD/kWh, constant efficiencies for charge and discharge (95%) and the consumption profile in Table 15P4.1.

c. Numerically check on the solution uniqueness/multiplicity. Hint: try employing different solvers or different starting points.

d. Now consider the capacity loss of the battery due to its usage. Assume that the capacity loss can be modeled as a second-order polynomial of the charge power with positive coefficients α_1 and α_2. Solve the problem for different values for α_1 and α_2 starting with $\alpha_1 = \alpha_2 = 10^{-4}$. Can you now find multiple solutions?

e. On the basis of your results for parts (c) and (d), can you conclude on the uniqueness of the solution?

P15.5 A wind generator of capacity X (X in MW) is installed to produce fuel H_2 in a remote location (no grid connection available). The average wind production factor for this location is 0.4 (mean power produced/installed power) and 200 kg H_2/day must be obtained from the facility.

a. Draw the ESS for this case, properly identifying the nodes, and formulate a suitable model.

b. On the basis of the previous model, estimate the minimum capacity X that is required to meet the H_2 production target.

c. For larger values of X, plot the minimum η_{el} versus X that is required to make the problem feasible, where η_{el} is the efficiency of the electrolyzer.

References

Adametz, P., Müller, K., Arlt, W., 2014. Efficiency of low-temperature adsorptive hydrogen storage systems. Int. J. Hydrogen Energy 39, 15604–15613.

Allman, A., Daoutidis, P., Tiffany, D., Kelley, S., 2017. A framework for ammonia supply chain optimization incorporating conventional and renewable generation. AIChE J. 63, 4390–4402.

AlZahrani, A.A., Dincer, I., 2017. Thermodynamic and electrochemical analyses of a solid oxide electrolyzer for hydrogen production. Int. J. Hydrogen Energy 42, 21404–21413.

Amores, E., Rodríguez, J., Carreras, C., 2014. Influence of operation parameters in the modeling of alkaline water electrolyzers for hydrogen production. Int. J. Hydrogen Energy 39, 13063–13078.

Andersson, J., Grönkvist, S., 2019. Large-scale storage of hydrogen. Int. J. Hydrogen Energy 44, 11901–11919.

Aneke, M., Wang, M., 2016. Energy storage technologies and real life applications – a state of the art review. Appl. Energy 179, 350–377.

Arenas, L., de León, C.P., Walsh, F., 2017. Engineering aspects of the design, construction and performance of modular redox flow batteries for energy storage. J. Energy Storage 11, 119–153.

ARES, 2020. <https://aresnorthamerica.com/> (accessed December 2020).

Baldsing, W., Hamilton, J., Hollenkamp, A., Newnham, R., Rand, D., 1991. Performance of lead/acid batteries in remote-area power-supply applications. J. Power Sources 35, 385–394.

Barthelemy, H., Weber, M., Barbier, F., 2017. Hydrogen storage: recent improvements and industrial perspectives. Int. J. Hydrogen Energy 42, 7254–7262.

Bender, D., 2016. Flywheels. Storing Energy. Elsevier, pp. 183–201.

Bents, D.J., Scullin, V.J., Chang, B., Johnson, D.W., Garcia, C.P., Jakupca, I.J., 2006. Hydrogen-oxygen PEM regenerative fuel cell development at NASA Glenn Research Center. Fuel Cell Bull. 2006, 12–14.

Bessarabov, D., Millet, P., 2018. Fundamentals of water electrolysis. PEM Water Electrolysis. Elsevier, pp. 43–73.

Bharathraj, S., Adiga, S., Mayya, K., Song, T., Kim, J., Sung, Y., 2020. Degradation-guided optimization of charging protocol for cycle life enhancement of Li-ion batteries with lithium manganese oxide-based cathodes. J. Power Sources 474, 228659.

Biegler, L.T., Grossmann, I.E., Westerberg, A.W., 1997. Systematic Methods of Chemical Process Design. Prentice Hall PTR, Chapter 15.

Bird, L., Cochran, J., Wang, X., 2014. Wind and solar energy curtailment: experience and practices in the United States. Tech. Rep. NREL.

Bird, L., Lew, D., Milligan, M., Carlini, E.M., Estanqueiro, A., Flynn, D., et al., 2016. Wind and solar energy curtailment: a review of international experience. Renew. Sustain. Energy Rev. 65, 577–586.

Bockris, J.O., Reddy, A.K., Gamboa-Aldeco, M.E., 2002. Modern Electrochemistry 2A. Kluwer Academic Publishers.

Brauns, J., Turek, T., 2020. Alkaline water electrolysis powered by renewable energy: a review. Processes 8, 248.

Caprio, S.S., Vudata, S.P., Bhattacharyya, D., Turton, R., 2020. Transient modeling and simulation of a non-isothermal sodium–sulfur cell. J. Power Sources 453, 227849.

Carmo, M., Stolten, D., 2019. Energy storage using hydrogen produced from excess renewable electricity. Science and Engineering of Hydrogen-Based Energy Technologies. Elsevier, pp. 165–199.

Chen, Q., Zhao, T., 2017. Heat recovery and storage installation in large-scale battery systems for effective integration of renewable energy sources into power systems. Appl. Therm. Eng. 122, 194–203.

Corengia, M., Torres, A.I., 2018a. Two-phase dynamic model for PEM electrolyzer. Computer Aided Chem. Eng. 44, 1435–1440.

Corengia, M., Torres, A.I., 2018b. Effect of tariff policy and battery degradation on optimal energy storage. Processes 6, 204.

Corengia, M., Torres, A.I., 2019. Quantification of battery degradation effects in optimal energy storage schedules. In: Foundations of Computer-Aided Process Design (FOCAPD 2019).

Corengia, M., Torres, A.I., 2020a. Operación óptima de baterías bajo el actual régimen tarifario en Uruguay. ENERLAC. Rev. de. energía de Latinoamérica y. el Caribe 4 (1), 56.

Corengia, M., Torres, A.I., 2020b. Exploiting wind/sun energy surplus for production of green-hydrogen: combining PEM and AE electrolyzers for cost efficiency. In: 2020 AIChE Annual Meeting.

Corengia, M., Estefan, N., Torres, A.I., 2020. Analyzing hydrogen production capacities to seize renewable energy surplus. Computer Aided Chem. Eng. 48, 1549–1554.

Cousins, K., Zhang, R., 2019. Highly porous organic polymers for hydrogen fuel storage. Polymers 11, 690.

Cui, Q., He, L., Han, G., Chen, H., Cao, J., 2020. Review on climate and water resource implications of reducing renewable power curtailment in China: a nexus perspective. Appl. Energy 267, 115114.

Dai, W., Wang, H., Yuan, X.-Z., Martin, J.J., Yang, D., Qiao, J., et al., 2009. A review on water balance in the membrane electrode assembly of proton exchange membrane fuel cells. Int. J. Hydrogen Energy 34, 9461−9478.

Davis, W., Martín, M., 2014. Optimal year-round operation for methane production from CO_2 and water using wind energy. Energy 69, 497−505.

Demirhan, C.D., Tso, W.W., Powell, J.B., Pistikopoulos, E.N., 2019. Sustainable ammonia production through process synthesis and global optimization. AIChE J. 65, e16498.

Denholm, P., O'Connell, M., Brinkman, G., Jorgenson, J., 2015. Overgeneration from solar energy in California: a field guide to the duck chart. Tech. Rep. NREL.

Ding, Z., Zhang, Z., 2019. A behind-the-meter battery control algorithm with the consideration of li-ion battery degradation. In: 2019 IEEE 28th International Symposium on Industrial Electronics (ISIE). IEEE.

Doddathimmaiah, A., Andrews, J., 2009. Theory, modelling and performance measurement of unitised regenerative fuel cells. Int. J. Hydrogen Energy 34, 8157−8170.

Ehsani, M., 2013. Hybrid energy storage hybrid energy storage systems for vehicle applications hybrid energy storage vehicle applications. Transportation Technologies for Sustainability. Springer, New York, pp. 614−626.

Estermann, T., Newborough, M., Sterner, M., 2016. Power-to-gas systems for absorbing excess solar power in electricity distribution networks. Int. J. Hydrogen Energy 41, 13950−13959.

Falcão, D., Pinto, A., 2020. A review on PEM electrolyzer modelling: guidelines for beginners. J. Clean. Prod. 261, 121184.

Fortenbacher, P., Andersson, G., 2017. Battery degradation maps for power system optimization and as a benchmark reference. In: 2017 IEEE Manchester PowerTech. IEEE.

Gahleitner, G., 2013. Hydrogen from renewable electricity: an international review of power-to-gas pilot plants for stationary applications. Int. J. Hydrogen Energy 38, 2039−2061.

Garvey, S.D., Pimm, A., 2016. Compressed air energy storage. Storing Energy. Elsevier, pp. 87−111.

Gil, A., Vicente, M.A., 2019. Energy storage materials from clay minerals and zeolite-like structures. Modified Clay and Zeolite Nanocomposite Materials. Elsevier, pp. 275−288.

Gong, Q., Wang, Y., 2020. Economic dispatching strategy of double lead-acid battery packs considering various factors. J. Electr. Eng. Technol.

Gravity Power, 2020. <http://www.gravitypower.net> (accessed December 2020).

Gu, W., Wang, C., Li, S., Geng, M., Liaw, B., 1999. Modeling discharge and charge characteristics of nickel−metal hydride batteries. Electrochim. Acta 44, 4525−4541.

Hunt, J.D., Zakeri, B., Lopes, R., Barbosa, P.S.F., Nascimento, A., de Castro, N.J., et al., 2020. Existing and new arrangements of pumped-hydro storage plants. Renew. Sustain. Energy Rev. 129, 109914.

Jiang, Y., Kang, L., Liu, Y., 2020. Optimal configuration of battery energy storage system with multiple types of batteries based on supply-demand characteristics. Energy 206, 118093.

Kelley, M.T., Baldick, R., Baldea, M., 2020. Demand response scheduling under uncertainty: chance-constrained framework and application to an air separation unit. AIChE J. 66.

Koskela, J., Lummi, K., Mutanen, A., Rautiainen, A., Jarventausta, P., 2019. Utilization of electrical energy storage with power-based distribution tariffs in households. IEEE Trans. Power Syst. 34, 1693−1702.

Li, L., Liu, P., Li, Z., Wang, X., 2018a. A multi-objective optimization approach for selection of energy storage systems. Computers Chem. Eng. 115, 213−225.

Li, Y., Kang, Z., Mo, J., Yang, G., Yu, S., Talley, D.A., et al., 2018b. In-situ investigation of bubble dynamics and two-phase flow in proton exchange membrane electrolyzer cells. Int. J. Hydrogen Energy 43, 11223−11233.

Li, T., Xu, J., Wang, C., Wu, W., Su, D., Wang, G., 2019a. The latest advances in the critical factors (positive electrode, electrolytes, separators) for sodium-sulfur battery. J. Alloy. Compd. 792, 797−817.

Li, Y., Yang, G., Yu, S., Kang, Z., Mo, J., Han, B., et al., 2019b. In-situ investigation and modeling of electrochemical reactions with simultaneous oxygen and hydrogen microbubble evolutions in water electrolysis. Int. J. Hydrogen Energy 44, 28283−28293.

Liu, H., Zhang, L., Ma, H., Lu, C., Luo, H., Wang, X., et al., 2021. Aluminum hydride for solid-state hydrogen storage: structure, synthesis, thermodynamics, kinetics, and regeneration. J. Energy Chem. 52, 428−440.

8. Integration of resources

Ma, T., Shen, L., Li, M., 2018. Electrical energy storage for buildings. Handbook of Energy Systems in Green Buildings. Springer, Berlin, Heidelberg, pp. 1079–1107.

Maia, L.K., Drünert, L., Mantia, F.L., Zondervan, E., 2019. Expanding the lifetime of li-ion batteries through optimization of charging profiles. J. Clean. Prod. 225, 928–938.

Martín, M., 2016. Optimal year-round production of DME from CO_2 and water using renewable energy. J. CO_2 Utilization 13, 105–113.

Martín, M., Grossmann, I.E., 2017. Towards zero CO_2 emissions in the production of methanol from switch-grass. CO_2 to methanol. Comput. Chem. Eng. 105, 308–316.

Mehrjerdi, H., Hemmati, R., 2019. Modeling and optimal scheduling of battery energy storage systems in electric power distribution networks. J. Clean. Prod. 234, 810–821.

Miao, D., Hossain, S., 2020. Improved gray wolf optimization algorithm for solving placement and sizing of electrical energy storage system in micro-grids. ISA Trans. 102, 376–387.

Miao, Y., Hynan, P., von Jouanne, A., Yokochi, A., 2019. Current li-ion battery technologies in electric vehicles and opportunities for advancements. Energies 12, 1074.

MidAmerican Energy Company, 2014. MidAmerican Energy Company electric tariff no. 2 filed with Iowa utilities board. <https://www.midamericanenergy.com>.

MIEM, ANCAP, UTE, 2020. Proyecto Verne. <https://www.ancap.com.uy/> (accessed December 2020).

Mohan, M., Sharma, V.K., Kumar, E.A., Gayathri, V., 2019. Hydrogen storage in carbon materials—a review. Energy Storage 1, e35.

Mongird, K., Fotedar, V., Viswanathan, V., Koritarov, V., Balducci, P., Hadjerioua, B., et al., 2019. Energy storage technology and cost characterization report. Tech. Rep. US Department of Energy.

Morabito, A., Hendrick, P., 2019. Pump as turbine applied to micro energy storage and smart water grids: a case study. Appl. Energy 241, 567–579.

NGK Insulators, 2020. <https://www.ngk-insulators.com/> (accessed December 2020).

Niermann, M., Beckendorff, A., Kaltschmitt, M., Bonhoff, K., 2019a. Liquid organic hydrogen carrier (LOHC)—assessment based on chemical and economic properties. Int. J. Hydrogen Energy 44 (13), 6631–6654.

Niermann, M., Drünert, S., Kaltschmitt, M., Bonhoff, K., 2019b. Liquid organic hydrogen carriers (LOHCs)—techno-economic analysis of LOHCs in a defined process chain. Energy Env. Sci. 12, 290–307.

Nouri-Khorasani, A., Ojong, E.T., Smolinka, T., Wilkinson, D.P., 2017. Model of oxygen bubbles and performance impact in the porous transport layer of PEM water electrolysis cells. Int. J. Hydrogen Energy 42, 28665–28680.

Okano, K., 2016. Development histories: fuel cell technologies. Green Energy and Technology. Springer, Japan, pp. 93–115.

Olabi, A., Wilberforce, T., Ramadan, M., Abdelkareem, M.A., Alami, A.H., 2020. Compressed air energy storage systems: components and operating parameters – a review. J. Energy Storage 102000.

Olabi, A., Onumaegbu, C., Wilberforce, T., Ramadan, M., Abdelkareem, M.A., Alami, A.H.A., 2021. Critical review of energy storage systems. Energy 214, 118987.

Ong, B., Kamarudin, S., Basri, S., 2017. Direct liquid fuel cells: a review. Int. J. Hydrogen Energy 42, 10142–10157.

Otashu, J.I., Baldea, M., 2019. Demand response-oriented dynamic modeling and operational optimization of membrane-based chlor-alkali plants. Computers & Chem. Eng. 121, 396–408.

Oudalov, A., Cherkaoui, R., Beguin, A., 2007. Sizing and optimal operation of battery energy storage system for peak shaving application. In: 2007 IEEE Lausanne Power Tech. IEEE.

Palys, M.J., Daoutidis, P., 2020. Using hydrogen and ammonia for renewable energy storage: a geographically comprehensive techno-economic study. Comput. Chem. Eng. 136, 106785.

Palys, M.J., Kuznetsov, A., Tallaksen, J., Reese, M., Daoutidis, P., 2019. A novel system for ammonia-based sustainable energy and agriculture: concept and design optimization. Chem. Eng. Process. Process Intensif. 140, 11–21.

Paul, B., Andrews, J., 2017. PEM unitised reversible/regenerative hydrogen fuel cell systems: state of the art and technical challenges. Renew. Sustain. Energy Rev. 79, 585–599.

Proost, J., 2019. State-of-the art CAPEX data for water electrolysers, and their impact on renewable hydrogen price settings. Int. J. Hydrogen Energy 44, 4406–4413.

REN21, 2020. Renewables 2020 global status report. Tech. Rep. Available from: <https://www.ren21.net>.

Robles, J.O., Almaraz, S.D.-L., Azzaro-Pantel, C., 2018. Hydrogen supply chain design: Key technological components and sustainable assessment. Hydrogen Supply Chains. Elsevier, pp. 37–79.

Saba, S.M., Müller, M., Robinius, M., Stolten, D., 2018. The investment costs of electrolysis — a comparison of cost studies from the past 30 years. Int. J. Hydrogen Energy 43, 1209—1223.

Sandeep, K.C., Kamath, S., Mistry, K., Kumar, M.A., Bhattacharya, S.K., Bhanja, K., et al., 2017. Experimental studies and modeling of advanced alkaline water electrolyser with porous nickel electrodes for hydrogen production. Int. J. Hydrogen Energy 42, 12094—12103.

Sarker, Mushfiqur R., Murbach, Matthew D., Schwartz, Daniel T., Ortega-Vazquez, Miguel A., 2017. Optimal operation of a battery energy storage system: Trade-off between grid economics and storage health. Electric Power Systems Research 152, 342—349. 10.1016/j.epsr.2017.07.007.

Schmidt, O., Gambhir, A., Staffell, I., Hawkes, A., Nelson, J., Few, S., 2017. Future cost and performance of water electrolysis: an expert elicitation study. Int. J. Hydrogen Energy 42, 30470—30492.

Seo, K., Edgar, T.F., Baldea, M., 2020. Optimal demand response operation of electric boosting glass furnaces. Appl. Energy 269, 115077.

Soha, T., Munkácsy, B., Harmat, Á., Csontos, C., Horváth, G., Tamás, L., et al., 2017. GIS-based assessment of the opportunities for small-scale pumped hydro energy storage in middle-mountain areas focusing on artificial landscape features. Energy 141, 1363—1373.

Solar Harmonics, 2020. <https://www.solarharmonics.com/top-5-best-tesla-powerwall-alternatives-for-2020/> (accessed November 2020).

Sone, Y., 2011. A 100-W class regenerative fuel cell system for lunar and planetary missions. J. Power Sources 196, 9076—9080.

Srinivasan, S., 2006. Fuel Cells. Springer-Verlag GmbH, Chapter 4.

Sundén, B., 2019. Hydrogen. Hydrogen, Batteries and Fuel Cells. Elsevier, pp. 37—55.

Syali, M.S., Kumar, D., Mishra, K., Kanchan, D., 2020. Recent advances in electrolytes for room-temperature sodium-sulfur batteries: a review. Energy Storage Mater. 31, 352—372.

Tesla, 2020. <https://www.tesla.com/powerwall> (accessed November 2020).

Thompson, A.W., 2018. Economic implications of lithium ion battery degradation for vehicle-to-grid (V2X) services. J. Power Sources 396, 691—709.

Tilton, James N., 1997. Fluid and particle dynamics, Perry's Chemical Engineers' Handbook., Seventh Edition McGraw-Hill.

Tong, Z., Cheng, Z., Tong, S., 2021. A review on the development of compressed air energy storage in China: technical and economic challenges to commercialization. Renew. Sustain. Energy Rev. 135, 110178.

Toyota, 2020. <https://www.toyota.com/highlander/> (accessed December 2020).

UTE, 2020. Pliego tarifario. <https://www.ute.com.uy>.

Wang, J., 2015. Barriers of scaling-up fuel cells: cost, durability and reliability. Energy 80, 509—521.

Wang, Y., Leung, D.Y., Xuan, J., Wang, H., 2017. A review on unitized regenerative fuel cell technologies, part B: unitized regenerative alkaline fuel cell, solid oxide fuel cell, and microfluidic fuel cell. Renew. Sustain. Energy Rev. 75, 775—795.

Wang, A., Kadam, S., Li, H., Shi, S., Qi, Y., 2018. Review on modeling of the anode solid electrolyte interphase (SEI) for lithium-ion batteries. NPJ Comput. Mater. 4.

Wang, H., Frisco, S., Gottlieb, E., Yuan, R., Whitacre, J.F., 2019. Capacity degradation in commercial li-ion cells: the effects of charge protocol and temperature. J. Power Sources 426, 67—73.

Wang, Y., An, N., Wen, L., Wang, L., Jiang, X., Hou, F., et al., 2021. Recent progress on the recycling technology of li-ion batteries. J. Energy Chem. 55, 391—419.

Wong, C., Wong, W., Ramya, K., Khalid, M., Loh, K., Daud, W., et al., 2019. Additives in proton exchange membranes for low- and high-temperature fuel cell applications: a review. Int. J. Hydrogen Energy 44, 6116—6135.

Wulf, C., Zapp, P., Schreiber, A., 2020. Review of power-to-X demonstration projects in Europe. Front. Energy Res. 8.

Yamchi, H.B., Shahsavari, H., Kalantari, N.T., Safari, A., Farrokhifar, M., 2019. A cost-efficient application of different battery energy storage technologies in microgrids considering load uncertainty. J. Energy Storage 22, 17—26.

Yan, X., Ozturk, Y., Hu, Z., Song, Y., 2018. A review on price-driven residential demand response. Renew. Sustain. Energy Rev. 96, 411—419.

Yartys, V., Lototskyy, M., Akiba, E., Albert, R., Antonov, V., Ares, et al., 2019. Magnesium based materials for hydrogen based energy storage: past, present and future. Int. J. Hydrogen Energy 44, 7809—7859.

Zhang, Q., Grossmann, I.E., 2016. Enterprise-wide optimization for industrial demand side management: fundamentals, advances, and perspectives. Chem. Eng. Res. Des. 116, 114–131.

Zhang, Q., Sundaramoorthy, A., Grossmann, I.E., Pinto, J.M., 2016. A discrete-time scheduling model for continuous power-intensive process networks with various power contracts. Computers & Chem. Eng. 84, 382–393.

Zhang, Q., Martín, M., Grossmann, I.E., 2019a. Integrated design and operation of renewables-based fuels and power production networks. Computers Chem. Eng. 122, 80–92.

Zhang, Z., Shi, J., Gao, Y., Yu, N., 2019b. Degradation-aware valuation and sizing of behind-the-meter battery energy storage systems for commercial customers. In: 2019 IEEE PES GTD Grand International Conference and Exposition Asia (GTD Asia). IEEE.

Zheng, W., Hennessy, J.J., Li, H., 2019. Reducing renewable power curtailment and CO_2 emissions in China through district heating storage. WIREs Energy Environ. 9.

Zoski, C.G. (Ed.), 2007. Handbook of Electrochemistry. Elsevier Science & Technology, Chapter 18.

Solutions

Chapter 2

P2.1. $W = (F) \times \frac{R \times k \times (T)}{((MW) \times (k-1))} \frac{1}{\eta_c} \left(\left(\frac{P_f}{P_{ini}} \right)^{(k-1)/k} - 1 \right)$

P2.2. $P = 2 \cdot \pi \cdot M \cdot N = P_o \cdot \rho \cdot N^3 \cdot T^5$

P2.3. yield $= 74.6301 + 0.4209 \times T(\text{reactor}) + 15.1582 \times \text{Cat} + 3.1561 \times \text{ratio\_met} - 0.0019 \times T(\text{reactor})^2 - 0.2022 \times T(\text{reactor}) \times \text{Cat} - 0.01925 \times T(\text{reactor}) \times \text{ratio\_met} - 4.0143 \times \text{Cat}^2 - 0.3400 \times \text{Cat} \times \text{ratio\_met} - 0.1459 \times \text{ratio\_met}^2$

P2.4. 58

P2.5. $x_1 = 3 + 1/3$; $x_2 = 2/6$

P2.6. $[x_1 - x_2 \le -1] \vee [-x_1 + x_2 \le -1]$

$0 \le x_i \le 5$

$big - M$

$x_1 - x_2 \le -1 + M(1 - y_1)$

$-x_1 + x_2 \le -1 + M(1 - y_2)$

Convex hull

$x_1 = \nu_1^1 + \nu_1^2$

$x_2 = \nu_2^1 + \nu_2^2$

$\nu_1^1 - \nu_2^1 \le -1 \cdot y_1$

$-\nu_1^2 + \nu_2^2 \le -1 \cdot y_2$

$0 \le \nu_1^1 \le 5y_1; 0 \le \nu_1^2 \le 5y_2; 0 \le \nu_2^1 \le 5y_1; 0 \le \nu_2^2 \le 5y_2$

P2.7. $y_1 = 1$; $y_2 = 3$.

P2.8. $a_1 x_1 + \cdots + a_n x_n = b$

$(\lfloor a_1 \rfloor + [a_1 - \lfloor a_1 \rfloor])x_1 + \cdots + (\lfloor a_n \rfloor + [a_n - \lfloor a_n \rfloor])x_n = \lfloor b \rfloor + [b - \lfloor b \rfloor])$

$[a_1 - \lfloor a_1 \rfloor] = f_1$

$(\lfloor a_1 \rfloor + f_1)x_1 + \cdots + (\lfloor a_n \rfloor + f_n)x_n = \lfloor b \rfloor + f$

$f_1 x_1 + \cdots f_n x_n - f = \lfloor b \rfloor - \lfloor a_1 \rfloor x_1 - \cdots - \lfloor a_n \rfloor x_n$

For a feasible x, RHD is integer

For a feasible x, LHD is integer

Since $0 \le f \le 1$, $x \ge 0$ and LHD is integer, it cannot be negative

$f_1 x_1 + \cdots f_n x_n - f \ge 0 \Leftrightarrow f_1 x_1 + \cdots f_n x_n \ge f$

Thus

$y_{Bi}^* + \sum_{j \in NB} \overline{a}_{ij} y_j^* = \overline{a}_{io}$

A cut is

$f_1 x_1 + \cdots f_n x_n \ge f$

P2.9. $x_1 = 41/9, x_2 = 26/9, u_2 = 2/9 Z = 0.056$

P2.10. : $-3 \times x_1 + x_2 = 0$

P2.11. 10.9;3.1;1.1

P2.12. Extreme points u^1 (0,1), u^2 (35/22, 1/22), Extreme rays v^1(1,3/2) y $v^2 = (1, 1/2)$

$$\min_{x,n} 3x_1 + 2x_2 - 4x_3 + n$$
$$\text{st} \quad n \geq 0(4 - 2x_1 + 4x_2 - x_3) + 1(-2 - x_1 - 2x_2 + 4x_3)$$
$$n \geq \frac{35}{22}(4 - 2x_1 + 4x_2 - x_3) + \frac{1}{22}(-2 - x_1 - 2x_2 + 4x_3)$$

$$1(4 - 2x_1 + 4x_2 - x_3) + \frac{3}{2}(-2 - x_1 - 2x_2 + 4x_3) \leq 0$$

$$1(4 - 2x_1 + 4x_2 - x_3) + \frac{1}{2}(-2 - x_1 - 2x_2 + 4x_3) \leq 0$$

$$x_j \leq 6, i = 1, 2, 3$$
$$x \in Z_+^3, n \in R^1$$

P2.13. The Dual: The concept of dual is interesting from the mathematical and physical point of view. It will be useful for the solution of other types of problems, for instance MILP, but it also provides useful information on the sensitivity of the solution of the LP problem at hand. It can be considered as an alternative point of view. The kind of information that brings to the table is that of the sensitivity of the solution, shadow prices, which in economics plays an important role.

$$\min c^T x$$
$$\text{st. } Ax \leq b$$
$$x \geq 0$$

The dual

$$\max b^T y$$
$$\text{st. } A^T y \leq c$$

The idea is to find the optimal bound to the solution, which is the tightest implied inequality that can be developed with the constraints of the original, primal, problem. Let's consider the following problem:

$$\text{Max} \quad 5x_1 + x_2 + 3x_3$$
$$\text{st.}$$
$$x_1 + 4x_2 \qquad \leq 1$$
$$2x_1 - x_2 + x_3 \quad \leq 2$$
$$x_1, x_2, x_3 \geq 0$$

Any feasible solution provides a lower bound for the maximization problem ζ^*. For instance, assuming (1,0,0), $\zeta^* \geq 5$ while if (0,0,2), $\zeta^* \geq 6$. By multiplying the first constraint by 2 and the second constraint by 3 and adding both, we develop an inequality whose coefficients for each one of the variables are larger or equal than the ones in the objective function, $8 \geq 5$, $5 \geq 5$, $3 \geq 3$. Thus we have created an upper bound for the objective function.

$$
\begin{aligned}
2(x_1 + 4x_2) \quad &\leq 2(1) \\
+ 3(2x_1 - x_2 + x_3) \quad &\leq 3(2) \\
\hline
8x_1 + 5x_2 + 3x_3 &\leq 8
\end{aligned}
$$

$$6 \leq 5x_1 + x_2 + 3x_3 \leq 8x_1 + 5x_2 + 3x_3 \leq 8$$

Thus $\zeta^* \leq 8$. We have bounded the objective between 6 and 8. The idea is to optimize the variables that multiply the constraints instead of selecting some by inspection.

$$
\begin{aligned}
y_1(x_1 + 4x_2) \quad &\leq y_1(1) \\
+ & \\
y_2(2x_1 - x_2 + x_3) \quad &\leq y_2(2) \\
\hline
(y_1 + 2y_2)x_1 + (4y_1 - y_2)x_2 &+ (0y_1 + y_2)x_3 \leq (y_1 + 2y_2)
\end{aligned}
$$

And now we make the coefficients to be upper bounds for the one sin the original objective function.

$$
\begin{aligned}
(y_1 + 2y_2) &\geq 5 \\
(4y_1 - y_2) &\geq 1 \\
y_2 &\geq 3
\end{aligned}
$$

The upper bound obtained $y_1 + 2y_2$ is to be minimized aiming at the best upper bound.

$$
\begin{aligned}
\text{Min} \quad & y_1 + 2y_2 \\
\text{st.} \quad & (y_1 + 2y_2) \geq 5 \\
& (4y_1 - y_2) \geq 1 \\
& y_2 \geq 3 \\
y_i \geq 0 &
\end{aligned}
$$

This procedure can be transformed in a series of rules to develop a dual optimization problem out of a primal as follows:

Primal (min)	Dual (max)	Primal (max)	Dual (min)
ith constraint $\leq$	ith variable ≤ 0	ith constraint $\leq$	ith variable ≥ 0
ith constraint $\geq$	ith variable ≥ 0	ith constraint $\geq$	ith variable ≤ 0
ith constraint $=$	ith variable unrestricted	ith constraint $=$	ith variable unrestricted
jth variable ≥ 0	jth constraint ≤ 0	jth variable ≥ 0	jth constraint ≥ 0
jth variable ≤ 0	jth constraint ≥ 0	jth variable ≤ 0	jth constraint ≤ 0
jth variable unrestricted	jth constraint $=$	jth variable unrestricted	jth constraint $=$

Using the rules for an example

$$\min Z = x_1 + x_2 - x_3$$
$$\text{st.}$$
$$2x_1 + x_2 \quad \geq 3$$
$$x_1 - \quad x_3 = 2$$
$$x_3 \geq 2$$

We asign y_1 to the first constraint and y_2 to the second one. Variables x_1 and x_2 are not restricted. Thus the dual constraints associated to them are equalities. x_3 is ≥ 0, thus the associated constraint is to be as ≤ 0. The first constraint of the primal is $\geq$, thus y_1 must also be $\geq$. The second is an equality constrain, thus y_2 is unrestricted.

$$\min Z = x_1 + x_2 - x_3$$
$$\text{st.}$$
$$y_1(2x_1 + x_2 \qquad \geq 3)$$
$$y_2(x_1 - \qquad x_3 = 2)$$
$$x_3 \geq 2$$

$\rightarrow$

$$\text{Max } Z = 3y_1 + 2y_2$$
$$\text{st.}$$
$$2y_1 + y_2 \qquad = 1$$
$$y_2 \qquad = 1$$
$$- y_2 \leq -1$$
$$y_1 \geq 0$$

P2.14.

$$\min \quad 2u_1 + 3u_2$$
$$\text{st.} \quad u_1 + 3u_2 \geq 5$$
$$u_1 + u_2 \geq 4$$
$$u_1 \geq 0, u_2 \geq 0$$

Chapter 3

P3.1. Solution: Cradle-to-gate scope will allow the comparison to other biodiesel alternatives assuming similar downstream life cycle activities. The LCA over the production enables a fair comparison since the use phase of any type of biodiesel is virtually equal.

P3.2.

Primary data	Flow	Value	Unit
FU	Biodiesel	1	t
Cultivation	Agricultural task (diesel burned in agricultural machinery)	3659.4	MJ
	Fertilizer	203.3	kg
	N_2O emission	0.81	kg
Separation	Electricity	15.2	kWh
	Transport	304	tkm[a]

(Continued)

(Continued)

Primary data	Flow	Value	Unit
FU	Biodiesel	1	t
Extraction	Electricity	136	kWh
Transesterification	Methanol	118.5	kg
	Sodium hydroxide	14.8	kg
	Heat, unspecific, in chemical plant	4650	MJ
	Electricity	0.4	kWh

[a]tkm is the abbreviation for tonne-kilometer, a unit of measure of transportation which represents the transport of one of goods over a distance of 1 km.

P3.3 The energy consumption for heating purposes in the transesterification plant is the more *significant contributor to the climate change impact in the studied system (32.1%)*.

P3.4 Given that the Climate Change impact scores a total value of 1.06 $kgCO_{2\text{-eq}}/kg$ biodiesel, it seems that the thistle-based biodiesel performs environmentally better than the soybean oil-based biodiesel. However, the scope, limitations, and general assumptions made in both studies must be thoroughly revised to assure a fair comparison. Moreover, to draw conclusions about sustainability other factors must be analyzed, such as technical and cost-effective feasibility and socioeconomic acceptance.

P3.5 For example, the application of heat integration techniques in the transesterification plant, the use of heat waste from other industrial facilities, the use of alternative technology or fuel for heat generation, the use of alternative fertilizer through the valorization of biowastes, the use of alternative means of freight transport, among others.

Chapter 4

P4.1. MEA: 1603 kW; DEA: 1062 kW; MDEA: 1254 kW
P4.2. $RelO_2 = 0.151$; $RelWa = 1.97$; $P = 11.4$ bar; $T = 950°C$
P4.3. See Fig. 4P3.1.
P4.4. See Fig. 4P4.1.
P4.5. 552.6°C, 38.4 kg tars/kg biomass

Chapter 5

P5.1. See Fig. 5P1.1.
P5.2. See Fig. 5P2.1.

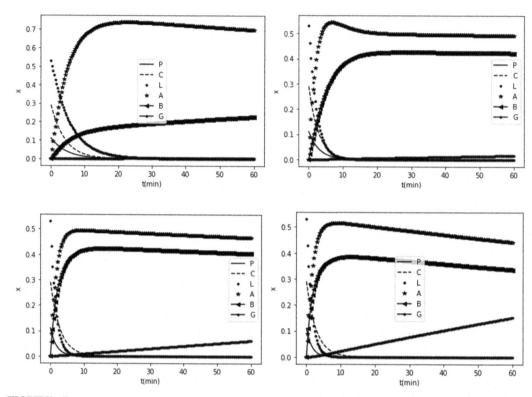

FIGURE 4P3.1   Profiles.

P5.3.

	Single	System
Heating (MJ/s)	23.0	12.5 + 6.9
Reboiler (MJ/s)	26.4	16.2

P5.4.

%	Duty (MJ/s) per 100 kg/s feed (HX + Reb)	Duty (MJ/kg etOH)	Reflux
4	23.8 + 30.7	13.6	6.7
6	23.0 + 26.4	8.2	4.5
8	22.2 + 31.3	6.7	3.6
12	21.0 + 39.5	5.0	2.6
15	20.2 + 45.7	4.4	1.1

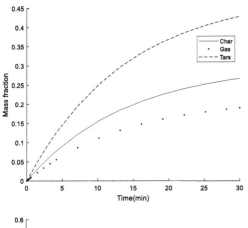

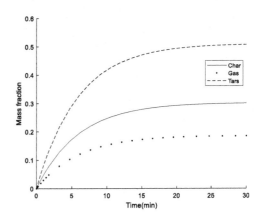

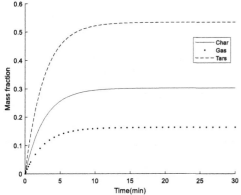

FIGURE 4P4.1   Profiles.

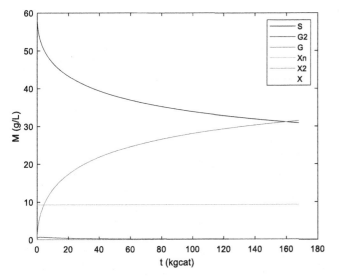

FIGURE 5P1.1   Sugar profiles.

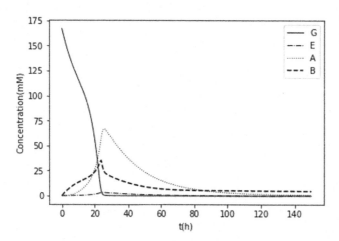

FIGURE 5P2.1 Substrate and product profiles.

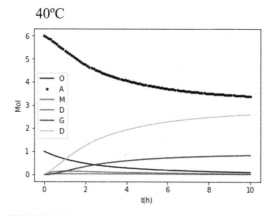 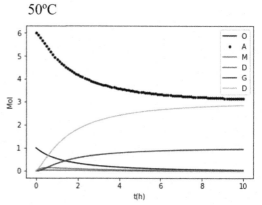

FIGURE 5P5.1 Biodiesel production profiles.

P5.5. See Fig. 5P5.1.

P5.6.

$$T \cdot \text{Cat}$$
$$(T\text{-}T^L) \cdot (\text{Cat-Cat}^L) \geq 0 \Rightarrow T \cdot \text{Cat} \geq T \cdot \text{Cat}^L + T^L\text{Cat- } T^L\text{Cat}^L$$
$$(T\text{-}T^U) \cdot (\text{Cat-Cat}^L) \leq 0 \Rightarrow T \cdot \text{Cat} \leq T \cdot \text{Cat}^L + T^U\text{Cat- } T^U\text{Cat}^L$$
$$(T\text{-}T^L) \cdot (\text{Cat-Cat}^U) \leq 0 \Rightarrow T \cdot \text{Cat} \leq T \cdot \text{Cat}^U + T^L\text{Cat- } T^U\text{Cat}^U$$
$$(T\text{-}T^U) \cdot (\text{Cat-Cat}^U) \geq 0 \Rightarrow T \cdot \text{Cat} \geq T \cdot \text{Cat}^U + T^L\text{Cat- } T^U\text{Cat}^U$$
$$T \cdot \text{RM}$$
$$(T\text{-}T^L) \cdot (\text{RM-RM}^L) \geq 0 \Rightarrow T \cdot \text{RM} \geq T \cdot \text{RM}^L + T^L\text{RM- } T^L\text{RM}^L$$
$$(T\text{-}T^U) \cdot (\text{RM-RM}^L) \leq 0 \Rightarrow T \cdot \text{RM} \leq T \cdot \text{RM}^L + T^U\text{RM- } T^U\text{RM}^L$$
$$(T\text{-}T^L) \cdot (\text{RM-RM}^U) \leq 0 \Rightarrow T \cdot \text{RM} \leq T \cdot \text{RM}^U + T^L\text{RM- } T^U\text{RM}^U$$
$$(T\text{-}T^U) \cdot (\text{RM-RM}^U) \geq 0 \Rightarrow T \cdot \text{RM} \geq T \cdot \text{RM}^U + T^L\text{RM- } T^U\text{RM}^U$$
$$\text{Cat} \cdot \text{RM}$$
$$(\text{Cat-Cat}^L) \cdot (\text{RM-RM}^L) \geq 0 \Rightarrow \text{Cat} \cdot \text{RM} \geq \text{Cat} \cdot \text{RM}^L + \text{Cat}^L\text{RM- } \text{Cat}^L\text{RM}^L$$
$$(\text{Cat-Cat}^U) \cdot (\text{RM-RM}^L) \leq 0 \Rightarrow \text{Cat} \cdot \text{RM} \leq \text{Cat} \cdot \text{RM}^L + \text{Cat}^U\text{RM- } \text{Cat}^U\text{RM}^L$$
$$(\text{Cat-Cat}^L) \cdot (\text{RM-RM}^U) \leq 0 \Rightarrow \text{Cat} \cdot \text{RM} \leq \text{Cat} \cdot \text{RM}^U + \text{Cat}^L\text{RM- } \text{Cat}^U\text{RM}^U$$
$$(\text{Cat-Cat}^U) \cdot (\text{RM-RM}^U) \geq 0 \Rightarrow \text{Cat} \cdot \text{RM} \geq \text{Cat} \cdot \text{RM}^U + \text{Cat}^L\text{RM- } \text{Cat}^U\text{RM}^U$$

```
Variable
Conversion
Positive variables
Temperature

RM
Cat;
Temperature.LO = 45;
Temperature.UP = 65;
RM.LO = 4.5;
RM.UP = 7.5;
Cat.LO = 0.5;
Cat.UP = 1.5

Equations
Yield;
Yield.. Conversion = E =      74.6302 + 0.4209*Temperature + 15.1582*Cat +
              3.1561*RM-0.0019*Temperature*Temperature-0.2022*Temperature*Cat
              -0.01925*Temperature*RM-4.0143*Cat*Cat-0.34*Cat*RM-0.1459*RM*RM;

Model Biodieselglobal /ALL/;
option NLP = BARON;
Solve Biodieselglobal Using NLP Maximizing Conversion;
```

−− **VAR Conversion**	−INF	99.465	+ INF	.
−− **VAR Temperature** ~	45.000	48.539	65.000	.
−− **VAR RM**	4.500	7.031	7.500	.
−− **VAR Cat**	0.500	0.500	1.500	−1.0

P5.7. Regions 1 ($81.5(g/m^2 d)$) and 9 ($80.5 (g/m^2 d)$)

P5.8. TotN = 5.04 g/L; TotP = 1.79 g/L

P5.9. % = 0.5; $T = 180°C$; $t = 11$ min: Enzyme = $50 \rightarrow 92\%$ y 62%

Chapter 6

P6.1.

	Cattle slurry
CO_2	0.3
O_2	0.05
N_2	0.02
H_2S	> 0.002
NH_3	5E−5
CH_4	0.675

P6.2. −0.79

P6.3. Use example 6.3 as reference

P6.4. Use model developed for example 6.5.

P6.5. RatioO$_2$ = 0.5; Steam = 31.3/60 (kg of biogas); Biogas as fuel 8.2/60, Pref: 8.2

Chapter 7

P7.1. Column C1: 27 stages; Distillate 25 kmol/h; Feed stage: 13; vapor stage; 27
Column C2: 59 stages; Reboiler duty 1500 kW; Reflux ratio: 5; Feed stage: 30; vapor stage; 31; Interconnecting vapor flow: 50 kmol/h

P7.2. Thermally coupled: Reboiler duty: 800 kW; 60 stages; 25 stages; Distillate rate 25 kmol/h; Reflux ratio: 4.
Thermally equivalent: 74 stages: Distillate rate: 25 kmol/h; Reflux ratio 4; Stages: 12 Reboiler duty 100 kW.

P7.3. See Fig. 7P3.1.

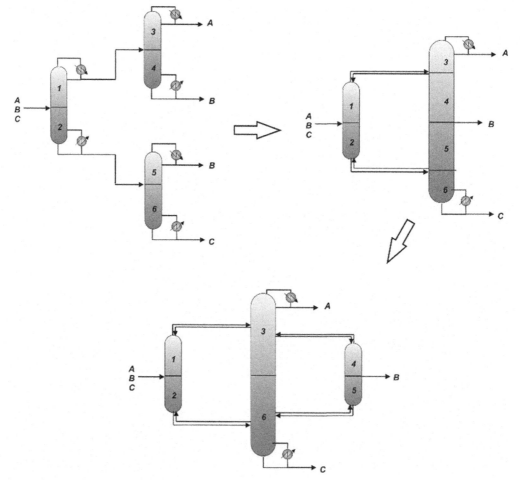

FIGURE 7P3.1   Equivalent columns.

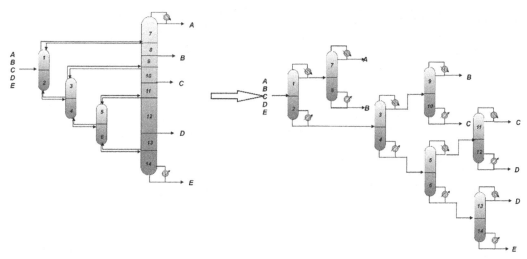

FIGURE 7P4.1   Equivalent distillation sequence.

P7.4. See Fig. 7P4.1.
P7.5. See Fig. 7P5.1.
P7.6. See Fig. 7P6.1.
P7.7. See Fig. 7P7.1.

Chapter 8

P8.1. Corn: F1 0.45; F2: 0.2; F3: 1.15; F4: 2.5; F5: 2.5
Wheat: F1 0.64; F3: 1.23; F4: 2; F5: 3
Corn: F2 1.31; F3: 2.45; F4: 22; F5: 4
P8.2. CS = 64; SW = 62 SW does not have enough lignin to be self-sufficient otherwise it will be the best one.
P8.3. $x_1 = 0.455$; $x_2 = 0.0$; $x_3 = 0.076$; $x_4 = 0.121$; $x_5 = 0.348$
Environmental = 4.0 py/kg
Cost = 21.06 €/kg
P8.4. MSW: 1000t; SLUDGE:0t; MANURE:1000t; LIGNO:500t; Profit:210000€
P8.5. (Baron, 5% of gap) (see Tables 8P5.1 and 8P5.2):

Chapter 9

P9.1. Isoeff:0.78. $W = 14{,}328$ kW; $Q = 24{,}640$ kW
P9.2. Average power 24.6 MW; $P = 40$ bar
P9.3. 168 MW
P9.4. Formulate the optimization problem for the selection of the allocation considering water and energy consumption.

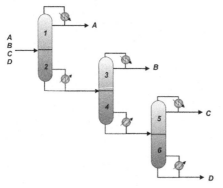

(a) Conventional Scheme 1

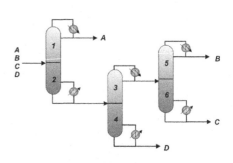

(b) Conventional Scheme 2

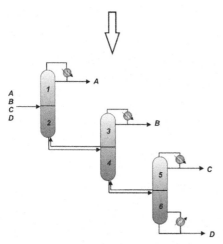

Thermally coupled 1 Conventional Scheme 1

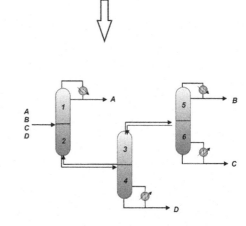

Thermally coupled 1 Conventional Scheme 2

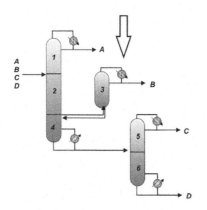

Thermally coupled 2 Conventional Scheme 1

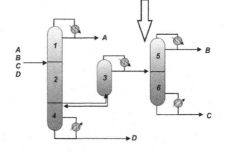

Thermally coupled 2 Conventional Scheme 2

FIGURE 7P5.1   Thermally coupled equivalent columns.

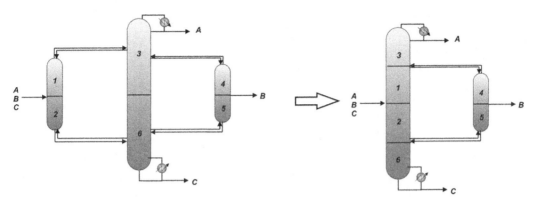

FIGURE 7P6.1   Equivalent sequences to thermally coupled system.

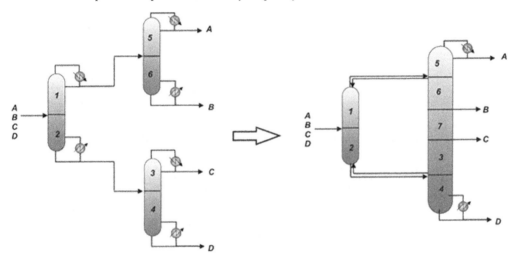

FIGURE 7P7.1   Equivalent Petlyuk column.

```
Scalar
Pressure     atm /1/
Ropt                /0.6/
Helios              /2000/
PowerR              /0.4/
RHelio              /0.9/
MinWater
MaxPower;

Binary variable
Region(loc)

Variable
Z;
```

TABLE 8P5.1 Solution of the problem 5 (part 1).

Ingredient	Supplier	Amount purchased	Product	Customer	Amount sold
1	1	40.19	1	1	300
1	2	159.81	1	2	300
2	1	122.3	1	3	300
2	2	127.7	2	2	45.875
3	1	142.5	**Location selected**	1	
3	2	32.5	**Profit (MM€)**	3,6	
4	2	100			
5	2	0.29			
5	3	0.65			
6	3	15			
7	2	0.94			
8	1	104.1			
8	2	99.9			

TABLE 8P5.2 Solution of the problem (part 2).

		Location	
Products	Ingredient	1	2
1	1	0.21099955	0.10714286
	2	0.2605	0.3273
	3	0.1884	0.0500
	4	0.1059	0.1000
	5	0.0010	0.0010
	6	0.0162	0.0291
	7	0.0010	0.0010
	8	0.2160	0.3844
2	1	0.2202	0.2500
	2	0.3381	0.3013
	3	0.1187	0.2500
	4	0.1030	0.1000
	5	0.0010	0.0087
	6	0.0099	0.0010
	7	0.0010	0.0010
	8	0.2081	0.0880

```
Parameter
WaterC(month,loc)
Fyield(month,loc)
Power(month,loc)
Waterloc(loc)
Powerloc(loc)

Waterobj(loc)
Powerobj(loc);
```

WaterC(month,loc) = -0.0002297*(Tempearture(month,loc))**2 + 0.798*(humidity(month, loc)/100)**2 + 7.09*Pressure**2 + 0.022*(Tempearture(month,loc))*(humidity (month, loc)/100)

         + 0.02993*(Tempearture(month,loc))*Pressure-0.515*(humidity(month, loc)/100)*Pressure-0.01533*(Tempearture(month,loc))-1.417*(humidity(month,loc)/ 100)-12.574*Pressure + 7.6256;

Fyield(month,loc) = Ropt-(0.01*(550-Tempearture(month,loc))/(DNI(month,loc)*1000));
Power(month,loc) = 1000*(DNI(month,loc))*Helios*120*Fyield(month,loc)/ 1000*PowerR*RHelio;

```
Waterloc(loc) = sum(month, WaterC(month,loc) )/12;
Powerloc(loc) = sum(month, Power(month,loc) )/12;

MinWater = smin(loc,Waterloc(loc));
MaxPower = smax(loc,Powerloc(loc));

Waterobj(loc) = 1-(Waterloc (loc)-MinWater)/MinWater;
Powerobj(loc) = Powerloc(loc)/MaxPower;

Equation
eq1, eq2;

eq1.. Z = E = sum(loc, Region(loc)*(Waterobj(loc) + Powerobj(loc)));
eq2..   sum(loc, Region(loc)) = E = 1;
```

P9.5. Rend_field = 0.565

P9.6. $W = 21$ MW $Q = -30$ MW. Regeneration = 22%. Flow of FTH = 122 kg/s; Flow steam = 16 kg/s

Chapter 10

P10.1. $T = 2857K$; $Si = 0.168$; $CO = 0.655$; $Sig = 0.04$; $Sic = 0.06$; $SiO = 0.09$

P10.2. $n = 1.2533T^{-0.01005}$

P10.3. Use example 10.3 as reference

P10.4. Sol Solve P1(Original) Fix Region("1") = 0 SolP2—Region = 2
Solve P2, Fix Region("9") = 0
Sol P3—Region = 11.

P10.5. Ropt: 0.183

Chapter 11

P11.1. 7D

P11.2. A subsidy of at least 270 €/MW is required

P11.3. $a' = a \cdot y_i;\; b' = b \cdot y_i;\; c' = c \cdot y_i$

P11.4. Use Example 11.4

P11.5. Use Example 11.4

P11.6.

Zone	1	2	3	4	**5**
V(ave) (m/s)	12.0	9.9	11.0	9.8	**10.3**
P_{WT} (kW)	1716	1348	1797	1353	**1690**
$P_{Windfarm}$ (MW)	85.8	67.4	89.9	66.7	**84.5**
Power (MW)	41.5	32.6	43.5	32.7	**40.8**
Cost (€/kW)	8588	8864	8455	8655	**8354**

Chapter 12

P12.1. 21.17 MPa

P12.2. $T_{Barcelona} = 119°C$; $T_{Galicia} = 119°C$; $T_{Aragon} = 171°C$; $T_{Lerida} = 128°C$; $T_{Salamanca/Caceres} = 119°C$; $T_{Madrid} = 137°C$; $T_{Sevilla/Cadiz} = 147°C$; $T_{Granada} = 109°C$; $T_{Almeria} = 105°C$; $T_{Canarias} = 208°C$

P12.3.

P (Flash) (bar)	Steam fraction	Turbine (MW)	Pump1 (MW)	Pump2 (MW)
0.5	0.17	4.55	0.0007	1.36
1	0.14	5.16	0.0013	1.37
2	0.10	4.89	0.002	1.38
5	0.043	2.63	0.002	1.39

P12.4.

Place	TWell (°C)	Power (MW)	Foc (kg/s)	T(HX1) (°C)	Tbrine(inj) (°C)
Aragon	175	6.1	87.1	87.9	73.0
Salamanca	120	2.0	28.8	87.9	86.3
Sevilla	150	4.3	60.6	87.9	79.1
Canarias	210	8.7	124.3	87.9	64.5

P12.5. $W = 1.37$ MW; 40% of the toluene evaporates

P12.6. The turbine produces 2.3 MW. Foc = 63.67 kg/s

P12.7. **37.2 kJ/m²**

P12.8. 137 m

P12.9. 1310 kg

Chapter 13

P13.1. $Q = 23.9$ m³/s; 3 pipes; $Q = 33$ m³/s; 3 pipes; $Q = 14.9$ m³/s; 2 pipes; $Q = 8.24$ m³/s; 1 pipe

P13.2. $Q = 85$ m³/s; 229 million m³

P13.3. 6.87% w/w.

P13.4. $X =: 99.1\%$. Heat consumption: 443.9 kW

Chapter 14

P14.1. Use model in Example 14.3.

P14.2. Use model in Example 14.3.

Chapter 15

P15.1. In this case, more than one type of storage may be available, so Eq. (15.39) does not apply. Instead, as C_3 and D_3 are different modes of the same equipment, if one is installed so does the other:

$$y_{C3} = y_{D3}$$

Also, just like the battery, this URFC cannot work simultaneously in charge and discharge mode, so a binary variable $(z_{S_{3,t}})$ is included: during charge $z_{S_{3,t}} = 1$; during discharge $z_{S_{3,t}} = 0$. Then, Eqs. (15.19) and (15.9) are replaced by:

$$0 \leq P^{in}_{C_{3,t}} \leq z_{S_{3,t}} IC_{C_3} \quad 0 \leq P^{out}_{D_{3,t}} \leq (1 - z_{S_{3,t}}) IC_{D_3}$$

P15.2. a) 3866 kWh for energy storage and 1955 kW for solar generation

b)

Time	$P^{max}_{G_t}$ (kW) Summer	Aut./Spring	$\Delta t \left(P^{max}_{G_t} - P^{dem}_t \right)$ (kWh) Summer	Aut./Spring	$\Delta t P^D_t$ (kWh) Summer	Aut./Spring
1–6	0	0	−250	−250	250	250
7	320	0	70	−250	0	250
8	860	400	610	150	0	0
9	1240	820	990	570	0	0
10	1420	980	1170	730	0	0

(Continued)

(Continued)

Time	$P^{max}_{G_t}$ (kW)		$\Delta t\left(P^{max}_{G_t} - P^{dem}_t\right)$ (kWh)		$\Delta t P^D_t$ (kWh)	
	Summer	Aut./Spring	Summer	Aut./Spring	Summer	Aut./Spring
11	1460	1020	1210	770	0	0
12	1480	1020	1230	770	0	0
13	1500	1040	1250	790	0	0
14	1460	980	1210	730	0	0
15	1400	940	1150	690	0	0
16	1340	880	1090	630	0	0
17	1240	720	990	470	0	0
18	940	280	690	30	0	0
19	200	0	−50	−250	50	250
20−24	0	0	−250	−250	250	250

Total surplus as 1.3 GWh.

c) Installed capacities are: ~ 1312 kWh for storage and ~ 1116 kW for generation. If solar generation capacity is set to 1150 kW, following (b) net surplus is 0.8 GWh/yr.

P15.3. a) **0.855; b) 0.63**

P15.4. a)

$$\min_{P^{grid}, P^{direct}_t, P^C_t, P^D_t, SoC_t} P^{grid}\$_{DC} + 30\left(\frac{d}{month}\right)\sum_t \Delta t_t \$_{EC}\left(P^{direct}_t + P^C_t\right)$$

s.t.

$$P^{direct}_t + P^C_t \le P^{grid}$$

$$SoC_t = SoC_{t-1} + \Delta t P^C_t \eta_C - \frac{\Delta t P^D_t}{\eta_D}$$

$$0 \le SoC_t \le C$$

$$P^{dem}_t = P^{direct}_t + P^D_t$$

b) 5913 USD/month

c) **It is not unique**

d) Two nlp solvers (Ipopt and Conopt) where used, both provide the exact same solution.

e) No. To assess whether or not the solution is unique, we should analyze if the problem is strictly convex. This is done by checking that all the eigenvalues of the Hessian matrix of the Lagrangian function for this problem are strictly positive.

P15.5. a)

See Fig. 15P5.1.

b) 1.03 MW

c)

See Fig. 15P5.2.

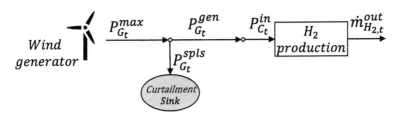

FIGURE 15P5.1   Energy Storage Systems of the example.

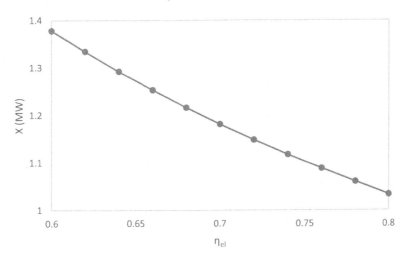

15P5.2   Figure Efficiency of the electrolyzer.

A

General nomenclature

a	Specific area (length^{-1})
a	Aperture of the parabola (length) (Chapter 9: Solar Thermal Energy)
A	Area (length units square)
A_i	Availability of species i (Chapter 8: Added-value Products)
A	Solar azimuth angle (Chapter 9: Solar Thermal Energy)
AM	Ratio of ammonia in unit of mass per mass of biomass
C	Concentration (mol per volume unit)
Cat	Mass fraction of catalyst added
CF$_i$, CP$_i$, CL$_i$	Corresponds to the crude fiber, protein, and lipids
CF	Capacity factor (Chapter 9: Solar Thermal Energy)
Cp	Heat capacity (energy per unit mass or mol and temperature)
d_{32}	Sauter mean diameter (m)
D	Distillate flow rate (molar per unit time)
D_i	Diffusion coefficient (length square per second)
D_p	Particle diameter (length units)
D_k	Demand of product k (Chapter 8: Added-value Products)
DM	Dry matter fraction
DNI	Direct normal irradiance (energy per unit time and area)
DQ	Digestibility coefficient
E	Reduction potential (V/mol)
E	Evaporated flow rate (Chapter 4: Thermochemical Processes) (mass per unit time)
E	Enzyme (Chapter 5: Biochemical-based Processes) (mass per unit volume)
f_i	Feed flow rate of component i (mass or molar per unit time)
f_i^o	Standard fugacity of species i (pressure units)
f	Friction factor
F	Flow rate (of the feed for distillation columns) (mass or molar per unit time)
F	Faraday constant
F_{12}	View factor
FFA	Free fatty acid
FAME	Fatty acid methyl ester
FAEE	Fatty acid ethyl ester
g_n	Phase lag
G	Gas flow rate (mass per unit time)
G	Gas flow rate per unit area (Ergun equation) (mass per unit time and area)
G	Growth rate in crystallization (length per unit time)

$\overline{G^o_i}$	Standard Gibbs free energy (energy)
h	Specific enthalpy (energy per unit mass or molar)
hi	Film coefficient (power per length and temperature)
h_g	Heat transfer coefficient (energy per unit length and temperature)
H	Henry constant (pressure units per molar fraction)
H	Enthalpy (energy per unit mass or molar)
H_{helio}	Height of the heliostat (length)
H_{tower}	Height of the tower(length)
HHV	High heating value (energy per mass)
I	Current intensity (Ampere)
I	Inhibition (Chapter 5: Biochemical-based Processes) (mass per unit volume)
J	Flux in membranes ($mol/m^2 s$)
K	Equilibrium constant (bar^p)
$K_L a$	Volumetric mass transfer coefficient (molar per unit time)
k_y	The mass transfer coefficient (mass per unit length)
k	Kinetic constant
k	Polytropic coefficient in compressors and expanders
K	Permeability (Chapter 12: Geothermal Energy)
l_i	Liquid flow rate of component i (mass or molar per unit time)
L	Liquid reflux (molar per unit time)
L_{yac}	Depth of the source (length units)
LOC	Limiting oxygen concentration
LHV	Low heating value (energy per mass)
LFL	Lower flammability limit
m/M	Mass flow (mass)
m_{TNT}	Equivalent mass of TNT (mass)
MW	Molar mass (mass per mol)
n	Molar flow (kmol per unit time)
n	Number of entities per unit volume (inverse of volume)
N	Number of trays
N	Moles (Chapter 5: Biochemical-based Processes)
N_i	Flux of species i (mass or molar units per unit time)
$p_{(l,\ k)}$	Composition in component (k) of pool (l)
$PQ_{(j,\ k)}$	Composition in component (k) of product (i)
P	Pressure (mass per unit length and time square)
P	Product (Chapter 5: Biochemical-based Processes) (mass per unit volume)
Pv	Vapor pressure (mass per unit length and time square)
P_T	Total pressure (mass per unit length and time square)
q	Specific energy (energy per unit mass)
q	Adsorption capacity (mass/mol per mass of adsorbent)
Q	Thermal energy (energy per unit time)
Q_c	Gas flow rate (volume per unit time)
R	Reflux ratio (L/D)
R	Gas constant (energy per unit mol and temperature)
RE	Molar ratio of ethanol to oil
RH	Relative humidity (Chapter 12: Geothermal Energy)
RM	Molar ratio of methanol to oil
r_i	Kinetic rate ($molar^q$ per unit time)
r_{yac}	Influence radius of the reservoir (length)
r_{well}	External radius of the well (length)
S	Entropy: energy per unit temperature and mass
S_i	Selectivity to species i
S	Substrate (Chapter 5: Biochemical-based Processes) (mass per unit volume)

SM	Solar multiple
t	Time
T	Temperature (K or °C)
u_G	Superficial gas velocity (length per unit time)
U	Heat transfer global transfer coefficient (energy per unit area and temperature)
U_b	Bubble rising velocity (length per unit time)
UFL	Lower flammability limit
v	Velocity (length per unit time) (Chapters 11: Wind Energy: Collection and Transformation; Chapter 13: Water as a Resource: Renewable Energies and Technologies for Brine Revalorization)
v	Specific volume of humid air (volume per mass of dry air)
V	Vapor flow rate (mass or molar per unit time)
V	Potential difference
V	Volume (Chapters 4: Thermochemical Processes; Chapter 8: Added-value Products) (volume units)
w_i	Mass fraction of component i
W	Residue flow rate in distillation columns (mass or molar per unit time)
W	Work (energy per unit time)
W_{helio}	Width of the heliostat (length)
x	Liquid molar fraction or mass fraction
$x_{i,L}$	Flows from raw material (i) to intermediate pool (l)
X	Conversion
y_i	Gas molar fraction
$y_{(l, j)}$	Corresponds to the flow from pool (l) to product (j)
y	Specific humidity (mass of vapor per mass of dried air)
$y_{(i, j)}$	Binary variable
$Y_{i/j}$	Yield from substrate j to product i (mass per mass)
z	Molar fraction
z	Polytropic coefficient (compression)
z	Distance downward from the top of the well (Chapter 12: Geothermal Energy)
$z_{(i, j)}$	The flow from raw material (i) to product (j)
z_e	Scaled distance
Greek letters	
α_{ij}	Relative volatility between species i and j
α	Solar altitude angle (Chapter 9: Solar Thermal Energy)
β_{ref}	Temperature coefficient (temperature^{-1})
δ	Layer thickness
δ	Solar declination angle (Chapter 9: Solar Thermal Energy)
ΔG	Gibbs free energy gradient (energy per unit mass or molar)
ΔH	Enthalpy gradient (energy per unit mass or molar)
ΔH_{comb}	Combustion enthalpy (energy per unit mass or molar)
ΔP	Pressure drop (mass per unit length and time square)
ΔR	Radial spacing (length)
ΔAz	The azimuthal spacing (length)
ε	Porosity
ε	Roughness (fluid flow)
ε_G	Gas hold up
ϕ	Sphericity factor (Ergun equation)
ϕ	Local latitude (Chapter 9: Solar Thermal Energy)
γ	Solar hour angle (Chapter 9: Solar Thermal Energy)
λ	Latent heat (energy per unit mass)
λ	Tilt angle (Chapter 9: Solar Thermal Energy)
λ_r	Tip speed ratio
μ	Viscosity (mass per unit length)
μ_i	Chemical potential of species

μ	Biomass growth (per unit time) (Chapters 5: Biochemical-based Processes; Chapter 6: Anaerobic Digestion and Nutrient Recovery)
φ	Relative moisture
φ	Loss of efficiency in turbine due to the load
Φ_L	Liquid fraction in the feed to a distillation column
Φ_V	Vapor fraction in the feed to a distillation column
η	Efficiency
Θ_i	Stoichiometric coefficient
π	Osmotic pressure (mass per unit length and time square)
θ	Pitch angle of rotor blades
θ_H	Azimuth angle of the heliostat relative to the tower base
θ	Loft angle
τ	Shear stress (pressure units)
τ	Residence time (time units)
ρ	Density (mass per unit volume)
σ	Surface tension (pressure length units)
v_n	Amplitude (Chapter 13: Water as a Resource: Renewable Energies and Technologies for Brine Revalorization)
ω	Absolute molar moisture (moles of water per moles of dry air)
ϕ	Azimuthal angle (Chapter 9: Solar Thermal Energy)

B

Thermodynamic data

B.1 Thermochemistry

	MW	ΔH_f (kcal/kmol)		Cp (kcal/kg K)	$a + bT + cT^2 + dT^3$ (T in K)		
				a	b	c	d
N_2	28	0	G	0.26614833	-0.00011594	2.28947E$-$07	-9.97949E$-$11
O_2	32	0	G	0.21012261	-2.7512E$-$08	1.30525E$-$07	7.96202E$-$11
H_2O	18	$-57,798$	G	0.42853535	2.5569E$-$05	1.40284E$-$07	-4.77937E$-$11
NH_3	17	$-10,960$	G	0.38439347	0.00033536	2.40276E$-$07	-1.6676E$-$10
H_2	2	0	G	3.24677033	0.00110931	-1.6519E$-$06	-4.30144E$-$10
CO	28	$-26,464.1148$	G	0.26374744	-0.00010979	2.38312E$-$07	-1.08681E$-$10
CO_2	44	$-94,203.3493$	G	0.10762832	0.00039928	-3.0459E$-$07	9.32634E$-$11
CH_4	16	$-17,877.2$	G	0.28756181	0.00077863	1.7886E$-$07	1.6909E$-$10
Cl_2	71	0	G	0.09073725	0.00011389	-1.304E$-$07	5.22272E$-$11

Aggregation state for ΔH_f: G, gas; L, liquid; S, solid.

B.2 Antoine correlation and phase change

$$Ln(Pv(\text{mmHg})) = A - \frac{B}{C + T(^\circ C)}$$

	A	B	C	λ (kcal/kmol)
H_2O	18.3036	3816.44	227.02	9723
CH_3OH	18.5875	3626.55	238.86	8431
C_2H_5OH	18.9119	3803.98	231.47	9266
NH_3	16.9481	2132.50	240.17	5583
H_2	13.6333	164.9	276.34	
N_2	14.9542	588.72	266.55	
CO	14.3686	530	260	
CO_2	22.5898	3103	272.99	
O_2	15.4075	734.55	266.7	1632
N_2	14.9542	588.72	266.55	1335
HCl	16.5040	1714.25	258.7	3866
FAME	17.4530	5003	122.13	20,181
$C_3H_8O_3$	17.2392	4487.94	132.80	14,623

B.3 Steam properties

$$\lambda(\text{kcal/kg}) = -0.0000043722 \cdot T(^\circ C)^3 + 0.00043484 \cdot T(^\circ C)^2 - 0.58433 \cdot T(^\circ C) + 597.48$$

$$H_{satvap}(\text{kcal/kg}) = -0.0000030903 \cdot T(^\circ C)^3 + 0.00022613 \cdot T(^\circ C)^2 + 0.42436 \cdot T(^\circ C) + 597.42$$

B.3.1 H and S for compressed liquid

$$H(\text{kJ/kg}) = 4.2921 \cdot (T) + 4.1269$$

$$S(\text{kJ/kg K}) = 1.1902 \cdot 10^{-5} \cdot (T)^3 - 3.7465 \cdot 10^{-3} \cdot (T)^2 + 4.5352 \cdot (T) + 0.64547$$

B.3.2 H and S for saturated liquid

$$H(\text{kJ/kg}) = 3.6082 \cdot 10^{-12}(T)^6 - 3.4120 \cdot 10^{-9}(T)^5 + 1.2303 \cdot 10^{-6}(T)^4 - 2.0306 \cdot 10^{-4}(T)^3 + 1.5552 \cdot 10^{-2}(T)^2 + 3.7216(T) + 3.0035$$

$$S(\text{kJ/kg K}) = 1.0372 \cdot 10^{-12}(T)^5 - 8.6494 \cdot 10^{-10}(T)^4 + 2.8965 \cdot 10^{-7} \cdot (T)^3 - 5.6730 \cdot 10^{-5}(T)^2 + 1.6802 \cdot 10^{-2}(T) - 2.1997 \cdot 10^{-2}$$

B.3.3 H and S for saturated vapor

$$H(\text{kJ/kg}) = -6.5690 \cdot 10^{-12}(T)^6 + 6.3049 \cdot 10^{-9} \cdot (T)^5 - 2.3080 \cdot 10^{-6}(T)^4 + 3.8339 \cdot 10^{-4}(T)^3$$
$$-3.0632 \cdot 10^{-2}(T)^2 + 2.7553(T) + 2.4957 * 10^3$$

$$S(\text{kJ/kg K}) = -2.0373 \cdot 10^{-12}(T)^5 + 1.8589 \cdot 10^{-9}(T)^4 - 7.1901 \cdot 10^{-7}(T)^3 + 1.6112 \cdot 10^{-4}(T)^2$$
$$-2.8904 \cdot 10^{-2}(T) + 9.1915$$

B.3.4 H and S for superheated steam (up to 10 bar)

$$H(\text{kJ/kg}) = \left(-6.3293 \cdot 10^{-6} \cdot (P(\text{bar})) + 3.3179 \cdot 10^{-4}\right) \cdot (T)^2 + (0.0124 \cdot (P(\text{bar})) + 1.8039)T$$
$$+ (-6.0707(P(\text{bar})) + 2504.6)$$

$$S(\text{kJ/kg K}) = 9.42 \cdot 10^{-10}(T)^3 - 3.09 \cdot 10^{-6}(T)^2 + 5.24 \cdot 10^{-3} \cdot (T) + \left(6.8171 \cdot (P(\text{bar}))^{(-0.069455)}\right)$$

B.3.5 H and S for superheated steam (10−150 bar)

$$H(\text{kJ/kg}) = (-1.1619 \cdot 10^{-13}(P(\text{bar}))^2 - 8.7596 \cdot 10^{-12}(P(\text{bar})) - 2.2611 \cdot 10^{-10})(T)^4$$
$$(4.298 \cdot 10^{-10}(P(\text{bar}))^2 + 3.276 \cdot 10^{-8}(P(\text{bar})) + 7.313 \cdot 10^{-7})(T)^3$$
$$+ (5.801 \cdot 10^{-7}(P(\text{bar}))^2 - 4.6 \cdot 10^{-5}(P(\text{bar})) - 5.009 \cdot 10^{-4})(T)^2$$
$$+ (3.383 \cdot 10^{-4}(P(\text{bar}))^2 + 0.02947 \cdot (P(\text{bar})) + 2.195)(T)$$
$$+ (-0.072042 \cdot (P(\text{bar}))^2 - 7.7877 \cdot (P(\text{bar})) + 2440.8)$$

$$S(\text{kJ/kg K}) = (1.5719 \cdot 10^{-11}(P(\text{bar})) + 7.4013 \cdot 10^{-10}) \cdot (T)^3$$
$$+ (-1.0074 \cdot 10^{-10}(P(\text{bar}))^2 - 3.0171 \cdot 10^{-8}(P(\text{bar})) - 2.8872 \cdot 10^{-6})(T)^2$$
$$+ (9.4914 \cdot 10^{-8}(P(\text{bar}))^2 + 2.9097 \cdot 10^{-5}(P(\text{bar})) + 5.0938 \cdot 10^{-3}) \cdot (T)$$
$$(4.1223 \cdot 10^{-5}(P(\text{bar}))^2 - 0.028841 \cdot (P(\text{bar})) + 5.9537)$$

B.4 Thermodynamic correlations

$$S(\text{J/mol} \times K); H(\text{kJ/mol}); T(K); P(\text{bar})$$

B.5 Toluene

B.5.1 Saturated liquid

$$H = 2.413 \times 10^{-13}T^6 - 5.986 \times 10^{-10}T^5 + 6.123 \times 10^{-7}T^4 - 0.0003303\,T^3 + 0.09922\,T^2 - 15.6T + 971.6$$

$$S = 4.583 \cdot 10^{-9}T^4 - 6.919 \cdot 10^{-6}T^3 + 0.003667T^2 - 0.3103T - 129.2$$

B.5.2 Saturated steam

$$H = -3.15523 \cdot 10^{-7}T^3 + 4.51006 \cdot 10^{-4}T^2 - 8.2643 \cdot 10^{-2}T - 16.3670$$

$$S = -8.01 \times 10^{-11}T^5 + 1.683 \times 10^{-7}T^4 - 0.0001422\,T^3 + 0.06072\,T^2 - 13.06T + 1208$$

B.5.3 Overheated steam

B.5.3.1 Pressure: 0.02–1.4 bar

$$H = 7.797 + 0.0007286T - 1.743P + 0.000174T^2 + 0.003393PT - 0.01305P^2$$

B.5.3.2 Pressure: 0.02–0.3 bar

$$S = -4.367 + 0.3199T - 323.9P + 4.004 \cdot 10^{-5}T^2 + 0.2449PT + 1156P^2$$
$$- 0.0003046T^2P - 0.03606TP^2 - 1774P^3$$

B.5.3.3 Pressure: 0.2–0.5 bar

$$S = 4.690 \cdot 10^{-6}P^3T^2 - 2.152 \cdot 10^{-3}P^3T - 74.800P^3 + 3.165 \cdot 10^{-5}P^2T^2 - 2.482 \cdot 10^{-2}P^2T + 120.891P^2$$
$$- 1.004 \cdot 10^{-4}PT^2 + 0.0859PT - 96.419P - 5.595 \cdot 10^{-7}T^2 + 3.5006 \cdot 10^{-1}T - 21.865$$

B.5.3.4 Pressure: 0.3–1.4 bar

$$S = 80.06 - 0.7629T - 53.46P + 0.004495T^2 - 0.2038TP + 119.7P^2 - 8.21 \cdot 10^{-6}T^3$$
$$+ 0.001046T^2P - 0.329TP^2 - 32.58P^3 + 5.706 \cdot 10^{-9}T^4 - 1.34 \cdot 10^{-6}T^3P + 0.0004433T^2P^2$$
$$- 0.01034TP^3 + 9.053P^4$$

B.5.4 Liquid

$$H = h_{ref} + (140140 - 152.3 \cdot T_{AV} + 0.695 \cdot (T_{AV})^2) \cdot 10^{-6} \cdot (T - T_{ref})$$

with:

$$h_{ref} = -14.55425 \text{kJ/mol}; \quad T_{AV} = \frac{T + T_{ref}}{2}; \quad T_{ref} = 298.15\text{K};$$

$$S = -228 + 0.1355P + 0.6739T - 0.002778P^2 - 3.308 \cdot 10^{-5}PT - 0.0002137T^2$$

B.5.4.1 Antoine equation

$$\ln(P_{sat} \cdot 750.064) = 16.0137 - \left(\frac{3096.52}{T_{sat} - 53.67} \right)$$

B.6 Benzene

B.6.1 Saturated liquid

$$H = -5.3293023 \cdot 10^{-14}T^6 + 1.3528752 \cdot 10^{-10}T^5 - 1.4069533 \cdot 10^{-7}T^4$$
$$+ 7.6948020 \cdot 10^{-5}T^3 - 2.3241987 \cdot 10^{-2}T^2 + 3.7956329T - 2.8223102 \cdot 10^2$$

$$S = 2.9008962 \cdot 10^{-11}T^5 - 5.6525126 \cdot 10^{-8}T^4 + 4.4218808 \cdot 10^{-5}T^3 - 1.7507080 \cdot 10^{-2}T^2 + 3.9400717T$$
$$- 4.3576804 \cdot 10^2$$

B.6.2 Saturated steam

$$H = -5.9672469 \cdot 10^{-12}T^5 + 1.0752354 \cdot 10^{-8}T^4 - 8.0309120 \cdot 10^{-6}T^3$$
$$+ 3.1637092 \cdot 10^{-3}T^2 - 5.6945064 \cdot 10^{-1}T + 56.601956$$

$$S = 1.75589 \cdot 10^{-15}T^6 - 3.20075 \cdot 10^{-11}T^5 + 6.31984 \cdot 10^{-8}T^4 - 5.42456 \cdot 10^{-5}T^3$$
$$+ 2.46331 \cdot 10^{-2}T^2 - 5.79659T + 640.791$$

B.6.3 Overheated steam

$$H = 9.847 + 0.01923T - 0.7586P + 0.0001165T^2 + 0.001124TP + 0.01733P^2$$

B.6.3.1 Pressure 0–0.8 bar

$$S = 6.739 + 0.3097T - 54.14P - 3.761 \times 10^{-5}T^2 + 0.003373TP + 32.55P^2$$

B.6.3.2 Pressure 0.8–2.0 bar

$$S = -10.06 + 0.3255T - 14.02P - 5.284 \times 10^{-5}T^2 + 0.001901TP + 2.361P^2$$

B.6.4 Liquid

$$H = h_{ref} + (162940 - 344.94 \cdot T_{AV} + 0.85562 \cdot (T_{AV})^2) \cdot 10^{-6} \cdot (T - T_{ref})$$

with:

$$h_{ref} = -7.6977 \text{kJ/mol}; \quad T_{AV} = \frac{T + T_{ref}}{2}; \quad T_{ref} = 298.15\text{K};$$

$$S = -173.8 + 0.5649T - 0.02493P - 0.0002062T^2 + 8.083 \times 10^{-6}PT + 0.001357P^2$$

B.6.4.1 Antoine equation

$$\ln(P_{sat} \cdot 750.064) = 15.9008 - \left(\frac{2788.51}{T_{sat} - 52.36} \right)$$

B.7 Cyclohexane

B.7.1 Saturated liquid

$$H = 5.64877 \cdot 10^{-11} \cdot T^5 - 1.10858 \cdot 10^{-7} \cdot T^4 + 8.61667 \cdot 10^{-5} \cdot T^3 - 0.0329151 \cdot T^2 + 6.32556 \cdot T - 510.124$$

$$S = 1.22384 \cdot 10^{-8} \cdot T^4 - 1.92074 \cdot 10^{-5} \cdot T^3 + 0.0110945 \cdot T^2 - 2.28934 \cdot T + 80.0381$$

B.7.2 Saturated steam

$$H = -4.47359 \cdot 10^{-13} \cdot T^6 + 1.06898 \cdot 10^{-9} \cdot T^5 - 1.05657 \cdot 10^{-6} \cdot T^4 + 0.000552199 \cdot T^3 - 0.160689 \cdot T^2 + 24.7614 \cdot T - 1565.25$$

$$S = -7.58895 \cdot 10^{-13} \cdot T^6 + 1.80099 \cdot 10^{-9} \cdot T^5 - 1.76164 \cdot 10^{-6} \cdot T^4 + 0.000905964 \cdot T^3 - 0.256920 \cdot T^2 + 37.9053 \cdot T - 2181.58$$

B.7.3 Overheated steam

$$H = 5.066 + 0.007424T - 0.8625P + 0.0001808T^2 + 0.001191PT + 0.02255P^2$$

$$S = -19.97 + 0.3682T - 94.6P + 7.489 \cdot 10^{-6}T^2 + 0.05242PT + 126.04P^2 - 4.627 \cdot 10^{-5}T^2P$$
$$- 0.02955P^2T - 100.4P^3 + 2.569 \cdot 10^{-5}P^2T^2 + 0.008158TP^3 + 40.39P^4 - 6.232 \cdot 10^{-6}T^2P^3$$
$$- 0.0003521TP^4 - 6.408P^5$$

B.7.4 Liquid

$$H = h_{ref} + (-220600 + 3118.3 \cdot T_{AV} - 9.4216 \cdot (T_{AV})^2 + 0.010687 \cdot (T_{AV})^3) \cdot 10^{-6} \cdot (T - T_{ref})$$

with:

$$h_{ref} = -9.3247 \text{kJ/mol}; \quad T_{AV} = \frac{T + T_{ref}}{2}; \quad T_{ref} = 298.15 \text{K};$$

$$S = -184.9 + 0.5331T + 0.1245P - 2.991 \times 10^{-5}T^2 - 0.0004279PT - 0.0008751P^2$$

B.7.4.1 Antoine equation

$$\ln(P_{sat} \cdot 750.064) = 15.7527 - \left(\frac{2766.63}{T_{sat} - 50.50} \right)$$

Index

ted States
ublisher Services